UNMANNED AERIAL VEHICLE SYSTEMS IN CROP PRODUCTION

A Compendium

UNMANNED AERIAL VEHICLE SYSTEMS IN CROP PRODUCTION

A Compendium

K. R. Krishna, PhD

AAP | APPLE ACADEMIC PRESS

Apple Academic Press Inc.
3333 Mistwell Crescent
Oakville, ON L6L 0A2
Canada

Apple Academic Press Inc.
1265 Goldenrod Circle NE
Palm Bay, Florida 32905
USA

Exclusive worldwide distribution by CRC Press, a member of Taylor & Francis Group

International Standard Book Number-13: 978-1-77188-756-4 (Hardcover)
International Standard Book Number-13: 978-0-42942-526-4 (eBook)

Library and Archives Canada Cataloguing in Publication

Title: Unmanned aerial vehicle systems in crop production : a compendium / K. R. Krishna, PhD.

Names: Krishna, K. R. (Kowligi R.), author.

Description: Includes bibliographical references and index.

Identifiers: Canadiana (print) 20190097248 | Canadiana (ebook) 20190097442 | ISBN 9781771887564 (hardcover) | ISBN 9780429425264 (eBook)

Subjects: LCSH: Agriculture—Remote sensing. | LCSH: Drone aircraft. | LCSH: Crops.

Classification: LCC S494.5.R4 K75 2019 | 630.208/4—dc23

Library of Congress Cataloging-in-Publication Data

Names: Krishna, K. R. (Kowligi R.), author.

Title: Unmanned aerial vehicle systems in crop production : a compendium / author: K. R. Krishna.

Description: Oakville, ON ; Palm Bay, Florida : Apple Academic Press, 2019. | Includes bibliographical references and index. |

Identifiers: LCCN 2019016571 (print) | LCCN 2019020458 (ebook) | ISBN 9780429425264 () | ISBN 9781771887564 (hardcover : alk. paper)

Subjects: LCSH: Agriculture--Remote sensing. | Drone aircraft. | Crops.

Classification: LCC S494.5.R4 (ebook) | LCC S494.5.R4 K75 2019 (print) | DDC 338.1/6--dc23

LC record available at https://lccn.loc.gov/2019016571

Apple Academic Press also publishes its books in a variety of electronic formats. Some content that appears in print may not be available in electronic format. For information about Apple Academic Press products, visit our website at **www.appleacademicpress.com** and the CRC Press website at **www.crcpress.com**

About the Author

K. R. Krishna, PhD

K. R. Krishna, PhD, is an agricultural scientist. His contributions deal with topics such as soil fertility, soil microbiology, crop production, agroecosystems, precision farming, robotics, and unmanned aerial vehicle (drones) techniques in agriculture. He is a member of the International Society for Precision Agriculture, American Society of Agronomy, Soil Science Society of America, Ecological Society of America, Indian Society of Agronomy, and Soil Science Society of India. He has authored several books on international agriculture, encompassing topics in agroecosystems, field crops, soil fertility and crop management, precision farming, and soil microbiology. His more recent titles deal with topics such as agricultural robotics and drones and satellite guidance to improve soil fertility and crop productivity. He is retired from the International Crops Research Institute for the Semi-Arid Tropics (ICRISAT) in India.

Contents

Abbreviations

2D	two-dimensional
3D	three-dimensional
ATLAS	The Advanced Light Acquisition System
AUAV	autonomous unmanned aerial vehicle
CIMMYT	International Maize and Wheat Center
CWSI	crop water stress index
DSM	Digital Surface Model
DSS	decision support systems
EMBRAPA	Empresa Brasileira de Pesquisa Agropecuaria
EO	electrooptical
FAA	Federal Aviation Agency
FAO	Food and Agriculture Organization
GCS	ground control station
HSE	Homeland Surveillance and Electronics LLC
IAI	Israel Aerospace Industries
ILS	Incident Light Sensor
INS	inertial navigation system
IR	infrared
ISTAR	intelligence, surveillance, target acquisition, and reconnaissance
KAP	kite aerial photography
KAT	kite aerial technology
LAI	leaf area index
LiDAR	light detection and ranging
LTA	lighter-than-air
MMC	MicroMultiCopter Aero Technology Co., Ltd.
MTOW	maximum take-off weight
NDRE	normalized difference red edge
NDVI	normalized difference vegetation index
NIR	near infrared
NIR	near-infrared radiation
PPK	postprocessed kinematic
R&D	research and development
RC	remote controlled

ROV	remotely operated vehicles
RPAS	remotely piloted aircraft system
TGI	triangulation greenness index
UAS	unmanned aerial system
UAV	unmanned aerial vehicle
UAVS	unmanned aerial vehicle system
VARI	visible atmospheric resistance index
VTOL	vertical take-off and landing

Preface

A new phenomenon generated through human ingenuity seems to have begun in the skies above agrarian regions of the world. It is called "unmanned aerial vehicle (UAV) technology." Agricultural UAVs are destined to fly and dominate the skies immediately above farm world. UAVs are being utilized frequently to obtain spectral images and digital data about natural resources, terrain, the soil, and cropping systems. It is a glaring fact that while agriculturists exploited soil and water tirelessly all through the past several millennia, the sky above crops was not utilized as efficiently. Aerial views and spectral analysis of a crop's canopy were not given great interest. Aerial data were not analyzed to any great extent to know what was happening to crops. At present, the agricultural UAVs, with their sharp optical and chemical sensors, can assess the crop's vigor, nutrient requirements, disease/pest attack, if any, and give the farmer an idea about the yield that he may harvest.

Historically, agrarian regions have been invaded with implements and gadgets. Mostly, they have been the gifts of human ingenuity. No doubt most of such introductions have aimed at reducing drudgery, making agronomic procedures easy and efficient, both physically and economically. Many have been perpetually adopted in crop fields, but several of them have faded or were discontinued for various reasons. The most recent gadget to enter the agricultural world is the agricultural UAV. The UAVs (also called drones) are flying and flocking over farms. Right now, we are learning the rudiments of UAV technology, with a clear intent to spread this technology. A final goal is to improve crop productivity and simultaneously reduce human drudgery in the fields. The agricultural UAVs are supposed to add accuracy and electronic sophistication to farming procedures.

We may note that agricultural UAVs are not a very costly proposition for farming companies and farmers with sizeable holding. Now, there is also a trend to teach farmers to make their own UAVs with indigenous materials. So, UAVs may make headway into agrarian regions of any economic level. The UAVs are versatile. Some of the models are well suited to an array of farming conditions and natural settings. There are highly specialized UAV models. So, UAV models that flood the market may be able to answer any sort of contingencies that farm experts or farmers may encounter.

So far, UAVs have been tried on several crop species cultivated in different agrarian belts of the world. Flight mission and image processing software are perhaps most important. The economics of purchasing a UAV system, using it, and obtaining worthwhile information for use on crop fields are also important. We may have to weigh out several technical and economic aspects attached with each UAV system, prior to purchasing it for regular use on farms.

Sensors on agricultural UAVs are the most important aspects of UAV technology. There are now several brands of the most commonly used sensors. Farmers have wide options based on the intended purpose. There are visual, multispectral, hyperspectral, infrared, near-infrared, and light detection and ranging (LIDAR) sensors. A few of them also have chemical sensors that detect greenhouse gas emission (e.g., Scentroid UAV). The UAV platform that provides vantage points and the sensors, together, are billed to initiate a revolution. They are expected to add electronic sophistication into the way crops are studied, analyzed, and assessed, and the way agronomic prescriptions are decided. Hard human labor and drudgery is expected to be avoided in a big way.

It is a fact that, right now, UAV aircrafts, that is, fixed-winged and copter UAVs, have gained popularity. However, a careful look at the literature shows that the agricultural UAV experts are also exploring, testing, and trying to assess the usefulness of low-cost balloons, blimps, and parachutes. Each of these contraptions has a certain clear advantage over aircraft versions. Who knows the future? A permanently floating balloon or a blimp above every farm may become the order of the day. These baloon or blimps could relay information about crops in a farmer's fields or experimental stations on a continuous basis. Parachutes may turn out to be cheaper and efficient too. The underlying competition among various types of UAVs will be interesting to watch.

There is stiff competition among UAV companies unfolding. This relates to selection of a UAV, say a fixed-winged aircraft, multirotor or a blimp, or even a kite. Next, an UAV company, to be viable, must compete or outcompete by offering the best suited UAV type and model. It should also offer accessories, and services, and make them all easy to operate and get results from the crop fields. This competition has just begun. Several UAV companies may foldup if their model is less preferred. Of course, a few may dominate the global agricultural UAV scenario. Right now, there are models such as RMAX helicopter, Hercules, eBee, PrecisionHawk's Lancaster, and few others that are more frequently accepted in farms worldwide. Sprayer UAVs seem to really dominate the plant protection procedures in farms.

So, we may expect them to be popular in the near future. UAV systems are facing certain constraints to their smooth acceptance into farm world. Many of them could get answered in due course through avionics and sensors research.

UAVs will replace farm labor that hitherto specialized in scouting and collecting data and, those involved in sprayer using Knapp sack or a tractor. Farmers must upgrade their skills and get trained in using UAVs and computer software. This aspect could be handled slowly and smoothly. Demographic data and skills need to be matched. Such adjustments in labor supply and need have occurred every time a new gadget, method, or a government legislation has entered the agrarian belt. UAVs are most recent ones to invade agrarian belts—literally.

The emphasis is on UAV aircrafts in this volume although there are other types of UAVs useful in agricultural operations such as parachutes, blimps, balloons, and kites. Over 200 models of UAV aircrafts have been listed along with their specifications. Currently, there has been a spurt in the number of UAV start-ups. New models with specific advantages are being churned out. Hence, firstly there is competition generated among drone models. A few models may eventually be very successful. There is already a certain preference shown by farmers for certain fixed-winged, helicopter, and multicopter models. They pick them based on the activity needed, such as aerial survey only or spraying or both. There is also an underlying competition about which type of UAV should be preferred by farmers, farming companies, and agricultural experimental stations. A good guess leads us to believe that blimps may take over crop monitoring, data procurement, and experimental evaluation of crops. They could be floating or hanging at a spot above for the whole season and could record every manifestation of crops in the field. Copters may vie for operations such as close-up analysis of crops and plantations. Rapid aerial imagery may be done better by fast-flying fixed flat-winged drones. So, we must wait and watch which type and model of UAVs dominate the skies above agrarian regions.

For the reader, anyway, this volume includes a wide range of models and types of UAVs, plus accompanying information on representative sensor (camera) types and software to control UAVs. The bottom line is that, we must be prepared to help farmers and researchers with software and to compare and select the best UAV for the said purpose. The UAV technology must be cheaper, efficient, and consistently profitable; otherwise, UAVs may end as one more intruder into agrarian regions that subsides after a while. Of course, the most important consequences of UAV technology are efficient monitoring and data collection on a crop's progress in the field.

The ability to manage farms with large land holdings but with reduced number of skilled farm workers is important. Elimination of farm drudgery is equally important.

In this volume, there are six chapters. An introductory chapter discusses the historical aspects in an abridged form. Further, it deals with different types of UAVs and their classification and uses in various sectors of human activity. Chapters 2, 3, and 4 form the centrepiece of this volume on UAVs. They deal with flat fixed-winged drones, versatile helicopter models, multicopters, parachutes, blimps, balloons, and kites. In Chapter 3, there is a special section on sprayer UAVs. Over 50 different sprayer copter UAVs have been described. At present, there is a surge in development of sprayer helicopters and multicopters. Their use has increased perceptibly in developed nations. Of course, it is related to their advantages in handling harmful pesticides, herbicides, and fungicides and in completing the tasks relatively very rapidly compared to human skilled workers. Chapter 5 deals exclusively with sensors (visual, multispectral, infrared, and near-infrared) and computer software needed to decide flight mission and image processing

There is no doubt that agricultural researchers, farm companies, and farmers must be conversant with a plethora of UAV models that are currently available in the market. In due course, many more UAV models will arrive in the market. Rather, too many more UAV models will flood the market. Farmers should pick the best using computer-aided selection process. Software to compare the UAVs for economic and agronomic efficiency are available.

Overall, this book should be most interesting to anyone dealing with agriculture, aeronautics, UAV technology, and artificial intelligence. It is a good book for the public, particularly to those inquisitive about improvements in food production tactics and UAVs. This volume is most useful to possess in public libraries, university libraries, research institutions, and commercial agencies dealing with UAVs. In educational institutions, professors, researchers, experts, and students in agriculture and engineering (aerial robotics) could use it as an excellent text or reference book.

— **Kowligi Krishna**
Bangalore, India, 2018

Acknowledgments

During the process of a literature survey, collecting important research papers and compiling the drafts of this treatise on unmanned aerial vehicles (UAVs), several persons have been helpful to me directly or indirectly. Several of them have offered permissions to use photographs, unpublished data, and sent information pertaining to a wide range of UAVs. I wish to thank them all. Any omissions are inadvertent. Following are the important industries, their CEOs, marketing and sales department officials, research scientists from universities and others, who have allowed me the use of images of UAVs.

CHAPTER 1

Dr. Clyde R. Beaver, Creative Service Manager, International Centre for Maize and Wheat (CIMMYT), El Baton, Mexico.

CHAPTER 2

Mr. Aleksandrs Manajenkovs and Konstantin Krivovs, Sales and Marketing Department, UAVFACTORY LTD. Jaunbridagi-1, Marupe, LV-2167, Latvia;

Dr. Alex Hardy, Director, Vulcan UAV Ltd., Gloucestershire, Great Britain;

Mr. Amaury Weist, Sunbirds, Business Development, 10, Avenure De Europe, 31 520 Ramonville-St-Agne, France;

Dr. Antonio Liska, Director, Robota LLC, Lancaster, TX, USA;

Mr. Anthony Hassal, Director Marketing and Ms. Neomi Tarancon, Sales Team, C/Jose Galan Merino 6, Edificio CREA, 41018 Sevilla, Spain;

Ms. Astrid Svara, Sales Team, C-Astral Aerospace Ltd., Adjovscina, Slovenia;

Mr. Atul Khosla, Director, OM UAV Systems, Jawahar Nagar, Delhi, India;

Mr. Benjamin Lehmann and Lehman Drone Team, Lehmann Aviation Inc., La Chapelle, Vendomoise, France;

Mr. Bill Nickerson, UASUSA Inc., Longmont, CO 80503, USA;

Dr. Blake Sawyer, Coordinator, Martin UAV LLC., 15455 North Dallas Parkway, Suite 1075, Addison, TX 75001, USA;

Ms. Céline Vergely, Corporate Communication Manager, Drone Volt, 14, Rue de la Perdrix – Lot 201, Roissy CDG Paris Nord 2, 93420 Villepinte, France;

Dr. Clyde R. Beaver, Creative Service Manager, International Centre for Wheat and Maize, El Baton, Mexico;

Dr. Dhruv Arora, TechnoSys Embedded Systems, SCO 66, FF, Sector 20-C, Chandigarh, India;

Dr. Fulvia Quagliotti, President and CEO, Mr. Gianluca Ristorto, Micro Aerial Vehicles Technology s.r.l., Torino, Italy, and Dr. Fulvia Quagliotti, President and CEO, Micro Aerial Vehicles Technology s.r.l., Torino, Italy'

Dr. Gene Robinson, President, RP Flight Systems, Wimberley, TX, USA;

Mr. Gianluca Ristorto, Micro Aerial Vehicles Technology—MAVTech, NOI Techpark & Business Incubator, Via Alessandro Volta 13/A, 39100, Bolzano, Italy;

Dr. Ian Freemantle, Founder and Director, Aerovision, PO Box 233, Noordhoek, Cape Town, South Africa;

Mrs. Lea Reich, Sr. Communications Director, Precision Hawk Inc., Noblesville, IN, USA;

Mrs. Lyudmila Piskunova, Marketing Manager, Autonomous Aerospace Systems, 3 Vuzuvsky Side Street, Krasnoyarsk, Siberia, Russia;

Mr. Matthew Wade, Marcomms Manager, senseFly Ltd—a Parrot company, Cheseaux-sur-Lausanne, Switzerland;

Mr. Mauricio Ortiz, Sales Support, Aeromapper, Aeromao Inc., Mississauga, Ontario L5N-2P3, Canada;

Dr. Michael Schmidt, Geschaftsfuhrer/Managing Director, Hanseatic Aviation Solutions GMBH, Herman Kohl Street 7. D-28199, Bremen, Germany;

Dr. Nigel King Director, Quest Technology Centre, Coquetdale Enterprise Park, Amble, Northumberland, UK;

Mrs. Ola Fristrom, Business Development Manager, SmartPlanes, Skelleftea, Sweden;

Mr. Phil Jones, Marketing Division, MartinUAV Inc., Austin, TX, USA; Blake Sawyer, VP, MartinUAV, Austin, TX, USA; and Robyn Olson, Executive Coordinator, Martin UAV LLC, 3335 Kifer Road, Santa Clara, CA, USA;

Mrs. Sarah Ritzen, Director Marketing and Communications Manager, Sentara LLC., 6636 Cedar Avenue South, Ste 250 Minneapolis, MN 55423, USA;

Dr. Sebastian Verling, Marketing Manager, Wingtra One, c/o ETH WEH H ETH Zurich (Wyss Zurich Project), Winberg Str., 35, 8092, Zurich, Switzerland;

Service Staff, Feiyu Technology, Qi Xiang, Guilin, China (Mainland);

Mr. Siim Juss, Project Manager of Marketing and Sales, Threod Inc., Kaare tee 3, Viimsi 74010, Estonia;

Mrs. Silvia McLachlan, PR and Marketing Communications, Trimble, Agriculture Division, Trimble Inc., Sunnyvale, CA, USA;

Support Service Team, Hubsan Intelligent Company Ltd., 13th floor Building 1C, Shenzhen Software Industry Base, Xuefu Road, Nanshan District, Shenzhen, China;

Mr. Terry Sanders, Vice President Marketing and Innovations, Homeland Surveillance and Electronics LLC. (HSE), USA;

Dr. Tom Nicholson, AgEagle LLC., Business Development, Neodesha, Kansas, KS, USA; Tel.: 620-330-2248;

Mr. Vick M. Sales and Marketing Division, Rise-Above Custom UAV Solutions Inc., Smeaton Grange, New South Wales, Australia.

CHAPTER 3

Mr. Adam Najberg, Marketing Department, DJI, Shenzhen, Guangdong, China;

Mr. Allen, J., Shenzhen Trump Aero Technology Co., Ltd., Donglong Zing Building, Huarong Rd., Dalang, Bao'an District, Shenzhen 518000, Guangdong, China;

Ms. Aya Kawakami, Marketing Division, Prodrone Ltd., Head Office, Naka-Ku, Nagoya-shi, Aichi 460-0004, Japan;

Mr. Brad Young, Aeryon Labs Inc., Kumpf Drive, Waterloo, Canada;

Ms. Céline Vergely, Corporate Communication Manager, Drone Volt, 14, Rue de la Perdrix—Lot 201, Roissy CDG Paris Nord 2, 93420 Villepinte, France;

Dr. Christoph Eck, Managing Partner (CEO), Aeroscout GmbH, Hengstrain 14, 6280 Hochdorf, Switzerland;

Mr. Donald Effren, CEO, Autocopter Inc., Raleigh, NC, USA;

Prof. Ing Fulvia Quagliotti, President and CEO, Micro Aerial Vehicles Technology—MAVTech, Via Vincenzo Gioberti 73 10128, Torino, Italy;

Mr. Jenny Wang, Shandong Joyance Intelligence Technology Co. Ltd., No. 2321, Beihai Road, Weifang, Shandong 261061, China;

Mrs. Kristy Hall, Advanced UAV Technology, Wellesley House, Duke of Wellington Avenue, Royal Arsenal, London SE18 6SS, UK;

Dr. Matthais Beldzik and Prof. Guido Morganthal, Ascending Technologies Gmbh, Konrad-Zuse Bogen, 482152, Kreiling, Germany;

Mrs. Meliha Boucher, Corporate PR and Marketing Director, ECA Group, L Gourde, France;

Ms. Mirjam Baumer, Marketing Manager-Europe, Microdrones Gmbh, Gutenbergstrasse, 86, Siegen, Germany;

Mr. Terry Sanders, Vice President Marketing and Innovations, HSE, USA;

Dr. Toru Tokushige , CEO and Founder, Terra Drone, Terra Motors, Shibuya-ku 5 chome Jingumae, Tokyo 53, 67 Cosmos Aoyama South Building 3F, Tokyo 150-0001, Japan;

Mr. Vick M. Sales and Marketing Division, Rise-Above Custom UAV Solutions Inc. Smeaton Grange, New South Wales, Australia;

Mr. Yohan Brochard, Marketing Manager and Designer, Aerialtronics, The Hague, Netherlands.

CHAPTER 4

Dr. Alexander Mijatovic, Structure and Aerodynamics, Aero Drum Ltd., Vojislava Ilica 99a, 11000 Belgrade, Serbia;

Dr. Gilles Dutray, Management, Anabatic LLC., Luins, Switzerland;

Mr. Ian Freemantle, Founder and Director Aerovision South Africa, Cape Town, South Africa;

Mr. John Edling, Tetracam Inc., Chatsworth, CA, USA;

Mrs Sandy Allsopp, Allsopp Helikites Ltd., Fordingbridge, Hampshire, United Kingdom

Dr. Thamm, H.L. CEO, Geo-Technics, Linz, Germany.

CHAPTER 5

Mr. Adam Svestka, Workswell s.r.o., Prague, Czech Republic; Mr. Jeroslov Bohm, Workswell Gmbh, Ibbenbüren-Laggenbeck, Germany;

Dr. Ardevan Bakhtari, President, Scentroid Inc., Stouffville, Ontario, Canada;

Mrs. Harriet Brewitt, Marketing Division, 3D Laser Mapping Ltd., Nottinghamshire, NG13 8gf, UKMr. Michael Barnes, CEO, Galileo Group, Inc., Research Triangle Park, NC, USA;

Mr. Micheal Zemlan Spectral Imagers, Advanced Spectroscopy Section, BaySpec Inc., San Jose, CA, USA;

Dr. Nolan Ramseyer, CEO, Peau Productions Inc., San Diego, CA 92108, USA.

CHAPTER 6

Dr. Clyde R. Beaver, Creative Service Manager, International Centre for Wheat and Maize (CIMMYT), El Baton, Mexico.

I wish to thank Dr. (Mrs.) Uma Krishna, Mr. Sharath Kowligi, and Mrs. Roopashree Kowligi, and offer my best wishes to Ms. Tara Kowligi.

CHAPTER 1

Introduction to Unmanned Aerial Vehicle Systems Utilized in Agriculture

ABSTRACT

This chapter introduces the concepts relevant to the use of unmanned aerial vehicles during crop production. Firstly, salient features of UAVs utilized in agriculture are described. Types of UAVs and their classification is briefly mentioned. Discussions cover a wide range of UAVs such as fixed-winged drone aircrafts, autonomous helicopters, fixed-winged VTOL hybrid UAVs, multi-rotor drones for aerial imagery and spraying plant protection chemicals, etc. Parachutes, parafoils, blimps, aerostats, and kites are other types of UAVs with potential to be adopted in farming in a big way. A few examples of successful use of UAVs in farming has been quoted briefly. Fixed-winged drones are efficient in supplying aerial imagery and spectral data about crops. Helicopters and multi-copter drones are useful for close-up surveillance of crops and during spraying pesticides/fungicides. UAVs such as aerostats, blimps, and parafoils are useful in aerial photography, surveillance, spectral analysis of crops, and detecting and reporting disasters. Tethered aerostats are used to collect crop's phenomics data on a continuous basis. Aerostats are also useful in rural telecommunication networks. Economic aspects of usage of UAVs are highly relevant for farmers, prior to their full-fledged adoption. Some facts about the extent of production of UAVs and forecasts have been listed.

1.1 HISTORICAL ASPECTS OF UNMANNED AERIAL VEHICLE TECHNOLOGY

The unmanned aerial vehicles (UAVs) have found a niche in the agrarian regions of the world. They are: flocking, flying, and at times hovering over the canopies of vast stretches of crops. Their entry into agrarian belts has been delayed compared to their usage in military and other civilian aspects of life.

Although belated, now UAVs seem to have simultaneously invaded different agrarian regions of the world. Agricultural UAVs are forecasted to have a very large impact on global crop production methods, the need for labor, inputs such as fertilizers, pesticides, fungicides, herbicides, and irrigation water. UAVs offer aerial images picked from vantage points above the crop canopies, that is, a "bird's-eye view" plus accurate spectral data for analysis. This aspect was never an easy possibility during past millennia (Krishna, 2018). Agricultural UAVs are among the latest of the techniques appreciated by farmers and agricultural researchers worldwide. These UAVs are expected to revolutionize crop husbandry methods. UAVs are here to flourish in the agrarian regions and offer us advantages for infinite period, in future.

Let us briefly consider the historical facts of UAVs, in general. It seems attempts to develop UAVs began with Nikola Tesla, who described a remote-controlled aircraft. It led to development of "tele-automation" which is crucial for UAV technology (Vroegindeweij et al., 2014). UAVs were first used in military campaign during World War I. Ruston Proctor Aerial Target of 1916 was the first pilotless aircraft (UAV). It was controlled, using Low's radio control techniques (Talon LPE, 2017). They were first developed in 1916; however, improvement in UAV technology related to military use occurred in the years between World Wars I and II (Nicole, 2015). Keane and Carr (2013) and Newcome (2004) offer reviews about the historical developments of UAVs. They state that development of UAV technology for military is now at least a century old. It seems British Navy in the Mediterranean region used it for the first time to hit enemy position. They launched military UAVs from an aircraft carrier (HMS Argus). Major initiatives to develop, test, evaluate, refine, and deploy UAVs as a defense arsenal occurred in the United States during 1950s and 1960s. It was done under a code name "Red Wagon." It began with the induction of *Ryan Firebees* series of military UAVs (Tetrault, 2014; Krishna, 2018). UAVs, therefore, have roots in military establishments of European nations and the United States (Krishna, 2018; Blake, 2017). It seems that development of UAV technology for warfare suffered a kind of stagnation and lack of interest, until it was deployed in a good scale in the Vietnam War during 1960s (Schwing, 2007; Kennedy, 1998). A pilot in the United States Air Force points out that, even as recently as half a decade ago, "UAVs" meant and conjured up idea of warfare, intelligence in battle fields, missile launches to destroy ground force, and air torpedoes. *It is not so now.* The enthusiasm to opt UAVs for civilian use, particularly in agricultural cropping belts has far surpassed the military importance of UAVs (Jacobsen, 2016; Gago et al., 2015). UAVs were utilized to study weather patterns by the meteorological services of the

United States in 1946. During past few decades, UAVs have offered excellent data about general weather conditions, tornadoes, storms, droughts, and floods in vast areas. Incidentally, Krishna (2018) has reviewed and listed several instances that exemplify the shift of focus of UAV technology, from solely military to civilian and agricultural uses, in the recent past. It has been pointed out that, so far, transition from predominantly military usage to agricultural aspects has occurred smoothly without a hitch. At times, it has also occurred in spurts. A few of the major military UAV-producing companies have abruptly shifted their focus. They have refined and modified their erstwhile UAV models to suit agricultural purposes, mainly aerial survey (e.g., R-Bat and Bat by Northrop Grumman Inc.). A few other UAV models have been modified to conduct spectral analysis of crops, procure digital data for precision farming vehicles, and to spray the crops.

Agricultural UAVs have a very short history. It spans for just past one decade. UAVs were introduced into farming zones relatively too recently. Reports suggest that UAVs were being evaluated for their efficacy in offering aerial photography sometime by 2007 (Dobberstein, 2013). UAVs were replacing piloted aircraft in many places. It was attributable to economic and logistical constraints related to human-piloted airplane campaigns. Forecasts done around 2007 suggested that UAVs will spread all through different agrarian belts. UAVs do affect the way crop husbandry procedures are conducted. To a certain extent, UAVs effects are evident. The UAV usage may get accentuated further in farm belts. UAVs have the potential to replace a large section of farm labor involved in scouting, collecting data, applying fertilizers, and spraying plant protection chemicals. This fact has induced large number of start-ups that produce a very wide range of UAV models. Specialized UAV models are also being flooded into the market since 2012–2015 (Krishna, 2018; Ottos, 2014; Bowman, 2015). In the United States, UAV models have crept into the airspace of maize, soybean, and wheat belts of the Northern Great Plains. Farmers in the Plains are promptly testing UAV models and utilizing them since past 3–4 years (United Soybean Board, 2014). Citrus groves in Florida are being surveyed for nutritional and water status and general health using UAVs. UAVs offer spectral data of orchards that can be analyzed to identify the occurrence and spread of weeds and Huanglongbing disease caused by a virus (Lee et al., 2008; Garcia-Ruiz et al., 2013).

The use of UAVs in general civilian and agricultural zones requires a set of few major regulatory guidelines, rules to be followed, registration of the equipment, pilot training, and license for pilot. Each nation seems to have already formulated regulations. For example, Canadian Aviation Agency has its own rules (Fitzpatrick and Burnett, 2014; Redmond, 2014). The

Federal Aviation Agency (FAA) of the United States is preparing its own well-discussed and detailed set of rules. The regulations are being prepared since 4–5 years. They were expected to be released for UAV technologists to follow, say, by end of 2016 (Krishna, 2018; Talon LPE, 2016; Dorr, 2014; Precision Farming Dealer, 2015).

Historically, the top few rankings of companies dealing with UAV (in general) and related accessories such as computers and hardware were garnered by those supplying military UAVs. However, now, the interest in military UAVs and demand for them has steadied and plateaued. Therefore, exchequer generated by them now seems to have stagnated for the past few decades. However, during past decade civilian and agricultural uses for UAVs have markedly improved. UAV companies producing small agricultural UAVs and related sensors to suit aerial imagery (visual and thermal) and sprayer equipment have taken the lead. The number of agricultural UAV companies has increased enormously. A recent market report states that UAV companies such as DJI of China, Parrot of France, Microdrones Gmbh of Germany, and 3D Robotics Inc., USA, offer a range of small agricultural UAVs. Similarly, Aeryon and SenseFly produce the popular models such as SkyRanger and eBee, respectively. The above few companies and Pix4D Inc. that produces image processing software are among the top 20 UAV and UAV-related companies (Drone Industry Insights, 2016). The report clearly mentions that ranking of UAV companies fluctuates. During recent times, there is a spurt in production and sales of agricultural UAVs. Therefore, companies dealing with agricultural UAVs top the rankings.

The major focus of this volume is on UAV. However, there are also other means and aerial vehicles that allow spectral analysis of crops from vantage points. It is intended to acquaint and discus about these other aerial vehicles, but only to a certain extent. Parachutes, blimps, balloons, and kites have all been used as UAVs. They are good enough to conduct aerial survey of ground features and agricultural belts. Parachutes were designed and tested by medieval intelligentsia. Sketches of parachutes by Leonardo da Vinci (1452–1519) clearly depict the existence of such contraptions (Bellis, 2017). Faust Vrančić, it seems, demonstrated parachute for the first time by jumping from the tower in Venice in 1617. First actual use of parachute was achieved by a Frenchman, Jean-Pierre Blanchard in 1785. It seems parachute was used to escape from disaster-prone balloons. There are art galleries and museums in European cities showing the use of parachutes during recent history, that is, 1700–1900. By 1890, a few improvements in parachute technology such as "fasteners" avoided any mishaps. Parachutes became easily portable since careful and accurate folding methods were devised. During early part of 20th century, parachutes were regularly used to jump from airplanes and to obtain images of ground features (Bellis, 2017). At present, parachutes with facility for inclusion of sensors, CPU, and even a pilot, if it happens to be semiautonomous parachute (glider), are available. Parachutes have been used to survey large patches of natural vegetation, agrarian regions, and individual farms. Parachutes have offered some excellent visual bandwidth data and multispectral signatures of crops. Parachutes have also been used to conduct regular evaluation of field crops in experimental farms. For example, SUSI 62 (Thamm, 2011) and "Pixy motorized parachute" (Lelong et al., 2016) have been adopted to collect data pertaining to crops. Parachutes have also been used to apply plant protection chemicals, provided, the wind interference is low. Parachutes offer one of the best endurance among aerial robots. They can stay afloat for days above the crop fields. They regularly relay data to ground station.

Now, let us consider a few historical facts about balloons, blimps, and kites used in wars for aerial reconnaissance, espionage, and offensive tactics. Let us list historical facts about these aerial contraptions and their role in agricultural farming. Montgolfier brothers of France were among the earliest to experiment with balloon. They were trying to use them as UAVs. These aerostats were used in warfare. These balloons (UAVs) were used by Austrians during their attack on Italian army during 1849 war. They used balloons to drop and detonate explosives exactly at different locations in Venice city (Talon LPE, 2017).

A search for the meaning of the word "blimp" suggested that, a blimp is a kind of airborne vehicle that levitates from the pressure of lifting gas. Blimps do not contain toughened airframe, unlike airships or Zeppelins (e.g., Hindenburg). Regarding the origin of the word blimp, it is said that in the military vernacular these vehicles are termed "Type B limp bags." So, the word blimp was derived out of this phrase. Another explanation states that, in 1915, when these air bag-like vehicles were examined, they made a sound whenever the pilots ran fingers over it. They referred it as blimp. A third explanation states that British had blimps in 1918 and Oxford dictionary traces the usage of word blimp to British military in 1916. The word "Zeppelin" was commonly used to refer to blimps and airships. This name is derived from its inventor Ferdinand Count Zeppelin of France.

Kites made using bamboo and silk were flown by Chinese and Japanese some 2000 years ago. In the Far East, kites have been used as part of religious activities since ancient period. Chinese (e.g., Han Hsin, 200 B.C.) used kites in military conquests. They conducted espionage by sending kites above the fort's walls. Kites were in vogue during medieval period. The British, French, Italian, and Russian armies have adopted kites during World War

I (NASA, 2016). Kites were used by the United States Navy during World War II. Samuel Cody, it seems, first drew attention of British military by crossing the "English Channel." He used kites to drag the boat. The "Cody kite" design is very popular even today (NASA, 2016). It seems, during later years British Army preferred regular blimps (airships).

Reports by kite enthusiasts and agricultural kite researchers point out that "kites" which we intend to deploy above crop fields is indeed a 100-year old technology (Table 1.2). Kites with multiple tails and camera were used to depict the vast destruction caused to San Francisco city due to earthquake of 1906. The famous photograph titled "San Francisco in Ruins" was taken from above. The camera was suspended from Conyne kites. Kite aerial photography (KAP) was done in 1888 by Batut (International Conference on Kite Aerial Photography, 2016). In North America, direct aerial photography was tested in 1839 using kites and balloons (Elsevier Ltd., 1987). A different report states that KAP was tried and practiced in 1880s. Kites with camera cradles were flown to obtain images (Conrad, 2017). A kite club was formed. It was known as "Franklin Kite Club." It mainly included people with interest in applying kites to conduct scientific experiments. The first recorded weather experiments with kites were conducted in 1749 at University of Glasgow in Scotland. William Eddy demonstrated the effectivity of using kites in obtaining weather data above the Boston region in the United States in 1894. During early 1900s, United States Department of Agriculture (USDA) and US Weather Observatory Department have used kites regularly (Millet, 1897; Robinson, 2003; USDA 1898). Kite-aided weather observatories were in vogue in Jutland in 1902, in India in 1905, and in Egypt in 1907.

Kites of different types and specifications are in vogue. They are used for recreation and fun. They are regularly used to obtain aerial images of ground features, particularly, crop fields and farms. There are regular shops that sell kites for aerial photography (Fosset, 2017). Cody kites are occasionally used to obtain aerial images of terrain, land surface features, crops, and other ground features. At present, we have homemade Cody kites and several variations of these box type kites made to specifications. Lutz Treczoks at Lüneburg in Germany since 1998 and Daniel Flintjer since 1983 at Buffalo in New York are examples of kite experts. They have produced Cody kites to conduct aerial photography in the recent times. Modern Cody kite's cost ranges from US$ 25–500. Kite-aided aerial photography should therefore be highly efficient and affordable to even farmers in the subsistence regions. As stated above, kites have offered meteorological data since early 1800s. An important point to note here is that kites are not entirely autonomous. They should be controlled and guided throughout their flight period. They need continuous vigilance.

However, there should be a possibility to develop remote controllers to keep kites in air safely for longer stretches and retrieve them in times of emergency.

1.2 WHAT ARE AGRICULTURAL UAVs

A simple definition would be that "agricultural UAVs" are aircrafts without a human pilot. Perhaps for most people interested in knowing about UAV technology, the word UAV hints about war and the various intelligence, reconnaissance, and destructive activities which these aerial robots accomplish (Jacobsen, 2016). The word UAV is basically a term derived from military jargons and names for their equipment, arsenal, and vehicles. It seems the word "drone" originates from remotely piloted De Havilland Aircrafts that were called "Queen Bees," in the 1930 (Vroegindeweij et al., 2014). At present, UAVs are part of most military establishments. They are also part of local administration. They are used in policing, traffic control, monitoring public events, etc. More recent trends suggest that agricultural UAVs are gaining in popularity in the farming belts of the world. Litchman (2015) states that the word UAV has now found common use with those which deal with small aerial robots and not just the military pursuits. Yet, use of the term "unmanned aerial vehicle" focuses our perceptions more toward civilian and agricultural uses of these machines.

Within the context of this book, an UAV is an unmanned aerial vehicle. It is an aircraft that could be small or big enough, but without a pilot on board. Earliest of the UAV models were regulated using remote controller (joystick), or sometimes using strings. Yet another definition states that UAV is a small or larger aircraft with computer software and decision support system for autonomous takeoff, navigation, and landing at predetermined location. "Unmanned aerial system (UAS)" usually refers to a full complement of UAV system. It includes the aircraft (platform) and ground station accessories such as telelink equipment, computer (iPad), and image processing unit. The word "system" denotes to the entire range of accessories along with platform (Krishna, 2018). In case of agricultural UAVs utilized for aerial sprays, UAS includes pesticide tank, sprayer bar attached with variable-rate nozzles, and computer processing unit (CPU). The CPU processes the digital data and sends commands to regulate variable-rate nozzles. Incidentally, the FAA of the United States prefers and recommends use of the acronym "UAVS," that is, unmanned aerial vehicle system (The UAV, 2015). In some European nations, the words UAV or AUAV (autonomous unmanned aerial vehicles) are used to denote a powered vehicle that does not carry a human operator on board. However, it uses aerodynamic forces to obtain lift and flies (navigates) autonomously, or it is piloted remotely. It is recoverable or expendable and can carry lethal or nonlethal payload

(Office of Secretary of Defence, 2005). We may also note that the word UAV is a misnomer. Agreed that, these aircrafts do not involve pilot (in the cockpit) (Stombaugh, 2016); however, an agricultural UAV is manned via remote control operator or through technicians who prepare flight path, pick images at the ground station (iPad or mobile), and process them, using appropriate computer software.

The list of small UAVs (aircrafts) that are suitable for flying, then, to collect aerial data about civilian and agricultural aspects on ground is large enough. It seems the FAA of the United States that examines and certifies the UAVs was provided with a list of 1085 UAV models. Several of them could be adapted to conduct agricultural tasks such as aerial survey, mapping of ground features and crops, and to spray crops (UAS Exemptions, 2017; Wikipedia, 2016, 2017; UAVGlobal, 2016). Right now, we have detailed information in the public domain for over 400 UAVs that could serve agriculturists. Here, in this book, specifications and details such as their use in farm and nonfarm situations have been listed for about 250 UAV models (Chapters 2, 3, and 4). They belong to different types such as: fixed-winged UAVs, fixed-winged vertical takeoff and landing (VTOL) UAVs, single-rotor helicopter UAVs, multirotor UAVs, parachutes, blimps, balloons, etc.

The theme of the book is to highlight the most recent developments and availability of numerous UAV aircraft models that suit farmers. However, to complete information about agricultural UAVs, in general, related topics such as parachutes, balloons, blimps, and kites have also been dealt upon, but feebly. These other types of aerial vehicles are also useful in obtaining aerial images and spectral data of crops. Most often, they could be semiautonomous or totally autonomous depending on model.

Parachutes have been used as UASs. Such autonomous parachutes fitted with petrol engines and sensors for aerial survey, imagery, and spectral analysis have been evaluated. They are in vogue in some areas of European agrarian region. For example, "SUSI 62" has been employed by German agricultural researchers to obtain aerial images. Such images could be eventually used to mark the "management blocks" necessary during precision farming (Thamm, 2011; Thamm and Judex, 2006). Balloons, blimps, and kites have also been adopted to act as autonomous and semiautonomous UAVs. They collect data about land and soil surface features, crops, and farm activity in progress. Powered parachutes with ability for autonomous navigation have been used to study water resources, vegetation, weather patterns above cropland, and even archaeological sites (Hailey, 2005).

Balloons and blimps could be classified under lighter-than-air (LTA) and tethered systems. LTA system includes balloons and blimps (dirigibles) filled with helium gas or any other LTA mixture. Balloons and blimps are large-sized UAVs. They are cumbersome to transport. They should be left floating for a long duration at a stretch to avoid repeated transport on ground and takeoff. These LTA systems also have constraints related to wind speed, atmospheric conditions (rain, storms, tornadoes, etc.), and flight stabilization (Stombaugh et al., 2016). Agricultural balloons and kites are not common. They are not yet standardized for farmers to adopt them frequently and commonly. These balloons and kites may have problem in avoiding farm features such as erect tall tree, an electric pole, a water tower, undulated or mountainous terrain, etc.

Blimps are also called "UAV airships." They are not made of tough metal frame, internally. But, they are soft and collapsible, if helium or LTA mixture that fills the vehicle is removed. There are innumerable uses attributed to blimps. They include military, civilian, and agricultural uses. Here, we are more interested with applications in surveying natural vegetation and monitoring crop production. One of the lists has 40 different types of uses for blimps. However, out of them, those related to farming are as follows: surveillance and inspection of farms, natural vegetation, encroachments; early warning of disasters such as large-scale drought, floods, disease/pest damage; round the clock farm security; general land survey and marking crop fields and "management blocks," precision farming; crop dusting; monitoring dams and irrigation projects; etc. Solar powered blimps could perpetually stay afloat and conduct aerial imagery and spectral analysis. They relay data on a continuous basis to the ground station computers (Mothership Aeronautics, 2016). Blimps have been used by agricultural experimental stations on a long-duration basis to collect data pertaining to local weather above crop fields, to conduct aerial imagery of crop, to obtain spectral data, to analyze growth (normalized difference vegetative index, NDVI), and to monitor boll maturity (Table 1.2; Plate 1.1) (Associated Press, 2004). At the International Maize and Wheat Center (CIMMYT), researchers are already using blimps to monitor and evaluate the growth traits of wheat. Blimp-derived data is then utilized to rank the large number of genotypes they grow each year/season. The blimp floats above the Norman E. Borlaug Experimental Station at Obregon to collect relevant data about wheat genotypes. It relays the spectral data (Plate 1.1; Table 1.1) (CIMMYT, 2012). In Australia, phenomics research groups have used blimps to collect data about crops grown in farms and experimental stations. They call them "*Phenoblimps*." The blimp hangs over the crop fields constantly and collects relevant data about performance of different genotypes of crops (The University of Adelaide, 2017).

Kites have offered excellent photography of ground surface. Kites are usually fixed with long tail to stabilize it while in flight. Sensors on the tail can provide good aerial pictures. For example, during an International Conference on Kite Aerial Photography, Ruijter (2016) displayed several clear images of agricultural landscape around Rotterdam, Netherlands. He used a 3.5 × 6 Fuji camera placed in the tail to obtain aerial images of farmland. There are indeed innumerable designs of kites. Each type has its specific advantages. For example, box-like and Cody kites can be very effective. They hold a couple of sensors. So, they offer good aerial images. Kites could be an excellent replacement in subsistence farming belts, particularly, if UAV aircrafts are costly. In fact, there are reports that kites could be used to provide aerial images of subsistence farms at very low costs to farmers. Reports suggest that kite-aided aerial photography has been effectively employed to study geological formations and other features on ground (Aber and Galazka, 2000). They flew kites across different regions of Poland. They used several types of kites (e.g., Sutton Flowforms) to suit different weather and wind patterns. They employed automatic Olympus camera to photograph ground features. Kites have been utilized to study soybean crop grown at Arkansas Agricultural Experimental Station, Fayetteville, USA (Miller, 2014). At least three different models of kites have been tested. Kites help to obtain useful atmospheric data such as air and canopy temperature of different genotypes of soybeans (Miller, 2014).

1.3 TYPES OF UAVS

We can consider a broader horizon of uses of UAVs. Then classify them into two major classes such as "military UAVs" and "civilian UAVs." Civilian UAVs include agricultural UAVs (Krishna, 2016, 2018). Military UAVs could be subgrouped into "target and decoy types." One group of military UAVs could be exclusively used for reconnaissance, surveillance, and stealth. Yet another group called "combat UAVs" could be used for offence. They could be used to bomb enemy positions, launch air-to-air or air-to-ground missiles. They are called "air torpedo." In the present context, we are concerned more with civilian UAVs and their usefulness to farming community worldwide.

Most agricultural UAV technologists agree that UAV models flooded into the market are ever increasing. Agricultural UAVs come with variations to suit a wide range of farmers (consumers). Agricultural UAVs are usually classified based on their exact use in the crop field or urban location. They are grouped, say, based on purposes such as aerial mapping, photography, surveillance, transport, sprayers, etc. Yet, we may have to accept that one of the best ways to classify agricultural UAVs is based on the characteristics of the platform and rotors. They are classified as fixed-winged UAVs, fixed-winged "hybrid" VTOL UAVs, single-rotor helicopter UAVs, and multirotor UAVs (Table 1.1). In addition to UAVs, parachutes, balloons, and blimps are also used for surveillance of agricultural regions. They are used to photograph crop fields and map the details on the ground.

Most of the UAVs could be grouped into fixed (flat)-winged UAVs (e.g., *Bramor gEO* by C-Astral Ltd., Slovenia; *eBee* by SenseFly Inc., Cheseaux-sur-Lausanne, Switzerland; *MTD UAV* by TechnoSys Embedded Systems Ltd., Chandigarh, India; *Nauru* by XMobots, Brazil; *UX 5* by Trimble Inc., California, USA; *Delta-M* by Autonomous Aerospace Systems, Krasnoyarsk, Siberia, Russia; and *VYOM 01*, *02*, and *08* by TechnoSys Embedded Systems, Chandigarh, India) or into copter (or rotor) UAVs (e.g., *RMAX* by Yamaha, Japan; *HEF 30* by High Eye Inc., Netherlands, *Vapor-55* by Pulse Aerospace Systems, Kansas, USA; Agras MG-1 by DJI, Shenzhen, China; Hummingbird by Ascending Technologies, Germany; and *Matrice-100* by DJI, Shenzhen, China). Fixed-winged UAVs could be small. They are light-weight machines with ability for rapid flight. But, their endurance could be small. Fixed-winged UAVs could also be large (e.g., Predator, Global Hawk). Fixed-winged types need to be hand launched or catapulted. Otherwise, a runway is to be used to gain liftoff and be airborne. Copter UAVs have single or multiple rotors. They are of VTOL type. Therefore, they do not require runway or catapult. They could be small quadcopters or large helicopters or octocopters. Fixed-winged UAVs with rotors that allow VTOL are available. These are called "fixed-winged VTOL UAVs" or "hybrid UAVs" (e.g., ALTI Transition, Quantrix, NASA GL-10, etc.). Incidentally, GL-10 is a large flat-winged VTOL hybrid. It has eight vertical propellers and horizontal stabilizer has two. It can be used for extended aerial surveys of agrarian regions (Atherton, 2015). Similarly, there are tilt-rotor UAVs developed based on larger standard military UAV models (e.g., tilt-rotor UAV based on V-22 Osprey) (Stone and Crandall, 2016).

TABLE 1.1 Types of UAVs Relevant to Agriculture.

Fixed-winged UAVs: Fixed-winged UAVs are designed just like the commercial airplane with flat wings. The fixed-winged UAVs do not use or require energy to stay afloat in air and overcome gravity. Wings provide the lift. They need energy only for forward thrust. They loiter for longer period (16–20 h). They rapidly fly using predetermined flight path and are supported by propulsion. For transit, flat-winged UAVs use rotors connected to batteries (e.g., eBee, PrecisionHawk's Lancaster, SmartPlane FFRAYE, SB4 Phoenix, SkyWalker CX8; Delta-M) or gasoline engine (e.g., TerraHawk). Fixed-winged UAVs are used for aerial survey and mapping, video imaging, and collecting data pertaining to crop characteristics. General inspection of crop fields can be done rapidly and repeatedly. Advantages stated often are: longer endurance, large area coverage during aerial survey, and higher flight speed. Disadvantages are that their launch may require runway or catapult. It is not a VTOL and hovering is not possible. Recovery of flat-winged UAV may at times need parachutes. Training is required to fly the UAVs and obtain aerial images. They may be expensive.

Flat-winged VTOL "hybrid" UAVs: They are a type of UAVs merging the benefits of fixed-winged UAVs with the ability to hover, just like a multirotor one. Fixed-winged hybrid VTOL may cost US$ 10–55,000 depending on sophistication and electronic accessories. Fixed-wing VTOL UAVs could be used for aerial survey, parcel transport, and crop growth analysis using spectral data. Advantages quoted are that it allows VTOL and runway supported takeoff, whichever is needed. It has long endurance. Disadvantages identified with VTOL hybrids are that they are not perfect in hovering or forward flight. There are relatively few VTOL hybrid small UAVs used in agriculture. But, their usage is expected to increase with their popularity, in the next few years. Training may be required to fly and manage a fixed-winged VTOL hybrid (e.g., TerraHawk VTOL, Quantix VTOL, F2VTOL, and ALTI Transition)

Helicopter UAVs: Single-rotor UAVs are like military and civilian helicopters. Except that they lack a human pilot. Helicopter UAVs can be powered by gasoline engines (e.g., RMAX, Chi-7, HEF-30, and RH2-Stern) or via electric batteries (RH3 Stern, Scout-B1, and Vapor 55). They have a single, big rotor placed above the cockpit. A small rotor at the tip of the tail (fuselage) is used to stabilize, guide, and set the direction of flight. They are guided remotely or by adopting predetermined pathway. Single-rotor helicopters are reportedly more efficient than multirotor UAVs. Helicopter crop sprayers and imagers may cost US$ 25–120,000 per unit. Advantages to consider are that they are robust VTOL and hovering-type UAVs. Endurance of helicopters could be long and extended further, if needed, by changing batteries/fuel tank. Helicopters carry relatively heavier payload. They are excellent for aerial survey, close-up images, and spraying pesticides. Disadvantages are that it needs thorough training to fly a helicopter UAV. Helicopters could be heavy and expensive to purchase and operate.

Multirotor UAVs: Multirotor UAVs are gaining in acceptance in farming regions. They are most frequently adopted to conduct aerial surveys, map the ground details, and to spray the crop with pesticides and liquid fertilizers. Multirotor UAVs may cost US$ 5000–65,000 per unit. Those used exclusively for quick aerial surveys may cost US$ 500–3000. Multirotor UAVs could be subclassified into tricopters (three rotors); quadcopters (Inspire and Matrice-100 by DJI, Shenzhen, China; MD-1000 and MD-3000 by Microdrones, Germany); hexacopter (six rotors) (e.g., Matrice-600 by DJI, Shenzhen, China; Raven by Vulcan UAV, United Kingdom), and octocopter (eight rotors) (e.g., Nemesis 88 by Allied UAVs, California, USA; AG-MRCD24 by HSE, Illinois, USA; AGRAS MG-1 by DJI, Shenzhen, China). At present, many multirotor UAV models are being flooded into market. Therefore, farmers may have easy access and pick suitable model for themselves. Multirotor UAVs are invariably VTOL types. They do not require runway space. They offer good control over the platform and the sensors, particularly cameras that obtain digital data and images. Multirotor UAVs fly close to crop canopy. They offer sharp and accurate aerial images (Tajar and Ahmad, 2013). Disadvantages mentioned are short endurance (20–30 min). Endurance is commensurate with the current battery technology. They may have only limited speed. Therefore, they allow small or moderate payload, restricted to 10–20 kg pesticide. Heavy-lift copters are possible but in exchange, they may have only shortened endurance. Gasoline engines with higher horsepower are not suitable. It is because, multirotor UAVs need high precision throttle changes to keep the UAV stabilized (Chapman, 2015). Hence, multirotor UAVs are restricted to electric batteries for energy source (e.g., DJI Phantom, DJI Inspire).

Sources: BAA Training, 2017; Chapman, 2015; ECA Group-Lima, 2017; Krishna, 2016, 2018; PrecisionHawk, 2014.

Note: Most popular classification of UAV aircrafts is to group them into fixed-winged and rotor UAVs.

TABLE 1.2 Parachutes, Balloons, Blimps, Aerostats and Kites that Are Adopted in Agriculture.

Parachutes: Parachutes have been used to conduct aerial surveys of crop fields and agrarian expanses (e.g., SUSI 62; PIXY) (Thamm, 2011; Thamm and Judex, 2006; Pudelko et al., 2008, 2012). They have been deployed to map the agrarian terrain, land resources, soil types, crop fields, and disease/pest attack if any. Regular aerial spectral analysis of crops can be conducted to assess canopy growth, leaf chlorophyll content, crop N status, water stress index, and grain maturity. Parachutes usually transit relatively slowly over the crop canopy. Usually, they float at relatively low altitude of 100–200 m above crop. So, imagery can be done with greater clarity. Parachutes could be programmed to fly predetermined path. Parachutes are usually energized using a gasoline/diesel engine. They float and keep moving for long duration. The endurance may reach over 1–2 days at a time. High-speed wind may cause them to drift. Hence, it is preferable to use parachute UAVs during the periods when winds are feeble. Parachute UAVs have been used in Germany to map the terrain and mark the "management blocks" for precision agriculture (Thamm, 2011). In Poland, they have been used to assess wheat crop growth, to judge nutrient status, and to decide fertilizer dosages and frequency of applications (Pudelko et al., 2008, 2012). Parachute UAV could also be used to spray foliar fertilizers or plant protection chemicals if wind-induced drift is not a factor. Recently, parachute type UAVs have also been used effectively to map farms and demarcate "management zones" during precision farming (Thamm, 2011; Thamm and Judex, 2006; Pudelko et al., 2008, 2012).

Balloons, blimps, and kites: A large balloon with helium or blimp can also serve as UAV. A modern blimp with its payload can be controlled using remote controller. However, the balloon is generally unstable, if the wind speed is beyond threshold (Yan et al., 2009). Sensors placed on balloons, particularly those useful for spectral analysis of crops have been utilized. They have offered data on seed germination, seedling establishment, crop canopy and growth, leaf chlorophyll, and water status (Inoue et al., 2000). Although not related directly to crops, balloons and kites have been adopted to study geological features, glacial movement, natural vegetation, and water resources (Boike and Yoshikawa, 2003). Blimps and kites have been utilized to conduct remote sensing of ground features by mounting suitable sensors on them. For example, blimps and kites have been used to prepare two-dimensional (2D) or three-dimensional (3D) maps of ground features and crops. The slow-moving blimps are known to provide good aerial images of surface soil erosion, gully formation, and loss of soil fertility (D'Oliere-Oltmans et al., 2012). Blimps are used to monitor wheat/maize genotypes and to collect phenomics data about thousands of germplasm lines and advance/elite lines (Plate 1.1).

There are several types of kites. They serve purposes such as fun flying, meteorological data collection, aerial imagery, etc. Kites have been made using paper, plastic, organic composites, textiles such as nylon, and plastic films. Kites are classified using a range of different characteristics. There are kites flown by multiple pilots. Multiple units of kites are also used (e.g., Cody, Conyne, etc.). A detailed knowledge about kites, their material, applications, and costs are available (Wikipedia, 2016).

Sources: Boike and Yoshikawa, 2003; Inoue et al., 2000; NASA, 2016; Pudelko et al., 2008, 2012; Ries and Marzolff, 2003; Thamm, 2011; Thamm and Judex, 2008; Wikipedia, 2016; Yan et al., 2009.

PLATE 1.1 A "blimp" at International Maize and Wheat Center's (CIMMYT) Experimental Station at Ciudad Obregon, Mexico, scouting the fields.
Source: Dr. Clyde Beaver, International Maize and Wheat Center (CIMMYT) Archives, Mexico.
Note: Soon, blimps with a range of sensors could become common above canopies of wheat/maize. In agricultural experimental stations, blimps could assess genotypes for growth, phenomics, grain yield, drought tolerance, cold tolerance, pest/disease resistance, etc. They are supposed to reduce the need for trained labor for manually scouting and maintaining crop yield data. *The process of agricultural experimentation itself could become easier and cheaper, if blimps are adopted.*

1.3.1 AGRICULTURAL UAVS AND THEIR CLASSIFICATION, USING DIFFERENT CRITERIA

Reports by sales staff of a popular UAV company state that basic classification used by consumers depend on how far the UAVs can fly from the ground station and for how long they stay in flight, that is, endurance. UAV types with ability for transiting 15 km or more and ability to stay afloat for lengthy periods are available. They are usually the military UAVs. In the realm of agriculture, fixed-winged or rotors with ability for endurance of 45–60 min are preferred. Short-range flight and capability for aerial imagery with high-resolution cameras are preferred characters. Such traits are enough to group them as agricultural UAVs (PrecisionHawk, 2014).

Based on the size of the UAVs, they could be classified into very small UAV (including micro or nano UAVs and mini UAVs), small-sized UAVs, medium-sized UAVs, and large UAVs.

- "Small UAVs" have a maximum takeoff weight of 20 lbs and altitude ceiling is <1200 ft above ground level. They may reach a speed of <100 knots. Small UAVs are usually hand launched or via catapult. Many of them come under the classification of fixed-winged UAVs. The small UAVs are most suited to carry out aerial survey of crops and obtain NDVI and crop growth status data. They are also used for spying and biological warfare. Examples for small UAVs are RQ-11 Raven by AeroVironment Inc.; RS-16 by American Aerospace; eBee by SenseFly, Switzerland; UX5 by Trimble Inc., California, USA.
- "Medium UAVs" have a maximum takeoff weight of 21–55 lbs. They are used when longer endurance and high-altitude flight is needed, say, to assess large agroecosystems, districts, or counties. The altitude ceiling is 3500 ft above ground level. They can reach only <250 knots. Examples of medium-sized UAVs are Eagle Eye by Boeing Aerospace; SkyEye R4E by BAE Systems; RQ-2 Pioneer, RS-20 by American Aerospace; etc.
- "Large UAV" has maximum takeoff weight of <1350 lbs. Its altitude ceiling is 18,500 ft above mean sea level and air speed attainable is <250 knots. Examples of large UAVs are Predator A and B and Northrop Grumman's Global Hawk.
- "Larger UAV" has a maximum takeoff weight of >1350 lbs. Its altitude ceiling is 18,000 ft above mean sea level and air speed attainable is about 200 knots.

- "Very large UAV" has a maximum takeoff weight of >1350 lbs. It flies at over 18,000 ft above mean sea level. Such UAVs perform long-distance travel and reach greater than 300 knots. The large and very large UAVs have found use in long-range transit, reconnaissance, and identification of targets (Arjomandi, 2013; Krishna, 2016).

We should note that, there are innumerable models produced that belong to each size-based class. During recent years, emphasis on small UAVs useful in agricultural farm is gaining. However, other size-based classes fit for defense- and transport-related activity are also in production.

UAVs have a range of characters related to platform (body) and its material, accessories attached, and ground control equipment. Equally so, there are large number of performance-related characters that could be utilized to classify UAVs. Arjomandi (2013) has made detailed effort to collect information about performance-related traits of an UAV. He has used them to class UAVs. They are weight, endurance, altitude, wing loading, engine type, and energy source. Let us now discuss the UAV's character, the terminologies used, and its meaning:

Weight: UAV's weight is an important trait. To a certain extent, it depends on the material used to construct the platform and payload. UAV's weight is easily a very good trait to classify the various models produced. Superheavy UAVs weigh over 2 t. They are mostly long-range UAVs. They hold high loads of fuel commensurate with endurance. They fly at high altitudes and are now operated for military reconnaissance and bombing. The "superheavy-weight UAVs" are retrievable and offer repeated use. They are used in lieu of missiles that are not retrievable (e.g., Global Hawk, Predator). Next class based on weight could be "heavy-weight UAVs." These UAVs weigh between 200–2000 kg per unit (e.g., Fire Scout). UAVs with a takeoff weight of 50–200 kg are classified as "medium-weight UAVs." They hold optimum levels of fuel and transit relatively longer distances compared to small UAVs. "Light-weight UAVs" are used to obtain aerial images of crops. They are more common in agrarian belts. They weigh between 5 and 50 kg. The RMAX copter belongs to this class of UAVs. It is currently among most preferred UAVs for aerial sprays and surveillance of standing crops. Other examples are Yintong's helicopter sprayer UAVs. UAVs that are smaller than one's described above are "micro UAVs." Micro UAVs are used to accomplish aerial survey of crops. They collect relevant and wide-ranging data about crop growth and grain formation trends. Micro UAVs weigh less than 5 kg. They are usually packed and carried in a suitcase or backpack.

Therefore, they are easily launchable from any point in a crop field (e.g., PrecisionHawk's Lancaster, Trimble's UX5, SenseFly's eBee).

UAV's attainable altitude: Here, altitude means the height above ground or crop canopy that an agricultural UAV takes while conducting aerial survey. UAVs, as such, are capable of flight at different altitudes. Based on UAV's stipulated altitude, they are classified into low-, medium-, and high-altitude UAVs. Low-altitude UAVs fly just a few meters above the crop canopy, say, 100–1000 m above crop canopy. Majority of agricultural UAVs are low- or medium-altitude types. Medium-altitude UAVs fly at 1000–5000 m above the crop canopy. They are used in aerial survey of large patches of crop land, natural vegetation, and riverine zones (Krishna, 2016, 2018). They are useful in collecting spectral data (near infrared, NIR; and thermal images) for entire county or district. High-altitude UAVs are used in military reconnaissance and when stealth and height are required. These UAVs lose height only to fix target and release bombs or missiles. High-altitude UAVs are used more frequently for aerial survey of county, district, or state's agrarian regions. They are also used to detect large-scale drought or flood-related devastation. In fact, altitude is a crucial characteristic for farm UAVs because of several reasons. First, the sharpness and resolution of images and area covered by cameras depend on the altitude. Close-up images are best when UAVs fly close to crop's canopy and at very low altitude. Higher range of altitude can make imagery hazy, low in resolution, and interference by cloud could also creep in. In case of a sprayer UAVs, they should invariably keep a close and low altitude. Otherwise, the pesticide may drift in the atmosphere. Low altitude allows accurate distribution of pesticides/fluid fertilizers during precision farming (Krishna, 2018).

UAV's endurance: Endurance of an UAV denotes its ability to stay afloat for a length of time and fly a distance, without refuel. Endurance largely depends on few characteristics of UAVs such as fuel/battery (its power storage capacity) or gasoline that fuel tank holds; and rate of fuel consumption. Endurance is perhaps a key trait that decides the period for which UAVs can keep themselves afloat and transit longer distance. This trait helps UAV to carry out aerial imagery, surveillance, or spray pesticides for longer period. UAVs are classified as those with long, medium and low endurance, depending on length of time they stay in flight. "Long-endurance UAVs" are common with military. Such UAVs keep afloat for over 24–48 h and cover 1500–20,000 km distance (e.g., Global Hawk, Predator, ScanEagle). "Medium-endurance UAVs" operate in flight for 5–24 h without refuel. They too travel long distance and conduct surveillance. They can be used when distance to be covered by the UAVs is around 100 km. UAVs used to fly over crop fields to collect precious spectral data and keep watch on crop growth rate are usually small and with short endurance period. The endurance of a fixed-winged UAV used for aerial survey of crops is about 15–45 min. The copters most frequently used to spray pesticides or fluid fertilizer also have a low endurance (usually 15–20 min) (Chapter 3). Agricultural UAVs have a short endurance. Therefore, they are used efficiently by predetermining the flight path. Even then, flight path is restricted to 50 km from the takeoff point (The UAV, 2015; Krishna, 2016, 2018).

UAV's range: Based on the range that UAVs transit, they are categorized into very close-range, close-range, short-range, and medium-range UAVs.

Very close-range UAVs: They are most frequently encountered in agrarian regions. They are used for aerial survey. They possess a range of 5 km from ground controls station, with an endurance time of 20–25 min per flight. The cost to procure a very close-range UAV is not high. It is most often affordable for a farmer or a crop production company. It costs about US$ 7–10,000. There are very close-range UAV models assembled at home and made of indigenous material and parts. They may cost the UAV engineers just about US$ 2–5,000 per unit.

Close-range UAVs: These UAVs maintain telelink with ground station for 50 km radius. They fly out to conduct aerial survey within that radius. The endurance ranges from 1 to 6 h. They are used for aerial surveys, reconnaissance, and regular surveillance of installations, farms, etc.

Short-range UAVs: These UAVs have a range of 150 km for flight and aerial survey. Their endurance is longer and extends to 8–12 h. Short-range UAVs are used for surveillance of farms, installations, crop fields, dams, irrigation systems, etc.

Medium-range UAVs: These are usually high-speed UAVs that loiter at high altitudes of 16,000 ft and transit at a speed of 650 km. They are used for military surveillance, target identification, and bombing missions. They are not common with farmers in agrarian regions. State agricultural agencies may, however, use them to obtain rapid data about large agroecosystems, riverine zones, and natural vegetation.

Wing loading: Wing loading is a technical term related to total weight of the UAV and area of the wings. Wing loading is a useful character to classify the UAVs. It is a value derived by dividing total weight of UAV by area of wings. Wing loading value for most agricultural UAVs ranges around 5 kg m^{-2}. They are low-wing load UAVs. UAVs with >100 kg m^{-2} are classified as "high-wing loading UAVs" (Global Hawk). Those with 50–100 kg m^{-2} are termed as UAVs with "medium wing loading" (e.g., Fire Scout, X-45).

Those with wing loading below 50 kg m^{-2} are termed as "low-wing loading UAVs" (Arjomandi, 2013, Krishna, 2016, 2018).

Engine and energy source: We may realize that UAV's engine and its source of energy (e.g., gasoline or electric batteries) are very important characters. Farmers may encounter UAVs with engine types such as turbo prop, turbo fans, two-stroke piston engine, push and pull type, or electric battery-powered motors connected to propellers. This criterion helps in the classification of UAVs. Farmers need UAVs of different endurance, fuel efficiency, and takeoff weight. First, they try to know the horsepower of the engine and its fuel source. Engines with better fuel efficiency are preferred. Those with low gasoline or battery power usage per unit time will show better endurance. Usually, UAVs with small- or medium-sized batteries (16,000–24,000 mAh) are preferred in farming sector. They are short-endurance UAVs. They fly rapidly past the crop and collect useful data or unload pesticides rapidly within 15–20 min.

We can also classify UAVs based on a couple or a combination of their characters. It is mostly based on importance bestowed to characters and the purpose. For example, Gogarty and Robinson (2012) have stated that during recent times, altitude and endurance are considered together. Particularly, when it comes to launching UAVs to do surveillance, attack (bombing), and destroy (bombing) long-range military targets. Such UAVs are grouped as HALE. It means high-altitude, long-endurance UAV. UAVs are grouped as medium altitude low endurance (MALE), when they aim at targets of medium range but possess long endurance.

There are UAVs that need periodic replacements of accessories, computer chips, propellers, etc. There are others with rugged bodies, toughened propellers and parts that last for longer durations. So, wear and tear and replacement of parts could be an important trait. Farmers may have to consider it while purchasing UAVs. Sometime, in future, we may also develop "use and throw" UAVs that are made of low-cost material (Krishna, 2018). They serve excellently when farmers want to do surveillance of their large fields rapidly, obtain images instantaneously, and monitor crop's condition. After their use, we are done with it. We do not retrieve the low-cost UAV. Such UAVs perhaps will be made of cardboard (degradable) and very low-cost sensors. We could then have two groups of agricultural UAVs, that is, "use and throw types" and "long reusable and durable types."

Based on takeoff traits, agricultural UAVs could be classified into "short runway" requiring UAVs. Such models possess undercarriage with wheels. UAVs can be launched by keeping them above a moving truck, pick-up van, or a transport vehicle. UAVs are also launched using a catapult. UAVs could be hybrids possessing moveable rotors. They could be used in vertical format for VTOL. There are regular VTOL UAVs, such as helicopters and multirotor UAVs. A vertical takeoff UAV is preferred when using copter to pick pesticides and spray it on a crop. They fly on a crop at slow speed and at low altitudes. Sometimes, farmers may even prefer hovering multirotor UAVs to spray larger quantities of pesticide. Short-runway (1000 m), rapid-speed (45–60 km/h) flat-winged UAVs are preferred during surveillance of crop fields, surveying for disease/pest or drought or flood affected regions, landslides and soil erosion, etc. Flat-winged UAVs launched by hand are perhaps most commonly used by farmers. Particularly, those who wish to collect data, rapidly.

In farming, like any other enterprise, cost of inputs and implements are thoroughly weighed out against the accrued profits. *UAV's cost and its impact on farmer's profit is an important aspect. The quantum of increase in exchequer is thoroughly analyzed prior to purchasing and adopting an UAV. Most importantly, agricultural UAVs could be grouped based on cost of the equipment (platform), its entire accessories, and the purpose it serves to the farmers. Depending on each task, it is most important to characterize UAVs into economically efficient or less efficient. Computer software that accepts a range of data about each UAV model, simulates, compares, and offers forecasts is necessary.* Farmers should invariably consider economic returns from different UAV models. Firstly, we need a ready reckoner or compendium of UAV models. It should list agricultural tasks conducted efficiently by each UAV model (Chapters 2, 3, and 4). The pool of UAV models available to farmer is ever increasing.

1.3.1.1 SPECIFICATIONS OF UAVS

A compendium that lists various UAV models available in the market and their specification is any day useful for farmers. They can consult it prior to procuring an UAV model. The specifications or criteria for selection of UAV model may vary with the purpose. The list of specifications commonly considered prior to using UAVs are not many. Most frequently used criteria (specifications) to assess suitability of UAV models for civilian and agricultural purposes are as follows:

(1) size of the UAV; (2) cost of the UAV; (3) payload capacity is important because it decides the number of sensors, fuel, and any other commodity or parcel the UAV can carry; (4) vehicle control using remote controller or

via preprogrammed flight path; (5) distance of operation from the remote controller or ground control station (iPad); (6) flight duration:endurance affects the type of aerial survey and sprays that the UAVs can conduct; (7) speed in the air affects the extent of area covered by the UAV. Speed, to a certain extent, decides amount of pesticides released per unit time, if it is a sprayer UAV; (8) fuel type (battery or gasoline) and consumption rate; (9) launch and recovery methods; (10) sensors and their capabilities—it ranges from visual, multispectral, thermal, and red edge bandwidth imagery; (11) sensor size—sensors could be as large as 20 kg if the UAV is large, or about 2 kg or just couple of hundred grams; (12) video resolution and image stabilization is key to clarity of aerial images; and (13) digital data that has to be utilized during precision farming.

The list of specifications that need to be considered while purchasing and using an UAV in a farm situation has been dealt, in greater detail (LGL Ltd., 2009; Krishna, 2016, 2018; eBee, 2017; Trimble Inc., 2017; PrecisionHawk, 2014; DJI, 2016). Also, it forms the focal point of this treatise in Chapters 2, 3, and 4. The list of specifications to be considered varies with the type of UAV. For example, fixed-winged UAVs meant exclusively to obtain aerial images, monitor ground features, mining activity, or traffic vehicles, or patrolling pipelines may have to be examined, for correctness of sensor accessories and other specifications. A copter UAV meant to spray pesticides needs to be checked for specifications related to payload (pesticide) tank, its capacity, nozzles, variable-rate techniques (software), and cost of the UAV equipment itself. Specifications related to aerial imagery and close-up shots may get greater attention, if the UAV is to be used for detection of diseases, pests, or weed flora and obtain close-up view of the ground situation.

Agricultural UAVs may gain further specialization regarding functions they perform from above the crop fields. UAV models and whole packages meant for a single or set of tasks may attract attention. UAV meant exclusively for rapid aerial survey of crops, or those exclusively meant to pick 2D and 3D images and obtain maps, to form "management blocks" may be available in future. Already, we know that, fixed-winged UAVs such as Precision Hawk's Lancaster or UX5 by Trimble Inc. are used predominantly to collect aerial data about crop growth, phenomics, leaf N status, water stress index, and grain maturity status (Perry et al., 2012a, 2012b). Next, the small, hand-launched, briskly flying *eBee* UAV by SenseFly Inc. is used more often to scout the crop and collect aerial images. This is used to assess seedling establishment, crop stand, phenomics, chlorophyll, and water status (SenseFly, 2016). So, we can classify these models as "aerial survey UAVs" or "mapping UAVs" or "phenomics UAVs." We have a couple of single-rotor helicopters such as RMAX, Hercules, or Yintong's U-5 (Yamaha, 2014; Yintong Aviation Supplies Company, 2012). These are predominantly used to spray the crop with fertilizers/pesticides. They can be grouped under "sprayer UAVs." Similarly, there are several multirotor sprayer UAVs (Chapter 3). UAVs to detect disease/pests based on spectral properties of healthy and disease/pest-afflicted crops are already a possibility. There are UAVs exclusively meant to spray pesticides, fungicides, or liquid fertilizer formulation (see Chapter 3 for details). There are UAVs meant to monitor farm activity on the ground and keep track of the progress of agronomic procedures such as plowing, seeding, interculture, plant protection sprays, and harvesting. Therefore, soon we may like to identify and classify small agricultural UAVs into "aerial survey UAVs," "precision farming UAVs," "sprayer UAVs" or "plant protection UAVs," "weed control UAVs," "irrigation UAVs," etc. Such a classification is based on specific tasks that the UAV models perform, of course, with better accuracy and efficiency. Such a classification of UAVs is highly recommended. It is useful to farmers and farming companies. Particularly, those who wish to buy most appropriate UAV model. It identifies the UAV with agricultural tasks better.

Land and soil management UAVs: These UAVs are manufactured specifically to conduct aerial survey of terrain, land resources, soil type, etc. (Santaana, 2016). The UAV offers 2D and 3D maps of surface features of land, soil color (organic matter), and textural classes. The images are used to demark the land into "management blocks" (Krishna, 2012; Thamm, 2011). A few good examples of small UAVs used in aerial survey of land in the agrarian belts are: SenseFly's *eBee*, Trimble's UX5, PrecisionHawk's Lancaster, DJI's Inspire Phantom and Delta-M). These are small UAVs. They are hand launched. They takeoff and fly quickly at 45–60 km/h. They obtain aerial imagery of land and soil surface features. They offer data about soil moisture status, need for plowing, and detect patches with soil maladies such as erosion, flooding, salinity, etc. Most importantly, the UAV's imagery of cropland is used to demarcate the "management zones." These UAVs could be used to monitor and take note of the progress of swarms of tractors (disc plows). UAVs monitor "autonomous tractors (driverless)" that are used to plow the land. They can also be used to monitor ridging, seeding, and seedling establishment.

Agronomy and crop husbandry UAVs: UAVs are expected to guide, monitor, and offer data for decision support systems relevant to crop husbandry operations. A few different UAV models, each suited to conduct an agronomic procedure may get deployed (Krishna, 2018). UAV models that are versatile and aid in the conduct of a series of agronomic procedures could be useful. UAVs could monitor agronomic operations conducted to raise a crop from seeding, seedling, and canopy establishment and until crop's maturity. So, we can opt for an "agronomy UAV," that is, a single model capable of several tasks relevant to crop husbandry. Alternatively, we can purchase a set of few different UAV models. A fixed-wing and a couple of quadcopters each geared up to conduct specific agronomic procedures. We can evaluate a series of UAV models and arrive at best suited ones for each agronomic procedure. They could be named accordingly (e.g., "irrigation UAVs," "plant protection UAVs," "weed control UAVs," "yield forecast and harvest monitoring UAV," etc.). Basically, the platform, set of sensors, and software to conduct the analysis and arrive at proper decisions are important. It will decide which model gets preferred for a specific agronomic procedure (Krishna, 2018). For example, an UAV with ability to obtain spectral data of crops, weeds, and locate weeds in interrow spaces could be selected. It could be fitted with computer software for identification of weed species, its intensity, and location (object based image analysis, OBIA). It will automatically be termed as "weed-detection UAV," "weed hunters," or "weed tracers." Add to it, if the UAV model is a copter with facility for herbicide spray and variable-rate nozzles then it is a full fledge "weed control UAV." We can think of "crop harvest UAVs." They would make aerial survey and detect panicle and grain maturity status, monitor the combine harvesters, and provide schedules for other harvest-related transport vehicles. A compendium on UAVs, in future, may have to identify which UAVs are termed what. It could be based on their specific functions during crop husbandry. Right now, we have already adopted the name "sprayer UAVs" when the single-rotor or multirotor UAVs are used exclusively to spray pesticides/herbicides or fertilizer formulations (Chapter 3).

Agricultural experimental station UAVs: These are essentially small, hand-launched, auto-takeoff UAVs. They collect data pertaining to crop growth, phenomics, leaf nutrient, and water status. They capture data at various stages of crop growth (Plates 1.2 and 1.3) (Dreiling, 2012; Knoth and Prinz, 2013; Krishna, 2016, 2018; Lumpkin, 2012; Mortimer, 2013; Tattaris and Reynolds, 2015; Perry et al., 2012a, 2012b). They are also used to keep an aerial watch on the happenings in the experimental station. They are used to prepare planting plans, fertilizer inoculation, spraying schedules, and also design irrigation and drainage. Since, crop breeders and agronomists could use such UAVs for routine evaluation of crops, we can also call them "crop breeding and evaluation UAVs." Agricultural experimental stations in the United States, in Mexico, in Zimbabwe, and at Rothamsted Experimental Station in England have utilized UAVs to evaluate wheat genotypes (Case, 2013; Farming Online, 2013). CIMMYT at El Baton in Mexico and its outreach stations in Zimbabwe are adopting UAVs and blimps to evaluate wheat genotypes (Plates 1.1–1.3). Specialized UAVs are used to assess drought tolerance of sorghum genotypes at Kearney Agricultural Experimental Station, California, USA (Vaan der Staay, 2017). They rapidly collect data and relay images relevant to drought tolerance of sorghum genotypes. We can evaluate thousands of sorghum genotypes for drought tolerance, in one go. UAVs collect data about them in a matter few minutes. We can also use a concept called "crop surface models" to assess canopy and growth traits of crop genotypes (Bendig et al., 2013a, 2013b, 2015). Overall, "experimental station UAVs" could be set of models that are swift, equipped with sensors for collecting wide array of data about crops and relay them to ground computers. In an agricultural experimental station, they help in assessing germplasm and elite lines.

PLATE 1.2 A fixed-winged UAV, *eBee*, being launched at International Center for Maize and Wheat (CIMMYT)'s Norman E. Borlaug Experimental Station at Obregon, Mexico.
Source: Dr. Clyde Beaver, Creative Service Manager, International Centre for Maize and Wheat, El Batan, Mexico.
Photo credit: Dr. Alphonso Cortes, CIMMYT, El Batan, Mexico.
Note: These fixed-winged UAVs swiftly scout and collect data. Such data helps to assess the performance of genotypes of wheat/maize. They could be easily called *experimental station UAVs.*

Now, there are already "vineyard UAVs" or some may call them "grapevine UAVs." Then, we have "citrus UAVs," etc. They are named so because they are equipped exactly to conduct aerial survey and obtain accurate data about the specific plantation crop. These UAVs assess crop canopies, foliage, and fruits. They relay the digital data to ground computer to process it. The aerial data could also be processed using in situ CPU that has necessary software to stitch and analyze the images immediately (e.g., PIX_4D). Such "plantation UAVs" conduct specialized tasks of assessing canopy, foliage, estimating nutrient status particularly, crop N status, and help in prescribing fertilizer N. They assess crop water stress index accurately and prescribe irrigation. The UAV machine (i.e., platform), computer software, processing unit, and iPads are all custom made. They are to be used on a specific crop. So, in future we may encounter UAVs that are specific to a crop species (e.g., grapevines, citrus, apple, wheat, rice, etc.) or a set of crops (e.g., cereals). Such specialized UAVs may work less efficiently, if used on other crops.

PLATE 1.3 A quadcopter above wheat genotypes.
Source: Dr. Clyde Beaver, International Centre for Maize and Wheat (CIMMYT), El Baton, Mexico.
Note: A quadcopter is being utilized to obtain aerial imagery and multispectral and thermal data of wheat genotypes grown at International Centre for Maize and Wheat (CIMMYT)'s Norman E. Borlaug Experimental Station, Obregon, Mexico
Photo credit: Dr. Peter Lowe, CIMMYT, El Batan, Mexico.

A step further, UAV technology and crop production methods may coevolve and get refined to a very great extent. Then, we may encounter a highly specialized way of treating crops using UAV technology. Such situations have been encountered with other farm vehicles. The guess is that, in future, farmers may have to buy a set of two or three UAVs and specific accessories and computer software. We can imagine a set of UAVs for rice production. It could include a small, flat-winged micro UAV (e.g., *eBee* or UX-5). Then, we may fly a different UAV model, midway in the crop season, to collect crop data in greater detail, using appropriate sensors. During seedling stage and until panicle development we may have to fly a copter UAV that flies low over the crop canopy and sprays few different pesticide formulations, weedicides, and fertilizer formulations as per prescriptions. We may again need a small UAV to rapidly assess panicle maturity and identify fields that are to be harvested. Overall, a package of UAV models for each crop species could be formulated and sold. In fact, they say in North America there are farm companies that own a series of UAVs. They are meant to perform specific tasks during a crop season.

In addition to UAVs and their models, there are elaborately and accurately prepared lists of UAV companies to watch and select. This aspect helps farmers and farming companies or those in other professions (e.g., military or transport) to select the most appropriate UAV company from where to buy the UAV. We ought to realize that UAV companies fluctuate regarding their popularity. The demand for their UAV models may also fluctuate. It is based on suitability of UAV model, usefulness, cost, ease of operation, etc. In 2016, about 70 UAV companies were short listed based on the various aspects of UAVs that they produce. The demand created for UAV types and models depends on aspects, for example, agriculture, transport, military, etc. (Perlman, 2016).

1.4 GENERAL USAGE OF UAVS: A FEW SALIENT EXAMPLES

Regarding excessive use of UAV surveillance, bombing, and missile attacks in the Middle East, Afghanistan, and Pakistan (Stanford Law School International Human Rights Group, 2012), Litchman (2015) points to a report that describes about the plight of civilians. He uses the phrase "living under fear of UAVs and their attacks." This information pertains to military use of UAVs. In this volume, we are concerned more with agricultural uses of UAVs. Now think of agricultural UAVs, their current popularity ratings, utility in farmland worldwide, reduction in farm/field drudgery and labor costs, easier crop scouting and spraying, and the economic impact. We should now use the phrase "agriculture under UAVs." After all, global agriculture under UAVs may be proportionately less tiring, more efficient, high yielding, and perhaps more profitable. We still need to conduct several field trials to prove UAV's profitability in agriculture. The shift in emphasis of UAV technology from military pursuits to food generation and reduction in farm drudgery seems real and imminent.

Jacobsen (2016) states that UAV-aided delivery of small packet-sized cargo is attracting attention of the civilian populations. Also, of those dealing with humanitarian situations (e.g., MatterNet, California, USA). For example, UAVs have been used to send out food packages, collect blood sample and deliver it to hospitals, and carry postal mail and parcels. However, there are bottlenecks. They are: endurance of small UAV, safety of these aerial robots, standardization of safe journey routes, landing sites, etc. Yet, UAVs may find acceptance because UAV-aided delivery seems to cost just US$ 1–6 per unit. It overcomes traffic problems and lands with parcels, direct at the destination.

Small and medium-sized UAVs with long endurance have been deployed to assess marine conditions off the coastline. Usually UAVs with ability to traverse at least 2000 km are selected. However, if the flight endurance is smaller, then UAV traverses only less than 200 km. In such a case, they are launched from a sea boat or streamer or a ship. For example, UAVs such as *Insight-20* (also called ScanEagle) produced by InSitu Group, USA; *Manta B* (Silver Fox) by Advanced Ceramics Research Inc, Tucson, Arizona, USA; *Fulmar* by Aerovision Vehiculos Aereos, S.L. (Sebastian, Spain); and *ZALA 421-16* by Aerosystems, Izhevsk, Russia are used to survey sea surface, for biological resources and weather parameters (LGL Ltd., 2009). There are indeed several other UAV models used to study sea, mountain ranges, marshland, volcanoes, etc. Again, a detailed knowledge about different UAVs and their utility for specific tasks in different geographical conditions is worthwhile. There is indeed a long list of humanitarian situations, such as disasters, floods, earthquakes, drought, famine, need for medical supplies, transport of food, etc. UAVs may accomplish many of these tasks with flying colors! (Krishna, 2018). UAV's usage in humanitarian and agrarian aspects may outstrip military needs (Litchman, 2015).

Small UAVs with appropriate sensor (e.g., Cannon 450D or Sony Alpha 7) and a flight endurance of 1–2 h has been utilized to detect the extent of landslides and crop loss (Rau et al., 2011; Chen et al., 2015). The UAV model used for aerial assessment, its sensors, and digital data processors are all important. Particularly, while conducting aerial surveys of land, detection of landslides and erosion if any. The UAV model and its ability to reach a stipulated altitude is important (Anders, 2014).

UAVs are versatile and there is indeed a long list of utilities attributed to them (Gogarty and Robinson, 2012; Frey, 2015). They have counted about 192 different uses for UAVs. Let us list a few uses. They are: (1) early warning systems; (2) emergency services; (3) news reporting; (4) delivery UAVs; (5) business activity monitoring; (6) gaming and entertainment; (7) farming and agriculture; (8) ranching UAV; (9) police UAVs; (10) real estate UAVs; (11) library UAVs; (12) military and spy UAVs; (13) educational UAVs; (14) healthcare UAVs; (15) travel and delivery UAVs; (16) sanctuary UAVs; and (17) lightning UAVs.

A compendium of UAV may incorporate details on uses of UAVs. In the present context, we are interested in only agricultural UAVs. In other words, UAVs that have utility in agricultural farms. Soon, UAVs may become as common as tractors. They could be aiding the conduct of different procedures, while flying in the atmosphere above the crop canopy. They are bound to find an excellent niche in the region closely above crop canopy. Farmers, it seems, have a grand list of several uses for these UAVs. These UAVs may conduct farm operations almost completely autonomously, if properly programmed. UAVs could also be guided, using remote controllers and ground station computers (iPad). Incidentally, a recent report by an UAV company clearly states that initially, UAVs with remote controllers were preferred. However, currently, the trend even with farm experts and growers is to procure UAV with iPad and software. We can then predetermine the flight path, way points, and camera shutter activity (PrecisionHawk, 2014). There are UAVs that specifically suit to conduct a farm task or group of similar tasks. For example, small flat-winged UAV may conduct aerial survey, surveillance of crops, map the crop canopy and provide NDVI data, measure water stress index, etc. There may be others, say, a few models of copters that are excellently suited to provide close-up shots of crops, and to detect pests/diseases. We may also find new added uses for UAVs in the realm of agricultural crop production. This aspect depends on research and innovation. Agricultural uses of UAVs may also depend on the region and crop species that farmers cultivate.

Each UAV company, currently, seems to advertise their UAV models by providing a list of agricultural usages. The list of agricultural uses provided by one such important UAV producing company is as follows: crop survey, crop growth monitoring, crop inventory tracing, nutrient deficiency detection and fertilizer management (aerial spray of liquid fertilizers), crop moisture status monitoring, disease and pest detection, aerial application of pesticides or fungicides, disease outbreak tracking, detection of soil erosion, flooding and drought affected zones, detection of weed infestation using spectral analysis, monitoring panicle/grain maturity, yield forecast and timing the harvest, farm asset inventory tracking, and farm surveillance (HSE, 2017).

Next, a different UAV company emphasizes on UAVs that make commercial assessments of crop fields. They provide information to insurance companies about flood, drought, soil erosion, and crop loss assessments. UAVs with fully automated and predetermined flight path are preferred. It helps to obtain aerial imagery and 2D/3D maps. Preparing 3D maps of agricultural fields is an important aspect. UAVs are among the main sources of such detailed maps (Nex and Remondino, 2013). This UAV company also stresses on accomplishing tasks such as: detecting weather-related damage to crops (e.g., cold front damage), need for fertilizer and herbicide usage, monitoring livestock and pasture, survey of irrigation structures, etc.

Many of the UAV companies producing small agricultural UAVs offer a list of uses. A few of them could seem somewhat similar. Yet, each UAV company emphasizes a certain set of uses. For example, a fixed-winged UAV producer company such as PrecisionHawk or *eBee* may emphasize rapid aerial imagery, crop scouting for canopy growth (NDVI), and phenomics, offering spectral maps and calculating need (quantum) for fertilizer or pesticides as their main uses (Shi et al., 2016; PrecisionHawk, 2014, 2017; SenseFly, 2016). While a company that predominantly produces copters may emphasize on spraying pesticides as the major usage. But, they may still quote several other uses as routinely conducted functions. For example, Yamaha's RMAX or Yintong's helicopter UAVs are suited best for aerial spraying. Microdrones Gmbh is a German UAV company. It manufactures a few popular brands of quadcopters of great utility in agriculture. They quote a range of important improvements and specialties related to the models. Some of the attractive specifications are longer endurance, weather resistant body, applications both in farm and industrial regions, low operating height, storage of flight and aerial data, and several more way points than many other models (Jung-Diveky, 2016). Such facts are important, because, when farmers purchase UAVs, they may find a certain model more efficient. They could be less costly to accomplish a certain set of tasks. No doubt, we need a computer software that evaluates and offers results, depicting the functions that, an UAV model can conduct most efficiently, in agricultural fields. Energy and economic consideration may decide the UAV that is to be selected. In such a situation, a ready reckoner or a compendium that lists as many UAV models available in market as possible, along with their salient features, would be useful.

At present, SenseFly's (Parrot Inc.) *eBee* and related advanced models of small, flat, fixed-winged UAVs are popular worldwide. They are swift in action and are among the best suited to offer aerial images and spectral data of ground features, soil conditions, crop growth, and phenomics data and to detect panicle/grain maturity. The company itself touts these small, relatively low-cost agricultural UAVs as well suited to conduct a series of tasks related to crop husbandry. Such functions range from monitoring land preparation, seeding, interculture, supply of inputs and right till harvest of grains (SenseFly Inc., 2016). These UAVs (*eBee*) are fit to conduct crop scouting, also set timetable to collect information on different crops (wheat, soybean, sorghum, etc.) and varieties sown at various locations, in the large farm. UAV usage may vary depending on the crop seasons and procedures to be conducted. Small UAVs (e.g., eBee) could be flown above crop fields to collect data about machinery required. For example, tillage equipment and extent of tillage required, fertilizer inoculators, herbicide sprayers and their activity, etc. During summer, UAVs could monitor application of fertilizers and spraying of pesticides, herbicides, and fungicides (SenseFly, 2016). During fall, eBee UAVs could be used to prepare harvest schedule and monitor combine harvesters' movement and grin transport vehicles. Therefore, a compendium, in future, may also have to identify and mention about tasks, that an UAV model can carry out in different seasons. For example, eBee may conduct a series of crop husbandry procedures but not spraying the crop fields with pesticides. A copter could be a good fit for spraying, but less opted to conduct aerial surveys in a particular season. There are also case studies about how certain UAV models suit, during the conduct of precision farming procedures, across different seasons (SenseFly Inc., 2016).

UAVs come in different types, models, and sizes. Their abilities related to aerial photography, spectral analysis, spraying, and to gather general information about the ground features may vary widely. *Matching UAVs and their specific model with purpose is required*. In fact, the series of agricultural uses attributable to UAV technology itself varies depending on the purpose, problem faced, and UAV model utilized. Set of agricultural uses attributed to UAV technology by private agricultural agencies in Northeast states of the United States are follows:

(a) UAVs can fly swiftly and create accurate (3 cm resolution or even less) 3D relief maps of fields, in a matter of minutes; they allow us to compare NDVI and its variations in a field. UAVs also offer high-resolution videos of crop fields. Such videos offer great details and offer better insights into happenings in the crop fields;

(b) farmers using three or four different crop species and their varieties may easily compare the relative performances of genotypes/crops and assess soil fertility trends, using high-resolution close-up images;

(c) UAVs equipped with thermal imaging sensors help in rapid detection of crop's water stress index. Such data helps in regulating water supply and in deciding quantum of irrigation. Thermal sensors also help us in detecting and monitoring cattle and other animals grazing the pastures;

(d) UAVs with videos help agricultural insurance agencies in assessing the need for rehabilitation, its quantum, and timing. Crop loan schemes too may utilize aerial images and crop videos derived using UAVs. For example, a real-time video of irrigation channels will help agricultural service and loan providing agencies with detailed knowledge about the functioning of existing irrigation setup, particularly leaks if any. Such information can be used to improve irrigation system and drainage system;

(e) UAV's imagery helps in detecting overwatered zones, flooded zones, and extent of soil erosion, if any;

(f) farmers can assess the damage to barns, silos, water storage units/ponds, and large farm vehicles such as tractors/combine harvesters. A set of uses described in great detail and classified into various aspects is offered by the UAV company Parrot Inc. (2017). It produces the popular brand known as *eBee* (Parrot Inc., 2017). Similar brochures are offered by several UAV companies.

UAV swarms composed entirely of fixed flat-winged UAV are possible. Multirotor UAVs can also be utilized in swarms. Even a mixture of flat-winged and multirotor UAVs working in cohesion, rhythmically, and in a well-coordinated fashion should be a possibility. UAV swarms should be explored, when the fields to be covered are vast. UAV swarms are useful when data pertaining to an agronomic operation must get accrued rapidly and in time. Swarms of UAVs could be a frequently adopted idea in large farms. The UAV models that suit to be included into a swarm or operated in groups must be identified. The inter-UAV communication needs emphasis. Excellent electronics and coordination of UAV movement in the air are required. Tasks that could be accomplished by swarms need special attention. Repetitions and collisions should be counted. The computer software, radio signals, and Global Positioning System (GPS) connectivity must be impeccable to operate UAV swarms (Gogerty and Robinson, 2012; Krishna, 2018). Not all UAV models could be flown in swarms and formations. A compendium, therefore, should identify UAV models that are amenable to be used in swarms above crop field. A compendium should include UAV models that could be operated in groups of 2, 3, 5, or 10 UAVs at a time.

Redmond (2014) opines that the value of UAV-derived aerial imagery is expected to get enhanced. Farmers will use it more frequently and routinely, in near future. It depends on how quickly we collect data regarding spectral signatures of disease/pest attacked crop patches and healthy ones. Multispectral and NIR images could pinpoint crop patches that need pesticide/fungicide application. Such digital data is transferred to autosteer sprayer systems on the ground. Also, a sprayer UAV could use spectral data immediately to dust the crop accurately. Such a procedure helps farmer to adopt plant protection measures quickly. Therefore, it thwarts disease/pest development at an early stage. It quantitatively reduces need for chemicals and repeated sprays. The reduction in requirement of pesticide/fungicide can be enormous. Sometimes it ranges about 80–90% reduction in chemical use compared to traditional blanket sprays of plant protection chemicals (Krishna, 2018). This strategy of using UAVs to obtain spectral data, detect disease/pest, and spray accurately, where required in apt quantities is efficient. It is expected to gain farmer's acceptance (Redmond, 2014). The need of the hour is to accumulate spectral data of as many plant diseases, pests, and weeds in the data bank as possible. Ultimately, it leads us to "integrated pest management" that includes aerial imagery and dusting/spraying by UAVs. We should match UAV model, spray equipment, fungicide formulation, and crop disease. At present, there are at least few hundred agricultural UAV models and several sprayer brands to choose (see Chapters 2 and 3 for details). Most urgently, we require a computer software that analyzes and offers us with best UAV models (ranking), sprayer equipment, pesticides/fungicide, and economic advantages accrued. We may even have to conduct computer-aided evaluation midway in the crop season. This is to keep pace with changes in disease/pest pattern as the crop grows. Overall, future for adoption of UAV technology per se during precision agriculture seems good. Companies that churn out UAV models with specific advantages could reap good profits (Redmond, 2014; Krishna, 2018).

1.5 ECONOMIC ASPECTS OF AGRICULTURAL UAVS

UAVs are gaining in popularity in agrarian regions. They are getting evaluated by farmers repeatedly. They are capable of accomplishing several different tasks relevant to agriculture. Many of them are small-sized UAVs that fit farmer's budgets. UAVs have been accepted because of their consistent and demonstrable use in assessing crop's status in the field. They are relatively inexpensive. Hence, they are preferred (Greenwood, 2016). Agricultural UAV start-ups have churned out an array of models. Further, Greenwood (2016) states that an agriculturally useful UAV can be purchased for as low as US$ 1000. It comes with necessary accessories for aerial survey and capture of spectral data. A multicopter with sophisticated sensors for visual and thermal imagery and with a sprayer device for pesticide application costs around US$ 5–10,000. In a few locations of China and Far East, farmers have used indigenous material to build UAVs, at perhaps, unbelievably low cost. They could be adopted to conduct aerial survey and obtain digitized spectral data (Krishna, 2016, 2018; Ministry of Agriculture, 2013).

There are indeed large numbers of small agricultural UAV models. They are mostly used for surveillance of crops. We also have moderately sized "sprayer UAVs" of great utility during pesticide dusting. Agricultural UAVs (e.g., DJI's MJ-Agras-1, PrecisionHawk, eBee) are capable of rapid aerial imagery. They cost much less at US$ 2000–3000. Of course, that includes full complement of accessories. An agricultural sprayer UAV costs US$ 5000–7500 (e.g., JMR-X1450; A5 Professional Agricultural sprayer quadcopter; see Chapter 3). A military UAV (e.g., MQ-9 Reaper) with 60 m wingspan that flies at 30,000 ft altitude, travels 2000 km, and delivers a few missiles may cost US$ 16 million per piece (Stuff, 2017). We can exchange a military UAV for few thousands of pieces of agricultural UAVs. The impact of agricultural UAVs on food crop generation could be impeccable. No doubt, government expenditure on agricultural UAVs may serve usefully, both, in terms of exchequer and total food grain generation. We should note that selection of most appropriate UAV models for the said agricultural zone, crops that flourish, and the problems that need to be corrected are essential. This means, we should develop software that compares military UAVs (optional), wide range of agricultural UAVs of contrasting features, and general purpose small UAVs. The economic evaluation could be centered around tasks such as aerial imagery, preparation of spectral maps depicting variations of nutritional status within a crop field, disease/pest attack, spraying pesticides, foliar application of fertilizers, etc.

In due course, our options for UAV models that exactly fit the situation, tasks to be performed, farmer's budget, and expected profits may increase. The number of UAV models flooded into the market is expected to be enormous. There are now at the least over 1300 UAV models that could perform useful tasks in crop fields worldwide. However, only 300–500 models have been deployed, tested, tried, and approved by farming community. We need to accumulate data pertaining to all models in one place. It helps us to compare them swiftly using appropriate software and assess crop productivity and economic consequences. In other words, we need software that evaluates all possible UAV models in one go. This way, farmers will not make mistakes while procuring UAV models.

Reports in 2014 suggested that agricultural uses of UAVs may take primacy among the set of civilian uses. Hence, they predicted that 80% of applications, usage, and market for UAVs will come from agricultural sector. Farmers need data about the crops rather too frequently. Therefore, it makes them depend on UAVs to get detailed aerial data, repeatedly and frequently. The need for sensors of different bandwidths will increase further. No doubt, software that processes the images and aids decision-making is the need of the hour. Farmers' interest in assessing nitrogen and water needs using aerial images will enhance UAV market (PrecisionHawk, 2014). Multispectral sensors and the range of spectral/digital data that UAVs offer is the game changer in case of UAV technology. It attracts our attention to UAV technology and its application in farming. They are inducing farmers, in plenty, to opt for UAV technology in North America and Europe and Far East. Scouting using UAVs will reduce need for human labor. Therefore, UAVs will become more profitable to farmers. These factors will further increase production and market for agricultural UAVs. Right now, the cost of an UAV unit is relatively high, but, as soon as production by the various start-ups all over the world increases, the cost of an UAV decreases. There is a common requirement related to all the above points mentioned. It is the need to know the UAV equipment and its specifications rather thoroughly. We need to match the UAV model and its capabilities with the agricultural tasks that need to be performed. Knowledge about profit margin is a basic requirement. Appropriate softwares that simulate and compare profits from UAV models are needed.

A few predictions based on authentic UAV sale data, current and expected rate of spread of UAV technology in different agrarian belts are available. "Agricultural UAVs" are deemed as the big winners. They are expected to spread into 60–65% of global agrarian belts. It could happen in a matter of few years, say 5–10 years from now (Blake, 2017). A few forecasts suggest that in the next decade, UAV sales, that is, just the platforms and simple accessories may cross US$ 85 billion and create 70,000 jobs a year. Jacobsen (2016) states that sales of UAVs have increased enormously during past 3–5 years. It is attributable to extraordinary interest of civilians to harness the various advantages of UAV technology. It is treated just like the smart phones, sensors, and nanotechniques adopted by the public. Incidentally, agriculture seems to garner most of the sales relevant to UAV technology. Consumers have spent over US$ 100 million during past 5 years to harness UAVs in farm world. It is expected to increase to US$ 1 billion within next 5 years. Clearly, we forecast an exponential use of UAVs in agrarian regions. It obviously triggers larger production and sales. Right now there are innumerable agricultural UAV models in the market. Knowledge about their use in crop fields such as aerial surveillance, spectral data accrual, and analysis by computers is mandatory. It helps to obtain best match between the various tasks, in this case, agricultural operations and UAV model. The best economic advantage also needs emphasis. We may need such software, more so, as the number of UAV models available in the market increases. A compendium or a dictionary of agricultural UAVs or a ready reckoner is the need of the hour. Farmers should evaluate a range of options before investing on a specific UAV model.

There are indeed innumerable reports that forecast the expansion of agricultural UAV market. The parameters emphasized while deriving such inferences may vary. A few forecasts may be exaggerations, while others may have underestimated. Knowledge about the number of UAV models flooded into agrarian belts and its fiscal consequences to UAV industry as well as individual farm production is essential (Krishna, 2018). Greenwood (2016) states that an estimate by Grand View Research suggested the UAV market worldwide to be at US$ 552 million in 2014. It was forecasted to grow to US$ 2.07 billion by the year 2022. A different study conducted by Markets and Markets Research Private Ltd. (2016) states that UAV-related market may reach US$ 4209.2 million by 2022. They have taken base year as 2015 and considered the rate of increase in UAV production, market, and usage by farmers. Right now availability of suitable software for processing the images, assessing crops, and devising input schedules seems to drive the UAV industry. Knowledge about demand for UAV models by farmers is important. It partially decides the future of UAVs in agriculture. Perhaps, by 2022, most farms would have possessed at least one or a few specialized UAV models to suit the farm operation. At present, North America and

Europe produces highest number of UAV units. They utilize them in civil and agricultural sectors. Far Eastern agrarian regions dominate the production and use of sprayer UAV models in their crop fields. These same reports suggest that we can also consider the UAV technologists trained and their designations. About 40% of UAV industries are engulfed by executives and administrators, technical staff account for 35% and rest 24%. North America garners 46% of global UAV market followed by European agrarian regions constituting 33% share. Asia-Pacific region accounts for 27% of global UAV market, while the rest of the area accounts for only 5–6% (Markets and Markets Research Private Ltd, 2016). They say markets for UAVs are expected to fluctuate depending on the major UAV industries, their models, and results that farmers obtain. From a different angle of thought, major UAV companies, models that they churn in big numbers, and their cost:benefit ratio will literally control adoption of UAV in farming regions.

In the present context, we may have to realize that both flat-winged and copter UAVs are being sought by farmers. A certain UAV with its salient features may get preferred more often than others. The efficiency of UAV-aided operations and economic advantages, of course, decides the UAV model that dominates an agrarian belt. Hence, we should supply computer software that evaluates a large number of different UAV models.

We may develop several UAV models. They could possess simple to complex capabilities that suit agricultural farms. Farmers may often be located away from cities that have UAV servicing agents. Greenwood (2016) rightly points out that UAV technology is adopted in villages and remote locations where crop production occurs. Villages need to be served well with electricity and/or lithium polymer batteries that suit the UAV model. As a precaution, lithium polymer batteries should be stored in quantities, if longer endurance UAVs in flight is needed. Timely availability of electric batteries is important. Electric batteries too could be a factor affecting UAV usage in villages.

Quiroz (2016) points out that UAV technology is being adopted in rural and agricultural farming zones. Further, considering the launch of Sentinel 1 and 2 by 2022, farmers and rural population could be using UAVs just like the mobile cell phones of today. This could happen within this decade. Most agricultural UAVs are being produced and utilized in North America, Europe, China, and Far East. Hence, many of the advances related to technology may show up first in these very regions.

During past couple of years, many agricultural research agencies have been toiling computing large amounts of data about crop production trends. Particularly, about use of UAVs and their impact on crop yield and farm economics. They say, in future, agricultural technology will be largely focused around UAVs and their various uses. Sensors, cloud computing, and software that process UAV-derived spectral and digital data will dominate the farm activity. This is because crop production in future years will be highly dependent on "data banks." It is driven by detailed and voluminous data the data banks hold. Economic advantages, therefore, will hinge on the extent to which farmers and companies adopt UAV technology. The introduction of UAVs itself will largely depend on the extent of venture capital and governmental support bestowed on UAV techniques. As stated above, there are many evaluations, reports, and guesses about the UAV-based market. They say venture capital bestowed on agricultural UAVs is ever increasing. It reached US$ 2.06 billion in the first half of 2015 (Winter Green Research, 2016). It could double to US$ 4.2 billion by the year 2021. Much of the interest and profits accruing due to use of UAVs will be derived through rapid aerial scouting and replacing farm human scouts. Aspects such as aerial mapping, detection of nutrient deficiencies, detecting water stress, if any, and prescribing irrigation quantum will get attention. Detecting pests/disease and weeds will also draw attention. Data driven agriculture will take over from the traditional prescriptions and blanket sprays, etc. (Winter Green Research, 2016). We must be alert to the fact that UAV technology spreads rapidly into agrarian belts. Consequently, it induces many start-ups that flood the market with innumerable agricultural UAV models. Therefore, farmers will need detailed information on as many UAV models as possible. Again, a software that compares and picks best UAV model is required. A compendium describing specifications of UAVs is essential. Although an abridged version, a big list of UAV models is provided in Chapters 2, 3, and 4 of this volume.

1.6 ADOPTION OF DIFFERENT MODELS OF AGRICULTURAL UAVS

Agricultural UAVs are versatile. Yet, evaluation of their performance in different agrarian regions is essential. The terrain, weather parameters, crops, and inputs may all influence the economic/energy efficiency of UAV technology. Let us consider a few examples. Wheat production in parts of Central Great Plains of the United States, in Australia, and elsewhere in other agrarian belts does depend, to a certain extent, on genetic resistance, plus plant protection measures. Surveillance for disease such as "yellow rust" or a pest such as "Russian aphid" is almost mandatory. Incidentally, Russian aphid is a vector for viral diseases. Adopting UAVs to regularly study the spectral changes of disease affected and healthy wheat crop is a good idea. Such a spectral investigation has been performed. The data accrued has been exchanged with several wheat-growing regions. It is believed that UAV techniques could play a great role in global plant protection. Initially, it may be focused to those maladies that are widespread. We may confine to diseases and pests that are amenable for detection using spectral reflectance data (Kansas State University, 2015). A project at Kansas Agricultural Experimental Station uses both flat-winged and copter UAVs for crop surveillance. The aim is to study crop diseases using spectral signatures.

In the Southern Great Plains, corn and sorghum are important cereals. UAV technology has been evaluated on these crops. It has been standardized to a certain extent for use on corn crop. There are reports suggesting UAV usage to obtain data about canopy growth, plant height, vegetative indices, and chlorophyll content. Nitrogen fertilizer and water requirements could also be assessed. In fact, UAV-aided evaluation of crop genotypes, particularly growth and yield traits have been possible (Plates 1.2 and 1.3) (Shi et al., 2016; Krishna, 2016, 2018). In most studies, correlation between UAV-derived data and ground reality data has been excellent. This aspect of UAV technology may induce farm researchers and institutions worldwide to depend more on UAVs. Particularly, to evaluate various crop cultivars and new farm procedures. In the experimental stations, UAVs could reduce cost of crop evaluation significantly when compared with skilled technicians or traditional human scouts.

In the Peruvian agrarian regions that show a dominance of potato culture, again, UAVs could have a great future. According to Allen (2016) of International Potato Centre, Lima, Peru, agricultural UAVs with simple sensors for aerial imagery and ground station facility (mobile or laptop) are good enough. UAVs collect useful spectral data. A data bank with spectral signatures of different genotypes of potatoes is proposed. Similarly, collection of spectral signatures of potato crop afflicted with disease and/or pests and healthy ones; those experiencing drought, nutrient deficiency, or retarded growth are contemplated. The data bank is being filled rapidly with relevant digital data. Such a data bank with spectral signatures is necessary for decision support computers. Allen (2016) states that the utility of small agricultural UAVs has been demonstrated on experimental farms of Centro Internacional De La Papa (CIP), at Lima, Peru. Therefore, it should not be difficult to replicate the effort in other potato-growing regions of Europe, North America, and hill tracts of Asia.

UAVs are expected to play a vital role in controlling locusts, particularly in subsistence farming regions of North Africa. For example, UAVs are used to provide early warning of locust swarms. UAVs could detect direction of movement of locusts. Then, relay it to farmers cultivating millets in Sudan (Cressman, 2016). Otherwise, Food and Agricultural Organization (Rome, Italy) depends mostly on tribes and nomads to collect local information. Sometimes, satellite imagery is also coupled to local information and utilized. UAVs are cost effective. They could be deployed at any time. Data, in the form of aerial photos could be relayed almost immediately. UAV models that suit best to conduct aerial surveys of disease/insect pests need to be identified. This will help farm agencies to scout and identify regions affected with locusts, fungal epidemics, or soil erosion. Regarding UAV models and capabilities, Cressman (2016) states that there is a need to sell and popularize UAVs that are small, but, with sufficient endurance, to cover over 100 km^2 area. Appropriate remote control gadgets, that is, telemetric link, GPS connectivity, and joystick or laptop are needed. UAVs should be able to locate and estimate locust density. They could also be adopted to spray pesticides at desired quantities. Hence, knowledge about the array of UAV models available and their specifications is a necessity. We need to match UAV models and the purpose, that is, locust control. In Nigeria, UAV technology is expected to provide farmers with a cost-effective method to monitor farm infrastructure. It has been utilized to design efficient irrigation systems for rice production (Le, 2016).

UAVs have begun hovering over crop fields in tropical Africa. In Kenya and Ghana, University of Lund and Swedish Agricultural University have aimed at introducing UAVs. Their aim is to do surveillance and obtain NDVI (crop vigor and health) of different crop species and natural vegetation (Hall, 2015). They hope to build a data bank of spectral signatures. They have also conducted infrared imagery of crop fields to ascertain water status of crops. Impending drought-like situations and need for water could be assessed. In the Sahelian region, UAVs have been demonstrated to agricultural agencies. They have been used to demarcate fields and plots, and then, decide on cropping pattern (Milla and Rolf, 2015). In this region, UAVs may soon play vital role in detecting seed germination trends, crop establishment, and growth pattern. UAVs could aid in developing forecasts about droughts or dust storms. Forage and grain productivity could be estimated using spectral reflectance and indices. There is a clear need to build a "spectral signature bank" for major crops of West Africa. This helps in detecting disease/pest attack very quickly in any subregion and to thwart its spread further. We also need a ready reckoner or software that suggests which UAV models suit best to perform such specialized tasks. We should also be aware of best sensors for capture of crop's spectral reflectance. In Southern Africa, agricultural UAVs are being evaluated by researchers from CIMMYT. UAVs are being evaluated for use in farms practicing rotations of wheat and legume (soybean). UAVs could be of great utility in detecting water deficits, nutrient deficiencies, and disease/pest attacked zones (Plate 1.4).

PLATE 1.4 A fixed-winged agricultural UAV is launched from crop fields at the International Maize and Wheat Center's experimental fields in Zimbabwe, Southern Africa.
Source: Dr. Clyde Beaver, Creative Service Manager, International Maize and Wheat Center, El Batan, Mexico.
Photo credit: CIMMYT Archives.

In the Philippines, project on UAVs concentrates on introducing them into cropping zones. Its aims are to conduct aerial survey and identify fields prone to disasters such as soil erosion, floods, and loss of crop stand. The intention is to map the disaster-prone zones. An early warning helps in rapid response to disasters in crop fields. There are about 25 UAV experts deployed by the Food and Agriculture Organization (FAO) of United Nations to alert the farmers with impending disasters. Farmers could be induced to take timely remedial measures (Plate 1.5) (FAO of the United Nations, 2016).

PLATE 1.5 Left: A fixed-winged UAV. Right: Launching an UAV to detect disaster-affected zones.
Source: Food and Agricultural Organization of the United Nations, Rome, Italy.
Note: FAO's UAV in Philippines helps to detect storm struck regions early. Therefore, farmers could overcome crop loss though timely remedial measures.

At present, United Nations (FAO, Rome) is making an effort to monitor pest/disease/weed epidemics that occur worldwide. They wish to forewarn and inform the regions that would get affected (FAO of the United Nations, 2016). No doubt, forecasts, alerts, and prophylaxis all are required, regularly. The FAO Plant Protection group should concentrate on informing farmers about impending disease/pest. Global Crop Protection Programs intend to take care of training farmers about UAV-aided sprays. They could do it through collaboration with local agencies. UAVs could act as watch dogs of disease/pest/weed infestation levels. They could also spray plant protection chemicals wherever feasible.

The Far East agrarian belts comprising Northeast China and Japan are among the most UAV savvy farming belts. They use UAVs to produce their major cereals such as rice, wheat, and maize. They are now trying to introduce UAVs and amalgamate them with "precision farming" procedures (Sakahito, 2016). Aerial data needed to prescribe variable-rate fertilizers, herbicides, and pesticides are collected using UAVs. They also automatically accrue data into an "agricultural data bank." This facilitates decision-making by the computers that prescribe fertilizer and pesticide dosages. The data bank service is available to farmers for a fee. UAVs could be rented. Thus, aerial imagery and canopy data required could be collected.

We select most appropriate tillage equipment and the tractor model. We may use, say, a disc or heavy tined plow for deep tillage. Otherwise, just a light plow can be used to obtain disturbance of surface soil. Clod crushers and hoes are used to break clods and get softer tilth. Tined hoes are used to mark the field for line sowing. Similarly, sprayers of different capacities, nozzle number, and spray rate are procured to achieve full drenching spray of pesticides. Further, simple sickles, small harvesters, or combines are used based on the crop and area to be harvested. No doubt, models of farm vehicles and implements/hitches are selected carefully by farmers. Now apply it to UAV technology. We already know that a few tasks such as aerial imagery, rapid monitoring, and surveillance of fields are done better by light, flat-winged UAVs (e.g., SenseFly's *eBee*, Trimble's UX5, PrecisionHawk's Lancaster). They fly rapidly at 50–60 km/h and offer ground station with detailed images of crop fields. They offer data about characteristics such as NDVI, leaf chlorophyll, etc. There are light quadcopters that can hover above the crops and pick detailed close-up and high-resolution images of canopy, leaves, insect species, or disease that has affected foliage/fruits, etc. (e.g., DJI's Inspire). Then, there are heavier hexa- and octocopters better suited to fly closely over the crop, lift a pay load of pesticide tank of 12–15 L and spray the crop carefully. UAVs spray at only spots that need attention (e.g., DJI's MJ Agras-1). *Overall, we should impart detailed knowledge to farmers and farming companies about array of UAV models available in the market. Then, suggest them about their suitability for specific tasks in farms. A thorough knowledge about different kinds of UAVs, their cost, agronomic and economic efficiency, reusability, etc., are needed, mandatorily.*

A recent update suggests that there are over 70 nations that produce UAV models. The UAV models are apt for military, civilian, and agricultural tasks. Total number of UAV models listed runs into over couple of thousands (Wikipedia, 2016). A different compilation states that, currently, there are over 1085 UAV models available. Those suitable for agricultural tasks are supposedly small. It is approximately 300–400 models. UAVs are expected to gain in popularity among farmers with sizeable land holdings and large farming companies. They may all prefer to own at least a few different types of UAVs. Each UAV model is meant to serve the farmer in conducting a specialized farm activity. For example, a couple of flat-winged small UAVs quickly survey, monitor, and get aerial data of crops. A slightly sophisticated UAV with longer endurance and a set of high-resolution cameras is utilized to gain spectral data about crops in detail. A copter UAV with pesticide tank is adopted to spray the crops with foliar (liquid) fertilizer or pesticides. Further, farmers may own a flat-winged UAV just to decide on which field has ripened and which farm activity should be monitored. Such practices will induce further production and sale of *specialized UAV* models. Therefore, farmers should be conversant about specifications and details of as many UAVs as available.

Australian farming companies intend to deploy UAVs in the agrarian regions. Their aim is to reap better harvests. Indeed, there are several different collaborations and initiatives from across different continents to test different UAV models for efficiency. A recent report states that a fixed-winged model such as SIRIUS (by MaVinci) and a rotory UAV, Aibotix (hexacopter), is being evaluated for their usefulness to farmers in Australia. SIRIUS is a small (1.6 m wingspan), compact, light-weight (2.7 kg) UAV. This UAV is finding acceptance to conduct crop scouting, 2D and 3D mapping, and preparing 3D models of terrain of the farm. SIRIUS, it seems, is a simple and low-cost flat-winged UAV to enter Australian farms. The rotor UAV AibotiX-6 is also being simultaneously evaluated. It is introduced to conduct different tasks in the crop fields. Some of the tasks are flying preplanned paths above crop fields. It picks accurate imagery of crop canopy and foliage. This is to estimate fertilizer and pesticide requirements, if any. A checklist of tasks conducted by both the UAVs, namely, the flat-winged and rotor UAV are being evaluated. The launch of a model of UAV is usually preceded by thorough tests and evaluations for accuracy, ease, and economic efficiency (Docherty, 2013). At this juncture, it is necessary to know the specifications, pros, and cons of each model that gets tested and deployed in each agrarian. We should be able to compute the characteristics of UAVs and simulate their functions, then arrive at best possible models. The software to simulate and compare the several UAVs in one go should be simple for farmers to adopt.

In New Zealand, ministry responsible for agriculture has initiated a series of UAV-related programs. They try to induce farmers to use UAVs to conduct several different types of farm activities. Some of the activities include general surveillance of crops, aerial survey for pests and disease, monitoring cattle, etc. There are a few models whose import and production has been suggested. At this juncture, we should note that a few UAV models have been selected by them. They are DJI's Phantom, SenseFly's *eBee*, and PrecisionHawk's Lancaster that are apt. They make aerial surveys and collect important spectral data. Their computers digitize spectral data and store/relay it to decision support systems. For example, an UAV such as DJI's Phantom with sensors flies at 10 m s^{-1} for 15 min. It collects relevant data for pest/disease control. It also collects data relevant to irrigation and fertilizer supply. The crux is to firstly gain a thorough knowledge of all UAV models available and verify their utility and economic efficiency. Then, use it to improve agricultural crop production (Stuff, 2017).

KEYWORDS

- **unmanned aerial vehicles**
- **fixed-winged UAVs**
- **multi-rotor UAVs**
- **aerostats**
- **sensors**
- **aerial photography**
- **crops**

REFERENCES

Aber, J. S.; Galazka, D. Kite Aerial Photography of Poland. *Geol. Q.* **2000,** *44*, 33–38.

Allen, W. Drones Detect Crop Stresses More Effectively. *ICT Update,* 2016; Vol. 82, pp 10–11. http://ictupdate.cta.Int (accessed Apr 2, 2017).

Anders, N.; Keesstra, S.; Cmmerat, E. *The Effect of Flight Altitude to Data Quality of Fixed Wing UAV Imagery: Case Study in Murcia, Spain*; 2014; p 9. http://adsabs.harvard.edu/abs/2014EGUGA. 1610295A (accessed Nov 9, 2017).

Arjomandi, M. *Classification of Unmanned Aerial Vehicles*; The University of Adelaide, Australia. http://www.personal.mecheng.adelaide.edu/mazair/arjoomani/aeronautical%20engineering%20projects/2006/group9.pdf (accessed Sept 1, 2015).

Associated Press. *Blimp, Infrared Camera Used to Fine-Tune Cotton Irrigation*; 2004; pp 1–5. http://chronicle.augusta.com/stories/2004/07/13/liv_422010.shtml (accessed Apr 4, 2017).

Atherton, K. D. Watch NASA's Greased Lightning Tilt-wing Take-off Like a Summer Time Fling (Video). *Popular Science* (A Bonnier Corporation Company), 2015; pp 1–4. http://www.popsci.com/ (accessed Nov 1, 2017).

BAA Training. *UAV Types: How to Choose Yours*; 2015; pp 1–7. http://www.baatrainging.com/uav-types-how-to-choose-yours/ (accessed Mar 22, 2017).

Batut, A. In *Earth from Above*, International Conference on Kite Aerial Photography, 1888; pp 1–3. https://www.wokipi.com/Kapined/art-va-html (accessed Apr 27, 2017).

Bellis, M. *History of the Parachute*; 2017; pp 1–22. http://theinventors.org/library/inventors/blparachute.htm (accessed Apr 26, 2017).

Bendig, J.; Bolten, A.; Bareth, G. UAV-based Imaging for Multi-temporal, Very High-resolution Crop Surface Models to Monitor Crop Growth Variability. *PFA* **2013a,** *13*, 551–562.

Bendig, J.; Willkomm, M.; Tilly, N.; Gnyp, M. L.; Bennertz, S.; Qiang, C.; Miao, Y.; Lenz-Wiedmann, V. L. S; Bareth, G. In *Very High-resolution Crop Surface Models (CSMs) from UAV-based Stereo Images of Rice Growth Monitoring in Northeast China*, International Archives of Photogrammetry and Remote Sensing, Spatial Information Science, 2013b; Vol. 40, pp 45–50.

Bendig, J.; Yu, K.; Assen, H.; Bolten, A.; Bennertz, S.; Broscheit, J.; Gnyp, M.; Bareth, G. Combining UAV Based Plant Height from Crop Surface Models, Visible and Infrared Vegetation Indices for Biomass Monitoring in Barley. *Int. J. Appl. Earth Obs. GeoInf.* **2015,** *39*, 79–87.

Blake, C. *Agriculture to Farm Two-thirds of UAV Drone Market*; 2017; pp 1–9. http://www.proag.com/News/Agriculture-to-Farm-twothirds-of-UAVDrone-market (accessed Apr 22, 2017).

Boike, A. K. M.; Yoshikawa, K. Mapping of Periglacial Geomorphology Using Kite/Balloon Aerial Photography. *Permafrost Periglac.* **2003,** *14*, 81–85.

Bowman, L. UAV Drones for Farmers and Ranchers. *Agriculture UAV Drones*, 2015; pp 1–2. http://www. agricultureuavs.com/uav_drones_for_farmers_ranchers.htm (accessed May 23, 2015)

Case, P. *Rothamsted Unveils Octo-copter Crop-monitoring Drone*; pp 1–2. http://www.fwi.co.uk.arable/rothamsted-unveils-octocopter-crop-monitoring-drone.htm (accessed June 25, 2013).

Chapman, A. Types of Drones: Multi-rotor vs Fixed Winged vs Single Rotor vs Fixed Winged VTOL Drones. *Australian UAV*, 2015; pp 1–9. http://www.auav.com.au/articles/drone-types (accessed Aug 12, 2016).

Chen, S. C.; Hsiao, Y. S.; Chung, T. H. Determination of Landslides and Driftwood Potentials by Fixed Wing UAV-borne RGB and NIR Images: A Case Study of Shenmu Area in Taiwan. *NASA Astrophys. Data Syst. (ADS)* **2015,** 8. http://adsaba.harvard.edu/abstract_service.html; http://adsabs.harvard.edu/bs/2015EGUGA..17.2491C/ (accessed Nov 9, 2017).

CIMMYT. *Obregon Blimp Airborne and Eyeing Plots*; International Maize and Wheat Centre: Mexico, 2012; pp 1–4 (accessed Apr 4, 2017).

Conrad, K. *Kite Aerial Photography*; 2017; pp 1–3. Brooxe.com; http://www.brooxes.com/newsite/HOME.html (accessed Apr 28, 2017).

Cressman, K. Preventing the Spread of Desert Locusts Swarms. *ICT Update,* 2016; Vol. 82, pp 8–9. http://ictupdate.cta.int (accessed Apr 1, 2017).

D'Oliere-Oltmans, S.; Marzolff, I.; Peter, K. D.; Ries, J. D. Unmanned Arial Vehicle (UAV) for Monitoring Soil Erosion in Morocco. *Remote Sensing* **2012,** *4*, 3390–3416. DOI: 10.3390/rs4113390 1-12 (accessed Nov 18, 2015).

DJI. Above the World. *UAS Magazine*, 2016; pp 238. http://www.uasmagazine.com/articles/1591/dji-explains-new-book-of-drone-captured-images (accessed Apr 20, 2017).

Dobberstein, J. Technology: Drones Could Change Face of No-tillage. *Precision Farm Dealer*, 2013; pp 1–7. http://www.no-tillfarmer.com/pages/Spre/SPRE-Drones-Could-Change-Faceof-No-Tillng-May-1-2-13 (accessed July 26, 2014).

Docherty, G. UAV: Fixed Wing or Rotary? *The Business of Drones*; sUAS News, 2013; pp 1–6. https://www.suasnews.com/2013/09/uav-fixed-wing-or-rotary/ (accessed Mar 22, 2017).

Dorr, L. Fact Sheet: Unmanned Aircraft System; 2014; pp 1–10. http://www.faa.gov/news/fact_sheets/news_story.cfm?newsid-14153 (accessed Aug 17, 2015).

Dreiling, L. New Drone Aircraft to Act as Crop Scout. Kansas State University, Salina, USA; pp 1–2. http://www.hpj.com/archives/2012/apr1216/apr16/springPlantingMACOLDsr.cfm (accessed May 30, 2015).

Drone Industry Insights. Top 20 Drone Company Rankings q1 2016; pp 1–24. http://www.drone11.com/top20-drone-company-ranking-q2-2016 (accessed Oct 30, 2016).

eBee. *The Professional Mapping Drone*; SenseFly Inc.: Switzerland; pp 1–19. https://www.sensefly.com/drones/ebee.html (accessed May 5, 2017).

ECA Group-Lima. Types of Drones: Explore the Different Models of UAVs. *Circuits Today,* 2017; pp 1–7. http://www.circuitstoday.com/types-of-drones (accessed Mar 22, 2017).

Elsevier Ltd. Aerial Photography. *Developments in Soil Science;* 1987; Vol. 15, pp 155–180. http://doi.org/10.1016/so166-2461(08)70033-3 (accessed Apr 28, 2017).

FAO of the United Nations. *Drones, Data, Food Security: How UAVs Offer New Perspectives on Agriculture*; Food and Agricultural Organization of the United Nations: Rome, Italy; pp 1–3. http://www.fao.org/news/podcst/drones/en/ (accessed Jan 1, 2017).

Farming Online. Using Drones to Monitor Crops. Rothamsted Experimental Agricultural Station: United Kingdom, 2013; pp 1–2. http://farming.co.news/article/9243 (accessed May 22, 2015).

Fitzpatrick, S.; Burnett K. *Regulation and Use of Drones in Canada.* The Canadian Bar Association, 2014. http://www.cba.org.org/CBA/section_irnd space/newsletter2013/drones.aspx (accessed Sept 14, 2014).

Fosset, R. Aerial Photography by Kite. KAP Shop, 2017; pp 1–2. http://www.kpshop.com/p35/Aerial-Photography-By-Kite/product_info.html (accessed Apr 28, 2017).

Frey, T. *Future Uses offor Flying Drones*; 2015; pp 1–37. http://www.futuristsspeakercom?2014, 2014/09/192-future-uses-for-flying-drones/ (accessed Apr 10, 2017).

Gago, J.; Douthe, C.; Coopman, R. F.; Gallego, P. P.; Ribas-Carbo, M.; Flexas, J.; Esclona, J.; Medrano, H. UAVs Challenge to Assess Water Stress for Sustainable Agriculture. *Agric. Water Manag.* **2015,** *153,* 1–14. DOI: 10.1016/j.agwat.2015.01.020 (accessed Mar 27, 2017).

Garcia-Ruiz, F.; Shankaran, S.; Maje, J. M.; Lee, W. S.; Rasmussen, J.; Ehsani, R. Comparison of Two Imaging Platforms for Identification of Huanglongbing-infected Citrus Orchard. *Comp. Electron. Agric.* **2013,** *91,* 106–115.

Gogarty, B.; Robinson, I. Unmanned Vehicles: A (Rebooted) History, Background and Current State of the Art. *J. Law Inf. Sci.* **2012,** *21,* 1–18.

Greenwood, F. Drones are on the Horizon: New Frontier in Agricultural Innovation. *ICT Update,* 2016; Vol. 82, pp 1–4. http://ictupdate.cta.Int (accessed Apr 1, 2017).

Hailey, I. T. The Powered Parachutes Archaeological Aerial Reconnaissance Vehicle. Plant Direct On-line Library, 2005; pp 1–12. http;//onlinelibrary.wiley.com/doi/10.1002/arp.247/full (accessed Apr 25, 2017).

Hall, O. 2015 The Challenge of Comparing Crop Imagery Over Space and Time. *ICT Update,* 2015; Vol. 82, pp 14–15. http://ictupdate.cta.int (accessed Apr 6, 2017).

HSE. *Agricultural and Conservation UAS*; Homeland Surveillance and Electronics LLC: Illinois, USA; pp 1–4 (accessed Apr 10, 2017).

Inoue, Y.; Moringa, S.; Tomita, A. A Blimp-based Remote Sensing System for Low Altitude Monitoring of Plant Variables: A Preliminary Experiment for Agricultural and Ecological Explanations. *Int. J. Remote Sens.* **2000,** *21,* 379–385.

International Conference on Kite Aerial Photography. *Kipned 10 Aerial Photography,* 2016; pp 1–3. http://www.wokipi.com/Kapined/art-va.html (accessed Apr 28, 2017).

Jacobsen, F. The Promise of Drone. *Harv. Int. Rev.* **2016,** *37* (3), 1–4.

Jung-Diveky, A. Microdrones MD4-3000 Pushing Boundaries Again. *Microdrones Gmbh,* 2016; pp 1–16. http://www.microdrones.com/en/products/md4-3000/ (accessed Nov 28, 2016).

Kansas State University. *Project Using Drones to Detect Emerging Pest Insects, Disease in Crops.* Department of Entomology, Kansas State University at Salina and Manhattan, USA, 2015; p 104. http://www.agprofessional.com/news/project-using-drones-detect-emerging-pest-insects-diseases-in-crops/ (accessed Apr 5, 2017).

Keane, J. F.; Carr, S. S. A Brief History of Early Unmanned Aircraft. *J. Hopkins Appl. Tech. D.* **2013,** *32,* 558–593.

Kennedy, M. W. *Moderate Course for USAF UAV Development*; National Technical Information Service, US Department of Commerce, 1998; pp 1–2. Info@ntis.goc (accessed Aug 10, 2015).

Knoth, C.; Prinz, T. In *UAV Based High Resolution Remote Sensing as an Innovative Monitoring Tool for Effective Crop Management*, Proceedings of a Meeting on "Crop Resource Use Efficiency and Field Phenotyping," Belton, North Grantham, UK; pp 1–4.

Krishna, K. R. *Precision Farming: Soil Fertility and Productivity Aspects*; Apple Academic Press Inc.: Waretown, NJ, 2012; p 188.

Krishna, K. R. *Push Button Agriculture: Robotics, Drones, Satellite-guided Soil and Crop Management*; Apple Academic Press Inc.: Waretown, NJ, 2016; p 452.

Krishna, K. R. *Agricultural Drones: A Peaceful Pursuit*; Apple Academic Press Inc.: Waretown, NJ, 2018; p 425.

Le, Q. A Bird's Eye View on Africa's Rice Irrigation Systems. *ICT Update,* 2016; Vol. 82, pp 6–7 http://ictupdate.cta.Int/ (accessed Apr 2, 2017).

Lee, W. S.; Ehsani, R.; Albrigo, L. G. In *Citrus Greening Disease (Huanglongbing Disease) Detection Using Aerial Hyperspectral Imaging,* Proceedings of the 9th International Conference on Precision Agriculture, Denver, Colorado, 2008; pp 32–39.

Lelong, C. C. D.; Burger, P.; Jubelin, G.; Roux, B.; Labbe, S.; Barer, F. Assessment of Unmanned Aerial Vehicles Imagery for Quantitative Monitoring of Wheat Crop in Small Plots. *Sensors* **2016,** *8,* 3557–3585.

LGL Ltd. *A Review and Inventory of Unmanned Aerial Systems for Detection and Monitoring of Key Biological Resources and Physical Parameters Affecting Marine Life During Off Shore Exploration and Production Activities*. Joint Industry Programs on Marine life. LGL Report No. TA4731-1; pp 1–12. http://www.soundand mrinelife.org (accessed Apr 15, 2017).

Litchman, A. Humanitarian Uses of Drones and Satellite Imagery Analysis: The Promises and Perils. *Am. Med. Assoc. J. Ethics*. **2015,** *17*, 931–937.

Lumpkin, T. *CGIAR Research Programs on Wheat and Maize: Addressing Global Hunger*; International Centre for Maize and Wheat (CIMMYT), Mexico. DG's Report, 2012; pp 1–8.

Markets and Markets Private Research Ltd. Agricultural Drones Market by Type (Fixed-wing, Rotary-blade, Hybrid, Data Analysis Software, Imaging Software), Application (Field Mapping, VRA, Crop Scouting, Livestock, Crop Spraying), Component and Geography (Global Forecast to 2022); 2016; pp 1–5. Markets and Markets.com; http://www.marketssand markets.com/Market-Reports/agriculture-drones-market-2370 (accessed Mar 23, 2017).

Milla, R. Z.; Rolf, A. D. Transforming Small-holder Farming. *ICT Update,* 2015; Vol. 82, pp 18–19. http://ictupdate.cta.int (accessed Apr 6, 2017).

Miller, F. Kites, *Balloon Collect Aerial Data for Soybean Drought Tolerance Research*; APS Crop Protection and Management Collection, Plant Management Network, 2014; pp 1–6. http://www.plantmanagment work.org/pub/crop/news/2014/AerialData/ (accessed Apr 28, 2017).

Millet, J. B. Scientific Kite Flying. *The Century Illustrated Monthly Magazine* 1897; Vol. 54, p 66.

Ministry of Agriculture. *Beijing Applies "Helicopter" in Wheat Pest Control*. Ministry of Agriculture of the Peoples Republic of China: A Report, 2013; pp 1–8. http://english,agri.gov.cn/news/dqnf/201306/t20130605 _19767.htm (accessed Aug 10, 2014).

Mortimer, G. *"Skywalker": Aeronautical Technology to Improve Maize Yields in Zimbabwe*; International Maize and Wheat Centre, Mexico, DIY Drones, 2013; pp 1–6. http;//www.ubedu/web/ub/en/menu_eines/notices/2013/04/006.html (accessed Feb 10, 2016).

Mothership Aeronautics. *100 Applications of Solar-Powered Blimps*; 2016; pp 1–5. http://www.mothership.aero/100-applications (accessed Apr. 4, 2017).

NASA. *Kites: A Background*; Aeronautics Research Mission Directorate-NASA. National Aeronautics and Space Agency: Washington DC, 2016; pp 1–65. http://www.nasa.gov/sites/default/files/atoms/files/kites_t4.pdf (accessed Apr 28, 2017).

Newcome, L. R. *Unmanned Aviation: A Brief History of Unmanned Aerial Vehicles*; American Institute of Astronautics Inc.: Reston, Virginia, 2004; pp 1–29.

Nex, F.; Remondino, F. UAV for 3D Mapping Applications: A Review. *Appl. Geo.* **2013,** *6*, 1–15.

Office of Secretary of Defence. In *Unmanned Aircrafts Systems Roadmap, 2005 to 2030*, Proceedings International Conference of Agricultural Engineering, Zurich, 2005; pp 1–8. http://www.eurageng.eu (accessed Apr 27, 2017).

Ottos, J. UAVs are Next Wave of Agricultural Technology. *Agrinews* **2014,** 1–3. http://agrinews-pubs.com/Content/ News/MoneyNews/Article/UAVs-are-next-wave / (accessed May 16, 2015).

Penner, N. *Sustainable Technology: Drone Use in Agriculture*; pp 1–7. https://wiki.usask.ca/display/~pdp177/Sustainable+Technology+Drone+Use+griculture (accessed July 6, 2015).

Parrot Inc. *Live the Intensity of Immersive Flight with Bebop 2 FPV and Disco FPV*; 2017; pp 1–3. https://www.parrot.com/ca/#drones-fpv (accessed May 5, 2017).

Perlman, A. *70 Drone Companies to Watch in 2016*; 2016; pp 1–55. http://uavcoach.com/drone-compnies-2016/ (accessed Nov 5, 2016).

Perry, E. M.; Band, J.; Kant, S.; Fritzgerald, G. J. *A Field Based Rapid Phenotyping with Unmanned Aerial Vehicles (UAV)*; 2012a; pp 1–5. http://www.regional.org.au/au/asa/2012/precision-agriculture/7933_perry.htm (accessed Aug 23, 2014).

Perry, E. M.; Fitzgerald, G. J.; Nutall, J. G.; O'Leary, M.; Schulthess, U.; Whitlock, A. Rapid Estimation of Canopy Nitrogen of Cereal Crops at Paddock Scale Using a Canopy Chlorophyll Content Index. *Field Crops Res.* **2012b,** *118*, 567–578.

Precision Farming Dealer. *Drone Owners Must Register Equipment with FAA: Starting Today*; 2015; pp 1–3. http://www.precisionfarming dealer.com/1881.htm (accessed Dec 22, 2015).

PrecisionHawk. *The Fundamentals of UAVs*; Precision Hawk Media, 2014; pp 1–8. http://www.precision hawk.com/media/topic/the-fundamentals-of-uavs/ (accessed Mar 22, 2017).

PrecisionHawk. *An Enterprise Drone Platform for Better Business Intelligence*; Precision Hawk Inc.: Indiana, 2017; pp 1–12. http://www.precisionhawk.com/ (accessed May 5, 2017).

Pudelko, J.; Kozyra, J.; Neirobca, P. Identification of the Intensity of Weeds in Maize Plantations Based on Aerial Photography. *Zembdirbyste* **2008,** *3*, 130–134.

Pudelko, R.; Stuzynski, T.; Borzeka-Walker, M. The Suitability of Unmanned Aerial Vehicle (UAV) for the Evaluation of Experimental Fields and Crops. *Zemdirbyste* **2012,** *99*, 431–436.

Quiroz, R. A. UAV-based Remote Sensing Will Be Like Using a Cell Phone Today. *ICT Update,* 2016; Vol. 82, p 13. http://ictupdate.cta.Int (accessed Apr 2, 2017).

Rau, J. Y.; Jhan, J. P.; Lo, C. F.; Lin, Y. S. *Landslide Mapping Using Imagery Acquired by a Fixed-wing UAV*; NASA Astrophysics Data System (ADS), 2011; p 5. http://dsabs.harvrd.edu/abstract_service.htm (accessed Nov 9, 2016).

Redmond, S. *The Future of UAVs for Agriculture*; pp 1–2. http://www.hdc.on.ca/grain-marketing/hdc-reports/29-grain-mrketing/253-hdc-future (accessed Mar 28, 2017).

Ries, J. B.; Marzolff, I. Monitoring of Gully Erosion in the Central Ebro Basin by Large Scale Aerial Photography Taken from a Remotely Controlled Blimp. *CATENA* **2003,** *50,* 309–328 (Under Experiments)

Robinson, M. *Meteorological Kites: Scientific Kites of the Industrial Revolution*; 2003; pp 1–5. http://kitehistory.com/Miscellneous/meteorological_kites.htm (accessed Apr 29, 2017).

Ruijter, G. D. In *Kites in Agriculture*, International Conference on Kite Aerial Photography, 2016; pp 1–3. http://www.wokipi.com/Kpined/art-va.html (accessed Apr 27, 2017).

Sakahito, S. "Drone Japan" Unveils Drone-based Field Data Analysis Service for Improved Rice Farming. *The Bridge*, 2016; pp 1–3. http://the bridge.jp/en/2016/ 10/drone-japan-announces-dj-agri-service-drone-rice/ (accessed Jan 9, 2017).

Santaana, R. Drone-based Land Management Meeting in Corpus Christi. *AgriLife TODAY*, 2016. https://today.grilife.org/2016/01/12/drone-land-managment-meet/ (accessed Mar 22, 2017).

Schwing, R. P. *Unmanned Aerial Vehicles-revolutionary Tools in War and Peace*. USAWC Strategy Research Project; United States Air Force; pp 1–22. http://fas.org/irp/program/collect/docs/97-6230D.pdf/A/Acsc/0230D/97-03 (accessed July 21, 2014).

SenseFly Inc. *Drones for Agriculture*; SenseFly Inc.—Parrot Company: Cheaseux-Louisanne, Switzerland, 2016; pp 1–9. http://www.sensfly.com/applictions/agriculure.html (accessed June 15, 2016).

Shi, Y.; Murray, S. C.; Rooney, W. L.; Vlassek, J.; Jeff J.; Ace, J. P.; Bowden, E.; Dongyan, Z.; Thomasson, J. A. *Corn and Sorghum Phenotyping Using a Fixed Wing UAV-based Remote Sensing System*; NASA Astrophysics Data System; p 1. http://adsabs.harvrd.edu/abstract_service.html (accessed May 1, 2016).

Stanford Law School International Human Rights Group. *Living Under Drones: Death, Injury, and Trauma to Civilians from US Drone Practices in Pakistan*; 2012; pp 1–38. http;//chgri.org/wp-content/uploads/2012/10/Living-Under-Drones.pdf (accessed Apr 3, 2017).

Stombaugh, T.; Smith, S.; Thamman, M. *The Use of Unmanned Aircraft Systems in Agriculture*. Biosystems and Agricultural Engineering Update; University of Kentucky Cooperative Extension Service, 2016.

Stone, B.; Crandall, Z. *Tilt: Rotor Drone*. Report Submitted to Electrical Engineering Department, California Polytechnic State University, San Luis Obispo, CA; pp 1–23.

Stuff. Drones Could Join Battle Against Pests and Diseases; pp 1–3. Stuff.co.NZ; http://www.stuff.co.nz/environment/64513141/drones-could-join-battle-against-persts- / (accessed May 5,2017).

Tajar, K. M.; Ahmad, A. An Evaluation on Fixed Wing and Multi-rotor UAV Images Using Photogrammetric Image Processing. *Int. J. Comp. Elec. Auto. Cont. Inform. Eng.* **2013,** *7,* 48–51.

Talon LPE. A Brief History of Unmanned Aerial Vehicles AKA Drones; pp 1–5. http://www.talonlpe.com/blog/a-brief-history-of-unmnned-aerial vehicles-aka-drones (accessed Apr 25, 2017).

Tattaris, M.; Reynolds, M. In *Applications of an Aerial Remote Sensing Platform*, Proceedings of the International TRIGO (Wheat) Yield Workshop; Reynolds, M., Mollero, G., Mollins, J., Braun, H., Eds.; International Maize and Wheat Centre (CIMMYT): Mexico, 2015; pp 1–5.

Tetrault, C. *A Short History of Unmanned Aerial Vehicles (UAVS)*; Dragon Fly Innovations Inc., 2014; pp 1–2. http://texasagrilife extension.edu/ (accessed July 24, 2014).

Thamm, H. P. In *SUSI A Robust and Safe Parachute UAV with Long Flight Time and Good Pay Load*, International Archives of the Photogrammetry, Remote Sensing and Spatial Information Sciences, 2011; Vol. 38, pp 1–6.

Thamm, H. P.; Judex, M. In *The Low-cost Drone. An Interesting Tool for Process Monitoring in a High Spatial and Temporal Resolution*, International Archives of Photogrammetry, Remote sensing, Spatial information Science, ISPs Commission 7th Mid-term Symposium, Remote Sensing: From Pixels to Process, Enchede, The Netherlands, 2006; Vol. 36, pp 140–144.

The UAV. *The UAV: The Future of the Sky*; 2015; pp 1–4. http://www.theuav.com (accessed June 26, 2016).

The University of Adelaide. *Australian Plant Phenomics Facility: Crop Plant Filed Module*; pp 1–3. http://www.plant phenomics.org.au/services/cropfield/?template=print (accessed Apr 29, 2017).

Trimble Inc. *The Complete Unmanned System*; Trimble Inc.: San Jose, CA, 2017; pp 1–10. http://www.us.trimble/ux5 (accessed May 5, 2017).

UAS Exemptions. *FAA Approved Unmanned Aircraft List*; Federal Aviation Agency: Washington, DC; p 83. http://usexpemptions.com/approved-unmanned-aircraft-list/ (accessed Apr 1, 2017).

UAVGlobal. *List of All Unmanned Systems and Manufacturers*. UAVGlobal Unmanned Systems and Manufacturers; pp 1–8. http://www.uavglobal.com/list-os-manufacturers/ (accessed Dec 10, 2016).

United Soybean Board. Agriculture Gives UAVs a New Purpose. *AG Professional*; p 1. http://www. agprofessional.com/news/agriculture-gives-unmanned-aerial-vehicles-a-new-purpose-255380931 (accessed Apr 23, 2014).

USDA. *The Use of Kites in the Exploration of the Upper Air;* Year Book of United States Department of Agriculture, Beltsville, 1898; p 344.

Vaan der Staay, L. *Drones Are Used for Research and Land Management: Do You Want to Learn How to Use Drones*? pp 1–2. Farms.com (accessed Apr 20, 2017).

Vroegindeweij, B. A.; Van Wijk, S. W.; Van Henten, E. J. In *Autonomous Unmanned Aerial Vehicles for Agricultural Applications*, International Conference of Agricultural Engineering, Zurich, Switzerland, 2014; pp 1–8. http://www.eurageng.eu (accessed Dec 17, 2016).

Wikipedia. *Kite Types*; pp 1–16. http://enwikipedia.org/wiki/Kite_types (accessed Apr 28, 2017).

Wikipedia. *List of Unmanned Aerial Vehicles*; pp 1–18. https://en.wikipedia.org.wiki/List_of-_unmnned_aerial_vehicles/ (accessed May 11, 2017).

Winter Green Research. *Agricultural Drones Market Shares, Strategies and Forecasts, Worldwide, 2016 to 2022*; 2016; pp 1–3. http://www.rnrmarketresearch.com/agricultural-drones-market-shares-strategies-and-forecsts-worldwide/ (accessed Jan 9, 2017).

Yamaha. *RMAX-History*; 2014; pp 1–4. http://www.rmax.yamaha-motor.com.all/history (accessed Sept 20, 2015).

Yan, L.; Gou, Z.; Duan, Y. A Remote Sensing System: Design and Tests. *Geospatial Technology for Earth Observation*; pp 27–38. DOI: 10.1007/978-1-4419-0050-0_2 (accessed July 11, 2009).

Yintong Aviation Supplies Company Ltd. *A Precision Agriculture UAV*; pp 1–3. http://www. china-yintong.com/en/productshow.asp?sortid=7&id=57 (accessed July 31, 2014).

CHAPTER 2

Fixed-Winged Unmanned Aerial Vehicle Systems Utilized in Agriculture

ABSTRACT

At present, fixed-winged unmanned aerial vehicles are popular with military establishments and civilian agencies. They are yet to gain popularity within agrarian regions. Their major usage is in obtaining aerial imagery and digital data about land, soil, and crops. Such data are utilized in farming. The spectral data and maps derived from fixed-winged UAVs are of utmost value, to those who practise precision farming. During past 5 years, agrarian regions are being flooded with innumerable models of small fixed-winged aerial UAVs (drone aircrafts). Large number of start-ups have appeared in North America, Europe, and Far East, particularly in China. The spurt in number of UAV companies and models they offer to agricultural enterprises is easily discernible. They have offered several different models of fixed-winged small drones useful in conducting aerial surveys of farm belts. This is actually in response to need for detailed aerial imagery and digital data, while we practice precision farming. Farm companies with large land holdings invariably prefer a set of fixed-winged drones, a few sprayer copter drones, and matching computer programs to process the data.

In this chapter, a wide range of fixed-winged UAVs of different weight classes have been discussed. In all about 103 fixed-winged UAVs models have been dealt in detail. The fixed-winged UAVs have been classified based on their weight into "Micro-weight fixed-winged UAVs (< 5 kg)' (e.g., *eBee*, Precision Hawk, AgEagle, Delta-M, and Phoenix 2), "light weight fixed-winged UAVs (5–50 kg) (ScanEagle)," medium weight fixed-winged UAVs (50–200 kg), heavy fixed-winged UAVs (200–2000 kg) (e.g., Gamma), and super heavy weight fixed-winged UAVs (>2 tons) (e.g., Global Hawk, Predator). In addition, there are some special types of fixed-winged UAVs discussed in this chapter. For example, detailed specifications for a set of "hybrid vertical take-off and landing (VTOL) fixed-winged UAVs" (e.g., Quantix, Aerosonde); "blended-winged UAVs" (e.g., Bramor, eBee, and UX5), and a few miscellaneous types are made available. Several models of each class have been discussed in detail and their specifications have been listed. Industrial production trends suggest that right now, agricultural regions are preferring micro- and light weight fixed-winged UAVs in greatest number. Knowledge about specifications and economic efficiency of fixed-winged drones is essential to farmers. Farmers have to select the most appropriate models(s) for their farm. In fact, there is a suggestion made to develop a computer program that helps farming enterprises while selecting the best possible fixed-winged UAV model. The information provided for each fixed-winged UAV model listed in this chapter pertains to firstly, the manufacturer's details and technical specifications. Technical specifications include material used to manufacture the UAV model, size, propellers, engine, payload allowed, launching information, ceiling altitude, flight speed, wind speed tolerance, remote control equipment, photographic accessories, agricultural and non-agricultural uses. Such details are needed by farmers envisaging to procure a fixed-winged UAV for their farm. A computer-based ready reckoner with detailed specifications and user guide/manual has also been suggested.

2.1 INTRODUCTION: FIXED-WINGED UNMANNED AERIAL VEHICLE SYSTEMS

Historically, the fixed-winged unmanned aerial vehicles (UAVs) were developed to suit the military pursuits, around 5–6 decades ago (Newcome, 2004; Nicole, 2015; Keane and Carr, 2013, Tetrault, 2014). At present, a range of fixed-winged UAVs with different size, weight, endurance, and capabilities for flight from a few kilometers to thousands of kilometers are available. They conduct surveillance, interception, and bombing tasks. During recent past, several models with wide range of utilities in surveillance, tracking, monitoring, aerial imagery, and spectral analysis of ground features were developed. Fixed-winged UAVs preceded the other types, such as helicopters, multicopters, vertical take-off and landing (VTOL) hybrids, and special designs, such as tilted-winged, etc. Within the context of this chapter, we are concerned more with fixed-winged UAVs relevant to conduct various aerial observations and draw useful data for agriculturists. Agricultural UAVs per se is a recent idea. Agricultural UAVs seem to have a great role in improving crop production tactics (Anderson, 2015; Greenwood, 2017; Krishna, 2018; National Geographic, 2016). The pace of development and use of agricultural UAVs, particularly, the small fixed-winged versions, have been marked during the past 5–8 years (Krishna, 2016, 2018).

Several hundreds of new companies have churned out, really a wide array of agricultural UAV models. Simultaneously, there are several military UAVs that have been modified and adapted to be useful to farm sector.

Fixed-winged UAV types have their set of advantages and lacunae about their usage above crop lands. They conduct aerial imagery with great pace. They can repeatedly conduct aerial imagery at rapid intervals. They fly at low heights over crop canopy, and pick aerial images and spectral data at rather rapid pace. They usually fly at 45–60 mi h^{-1} and conduct the aerial imagery. The spectral data are highly useful to farm experts and farmers. The crop data is utilized to prescribe fertilizer, irrigation, and other agronomic procedures. This is in addition to the regular surveillance and monitoring tasks in the farms. The small fixed-winged UAVs are among the most preferred models in agricultural operations. Usually, the larger, heavier, and long-distance (endurance) models are preferred for military or other high-altitude observations. Right now, we have a wide range of fixed-winged UAV models. They can be grouped and identified based on size, weight, engine type, source of power, endurance, take-off characteristics, etc. (Arjomandi, 2013; ECA Group-Lima, 2017; see Krishna, 2016, 2018). Fixed-winged UAVs with their relatively lightweight and small payloads are not preferred for spraying pesticides or applying foliar fertilizers. Sprayer UAVs, for now, have been mostly picked from helicopter or multirotor types. Rotor UAVs, because of their versatility in flight, hovering ability, and better payload (pesticide tanks) are used more commonly, for spraying. Perhaps, only a couple of fixed-winged UAVs are known to have been used for spraying. The rapid transit above the crop canopy creates problems related to drift of harmful chemicals and fertilizers. There are several characteristics, based on which the fixed-winged UAVs can be classified. In this chapter, however, for convenience, fixed-winged UAVs are categorized first using their weight. They are grouped as micro-UAVs (<5 kg), light-weight UAVs (5–50 kg), medium-weight UAVs (50–200 kg), heavy UAVs (200–2000 kg), and superheavy UAVs (>2 t). Nano-UAVs have not been dealt. They may have an important role to play in small farm, but there are yet many modifications required. Fixed-winged UAV classes (groups) such as those with wings and fuselage integrated or blended, those with vertical take-off facility, also known as "hybrids," and a few special types have been treated separately in the chapter. Most of the fixed-winged UAV models are best suited for rapid aerial imagery, collection of spectral data of crops and cropping systems, detecting disease/pests, and need for fertilizer and water. There are a couple of UAV models such as Hubsan's Spy Hawk that are small and meant currently for recreation. However, with slight modification, it can deliver good aerial imagery of crops and the happenings in a large farm.

Knowledge about various fixed-winged UAV models, their specifications, and their suitability to conduct different tasks related to crop production (i.e., agricultural uses) is essential. We need a computer-based data bank or at least a compendium of agricultural UAV models. Such a compilation is useful to farm experts/farmers while they decide to procure a UAV. They should opt for the best fixed-winged UAV, depending on the tasks that they wish to accomplish. In the following pages, specifications and uses of about 100 plus different fixed-winged UAVs of relevance to farming sector have been listed. There are also a few models that are not entirely suited to farm sector, but they offer useful insight to the reader. A couple of them listed here are basically military UAV models (e.g., Scan eagle, Aerosonde, and Arcturus), but they can be switched to conduct aerial imagery of large cropping expanses, etc. The sensors used for military or agricultural purposes are same. Of course, the end use of spectral data is different. The UAV models and specifications listed below are arranged in alphabetical order within a group.

2.2 FIXED-WINGED UNMANNED AERIAL VEHICLE SYSTEMS IN AGRICULTURE: MICRO-UNMANNED AERIAL VEHICLES (<5 kg)

Micro-UAVs are small and very light in weight. Hence, they are easily packed in a suitcase or backpack. They are portable to any corner of the crop field. They are most commonly adopted in farms to conduct aerial surveillance of crops and agronomic procedures being conducted by the farm crew. They usually carry payload of compact gimbal with few different types of sensors. They are not amenable for heavy payloads of pesticide tanks. Hence, they are not known to conduct aerial spraying sprees. These microweight UAVs allow easy hand-held take-off. We should hold them at shoulder height and release. Landing could be on belly, or via parachute (e.g., AeroMapper, Delta M) or even utilizing a landing gear. Micro-UAVs show excellent versatility regarding flight path and waypoints covered within a short spell of flight. Most importantly, they are supported by lithium polymer electric batteries and have short endurance period in the air. However, repeated flights at short intervals and adopting the same flight path are also possible. In the following paragraphs, detailed description of the UAV model and address of the company that produces it are provided. Specifications useful to farmers and farm companies are listed for UAVs that are relevant to agriculture.

Name/Model: Aero-M fixed-wing UAV (3DR Aero-M Drone)

Company and address: 3D Robotics, Berkeley, California, USA

Sales agent: Unit 2, 40 Dun Road, Smeaton Grange, New South Wales 2567, Australia. Phone: + 61 (02) 4647 3450; e-mail: sales@riseabove.com.au; technical support e-mail: support@riseabove.com.au, http://www.riseabove.com.au

This is an UAV company started during early 2003. It manufactures small, light-weight micro-UAVs. UAVs produced by this company are modestly costly at Australian $ 8–10,000 (Plate 2.1.1). UAVs produced by this company serve both agricultural and nonagricultural purposes. This UAV is packed in a small briefcase. Therefore, it could be transported anywhere in the farms and cities to launch. It is a versatile UAV capable of aerial imagery of terrain, soils, cropping belts, irrigation sources, farm vehicles, their operation, etc. Detection of crop disease/insects, etc. is also possible (Plates 2.1.1 and 2.1.2).

TECHNICAL SPECIFICATIONS

Material: UAV has a robust frame made of foam. Yet, the platform is light in weight. The frame withstands belly landing on rough surface, if any.

Size: Length: 51 in (129 cm); width (wingspan): 74 in (188 cm); height 20 in (49 cm); weight: 6.8 lb (3 kg)

Propellers: Gemfan 11 × 7 cm. It has one propeller with two blades placed at the center of wingspan, just above the fuselage (Gemfan 11 × 7). The propeller faces backward. Propeller is connected to a tiger motor (2820–2830 kV)

Payload: 1.1 lb or 500 g

Launching information: It is hand-launched at shoulder height. It can also be launched using a catapult.

Altitude: UAV is operated at low altitude just above the crop canopy for close-up shots. It reaches up to 400 m a.g.l. for farm-related operations. However, it can reach 4000 m a.g.l. for studying large areas of crops and vegetation.

Speed: 40 km h^{-1} (9.8 cm s^{-1})

Wind speed tolerance: It withstands <25 km h^{-1}

Temperature tolerance: −20 to +60°C

Power source: Electric batteries

Electric batteries: 4S 6000 mA h 35C lithium–polymer batteries for flight. It has two flight batteries.

Fuel: Not an IC engine-supported UAV

Endurance: 40–60 min per flight

Remote controllers: UAV's path is preprogramed, using Pixhawk autopilot system. Ground control station (GCS) has a 915-MHz connectivity (Spektrum DX7s radio-control). It helps in regulating UAV's flight midway and relay communications. The radioconnectivity is functional for up to 1 km radius from ground station.

Area covered: About 250 ac (1.0 km^2) per flight of 40 min could be surveillanced and imaged.

Photographic accessories: Cameras: Canon SX260 and S100. These Canon cameras are of high resolution (12 MP). This UAV is not compatible with GoPro camera.

Computer programs: This UAV provides 3D imagery. The 3D imagery is stitched using Pix4Mapper LT 3DR version. It is specific to the UAV model. It works with Windows operating system only. UAV's cameras provide accurate high-resolution images. Each image is geo-referenced using GPS coordinates. Ground sampling distance is 2 in. pixel^{-1} (5 cm pixel^{-1}). The orthomosaic accuracy is 3–16 ft (1–5 m).

Spraying area: Not a spraying UAV.

Volume: Not applicable

Agricultural uses: It is supposedly a perfect small flat-winged UAV. It is meant to conduct aerial surveys, scouting, and imagery that help a series of agricultural operations (Plate 2.1.2). It is also useful in monitoring farm installations, dairy cattle, farm vehicles and their movements, etc. Crop scouting, crop phenotyping, measuring normalized difference vegetation index (NDVI), and crop-N status are also possible. The natural vegetation monitoring is also done using "Aero-M UAV."

Nonagricultural uses: Monitoring construction activity, industrial installations, rail roads, highway, and city traffic; surveillance of mines, mining activity, and ore-dumping zones.

PLATE 2.1.1 3DR Aero-M fixed-wing UAV.
Source: Mr. Vick M. Sales and Marketing Division. Rise-Above Custom UAV Solutions Inc. Smeaton Grange, New South Wales, Australia; Mr. Chris Anderson and Mr. Jordi Munoz, 3D Robotics Inc., Berkeley, CA, USA.

PLATE 2.1.2 Left: An aerial map of farm. It was obtained using visual bandwidth sensor on aero-M flat-winged UAV. Right: Ground-station desk computer. It is used to collect images from UAV. Such images could be scanned and relayed to other computers for 2D and 3D processing.
Source: Mr. Vick M. Sales and Marketing Division, Rise-Above Custom UAV Solutions Inc. Smeaton Grange, New South Wales, Australia; Mr. Chris Anderson and Mr. Jordi Munoz, 3D Robotics Inc. Berkeley, CA, USA.

Useful References, Websites, and YouTube Addresses

http://www.riseabove.com.au/3dr-aero-m-915 (accessed June 25, 2017).
https://www.youtube.com/watch?v=xl91tKg126Q (accessed June 25, 2017).
https://www.youtube.com/watch?v=KjNTyFs6tD0 (accessed Nov 3, 2017).
https://www.youtube.com/watch?v=8bwtrYzsBd4 (accessed June 25, 2017).

Name/Models: AeroMapper EV2

Company and address: AeroMao Inc. 33-7030 Copenhagen Road, Mississauga, ON, Canada. E-mail: info@aeromao.com; phone: +1 647 928 4747; website: http://www.aeromao.com/home

AeroMao Inc. located in Ontario, Canada, manufactures the "Aeromapper" series of UAVs. They are meant for acquisition of images for mapping and surveying applications. The "AeroMappers" are currently in use, in more than 45 countries. Clients, include government institutions, private companies, universities, and research organizations. They adopt "AeroMappers" in a wide variety of applications. These UAVs are said to be efficient and accurate, in performing the tasks. Therefore, these UAV models outcompete other brands. AeroMapper is usually packed in a suitcase and transported (Plates 2.2.1 and 2.2.3).

The AeroMapper EV2 is a versatile and affordable UAV. It is designed for remote-sensing and mapping applications. It features a 24-MP camera. It is easy to hand-launch an "AeroMapper." It is fully autonomous but has manual override. It is recovered using a parachute. It has been built to last and is made of reinforced carbon fiber. AeroMapper offers 60-min endurance with full long-range datalink and control. AeroMapper EV2 has wide range of applications in farming, commercial, and even military realms.

TECHNICAL SPECIFICATIONS

Material: The airframe is made of fully reinforced carbon fiber. The payload area is made of kevlar and fiber glass.

Size: length: 120 cm, width (wingspan): 200 cm; height: 35 cm; weight: take-off weight is 4.5 kg

Propellers: One propeller placed in front at the tip of the nose on fuselage. The propeller has two plastic blades.

Payload: Payload allowed is 1.2 kg. It includes a series of cameras, Lidar, and radar equipment.

Launching information: This UAV is hand-launched. It adopts parachute recovery system (Plate 2.2.2). It has autonomous take-off mode and flight. It uses Pixhawk (by 3D Robotics Inc.) software during autonomous navigation. Flight mode can be either manual, automatic, or return to home or loiter. The parachute is made of kevlar linings. It is popped using a single switch, when the UAV is at least at 40 m a.g.l. Belly landing is also possible, but with the assistance of pilot.

Altitude: 4500 m above sea level is the ceiling.

Speed: Cruise speed is 60 km h^{-1}. Maximum speed is 120 km h^{-1}. This UAV is capable of roll and pitch, while on flight. Yet, it offers great stability about photographic equipment and sharpness and accuracy of images. Gimbals are not needed. It has dihedral, small lateral profile, and rudder control.

Wind speed tolerance: It withstands 45 km h^{-1} wind speed during flight and 25 km h^{-1}, if landing, using a parachute. The AeroMapper EV2 can fly even in slight drizzle, since all the electronics are tightly enclosed.

Temperature tolerance: −20 to 60°C

Power source: Electric batteries

Electric batteries: The UAV is supplied power by lithium polymer batteries 28 A h high capacity with four cells.

Fuel: Not an UAV fitted with IC engine.

Endurance: 60 min per flight. Endurance could be increased by enhancing battery power generation and storage.

Remote controllers/Preprogramed flight paths: The ground-station radiolink for flight commands and data download lasts for up to 20 km radius. The telemetry instrument uses 915 or 868 MHz frequency. Flight mission planning is done at GCS, using computers (iPad). Usually, aspects such as altitude, side and length-wise overlaps, orientation of lines, starting point and end, with over 200 way points are decided, using the touchscreen.

Area covered: This UAV covers 3 km^2 from 85 m altitude and offers 1.6 cm pixel^{-1} resolution. It covers 9 km^2 from 250 m a.g.l. and offers 7.8 cm pixel^{-1} resolution. It covers 21 km^2 area from 600 m a.g.l. and offers 11.7 cm pixel^{-1} resolution. The area calculated is based on a 30° side overlap. The area covered is per 60-min flight. Obviously, if more area should be covered, multiple flights are necessary or a few UAVs (swarms) must be launched.

Photographic accessories: Cameras: Sony A 6000 with 20 mm lens for visual R, G, and B aerial imagery. The AeroMapper EV2, it seems that it can provide excellent maps of high accuracy and clarity with its 24-MP camera. The camera has survey-grade lenses and 1.6 cm pixel^{-1} resolution. In addition to multispectral cameras, it is fitted with near infrared (NIR) Sony A 5100 camera. This camera picks reflectance of leaf chlorophyll and also spectral signatures of natural vegetation and crops. It is also fitted with a Parrot Sequoia camera to obtain data of crops such as NDVI, canopy growth, leaf area. The cameras allow estimation of crop-N status. Hence, it helps in calculating fertilizer-N need of the crop at different stages.

Computer programs to process orthomosaic: The ground-station computers utilize Pix4D Mapper to process and produce well-tailored and highly accurate information. The images are marked with GPS coordinates on each frame. Software such as Agisoft or Pix4D is used to develop 3D models and imagery of farm, crop fields, and other features. The Agisoft Photoscan, in fact, helps in producing high-resolution images with geo-reference data.

Spraying area: Not a spraying UAV.

Volume: Not applicable

Agricultural uses: AeroMapper EV2 is utilized to conduct a range of agricultural and related operations from the air. It primarily offers images of the land and its features, natural vegetation and its diversity, crops species cultivated, etc. The sensors are used to collect detailed aerial data about crop, it growth stage, leaf area, leaf-N/crop-N status, disease/pest attack, if any. Such data can be utilized directly, in the variable-rate applicators to supply fertilizers or pesticide. The aerial survey by this UAV also offers information and accurate GPS coordinates of areas affected by drought, floods, nutrient deficiencies, etc. Crop scouting is performed regularly during a crop season, to develop "Digital Crop Surface Models."

Nonagricultural uses: Monitoring and relaying information/images of large industrial installations, buildings, highways, rail roads, oil pipelines, electric lines/poles, etc.; surveillance of mines, ore movement and ore-dumping zone.

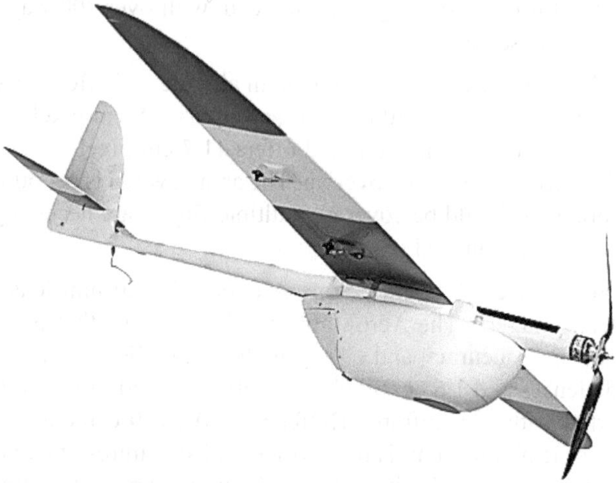

PLATE 2.2.1 AeroMapper EV2.
Source: Mr. Mauricio Ortiz, AeroMao Inc., Mississauga, Canada.

PLATE 2.2.2 AeroMapper EV2 UAV being guided to land, using a popped parachute.
Source: Mr. Mauricio Ortiz, AeroMao Inc., Mississauga, Canada.

PLATE 2.2.3 An AeroMapper EV2 system packed in a suitcase for portability.
Source: Mr. Mauricio Ortiz, AeroMao Inc., Copenhagen Road, Mississauga, Canada.
Note: An AeroMapper EV2 system consists of following items: AeroMapper EV2 platform in carrying case; flight batteries for 60 min of flight; additional batteries can be purchased separately; ground telemetry module (915 MHz, +20 km range); hand-held controller with long-range system (+40 km range) and accessories in carrying case; Sony A-6000, 24 MP camera; survey-grade wide-angle 15-mm lens with adaptor; lithium polymer battery charger; full user's manual with detailed step-by-step instructions; mission planner with instructions to create survey missions and use as GVS and a laptop of choice.

Useful References, Websites, and YouTube Addresses

www.aeromao.com/ (accessed *June 3, 2017*).
www.aeromao.com/aeromapper_uav (accessed *June 3, 2017*).
https://www.unmannedsystemssource.com/shop/vehicles/aeromapper-ev2/ (accessed *June 3, 2017*).
http://www.aeromao.com/aeromapper_uav/ (accessed *Nov 12, 2017*).
https://www.youtube.com/watch?v=gvQeWJAs_sY (accessed *June 17, 2017*).
https://www.youtube.com/watch?v=avrut9DneGw (accessed *June 3, 2017*).

Name/Models: AeroMapper Talon UAV

Company and address: AeroMao Inc. 33-7030 Copenhagen Road. Mississauga, ON, Canada; e-mail: info@aeromao.com; phone: +1 647 928 4747; website: http://www.aeromao.com/home

The AeroMappers are systems that offer the best combination of optics, swappable multisensors capabilities, parachute recovery, strength, reliability, and ease of use (Plates 2.3.1 and 2.3.2). Each UAV manufactured is individually flight tested and delivered. The UAV comes with a detailed step-by-step manual for users with zero experience. The AeroMappers are turnkey systems. They are delivered in ready to use condition with no assembly or set-up required.

Talon offers the easiest way to achieve professional geo-referenced DEMs and orthomosaics. Talon has a very strong epoxy (EPO) foam body. It is made of internal carbon fiber reinforcements and a parachute landing system. It has 24-MP camera with survey-grade lens, fully autonomous flight, hand-launch (no launcher required). The package also includes long-range control and long-range datalink (20 km). All units are delivered flight tested and in ready-to-fly condition. **The camera is installed in panoramic orientation for most optimal area coverage with maximum side overlap.** The AeroMapper Talon complies with all the design standards for UAVs set by "Transport-Canada."

TECHNICAL SPECIFICATIONS

Material: EPO foam reinforced with carbon fiber

Size: length: 140 cm; width (wingspan): 200 cm; height: 30 cm; weight: 3.5 kg

Propellers: One propeller at the rear end of fuselage. It has two blades made of plastic.

Payload: It consists of a range of cameras such as visual, multispectral, hyperspectral, NIR, and infrared (IR) bandwidth.

Launching information: AeroMapper Talon is hand-launched by holding it at the shoulder height. This UAV adopts Pixhawk software produced by 3D Robotics Inc. However, flight modes can be manual, fly-by-wire or autonomous. The flight is autostabilized. The UAV returns to launch spot, if errors occur in navigation or radiolink is lost for any reason. Landing is accomplished using parachute. The parachute is released at least 40 m height above from ground. The parachute is effectively released by just pressing a switch (button). Preprogramed automatic release of parachute is also possible. Parachute is made of kevlar linings. It is easily packed into a UAV, in just under 2 min, before flight. Belly landing is also possible, but with the aid of pilot or preprograming.

Altitude: It is flown at 50–400 m a.g.l. for aerial imagery of land, natural resources, and crops. However, ceiling is 3000 m a.g.l.

Speed: The cruise speed is 50 km h^{-1}; maximum speed attainable is 85 km h^{-1}

Wind speed tolerance: It resists 35 km h^{-1}, if in flight; 25 km h^{-1}, if descending using a parachute.

Temperature tolerance: −20 to +45°C

Power source: Electric batteries

Electric batteries: 16 mA h, two cells

Fuel: It is not an IC engine-fitted UAV, so no need for petrol/diesel

Endurance: 2 h with full payload

Remote controllers: The radiolink at the GCS keeps contact with UAV, for up to 20 km radius, if it is aerial imagery and up to 50 km, if just for flight (loiter). Telemetry is supported by electric batteries. The GCS uses 20 Hz dual frequency 1.1/1.2 radiolink-GNSS (GPS plus GlONASS) receiver.

Area covered: At an altitude of 85 m a.g.l., this UAV covers about 5.2 km^2 area; at 125 m a.g.l., it covers 8 km^2; and at 600 m, it covers 35 km^2. It picks detailed digital data and images of earth's surface features. This is if calculated at 30° overlap. For 3D imagery, 60° side overlap is needed; so, it covers less area.

Photographic accessories: Cameras: This UAV is primarily fitted with Sony 6000 with survey-grade wide-angle rectilinear 20-mm lens. Other sensors include multispectral, hyperspectral, NIR, IR, and red-edge (thermal) sensor. It is recommended to use Parrot Sequoia + RGB Sony 24 MP cameras to get NDVI data and orthomosaics in a single flight. Additional cameras are mica sense–red-edge camera, thermal FLIR, FPV. The cameras are controlled from GCS or preprogramed. They all offer geo-referenced images and videos.

Computer programs: The imagery is processed at the UAV using Pix4D Mapper or Agisoft for 3D imagery. The GCS is equipped with iPads that process images swiftly and relay for analysis.

Spraying area: Not a spraying UAV.

Volume: Not applicable

Agricultural uses: AeroMapper Talon (Plate 2.3.1) could be utilized to survey agricultural crops aerially. Its aerial imagery can provide data such as NDVI, canopy-growth characteristics, leaf chlorophyll (crop-N status), nutrient deficiency if any, water stress index, and grain maturity. The UAV can scout for disease, pest, weeds, drought, and flood effects. Natural vegetation monitoring is also possible (Plate 2.3.3).

Nonagricultural uses: AeroMapper could be used to surveillance buildings, rail roads, city traffic, mines, and mining activity, etc.

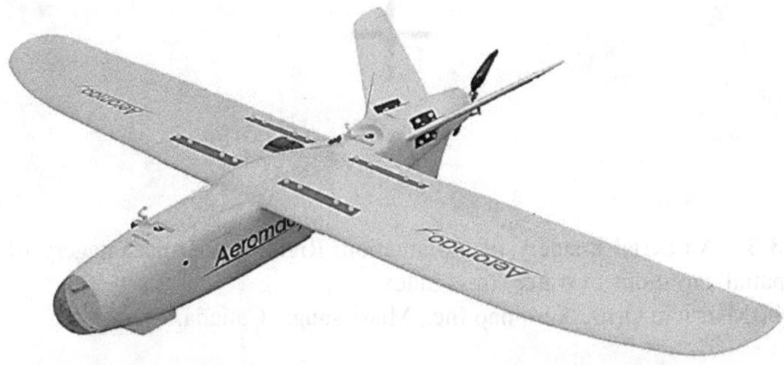

PLATE 2.3.1 AeroMapper TALON UAV.
Source: Mr. Maurico Ortiz, Aeromao Inc. Mississauga, Canada.

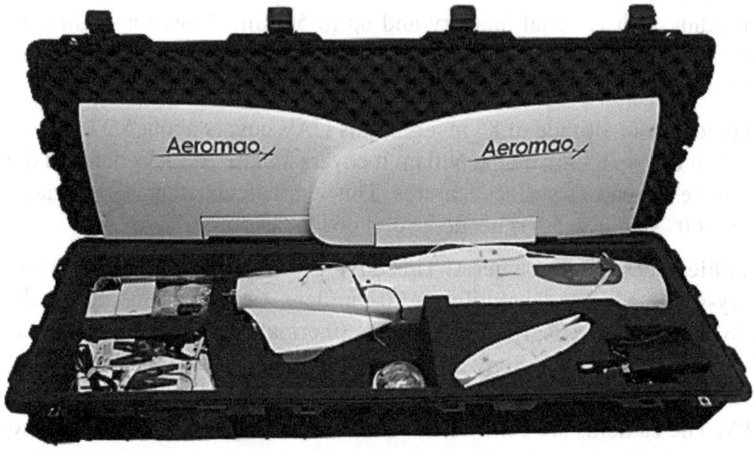

PLATE 2.3.2 A complete package of AeroMappr Talon UAV in a suitcase.
Source: Mr. Mauricio Ortz, Aeromao Inc., Mississauga, Canada.
Note: A complete package includes the following: AeroMapper Talon UAV ready to fly and flight tested; parachute system installed and ready to use; Sony a6000 24 MP camera and 20 mm lens; hand-held controller with long-range system (+50 km range); flight batteries for 160 min of flight. Additional batteries can be purchased separately; telemetry modules (915 MHz, +20 km range); lipo-battery charger and lipo-monitor; carrying case with custom cut foam; full instructions manual with tons of pictures, examples and step-by-step instructions for your first flight and survey missions. No training is necessary; mission planner and full instructions to create survey missions and use as GCS and laptop of choice. Such a complete system of AeroMapper Talon UAV cost US$ 9985 on December 15, 2016).

PLATE 2.3.3 An aerial imagery of a plantation. Right: A thermal imagery of field crop showing spatial variations in water stress index.
Source: Mr. Mauricio Ortiz, Aeromao Inc., Mississauga, Canada.

Useful References, Websites, and YouTube Addresses

https://www.thedroneproshop.com/products/aeromao-aeromapper-talon (accessed June 24, 2017).

https://www.suasnews.com/2016/11/aeromapper-talon-two-hour-endurance/ (accessed June 24, 2017).

http://www.aeromao.com/aeromapper_talon_uav (accessed July 10, 2017).

http://diydrones.com/profiles/blogs/aeromapper-talon-mapping-drone (accessed Nov 3, 2017).

https://www.youtube.com/watch?v=yd8JUzGGgVc (accessed Nov 3, 2017).

https://www.youtube.com/watch?v=9NJXPt2K2aA (accessed Nov 3, 2017).

https://www.youtube.com/watch?v=FgSn0sxcX9A (accessed Nov 3, 2017).

Name/Models: AgDrone System

Company and address: HoneyComb Corporation, 7929 SW Burns Way STE A, Wilsonville, OR 97070, phone +1 503 563 6382; e-mail: not available; Website: http://www.honeycombcorp.com/agdrone-system/

HoneyComb Inc. is a UAV company involved predominately with agriculture and forestry-related aerial survey and growth analysis. The above UAV model (AgDrone) was ranked among the top seven best agricultural UAVs during 2015. HoneyComb Inc. concentrates on aspects related to precision agriculture, such as crop imaging, yield forecasting, yield mapping, NDVI, etc. Aerial imagery is used to mark "management blocks." They also provide digital data for variable-rate applicators. The company is well focused on UAV-based aerial imaging solutions for precision agriculture. HoneyComb Inc. strives to provide robust UAV-based sensing and imaging technologies for agriculture. It also tries to offer accessible and effective data processing solutions. Overall, they intend to increase crop yield and reduce costs.

TECHNICAL SPECIFICATIONS

Material: AgDrone's airframe is made of graphite reinforced plastic. It is toughened to withstand harsh weather conditions.

Size: length: 64 cm, width (wingspan): 124.5 cm; height: 34 cm; weight: 4.95 lb. The UAV comes with a rugged lightweight, waterproof case. The suitcase holds UAV platform and all accessories such as cameras, remote controls, and image-processing material.

Propellers: One on the fuselage at the confluence of wings. The propeller is connected to 575-W electric motor.

Payload: 800 g

Launching information: AgDrone is launched by hand. Computer guides the flight. The UAV takes off at 15° angle. Landing is usually achieved by belly skids on grassy terrain. It needs at least 400 ft clear space to land.

Altitude: UAV operates at 400 m a.g.l. for agriculture-related procedures, like crop survey, monitoring, and analysis. It can reach about 4000 m a.g.l.

Speed: Average speed is 47 km h^{-1}. However, its maximum speed is 82 km h^{-1}.

Wind speed tolerance: 8–16 km h^{-1}

Temperature tolerance: −20 to +60°C

Power source: Electric batteries

Electric batteries: 8000 mA h lithium–polymer batteries

Fuel: It is not an IC engine-fitted UAV, so, no need for petrol or diesel

Endurance: 55 min per flight + 11 min as safety factor

Remote controllers: Communication control for flight path and imagery works up to distance of 4.5 km from the ground station. The UAV can be preprogramed by adopting suitable software.

Area covered: 858 ac h^{-1} or 344 ha h^{-1} at 400 ft altitude above crop fields

Photographic accessories: Cameras: It has visual, red, green, and blue and NIR cameras. Simultaneous exposures are possible. It is done using predetermined programs. Cameras are triggered automatically, using computer programs. The CPU processes and transmits images instantaneously to ground-station computers. Crop-related decisions could be made by visualizing the fields on a computer screen/iPad.

Computer programs: The UAV package comes with "Windows surface Pro 3" tablet. It exports imagery and processed maps instantaneously. It is highly useful for scouting, since images are relayed instantaneously to computer screens. The data could be shared with several computers in the ground station. Field markers and documentation is possible instantaneously. All images are tagged with GPS coordinates. The computer software helps in mission planning, flight path, autonomous landing, safe routines, and in automatic triggering of cameras. Tablet is charged with HiTec x4 plus multicharger (AC/DC)

Spraying area: This is not a sprayer UAV with tanks and spraying nozzles.

Volume: Not applicable

Agricultural uses: It includes crop scouting, phenotyping, monitoring farm activity, and vehicles. Estimating crop NDVI, growth pattern, and maturity are other functions. Detecting drought and flood-affected spots in the crop fields is possible. Devising irrigation schedules based on crop water stress index is possible. Providing images of grain maturity to devise harvest schedules is done routinely.

Nonagricultural uses: Surveillance of large industrial zones, buildings, detecting disasters, and rapid relay of information from disaster zones; monitoring mines, rail roads, and city traffic.

Useful References, Websites, and YouTube Addresses

http://www.honeycombcorp.com/about/ (accessed July 28, 2017).
http://www.honeycombcorp.com/agdrone-system/ (accessed July 28, 2017).
https://www.youtube.com/watch?v=3aVUbcPnTM4 (accessed July 28, 2017).
https://www.youtube.com/watch?v=8djQNiQPFig (accessed July 28, 2017).
https://www.youtube.com/watch?v=-ddCkCkiGfk (accessed July 28, 2017).

Name/Models: AgEagle

Company and address: AgEagle Aerial Systems Inc. 117 South, 4th Str., Neodesha, KS, USA

Phone: +1 620 325 6363; + 620 625 4626; +1 800 625 4626; e-mail: nicolef@ageagle.com; Website: http://ageagle.com/resources/image-gallery/; http://ageagle.com/contact/

AgEagle is a UAV company that caters primarily to farm operations in the Central Plains of North America, and also elsewhere in other crop production zones. It was started in 2008. They supply UAVs (e.g., AgEagle RX60) and related accessories needed to obtain aerial imagery of crop. They also offer digital data that can be used during precision farming methods (Plate 2.4.1; Cheslofska, 2017; M3 Productions, 2017).

TECHNICAL SPECIFICATIONS

Material: The UAV's airframe is made of light-weight carbon fiber, toughened plastic, and foam.

Size: length: 45 cm; width (wingspan): 135 cm; height: 30 cm; weight: 1.8 kg

Propellers: One propeller with two blades. Propeller is located at the rear end of the fuselage.

Payload: 550 g

Launching information: AgEagle is launched using catapult. Autopilot and landing facility is available.

Altitude: 50–150 m for crop surveillance. However, the UAV can reach its ceiling altitude of 5000 m a.g.l.

Speed: 45–60 km h^{-1}

Wind speed tolerance: 20 km h^{-1}

Temperature tolerance: −20 to +50°C

Power source: Electric batteries

Electric batteries: Lithium–polymer batteries 8 mA h

Fuel: This is not an IC engine-fitted UAV, so does not require diesel or petrol.

Endurance: 45 min

Remote controllers: Ground station is usually equipped with 4G LTE and high-speed *Wifi* connectivity. UAV relays images instantaneously to ground-station computers. Images can be processed at the ground station. Processed images are drawn from the CPU on UAV, that is, when processors in the fuselage are used.

Area covered: 150–300 ha per flight

Photographic accessories: Cameras: The UAV carries a set of lightweight visual range sensors that operate at R, G, and B wavelength bands. It also has NIR and IR thermal sensors for detecting crop water stress.

Computer programs: The images are processed using software in the UAV or at the ground-station computers/mobile. Aerial images are instantaneously transferable to any other computer via internet or mobile connectivity. Aerial images are accurate due to GPS coordinates and of high resolution. UAVs are flown low above the crop to achieve high-resolution images of the crop.

Spraying area: AgEagle is not a spraying UAV.

Volume: Not applicable

Agricultural uses: Major uses of AgEagle RX 60 are aerial mapping of farm and its terrain. Further, UAV's images also depict field topography, land, and water resources. Such data showing variations in soil characteristics are utilized to mark the "management blocks" necessary for precision farming. UAV-derived images help in deciding cropping systems, planting density, and scheduling crop husbandry procedures. Crop scouting, measurement of NDVI, mapping leaf chlorophyll, and leaf-N variations in the field are done routinely (Plate 2.4.1), providing digital data and imagery for variable-rate applicators of fertilizer-N. Thermal images to understand variations in crop water stress index. It also supplies data to GCS to regulate center-pivot irrigators with variable-rate nozzles. In other words, it helps to irrigate crops using precision farming principles. Arial survey of weeds, using spectral signatures specific to weeds and crops is a possibility. This UAV also used to monitor farm installations, dairy cattle, and storage facilities (Plate 2.4.2). Natural vegetation monitoring is also a clear possibility.

Nonagricultural uses: AgEagle RX 60 is used in monitoring mining activity, transport vehicle and their movement, city traffic and policing of public events, oil pipeline surveillance, industrial installations, etc.

PLATE 2.4.1 An AgEagle—a fixed-winged UAV above a maize crop.
Note: The UAV is flying at a low altitude above the crop. It allows the farmers to obtain high-resolution images of the crop. It helps in accruing accurate data about NDVI, leaf chlorophyll content, disease/pest attack, drought, and flood effects on crops/fields.
Source: Mr. Tom Nicholson, CEO, AgEagle Aerial Systems Inc., Neodesha, KS, USA.

PLATE 2.4.2 An aerial image of a farm obtained, using visual range sensors on the AgEagle UAV.
Note: Farm installations could also be monitored regularly, using this small fixed-winged UAV.
Source: Mr. Tom Nicholson, CEO, AgEagle Aerial Systems Inc., Neodesha, KS, USA.

Useful References, Websites, and YouTube Addresses

Cheslofska, D. 2015 *The 7 Best Agricultural Drones on the Market*; DANICA, pp 1–3. http://dronelife.com/2015/10/14/7-best-agricultural-drones-market/ (accessed June 19, 2017).

M3 Productions. *2017 M3 Aerial Productions*; pp 112. http://www.m3aerial.com/ (accessed June 19, 2017).

http://ageagle.com/technology/ (accessed June 24, 2017).

http://ageagle.com/resources/image-gallery/ (accessed Dec 10, 2017).

http://www.falconuav.com.au/uav-products (accessed June 24, 2017).

https://www.youtube.com/watch?v=ikU39yitmYk/ (accessed June 24, 2017).

https://www.youtube.com/watch?v=dJWxQpdjgp4/ (accessed June 19, 2017).

https://www.youtube.com/watch?v=kvyh-fxpeSE/ (accessed June 19, 2017).

https://www.youtube.com/watch?v=2sND-jnEHbQ/ (accessed June 19, 2017).

Name/Models: AGPlane Arara

Company and address: AGX Tecnologia Ltd.; A, R. José Rodrigues Sampaio, 361, Centreville, São Carlos, SP CEP 13560-710, Brazil; Tel./Fax: +55 16 3372 8185; e-mail: commercial@agx.com.br; contato@agx.com.br; website: www.agx.com.br

AGX Technologia initiated an agricultural UAV project in 2002. It was done in collaboration with the Computer and Mathematical Science Department, of Instituto Ciencias and Computaceo, University of Sao Paolo and EMBRAPA (Agricultural Research Organization of Brazil). The project developed light-weight UAV that aids accurate studies via imagery of crops (Pelizzoni et al., 2011). The UAV is specially developed to study crops, and their growth pattern, using aerial imagery. It helps to collect data on NDVI, crop water stress index, develop irrigation schedules, and forecast yield (Trindade and Jorge, 2004). It also helps in assessing environmental parameters. It offers generalized aerial imagery of natural resources, large forest plantations, and riverine zones. This company's other UAV projects involve development of heavy UAVs with medium and long-range surveillance capabilities for military and civilian uses. A few other UAV models by AGX help in geospatial information. AGX Technologia Ltd. also specializes in producing sensors and software, for rapid and accurate aerial imagery and relay to ground stations (Year Book, 2012). *Arara* means parrot.

TECHNICAL SPECIFICATIONS

Material: The platform is made of composite developed using carbon fiber, toughened plastic and foam.

Size: length: 115 cm; width (wingspan): 140 cm; height: 45 cm; weight: 3.2 kg

Propellers: One propeller with two blades. It is placed at the center of the wingspan, just above the fuselage.

Payload: 1.5 kg

Launching information: The UAV is launched via a catapult or by using a pick-up van. The pick-up van carries the UAV on top and moves at a high speed for a distance to get the thrust. It has autonavigation software and flies fully autonomously. Landing gear with three wheels helps in landing. Parachute landing is also possible. The parachute should be popped about 40 m above the ground.

Altitude: It gains altitude as stipulated by ground computers and changes are automatic based on preprograming.

Speed: 45–60 km h^{-1}

Wind speed tolerance: 20 km h^{-1}

Power source: Electric and internal combustion engine (40 cm^3, 5 HP engine)

Electric batteries: 14,000 mA h to support electronic circuits and CPU computers

Fuel: AVGAS plus 2T lubricant

Endurance: 4 h per flight

Remote controllers: The GCS computers and antenna are packed in a portable suitcase for transit. The GCS has automatic videolink with UAV for up to 15 km radius.

Area covered: 400 to 600 ha per flight at 400 m altitude

Photographic accessories: Cameras: The gimbal and photo cameras are placed under the fuselage. It includes visible range (R, G, and B), multispectral, hyperspectral, NIR, IR, and red-edge sensors. It may also carry a Lidar pod.

Computer programs: The raw images relayed to GCS computers are processed, using software such as Pix4D Mapper or Agisoft's photoscan.

Spraying area: This is not a spraying UAV.

Volume: Not applicable

Agricultural uses: Major emphasis is on-field scouting for seed germination, seedling establishment, NDVI, leaf area, leaf-N, grain maturity, etc. It is also used to map disease/pest-attacked zones, drought, and flood-affected patches. Natural vegetation monitoring is also a possibility.

Nonagricultural uses: Surveillance of mines, mining activity, industrial sites, buildings, oil and gas pipelines for leakage and pilferage, electric transmission lines, city traffic and tracking vehicles, etc.

Useful References, Websites, and YouTube Addresses

Pelizzoni, J. M.; Neris, L. O.; Trindade, O., Jr.; Osório, F. S.; Wolf, D. F. Tiriba—A New Approach of UAV Based on Model Driven Development and Multiprocessors. *ICRA Commun.* **2011**, 1–9.

Trindade, O. and Jorge, A. C. Using UAVs for Precision Farming. *Unmann. Syst. Mag.* **2004,** 35–39.

Year Book. Brazilian UAS Community—Country Reports. *UAS—The Global Perspective*, 9th ed.; Byenburgh and Co., 2012, pp 85/216. http://www.usv-info.com (accessed Nov 26, 2016).
https://www.youtube.com/watch?v=Wyecwkmb67A (accessed Nov 12, 2016).
https://www.youtube.com/embed/Wyecwkmb67A?fs=1 (accessed Nov 12, 2016).

Name/Models: AGRI 2000 RPAS

Company and address: Micro-Aerial Vehicles Technology (MAVTech s.r.l.), MAVTech s.r.l.—corso Galileo Ferraris 57, 10128 Torino, Italy; Tel.: +39 011 5808482; fax: +39 011 5808579; e-mail: mavtech@mavtech.eu; website: http://www.mavtech.eu/it/contacts/

MAVTech s.r.l. was founded as a spin-off company of Politecnico di Torino (2005–2014). Currently, it is in Bozen, as a Technology Company of TIS Innovation Park. MAVTech s.r.l. is involved in designing and developing technical solutions related to UAVs. MAVTech s.r.l. is a company with a primary interest in the development of innovative solutions for aerial surveillance and tactical operation support for civil applications, transferring new aerospace technologies from research to the operational domain. Their UAV is also used in judging atmospheric quality (Malaver et al., 2015).

AGRI2000 RPAS (remotely piloted aircraft system) is a fixed-wing-powered aircraft. It has a wider application, because of better payload capabilities and flight performances. It has superior mission range and endurance, when compared with few other existing concepts. The design is compact with an adequate aerodynamic efficiency. Structures are light; the vehicle is spin resistant and stable in flight. The entire UAV and small accessories are easily foldable. They are enclosed in a portable suitcase. Therefore, it can be launched from anywhere in the crop field. It costs moderately (see Guglieri and Ristorto, 2016).

TECHNICAL SPECIFICATIONS

Material: Agri 2000 UAV is made of EPO and carbon fiber.

Size: length: 140 cm; width (wingspan): 200 cm; height: 30 cm; weight: 4.0 kg

Propellers: One propeller at the rear end of fuselage. Propulsion is achieved using a battery-powered brushless motor.

Payload: <1.0 kg

Launching information: Agri 2000 UAV's take-off could be accomplished by hand. We must hold the airframe at shoulder height. It can be launched using catapult. The flight is autonomous, dictated by a preprogramed flight path. It can also be recovered (landed) by guiding it, using radiolink or via autonomous landing instructions. Emergency landing via parachute is also possible. The parachute should be ejected at least 40 m a.g.l.

Altitude: Operating altitude is 150 m a.g.l.

Speed: 7.5–20 m s^{-1}

Wind speed tolerance: It tolerates 20 km h^{-1} disturbing wind. It also withstands slight drizzle during flight.

Temperature tolerance: −20 to +60°C

Power source: Electric batteries

Electric batteries: Lithium–polymer batteries 28 A h; four cells.

Fuel: It does not possess an IC petrol engine.

Endurance: 60 min at a speed of 15 m s^{-1}

Remote controllers: The ground station uses a 2.4-GHz in-flight mode. Telemetric link is available via radiomodem at 433 MHz override. MAVTech s.r.l has released a new software package (*JavaCube*) that focuses on the specific needs and requirements of flight planning and passive collision avoidance (including grid waypoint planning for optimized photogrammetric surveys).

Area covered: 50–200 ac h^{-1}

Photographic accessories: Cameras: A standard video system is provided as default. It relays image and digital data to ground station instantaneously. The assortment of cameras for agricultural crop survey by AGRI 2000 UAV includes visual (R, G, and B), high-resolution camera (Sony a5100); multispectral camera (Canon); thermal-imaging camera; and mica-sense red-edge camera. These cameras offer a wide range of digital information for use in variable-rate applicators.

Computer programs: Photographic details and digital data processing post flight is supplied as default. It usually consists of software such as Pix4D Mapper, or Agisoft Photoscan.

Spraying area: Agri 2000 not a spraying UAV.

Volume: Not applicable

Agricultural uses: The UAV company lists following as major agriculture-related uses of AGRI 2000: They are crop health and disease monitoring; vegetation identification and mapping; terrain mapping; invasive species identification and analysis (invasive plants); land and forestry research; vigor mapping and frost mitigation; agricultural insecticide spraying; agricultural fertilizer dispensing, etc.

Nonagricultural uses: It includes surveillance and monitoring of large industries, buildings, dams, highways, rail roads, vehicle traffic, vehicle tacking using tracking devices, monitoring mining activity and ore transport, etc.

Useful References, Websites, and YouTube Addresses

Guglieri, G.; Ristorto, G. Safety Assessment for Light Remotely Piloted Aircraft Systems. In *Proceedings of International Conference on Air Transport*; 2016; pp 1–7.

Malaver, A.; Motta, N.; Corke, P.; Gonzalez, F. Development and Integration of a Solar Powered Unmanned Aerial Vehicle and a Wireless Sensor Network to Monitor Greenhouse Gases. *Sensors* **2015,** *15*, 4072–4096. DOI:10.3390/s156204072 (accessed June 13, 2017).
http://www.mavtech.eu/it/contacts/ (Dec. 7, 2016).
http://www.mavtech.eu/it/applicazioni/agricoltura-di-precisione/ (accessed Dec. 7, 2016).
http://www.mavtech.eu/it/applicazioni/mappatura-aerea/ (accessed Dec. 7, 2016).

Name/Models: Agriculture UAV

Company and address: Aerial Agriculture LLC, 320 North Street, West Lafayette, IN 47906, USA. Phone: 765 807 2585; e-mail: info@AerialAgricultureUSA.com; website: http://www.AerialAgUSA.com

Technical specifications

Material: The UAV is made of ruggedized foam, rubber, carbon fiber, and steel (rods) (Plate 2.5.1)

Size: length: 75 cm, width (wingspan): 125 cm; height: 35 cm; weight: 2.3 kg

Propellers: It has one propeller with two blades. The propeller is fixed to the center of the wing span and facing backward.

Payload: 500–750 g. It includes visual and IR cameras, CPU, and batteries.

Launching information: It is launched by hand or using general catapult.

Altitude: It reaches 400 m height above crop's canopy for agricultural survey. But, it can attain about 11,000 ft a.g.l.

Speed: 60–70 km h^{-1}

Wind speed tolerance: 35–40 km h^{-1}

Temperature tolerance: −17 to +55°C

Power source: Electric batteries

Electric batteries: Lithium-polymer batteries, 24,000 mA h.

Fuel: Not an IC engine-fitted UAV; so, no requirement for petroleum. No need for petrol or diesel.

Endurance: 45 min per flight

Remote controllers: UAV is controlled using a radioconnectivity from ground controls station, for up to 3 km. Flight path can be altered using override facility at the GCS iPad/mobile app.

Area covered: This UAV surveys and images covering about 200 ac of crop land per flight of 45 min to 1.0 h

Photographic accessories: Cameras: The UAV has usual set of visual, multispectral, hyperspectral, NIR, IR, and red-edge cameras.

Computer programs: Aerial imagery is processed using software such as Pix4D Mapper or Agisoft Photoscan.

Spraying area: This is not a sprayer UAV.

Volume: Not applicable

Agricultural uses: This UAV is useful in scouting land, soil, and water resources. It is used to obtain aerial imagery of crops, canopy growth, leaf chlorophyll content (green index via NIR camera), crop's N status, and water stress index, etc. Natural vegetation monitoring is also possible.

Nonagricultural uses: They are surveillance of industries, large buildings, public events, city traffic, rail roads, mines, convoy of transport vehicles, etc.

PLATE 2.5.1 An agricultural UAV flying above the crop fields in Indiana.
Source: University of Purdue, Lafayette, USA; Aerial Agriculture LLC, West Lafayette, IN, USA.

Useful References, Websites, and YouTube Addresses

https://www.purdue.edu/newsroom/releases/2016/Q3/purdue-students-launch-agricultural-drone-startup-to-help-reduce-farming-costs.html (accessed Jan 3, 2017).

Name/Models: AI 450-ER

Company and address: Aeroterrascan Inc. Jalan Haji Wasud No. 17, Dipatikur, Bandung West Java, Indonesia, ID 40132; phone: +62 22 250 2323; e-mail: contact@aeroterracsan.com; Website: http://www.aeroterrascan.com

Aeroterrascan Inc., situated at Bandung in Indonesia, is a relatively recent UAV company. It was started in 2010. It aims at developing a range of UAV models that suit the need of farmers, foresters, and commercial companies. They try to accomplish aerial survey and collect digital data, swiftly. UAV, such as AI 450-ER helps in monitoring the crop fields, farms, farm animals, and vehicles. This UAV also helps in surveying and monitoring geologic sights. These are low-cost, small UAVs that are launched from anywhere.

AI 450-ER series is considered the flagship UAV model of the company. This UAV serves the farmers and other commercial clients with accurate aerial survey, aerial mapping, monitoring, and collecting digital data about land features.

TECHNICAL SPECIFICATIONS

Material: This UAV is made of a composite of lightweight, carbon fiber, and fiber glass.

Size: length: 105 cm; width (wingspan): 145 cm; height: 48 cm; weight: 2.4 kg

Propellers: Propulsion is achieved by linking battery-powered motors to propeller.

Payload: Payload consists of a series of visual, multispectral, thermal, and mica-sense red-edge wave band cameras.

Launching information: It needs a launcher to take-off. It is autonomous regarding launch, flight and navigation, and landing. A software such as Pixhawk autopilot is utilized by the CPU in the UAV for autopilot. It uses satellite guidance/GPS to navigate.

Altitude: It operates at 100–400 m height for aerial imagery of crop fields. However, it can reach 1200 m, if needed, particularly, while on surveillance flight over civil land scape.

Speed: 60 km h^{-1}

Wind speed tolerance: 20–45 km h^{-1}

Temperature tolerance: −20 to +55°C

Power source: Electric batteries

Electric batteries: 16,000 mA h

Fuel: This is not an IC engine-fitted UAV; so, no need for petrol or diesel.

Endurance: 60–70 min per flight

Remote controllers: The communications between the UAV and GCS is established, using 2.4-GHz remote 900-MHz telemetry. Radiolink keeps the UAV/GCS in contact for up to 10 m radius. There are options to extend telemetric link to 25 km.

Area covered: This UAV covers about 60 km linear length about 1000 ha of farmland per flight.

Photographic accessories: Cameras: It is fitted with gyro-stabilized gimbal that houses R, G, and B Sony A 5011 visual camera. The cameras offer 24 MP frame resolution. It has a videocamera "Ya" (1080, p60). It also has a thermal camera to detect canopy and air temperature. It helps in finding water stress index of the crops.

Computer programs: Software such as Pix4D Mapper or Agisoft Photoscan is used for processing the orthoimages.

Spraying area: This is not a spraying UAV.

Volume: Not applicable

Agricultural uses: This UAV can be utilized to conduct aerial survey of land and its features, general topography, to study cropping systems, crop canopy characteristics, leaf chlorophyll (leaf-N), crop-N status, water stress index, and grain maturity status. Crop scouting for disease, pest attack, drought, or flood damage can be conducted with great accuracy and at a very fast rate.

Nonagricultural uses: They are surveillance of industrial sites, buildings, city traffic, public events, oil pipelines, highways, rail roads, and mines are all possible with this UAV.

Useful References, Websites, and YouTube Addresses

http://www.aeroterrascan.com (accessed Dec 30, 2106).
https://www.youtube.com/watch?v=0g5dAuhMYjs (accessed Dec 30, 2106).
https://www.youtube.com/watch?v=sruMqow7qPE (accessed Dec 30, 2106).
https://www.youtube.com/watch?v=pf92N5F7j4k (accessed Dec 30, 2106).

Name/Models: AI 600

Company and address: Aeroterrascan Inc. Jalan Haji Wasud No. 17, Dipatikur, Bandung West Java, Indonesia, ID 40132; phone: +62 22 250 2323; e-mail: contact@aeroterracsan.com; website: http://www.aeroterrascan.com

AeroTerrascan Inc. is a small UAV manufacturing company. It is situated in Bandung, Indonesia. They offer a series of different models of fixed-winged UAVs. These models are useful in civilian and agricultural aspects.

The AI 600 is a sleek, small fixed-winged UAV. It is one of the more recent models released by Terrascan Inc. It has a set of visual still and videocameras to help in obtaining aerial images of crop fields and forests.

Technical specifications

Material: It is made of carbon fiber, plastic, and fiber glass.

Size: length: 65 cm, width (wingspan): 90 cm; height: 35 cm; weight: 1.3 kg

Propellers: It has one propeller attached to tip of the sleek fuselage.

Payload: 750 g

Launching information: It is hand-launched. It can also be launched using a catapult.

Altitude: It operates at 100–400 m height for aerial imagery of crop fields. However, it can reach 1200 m, if needed, particularly, while on surveillance flight over civil land scape.

Speed: It transits at 40–80 km h^{-1}

Wind speed tolerance: 20 km h^{-1} disturbance

Temperature tolerance: −20 to +55°C

Power source: Electric batteries

Electric batteries: 24 mA h lithium polymer battery

Fuel: No need for petrol or diesel.

Endurance: 60–70 min per flight. The GCS telemetry uses 2.4 GHz, 900 MHz

Remote controllers: The GCS has radiolink with UAV for up to 10 km radius. There is an option to extend this to 25 km radius.

Area covered: It is 60 km in linear distance or about 1000 h, depending on the flight path planned.

Photographic accessories: Cameras: It has visual still camera Sony-5100 (R, G, and B). It also has a videocamera "Ya" (1080, p60).

Computer programs: The images are processed using software such as PIX4 D Mapper or Agisoft's Photoscan.

Spraying area: This is not a sprayer UAV.

Agricultural uses: Major functions of this sleek UAV are crop scouting and monitoring natural vegetation. It collects spectral data useful to be used in precision farming.

Nonagricultural uses: They are surveillance of mines, mining regions, ore movement, oil and gas pipelines, rods, city traffic, tracking vehicles, etc.

Useful References, Websites, and YouTube Addresses

http://www.aeroterrascan.com (accessed Jan 7, 2017).
https://www.youtube.com/watch?v=0g5dAuhMYjs (accessed Jan 7, 2017).

Name/Models: ALTiMapper AV-01

Company and address: Aerovision Unmanned Arial Solutions Pty. Ltd., 43, Sea Cottage Drive, Noordhoek, Cape Town, 7979, South Africa; info@aerovision-sa.com; Website: http://www.aerovission-sa-com/contact; Tel.: 0825641809 (internal), +27825641809 (international)

Aerovision South Africa was founded in 2012. This UAV company aims to satisfy the demand for UAV imagery and remote sensing in South Africa. It also strives to be an industry leader for Unmanned Aerial Solutions in Africa. Aerovision is currently focusing on providing cost effective, yet, professional aerial videographic services.

Aerovision is also an agricultural services agency dealing with UAVs. It aims to supply crop data to farmers in South Africa and elsewhere. AltiMapper AV 01 was launched recently in 2016. They use ALTiMapper flat-winged UAV and a few other copters, to collect data about topography, land, soil, and cropping systems.

ALTiMapper is used for aerial mapping of agricultural regions, particularly, crop fields. In addition, their aerial solutions include pilot training, service and maintenance of UAVs, and GCS equipment (Plates 2.6.1 and 2.6.2). They also use helioballoons systems to survey agricultural farms, scout them for disease/pests, water status and grain maturity.

TECHNICAL SPECIFICATIONS

Material: Tough EPO foam fuselage and wings reinforced with carbon fiber.

Size: length: 15 cm; width (wingspan): 195 cm; height: 12 cm; weight: 2.6 kg

Propellers: A single propeller with two blades. Propeller is paced at the rear tip of the fuselage facing backward.

Payload: <3 kg payload. It includes cameras, CPU, and batteries. The cameras and CPU are placed tightly in the fuselage and well protected against vibrations.

Launching information: It has a fully automated take-off and landing facility. It is also amenable for Bungee and hand launching. The UAV returns to point of launch, in case of emergency or at the end of flight mission. The autopilot mode is supported by Pixhawk Autopilot software, mainly to fix launching, flight pattern, and landing.

Altitude: It operates at <400 m a.g.l. for agricultural survey, scouting, and mapping. However, the UAV can reach 3–4000 m a.g.l.

Speed: The UAV flies at 45–60 km h^{-1}

Wind speed tolerance: 30 km h^{-1}

Temperature tolerance: −20 to +55°C

Power source: Electric batteries

Electric batteries: Lithium–polymer batteries (8000 mA h)

Fuel: This is not an IC engine-powered UAV. There is no requirement for petrol or diesel.

Endurance: 60 min per flight

Remote controllers: ALTiMapper UAV has 2 km radiocontrol range from the ground station

Area covered: It covers an area of 50–150 ac per flight. It has computerized mission planning facility. Flight plans could be altered midway using iPads.

Photographic accessories: Cameras: UAV supports the usual four different visual cameras, the R, G, B, and NIR bandwidth. It can also support a thermal sensor. Usually, this UAV is fitted with Sony a5100 (24 MP) visual camera for high-resolution imagery. A modified Sony a5100 is used at NIR bandwidth for chlorophyll and IR for thermal imagery. Cameras are integrated with ground-station computers and are triggered automatically. Cameras provide 24 MP images of ground and crops (Plate 2.6.2). They offer GPS-tagged images so that images can be swiftly used in ground vehicles, during variable-rate supply of inputs like fertilizers, water, and plant protection chemicals.

Computer programs: UAV's images are processed using Pix4D Mapper, Agisoft, or similar custom-made software.

Spraying area: This is not a spraying UAV.

Volume: Not applicable

Agricultural uses: They are aerial imagery of general topography, land, soil type, crops, and natural vegetation monitoring; crop scouting for seed germination, crop stand, NDVI, and growth measurements. It offers data about crop water stress index (thermal cameras), leaf area index, leaf chlorophyll content, crop-N status, and grain maturity, etc. It could be used to monitor progress of agronomic procedures in large farms. It records positions of different farm vehicles and robotic tractors. It can be used to surveillance irrigation equipment, note water level in reservoirs, and functioning of irrigation channels. It can be used to report drought/flood-affected locations, within a large farm.

Nonagricultural uses: They are mapping construction areas and reporting activity, monitoring buildings, large industrial installations, pipelines and electric lines, monitoring golf estates, and detecting and reporting fire damage in industries and civil locations. Reporting on mining progress and observation of mine vehicle movement are also possible.

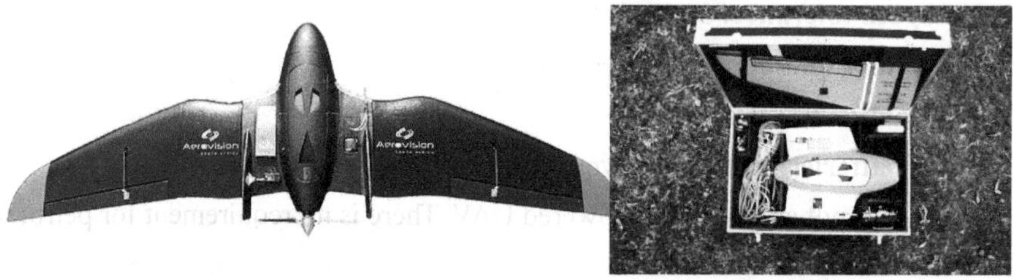

PLATE 2.6.1 ALTiMapper AV-01.

Left: The UAV platform. Right: Attach case for packing and transporting the UAV plus the ground-station accessories (remote controller).

Source: Mr. Ian Freemantle, CEO, Aerovision Unmanned Aerial Solutions Pty. Ltd., Noordhoek, Cape Town, South Africa.

Note: UAV's cameras allow high-precision imagery at a wide range of bandwidths including visual, NIR, and red-edge wavelengths. Images and data are useful during precision farming operations.

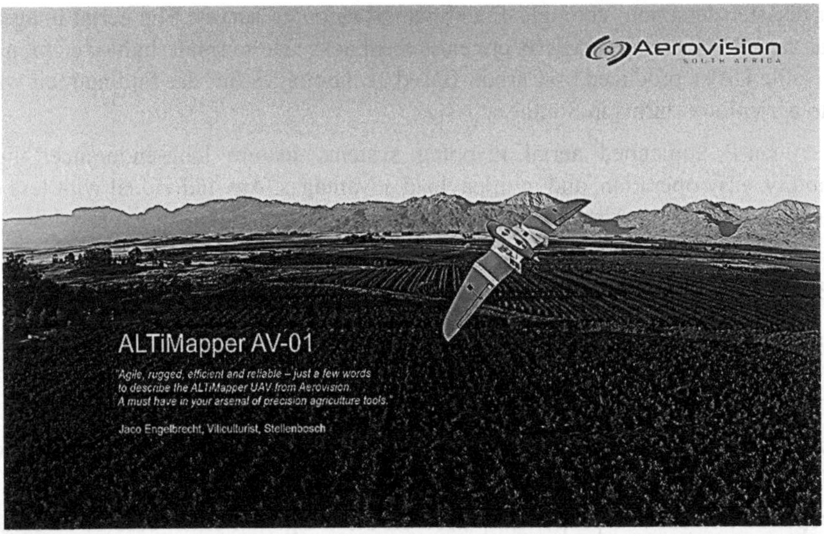

PLATE 2.6.2 An ALTI Mapper UAV on a flight mission over crops in South African farmland.
Source: Mr. Ian Freemantle, CEO, Aerovision Ltd., Noordhoeck, Cape Town, South Africa.

Useful References, Websites, and YouTube Addresses

http://www.aerovision-sa.com/altimapper (accessed Mar 12, 2017)

http://www.aerovision-sa.com/unmanned-aerial-solutions (accessed Mar 12, 2017).

http://www.aerovision-sa.com/altimapper?lightbox=dataItem-iqnkkbmj/ (accessed Mar 12, 2017).

http://www.suasnews.com/2016/09/altimapper-av01-fixed-wing-mapping-platform-launched/ (accessed Mar 12, 2017).

https://www.youtube.com/watch?v=swodxuPITek/ (accessed Sept 12, 2017).

https://www.youtube.com/watch?v=AE-pJd5QL1w/ (accessed Sept 12, 2017).

https://www.youtube.com/watch?v=2qtm-ZuRUYU (accessed Sept 12, 2017).

Name/Models: Avian-P

Company and address: Carbon-Based Technology Inc., Central Science Park/3f, No 30, Keya Road, Daya District, Taichung City, 42881, Taiwan; phone +886 4 2565 8558; e-mail: info@uaver.com; website: http://www.uaver.com

Carbon-Based Technology Inc. was established at Taichung City in Taiwan around March 2007. This company concentrates on development of design, integration, manufacturing, and global marketing of unmanned aircraft system. It offers high reliability unmanned aerial system (UAS) useful for "intelligence gathering." These UAVs also conduct aerial surveys of commercial and agricultural regions and various other aspects. The UAS can integrate with techniques of data processing and analyzing at GCS. The team at Carbon-Based Technologies Inc. offers solutions for disaster detection, geological, and natural resource survey. The aerial imagery that these UAVs relay serves as priceless feedbacks. These small, light-weight, and versatile UAVs produced by Carbon-Based Technologies Inc. are finding their way into agricultural farms in Southeast Asia.

Avian-P unmanned aerial mapping systems feature long-endurance, high stability, easy operation, and multipayload advantages. Any individual with less or even no aviation experience can easily accomplish and aerial mapping mission.

TECHNICAL SPECIFICATIONS

Material: The airframe is made of evelar, carbon fiber, and foam (honeycomb structure). The body is sturdy and toughened to withstand harsh weather and rough landing.

Size: length: 120 cm; width (wingspan): 160 cm; height: 35 cm; weight: 4.7 kg (10.3 lb)

Propellers: It has two propellers placed on the wings, on either side of fuselage. They have two plastic blades each.

Payload: 1.8 kg. It includes cameras such as visual, NIR, IR, and red edge.

Launching information: It is launched by placing it on a launch rack and using a "Bungee cord" to get the thrust. It has autolaunch, navigation, and landing mode. The autonomous flight is regulated, using Pixhawk autopilot software. Landing is achieved using a parachute.

Altitude: Agricultural crop survey is done at 50–100 a.g.l. However, ceiling is 4000 m a.g.l.

Speed: Cruise speed 63–81 km h^{-1} (35–45 knots).

Wind speed tolerance: Beaufort scale 6 (39–45 km h^{-1} or 22–27 knots)

Temperature tolerance: −20 to +60°C

Power source: Electric batteries

Electric batteries: 16 A h high capacity, Li–polymer batteries.

Fuel: It is not an IC engine-fitted UAV. It does not require petrol or diesel.

Endurance: 80–90 min

Remote controllers: The GCS has a radiolink that operates at 433/868/915 MHz (hopping). The radiocontrol has limit for datalink and download. It functions within 12 km radius.

Area covered: The UAV's cameras cover an area of 6 km^2.

Photographic accessories: Cameras: The main cameras included are visible range, multispectral, hyperspectral, NIR, IR, and red-edge bandwidth. It has a Sony α5100 camera with 16 Mega-pixel resolution (or 5 cm resolution).

Computer programs: The aerial images are processed using the software such as Pix4D Mapper or Agisoft Photoscan.

Spraying area: This is not a spraying UAV.

Volume: Not applicable

Agricultural uses: Avian-P has a range of applications in the realm of agricultural crop production. The UAV could be effectively used to scout the crop fields for seed germination, seedling stand, crop canopy growth and surface features, leaf-N, and crop-N status. Crop scouting is also done to detect disease, pest-attacked zones, drought or flood-affected zones, etc. Natural vegetation monitoring could be done regularly and at rapid intervals, using this UAV.

Nonagricultural uses: Avian-P is useful in aerial surveillance, imaging, mapping, and scouting. This UAV is used to monitor large industries, buildings, mines, highways, rail roads, oil pipelines, electric lines, etc.

Useful References, Websites, and YouTube Addresses

http://www.uaver.com/avian-p-aerial-mapping.html (accessed Mar 24, 2017).

http://www.manufacturers.com.tw/showroom.php?c=9367&l=4&f=11&p=wxFG0-BCmKw (accessed Mar 24, 2017).

https://www.youtube.com/watch?v=BVFminpiv38 (accessed Mar 24, 2017).

https://www.youtube.com/watch?v=sYRMrKWvjG4 (accessed Mar 24, 2017).

Name/Models: AW1 E-STAR

Company and address: Allied Drones Inc., 3480 W. Warner Avenue, Santa Ana., CA 92704, USA; phone: +1 844 832 8412; e-mail: info@allieddrrones.com, media@allieddrones.com; website: https://allieddrones.com/contact-us/

Allied Drones Inc. is a company specializing in fixed-wing UAVs. They cater different models of small UAVs meant for civilian, agricultural, and military reconnaissance.

AW1 E-STAR is a lightweight fixed-wing UAV. This is a wide area coverage UAV. It has been designed to be useful in aerial survey, mapping, and precision agriculture. It offers coverage of wide area during imagery. This UAV is useful in collecting information, during disasters and floods, etc. This is an easily portable, small UAV of 1.5 kg. It has easily detachable and foldable wings and fuselage portion. This allows quick transport to anywhere in a crop field for relaunch.

TECHNICAL SPECIFICATIONS

Material: The UAV's body is made of durable high-density foam, ruggedized plastic, and carbon fiber composite.

Size: length: 120 cm, width (wingspan): 180 cm; height 45 cm; weight: 1.5 kg without batteries

Propellers: It has one propeller at the rear end of fuselage.

Payload: It allows 1.5 kg payload inside two internal bays in the fuselage. Payload bay of 40 × 8 × 8 cm is situated in the center and the other of 40 × 5 × 8 cm in front.

Launching information: This UAV can be launched using hand (at shoulder height) or using a bungee catapult system. It has autopilot mode supported by 3DR Pixhawk software. Landing is on belly.

Altitude: Generally, it is flown at 50–400 m a.g.l. for precision agriculture purposes. However, ceiling is 3000 m.

Speed: 45–60 km h^{-1}

Wind speed tolerance: 15 km h^{-1}

Temperature tolerance: −20 to +60°C

Power source: Electric batteries

Electric batteries: Lithium–polymer batteries 8–14,000 mA h; two cells

Fuel: Not UAV fitted with IC engine

Endurance: 1 h

Remote controllers: Ground station is made of iPads or PCs. They regulate flight path. UAV is capable of preprogramed flight path and autolanding. This UAV has been designed for belly landing.

Area covered: 150–200 ha per flight

Photographic accessories: Camera: This UAV comes with a preconfigured Sony QX1 24 MP, NIR, IR, red edge, and Lidar sensors. The fixed wing can also accommodate an extra camera each for hyperspectral imagery of crops with greater accuracy.

Computer programs: UAV's images are processed using Px4D Mapper or Agisoft software. Preprocessed images could also be transmitted from UAV's CPU, directly to ground iPads.

Spraying area: This is not a spraying UAV.

Volume: Not applicable

Agricultural uses: Agriculture-related uses for this UAV are in precision methods. It is utilized to survey the crop growth trends, canopy characteristic, determine crop-N status and fertilizer needs, etc. Crop scouting for nutrient deficiencies, disease, and pest attack is done. This model is used to identify flood or drought-affected spots. This UAV is also useful for natural vegetation monitoring.

Nonagricultural uses: They are surveillance of military installations, borders, civilian buildings, highway traffic and vehicle movement, city and public events, etc. It could also be used for surveillance of mines and direct mining activity.

Useful References, Websites, and YouTube Addresses

https://www.allieddrones.com/ (accessed Nov 10, 2016).
https//allieddrones.com/portfolio-itm/aw-1e-star/ (accessed Nov 10, 2016).
https://www.youtube.com/user/allieddrones/ (accessed Nov 10, 2016).

Name/Models: Baaz Flying Wings

Company and address: OM UAV, 60-UA, Jawahar Nagar, Delhi 110007, India. E-mail: atul@omuavsystems.com; mobile: +91 9873690212; Website: http://www.omuavsystems.com/guardian.html.

Baaz is a small UAV produced by a company located in Delhi, India. This UAV company has a few other fixed, flat-winged UAVs to offer, such as Guardian UAV, etc. This company was initiated in 2009. The UAV models have been adopted during survey and aerial imagery of natural resources. They may make a mark in the agricultural belt of India soon.

Baaz is relatively a small UAV. It is swift to assemble and launch. It is autonomous during launch, navigation and landing. It has a range of uses related to agriculture, commercial and even military aspects. A standard package of Baaz, which is not very costly has a Baaz aircraft (platform), an Android tablet (GCS), video monitor, Yagi antenna for data download, propulsion batteries, microprocessor to fit the fuselage cockpit, radiocontrol transmitter and a minitool kit. This UAV could be assembled, started, and operated to get aerial images and data by a single-trained person (Plate 2.7.1).

TECHNICAL SPECIFICATIONS

Material: Baaz UAV is made of a composite material having foam, plastic, and carbon fiber.

Size: length: 45 cm, width (wingspan): 120 cm; height: 12 cm; weight: 800 g

Propellers: It has one at the rear end of fuselage. Propulsion is via motors that generate 11 V.

Payload: 100 g

Launching information: Baaz is hand-launched by holding it at shoulder height. It has autolaunch, navigation, and landing facility. It is autopilot UAV. It adopts ARM Cortex M4 32-bit processor. It has software for autolanding on belly. It lands with an accuracy of 10 m from the programed location. Parachute landing is possible in times of emergency.

Altitude: 50–1000 m a.g.l. is the ceiling. The UAV has a high-resolution barometer.

Speed: Maximum speed is 68 km h^{-1}. Stall speed is 35 km h^{-1}

Wind speed tolerance: 20 km h^{-1}

Temperature tolerance: −20 to +60°C

Power source: Electric engine

Electric batteries: 6800 mA h.

Fuel: No requirement for petroleum fuel.

Endurance: 40 min for regular flight + 10-min reserve in case of emergencies and for fail safe.

Remote controllers: Flight path can be programed for 300-way points. It can also be put on loiter mode. This UAV can also be guided midway by the iPads located at the GCS. The GCS has radiocontrol equipment, such as transmitter with 400–455 MHz, with eight channels. The range of telemetric control extends to 10 km radius from the station. The radiocontrol is supported by lithium–polymer battery of 11.1 V. It operates for 12 h at a stretch. Datalink operates up to an altitude of 1000 m a.g.l. It operates at a frequency 900 MHz and transmits 57,600 kb s^{-1} to ground-station iPads. The range for datalink is also 10 km radius using Yagi antenna.

Area covered: 200–300 ha per flight

Photographic accessories: Cameras: Visual range cameras (Sony 7A), NIR, IR, and thermal camera (thermal 320 × 270 pixels photos). A color videographer with 32 GB memory of 1080 × 270-pixel photos is also possible on the UAV.

Computer programs: This UAV has gyro-stabilized gimbal that is compatible with GPS, GLONASS, Galileo, and Biedou systems. Images and videos are all tagged with GPS coordinates accurately. The gimbal also accommodates a magnetometer for magnetometer mapping.

Spraying area: This is not a spraying UAV.

Volume: Not applicable

Agricultural uses: Land-resource mapping, crop scouting for NDVI, phenomics, and grain maturity. Detection of disease/pest attacks, drought and flood effects, if any, in the field. It is also used for natural vegetation monitoring.

Nonagricultural uses: Surveillance of industrial sites, buildings, oil pipelines, electric lines, dams, river banks, etc.; monitoring mines and mining activity, ore movement, and dumping; monitoring and guiding transport vehicles, etc.

PLATE 2.7.1 Bazz—a micro-UAV.
Source: Mr. Atul Khosla, Director, OM UAV Systems, Delhi, India.

Useful References, Websites, and YouTube Addresses

http://www.omuavsystems.com/baaz.html (accessed Feb 4, 2017).
https://www.scribd.com/document/316689891/Bazz-Mini-UAV (accessed Feb 4, 2017).
https://www.pinterest.se/pin/103019910205416348/ (accessed Feb 4, 2017).

Name/Models: Besra

Company and address: Carbon-Based Technology Inc., Central Science Park/3f, No. 30, Keya Road, Daya District, Taichung City, 42881, Taiwan; phone: +886 4 2565 8558; e-mail: info@uaver.com; website: http://www.uaver.com

Carbon-Based Technology Inc. was established at Taichung City in Taiwan around March 2007. This company concentrates on development of design, integration, manufacturing, and global marketing of unmanned aircraft system. It offers high-reliability UAS useful for "intelligence gathering." These UAVs also conduct aerial surveys of commercial and agricultural regions. The UAS can integrate with techniques of data processing and analyzing. The team at Carbon-Based Technologies Inc. offers solutions for disaster detection, geological, and natural resource survey. The aerial imagery that these UAVs relay, serve as priceless feedbacks. These small, light-weight, and versatile UAVs produced by Carbon-Based Technologies are finding their way into agricultural farms in Southeast Asia.

Besra is a small, light-weight UAV suitable for use in agricultural farms and other civilian locations. It has a major role in aerial mapping of crop belts, spectral analysis of crops, monitoring and surveillance of infrastructure, industrial assets, highway vehicular traffic, mines, oil pipelines, etc. BESRA can fly for more than 40+ min. It is equipped with professional flight control system. BESRA can work with multiple payloads as options. Its real-time transmission image allows us to see first-hand aerial view, easily.

TECHNICAL SPECIFICATIONS

Material: EPP plus carbon fiber and ruggedized plastic.

Size: length: 120 cm; width (wingspan): 150 cm; height: 35 cm; weight: 2.1–2.3 kg

Propellers: BESRA has one propeller placed at the tip of the fuselage (nose). It has two plastic blades.

Payload: 1.2 kg includes gyro-stabilized gimbal with cameras, radar, and Lidar.

Launching information: It is hand-launched by holding the frame at shoulder height.

Altitude: It is flown at 50–100 m a.g.l. when surveying agricultural crops. Its altitude ceiling is 3000 m a.g.l.

Speed: 54–63 km h^{-1}

Wind speed tolerance: Beaufort scale 4 (i.e., 28–33 km h^{-1})

Temperature tolerance: −20 to +55°C

Power source: Electric batteries

Electric batteries: 4 A h high-capacity lithium–polymer battery

Fuel: No requirement of petroleum requirement

Endurance: 40 min per flight

Remote controllers: The ground control equipment includes radiooperative at 2.4 GHz. It keeps data download link for 3–5 km radius.

Area covered: Besra covers about 2 km^2 photo area per flight.

Photographic accessories: Cameras: Besra is attached with cameras such as RICOH WG-5 (or equivalent) for real -time aerial mapping. It provides video/still images of 16 MP resolution. Other cameras include hyperspectral, multispectral, NIR, IR, red edge, and Lidar.

Computer programs: The images are processed in situ, at the UAV's CPU or at the ground station, on iPads. The computers utilize software such as Pix4D Mapper or Agisoft Photoscan.

Spraying area: Not a spraying UAV.

Volume: Not applicable

Agricultural uses: Besra is a lightweight UAV best suited for swift and repeated aerial imagery of crops. It is utilized to scan the crops for seedling density, plant stand, foliage, leaf area, leaf chlorophyll, and crop-N status. This UAV could be utilized to monitor crop growth regularly and identify grain maturity. Scouting for disease, pest, and weeds in the field is a clear possibility. Aerial imagery that depicts drought or flooded–affected locations or soil erosion could be obtained swiftly, using Besra UAV.

Nonagricultural uses: In general, Besra UAV could be utilized to surveillance large buildings, industries, mines, rail roads, highways, and track vehicles, if tracking device is also attached to it.

Useful References, Websites, and YouTube Addresses

http://www.uaver.com/product-Besra-Besra.html (accessed Jan 17, 2017).

http://www.uaver.com/product-Besra-Besra.html (accessed Jan 17, 2017).

http://www.dronepeople.club/drone-video/drone-uaver-besra-introduction (accessed Jan. 17, 2017).

http://dronecademy.org/uaver-besra-aerial-surveying-and-mapping/ (accessed Jan 17, 2017).

https://www.youtube.com/watch?v=N1iM0PO4rqc (accessed Jan 17, 2017).

https://www.youtube.com/watch?v=8X3NeuSztQ4 (accessed Jan 17, 2017).

https://www.youtube.com/watch?v=H4CYoc2pf8s (accessed Jan 17, 2017).

Name/Models: CREX-B

Company and address: Selex ES: A Finmeccanica company, registered head office—Pizza Monte Grappa no. 4, 00195, Rome, Italy, e-mail: info@leonardocompany.com; or

Selex ES S.p.A—A Finmeccanica company, Via M. Stoppani, 21-34077 Ronchi dei Legionari GO, Italy; phone: +39 0481 478111; website: http://www.selex-es.com

Selex ES is a subsidiary company of Finmeccanica, Rome, Italy. It specializes in the production of a wide range of aircrafts and UAVs meant for military reconnaissance, commercial purposes such as city traffic surveillance and even agricultural/natural vegetation aerial mapping, etc. CREX-B is one of the few recently released models of UAVs, by the Selex ES Company. Others are Selex ES SPYBALL-B, Selex ES ASIO-B, etc.

CREX-B is a microelectric fixed-wing UAS. It is powered by electric batteries to support a flight endurance of 75 min per flight. It is a hand-launched autonomous navigation UAV. The UAV comes in a suitcase. It can be assembled in farm, crop field, or a border post in about 10 min. It is a moderately priced UAV. It is expected to serve a variety of functions required by agricultural farms, commercial units, and military.

TECHNICAL SPECIFICATIONS

Material: It is a fixed-winged UAV made of EPO, kevlar, carbon fiber, and ruggedized plastic.

Size: length: 45 cm; width (wingspan): 170 cm; height: 30 cm; weight: <2.1 kg (±5%)

Propellers: One propeller at the rear end of fuselage

Payload: Payload includes day/night electrooptical (EO) stabilized gimbal. It also allows 1–4 sensors for aerial imagery.

Launching information: This UAV is small and light in weight. Therefore, it is easily handled by a single person. It is hand launched. It has autopilot mode and navigation as decided by the software, such as "Pixhawk Autopilot" or other custom-made software. This UAV has low-noise signature during flight.

Altitude: The operational altitude for crop scouting is 30–500 m a.g.l. Maximum rate of climb is 7.2 km h^{-1}

Speed: Maximum speed is 110 km h^{-1}, but operational speed is 50–60 km h^{-1}. Cruise speed is 36 km h^{-1}.

Wind speed tolerance: It withstands disturbing wind speed up to 46.8 km h^{-1}. It is a UAV with rainproof body and works in slight drizzle.

Temperature tolerance: −20 to +49°C.

Power source: Electric batteries

Electric batteries: 14 A h; two cells.

Fuel: Not an IC engine-fitted UAV. It does not require petroleum fuel.

Endurance: Maximum endurance is 75 min per flight

Remote controllers: The ground control uses Tx/Rx digital communication links with the UAVs, CPU. The communication works within 10 km radius range—that is, line of sight. This UAV has facility for communication at different radiofrequencies. It has at least two different channels. The ground data terminal houses, both, RF terminals and the central CPU, along with storage memory. It is also equipped with GPS facility, to locate the UAV, at any time. The ground computers are interfaced to provide videoimages and still images, if needed. The ground computer is an iPad of 7 in. touchscreen.

The ground control is used mainly to plan flight mission, map the images, and build a database, command the aerial transit of UAV through telemetric link, receive videoimages and digital data, video processing, tracking, and storage of general digital data. The GCS contains both commercial software such as Pix4 D Mapper and Agisoft Photoscan and a few other custom-made software. These are provided along with the UAV unit at the time of purchase.

Area covered: It covers about 50–60 km linear distance or 200–250 ha of land surface filled with crops or other features.

Photographic accessories: Cameras: The sensors include visual bandwidth R, G, and B cameras, also NIR and IR cameras. The IR camera is placed in a gyro-stabilized gimbal.

Computer programs: The aerial images obtained at visual and IR bandwidth are processed using the usual commercial image-processing software (e.g., Pix4D Mapper).

Spraying area: This is not a spraying UAV.

Volume: Not applicable

Agricultural uses: They are aerial mapping of terrain, soil types, crops, and water resources. Crop scouting for seed germination, seedling establishment, plant stand, growth and canopy traits, crop-N status, and water stress index could be accomplished. We can use the same set of sensors and image-processing facility and few programs for crop growth analysis.

Nonagricultural uses: It is a typical close-range tactical UAV. It is useful in surveillance of military establishments and vehicles, mines and mining activity, large industries, city traffic, tracking vehicles, monitoring events, oil pipelines, etc.

Useful References, Websites, and YouTube Addresses

http://www.selex-es.com (accessed May 23, 2017).

http://www.leonardocompany.com/en/-/crex-b-1 (accessed May 23, 2017).
http://www.leonardocompany.com/product-services/sistemi-avionici-spaziali-airborne-space-systems-2/capabilities/uas (accessed May 23, 2017).

Name/Models: CropCam Unmanned Aerial Vehicle

Company and address: CropCam Ltd., PO Box 720, 72067 Rd 8E, Sturgeon Road, Stony Mountain, Canada R0C 3A0. Phone: +1 204 344 5617, +1 (204) 344 5706; e-mail: info@cropcam.com

CropCam is a North American UAV company. It manufactures lightweight agricultural micro-UAVs. These UAVs are versatile and low-cost units. CropCam also produces a few other models that serve specific purposes in farming and general civilian usage. CropCam fixed-winged UAVs are already operative for over 12 years now, with thousands of trouble-free flight hours and several thousand units sold in different regions of the world (Plate 2.8.1).

TECHNICAL SPECIFICATIONS

Material: CropCam is made of carbon fiber reinforced plastic. It is toughened to withstand harsh weather conditions.

Size: length: 4 ft, width (wingspan): 6 or 8 ft; height: 1.0 ft; weight: 6 lb (2.7 kg)

Propellers: It has one propeller at the tip of nose on the fuselage
Payload: 700 g

Launching information: It is hand-launched. It possesses automatic navigation and landing, using Pixhawk Autopilot software.

Altitude: It is alterable it ranges from low altitude above crop at 50–3800 m a.g.l.

Speed: It flies at an average speed of 60 km h^{-1}

Wind speed tolerance: 30 km h^{-1}

Temperature tolerance: Withstands a range from −20 to +50°C

Power source: Electric batteries

Electric batteries: One set of lithium–polymer batteries (four packs); four thunder power batteries lithium–polymer 2100 mA h/11.1 V (three cells). This UAV consumes about 100 MA per 10 min flight.

Fuel: Not a petrol engine

Endurance: 55–60 min per flight without brake

Engine: Electric engine Axi Brushless

Remote controllers: Remote controllers operate up to 3 km from the UAV/ground station. Remote controllers include laptop and radiocontrol system.

Area covered: It covers 50–200 ha of land surface per flight.

Photographic accessories: Cameras: It has one set of CropCam cameras for R, G, and B bandwidth imagery, a multispectral camera, and one for NIR, IR, and red-edge imagery. It provides GPS-tagged color images. Cameras provide 640 × 480 mm pictures shot at 30 frames s^{-1}. Images are transmitted instantaneously for up to a range on 1.6 km from the UAV (Plate 2.8.1).

Computer programs: Package includes CropCam image-processing software such as Pix4D Mapper and Agisoft Photoscan. Images can also be processed right at the UAV using CPU in the fuselage. The processed images could be relayed to ground-station computers/iPad for analysis. It helps us in revising flight paths and imaging spree as needed.

Spraying area: This is not a sprayer UAV.

Volume: Not applicable

Agricultural uses: Excellent images of crop fields, irrigation lines, natural vegetation, and farm equipment are provided. It scouts the progress of agronomic procedures. Images of farm activity are relayed to ground station or provided as digital data. Surveying and mapping agricultural zones and depicting land-use trends; monitoring irrigation and drainage systems, crop phenotyping, collecting data on grain maturity. Yield forecasting is also possible, using digital data derived by the UAV's cameras.

Nonagricultural uses: CropCam UAV is also used to surveillance industrial units, large buildings, rail roads, city traffic, oil/gas pipelines, environmental parameters, terrain, and natural vegetation.

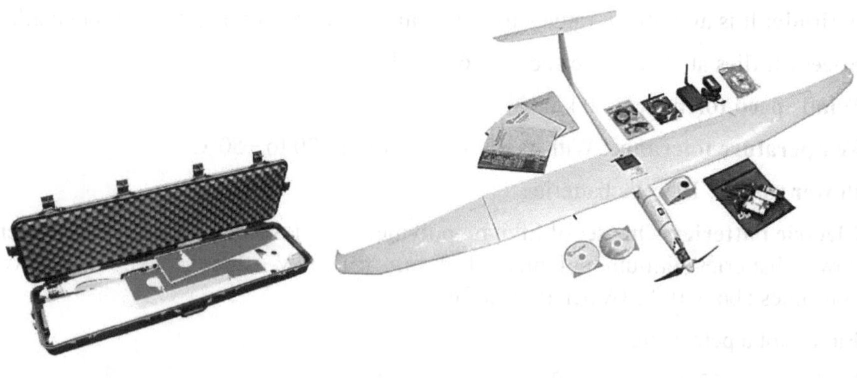

PLATE 2.8.1 CropCam—an agricultural UAV.
Source: CropCam Inc., Stony Mountain, Canada; Robot Shop Inc., RobotShop, Canada.
Note: CropCam is a very light and low-cost fixed-winged UAV. It is useful in scouting crops and in accumulating wide range of data about crops.

Useful References, Websites, and YouTube Addresses

http://www.cropcam.com/products.htm (accessed June 7, 2016).
http://www.cropcam.com/samples.htm (accessed June 7, 2016).
http://www.robotshop.com/en/cropcam-en.html (accessed June 7, 2016).
https://www.youtube.com/watch?v=Jwq-EPxkvY0 (accessed June 7, 2016).
https://www.youtube.com/watch?v=wHKcA74ujPE (accessed June 7, 2016).
https://www.youtube.com/watch?v=qSeVP0oWWm8 (accessed June 7, 2016).
https://www.youtube.com/watch?v=CrQxYY1MozE (accessed June 7, 2016).

Name/Models: Delair-Tech's UAVs DT-18 and DT-26X

Company and address: Delair Tech, 395, Route de Saint Simon, 31100 Toulouse, France; phone: +33 5 82 95 44 06; e-mail: hotline@delair-tech.com; Website: http://www.delair-tech.com/our-company/

Delair-Tech is a European UAV company located in southern France. It was inaugurated in 2011. It has outlets for UAV sales and servicing, including processing of images and prescriptions for efficient farming. Delair-Tech's services include aspects such as geospatial techniques, mining, security of oil and gas pipelines, electric powerlines, and utility services. Delair-Tech is a company with over 100 personnel dealing with UAV production, sales, and servicing. They are present in over 80 nations worldwide. Their UAVs, DT 18 and DT 26X and M are operative in many agrarian regions, including Australia.

TECHNICAL SPECIFICATIONS

Material: DT-18 is made of foam, carbon fiber, and plastic composite. The material is ruggedized to withstand rough weather and landing on belly.

Size: length: 35 cm; width (wingspan): 110 cm height; weight: 2 kg (DT 18); 8–11 kg plus 4 kg payload (DT 26X and 26M)

Propellers: DT-18 has one propeller at the front tip of the fuselage.

Payload: 2 kg (DT-26); 300 g (DT-18)

Launching information: Take-off is possible both via catapult and as hand-held UAV at shoulder height. This UAV can be deployed in less than 5 min.

Landing: It lands on belly. Otherwise, net landing is also possible. It has autolaunch, flight, and landing facility, through Pixhawk software.

Altitude: 50–150 m above crop, but it can reach up to 2500 m a.g.l.

Speed: 45–50 km h^{-1}

Wind speed tolerance: 70 km h^{-1} (DT-26); 45 km h^{-1} (DT-18)

Temperature tolerance: −20 to +50°C

Power source: Electric batteries operate for over 60 min.

Electric batteries: 14,000 mA h; two cells.

Fuel: No requirement for petroleum fuel.

Endurance: 3 h (DT-26); 2 h (DT-18)

Remote controllers: GCS signals operate on UAVs for up to 50 km radius. For long-range missions, 3G/4G abilities and a sitcom C2 Pix4D processing is essential.

Area covered: 100–150 km per flight or 1200 ha per flight of 2 h endurance

Photographic accessories: Cameras: 1080 p HD camera

Computer programs: Postflight image processing is done, using software such as Pix4D Mapper or Agisoft Photoscan. The images are also transmitted instantaneously, after being processed at the UAV itself, using CPU software. Images could also be stored for analysis later, by the ground-station computers.

Video transmission: Instantaneous video transmission is possible for up to 15 km from the UAV.

Spraying area: This is not a sprayer UAV.

Volume: Not applicable

Agricultural uses: Imaging terrain, topography, and facilities in the farm and crop fields; land and field mapping and collecting NDVI data; estimation of leaf chlorophyll, leaf-N content, plant water status, etc.; monitoring weed and disease/insect pest incidence.

Nonagricultural uses: The above UAVs are useful in security and defense departments. The UAV offers emergency services, monitoring of oil pipeline, power lines, public safety monitoring, traffic monitoring, mapping terrain, surveillance of geological and mining site.

Useful References, Websites, and YouTube Addresses

http://www.delair-tech.com/en; http://www.delair-tech.com/uavs/ (accessed May 4, 2017).
http://www.dronelife.com/product/DT-26 (accessed May 4, 2017).
https://www.youtube.com/watch?v=JRZf0-2gLUM (accessed May 4, 2017).
https://www.youtube.com/watch?v=czixY4bGfuw (accessed May 4, 2017).
https://www.youtube.com/watch?v=W-ylWviFi3c (accessed May 4, 2017).

Name/Models: Delta-FW 70

Company and address: The Homeland Surveillance and Electronics LLC (HSE), Morton, Illinois, USA; phone: +1 309 361 7656; e-mail: tsanders@hse-uav.com; website: http://www.hse-uav.com/

Homeland Surveillance and Electronics LLC (HSE) is a UAV and UAV-related service provider. It produces several models of UAVs. The products comprise both fixed-wing and rotor copters. These UAV models, it seems were results of over 10 years of research and development by the company. The HSE was started in 2009 by Mr. Dave Sanders and his entourage. They have industrial facility in Denver, Colorado, USA, that produces UAVs. The flat-wing UAV-Delta-FW-70 is among premier products of the company. It serves as a good agricultural UAV. Delta-FW-70 is large enough to hold sensors. It has four separate payload bays. HSE LLC deals with several aspects of agriculture. For example, they also produce autonomous "Sprayer UAVs." The sprayer assembly with precision techniques imbibed can also be attached to Delt-FW-70. Delta FW-70 is an excellent UAV to obtain aerial imagery of land resources, topography and crop fields (Plate 2.9.1). It helps to assess various aspects of crops, such as NDVI, water stress index, etc. These UAVs can also be used, to surveillance natural resources, industrial installations, civil, military and public infrastructure, etc. (HSE, 2016).

TECHNICAL SPECIFICATIONS

Material: It is made of graphite-reinforced plastic. It is toughened to withstand harsh weather conditions.

Size: length: 60 cm; width (wingspan): 75 cm; height: 35 cm; weight: 2 lb, in case, the body is made of dense foam. Otherwise, 1 lb, if made of lightweight foam. Dense foam body resists landing impact better.

Propellers: It has one propeller at the nose (tip) of the fuselage. It is made of light reinforced plastic.

Payload: volume: 375 sq. in.; and weight 6 lb

Launching information: It is hand-launched by holding the airframe at shoulder height. It has autolaunch, predetermined flight path and autolanding software (e.g., Pixhawk Auotpilot)

Altitude: The operational height for agricultural aspects is 50–150 m a.g.l.

Speed: 59 km h^{-1}

Wind speed tolerance: <25 km h^{-1}

Temperature tolerance: −20 to +50°C

Power source: Electric batteries

Electric batteries: 28,000 mA h; two cells

Fuel: It is not powered by IC engines; so, no requirement for petroleum engine.

Endurance: Up to 2 h at one stretch of flight

Remote controllers: The ground control computers keep contact with UAV, for up to 6 km radius. GCS computers have override facility. They can alter the flight path midway.

Area covered: Flight distance covered depends on weather and windy conditions. Under humid windy condition, it covers 32 km distance per flight. The imagery is securely stored in the servers through rapid and instantaneous transmittance of digital data.

Photographic accessories: It has DSLR EO and NIR cameras, Lidar, and multiple sensors for multispectral analysis.

Cameras: Canon EOS SL1 DSLR camera body modified to conduct NDVI imagery; Canon EF-S Pancake lens; External Canon GP-E2 GPS receiver; Canon hot shoe/ GPS extension cable made by Vello (cameras have a 30-day warranty) (HSE, 2016).

Computer programs: The UAV's cameras relay images instantaneously to ground-station PCs and iPad. Image processing can be done right at the UAV using CPU. Otherwise, it could be processed at GCS iPad, using Pix4D Mapper or Agisoft's Photoscan or any of the similar software.

Spraying area: This is not a sprayer UAV.

Agricultural uses: The aerial imagery by Delta-FW-70 is helpful in detecting weather-related damage to crops and soil surface (e.g., erosion). It helps in reducing use of fertilizer and pesticides, by showing up spots that need attention. UAV's images help in scouting crops at grain maturity and arriving at forecasts for crop yield. Aerial images can show up weeds and invasive species in the crop fields and surrounding natural vegetation. Overall, this UAV aids detection of weather-related damage to fields, monitoring disease, survey for irrigation needs in different regions of the field, and locate invasive species of plants. It is also used to monitor livestock and pasture conditions.

Nonagricultural uses: Delta-FW-70 has been designed by HSE, primarily, to function as a useful agricultural UAV. However, its uses include surveillance of industrial installations, dams, bridges, buildings, oil pipelines, electric transmission lines, rail road, and highway traffic. These UAVs can be used as border patrols.

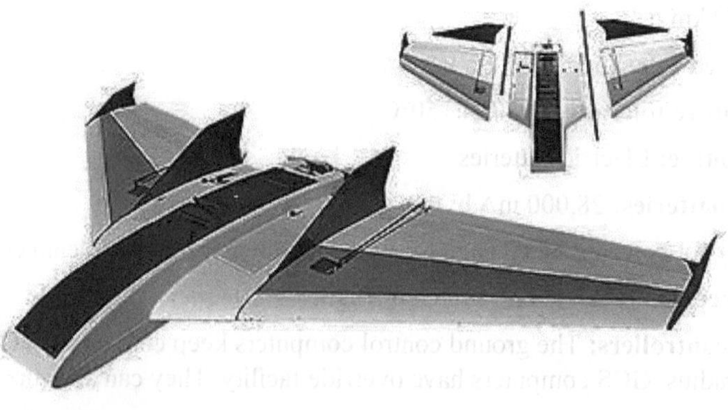

PLATE 2.9.1 Delta-FW-70.
Source: Dr. Terry Sanders, Homeland Surveillance and Electronics Inc., Denver, Colorado, USA.

Useful References, Websites, and YouTube Addresses

HSE. *Delta FW 70 Fixed Wing UAVs*; 2016; pp 1–14. http://www.hse-uav.com/ (accessed Nov 30, 2016).

http://www.uavcropdustersprayrs.com/agricultuure_delta_fw70_fixed_wing_uav.htm (accessed Oct 20, 2016); pp 1–10.

http://www.hse-uav.com/ (accessed Nov 30, 2016).

http://www.hse-uav.com/about_us.htm (accessed Nov 30, 2016).

http://www.hse-uav.com/delta_fw70_uav_package.htm (accessed June 15, 2017).

Name/Model: Delta-M UAV System

Company and address: AVASyS LLC—GeoService Autonomous Aerospace Systems; 3 Vuzovsky side street office 223, Krasnoyarsk, Russia 660025; phone: +07 391 286 61 09, +07 953 5866109; e-mail: info@uav-siberia.com; website: http://uav-siberia.com

AVASyS LLC is a Russian UAV company located at Krasnoyarsk in Siberia. The company is operated by the Siberian Federal University, Institute for Geology and Geo-technology, with the support of finance from Prognoz Group LLC. AVASyS LLC aims at manufacture of high numbers of autopilot UAVs. They aim at developing UAV systems that suit different purposes, such as mining, geology, road construction, inspection of buildings, and importantly natural vegetation and agricultural crop production (AVASyS LLC, 2016).

Delta-M model is a small UAV. It is launched using catapult. It has autopilot mode during flight. It lands using a parachute. This is supposedly an excellent UAV useful in aerial imagery, storage of digital data, and relay to ground stations. It offers excellent GPS (GNSS)-tagged imagery, 3D images, and videography about natural resources, buildings, crop land, and farm installations. It uses Sony RX-1 and Canon EOS-M cameras (Plate 2.10.1).

TECHNICAL SPECIFICATIONS

Material: Delta-M is made of toughened foam, plastic, and carbon fiber composite. It is a lightweight material. The airframe lasts at least over 50 hard landings and it is manually replaced, later.

Size: length: 45 cm; width (wingspan): 105 cm; height: 30 cm; weight: 2.3 kg

Propellers: It has one propeller located at the rear end of fuselage. It is made of two plastic blades.

Payload: 800 g

Launching information: It is launched using a catapult (see Plate 2.10.1). It has autopilot mode that uses Pixhawk software. The autopilot also uses Hi-Rel aircraft control system. It automatically checks subsystems and devices. It prevents errors generated and calibrates the flight path based on obstacles. It lands using a parachute.

Altitude: 100–3000 m a.g.l.

Speed: 65–85 km h^{-1}

Wind speed tolerance: It tolerates 15 m s^{-1} wind disturbance

Temperature tolerance: −35 to +40°C

Power source: Electric batteries

Electric batteries: Lithium Polymer Batteries 14,000 mA h.

Fuel: Not an IC engine-fitted UAV. Batteries are replaced after 80% of the stored power has been expended.

Endurance: 40−60 min per flight

Remote controllers: The GCS computers are effective up to 12 km radius. The preprogramed flight path can be modified midway, using override.

Area covered: 50−200 ac per flight.

Photographic accessories: Cameras: Sony RX-1 full-sized matrix 35 mm with a resolution of 24 MP. Alternative option is Canon EOS-M with 50-mm focal length USM lens. Photography is connected to GNSS receiver and RTK. The UAV comes with a three-axis camera gimbal stabilization. It allows high flexibility during aerial imagery. The gimbal guarantees safety of cameras, despite rapid change of speed, angle, and even roles conducted, by the UAV as per directions of the ground control. Free orientation of cameras in the gimbal allows rapid imagery of sides of buildings, lateral shots of crop land and natural features, etc.

Computer programs: Image processing is done by Agisoft's "Photoscan" software for 3D and mapping. Pix4D Mapper is used to show ground relief and crop imagery. Racurs Photomod software is also used to process images.

Spraying area: It is not a spraying UAV.

Volume: Not applicable

Agricultural uses: Aerial imagery of farm installations, land resources, crops, and cropping pattern. Crop scouting for phenomics data, NDVI, leaf chlorophyll, crop-N status, canopy temperature (water stress index) are other major functions. Natural vegetation monitoring is routinely done using Delta-M UAV.

Nonagricultural uses: It includes surveillance of industries, oil pipelines, electrical installations, rail roads, city traffic, vehicle tracking, etc.

PLATE 2.10.1 Top: A Delta-M UAV aircraft. Middle: Cameras and flight plan; bottom: Catapult and Parachute aided launch of the UAV.
Source: Mrs. Lyudmila Piskunova, Autonomous Aerospace Systems, AVASyS LLC, Krasnoyarsk, Siberia, Russia.

Useful References, Websites, and YouTube Addresses

AVASyS LLC. *Delta-M UAV Systems. Autonomous Aerospace Systems-Geoservice*; Krasnoyarsk, Siberia, Russia, 2016; pp 1–8. http://uav-siberia.com/en/content/delta-m-uav-system?gclid=CNKdmZ3-0NACFdi. (accessed Dec 1, 2016).

http://uav-siberia.com/en/content/delta-m-uav-system?gclid=CNKdmZ3-0NACFdi. (accessed Dec 30, 2017).

Name/Models: DV Wing

Company and address: UAV Volt Inc., UAV Volt 14, rue de la Perdrix, Lot 201, Villepinte 95934, Roissy, Charles de Gaulle Cedex, Paris, France; phone: +33 0180 894444; e-mail: contact@drronevolt.com; website: http://www.dronevolt.com

DV Wing is a micro-UAV of under 1 kg weight. This UAV is easy for package and transport to any location for launch (Plate 2.11.1). It could be launched in a matter of 5–10 min. It is made of high-quality EPO and foam. It sends images to GCS rapidly and covers larger area of crop fields (Plate 2.11.2). It is easy to fix flight path and control it, using remote controller. It does not need special flying training or skills. It costs about US$ 3500 (2017 price level).

TECHNICAL SPECIFICATIONS

Material: It is made of a composite derived, using carbon fiber, evelar, and rugged foam. The material withstands rough belly landing. The entire UAV is packed in a suitcase of the size 103 × 58 × 23 cm.

Size: length: 45 cm; width (wingspan): 90 cm; height: 39 cm; weight: 0.98 kg

Propellers: It has electric pusher propeller. One propeller located at the rear end of the fuselage.

Payload: 0.45 kg

Launching information: It is hand-launched by holding it at shoulder height. It has autolaunch, navigation, and landing software, namely, Pixhawk. Otherwise, it could be included with similar custom-made software such as "UAV VOLT" software. It lands by circling against wind and then does belly landing.

Altitude: The operational altitude is 400 m above crop canopy. Ceiling is 4000 m above ground surface

Speed: The cruising speed is 14 m s^{-1}

Wind speed tolerance: 15 km h^{-1} wind disturbance

Temperature tolerance: −10 to +60°C

Power source: Lithium polymer electric batteries

Electric batteries: Lithium–polymer batteries of 5500 mA h; 11.1 V.

Fuel: No requirement for petroleum fuel.

Endurance: It has a slightly longer flight endurance of 85 min per flight

Remote controllers: The telemetric link operates with UAV for up to 2–3 km radius.

Area covered: It conducts aerial survey or imagery of 300 to 500 ha of crop field per flight.

Photographic accessories: Cameras: This UAV is fitted with a stabilized gimbal (three-axis). It holds high-resolution visual sensor. It produces high quality photographs of land surface and crop stand. It offers 1:500, 1:1000, and 1:2000 orthophotos for transmission and processing.

Computer programs: The orthoimages are processed, using software such as Pix4D Mapper/Agisoft Photoscan/Trimble's Inpho or UAV Deploy.

Spraying area: This is not a sprayer UAV.

Volume: Not applicable

Agricultural uses: They are aerial survey of land resource, soil type, and cropping systems. Crop scouting for canopy growth, NDVI values, leaf-chlorophyll status (crop-N status), canopy temperature and water stress index, panicle, and grain maturity status. The data from UAV's imagery could be utilized to decide on fertilizer supply, using precision techniques. It could be used to suggest irrigation timing, frequency, and quantity, based on variations in crop's water stress index. Natural vegetation monitoring is also possible (see Plate 2.11.2).

Nonagricultural uses: They are surveillance of industrial sites, buildings, public places, mines and mining activity, highways and transport vehicles, etc.

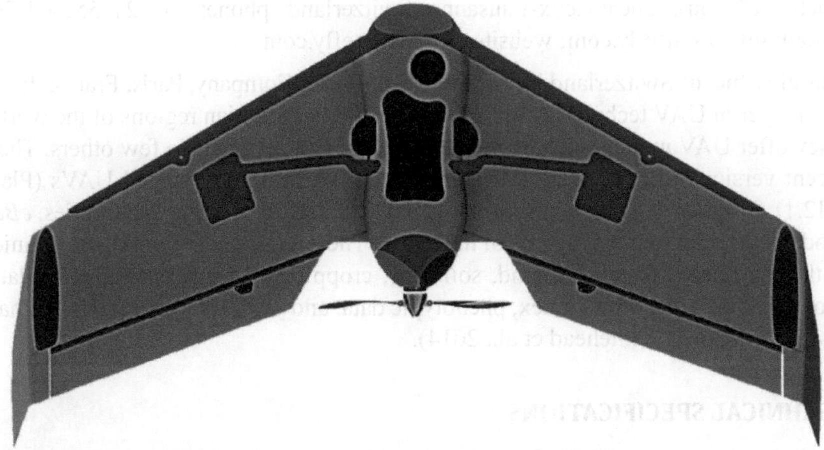

PLATE 2.11.1 DV Wing—a micro-UAV for use in agricultural fields.
Source: Mrs. Celine Vergeley, Drone volt, Roissy, Paris Nord, France.

PLATE 2.11.2 A DV fixed-winged UAV monitoring sprayer movement in the crop field.
Source: Mrs. Celine Vergeley, Drone Volt, Roissy, Paris Nord, France.

Useful References, Websites, and YouTube Addresses

http://www.expouav.com/news/latest/dv-wing-drone-dedicated-precision-agriculture/ (accessed Apr. 20, 2017).

http://www.dronevolt.com/en/expert-solutions/dv-wing/ (accessed Apr. 20, 2017).

http://www.drone.volt.com/en/expert-solutions/dv-wing/?gclid=CJTNuZ2MjNICFdKJ (accessed Apr. 20, 2017).

https://www.youtube.com/watch?v=g2mmevhvzjy&autoplay=1)dvwingbydronevolt (accessed Apr. 20, 2017).

https://www.youtube.com/watch?v=jllOLZiB63s (accessed Apr. 25, 2017).

https://www.youtube.com/watch?v=ZglSMaG5HQo (accessed Apr. 25, 2017).

Name/Models: *eBee*

Company and address: SenseFly—A Parrot Company, 1033 Route Geneva, (Z.I. Chaterlard Sud), Cheassaeux-Lausanne, Switzerland; phone: +41 21 552 04 20; e-mail: info@sensefly.com; website: www.sensefly.com

SenseFly Inc. of Switzerland is a subsidiary of Parrot Company, Paris, France. It is a key player in UAV technology in European and other agrarian regions of the world. They offer UAV models such as Swinglet Cam, *eBee* series and a few others. Their recent versions of *eBee* are lightweight, highly versatile flat-winged UAVs (Plate 2.12.1). They carry a range of sensors with wide range of imaging capabilities. *eBee* model is popular in most regions of the world. These UAVs help farming companies with data related to terrain, land, soil type, cropping systems, crop growth data, drought, crop water stress index, phenotypic data, and yield forecasting (Haghighattalab et al., 2015; Whitehead et al., 2014).

TECHNICAL SPECIFICATIONS

Material: It is made of EPP, foam, and graphite composite.

Size: length: 35 cm; width (wingspan): 96 cm; height: 8 cm; weight: 750 g (1.5 lb)

Propellers: It has one propeller at the nose of fuselage. It is an electric pusher propeller connected to motor that works at 160 W brushless DC motor.

Payload: 0.15 kg per flight.

Launching information: It is launched at shoulder height, using hand. Landing is on belly and its accuracy is 5 m.

Altitude: <150 m from earth's surface for agricultural operations.

Speed: The cruise speed during agricultural operations is 40 km h^{-1}, but it could reach up to 90 km h^{-1}.

Wind speed tolerance: 12–15 km h^{-1}

Temperature tolerance: −20 to +45°C

Power source: Electric batteries

Electric batteries: Lithium Batteries 11.1 V, 2150 mAG.

Fuel: Operates on electric power, not on petrol/diesel

Endurance: 45–50 min per flight without a relaunch

Remote controllers: Radiolink from UAV to ground station is up to 3 km radius. UAV flight path is regulated via remote controller or could be preprogramed. Software allows automatic 3D flight path determination. UAV is controlled using a software known as *eMotion.*

Area covered: New *Ebee*⁺ model covers 220 ha or 540 ac per flight.

Photographic accessories: Cameras: *Ebee* is usually fitted with Sony G9X, S110, NIR/RE, Sequoia, and *Thermo*Map.

Computer programs: Image processing is done using Pix4D Mapper software. UAV's cameras relay images, instantaneously, to ground-station PCs and iPad. Image processing can be done at the UAV, using CPU or could be processed at GCS iPad, using Pix4D Mapper or Agisoft Photoscan or similar software. The images are all GPS tagged. They offer high accuracy while analyzing spatial effects of factors such as fertilizers, crop genotypes, or irrigation.

Spraying area: This is a small UAV. This is not known to be attached with chemical tanks or used for spraying pesticides, herbicides or liquid fertilizers.

Volume: Not applicable

Agricultural uses: *eBee* is among the most popular agricultural UAVs. *eBee* is an excellent field/crop mapping UAV. It is useful in scouting, aerial imagery, detection of disease/pest incidence, study of field topography, and soil conditions. Surveillance of farm vehicles and their progress in fields is also possible (Plate 2.12.1).

Nonagricultural uses: Mineral exploration, geophysical studies, mine-pit, and ore-dump management; ore haul route management, mapping mining zones, and imaging steep and inaccessible routes in the mining zones are important uses of *eBee*. Monitoring rivers and irrigation projects is possible (Whitehead et al., 2014). Locating and imaging erosion and flooding are done with ease by *eBee*. Monitoring pipeline and electric lines, industrial installations, monitoring, and managing city traffic and public events could be achieved routinely by *eBee*.

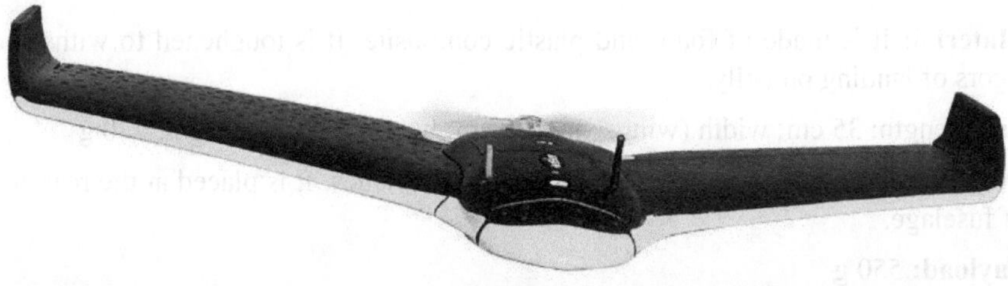

PLATE 2.12.1 Top: An *eBee* Agricultural UAV.
Source: Dr. Mathew Wade, SenseFly- A Parrot Company, Cheseaux-Lausanne, Switzerland.

Useful References, Websites, and YouTube Addresses

Haghighattalab, A.; Gonzlez Perez, C.; Mondl, S.; Singh, D.; Schomstock, D.; Rutkoski, J.; Ortiz-Monsteria, I.; Singh R. P. Goodire, D.; Poland, J. Applications of Unmanned Aerial Systems for Phenotyping of Large Wheat Breeding Nurseries. *Plant Methods* **2016,** 1–8. DOI:10.1186/s13007-016-016346 (accessed June 17, 2017).

Whitehead, K.; Hugenholtz, C. H.; Myshalk, S.; Brown, O.; LeClair, A.; Tamminga, A.; Brycha, T. E.; Moorrman, B.; Eton, B. Remote Sensing of the Environment with Small Unmanned Air Craft (UAVs), Part B: Scientific and Commercial Applications. *J. Unmann. Vehicle Syst.* **2014,** *2,* 1–9. http://www.nrcresearchpress.com/doi/full/10.1139/juvs-2014-0007#.WUU-2dvrbmQ (accessed June 17, 2017).

https://www.sensefly.com/drones/ebee.html (accessed Feb. 12, 2017).

https://www.sensefly.com/drones/ebee-plus.html (accessed Feb. 12, 2017).

https://www.sensefly.com/drones/ebee-sq.html (accessed Mar. 2, 2017).

https://www.sensefly.com/home.html (accessed Feb. 12, 2017).

https://www.youtube.com/watch?v=yybxZuuWX-Y (accessed Feb. 12, 2017).

https://www.youtube.com/watch?v=wtjqFtOvtIs (accessed Feb. 12, 2017).

https://www.youtube.com/watch?v=AyPQ6j9prP8 (accessed Mar. 2, 2017).

https://www.youtube.com/watch?v=eYotEDFOcdI (accessed Mar. 2, 2017).

Name/Models: Echar 20 A

Company and address: XMobots, Rua Santa Cruz, 979 Sao Carlos, SP, Brazil; phone: +55 (16) 3413 0655; website: http://www.xmobots.com; Phone +55 (16) 3413 0655; e-mail: contato@xmobots.br

Xmobots is a UAV company initiated in 2007 in the Sao Paulo region of Brazil. They produce UAVs such as Apoena 1000B, Nauru 500A, Echar 20A, and CAVE/ ANAC. They are all flat, fixed-winged UAVs. These UAVs have been developed in collaboration with the Universidade de Sao Paulo and Empresa Brasileira de Pesquisa Agropecuaria (EMBRAPA) research scientists. They cater to variety of purposes such as aerial reconnaissance, aerial imagery of land resources, agricultural farms, crops, and crop analysis using visual and NIR sensors (Nunes et al., 2015). This UAV also used for industrial surveillance, rail-road monitoring, public surveillance, security, traffic, etc.

TECHNICAL SPECIFICATIONS

Material: It is made of foam and plastic composite. It is toughened to withstand rigors of landing on belly.

Size: length: 35 cm; width (wingspan): 90 cm; height: 10 cm; weight: 780 g

Propellers: It has one propeller with two plastic blades. It is placed at the rear end of fuselage.

Payload: 550 g

Launching information: It is a hand-launched fixed-winged UAV.

Altitude: It operates at 50–400 m a.g.l. during aerial survey and imagery. Its ceiling height is 3000 m a.g.l.

Speed: 48–55 km h^{-1}

Wind speed tolerance: It tolerates a 20 km h^{-1} wind disturbance.

Temperature tolerance: −20 to +45°C

Power source: Electric batteries

Electric batteries: 24 mA h lithium–polymer batteries

Fuel: No requirement for petroleum

Endurance: 45 min

Remote controllers/Preprogramed flight paths: The UAV keeps contact with GCS for up to 20 km radius using radiolinks. Flight path is regulated using Pixhawk software or a similar custom-made software.

Area covered: It covers about 50–100 ha per flight

Photographic accessories: Cameras: It has a series of five cameras. They are visual, multispectral, hyperspectral, IR, NIR, and width cameras.

Computer programs: The images derived by the UAV is processed using Pix4D Mapper or Agisoft Photoscan

Spraying area: Not a spraying UAV

Agricultural uses: It is used to conduct aerial survey of terrain, land resources, rivers, crop fields, cropping systems, etc. The imagery is also used collect data about canopy growth, leaf chlorophyll, leaf-N status, water stress index, etc. Natural vegetation monitoring is also possible.

Nonagricultural uses: They are surveillance of mines, mining activities, ore dumps, transport vehicles, etc.

Useful References, Websites, and YouTube Addresses

Nunes, G. M.; Esteves Vieira, E. V.; Carvlho, S. C. Preliminary Data Obtained by UAV into Clonal Forest Stands of *Eucalyptus urograndis* H13 in the State of Mato Grosso. *Geográficas Aplicados à Engenharia Florestal. SENGEF—XI Seminário de Atualização em Sensoriamento Remoto e Sistema de Informações* **2015,** *26,* 1–5.

http://www.xmobots.com.br/ (accessed Nov 20, 2016).

http://www.xmobots.com.br/index.php?sec=blog&funcao=post&id=26 (accessed Nov 20, 2016).

https://www.youtube.com/watch?v=NsrpUrAL0To (accessed Nov 20, 2016).

https://www.youtube.com/watch?v=rccb-lv5u08 (accessed Nov 20, 2016).

https://www.youtube.com/watch?v=rccb-lv5u08 (accessed Nov 24, 2016).

https://www.youtube.com/watch?v=61qrs54Z7vg (accessed Nov 24, 2016).

Name/Models: ECLIPSE

Company and address: Robota LLC, 820 Ferris Rd Hangar #28 Lancaster, TX 75146 Lancaster, TX, USA; Phone: +1 925 388 6267; e-mail: info@robota.us; website: http://www.robota.us/

Robota LLC was founded at Lancaster, Texas, in 2006. The company aims at developing and producing UAVs for commercial and agricultural markets in USA and elsewhere in other continents. The engineers were originally trained in military UAVs. However, they shifted focus to civil and agricultural UAVs. Robota LLC produces several of the accessories required for adoption of UAV technology in crop fields. The "Goose autopilot" and "Triton mapping system" were developed and sold starting from 2011. Eclipse UAV is frequently used in industrial surveillance, precision agriculture, geological surveys, imaging topography, photogrammetry, in monitoring forest plantation, and in live relay of images from public events (Plate 2.13.1; Robota LLC, 2015).

TECHNICAL SPECIFICATIONS

Material: It is made of foam and plastic composites. It is toughened to withstand harsh weather conditions.

Size: length: 45 cm; width (wingspan): 95 cm; height: 30 cm; weight: 3 lb

Propellers: One at the rear end of fuselage

Payload: 3 kg

Launching information: This UAV is hand-launched, manually, by holding it at shoulder height. Landing is smooth due to reverse thrust capability.

Altitude: It is 400 ft for imagery, but the UAV can reach 11,000 ft a.s.l.

Speed: 38–75 km h^{-1}

Wind speed tolerance: 28 km h^{-1}

Temperature tolerance: −20 to +50°C

Power source: Electric batteries

Electric batteries: Lithium batteries

Fuel: Not a UAV with IC engine. No need for petrol/diesel

Endurance: 50 min per flight

Remote controllers: UAV stays in contact with ground-station computers for up to 3 km radius. Ground-station iPads receive either orthomosaics or preprocessed images (see Plate 2.13.1)

Area covered: It covers 400 ac per flight; mapping area is 200 ac, if red edge is used.

Photographic accessories: Cameras: Sony RX100 (1.1 in./pixel at 400 ft altitude) and mica-sense red edge (3.4 in./pixel at 400 ft altitude) both take images simultaneously (see Plate 2.13.1). The visual range camera is used for high-resolution imagery of crops, their growth, and maturity stages. Red-edge is used for thermal imagery and in detection of crop water stress index.

Computer programs: Robota's ECLIPSE includes accessories such as "The Goose" a software for autopilot, image storage and relay. The goose allows farmers to control and fix flight path of the UAV. ECLIPSE has a complete mapping system that offers farmers, with instantaneous maps of fields, maps depicting NDVI, leaf chlorophyll, and plant-N status (Plate 2.13.1).

Spraying area: This is not a sprayer UAV.

Volume: Not applicable

Agricultural uses: This flat-winged UAV is used to monitor agricultural fields, study soil, its variations in color, thermal properties, and water content. It is used to estimate and map NDVI variations in the crop fields. It is also used to estimate leaf chlorophyll, leaf-N, and plant water status. It is used to monitor work progress using visual cameras and GCSs. It surveillances crop fields for diseases, pests, drought, and/or flood-affected regions.

Nonagricultural uses: ECLIPSE is attached with visual and IR cameras. They are used to study geological aspects and natural resources distribution, general topography and features on the ground, that is, surface of earth. It is used to surveillance industrial installations, buildings, pipelines, mining zones, etc.

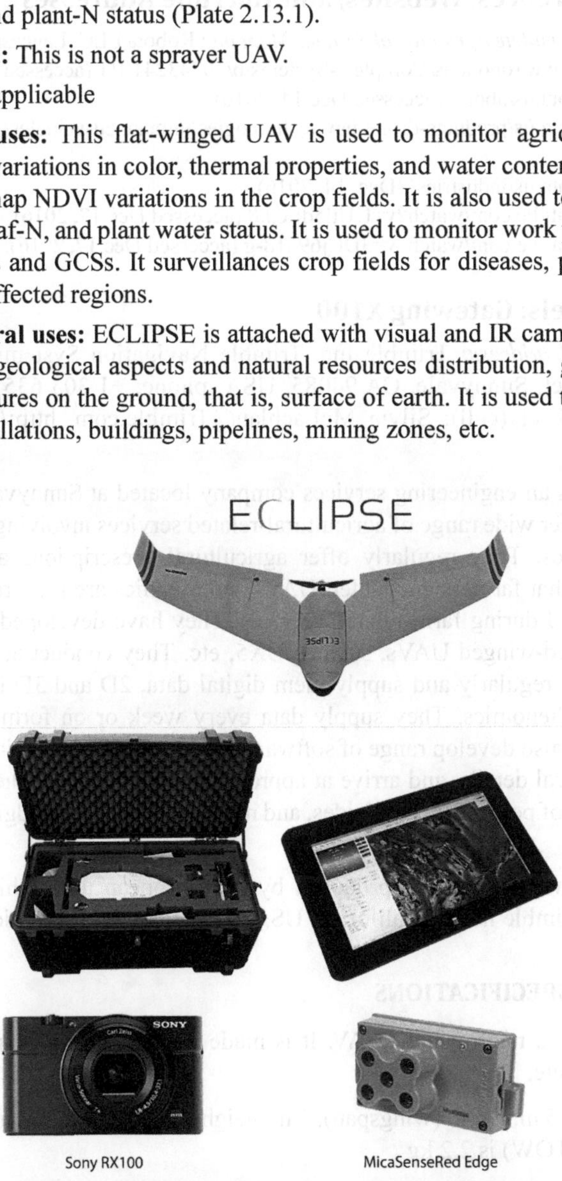

PLATE 2.13.1 Top: The UAV (ECLIPSE); middle: its case, iPad, that is, remote controller; bottom: visual range bandwidth sensor and red-edge bandwidth sensor.

Source: Dr. Antonio Liska, Robota LLC, Lancaster, Texas, USA.

Useful References, Websites, and YouTube Addresses

Robota LLC. *Complete Agricultural Systems Mapping*; Robota LLC: Lancaster, TX, 2015, pp 1–17. http://www.robota.us/Complete-Systems/b/10743241011 (accessed May 16, 2015).

http://www.robota.us/about/ (accessed Dec 14, 2016).

http://www.robota.us/products-drone-uav-technologies/unmanned-vehicles/ (accessed Dec 14, 2016).

http://www.robota.us/industries/ (Dec. 14, 2016).

https://www.youtube.com/watch?v=UUtl5dEcjk8 (accessed Dec 17, 2016).

https://www.youtube.com/watch?v=3DCj69q18-g (accessed Dec 17, 2016).

Name/Models: Gatewing X100

Company and address: Trimble Inc. Trimble Navigation Systems, Agricultural Division, Street, Sunnyvale, CA 94085, USA; phone: +1 303 635 8707 (office), +1 303 803 5711 (cell); Silvia_McLachlan@Trimble.com, http://www.trimble.com

Trimble Inc. is an engineering services company located at Sunnyvale, California, USA. They offer wide range of agricultural-related services involving farm vehicles with electronics. They regularly offer agricultural prescriptions after analyzing the problems that farmers encounter. UAVs and avionics are their recent specialty gadgets utilized during farm-related services. They have developed and produced a range of fixed-winged UAVs, such as UX5, etc. They conduct aerial surveys of farmers' fields regularly and supply them digital data, 2D and 3D images of crop growth, and phenomics. They supply data every week or on fortnightly basis to farmers. They also develop range of software necessary to stich the images, analyze them for spectral details, and arrive at appropriate prescriptions. Their suggestions relate to spray of pesticides, fungicides, and most importantly in judging fertilizer-N requirements.

The "Gatewing X100" is produced by the European unit, which is a recent takeover by Trimble Inc. of California, USA (Plate 2.14.1; UAV Insider, 2013).

TECHNICAL SPECIFICATIONS

Material: It is a microweight UAV. It is made of ruggedized foam, carbon, and plastic composite.

Size: length: 0.5 m; width (wingspan): 1 m; height: 0.35 m; weight: maximum take-off weight (MTOW) is 2.2 kg

Propellers: It has one propeller placed on the fuselage at the tip of rear end. Propeller is powered by 0.25-kW brushless electric motor.

Payload: 700 g

Launching information: It is launched using a standard pneumatic catapult; landing on belly needs, at least 150 × 30 m open land.

Altitude: The ceiling is <2500 m

Speed: The cruising speed is 80–95 km h^{-1}.

Wind speed tolerance: 30–45 km h^{-1}

Temperature tolerance: −17 to +55°C

Power source: Electric batteries

Electric batteries: Lithium–polymer batteries

Fuel: Not an internal combustion engine. It does not require petrol/diesel.

Endurance: 45 min

Remote controllers: Ground-station computers can preprogram the flight path. Otherwise, the UAV could be controlled, through a remote controller. UAV keeps radiocontact for 5 km radius from the ground station. Farmers can reorganize the flight path over the field, midway through the flight.

Area covered: Maximum flight distance is 53 km.

Photographic accessories: Cameras: The payload area accommodates a set of visual cameras. The cameras operate at R, G, and B bandwidths. It also carries IR and NIR cameras, plus thermal imaging facility. UAV relays 3D images instantaneously to ground station.

Computer programs: The images or digital data collected are instantaneously relayed to Trimble ground station for rapid processing and storage. Images are analyzed for a range of crop characteristics.

Spraying area: This is not a sprayer UAV.

Volume: Not applicable

Agricultural uses: Gatewing X100 was designed by Belgian Engineers to serve low level photogrammetry, imagery, and agricultural scouting. It is frequently utilized to scout crop and natural vegetation. This UAV is used to monitor crop growth periodically. It is used to estimate NDVI, leaf chlorophyll, leaf and plant-N status, crop water stress index, and soil moisture in the surface layers. It is also useful to obtain 3D images to mark "management zones," during precision farming (Plate 2.14.1).

Nonagricultural uses: They are surveillance of mines, earth works and vehicles, industrial and construction sites, transport vehicles, etc.

PLATE 2.14.1 A Gatewing-X 100 UAV.
Note: The fixed-winged UAV has been readied for take-off using catapult.
Source: Mrs. Silvia Mclachlan, Trimble Inc. Sunnyvale, California, USA (European unit at Antwerp, Belgium, produces this model).

Useful References, Websites, and YouTube Addresses

UAV Insider. *Trimble Gatewing X100*; 2013; pp 1–3. http:/www.uavinsider.com/trimble-gatewing-x100 pp 1–3 (accessed Apr 12, 2016).

http://en.avia.pro/blog/tromble-gatewing-x100-technischeskie-harakteristiki-foto (accessed Nov 3, 2016).

http://www.uavinsider.com/trimble-gatewing-x100/ (accessed Dec 20, 2017).

http://www.uavinsider.com/trimble-gatewing-x100/ (accessed Dec 20, 2017).

https://www.youtube.com/watch?v=NxnWmc0ZTvA (accessed Dec 20, 2017).

https://www.youtube.com/watch?v=MOucJcdFN5c (accessed Dec 20, 2017).

https://www.youtube.com/watch?v=XkqK8oeQ7Fc (accessed Nov 3, 2016).

https://www.youtube.com/watch?v=NxnWmc0ZTvA (accessed Nov 3, 2016).

https://www.youtube.com/watch?v=Sz_T-VKTXs8 (accessed Nov 3, 2016).

Name/Model: Gemini V2

Company and address: Baerospace Industries, http://www.bearospaceindustries.com/aircraft.html

This UAV is easily dismantled into different parts and loaded in a box. The box with complete set of parts for UAV fits into the boot of a car. The combination of a low-cost simple construction, toughened body, and long-range flight capability is highly preferred.

TECHNICAL SPECIFICATIONS

Material: It is made of hardened foam, plastic, carbon fiber, and air foil.

Size: length: 90 cm, width (wingspan): 320 cm (126 in.); height: 35 cm; weight: airframe weight is 4.9 kg (10.8 lb), with payload the weight is 9.4 kg (20.75 lb)

Propellers: It has 13 × 8 P and 13 × 8E APC thin electric two-blade propellers

Motors: 2X E-Flite "Power 15" 950 kV outrunner electric motors. These are connected to propellers.

Payload: 2.8 kg

Launching information: It is launched using a small runway. Otherwise, it can be carried on a pick-up van to get the lift-off and thrust. It has autopilot mode during flight. Landing is on wheels fitted to a foldable landing gear. It has three nose wheels.

Altitude: The ceiling 13,000 ft (4000 m)

Speed: The raising speed is 40–70 speed. Best loitering speed is 42 km h^{-1}.

Wind speed tolerance: 15 mi h^{-1}

Temperature tolerance: −20 to +60°C

Power source: Electric batteries

Electric batteries: 50,000 mA h, three lithium–polymer or tesla-type 64 A h lithium–polymer.

Fuel: No petroleum fuel required

Endurance: 4 h with payload

Remote controller: Ground control keeps contact with UAV for 12 km radius. GCS iPad can override previous program and alter flight path.

Area covered: 120 km linear distance

Photographic accessories: Cameras: It has the usual complement of visual, multi-spectral, NIR, IR, and red-edge bandwidth sensors.

Computer programs: Images are relayed instantaneously. They can be processed at the UAV's CPU. Images could also be processed at GCS iPad. Software such as Pix4D Mapper or Agisoft Photoscan is utilized.

Spraying area: This is not a sprayer UAV.

Volume: Not applicable

Agricultural uses: This UAV is used to scout the crop for seedling establishment, phenomics, weeds, disease/pest tack floods, drought, and soil erosion. It is also used to fix fertilizer-N and irrigation schedule, natural vegetation monitoring is possible.

Nonagricultural uses: They are surveillance of industries, buildings, rail roads, oil pipelines, boundaries, mines, mining activity, ore dumps, etc.

Useful References, Websites, and YouTube Addresses

http://diydrones.com/forum/topics/gemini-v2
https://www.youtube.com/watch?v=mU0ZGcYdCjk (accessed Dec 12, 2017).
https://www.youtube.com/watch?v=19RDaBTwC6I (accessed Dec 14, 2017).
https://www.youtube.com/watch?v=KBhZNQpeXeE (accessed Nov 29, 2017).
https://www.youtube.com/watch?v=NV8SnrK3kQo (accessed Nov 29, 2017).
https://www.youtube.com/watch?v=wDcc3RYw_b8 (accessed Nov 29, 2017).

NAME/MODELS: GUARDIAN UAV

Company and address: OM UAV, 60-UA, Jawahar Nagar, Delhi 110007, India; mobile: +91 9873690212; e-mail: atul@omuavsystems.com; website: http://www.omuavsystems.com/guardian.html

Guardian UAV (mini-UAV) is produced by a company located in Delhi, India. This UAV company has a few other fixed-winged UAV models to offer, such as Baaz UAV, etc. This company was initiated in 2009. The UAV models have been adopted to survey and obtain aerial imagery of natural resources. They may make a mark in the agricultural belt of India soon.

Guardian UAV is a small aerial robot. It is also called as "mini-UAV." It is easily portable since all the accessories and airframe can be packed and transported to any location for hand launch. It needs small area (opening) to launch and land (Plate 2.15.1). The airframe and software for navigation and landing are custom-made by the company. Its propulsion is achieved, using an electric motor. The motor is energized using lithium-polymer batteries 11.1 V. The ground-control equipment has telemetric (radio) contact and internet connectivity. It is used to relay commands and for receiving videoimages, still images, or digital data about earth's surface features. This small UAV is highly useful to farmers who wish to get aerial imagery, thermal images, and 3D images (Plate 2.15.1).

TECHNICAL SPECIFICATIONS

Material: A composite of very light material that includes carbon fiber, ruggedized foam, rubber, and plastic is used to make the UAV.

Size: length: 130 cm; width (wingspan): 176 cm; height: 20 cm; weight: 300 g

Propellers: It has one propeller with two blades of 9 × 6 cm and made of carbon fiber.

Payload: It carries a payload of 800 g. The payload includes photographic equipment such as color videocameras, high-resolution still cameras, multispectral cameras, thermal sensor, etc.

Launching information: Guardian is a small UAV. It is easily hand-launched by holding the frame at shoulder height. It has autonomous take-off and navigation facility. The flight can be programed for up to 300-way points. It has manual override option, using an iPad (GCS). The GCS has control over photographic schedule. It can activate or deactivate cameras. Navigation can also be altered, using telemetric connection. Manual landing facility exists, but it depends on the wind, weather conditions, and size of the landing site. For landing, this UAV needs 100 × 20 m obstacle-free location. Landing is accurate to 10 m from the programed spot.

Altitude: It is generally operated at 50–300 m a.g.l. Ceiling is 2000 m a.g.l.

Speed: Cruise speed is 48 km; maximum speed is 60 km h^{-1}

Wind speed tolerance: 20 km h^{-1}

Temperature tolerance: −20 to +60°C

Power source: Electric batteries

Electric batteries: 1240 mA h batteries; two cells.

Fuel: It does not require petroleum fuel.

Endurance: 80 + 10 min for fail safe.

Remote controllers: The GCS consists of mission control software in iPads/laptop. It has video-monitoring facility. The power supply, tracker antenna, and Yagi antenna are also situated in the GCS. Videocontact is established using analog audio and video. It works at a frequency of 5.8 GHz. It has 32 channels. Ground control systems are supported by a 300-mA@11.1-V power source. The GCS keeps contact with the UAV via telemetric and internet connectivity, for up to 10 km radius. It operates at a frequency of 900 (885–915 MHz). Baud rate is 19,200 kb s^{-1}.

Area covered: It covers a linear length of 50 km and an area of 300–500 ha of farmland depending on flight plan.

Photographic accessories: Cameras: HD color video recorder with 32 GB memory card. It provides images of 1270 × 720 pixel. Maximum resolution of videocamera is 2700 × 720 pixel. The UAV has on board a video-recording and storing facility with a 32 GB memory card. It has day light and low-light zoom lens. Photographic cameras can be commanded, using telemetric as well as internet connectivity. Thermal camera (IR bandwidth) has 640 × 480-pixel resolution at 40°. It also has a range of selectable thermal range lenses.

Computer programs: Images are processed using PIX4 D Mapper, Agisoft Photoscan or a custom-made software for image processing.

Spraying area: This not a sprayer UAV.

Volume: Not applicable

Agricultural uses: The UAV is used to scout crops and monitor farm vehicles. It is useful in collecting spectral data pertaining to crop growth (NDVI), phenomics, grain maturity, disease/pest attacks, drought/flood, etc. It is also used for natural vegetation monitoring.

Nonagricultural uses: They are surveillance of industrial sites, buildings, oil pipelines, mines, mining activity, ore dumps, city traffic and tracking transport vehicles, etc.

PLATE 2.15.1 Guardian UAV.
Source: Mr. Atul Khosla, Director, OM UAV Systems Ltd., Delhi, India.

Useful rReferences, Websites, and YouTube Addresses

http://www.omuavsystems.com/ (accessed May 12, 2017).
https://www.linkedin.com/company-beta/2988522 (accessed May 12, 2017).
https://www.youtube.com/watch?v=gevVZ3a2o1E (accessed May 12, 2017).

Name/Models: Hubsan-H301S-SPY HAWK

Company and address: Hubsan Intelligent Company Limited; office address: 13th Floor, Building 1C, Shenzhen Software Industry Base, Xuefu Road, Nanshan District, Shenzhen, China; Tel.: 0769 87923231/0755 83947816; fax: +86 769 82063083/0755 83947819; e-mail: Service@hubsan.com

TECHNICAL SPECIFICATIONS

Material: It is made of foam and plastic, ruggedized to withstand rough landing on belly (Plate 2.16.1)

Size: length: 75 cm, width (wingspan): 100 cm; height: 20 cm; weight: 355 g

Propellers: It has one propeller placed on the fuselage facing backward. It is connected to an electric motor: 1812 Brushless KV3200.

Payload: 150 g

Launching information: Spy Hawk is hand-launched at shoulder height. It has software for automatic landing.

Altitude: It flies at 400 m a.g.l. during surveillance of crop fields, but its ceiling altitude is 3700 m a.g.l.

Speed: 45 km h^{-1}

Wind speed tolerance: 15–20 km h^{-1}

Temperature tolerance: −20 to +45°C

Power source: Electric batteries

Electric batteries: 7.4 V, 1300 mA h

Fuel: No requirement for petroleum fuel

Endurance: 30 min per flight

Remote controllers: Radiocontrol range is 2 km from ground station/remote controller. It is an autopilot UAV. It has software such as Pixhawk for autonomous navigation. It autonomously returns to launch point. The GCS computers (iPad) have override facility.

Area covered: 100–200 ha per flight

Photographic accessories: Cameras: The UAV carries a high-resolution visual (R, G, and B) camera. Images are collected, using 2.4 GHz control system. Images are accurately tagged using GPS. Images are transmitted immediately, using 5.86 image transmission function. Cameras have video-recording facility.

Computer programs: Images relayed could be processed, using the usual Pix4D or Agisoft's Photoscan software. UAV kit comes with a memory card to store data and images.

Spraying area: This is not a sprayer UAV.

Volume: Not applicable

Agricultural uses: This UAV requires certain modifications before it is adapted for use in agricultural regions; aerial survey of land, crops, and cropping systems; crop scouting for growth, phenomics, and grain maturity should be possible; and monitoring natural vegetation is also a possibility (Plate 2.16.1).

Nonagricultural uses: It is a good-hobby UAV. Using certain modifications in cameras, remote control, and software, we can adopt it to conduct surveillance of industrial units, buildings, oil pipelines, electric installations, dams, rivers, mines, mining activity, ore movement, etc.

PLATE 2.16.1 Hubsan-H 301S-SPY HAWK flat-winged lightweight UAV.
Source: Support Service Team, Hubsan Intelligent Company Limited, Nanshan District, Shenzhen, China.

Useful References, Websites, and YouTube Addresses

http://n.stuccu.com/s/Hubsan%20Spyhawk (accessed Jan 1, 2017).

https://www.alibaba.com/product-detail/Hubsan-H302S-SPY-HAWK-58G_6014405 (accessed Jan 1, 2017).

https://www.youtube.com/watch?v=GsWetEMgmos (accessed Jan 1, 2017).

https://www.youtube.com/watch?v=DLnX0YukJfs (accessed Jan 1, 2017).

https://www.youtube.com/watch?v=DCgKKU0_Xug (accessed Mar 31, 2017).

Name/Models: KAHU

Company and address: Skycam UAV 2014 Ltd., Palmerston North, New Zealand; phone +64 21 110 1099; e-mail: info@skycam.co.nz; Website: http://www.kahunet.co.nz/

Skycam-2014 Ltd. is a UAV-manufacturing company. It concentrates on offering small, light-weight, and highly useful UAVs. They are meant for aerial surveillance and imagery. Skycam UAV Ltd. also offers accessories such as ground control equipment for telemetry (radiocontrol), iPads for programing UAV flight-path and movement, etc. Skycam's UAVs are being used in New Zealand and adjoining nations, mainly, to monitor natural resources and agricultural regions.

The Kahu UAV has a small fully composite airframe. It has been designed to achieve better endurance. The aircraft incorporates an electric motor (1200 W) and novel cooling mount. It has DTA pilot static sensor for measurement of air data. It can fly even in light rain. Kahu airframes are constructed using aluminum molds. Fuselage has large camera port. Wings are detachable. Wings store batteries and other avionics-related parts. Tail fins are also detachable. Kahu has a series of applications in the realms of agriculture, commercial, and civilian activities.

TECHNICAL SPECIFICATIONS

Material: The airframe is made of light-weight aluminum alloy, and a composite of S-glass, E-glass, carbon fiber, and plastic. This combination is supposed to offer better strength, rigidity, and minimize flight weight.

Size: length 2.2 m; width (wingspan): 2.8 m; height: 0.3 m; weight: 3.9 kg MTOW.

Propellers: It has one propeller placed at the tip of the nose, on the fuselage. It has two blades made of plastic. Propeller is connected to 1200-W brushless electric motor system.

Payload: 1.4 kg. It includes a gyro-stabilized gimbal. A series of cameras for visual, NIR, IR, and thermal imagery are fitted.

Launching information: It is launched using hand or catapult or with a bungee catapult. It has autopilot mode. It adopts "hawk-eye autopilot system" for navigation and landing. Landing is on belly and with an accuracy of 5–10 m from the programed spot. It can be recovered using a parachute. Parachute should pop-up at 100 m a.g.l.

Altitude: It operates at 50–300 m for surveillance of agricultural crops and civilian installations. Its ceiling height is 5000 m a.g.l.

Speed: The cruise speed is 60 km h^{-1}, but it can reach up to 100 km h^{-1}.

Wind speed tolerance: 20 km h^{-1}

Temperature tolerance: −20 to +55°C

Power source: Electric batteries

Electric batteries: It has four cell lithium–polymer batteries (16.8 V, 8000 mA h).

Fuel: This is not an IC engine-fitted UAV. There is no requirement for petroleum fuel.

Endurance: 2 h per flight

Remote controllers: The GCS is equipped with PCs or iPads to monitor UAV's flight and decide its path. Flight plan can be either to preprogramed or modified midway, if needed. The telemetric connectivity functions within a range of 25 km radius from GCS. Flight path is decided using Pixhawk autopilot software or similar custom-made software.

Area covered: It covers 60–70 km linearly. Based on flight program and path, it can cover an area of 100–200 ha.

Photographic accessories: Cameras: Sensors include visual cameras (R, G, and B), IR, NIR, and red-edge waveband cameras. Videography is also possible using appropriate camera. The UAV is usually fitted with a choice from a range of cameras such as Pentax, Sony NEX 5, or Canon series.

Computer programs: The images derived using UAV cameras are processed, at the CPU on board. Otherwise, it is relayed to GCS. The images are processed using Pix4D Mapper or Agisoft Photoscan.

Spraying Area: This is not a sprayer UAV.

Volume: Not applicable

Agricultural uses: This UAV is capable of rapid and repeated deployment to monitor crops, surveillance them specifically to get data such as NDVI, canopy growth pattern, biomass accumulation, nutritional status, and fertilizer needs. Crop scouting is also done to assess drought or flood damage to crops, diseases, pests, and weeds. When flown at low altitude, this UAV provides excellent data about spectral signatures of disease/pest-attacked zones and weed species. Monitoring natural vegetation frequently is a clear possibility.

Nonagricultural uses: Aerial photography, mapping, and surveillance of city installations. These small and low-cost UAVs are also useful to monitor disaster-affected zones, mines, hazardous locations such as volcanoes, etc.

Useful References, Websites, and YouTube Addresses

http://www.kahunet.co.nz/ (accessed Apr 1, 2017).
http://www.kahunet.co.nz/kahu-uav.html (accessed Sept 5, 2017).
http://www.kahunet.co.nz/image-processing.html (accessed Sept 5, 2017).
https://www.youtube.com/watch?v=FKyeI3l3w34 (accessed Sept 5, 2017).

Name/Models: L-A series (LA 500-AG for Precision Agriculture)

Company and address: Lehmann Aviation Inc. La Chapelle, Vendomoise, France; e-mail: info@lehmannaviation.com; websites: info@lehmannaviation.com, http://www.lehmannaviation.com/

Lehmann Aviation Inc. specializes in small UAVs. It was initiated in 2005 by Mr. Benjamin Lehmann, founder and CEO of Lehmann Aviation Inc. They have announced launch of a totally new L-A Series line of professional mapping UAVs for operators around the world (Ridgeline, 2016). Lehmann Aviation also offers the new Operation Centre v2, the most advanced UAV Management System. The L-A series has major mapping applications such as LA500 for mining and construction, "LA500-AG" for precision agriculture and LA500-RTK for accurate mapping.

TECHNICAL SPECIFICATIONS

Material: The platform is made of graphite-reinforced plastic. It is ruggedized and toughened to withstand harsh weather conditions (Plate 2.17.1).

Size: length: 14 in (or 35 cm); width (wingspan): 47.2 in. (or 116 cm, including detachable winglets); height: 14 in. (or 35 cm); weight: 1250 g including cameras and batteries

Propellers: Electric propellers with brushless motors. Propeller is placed on fuselage between wings at the rear end.

Payload: 250 g but could be upgraded to 400 g with optional wings

Launching information: It is a hand-launched UAV. It has autolaunch, navigation, and landing software, such as Pixhawk.

Altitude: Its operational altitude is 50–300 m a.g.l. during crop survey, with maximum ceiling 3500 m a.s.l. or 11,480 ft a.s.l.

Speed: The cruise speed usually ranges from 20 to 80 km h^{-1} (12–50 mi h^{-1})

Wind speed tolerance: 37 km h^{-1}

Temperature tolerance: −25 to +60°C

Power source: Electric batteries

Electric batteries: Lithium batteries

Fuel: No requirement for petroleum fuel.

Endurance: 45 min without brake in flight

Remote controllers: The remote-control equipment at GCS operates up to 3 km from the UAV.

Area covered: The UAV covers around 1000 ha per flight, picking digital imagery of land surface, crops, and other features. It operates at 1000 m altitude during imaging. Oblique imagery is also possible. It is done by flying the UAV at low altitude and manipulating camera angles. The GCS iPad has override facility to alter the flight path and shutter activity, midway during flight.

Photographic accessories: Cameras: GoPro Hero-4, Sony α6000; options include red, green, and blue, NIR, resolution reaches up to 24.3 MP.

Computer programs: Flight planning and paths are decided using Operation Centre v2 (a windows touchscreen device). Image processing is done using Agisoft photoscan or Mapper native files (software). Digital images could be processed immediately and downloaded at a ground station.

Spraying area: This is not a sprayer UAV.

Agricultural uses: UAV models such as LA 500-Ag are used to develop digital maps of crop fields, detect drought, pests, and disease, at an early stage of the crop. They are also used to analyze natural vegetation, its species diversity, and study floral changes on a broad scale. LA-500 is used to develop 3D models of terrain and vegetation distribution.

Nonagricultural uses: LA-500 is utilized to monitor and surveillance construction activity in large areas, buildings, dams, etc. It is used to keep a close watch of mining region, its work progress, mine dumps, and transshipment of ores. LA-500 is also used to monitor traffic in large metropolis and policing events, monitor oil pipelines, rail roads, and boundaries (Frey, 2014).

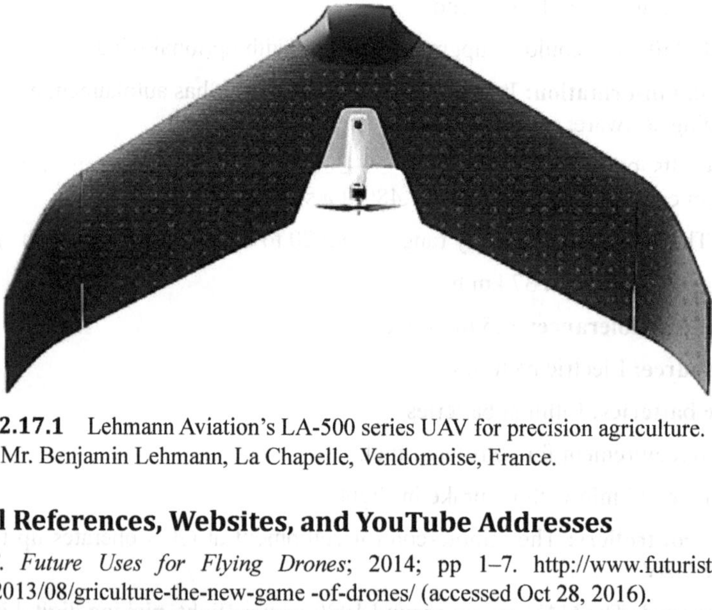

PLATE 2.17.1 Lehmann Aviation's LA-500 series UAV for precision agriculture.
Source: Mr. Benjamin Lehmann, La Chapelle, Vendomoise, France.

Useful References, Websites, and YouTube Addresses

Frey, T. *Future Uses for Flying Drones*; 2014; pp 1–7. http://www.futuristspeaker.com/2013/08/griculture-the-new-game -of-drones/ (accessed Oct 28, 2016).

Ridgeline. *Lehman Aviation L-M 500*; 2016; p 17. http://www.ridgeline.com/products/lehman-aviation-l-m-500-series-drone (accessed Oct 2016).

http://www.lehmannaviation.com/ (accessed Nov 25, 2016).

http://www.lehmannaviation.com/lasiers?gclid=CITDw6Ku8c8CFdWJaAodQkMB2g. (Oct. 25, 2016).

http://www.lehmannaviation.com/contact (accessed Nov 25, 2016).

https://www.youtube.com/watch?v=9aUyrbJ-S6I (accessed Nov 25, 2016).

https://www.youtube.com/watch?v=9aUyrbJ-S6I (accessed June 15, 2017).

Name/Models: Ma 410

Company and address: Aeroterrascan Inc. Jalan Haji Wasud No. 17, Dipatikur, Bandung, West Java, Indonesia, ID 40132; phone: +62 22 250 2323 e-mail: contact@aeroterracsan.com; website: http://www.aeroterrascan.com

Ma 410 is a small fixed-winged UAV. It is hand launched and easily packed in a suitcase and transported to any place in a large farm. It is commonly used for surveillance of natural vegetation, crops, and mining activities. They have also used it for surveillance of public places and sanctuaries.

TECHNICAL SPECIFICATIONS

Material: It is made of carbon fiber, foam, and plastic foam.

Size: length: 45 cm; width (wingspan): 90 cm; height: 35 cm; weight: 1.8 kg

Propellers: It has one propeller placed at the rear end of the fuselage.

Payload: 800 g

Launching information: It is an autolaunch, autonavigation, and autolanding UAV.

Altitude: Its operative altitude is 100–400 m above the crop canopy. However, it can reach 3200 m a.g.l.

Speed: 40–60 km h^{-1}

Wind speed tolerance: 15–20 km h^{-1} disturbance

Temperature tolerance: −25 to +60°C

Power source: Electric batteries

Electric batteries: 14 or 28 mA h; two cells

Fuel: No need for petrol or diesel.

Endurance: 40–60 min per flight

Remote controllers: The UAV keeps contact with GCS via radiocontrol for up to 22 km radius. The GCS iPad can program the flight path for autonavigation. The UAV returns point of launch in times of emergency.

Area covered: 50–250 ha

Photographic accessories: Cameras: The UAV is fitted with a usual complement of visual (R, G, and B), multispectral, hyperspectral, NIR, IR, and mica-sense cameras. It also has a Lidar pod.

Computer programs: The images transmitted are processed using Pix4D Mapper or Agisoft's Photoscan or any other custom-made software.

Spraying area: This is not a sprayer UAV. Not applicable.

Agricultural uses: They are aerial survey of land resources, crop-scouting mica and natural vegetation monitoring are the major functions related to agriculture.

Nonagricultural uses: They are surveillance of industries, buildings, electric power lines, oil and gas pipelines, city traffic, tracking transport vehicles, etc.

USEFUL REFERENCES, WEBSITES, AND YOUTUBE ADDRESSES

http://www.aeroterrascan.com (accessed Oct 25, 2016).
https://www.youtube.com/watch?v=0g5dAuhMYjs (accessed Oct 25, 2016).

Name/Models: MTD UAV

Company and address: TechnoSys Embedded Systems, SCO 66, 1st floor, Sector 20C, Chandigarh 160020, India; phone: +91 172 5085909, +91 172 4671082; e-mail: info@technosysind.com, sales@technosysind.com, support@technosysind.com; Website: http://www.technosysind.com

TechnoSys Embedded Systems (P) Ltd. is a UAV company. It was started in 2003. Their products relate mainly to electronic control systems for various gadgets, including UAV. They develop software for UAV control, flight path determination, imagery, its frequency, processing orthomosaics, etc. MTD is a twin-propeller UAV. It was developed recently by the company (Plate 2.18.1). This fixed-wing UAV is useful in aerial imagery of agricultural regions, land, soil, crops, and water resources. The emphasis is on high-resolution mapping of crop production zones. The focus is on application of digital data to conduct agronomic procedures efficiently.

TECHNICAL SPECIFICATIONS

Material: The UAV is made of rugged foam, plastic, and carbon fiber. It is kept in a rugged box, so that UAV can be transported, to any location in a farm and launched (Plate 2.18.1).

Size: length: 60 cm, width (wingspan): 180 cm; height: 18 cm; weight: 1.8 kg

Propellers: It is twin-propeller UAV. Each propeller is powered by an electric motor.

Payload: 1200 g

Launching information: It is launched by hand. It can also be launched using the catapult. Its flight path and landing are autonomous.

Altitude: MTD UAV flies at around 400 m a.g.l. for agricultural survey, mapping, scouting, phenotyping, and imagery. It can reach 4200 m a.g.l.

Speed: 45–60 km h^{-1}

Wind speed tolerance: 20 km h^{-1} disturbance

Temperature tolerance: It endures hot tropical weather and tolerates −20 to +60°C

Power source: Electric batteries

Electric batteries: 1000 mA h.

Fuel: No requirement for petroleum fuel.

Endurance: 65 min per flight

Remote controllers: The UAV stays in contact with ground-station computers, for up to 10 km radius. Flight path is created on an iPad or computer screen. Flight paths may be edited midway through the flight. The UAV automatically returns to the point of launch, in case of failure of telemetry or errors in flight path design. The CPU in the UAV helps in mission planning and alterations, if any.

Area covered: 200 ha per flight. It picks imagery and relays it to GCS immediately.

Photographic accessories: Cameras: Sony α7, NIR, and red-edge bandwidth cameras. These are a set of high-resolution and multispectral cameras. Photographic accessories commonly attached to MTD UAV are Sony Alpha a5100, Mirrorless Digital Camera with 16–50 mm lens, 24.3 MP; APS-C Exmoor HD CMOS Sensor; BIONZ X Image Processor

Computer programs: This UAV carries a 24-MP camera. It is capable of imaging at R, G, B, and NIR wavelength bands. It can also be fitted with red-edge camera for thermal imagery and detection of crop water status. All the still images and videographics are tagged with GPS coordinates. This helps in easy stitching of orthomosaics. Images can be easily processed, at the ground station, using software such Pix4D Mapper or Agisoft Photoscan.

Spraying area: MTD UAV is not a spraying UAV.

Volume: Not applicable

Agricultural uses: MTD UAV is adopted for survey of agricultural terrain, soils, and irrigation channels. Crop scouting, crop phenotyping, accruing data on NDVI, leaf area, leaf chlorophyll content are other functions. Weed infestation detection is possible. Detection of disease/insect pest attacks, if any, on the crops, drought-affected areas, soil erosion, flooding and any other damage due to fire is a clear possibility. It is useful in forest management, particularly, in the surveillance of forest stands, growth monitoring, and logging.

Nonagricultural uses: They are surveillance of mines, traffic, and transport vehicle movement. It is useful in defense services. It is operated during detection and first-response coordination in a disaster zone, etc. This UAV has also been used in geological survey, urban planning and environmental engineering. It has been used for surveillance of industrial installations.

PLATE 2.18.1 A MTD UAV – A flat fixed-winged, ready-to-fly UAV.
Note: It is a rugged bodied small agricultural UAV. It is useful in surveillance of agricultural zones.
Source: Mr. Dhruv Arora, TechnoSys Embedded Systems (P) Ltd, Chandigarh, India.

Useful References, Websites, and YouTube Addresses

http://www.technosysind.com/tes.php.
https://www.youtube.com/watch?v=6WvU45J9REg (accessed Dec 26, 2016).
https://www.youtube.com/watch?v=OZPuDYKX9BY (accessed Dec 26, 2016).
https://www.youtube.com/watch?v=2WGrnwFSzYA (accessed Dec 26, 2016).
https://www.youtube.com/watch?v=WR8b7TwYblg (accessed Dec 26, 2016).

Name/Models: Multirole M-011

Company and address: AeroTekniikka UAV Ltd., Jouni Jalkanen, Managing Director, Tuomaalantie 58, FI-41500 Hankasalmi, Finland; phone: +358 45 321 9018; e-mail: firstname.lastname@aerotekniikka.fi

Aeroteknikka is a UAV company located in Finland. It has developed a few models of small UAVs. These UAVs such as "Multirole M-011" have a big role, to play in agricultural crop production, in future years. These UAVs help in obtaining aerial imagery and spectral data of crops. They help in conducting precision farming procedures by detecting variations in land, soil, and crop. AeroTekniikka UAV Ltd. offers training in the control of UAV (UAVs). This company believes that in addition to production, maintenance and training of UAVs, the use of UAV in farms and civilian locations is important. This company offers quick supply of spare parts produced in its Finland factory.

Multirole M-011 was developed and first tested in 2009. Multirole M-011 is easily managed by one person. The UAV is energized by lightweight electric batteries. It avoids use of fuel oil and need for larger fuselage. Compared with previous models, Multirole M-011 is economical to buy and operate. This UAV carries several cameras, Lidar, and radar equipment in its gimbal. Payload can be packed onto the aircraft, such as various cameras or other electronics. This UAV withstands harsh weather conditions. It does not need a runway to take-off or land.

TECHNICAL SPECIFICATIONS

Material: It is made of a composite of lightweight carbon fiber, plastic, ruggedized foam, and fiber glass sheets.

Size: length: 120 cm; width (wingspan): 90 cm; height: 30 cm; weight: 3.0 kg MTOW

Propellers: It has one propeller at the rear end of fuselage near tail fins.

Payload: 1.2 kg

Launching information: This UAV is hand-launched. It has autolaunch mode and autonomous navigation. It lands using belly landing. Otherwise, it can be recovered using parachute.

Altitude: 50–300 m for agricultural crop survey and general monitoring and aerial imagery; ceiling is 3500 m a.g.l.

Speed: 40–60 km h^{-1}

Wind speed tolerance: 20 km h^{-1}

Temperature tolerance: −25 to +45°C

Power source: Electric batteries

Electric batteries: lithium–polymer batteries 14,000 mA h.

Fuel: No requirement of petroleum fuel.

Endurance: 2 h of aerial still and videoimagery

Remote controllers: The GCS (radio) contact covers 12–15 km radius. The preprogramed flight path can be altered, if needed, using computers at the GCS. From the ground station, it is possible to share real-time video with others. The images are transmitted swiftly.

Area covered: 50–300 ha per flight

Photographic accessories: It has a set of visual, NIR, IR, and mica-sense red-edge camera. This UAV transmits high-resolution videoimages. The UAV cameras offer images from oblique angles, elevation, and other desired perspectives. The images are derived from steady cameras located in a gyro-stabilized gimbal.

Computer programs: The UAV carries a set of software in the CPU located in the fuselage. The software is utilized to store images, process them in situ, or transmit them to GCS (iPads). The GCS iPads have software such as Pix4D Mapper, Agisoft Photoscan, and other similar custom-made software for image processing.

Spraying area: This is not a spraying UAV.

Volume: Not applicable

Agricultural uses: Advanced research in crop production technology is supported by UAVs with their ability for aerial imagery. The UAVs such as "Multirole 011" are useful in surveillance and aerial mapping of land resources, soil type, crops, and cropping patterns. It is used to estimate canopy growth, leaf-N, estimation of fertilizer-N requirement, crop water stress index, etc. This UAV also provides detailed information about disease and pest attack, through its ability for collecting spectral signatures of healthy and affected crops/canopies and leaves. Crop scouting is also done to detect drought or flood-affected regions in the fields. In general, this UAV is useful in natural vegetation monitoring.

Nonagricultural uses: They are surveillance of industries, buildings, highway traffic, public areas, events and vehicle movement is possible, using aerial imagery from this UAV.

Useful References, Websites, and YouTube Addresses

http://aerotekniikka.fi/english/entrepreneur.html (accessed Jan 23, 2018).

Name/Models: Nauru 500 A

Company and address: XMobots, Rua Santa Cruz, 979 Sao Carlos, SP; Brazil; phone: +55 (16) 3413 0655; e-mail: Not available; website: http://www.xmobots.com

Xmobots is a UAV company initiated in 2007. It is situated in the Sao Paulo region of Brazil. They produce UAVs, such as Apoena 1000B, Nauru 500A, Echar 20 A, and CAVE/ANAC, all flat-winged UAVs. Nauru is an important small UAV. It is operative in different sectors such as civil and agriculture in Brazil. It has about 5.5-h endurance and can travel at a highest speed of 108 km h^{-1}. This is an excellent UAV for agricultural purposes. It can offer high-resolution images of crops at 2.5 cm $pixel^{-1}$.

TECHNICAL SPECIFICATIONS

Material: It is made of foam, carbon fiber, and steel.

Size: length: 95 cm; width (wingspan): 128 cm; height: 22 cm; weight: 3.4 kg

Propellers: It has one propeller placed at the rear end of fuselage.

Payload: 1.1 kg

Launching information: It is launched using a runway or a catapult

Altitude: 50–200 m above crop's canopy for aerial survey of agricultural crops. Ceiling is 2000 m a.g.l.

Speed: 35–45 km h^{-1}

Wind speed tolerance: 20 km h^{-1}

Temperature tolerance: −15 to +45°C

Power source: Electric batteries

Electric batteries: 24 mA h; two cells

Fuel: No requirement for petroleum

Endurance: 25–45 min

Remote controllers/Preprogramed flight paths: The GCS computers keep contact with UAV for up to 15 km radius. It has direct radiolink to monitor the flight or to modify the path midway.

Area covered: 50–200 ha per flight depending on the altitude stipulated.

Photographic accessories: Cameras: It has a full complement of sensors that function at visual (R, G, and B), multispectral, hyperspectral, and NIR bandwidth.

Computer programs to process orthomosaic: The images are processed using Pix4D Mapper or Agisoft Photoscan or Trimble's Inphos. The images are relayed instantaneously to GCS.

Spraying area: Not a spraying UAV

Agricultural uses: It is useful in crop scouting leaf chlorophyll distribution, leaf-N status, canopy temperature, disease/pest attack, flood, and drought image if any. Monitoring of natural vegetation is also possible.

Nonagricultural uses: Surveillance of mines, mining activity, ore transport, industrial sites, large buildings, transport vehicles, etc.

Useful References, Websites, and YouTube Addresses

http://www.xmobots.com.br/ (accessed May 25, 2017).

http://mundogeo.com/blog/2015/07/17/xmobots-lancara-nova-versao-do-drone-nauru-na-alemanha/ (accessed May 25, 2017).

https://www.youtube.com/watch?v=rccb-lv5u08 (accessed May 25, 2017).

https://www.youtube.com/watch?v=y16WFFiFirQ (accessed May 25, 2017).

https://www.youtube.com/watch?v=y16WFFiFirQ (accessed June 10, 2017).

https://www.youtube.com/watch?v=mAvhkQgwr70 (accessed June 10, 2017).

https://www.youtube.com/watch?v=y16WFFiFirQ (accessed June 10, 2017).
https://www.youtube.com/watch?v=y16WFFiFirQ (accessed June 10, 2017).

Name/Models: NOMAD

Company and address: Mr. Calle Jose Gallan, NOMAD AGR, Merino, 6. EDIFICIO CREA-Mosulo 13, 41015 Sevilla, Spain; phone: +34 954 871 782; e-mail: contactus@novadrone.com; Website: http://www.novadrone.com

Novadrone is a UAV-manufacturing company located in Spain. Two important models, namely NOMAD and NOUR, both are well adopted to conduct aerial survey and imagery of farmland, soils and crops. It conducts aerial imagery at different wavelengths and different altitudes. These UAVs specialize in farmland surveillance, measurement of crop growth, NDVI, phenotyping, and yield forecasting.

TECHNICAL SPECIFICATIONS

Material: UAV is built using kevlar and carbon fiber composites to avoid breakages. The engine, batteries, and electronic components are enclosed tightly, inside an alumina reinforced vault, to avoid shocks and breakage.

Size: length: 140 cm; width (wingspan): 200 cm; height: 35 cm; weight: 3.5–5 kg depending on accessories

Propellers: It has one propeller placed at the tip of fuselage.

Payload: 2 kg

Launching information: It is hand-launched. It has facility for automatic take-off, flight mission, and landing (on belly).

Altitude: Usual operating altitude for agricultural activities is 400 ft, but maximum attainable altitude is 18,000 ft a.g.l.

Speed: 18 m s^{-1} or 45 km h^{-1}

Wind speed tolerance: 70 km h^{-1} along with light rains/drizzle

Temperature tolerance: −20 to +60°C

Power source: Electric batteries

Engine: Axi brushless motors

Electric batteries: Electric batteries—Lipo/Li ion. The autopilot operates at 6.5–30 V.

Fuel: No requirement for petroleum fuel.

Endurance: 1–5 h without brake in flight.

Remote controllers: UAV keeps contact with ground station for 67 km radius. The UAV is also linked to ground station via 2.4 GHz/900 MHz. Ground control software includes encrypted telemetry and transmission of commands.

Area covered: UAV may cover up to 300 ac h^{-1} per flight, depending on batteries

Photographic accessories: Cameras include R, G, B, multispectral, and NIR. All cameras are loaded into fuselage with antivibration devise. It helps in accurate photography. Cameras operate at 20.1 MP resolution. Cameras relay pictures formatted as .JPG to the ground station. Cameras include Sony α7, and Sony series with 20 MP, Exmore APS-C type CMOS, and 16-mm fixed lens. It also carries Sony NIR and IR range cameras for thermal imagery of crops.

Computer programs: The UAV system includes several of the software required for imaging crops, collecting data, and other agricultural aspects. For example, Pix4D Mapper or Agisoft's Photoscan is used for processing orthomosaics. Other software include Global Mapper, ArcGIS, Quick terrain reader, Google maps, etc. The software allows UAV operators, to collect both 2D and 3D images of terrain, crops, irrigation lines, etc.

Spraying area: This is not a sprayer UAV.

Volume: Not applicable

Agricultural uses: NOMAD is useful to farmers/companies during all the four seasons.

During autumn: It is used to study and analyze preharvest crop.

During spring: It helps with images for soil analysis, tillage decisions, and planting.

During summer: This UAV is used to study crop water stress index, decide on irrigation schedules, and to observe, center-pivot sprinklers in the fields. Locations prone to stagnation and needing drainage are imaged swiftly.

During winter: This UAV is used to observe farm operations and vehicles. NOMAD is used to assess planting and seedling emergence. NOMAD is also used to study disease/pest occurrence in the crop fields, plus water stress effects, if any. NOMAD is also apt to phenotype the crop. It helps in irrigation scheduling. The UAV's imagery is also used to prepare "management blocks." It offers digital data for precision applicators (Fig. 2.1; Tarancon, 2016).

Nonagricultural uses: NOMAD model is used in mapping terrain, mining regions, water resources, surveillance of pipelines, electric and industrial installations, traffic, etc.

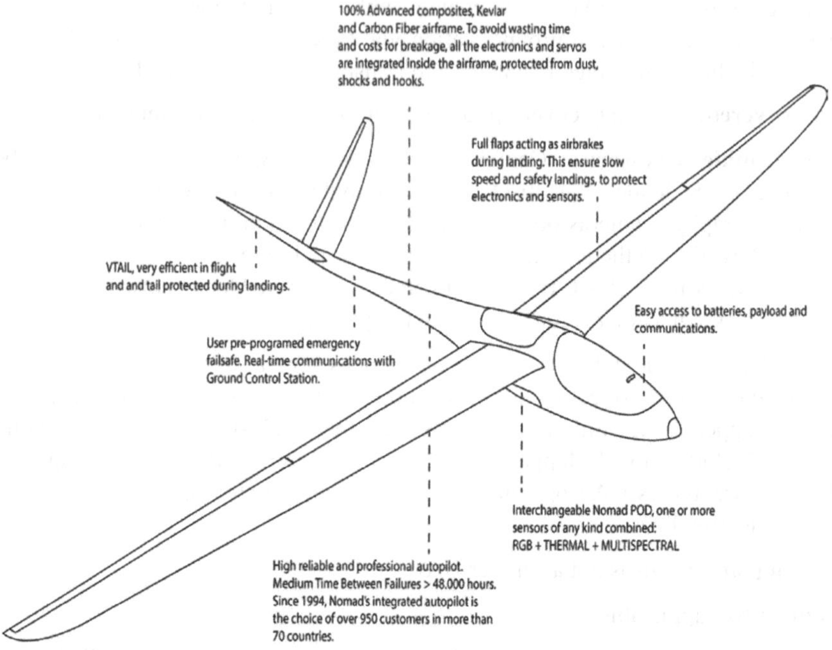

FIGURE 2.1 Diagram of Novadrone and its accessories.
Source: Mr. Anthony Hassal and Ms. Neomi Tarancon, Nomad AGR, Sevilla, Spain.

Useful References, Websites, and YouTube Addresses

Tarancon, N. *Novadrone NOUR: The Most Versatile UAV Is Its Class*; Novadrone Inc., 2016; pp 1–12. https://app.hubspot.com/presentations/2560331/veiw/2422218?acessId=50f6c8 (accessed Nov 10, 2016).

http://nomadgcs.com/industry-solutions/uas/ (accessed Oct 12, 2017).

https://novadrone.com/en/drones/nomad/ (accessed Oct 17, 2017).

https://novadrone.com/en/applications/drone-for-agriculture/?gclid=CPjOoY2A_88C (accessed Oct 17, 2017).

https://novadrone.com/en/?utm_source=TOPOGRAFOS%20DE%20CANADA&utm_campaign=49f1644165-EMAIL_CAMPAIGN_2017_06_14&utm_medium=email&utm_term=0_e4ec7dd30b-49f1644165-281688697 (accessed Oct 17, 2017).

https://www.youtube.com/watch?v=71XAXaEdIGY (accessed Oct 17, 2017).

Name/Models: Parrot's Disco Drone

Company and address: Parrot Inc., Paris, France.

Parrot Inc. is an important UAV manufacturing company in Europe. It is situated in Paris, France, and several other countries. They manufacture UAVs of different sizes and capabilities. They also govern and own several other companies that produce accessories for UAVs. For example, mica sense offers excellent sensors, etc.

TECHNICAL SPECIFICATIONS

Material: It has toughened (ruggedized) foam and graphite reinforced plastic, to withstand harsh weather conditions and landing. Wings are detachable, if needed.

Size: length 58 cm; width (wingspan): 115 cm; height: 12 cm; weight: 750 g (1.5 lb or 26 oz).

Propellers: It has one propeller at the center of wingspan and located on the fuselage. The propeller faces the rear end of UAV.

Payload: 450 g (usually contains R, G, B, and multispectral camera)

Launching information: It is hand-launch (automatic). This light UAV has just to be tossed in the air. It finds its altitude and circles until further command from remote control (iPad). Landing is automatic. It lands on belly, exactly at the place where it was launched.

Altitude: It reaches up to 1500 m a.g.l.

Speed: 47–80 km h^{-1}

Wind speed tolerance: 25 km h^{-1}

Temperature tolerance: −20 to +45°C

Power source: Electric batteries

Electric batteries: 2700 mA h, 25-A three cell batteries; lithium–polymer batteries.

Fuel: Not an IC engine

Endurance: 45 min

Remote controllers: UAV is connected and controlled, using *WiFi* AC type 2 bi-band antennas (2.4 and 5 GHz. UAV can be controlled, for up to 1.24 miles or 21 km from the ground station.

Area covered: 30 km^2 per flight

Photographic accessories: Cameras: It has 1080p full HD camera; with video streaming 360p/720p. Sensors include R, G, B, optical, and thermal bandwidths. Images are coded with GPS and altitude information. This UAV is also connected with GLONASS.

Computer: UAV has a CPU with a flash memory of 32 GB; computer processing is done by Linus and open source SDK.

Spraying area: This is not a sprayer UAV.

Volume: Not applicable

Agricultural uses: Field crop and pasture monitoring, collection of NDVI data; measuring leaf chlorophyll content. It helps in surveillance of irrigation channels and farm installation. It helps in monitoring progress of agronomic operations in the field.

Nonagricultural uses: They are surveillance of natural resources, mining sites, pipelines, industrial installation and imaging city traffic, and policing of public event.

Useful References, Websites, and YouTube Addresses

http://www.theverge.com/2016/1/4/10711178/new-parrot-disco-drone-announced-ces (accessed Nov 10, 2017).
https://www.parrot.com/us/drones/parrot-disco-fpv (accessed Nov 10, 2017).
https://www.youtube.com/watch?v=2asA1dDOvfI (accessed Nov 10, 2017).
https://www.youtube.com/watch?v=pxS6OjU-OWU (accessed Nov 17, 2017).
https://www.youtube.com/watch?v=2j660YoJfrY (accessed Nov 17, 2017).

Name/Models: Phoenix 2

Company and address: Sentera Inc. 6636 Cedar Avenue, Minneapolis, Minnesota 55423, USA; phone: +1 844 736 8372; e-mail: info@sentera.com; www.sentera.com

Sentera Inc. is a company situated at Minneapolis, Minnesota, USA. They specialize in UAV models such as Phoenix 2 and Omni UAV. They also supply accessories such as sensors, software such as AgVault.

They aim at developing simple and elegant UAV imagery solutions. They strive to improve the performance and efficiency of UAVs, the camera, hardware, and software.

"Phoenix 2," that is, Sentera's UAVs and sensors are changing the way industries approach remote-sensing technology. In agricultural fields, UAVs capture high-precision, high-resolution color, NIR, and NDVI data. Captured data may be used to create detailed 3D maps, assisting geospatial analysts, engineers, inspectors, and first responders with measuring geographical volumes, analyzing surface data, and planning for contingencies (Plate 2.19.1).

TECHNICAL SPECIFICATIONS

Material: It is made of foam and plastic toughened to withstand harsh weather conditions.

Size: length: 2.5 ft; width (wingspan): 4.5 ft; height: 1.2 ft; weight: 1.8 kg (or 4 lb)

Propellers: It has single propeller placed at the tip of nose on the fuselage.

Payload: 450 g. It comprises a set of at least four sensors and CPU placed in the fuselage.

Launching information: It is hand-launched. The UAV gains height at a rate of 1000 ft min^{-1}

Altitude: Typical altitude for flight is <400 ft which is stipulated for agricultural UAVs. However, it can reach up to 12,000 ft a.g.l.

Speed: It cruises at 48–50 km h^{-1} (26 knots); maximum speed is 75 km h^{-1}.

Wind speed tolerance: It withstands a wind speed of 40 km h^{-1} (or 21 knots) without any distraction to flight course.

Temperature tolerance: −20 to +50°C

Power source: Electric batteries

Electric batteries: 14,000 mA h lithium–polymer batteries; two cells.

Fuel: No requirement for petroleum fuel

Endurance: 60 min

Remote controllers: UAV's flight and imagery can be controlled for up to 1.6 km radius from the ground station (iPad).

Area covered: The UAV images and covers over 100 ac per flight of 45 min endurance.

Photographic accessories: Cameras include single visual sensor, Quadsensor (multispectral and NIR sensors for thermal imagery (see Plate 2.19.2).

Computer programs: The UAV systems comes with computer software called "AgVault." AgVault 2.0 is a mobile compatible software. It allows the UAV operator to collect images both at visual and NIR bandwidth. It offers NDVI data and HD color photos of terrain, land, and crops. AgVault has options for Quick titles. "Quick titles" provides accurate ID for storage of imagery and digital data. It allows rapid stitching of orthomosaics, using the titled photos. It automatically matches fields with photos and GPS-tagged locations. Images are accurately catalogued and stored for future use. This software also allows operator to prepare files for export. Data export along with popular image stitching software such as Pix4D Mapper is possible. The system comes with facility for timely upgradation of software for collection, storage, stitching, and export of images and digital data (see Grassi, 2017).

Spraying area: This is not a spraying UAV.

Spray volume: Not applicable

Agricultural uses: UAV system offers color images of terrain, soil surface, and crop fields. It relays NDVI data rapidly. It is used for general survey of crop fields in large farms. Imagery includes both visual range and NIR (Plate 2.19.2). The UAV is used to surveillance farms, watch the progress of agronomic procedures, monitor movement of farm vehicles in the field and outside. UAV is also used to monitor irrigation of large farms (Grassi, 2017).

Nonagricultural uses: This UAV is light and swift to take-off. It is used to surveillance mining zones, natural vegetation, terrain, pipelines, electric lines, industrial installations, traffic, and public events.

PLATE 2.19.1 Sentera's Phoenix 2 UAV.
Source: Mrs. Sarah Ritzen, Director Marketing Division, Sentera Inc. Minneapolis, Minnesota, USA.

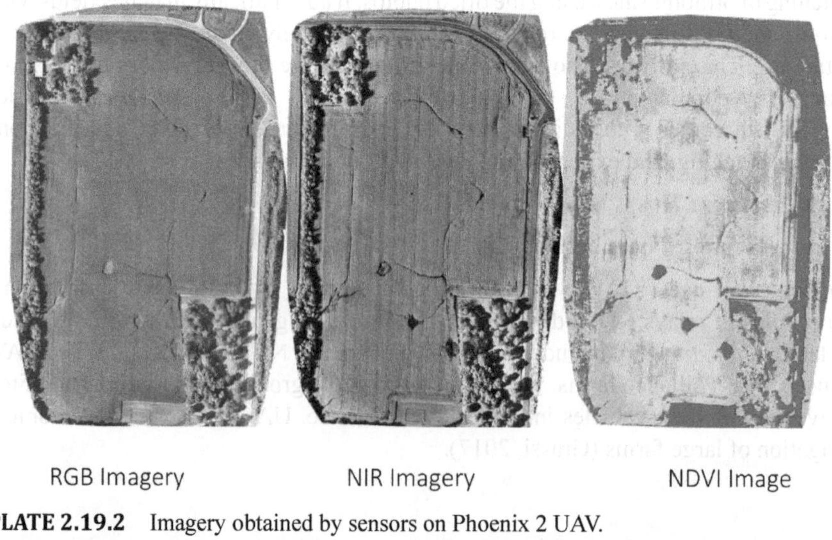

PLATE 2.19.2 Imagery obtained by sensors on Phoenix 2 UAV.
Source: Mrs. Sarah Ritzen, Director Marketing Division, Sentera Inc. Minneapolis, Minnesota, USA.

Useful References, Websites, and YouTube Addresses

Grassi, M. J. *Sentara Phoenix Drone Nears Coveted 1000-Acre Coverage Ceiling*; 2017; pp 1–2. http://www.precisionag.com/systems-management/sentera-phoenix-drone-nears-coveted-1k-acre-coverage-ceiling/ (accessed June 18, 2017).

https://sentara.com (June 1, 2017).

https://sentara.com/phoenix-2-uav (accessed June 1, 2017).

https://sentara.com/agriculture-solutions/ (accessed June 1, 2017).

http://michaelleeunmannedsystems.blogspot.in/2016/06/sentaras-phoenix-2-uav.html (accessed June 1, 2017).

https://www.youtube.com/watch?v=UFk15XHqww0 (accessed June 1, 2017).

https://www.youtube.com/watch?v=fCZioRW_C9Q (accessed May 30, 2017).

https://www.youtube.com/watch?v=8v3D-oXOnJQ (accessed May 30, 2017).

https://www.youtube.com/watch?v=oUxd6ywfyLI (accessed May 30, 2017).

https://www.youtube.com/watch?v=XBTA4vUUVy4 (accessed May 30, 2017).

Name/Models Precision Hawk's Lancaster

Company and address: Precision Hawk Inc. Street, etc. Noblesville, IN, USA; phone: +1 844 328 5326; e-mail:info@precisionhawk.com

Precision Hawk is a major UAV company that caters to agriculture worldwide. Precision Hawk's Lancaster is among the most popular agricultural UAVs. Lancaster is used predominately to survey, monitor, and schedule agricultural operations (Plate 2.20.1). The UAV is also used to prescribe fertilizer dosages. The UAV's thermal cameras help in assessing crops' water needs. Hence, it helps in fixing irrigation schedules.

TECHNICAL SPECIFICATIONS

Material: The platform is made of carbon fiber with reinforced plastic. It is toughened to withstand harsh weather conditions.

Size: length: 0.75 m; width (wingspan): 1.5 m (4.9 ft); height: 0.3 m; weight: 3.55 kg (7.8 lb)

Propellers: It has one propeller at the nose of fuselage. It is supported by a single electric motor.

Payload: 2.2 lb

Launching information: It is hand-launched at shoulder height. It has autolaunch, navigation, and landing software such as Pixhawk (Plate 2.20.2)

Altitude: Generally, it operates at 50–300 m height above crop canopy, for agricultural purposes. It can reach 2500 m (7000 ft a.s.l.).

Speed: The cruise speed 12–16 m s^{-1} (43.2–57.6 km h^{-1}). It attains a maximum speed of 79 km h^{-1}

Wind speed tolerance: 20–25 km h^{-1}

Temperature tolerance: −17 to +45°C

Power source: Power management is done by Smart Reflex TM 2 technology.

Electric batteries: Lithium–polymer batteries 7000 mA h

Endurance: 45 min per flight without relaunch

Remote controllers: The UAV package includes ground-kit with a list of accessories. It includes laptop, tablet flight designs to regulate flight path, and remote controller (hand held) to control speed and direction; a central processing unit (Linux computer) of 4 × 750 MHz. Ground control computers keep contact with UAV for up to 2 km radius (Plate 2.20.2).

Area covered: 150 to 200 ha h^{-1} at an altitude of 328 ft

Photographic accessories: Cameras: UAV's fuselage accommodates four different cameras. They are visual (R, G, and B); 18.4 MP, 0.7-cm pixel resolution, multispectral, thermal, hyperspectral, and Lidar.

Computer programs: It includes Pix4D to construct sharp images, using orthomosaics relayed by the CPU on the UAV. Several other programs such as Agisoft's Photoscan are also adopted by the ground-station computers.

Spraying area: This is not a sprayer UAV.

Volume: Not applicable

Agricultural uses: Precision Hawk's UAV is utilized in several ways during agricultural crop production. Visual range cameras are utilized for aerial mapping, photogrammetry, high-resolution videos of farm activity, scouting for seed germination, seedling establishment, plant counts, etc. It is used for surveillance of crop for growth, disease/pests, drought and floods (Plate 2.20.3). Thermal cameras on the UAV help in detecting heat signatures of crops. Determination of water stress index is possible. Multispectral cameras aid in detecting plant health, growth rates, NDVI, leaf chlorophyll, and crop-N status. Fertilizer-N supply could be decided, using leaf chlorophyll data. Lidar helps in obtaining 3D images of terrain, crop height and surface features, soil erosion-affected sites, etc. (Shi et al., 2016).

Nonagricultural uses: They are surveillance of oil wells, pipelines, and storage tanks, and aerial imagery to assess damages to large buildings, industries, and other installations. Such assessments are helpful to insurance companies as well as surveillance of mining activity, transport of ores, and dumping yards. It is used in rail road and highway traffic maintenance. Monitoring and detecting faults in electric transmission installations and lines. Urban policing is another function performed by the UAV (Frey, 2014).

PLATE 2.20.1 A Precision Hawk's Lancaster UAV.
Source: Mrs. Lea Reich, Marketing Division, Pecision Hawk Inc., Noblesville, Indiana.

PLATE 2.20.2 Flight plans on a mobile application.
Note: In-Flight software mobile app; Android app on Google Play.
Source: Ms. Lea Reich, Marketing division, Precision Hawk Inc., Noblesville, Indiana.

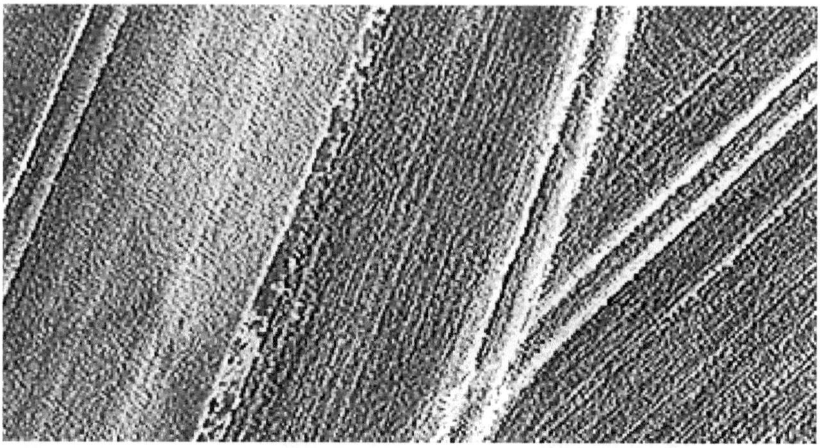

PLATE 2.20.3 An aerial view of crop field.
Source: Mrs. Lea Reich, Director Communication and Marketing, Precision Hawk Inc., Noblesville, Indiana, and North Carolina, USA.

Useful References, Websites, and YouTube Addresses

Frey, T. *Future Uses for Flying Drones*; 2014; pp 1–7. http://www.futuristspeaker.com/2013/08/griculture-the-new-game-of-drones/ (accessed Oct 28, 2016).

Shi, Y.; Thomasson, J. A.; Murray, S. C.; Puch, N. A.; Rooney, W. L.; Shafian, S.; Rajan, N.; Rouze, G.; Morgan, C. L. S.; Neely, H. L.; Rana, A.; Bagvthiannan, M. V.; Herrickson, J.; Bowden, E.; Vlseck, J.; Olsenholler, J.; Bishop, M. P.; Shridan, R.; Putman, E. B.; Popescu, S.; Burks, T.; Cope, D.; Ibrahim, A.; McCutchen, B. F.; Baltensperger, D. D.; Vrunt, R.; Vidrine, M.; Yng, C. Unmanned Aerial Vehicles for High Throughput Phenotyping and Agronomic Research. *PLoS One* **2016,** *11*, 1–15. http://dx.doi.org/10.1371/journal.pone.0159781 (accessed Nov 9, 2017).

http://www.precisionhawk.com/lancaster (accessed Nov 1, 2017).

http://www.precisionhawk.com/lancaster (accessed Nov 1, 2017).

https://www.youtube.com/embed/lUCYuvQ3ccA?rel=0&showinfo=0&autoplay=1 (accessed Nov 1, 2017).

https://www.youtube.com/watch?v=2SI1oLN05Io (accessed Nov 1, 2017).

https://www.youtube.com/watch?v=pACnG7b3FCM (accessed Nov 1, 2017).

https://www.youtube.com/watch?v=KHld29tBNYQ (accessed Nov 1, 2017).

https://www.youtube.com/watch?v=7d3rNk01nOM (accessed Nov 1, 2017).

Name/Models: Pteryx Pro UAV

Company and address: Trigger Composites, DGrodzisko, Dolne 800, Poland; phone: +48 608 683 984, +48 17 243 0037; website: http://www.trigger.pl/pteryx/

Trigger Composites Inc., a Polish aeronautics and UAV company was founded in 2014. It has developed a series of fixed flat-winged and copter UAVs. The UAVs such as Pteryx are adapted to conduct agricultural surveillance and aerial imagery in the European cropping zones. Trigger composites also manufactures EasyMap UAV and MasterFly. It is based in Grodzisko Dolne, Poland. Since its beginnings, the company has executed many projects related to UAVs, for both Polish and foreign clients.

Pteryx is a relatively small UAV that suits to conduct aerial survey and spectral analysis of crop fields. It has a range of functions useful to farmers during precision farming. This UAV has demonstrated utility in general commercial operations such as surveillance of installations, highways, etc. Forecasts suggest that such UAV models may become more common, in the farming belts.

TECHNICAL SPECIFICATIONS

Material: The UAV's frame is made of carbon fiber and kevlar composite. Its wings are made of fiber glass.

Size: length: 90 cm; width (wingspan): 150 cm; height: 30 cm; weight: 4.8 kg

Propellers: It has one propeller attached on fuselage near the wing. Propeller is energized by a high efficiency brushless motor.

Payload: Pteryx can lift up to 1 kg of cargo: It includes cameras, camcorders, or other research equipment. Moreover, the whole payload is housed in roll-stabilized head in the front of the fuselage.

Launching information: It is launched manually at shoulder height. It could also become air borne automatically, using a Bungee catapult. It has autopilot mode using software such as Pixhawk. Landing is on belly. Otherwise, the UAV can be recovered, using parachute. It returns to launching point, in times of emergency or if errors occur in flight programing.

Altitude: It reaches up to 50–400 m a.g.l. for land surface/crop's aerial imagery and spectral analysis. However, ceiling is 3500 m a.g.l.

Speed: 50 km h^{-1}. This is an optimum speed that takes care of resistance by wind, safety of machine, and energy efficiency.

Wind speed tolerance: 10 m s^{-1}. It also tolerates rain and snow to a certain extent.

Temperature tolerance: −20 to +60°C

Power source: Electric batteries

Electric batteries: 8–14,000 mA h, four cells.

Fuel: No requirement for petroleum fuel.

Endurance: 2 h. It is said that this length of flight endurance is enough to plan a longer flight and mission. It suits this class of UAV in size and weight.

Remote controllers: The GCS has telemetric connectivity. The autonomous flight path can be reprogramed, using GCS iPad. The telemetric connectivity and computer commands are effective, up to 10 km radius. GPS update rate is 10 Hz.

Area covered: 80–100 km in linear distance per flight or 50–300 ha per flight.

Photographic accessories: Cameras: Pteryx can accommodate wide variety of sensors. Moreover, sensors are installed in a camera socket and they stay fully enclosed. Camera shutters are fully integrated with software and can be operated, using computer generated commands. Locations and timing of aerial imagery can be preprogramed. We can also alter it manually. Sensors may be changed for each flight. The ground-station telemetry is interfaced and it allows rapid download of aerial imagery. "Pteryx Pro" UAV unit is usually delivered with 16 MP APSC daylight camera and wide lens. Sensors include two camera set (one daylight and one NIR), multispectral camera (e.g., Tetracam Mini MCA), and IR camera.

Computer programs: Pix4D Mapper and other postflight processing software are utilized to develop sharp images and videos. Pteryx Agro Mapper video is utilized. All the images are GPS tagged.

Spraying area: This is not a sprayer UAV.

Volume: Not applicable

Agricultural uses: UAV offers aerial imagery useful during precision agriculture. It offers farmers with aerial images that depict variations in crop growth and soil fertility effects. Crop scouting for canopy traits, leaf area, leaf-N, crop-N status, water stress index, and grain maturity is possible. Natural vegetation monitoring is accomplished with greater ease using this UAV.

Nonagricultural uses: Pteryx is used to surveillance industrial installations, buildings, rail roads, highways, mines, transport vehicles, etc. It is used in environmental survey and to assess climate change, etc.

Useful References, Websites, and YouTube Addresses

http://www.trigger.pl/pteryx/ (accessed Apr 24, 2017).

http://www.easymapuav.com/contact-us.html (accessed Apr 7, 2017).

https://www.youtube.com/watch?v=9LwBScnRZcU (accessed Apr 24, 2017).

https://www.youtube.com/watch?v=Q2sUuFABblw (accessed Apr 24, 2017).

https://www.youtube.com/watch?v=uoVtFUpm3X8 (accessed Apr 24, 2017).

Name/Models: QuestUAV Q200 Agri-Pro

Company and address: Quest Technology Centre, Coquetdale Enterprise Park, Amble, Northumberland, United Kingdom. Phone: +44 (0) 191 495 7340, +44 (0)

191 495 7341; e-mail: info@questuav.co.uk; sales@questuav.co.uk; http://www.questuav.com/enquirey-form

Quest Technology is a recently initiated UAV and UAV-related services company. They produce a few different types of UAVs. These models are apt to conduct survey and collect data on terrain, localities in urban areas industries, etc. (e.g., Datahawk). They produce a highly sophisticated model for agricultural services. It is utilized for collecting orthomosaics of farms, their crops, irrigation, farm vehicles, etc. (e.g., QuestUAV Q-200 Agri-Pro) (Plates 2.21.1 and 2.21.2). They offer full package of UAV, including the ground-station accessories, CPU on the UAV, cameras that include Sony NEX, Sony a5100, Sony IR and NIR range cameras, etc. They also cater UAV-aided crop analysis and prescriptions of fertilizer-N supply, irrigation supply, harvest scheduling, etc. Quest center also produces copter UAVs. The copter UAVs are specifically useful in monitoring wide range of farm, urban and industrial activities (e.g., Quest Oktocopter). In addition to United Kingdom, Q200 Agri-Pro and other related models are operated, in countries such as Canada, United States of America, Central Asian Nations such as Kazakhstan, Tadzhikistan and Far-East nations such as South Korea and Japan (QuestUAV, 2016a; Themistocleous, 2016).

TECHNICAL SPECIFICATIONS

Material: This fixed-winged agri-UAV is made of lightweight carbon fiber, plastic, and foam composite.

Size: length: 75 cm; width (wingspan): 110 cm; height: 15 cm; weight: 1.2 kg

Propellers: It has one propeller with two plastic blades. The propeller is attached at the rear end of the fuselage.

Payload: 500–800 g

Launching information: The UAV is launched, using bungee launch facility or it could be hand launched. It has an autolaunch, fully autonomous flight mode, and landing software. It also has autonomous landing facility, in times of emergencies or wrong instructions sent from ground station. Autolanding sends the UAV, back to location of launch. Q200 Agri-Pro UAV can also be recovered, using parachute-aided landing.

Altitude: It reaches 50–150 m a.s.l. for agricultural survey and crop analysis. However, this UAV can reach 4200 m a.s.l.

Speed: 45–60 km h^{-1}

Wind speed tolerance: 20–25 km h^{-1}

Temperature tolerance: −17 to +55°C

Power source: Electric batteries

Electric batteries: UAV is provided with lithium–polymer batteries of 10,000 mA h. But, it can be upgraded to 25,000 mA h, if long duration flight is envisaged.

Fuel: No requirement of petroleum fuel

Endurance: 45–60 min. Endurance depends on battery options

Remote controllers: Ground station need not have bulky PCs or even laptop that need to be formatted. Ground station is usually an iPad that can be operated anywhere in the crop fields or on the bund. Ground-station telemetric connections work up to a range 3 km. Telemetric connectivity is used to guide UAV's flight and instruct imaging spree.

Area covered: This UAV can survey and obtain imagery of 200–1000 ha

Photographic accessories: Cameras: A simple Datahawk UAV comes with visual range Sony a5100 camera. The camera set has Exmor APS HD CMOS sensor. It provides 2D and 3D visual range photos. These photos are helpful in preparing surface models of crop land. QuestUAV Q200 Agri-Pro comes with Agri-Q-Pod having NDVI sensors for obtaining NDVI data, measuring biomass and growth pattern. Thermal Q-Pod has OPTRIS P 1450 sensor. Multispectral Pod with tetracam Mini MICA 6 provides images, using 4–6 bandwidths. A single gimbal helps to stabilize the sensors and offers sharp imagery.

Computer programs: Postflight image processing is usually done by adopting Pix4D Mapper software or Agisoft's Photoscan. In-flight relay of processed images using the CPU in the fuselage is also a possibility.

Spraying area: This is not a sprayer UAV.

Volume: Not applicable

Agricultural uses: Survey of terrain, farmland features, topography (2D and 3D); obtaining imagery of soil type, crops, and cropping system. The UAV's imagery is used to assess planting density, seedling emergence and gap-filling needs. Determination of crop NDVI, growth stage, leaf chlorophyll, and crop-N status are possible. Crop scouting for disease/insect pest attack and weed infestation is done routinely. Determining both weed species and intensity of infestation is possible, using spectral signatures and survey of crop field to detect drought-affected zones, soil moisture dearth, and flood-affected regions, if any. Yield forecasting is done using panicle data, and farm vehicle monitoring and assessment of progress of agronomic procedures in the field (Plate 2.21.2; QuestUAV, 2016b).

Nonagricultural uses: Primary uses are surveillance of large buildings, industrial installations, rail roads, city traffic, oil pipelines, and electric transmission lines and surveillance of mining activity, ore-dumping regions, and movement of vehicles is also possible.

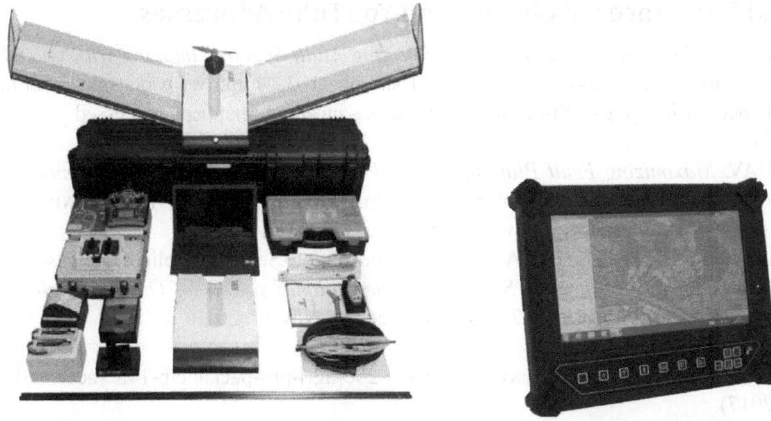

PLATE 2.21.1 QuestUAV agricultural UAV, its accessories, and iPad ground station.
Note: The QuestUAV Q200 Agri-Pro or Surveyor comes with accessories. It is a package that makes it quicker to set up the UAV and fly it. The list is as follows: sensor ready UAV; QuestUAV Platform-A frame with "autopilot system"; ruggedized wingset; laptop ground station; flight planning control software; Pix4D Mapper, discovery postprocessing software; Sony A6000 RGB Sensor System; Spectrum DX9 2.4 GHz control transmitter; programmable battery charger; transmitter case; Safe Launch Bungee System; Ruggedized Sensor Gimbal; 600 W brushless motor system; 5000 mA h 11.1 V lipo-battery × 6; Watt meter; servo tester; windmeter; wind sock; ground station stand; rugged transport case; spare wingtip set × 1; spare carbon spars × 2; spare electronic speed controller × 1; spare propeller × 6; and spares and repairs tool kit.
Source: Mr. Nigel King, Director, Quest Technological Centre, Northumberland, United Kingdom.

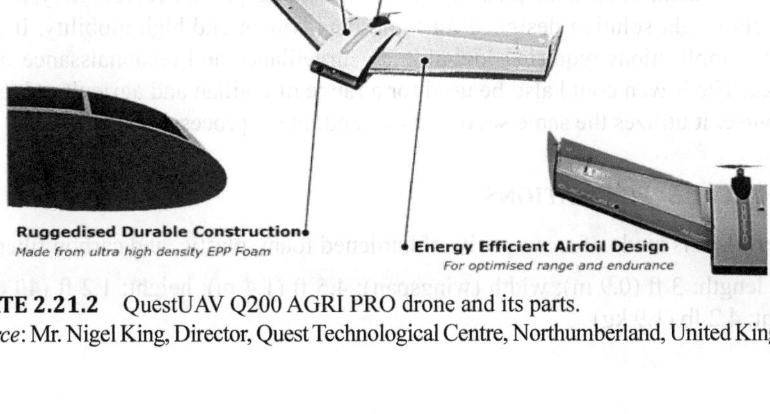

PLATE 2.21.2 QuestUAV Q200 AGRI PRO drone and its parts.
Source: Mr. Nigel King, Director, Quest Technological Centre, Northumberland, United Kingdom.

Useful References, Websites, and YouTube Addresses

QuestUAV. A *Breakthrough in Precision Farming Multi-Spectral Imaging with QuestUAV, Mica Sense and Pix4D Mapper*; 2016a; pp 1–3. http://www.questuav.com/news/multispectral-imaging-with-questuav-micasense-and-pix4dmapper (accessed Nov 17, 2016).

QuestUAV. *Maximizing Fruit Plantations Yield Using UAV Imagery: Plant Counting and Gap Filling Techniques*; 2016b; pp 1–3. http://www.questuav.com/news/maximizing-a-fruit-plantations-yield-using-uav-imagery (accessed Nov 17, 2016).

Themistocleous, K. The Use of UAV Platforms for Remote Sensing Application: Case Studies in Cyprus. In *Proceedings of SPIE—The International society of Optical Engineering*; 2016, pp 1–12. DOI:10.1117/12.2069514 (accessed Nov 17, 2016).

http://www.questuav.com/

http://www.questuav.com/store/uav-packages/q-200-agri-pro-package/p-148 (accessed Nov 10, 2017).

http://www.geosystems.fr/images/PDF/20140613_Brochure_QPod_complete.pdf (accessed Nov 10, 2017).

https://www.youtube.com/watch?v=6ExystDyhpU (accessed Nov 10, 2017).

https://www.youtube.com/watch?v=RYWwGGCjlXI (accessed Nov 10, 2017).

https://www.youtube.com/watch?v=yw1cSFjpno0 (accessed Nov 10, 2017).

https://www.youtube.com/watch?v=wA_eUCt0Yds (accessed Nov 10, 2017).

Name/Models: Raven

Company and address: AeroVironment Inc., 800 Royal Oaks Drive, suite 210, Monrovia, CA 91016 6347, USA; phone: +1 626 357 9983; e-mail: ir@avinc.com; website: https://www.avinc.com/contact.

AeroVironement is a Southern California aviation company. It was initiated some 40 years ago. They found a breakthrough with the production of a series of UAVs (drones), such as Puma, Raven, Wasp, Shrike, etc. They cater to military, farm, and civilian sectors with UAV models. Their UAVs are used in different continents. Most recent model touted by the company is named Quantix (VTOL) (AeroVironement Inc., 2016).

The "Raven" is the most widely used unmanned aircraft system. The "Raven B DDL" system is an enhanced version of the battle proven Raven B system. It is a lightweight solution designed for rapid deployment and high mobility. It suits military applications requiring low-altitude surveillance and reconnaissance intelligence. The Raven could also be used for a range of civilian and agricultural tasks. Of course, it utilizes the same set of sensors and image processing facilities.

TECHNICAL SPECIFICATIONS

Material: It is made of a composite of hardened foam, plastic, and carbon fiber.

Size: length: 3 ft (0.9 m); width (wingspan): 4.5 ft (1.4 m); height: 1.2 ft (40 cm); weight: 4.2 lb (1.9 kg)

Propellers: It has one propeller with two plastic blades. It is placed at the center of wing span, just above the fuselage. It faces backward.

Payload: It allows 1.3 kg payload. Payload is usually a gimbal and a set of EO/NIR/IR cameras with illuminator. The cameras are digital stabilized. The gimbal allows continuous +10 to 90° tilt, for imaging ground spots.

Launching information: It is a small UAV and light in weight at 2.0 kg. It is hand-launched by holding the frame at shoulder height. It has autonavigation and landing software Pixhawk.

Altitude: Routinely, this UAV is flown at 100–500 ft (30–152 m) a.g.l. Maximum height attainable is 14,000 ft a.g.l.

Speed: 32–81 km h^{-1} (17–44 knots)

Wind speed tolerance: 15–20 km h^{-1}

Temperature tolerance: −20 to +60°C

Power source: Electric batteries

Electric batteries: 14,000 mA h; two cells.

Fuel: No requirement for petroleum fuel.

Endurance: 60–90 min

Remote controllers: The GCS computers and radiolink keep contact with the UAV for 10 km radius. The UAV's path is usually predetermined, if it is for civilian and agricultural survey. Flight path changes can be done during both, day and night, while on military surveillance spree. Pixhawk autopilot or similar software is used to control UAV's flight path. It covers up to 80 km h^{-1} in linear length in one uninterrupted flight. UAV detects obstructions to flight path and alters course, automatically. It lands at the launch spot in case of emergency or errors in electronics.

Area covered: This UAV is quick and covers 50–200 ha of cropland or ground surface in one flight.

Photographic accessories: Cameras: It carries a full set of sensors in the gimbal. It has a set of visual cameras, NIR, and IR cameras. It has Lidar. For nighttime surveillance, the UAV's IR and night vision mode is utilized.

Computer programs: The photographic images sent to ground station are processed, using Agisoft Photoscan or Pix4D Mapper software. The videoimages could be relayed in processed condition, from the UAV to ground-station computers and iPads. The images could be processed right at the CPU of the UAV.

Spraying area: This is not a sprayer UAV.

Volume: Not applicable

Agricultural uses: Major uses are crop scouting for assessing seed germination, seedling establishment, canopy stand, leaf chlorophyll, crop-N status, canopy temperature, water stress index, disease/pest attacks, drought, and flood effects.

Nonagricultural uses: They are surveillance of military installations, industries, buildings, oil pipelines, rail roads, city traffic, tracking vehicles, etc. We can monitor mines, mining activity, and ore movement. Natural vegetation monitoring and tracing spectral signatures of diverse species of plants is possible.

Useful References, Websites, and YouTube Addresses

Aerovironment Inc. *UAS: RQ-11B Raven*; 2016; pp 1–3. https://www.avinc.com/uas/view/raven (Jan. 20, 2017).

https://www.avinc.com/uas (accessed Sept 20, 2017).

https://www.avinc.com/about (accessed Sept 20, 2017).

https://www.avinc.com/uas/view/raven (accessed Sept 20, 2017).

https://www.guavas.info/drones/Aerovironment,%20Inc.-Raven%20UAV-334 (accessed Sept 20, 2017).

https://www.youtube.com/watch?v=BcerPCn3MlA (accessed Sept 20, 2017).

Name/Models: SB4 Phoenix—a solar drone

Company and address: Sunbirds, 10, Avenue de l'Europe, 31 520 Ramonville-St-Agne, France; Tel.: +33 5 34 32 03 25; contact@sunbirds-uas.com; website: http://www.sunbirds-uas.com

Sunbird is a UAV-producing company located in France. They offer a few different types of UAVs. The UAV, "SB4 Phoenix" is a specialty type, since it is solar-cell supported. This UAV has been tested thoroughly for its utility in agricultural and civilian sector.

"SB4 Phoenix" is a solar UAV. It operates throughout the day, using power from sunlight, plus an hour in the dark, using stored energy. It is totally noiseless. It leaves no sound signatures for identification. So, it is a silent UAV. By classification, it is microweight fixed-wing UAV. It has autolaunch capability. It flies using solar energy but standby batteries takeover in case of emergency or failure of solar energy supply. It is a low-cost UAV. So, it is good even for farmers with smaller farm holdings and cropping area. It is packaged in a box and transported to any location for quick launch (Plate 2.22.1; Malaver et al., 2015).

TECHNICAL SPECIFICATIONS

Material: It is made of a composite derived from light carbon fiber, rugged foam and plastic.

Size: length: 130 cm, width (wingspan): 300 cm height: 45 cm; weight: 3.0 kg

Propellers: There are two propellers made of plastic blades. They are powered by brushless electric motors

Payload: The payload is 0.8–1.2 kg. It includes both, still and videocameras (21 MP and 4 K video).

Launching information: It is an autolaunch UAV. It is launched at shoulder height into the air, manually. The UAV's flight could be made entirely autonomous, using a software such as Pixhawk. Otherwise, it could be handled manually using a remote controller (joystick) (Plate 2.22.1).

Altitude: Operating height is 300–500 m a.g.l. However, the ceiling is 1000 m a.g.l.

Speed: 30 km h^{-1} (8 m s^{-1})

Wind speed tolerance: 15 to 20 km h^{-1} disturbance

Temperature tolerance: −10 to +50°C

Power source: Solar energy via solar cell panels.

Electric batteries: Solar batteries plus stand by lithium–polymer batteries.

Fuel: Not an IC engine-fitted UAV. So, no requirement for petroleum fuel.

Endurance: All through the day when sun shines. Plus, 1 h using the stored energy in the batteries.

Remote controllers: Remote controller is a joy stick. It could be used manually to guide the UAV. The flight could also be predetermined. The GCS keeps telemetric link with UAV, for up to 15 km distance. The connectivity with the UAV is maintained, using Wi-Fi, 3G/4G and blue tooth.

Area covered: It ranges from 50 to 100 ha of aerial imagery per day of 6–8 h. Area covered by the UAV depends on flight path and frequency of photography, as well as UAV's set speed and altitude.

Photographic accessories: Cameras: The UAV carries a gyro-stabilized gimbal with visual (R, G, and B), a multispectral high-resolution camera, NIR, IR, and red-edge cameras.

Computer programs: The orthomosaics are transmitted to GCS computers. Images are processed, using software such as Pix4D Mapper or Agisoft Photoscan. The processed digital data can be transmitted, to precision agricultural vehicles (e.g., variable-rate applicators, GPS-RTK tractors, combine harvesters, etc.

Spraying area: This is not a sprayer UAV.

Volume: Not applicable

Agricultural uses: Major applications are in aerial survey of land resources, soil type and crops. It helps in crop scouting for NDVI variations, leaf chlorophyll status (crop-N), canopy temperature (thermal imagery), grain maturity and senescence. It could be used to monitor movement of farm vehicles, such as combine harvesters. Monitoring natural vegetation for NDVI could be done regularly, using this UAV. Detection of plant diversity, using spectral data is also a possibility.

Nonagricultural uses: They are surveillance of industrial sites, buildings, rail roads, highways, international borders, movement of military and civilian vehicles, etc. It is also used in environmental monitoring, particularly, emission of greenhouse gases (Malaver et al., 2015).

PLATE 2.22.1 SB4 Phoenix UAV platform and its accessories.
Source: Mr. Amaury Wiest, Sunbirds, Ramonville-St-Agne, France.
Note: SB4 Phoenix is a solar-powered small UAV.

Useful References, Websites, and YouTube Addresses

Malaver, A.; Motta, N.; Corke, P.; Gonzalez, F. Development and Integration of a Solar Powered Unmanned Aerial Vehicle and a Wireless Sensor Network to Monitor Greenhouse Gases. Sensors **2015,** *15*, 4072–4096. DOI:10.3390/s156204072 (accessed June 13, 2017).

http://www.sunbirds-uas.com/en/drones/drones/sb4phoenix.html (accessed May 26, 2017).

http://www.phoenixsolar-group.com/business/us/en/about-us/Our-Company/Our-Leadership-Team.html (accessed May 26, 2017).

http://www.sunbirds-uas.com/en/drones/drones/sb4phoenix.html (accessed June 24, 2017).

https://www.youtube.com/watch?v=ecUkAZo4a8g (accessed June 24, 2017).

https://www.youtube.com/watch?v=ptDPoYTHywg (accessed June 24, 2017).

https://www.youtube.com/watch?v=ecUkAZo4a8g (accessed June 24, 2017).

http://www.aeronewstv.com/en/industry/drones/3183-a-solar-drone-for-eight-hour-assignments.html (accessed June 24, 2017).

Name/Models: SkyWalker X8

Company and address: TechnoSys Embedded Systems Ltd., SCO 666, Sector 20 C, Chandigarh, India; phone: +91 172 508 5909; +91 172 467 1082; e-mail: info@technosysind.com, sales@technosysind.com, support@technosysind.com; website: http://www.technosysind.com

TechnoSys Systems Ltd. is a company that offers few different fixed-winged UAV models. Each has its suitability for different agricultural and nonagricultural functions. "Sky Walker X8" is again a fixed-wing UAV of relatively small size. It suits best to surveillance and monitor the agricultural belt. It is a low-cost, ready-to-fly UAV model that suits the small holder farmers. It has been designed to replace the noisy, polluting, and costly manned aircraft missions. The "Sky Walker X8" is versatile and can be launched, at any time, during the day and as many times (Plate 2.23.1).

TECHNICAL SPECIFICATIONS

Material: The UAV is built using EPO, foam, plastic, and carbon fiber. The body is tropicalized and ruggedized to withstand harsh weather conditions and rough landing on the belly.

Size: length: 99 cm; width (wingspan): 212 cm; height: 35 cm; weight: 3.2 kg

Propellers: It has one propeller placed at the rear end of the fuselage. Propeller has two plastic blades. Propeller is attached to 750 Kc brushless motor.

Payload: 1.2 kg

Launching information: It can be hand-lunched since it is lightweight UAV. It can also be easily launched using a catapult. It is an autolaunch and navigation UAV. It adopts PX4 advanced 32-bit ARM cortex M4 processor running NuttX RTOS to decide flight path and landing location.

Altitude: It is operated at 50–400 m a.g.l. for crop scouting, mapping and other agriculture-related activity. However, the UAV can reach a ceiling height of 3000 m a.g.l.

Speed: 40–60 km h^{-1}

Wind speed tolerance: 20 km h^{-1} disturbance

Temperature tolerance: −10 to +50°C

Power source: Electric batteries

Electric batteries: 25,000 mA h, 4S battery, two cells. There are battery current sensors to detect fluctuations, if any, in the flow of energy.

Fuel: No need for petrol or diesel.

Endurance: 55 min

Remote controllers: The GCS has iPads and computer software to control the UAV and imaging schedules. The UAV is connected to GPS. Telemetry link is maintained for 20 km radius using RFD900.

Area covered: 300–400 ha per flight

Photographic accessories: Cameras: The UAV is fitted with usual complement of visual (R, G, and B), NIR, IR Red-edge, and a Lidar pod. These cameras should be procured separately. They are not part of UAV's package.

Computer programs: The images are processed using Pix4D Mapper or Agisoft's Photoscan.

Spraying area: This is not a sprayer UAV.

Volume: Not applicable

Agricultural uses: They are scouting cropland from seed germination to crop-stand establishment, identification of gaps in canopy, collecting data about phenomics, need for fertilizer-N, water stress index, and grain maturity. Natural vegetation monitoring and identification of plant species diversity is a clear possibility.

Nonagricultural uses: The utility of this UAV has been demonstrated in territory survey, defense, and police services, geological survey, urban planning, industrial site monitoring, environmental engineering, and disaster-site monitoring.

PLATE 2.23.1 Sky Walker X8.
Note: UAV is readied to be catapulted.
Source: Dhruv Arora, TechnoSys Embedded Systems (P) Ltd. Chandigarh, Punjab, India.

Useful References, Websites, and YouTube Addresses

http://www.technosysind.com/ (accessed Sept 12, 2017).
https://www.youtube.com/watch?v=a4togifypBc (accessed Sept 12, 2017).
https://www.youtube.com/watch?v=PoJLgAlkABE (accessed Sept 12, 2017).
https://www.youtube.com/watch?v=a4togifypBc (accessed Sept 12, 2017).
https://www.youtube.com/watch?v=PoJLgAlkABE (accessed Sept 12, 2017).

Name/Models: SmartPlane FRAYE

Company and address: SmartPlanes, Gymnasievägen 16, S-931 57 Skellefteå, Sweden; phone: +46 (0) 910 36260; e-mail: contact@smartplanes.se; website: www.smartplanes.se; http://smartplanes.se/our-system/

SmartPlanes, a company from Sweden, produces professional commercial UAVs. This company was inaugurated in 2004. It aims to offer UAV technology to populations in Nordic countries. Their UAVs perform missions of surveying, mapping, city-planning, mining, forestry, agriculture, volume calculations, science, wildlife protection, and much more. Smartplanes are also used, in aerial geomatics, geospatial technology, for geo-referenced orthophotography, and 3D modeling (Plate 2.24.1).

Smart plane complies with strict aviation-safety requirements. It has received operational permits in Sweden and many other countries around the world. These UAVs are designed to last many years of wear and tear. The flexible materials used in the shell of the UAV absorb shocks. Regarding the UAV's usage in agriculture, it is said innovation is the focus for the company. Right now, they are developing a unique world-class camera that in the future will give even more knowledge about a wide range of crops.

TECHNICAL SPECIFICATIONS

Material: It is made of ruggedized foam, light plastic, and carbon-fiber composite. SmartPlane can be assembled together and launched in less than 10 min. Smart plane's design makes it easy to put together and to take apart once flight mission is over.

Size: length: 40 cm at the fuselage; width (wingspan): 95 cm; height: 30 cm; weight: 1.1–1.5 kg

Propellers: It has one propeller placed at the rear end of the fuselage. It has two plastic blades.

Payload: It carries 200–800 g. The payload consists of the usual visual, multispectral, hyperspectral, NIR, IR, and red-edge cameras. The payload bay of the UAV has extraspace, if you would like to replace the original camera with another sensor. As an example, you can add an additional smaller camera, facing sideways.

Launching information: Smart plane is small and very lightweight UAV. It is packed in a briefcase. Therefore, it can be easily transported to any spot in the field or other places. It can be launched quickly. It is hand-held at shoulder height to launch into air. It has autolaunching mode and autonavigation, using software such as Pixhawk. Manual override facility is also available, if one decides to guide the UAV, all through the flight. It has fail safe-landing software. It returns to the spot where it was launched. It lands by skidding on the belly (Plate 2.24.1).

The UAV is launched by simply throwing it into the air. We can control it manually or let the autopilot fly it for us. Plus, the easy-to-use mission planning software can be run on any standard laptop.

Altitude: Normally, it flies at 50–300 m a.g.l. However, ceiling is 4000 m a.g.l.

Speed: 40–60 km h^{-1}

Wind speed tolerance: It is designed to withstand 13–15 m s^{-1}

Temperature tolerance: −25 to +45°C

Power source: Electric batteries

Electric batteries: Lithium–polymer batteries; four cells. Lithium–polymer batteries generate 14,000 mA h.

Fuel: Not an IC engine-fitted UAV; so, no requirement of petroleum fuel.

Endurance: 45–75 min per flight

Remote controllers: GCS consists of software and ground radiomodem (868, 900 MHz or 2.4 GHz) R/C radio transmitter for flying in assisted mode. Ground control radios keep contact with UAV for 15–20 km radius.

Tracker: "SmartPlane Freya" is utilized in difficult situations, such as strong winds, in mountainous areas, out of line of sight, or over very large areas. The tracker is a useful safeguard against unexpected aircraft loss. It has a portable hand-held receiver plus small transmitter. They fit nicely into the cargo bay. A field mobility kit, a backpack, and a foldable stool make the work in remote areas comfortable.

Area covered: It can transit for 60 km in linear distance and offer aerial imagery. In crop fields, it can cover 200–300 ha per flight and offers orthomosaics or processed still images or videos.

Photographic accessories: Cameras: The UAV carries digital camera customized for aerial mapping. For example, Ricoh GR, 16.2 MP with large APS-C CMOS sensor, spare parts, and field-repair kit support services.

Ricoh GR: The Ricoh GR is equipped with a high-sensitivity, wide dynamic range APS-C large-size CMOS sensor for image-quality equivalent to much larger DSLR cameras. Cameras are fixed with high-performance GR LENS 18.3 mm F2.8 (equivalent to 28 mm in the 35 mm format). The camera provides a clear depiction, all the way to the edges of the image. By designing the sensor and lens to be mutually optimized, and inclusion of a large-format sensor, the GR achieves outstanding results in the CPU of UAV or ground-station computers.

Ricoh GR NIR: This is a specially modified version of the Ricoh GR camera. It can acquire image data also in the NIR spectrum. It is designed for applications within the fields of precision agriculture and vegetation health monitoring. The spectral information can be used to compute the NDVI and other spectral indices. The high-resolution optics (APS-C size), combined with very good signal-to-noise ratio, makes it ideally suited for accurate measurement of spectral properties. It also features manual focus and exposure settings. Image data can be acquired in the uncompressed RAW format preserving the integrity of the radiometric data.

Computer programs: A standard package for processing the images procured by the sensors includes Smart4D advanced photogrammetric software, powered by Pix4D. It automatically processes aerial imagery into orthomosaics, DSMs, DTMs, Point clouds, and much more. Smart4D is only sold as a part of SmartPlanes Advanced Package.

Some of the options for processing software are as follows:

Agisoft PhotoScan is a stand-alone software product. It performs photogrammetric processing of digital images and generates 3D spatial data.

Septentrio-based postprocessed kinematic (PPK): The GPS PPK is a GPS position location process, whereby signals received from a mobile location receiving device stores the position data. It can be adjusted using corrections from a reference station after the data has been collected.

Spraying area: This is not a spraying UAV.

Volume: Not applicable

Agricultural uses: SmartOne UAV covers large areas of cropland. The aerial imagery it sends, usually gives an excellent overview of crops and plantations. The images can be used to prepare terrain development plans. Multispectral, NIR and IR sensors allow us to make high-precision analysis of crops, detect pests or stress. In case, the UAV lands in crop field where it is hard to trace (e.g., sugarcane or cornfield), an electronic **tracker** included in the system will guide to the exact location of the plane (Plate 2.24.1).

The NIR, IR, and red-edge bandwidth cameras can be used, for surveying crops and plants. Since chlorophyll reflects IR, a healthy plant with more chlorophyll will, for example, reflect more IR than an unhealthy plant. This provides us with valuable information about crop's nutrient status, particularly nitrogen. Fertilizer-N application schedules could be decided, accurately. The camera will also help to detect signs of stress, ripeness, and overall development of the crop (Plate 2.24.1).

Nonagricultural uses: Smart-plane FRAYE is utilized to surveillance large infrastructure, buildings, industries, highways, natural vegetation zones and sanctuaries, mines, oil pipelines, highways, rail roads, and vehicle movement.

PLATE 2.24.1 Top: SmartPlane Fraye—a fixed-winged UAV; middle left: a smart-plane landing on the belly with skid; middle right: iPad to mark the flight path at the GCS; bottom: an aerial imagery of crop field.
Source: Ms. Ola Fristrom, SmartPlanes, Skelleftea, Sweden.

Useful References, Websites, and YouTube Addresses

http://smartplanes.se/our-system/ (accessed July 20, 2017).
https://www.youtube.com/watch?v=jAX3Shl5Ylk (accessed July 20, 2017).
https://www.youtube.com/watch?v=ZFV2LHH7GWI (accessed Aug 1, 2017).
https://www.youtube.com/watch?v=ZFV2LHH7GWI (accessed Aug 7, 2017).

Name/Models: Spectra C

Company and address: RP Flight Systems Inc., Wimberley, Texas, USA; phone: +1 512 665 9990, +1 408 832 11195; e-mail: info@rpflightsystems.com, sales@rpflightsystems.com; website: http://www.RPFlightSystems.com

TECHNICAL SPECIFICATIONS

Material: UAV's body is made of fully composite carbon fiber and fiber glass (Plate 2.25.1).

Size: length 55 cm; width (wingspan): 58.5 cm; height: overall height 7 in.; weight: take-off weight 5 lb (2.3 kg)

Propellers: It has one propeller (11 × 7 in.) placed at the rear end of fuselage. Propeller has two blades. Propellers are energized using 4S motor.

Payload: 1.2 kg that includes gyro-stabilized gimbal with visual and multispectral sensors.

Launching information: This UAV is launched by hand, holding it at shoulder height. The UAV has autopilot software, Pixhawk. It has autonavigation and landing facility. Landing is on belly.

Altitude: It reaches 50–150 m for aerial survey and mapping of crops. However, the ceiling is 2500 m a.g.l.

Speed: 60 km h^{-1}

Wind speed tolerance: 20 km h^{-1}

Temperature tolerance: −20 to +60°C

Power source: Electric batteries

Electric batteries: 14,000 mA h

Fuel: No requirement for petroleum fuel.

Endurance: 1 h

Remote controllers: This UAV adopts "Robota Goose Autopilot System." It helps in programing flight path and photography events. The images are all tagged with GPS coordinates for accuracy.

Area covered: 60 km linear distance or 150–200 ha per flight

Photographic accessories: The cameras include 3D Sony RX-100, FLIR, red-edge, Sequoia sensors, and Lidar. It also has a videocamera. The cameras could be operated using remote control.

Computer programs: The images could be processed in situ using UAV's CPU software or they could be relayed to GCS iPad. The images are usually processed using Pix4D Mapper, or Agisoft Photoscan or similar custom-made software.

Spraying area: This is not a sprayer UAV.

Volume: Not applicable

Agricultural uses: They are aerial survey and mapping of land resources, soils, cropping pattern, and irrigation channels. Other functions are crop scouting for nutrient deficiencies, growth pattern (NDVI), crop-N status, and water stress effects, if any. This UAV is also used for natural vegetation monitoring (Plate 2.25.1).

Nonagricultural uses: They are surveillance of Industries, buildings, dams, mines, transport vehicles, public events, etc.

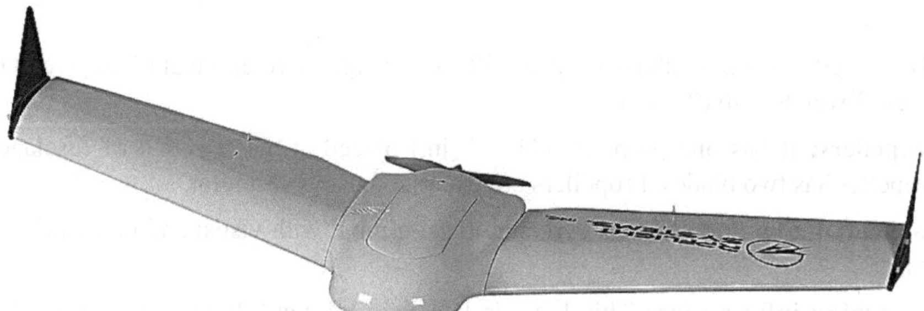

PLATE 2.25.1 Spectra C.
Source: Dr. Gene Robinson, President, RP Flight Systems, Wimberley, Texas, USA.

Useful References, Websites, and YouTube Addresses

www.rpflightsystems.com/ (accessed June 15, 2017).
http://www.asiatechdrones.com/ (accessed June 15, 2017).
http://www.asiatechdrones.com/_p/prd1/4083679471/product/spectra-c---ag%2C-sar%2C-mapping%2C-survey-platform (accessed June 15, 2017).
https://www.youtube.com/watch?v=CTLZ-DmclFc (accessed May 12, 2017).
https://www.youtube.com/watch?v=72vj69Hluwk (accessed May 12, 2017).

Name/Models: Swinglet Cam

Company and address: SenseFly—A Parrot Company, 1033 Route Geneva, (Z.I. Chaterlard Sud), Cheassaeux-Lausanne, Switzerland; phone: +41 21 552 04 20; e-mail: info@sensefly.com; website: www.sensefly.com

The Swinglet Cam is a small, fixed-winged UAV. It is a professional tool that offers GPS-tagged aerial images to farmers and other clients. This UAV offers quick orthomosaics that can be processed at ground station. This UAV could be used extensively in monitoring farms, crop fields, and agronomic procedures. However, we should note that Swinglet Cam is also used in a series of other activities such as in industry, policing city traffic, monitoring mines, pipelines, etc. We can think of many more uses for this versatile UAV (Frey, 2014).

TECHNICAL SPECIFICATIONS

Material: Swinglet Cam is made of EPP foam, carbon fiber, and composite parts made of plastic. The material is toughened to withstand harsh weather conditions (Plate 2.26.1).

Size: length: 75 cm; width (wingspan): 80 cm height; weight: 1.2 lb (700 g) the carry case dimensions are 83 × 135 × 56.5 cm.

Propellers: It has electric pusher propeller. It is connected to 100 W brushless motor.

Payload: 800 g

Launching information: It is hand-launched by releasing at shoulder height. It can also be launched into air, using catapult. Landing is possible with an accuracy of 5 m.

Altitude: It reaches 100–400 m above crop canopy for collection of crop's spectral data. However, the ceiling is 1100 m a.g.l.

Speed: 10 m s^{-1} (36 km h^{-1})

Wind speed tolerance: 25 km h^{-1}

Temperature tolerance: −20 to +60°C

Power source: Electric batteries

Electric batteries: 1350 mA h, 11.1 V

Fuel: No need for petrol or diesel.

Endurance: It is 30–35 min at a stretch without relaunch.

Remote controllers: The GCS computers keep radiocontact with UAV for up to 1 km length. The GCS can control many Swinglet Cam UAVs, that is, multiple UAV control is possible. UAV swarms or multi-UAV operation is possible because the package has software for avoidance for mid-air collision.

Area covered: 36 km distance per flight or covers 6 m^2 per flight

Photographic accessories: Cameras: 16-MP camera is electronically integrated to transmit images to ground station. It relays both 2D and 3D images depending on the cameras. The images offered are sharp and high-resolution ones (1.5 cm/pixel). The UAV is useful in planning, using 3D relief maps.

Computer programs: The image-processing software such as Pix4D Mapper or Agisoft's Photoscan is used.

Spraying area: This is not a sprayer UAV.

Agricultural uses: This is a quick and rapid action small UAV (Plate 2.26.1). It can be used repeatedly over the crop canopy to conduct aerial survey, obtain images, and send digital data to GCS computers. This UAV is used to scout crops to assess germination, identify gaps in seedling establishment, monitor growth and phenomics data, and identify and map locations affected by disease, pest, drought, or floods. This UAV could be used to monitor progress of agronomic procedures in large farms. It can track farm vehicles (Saberioon and Gholizadeh, 2016).

Nonagricultural uses: They include surveillance of industrial installations, buildings, city traffic, public events, oil and gas pipelines, rail roads, and mines and mining activity.

PLATE 2.26.1 Swinglet CAM.
Source: Dr. Mathew Wade, SenseFly Inc., Cheseaux-Laussanne, Switzerland.
Note: Swinglet CAM is a small fixed-winged UAV. It is useful for mapping natural resources and scouting agricultural crops. The entire UAV and accessories fit into a suitcase. So, they are easily portable to any location.

Useful References, Websites, and YouTube Addresses

Frey, T. 2014; pp 1–7. http://www.futuristspeaker.com/2013/08/griculture-the-new-game-of-drones/ (accessed Oct 28, 2016).

Saberioon, M. M.; Gholizadeh, A. Novel Approach for Estimating Nitrogen Content in Paddy Fields Using Low Altitude Remote Sensing System. *The International Archives of the Photogrammetry, Remote sensing and Spatial Information Sciences XLI-B1*, 2016; pp 12–19. DOI:10.5194/isprsarchives-XLI-B1-1011-2016 (accessed Aug 23, 2017).

https://www.sensefly.com/drones/overview.html (accessed June 6, 2017).

https://www.youtube.com/watch?v=OzlaNgKVX8o (accessed Jan 12, 2017).

https://www.youtube.com/watch?v=h5NeH8lEsQc (accessed Jan 12, 2017).

https://www.youtube.com/watch?v=B-xVy2E1sT4 (accessed Jan 12, 2017).

Name/Models: Swallow-P

Company and address: Carbon-Based Technology Inc., Central Science Park/3f, No. 30, Keya Road, Daya District, Taichung City, 42881, Taiwan; phone +886 4 2565 8558; e-mail: info@uaver.com; website: http://www.uaver.com

Carbon-Based Technology Inc. was established at Taichung City, Taiwan, around March 2007. This company concentrates on development of design, integration, manufacturing, and global marketing of unmanned aircraft system. It offers high reliability UAS, useful for "intelligence gathering." These UAVs also conduct aerial surveys of commercial and agricultural regions, and various other aspects. The UAS can integrate with techniques of data processing and analyzing. The team at Carbon-Based Technologies Inc. offers solutions for disaster detection, geological, and natural resource survey. The aerial imagery that these UAVs relay serve as priceless feedbacks. These small, light-weight, and versatile UAVs produced by Carbon-Based Technology Inc. are finding their way into agricultural farms in Southeast Asia.

Swallow-P unmanned aerial mapping system features lightweight, portable, and easy-operation advantages. Swallow-P UAVs are getting popular in some farming zones. This model is adopted to get aerial imagery of variety of features on earth. It has a big role to play in surveillance, aerial imagery, and data gathering, particularly, spectral variations and signatures of crops. It is easy to launch and operate. It seems any individual with less or even no aviation experience can easily accomplish an aerial mapping mission.

TECHNICAL SPECIFICATIONS

Material: It is made of ruggedized foam (honeycomb structure) and thoroughly weaved carbon fiber. The airframe is rigid. It withstands harsh weather and rough landing.

Size: length: 120 cm; width (wingspan): 1.0 m (3.28 ft); height: 35 cm; weight: 2.5 kg (5.5 lb)

Propellers: It has one propeller placed at the rear end of the fuselage of the UAV. The propeller has two blades made of tough plastic. Propellers are connected to brushless motors.

Payload: 1.2 kg. It includes gyro-stabilized gimbal that accommodates at least 4–5 sensors. Sensors such as visual, multispectral, IR, red edge, and Lidar are fitted.

Launching information: The system consists of three main components, that is, GCS, UAV, and launch system. The whole system can be set up in 10 min. The system is packed in a hard case. It can be easily transported within the boot of a sedan. The launch pad consists of a launch rack and bungee to get the thrust. This UAV has autolaunch, autonomous navigation and landing ability, using software such as "Pixhawk."

Altitude: 2000 m a.g.l. Agricultural imagery is done at 50–200 m a.g.l.

Speed: The cruising speed is 72–90 km (45–50 knots)

Wind speed tolerance: It withstands weather parameters at Beaufert scale 4 (20–28 km h^{-1} wind speed) or 11–16 knots wind speed).

Temperature tolerance: −20 to +60°C

Power source: Electric batteries

Electric batteries: It is supported by 8 A h high-capacity lithium polymer battery.

Fuel: No requirement for petroleum fuel.

Endurance: 45 min per flight

Remote controllers: The ground control equipment includes radiotransmitter that works at 433/868/915 MHz (hopping). It has a laptop to preprogram flight path and alters the path, midway, if necessary.

Area covered: 3 km^2 per flight with facility for continuous or programed aerial imaging events.

Photographic accessories: Cameras: Swallow-P carries a Sony α5100 (16 mm lens) with 24.3 MP resolution. It offers images of crop land at 5 cm resolution.

Computer programs: Swallow-P UAV and the ground-station computers possess useful software, such as Pix4D Mapper or Agisoft Photoscan. Software provides final images for farmers to analyze and take appropriate decisions.

Spraying area: It is not a sprayer UAV.

Volume: Not applicable

Agricultural uses: Swallow-P is a small UAV. It suits best to obtain aerial scans of crops rather swiftly. It offers aerial imagery of land resources, soil type, and cropping patterns. It also offers imagery useful to detect crop growth stage, growth rate and its variations in fields, crop's canopy structure, crop surface features (CSM), leaf area, leaf chlorophyll, and crop nutrient status (nitrogen). Crop scouting for drought or flood damaged spots, disease/pest-attacked zones, and land surface affected by erosion or salt accumulation and other maladies is possible. Monitoring natural vegetation and forest is done with greater efficiency using the UAV.

Nonagricultural uses: This is a small UAV capable of reaching locations not usually amenable for other vehicles or humans. Hence, it suits well to scan and obtain aerial imagery of areas affected by disaster, fire, floods, earthquake, high mountains, deep mines, etc. Swallow-P can also help in monitoring industrial infrastructure, buildings, highways, railroads, electric lines, and vehicle movement. This UAV has been tested and certified across different locations and for different purposes.

Useful References, Websites, and YouTube Addresses

http://www.uaver.com/swallow-p-aerial-mapping.html (accessed Aug 20, 2017).

http://www.manufacturers.com.tw/showroom-9367-4-5-0-0.php (accessed July 7, 2017).

https://www.taiwantrade.com/company/carbon-based-technology-inc-277185.html (accessed July 7, 2017).

https://www.youtube.com/watch?v=oZ0CZX5-HNA (accessed Aug 20, 2017).

https://www.youtube.com/watch?v=TBJXiGMZWhA (accessed Aug 20, 2017).

Name/Models: Swampfox

Company and address: Skycam UAV 2014 Ltd. Palmerston North, New Zealand; phone +64 21 110 1099; e-mail: info@skycam.co.nz; website: http://www.kahunet.co.nz/

Skycam UAV Ltd. was initiated as a UAV company. Originally, this company intended to produce small (micro) weight UAV models that suits New Zealand's military needs. The project on Swampfox was initiated in 2006. In addition, Skycam Ltd. offers a few other fixed, flat-winged, and rotocopter UAVs.

Swampfox was developed by Skycam UAV Ltd. in response to New Zealand government's needs, for a lightweight dual purpose UAV, capable of military and commercial aerial survey and imagery. Swampfox is portable. It can be assembled and launched in just a few minutes from any location in a farm or other place. It has been used by New Zealand's scientists to survey crops, natural vegetation, and algal blooms. It has been used to study the changes in rivers, swamps, and adjoining vegetation. This UAV model has also been utilized to study natural geographical features in the icy terrain of Antarctica.

TECHNICAL SPECIFICATIONS

Material: This UAV has a light-weight aluminum frame. The body is made of a composite of foam, carbon fiber, and plastic.

Size: length: 90 cm; width (wingspan): 180 cm; height: 35 cm; weight: 4.5 kg MTOW

Propellers: It has one propeller. It is placed at the rear end of the fuselage. The propeller has two blades made of tough plastic. Propellers are powered by a brushless electric motor.

Payload: It holds up to 2.2 kg. Payload includes cameras such as visual, NIR, IR thermal, and red-edge cameras. It also has two videocameras for surveillance activity.

Launching information: It is hand-launched or using a catapult. It has autopilot launch and navigation mode. It has autolanding software (e.g., Pixhawk). This UAV can be recovered using parachute landing. Parachute may be triggered using GCS computers (or joystick) or it could be preprogramed.

Altitude: Generally, it is flown at 50–450 m a.g.l. for surveying land resources, crops and farm vehicles.

Speed: The cruise speed is 60 mi h^{-1}. It can reach 80–100 km h^{-1}. Stall speed is 32 km h^{-1}.

Wind speed tolerance: 20 km h^{-1}

Temperature tolerance: −20 to +60°C

Power source: Electric batteries

Electric batteries: lithium–polymer batteries with 14.8 V capacity

Fuel: Not an IC engine-fitted UAV

Endurance: 50 + 10 min for fail safe

Remote controllers: GCS has telemetric equipment like radio, antenna, iPads, and PCs to track the UAV in its flight. They are also used to command the UAV's cameras regarding imaging points. Ground control keeps connectivity with UAV for 15 km h^{-1}.

Area covered: 200–300 ha per flight

Photographic accessories: Cameras: It has Twin Sony NEX-5 stills camera; Single Sony NEX-7 stills camera; GoPro Hero 2 or Hero 3; Mintron 10× optical zoom videocamera; Flir Tau Thermal imaging camera; Full digital video downlink.

Computer programs: Aerial imagery is processed using Pix4D Mapper. Onboard CPU also has facility for processing aerial imagery.

Spraying area: This is not a sprayer UAV.

Volume: Not applicable

Agricultural uses: This small UAV is apt for aerial survey of land resources, fields, soil type, cropping systems, and to subject crops to spectral analysis. Crop scouting for nutritional status, leaf chlorophyll content, NDVI values, and fertilizer needs (fertilizer-N) could be done, using Swampfox. The UAV's cameras can provide spectral data about disease/pest attacked zones, weed species, etc. Natural vegetation monitoring is also done efficiently, using Swampfox. It has also been used to study algal blooms, cyanobacterial patches on water bodies and swamps.

Nonagricultural uses: Swampfox was originally developed as a dual-purpose small UAV that suits New Zealand's military needs. At the same time, it had to be capable of commercial and agricultural functions. It is used to surveillance military installations, international boundaries, large industries, buildings, natural sanctuaries, rivers, dams, etc. It is used in city-traffic patrolling, highway vehicle tracking, rail roads, electric lines, and oil pipelines.

Useful References, Websites, and YouTube Addresses

http://www.kahunet.co.nz/swampfox-uav.html (accessed Apr 13, 2017).
http://www.kahunet.co.nz/ (Apr. 13, 2017). (accessed May 10, 2017).
https://www.youtube.com/watch?v=7EdAqZnL97c (accessed May 10, 2017).
https://www.youtube.com/watch?v=mX1wGTsj9B8 (accessed May 10, 2017).
https://www.youtube.com/watch?v=xpWFOy0S1eo (accessed Apr 13, 2017).
https://www.youtube.com/watch?v=ANIFZE-S-w0 (accessed Apr 13, 2017).

Name/Models: Tempest

Company and address: UASUSA, 229 Airport Road, East Hangar, Longmont, CO 80503, USA; phone: +1 720 608 1827; e-mail: info@uasusa.com; website: http://www.uasusa.com/

UASUSA is originally a small aircraft company. It was founded by a gentleman named Mr. Skip Miller, at Boulder in Colorado. They produced a range of piloted small aircraft. They have cumulative of 85 years of experience in producing small aircraft and unmanned UAVs for various purposes. Their moto about small UAV states that "small UAVs are like civil servants in the air." It means that they cater to a wide range of civilian tasks. The Tempest UAV is a small, sleek UAV useful to farmers and ranchers (Plate 2.27.1). They offer excellent and useful data about crops, their growth status, nutritional requirements and grain maturity, at a very rapid pace. They are also used to detect weeds in crop fields, disease/pest infestation, droughts, floods, erosion effects, etc.

TECHNICAL SPECIFICATIONS

Material: It is made of lightweight composite that has reinforced plastic. It is toughened to withstand harsh weather conditions.

Size: length 5.2 ft; width (wingspan): 10 ft; height 20 cm; weight: 4.6 kg

Propellers: Propeller is made of reinforced plastic, one fixed at the tip of the nose in front.

Payload: 3.6 lb (1.7 kg)

Launching information: It is hand launched by holding the plane at shoulder height. It could also be launched using an open pick-up van. This UAV has autopilot and landing facility, using the software Pixhawk. Pixhawk technology is a superior autopilot system featuring advanced processor, sensor technology, and a real-time operating system. This system delivers incredible performance, flexibility, and reliability, for controlling any autonomous vehicle.

Altitude: It reaches up to 400 ft for agricultural purposes, but it can reach 28,000 ft height.

Speed: 70–85 km h^{-1} (max speed 175 km h^{-1})

Wind speed tolerance: 50 km h^{-1}

Temperature tolerance: −60 to +50°C

Power source: Electric batteries

Electric batteries: Lithium batteries

Fuel: It does not possess an IC engine. No need for petrol or diesel.

Endurance: 1.5 h.

Remote controllers: Radiocontrol operates for up to 115 km from ground station. Flight paths can be altered midway using GCS iPad. The UAV returns to point of launch in time of emergency.

Area covered: 3.6 sq. miles (or 2000 ac) per flight of 1.5 h

Photographic accessories: Cameras include visual, multispectral, hyperspectral, thermal, and Lidars. They offer high-resolution aerial images of land, crops, and vehicles movement in farms. The images are processed right at the CPU on the UAV. Otherwise, it is relayed as orthomosaics and instantaneously processed into sharp images of crops. Images show plant stand, disease/pest infestation, grain maturity, etc.

Computer programs: The commonly used software for processing orthomosaics relayed by the UAV, that is, Pix4D Mapper or Agisoft's Photoscan is appropriate.

Spraying area: This is not a spraying UAV.

Volume: Not applicable

Agricultural uses: It includes aerial mapping of crop fields. Aerial imagery is conducted for canopy characters, such as leaf area, NDVI, plant height, weed maps, crop counting, leaf area index, chlorophyll maps, and leaf nitrogen, using UAV and its sensors.

Nonagricultural uses: It is employed in environmental studies, geophysical studies, terrain mapping, geology and mining, oil pipelines, and electric transmission line surveillance, etc.

PLATE 2.27.1 Tempest.
Source: Mr. Bill Nickerson, UASUSA, Longmont, Boulder, Colorado, USA.

Useful References, Websites, and YouTube Addresses

http://uasusa.com (accessed May 27, 2017).
http://blackswifttech.com/ (accessed May 27, 2017).
http://blackswifttech.com/ (accessed May 27, 2017).
https://www.youtube.com/watch?v=qjcLquGpVhM (accessed June 10, 2017).
https://www.youtube.com/watch?v=l5Zc_Q8nXfQ (accessed June 10, 2017).
https://www.youtube.com/watch?v=V3ZvcWtxT5Y (accessed June 10, 2017).

Name/Models: TerraHawk T-32 (fixed flat winged)

Company and address: Phoenix Aerial Systems Inc., 3364 Robertson Pl Suite 102, Los Angeles, CA, USA; phone: +1 323 577 3366; e-mail: info@phoenix-aerial.com; website: http://phoenix-aerial.com

Phoenix-Aerial Systems Inc. is a UAV manufacturing company located near Los Angles, California, USA. It specializes in small- to medium-sized agricultural UAVs. These UAVs are versatile and offer wide range of advantages to both farming companies and large clientele opting more generalized use of such flat-winged UAVs. This UAV helps in surveillance of large industrial installations, public buildings, localities, city traffic, transport vehicles, etc. This UAV can be used to get pictures of disaster regions, volcanoes, and locations difficult for humans to reach. These UAVs are also used to image and relay information of public events, sports events, etc. to ground station.

TECHNICAL SPECIFICATIONS

Material: Terrahawk T-32 is made of ruggedized foam, plastic, and carbon fiber composite.

Size: length: 225; width (wingspan): 300 cm; height: 45 cm; weight: 8 kg (MTOW), empty weight, that is, without payload and fuel is 4 kg.

Propellers: It has one at the front tip of the fuselage.

Payload: Maximum of 2.5 kg

Launching information: It is launched using a catapult. Launch, flight, and landing is regulated by autopilot software.

Altitude: Recommended altitude over the crop canopy is 20–60 m for sharp high-resolution imagery of crops.

Speed: Its cruise speed is 42 km h^{-1}

Wind speed tolerance: 20–25 km h^{-1} disturbance

Temperature tolerance: −10 to +60°C

Power source: Electric batteries. Power consumption is 12 W.

Electric batteries: Lithium–polymer 24,000 mA h; four cells

Fuel: No requirement for petroleum fuel

Endurance: 60–80 min

Remote controllers: The UAV's flight is regulated using an iPad or mobile app. Radiolinks operate for up to 6 km from ground station.

Area covered: 50–150 ha per flight

Photographic accessories: Cameras: The UAV carries a Sony A 6000 or Basler Ace cameras. The sensors operate at R, G, and B and multispectral bandwidths. Cameras with NIR and IR are also present for thermal imagery. Sensors offer highly sharp images, up to a height of 75 m from the crop canopy. This UAV also has Lidar. Laser operates at 905 nm.

Computer programs: Postflight processing is done, using the usual Pix4D Mapper or Agisoft otoscan or NovAtel's Inertial Explorer. Sensors offer high-resolution images of great utility to farmers and farming companies.

Spraying area: This is not a sprayer UAV.

Volume: Not applicable

Agricultural uses: TerraHawk T-32 has a range of uses for farming companies, particularly, those with large acreage. This UAV offers excellent aerial imagery of land, topography, crops, and growth status. Crop scouting is done, usually, to check seedling emergence, crop stand, leaf area, and crop maturity. It is also used to collect data like NDVI, biomass accumulation pattern, crop water stress, drought, and flood effects, if any and disease/pest attack. This UAV provides data useful in forecasting crop yield. Natural vegetation monitoring is done regularly using Terra hawk.

Nonagricultural uses: Major uses are in observation of industrial installations, large building complexes, rail roads, city traffic, surveillance of mines, mining activity, and ore movement, etc.

Useful References, Websites, and YouTube Addresses

http://www.phoenix-aerial.com/contact-us/contact/ (accessed June 27, 2017).
https://www.youtube.com/watch?v=duLoZ7JUT5E (accessed July 3, 2017).
https://www.youtube.com/watch?v=duLoZ7JUT5 (accessed July 3, 2017).

Name/Models: Tiriba

Company and address: AGX Tecnologia Ltd, R. José Rodrigues Sampaio, 361. Centreville, São Carlos, SP CEP 13560-710, Brazil; Tel./Fax: +55 16 3372 8185; comercial@agx.com.br; contato@agx.com.br

Tiriba means a small parrot. Tiriba is a sleek, microweight UAV. It has carbon fiber composite body that is rugged. Its services are mostly related to civilian and agricultural aspects. The sensors offer detailed multispectral data to farmers. It is predominantly used for surveillance of civilian population and its activities. It is a low-priced UAV produced for Brazilian famers (Joaquim, 2013; Master, 2013; Pelizzoni et al., 2011).

TECHNICAL SPECIFICATIONS

Material: It is made of carbon fiber, toughened plastic, and foam. It withstands rough belly landing.

Size: length: 105 cm; width (wingspan): 230 cm; height: 22 cm; weight: 4.0 kg

Propellers: It has one propeller placed at the nose tip of the fuselage.

Payload: 1.2 kg

Launching information: It is launched manually. It is done by hand, holding the UAV at shoulder height and releasing it swiftly into the air.

Altitude: It functions at 200–400 m a.g.l. for aerial survey and imagery of crops.

Speed: 70–100 km h^{-1}

Wind speed tolerance: 20 km h^{-1}

Temperature tolerance: −10 to +60°C

Power source: Electric batteries

Electric batteries: 14,000 mA h; two cells

Fuel: No need for petrol or diesel

Endurance: 35 min

Remote controllers: The GCS has iPad and radiolink with UAV for up to 20 km radius. The flight path is usually predetermined using Pixhawk software. The flight path can be altered midway through the flight by the GCS iPad.

Area covered: 50–250 ha depending on the flight path pattern

Photographic accessories: Cameras: The UAV carries a full complement of sensors that it includes high-resolution visual cameras, multispectral camera, videographic equipment, and thermal camera that operates at NIR, IR, and red-edge wave bandwidths.

Computer programs: The images procured by the UAV are processed using Pix4D Mapper or Agisoft Photoscan or any other custom-made processer.

Spraying area: This is not a sprayer UAV.

Volume: Not applicable

Agricultural uses: This UAV is useful in scouting land resources, crops, and irrigation facilities. Scouting of stand is usually done to procure NDVI data, assess canopy growth, obtain data about leaf chlorophyll (crop's N status), and water stress index. Natural vegetation monitoring is also possible.

Nonagricultural uses: It is useful in surveillance of industrial sites, mines, transport vehicles, rail roads, dams, irrigation channels, etc.

Useful References, Websites, and YouTube Addresses

Master, W. Brazil Leads the Way in Global Commercial UAVs. In *Unmanned Cargo Aircraft Conference*, Maspoort, Venio, The Netherlands, 2013; pp 1–3.

Joaquim, R. The Flight of the Falcon. *Pesquisa FAPESP*; 2013; pp 1–5. http://revistapesquisa.fapesp.br/en/2013/10/23/the-flight-of-the-falcon/ (accessed June 9, 2017).

Pelizzoni, J. M.; Neris, L. O.; Trindade Junior, O.; Osório, F. S.; Wolf. D. F. Tiriba—A New Approach of UAV Based on Model Driven Development and Multiprocessors. *ICRA Commun.* **2011,** *23*, 45–48.

https://unmannedcargoaircraftconference.com/tag/tiriba/ (accessed June 9, 2017).

http://revistapesquisa.fapesp.br/en/2013/10/23/the-flight-of-the-falcon/ (accessed June 9, 2017).

https://dronesdelsur.org/industria/brasil/ (accessed June 9, 2017).

http://www.agx.com.br/n2/pages/index.php?opt=video_tiriba (accessed June 9, 2017).

https://www.youtube.com/watch?v=zYuOSIQ_F9Q (accessed June 9, 2017).

Name/Models: UX5 Arial Image Rover

Company and address: Trimble Inc. Trimble Navigation Systems, Agricultural Division, Street, Sunnyvale, CA 94085, USA; phone: +1 303 635 8707 (office), +1 303 803 5711 (cell); Silvia_McLachlan@Trimble.com, http://www.trimble.com

Trimble Inc. is an engineering services company located at Sunnyvale, California, USA. They offer wide range of agriculture-related services involving farm vehicles with electronics. They regularly offer agricultural prescriptions after analyzing the problems that farmers encounter. The UAVs and avionics are their recent specialty gadgets utilized, during farm-related services. They have developed and produced a range of fixed-winged UAVs such as UX5, etc. They conduct aerial surveys of farmers' fields regularly and supply them digital data, 2D and 3D images of crop growth and phenomics, every week or on fortnight basis to farmers. They also develop range of software necessary to stich the images, analyze them for spectral details, and arrive at appropriate prescriptions, regarding spray of pesticides, fungicides, and most importantly in judging fertilizer-N requirements (Beveridge and Russell, 2015).

The Trimble's UX5 Imager is among the most popular models of agricultural UAVs. They are small, microweight, and highly versatile UAVs. They can be assembled quickly in less than 5 min and hand launched. They are swift to fly and pick images at a rapid pace based on the predetermined programs. They are low-cost UAVs (Plate 2.28.1; Cosyn and Miller, 2013).

TECHNICAL SPECIFICATIONS

Material: It is made of graphite-reinforced plastic and toughened to withstand harsh weather conditions.

Size: length: 65 cm; width (wing span): 100 cm, and wing area is 34 dm^2; height: 10 cm; weight: 2.5 kg (5.5 lb).

Propellers: It has one propeller placed centrally, at the tip of fuselage. It is powered by electricity from batteries.

Payload: 1200 g. It includes gimbal and a set of cameras.

Launching information: Hand-held launching is possible. Landing is smooth due to reverse thrust.

Altitude: This UAV operates at 50–400 m a.g.l. for agricultural survey and imagery. However, it can reach a ceiling height of 2500 m a.g.l.

Speed: 80 km h^{-1}

Wind speed tolerance: 37 km h^{-1}

Temperature tolerance: −20 to +60°C

Power source: Electricity

Electric batteries: 2 Electric batteries of 14.8 V.

Fuel: No requirement of petroleum fuel.

Endurance: 50 min without a relaunch

Remote controllers: The GCS computers keep contact with the UAV via radiolink and iPads for up to 20 km radius.

Area covered: 180 ac per flight or 60 km distance per flight or 37 miles in one stretch and imagery

Photographic accessories: Cameras: Sony 16-MP APSc camera and NIR camera.

Computer programs: The imagery derived by UAVs is processed using Pix4D Mapper or Agisoft Photoscan or Trimble's Inphos or any other custom-made software.

Spraying area: This is not a sprayer UAV.

Volume: Not applicable

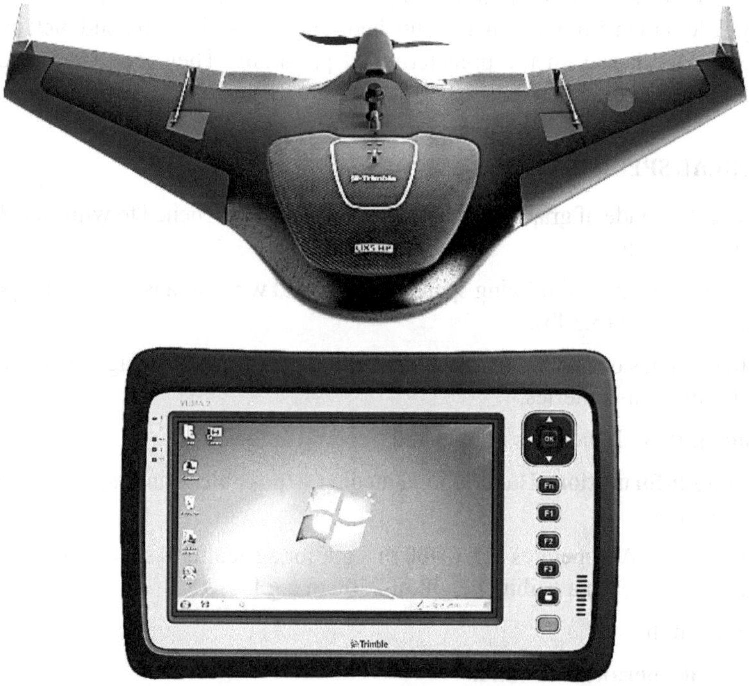

PLATE 2.28.1 Trimble UX5 Aerial Imaging Solution along with a remote controller tablet.
Source: Ms. Silvia McLachlan, Coordinator Agricultural Division, Trimble, Sunnyvale, California 94085, USA.
Note: Trimble UX5 Imaging Solution is a fixed-wing, fully autonomous UAV (UAS). It is capable of aerial imagery of natural resources, farm topography, soils, crops, crop growth, and nutrient status. The flight path of Trimble-UX5 can be predetermined using the tablet.

Useful References, Websites, and YouTube Addresses

Beveridge, M.; Russell, A. Using UAVs for the Detection of Crop Pest Damage in Canola. *Grain Research and Development Centre Southern Farming Systems: A Report*, 2015, pp 1–8. http://www.farmtrials.com.au/trial/18829 (accessed July 27, 2017).

Cosyn, P.; Miller, R. *Trimble UX5 Aerial Imaging Solution: A New Standard in Accuracy, Robustness, and Performance for Photogrammetric Aerial Mapping. A White Paper.* Trimble Navigation Ltd.: Westminister, CO, USA, 2013; pp 1–7.
http://uas.trimble.com/Agriculture/index.aspx (accessed Aug 3, 2017).
http://uas.trimble.com/Agriculture/field-iq.aspx (accessed Aug 3, 2017).
http://uas.trimble.com/ux5 (accessed Aug 3, 2017).
http://www.trimble.com/survey/ux5.aspx (accessed Aug 3, 2017).
http://www.trimble.com/agriculture/ux5 (accessed Aug 3, 2017).
https://www.youtube.com/watch?v=DZZcq2ofY-c (accessed July 30, 2017).
https://www.youtube.com/watch?v=-7VSDNqEh44 (accessed July 30, 2017).
https://www.youtube.com/watch?v=0D9-scSs0tc (accessed July 30, 2017).
https://www.youtube.com/watch?v=UbVdI_odRYY (accessed July 30, 2017).

Name/Models: UAS: Wasp AE

Company and address: AeroVironment Inc., 800 Royal Oaks Drive, Suite 210, Monrovia, CA 91016 6347, USA; phone: +1 626 357 9983; e-mail: ir@avinc.com; Website: https://www.avinc.com/contact

AeroVironement is a Southern California aviation company. It was initiated some 40 years ago. They found a breakthrough with production of a series of UAVs (drones) such as Puma, Raven, Wasp, Shrike, etc. They cater to military, farm, and civilian sectors with UAV models. Their UAVs are used in different continents. Most recent model touted by the company is named "Quantix (VTOL)" (AeroVironment Inc., 2016).

Wasp is a small, portable, reliable, and rugged unmanned aerial platform. It is designed for frontline day/night reconnaissance and surveillance. The Wasp AE Micro Air Vehicle (MAV) is the all environment version of AeroVironment's battle proven Wasp III. Wasp AE has special design considerations for maritime and land operations. Wasp AE is delivered in a package. It has exceptional features of superior imagery, increased endurance, encrypted video, and ease of use. These aspects are inherent in all AeroVironment UAS solutions. This UAV could also be employed to conduct a series of functions relevant to crop production, particularly precision farming.

TECHNICAL SPECIFICATIONS

Material: This small UAV is made of a composite derived from carbon fiber, plastic, and foam.

Size: length: 2.5 ft (75 cm); width (wingspan): 3.3 ft (102 cm); height: 1.0 ft (30 cm); weight: 2.85 lb (1.3 kg)

Propellers: It has one propeller placed at the tip of the nose. It has two plastic blades. Propellers are fitted with low acoustic propulsion system. This trait is useful when the UAV is used in battle fields or during reconnaissance.

Payload: It supports gimbaled payload with pan and tilt-stabilized, high-resolution EO and IR camera.

Launching information: It is hand-launched. It adopts a deep-stall landing in confined areas on land or water. Wasp AE can be operated manually. It can also be preprogramed for autonomous operation, utilizing the system's advanced avionics and precise GPS navigation. It has autolanding facility, using Pixhawk software.

Altitude: Normally, this UAV is flown at 500 ft (152 m) a.g.l.

Speed: 20 knots at cruise.

Wind speed tolerance: 20 km h^{-1}.

Temperature tolerance: −20 to +60°C.

Power source: Electric batteries

Electric batteries: The motor and motor controller battery, payload, and control surface actuators are all replaceable in the field within minutes. Replacements can be done without the use of special tool. This increases the operational availability of each UAS.

Fuel: This is not an IC engine-fitted UAV. So, petroleum fuel is not required.

Endurance: 50 min

Remote controllers: The UAV is operated from AeroVironment's battle-proven ground control system (GCS). The communication range is 5 km. It can be operated manually or programed for autonomous operation (AeroVironment Inc., 2016).

Area covered: 50 km linear distance. The UAV can conduct imagery of 50–150 ha of crop land per flight.

Photographic accessories: Cameras: It carries the usual component of three visual cameras (R, G, and B), multispectral camera, NIR, IR, and red-edge sensors.

Computer programs: The UAV operates virtually undetected. Wasp AEs have mechanically stabilized EO/IR gimbal payload that transmits advanced imagery, even in high wind conditions.

Spraying area: This is not a sprayer UAV.

Volume: Not applicable

Agricultural uses: This small UAV is swift and can conduct aerial survey of land, its topography, soil type, cropping system, disease/pest incidence, drought, floods, or soil erosion effects. This UAV can be utilized, to image the crop at different growth stages, record its NDVI (biomass), leaf area, and leaf chlorophyll, that is, crop-*N* status. Using IR and red-edge sensors, this UAV can provide useful data on soil moisture and crop water status. Irrigation can be scheduled most accurately, using the data from the UAV's sensors. The UAV is highly useful, in farms adopting precision techniques. It has been stated that AeroVironment's UAV is equipped with optical and multispectral sensors linked to cloud-based data processing. They are poised to become an important equipment of the farm. With their sophisticated eyes-in-the-sky and extensive monitoring capabilities, farmers can grow crops with greater precision.

Nonagricultural uses: This UAV is useful to surveillance mines, mining activity, ore dumps and ore movement. It is useful in keeping watch on industrial infrastructure, buildings, city traffic public events, rails rods, oil/gas pipelines, borders, etc.

Useful References, Websites, and YouTube Addresses

Aerovironment Inc. *UAS: Wasp AE*; 2016; pp 1–3. https://www.avinc.com/uas/view/raven (accessed Jan. 20, 2017).

https://www.avinc.com/uas/adc/wasp (accessed Jan 20, 2017).

https://www.avinc.com/uas/view/wasp (accessed Jan 20, 2017).

https://www.youtube.com/watch?v=vtaqPV7M7bc (accessed Jan 20, 2017).

https://www.youtube.com/watch?v=8RMB9WMdG6g (accessed Jan 20, 2017).

Name/Models: VYOM 08 UAV

Company and address: TechnoSys Embedded Systems, SCO 66, 1st floor, Sector 20 C, Chandigarh 160020, India; phone: +91 1725085909; +91 1724671082; e-mail: info@technosysind.com; sales@technosysind.com; support@technosysind.com Website: http://www.technosysind.com

Technosys Ltd. produces a few different UAV models. Some of them are specifically meant for rapid aerial imagery of terrain, crops, and monitoring farm vehicles. VYOM 08 is churned out by the company, to specifically help in repetitive flights at short notice in agricultural farms and urban departments. VYOM 08 is a low-cost UAV (Plate 2.29.1). It is destined to help the farmer in the North Indian Plains. It is a fail-safe UAV, since it returns to starting point, if communications or preprograms on computer go awry.

TECHNICAL SPECIFICATIONS

Material: It is made of honeycomb-structured composite made of foam, plastic, and steel.

Size: length: 170 cm; width (wingspan): 240 cm; height: 35 cm; weight: 2.2 lb or 1 kg

Propellers: It has one propeller at the tip of the nose, on the fuselage. It has two plastic blades. Propeller is connected to servos (Hitec645MG).

Payload: 800 kg

Launching information: It is hand-launched. It can be launched using a catapult.

Altitude: It operates at <400 m a.g.l. for crop surveillance and urban monitoring. It can reach 5000 m.a.gl, if needed

Speed: 100 km h^{-1}

Wind speed tolerance: 20 km h^{-1}

Temperature tolerance: −17 to +55°C

Power source: Electric batteries and gasoline. UAV is powered by an IC engine DLE 60.

Electric batteries: Lithium–polymer 8 mA h

Fuel: Gasoline

Endurance: 5 h

Remote controllers: Ground station includes a laptop. Laptop helps operator in programing flight path and landing. UAV keeps radio contact with remote controller (JR 8C). Autopilot is made of DZY025. Remote controllers operate for 30–50 km range (radius).

Area covered: VYOM covers 200–250 ac of crop land per flight. The UAV rapidly picks images that can be stored in the CPU and later processed. The timing and frequency of imaging events could be regulated using remote control or preprogramed.

Photographic accessories: Cameras: VYOM 08 comes with a sensor (Sensor: Canon 5D Markil). For thermal imagery, cameras that operate at NIR and IR bandwidth ranges can be added to gimbal.

Computer programs: The aerial imagery is relayed instantaneously to ground station, for processing. The orthomosaics are processed using Pix4D software. The images come with GPS data.

Spraying area: This is not a sprayer UAV.

Volume: Not applicable

Agricultural uses: It includes a wide range of farm activities. They are surveillance and mapping of terrain, soil types, seeding, monitoring seedling emergence, crop growth, NDVI estimation and mapping, leaf area determination, leaf chlorophyll, and crop-N estimation. Determination of fertilizer-N requirement is an important aspect where UAV technology could be applied. Detection of climate change effects such as soil erosion, floods, and droughts is possible. It monitors natural vegetation monitoring.

Nonagricultural uses: They are monitoring of territory, geological features, natural resources, rivers, dams, etc. This UAV offers imagery useful in urban planning; surveillance of industrial installations, pipelines, and electric power lines; and surveillance of mines, transport vehicles, etc. UAVs help in first response in times of disaster and accruing information about extent of damage.

PLATE 2.29.1 VYOM 08 UAV.
Source: Dr. Dhruv Arora, TechnoSys Embedded Systems (P) Ltd. Chandigarh, India.

Useful References, Websites, and YouTube Addresses

http://www.technosysind.com/product-details.php=VYOM08UAV (accessed Nov 11, 2016).
https://www.youtube.com/watch?v=AUPuSmF9-QU (accessed June 15, 2017).

Name/Models: X8 Fixed-wing UAV

Company and address: GuiLin Feiyu Electronic Technology Co Ltd., Feiyu Tech, 3rd floor B, Guilin Electric Valley, Innovation Building, Information Industry Park, Chao Yang Road, Qi Xiang District, Guilin 541004, China; Tel.: +86 0773 2320861; e-mail: service@feiyu-tech.com; sales@feiyu-tech.com; service@feiyu-tech.com

Guilin FeiYu Electronic Technology Company Ltd. (Feiyu Tech) is a company established in 2007. It specializes in production of UAV platforms, gimbals, and autopilot software. It is a pioneer company in South China. It caters to both indigenous and foreign market, regarding UAVs and their accessories. They produce unmanned aerial robots (UAVs), unmanned boats to navigate river waters in China, and electronic accessories for UAVs. They also manufacture software, to control launching, flight, and autolanding. X8 UAV is an important UAV model produced by this company.

TECHNICAL SPECIFICATIONS

Material: X8 (UAV) has a light but strong body made of ruggedized foam, rubber, carbon fiber, and metal alloy. It withstands vagaries of weather conditions and rough belly landing. The UAV is carried in small briefcase. All the parts are placed in a suitcase so that it can be carried to any location and launched. It is easy to assemble and disassemble UAV (Plate 2.30.1).

Size: length: 100 cm; width (wingspan): 212 cm; height: 2018 cm; weight: 4.0 kg

Propellers: It has one propeller made of two blades (toughened plastic). It is placed at the tip of the rear end of fuselage and facing backward. Propeller is powered by electric motor.

Payload: 600 g. It is made of cameras and CPU; long-range version of the UAV X8 can carry a payload of up to 2 kg

Launching, flight, and landing information: This is a ready-to-fly model. It is launched using a catapult launching device. It has auto take-off and autolanding system. UAV is supplied with "Panda 2 Auto Pilot system" for effortless launch, flight control, and landing. The long-range version is guided by "Pixhawk Flight Controller." It operates until the UAV covers about 3.2 miles per flight time. Landing instructions can be transmitted, using computers. Computers at GCS are operated, using mouse, or it could be done using touchscreen iPad. Manual controls are also possible, but they are cumbersome. It has intelligent return system. In case of emergency, X8 UAV returns to the location where it was launched. Parachute landing is also possible with X8 UAV. Parachute placed in fuselage needs to be triggered, using remote control.

Altitude: The ceiling is <1000 m

Speed: It transits at 65–70 km h^{-1}

Wind speed tolerance: 5–15 km h^{-1}

Temperature tolerance: −17 to +55°C

Power source: Electric batteries

Electric batteries: Lithium–polymer batteries.

Fuel: No requirement of petroleum fuel.

Endurance: It is 90 min, but this gets extended in the long-range version of the UAV by a further 30 min.

Remote controllers/Preprogramed flight paths: The ground station is interfaced with "Google Earth Platform." Remote control allows operator to place the UAV in multiple modes of flight, such as linear rapid transit, circular surveillance mode, autoreturn after fixed time of flight, auto take-off, and landing. The operator is also provided with information on battery life consumption, power consumption, and current voltage. "Pixhawk flight controller" and "Mission planner software" are also supplied with the UAV kit. Mission planer displays the flight range, location (GPS tagged), battery information, voltage, altitude, and speed of the UAV.

Area covered: It covers 2 sq. mi. of land surface in a single flight.

Photographic accessories: Cameras: This UAV can be fixed with the usual complement of cameras that operate at visual bandwidths (e.g., Sony α7 series; Sony NEX SR), NIR, and IR for thermal imagery. UAV is supplied with videographic cameras. Computer programs allow fixed timely photographic exposures by the multispectral sensors. The images come with accurate GPS coordinates, latitude, longitude, altitude, and speed (shutter speed).

Computer programs: The GCS adopts software such as Agisoft Photoscan or Pix4D Mapper for processing the images relayed by the UAV.

Spraying area: This is not a sprayer UAV.

Agricultural uses: It includes aerial survey of terrain, topography, cropping systems, soil characteristics, etc. and crop scouting for seedling establishment, canopy characteristics, nutritional deficiencies, leaf chlorophyll (crop-N) mapping, canopy temperature (water stress index) etc. It is also used to detect disease/pest-attacked regions in the fields. Natural vegetation monitoring is possible.

Nonagricultural uses: To study surface geology, geographical features, volcanoes, coastline, rivers, and lakes. It is used in first response teams during disaster clearance. Surveillance of mines, industrial installations, highways, and traffic control is possible. Monitoring transport of goods and scheduling is also done using X8 UAV. A few other functions are environmental analysis through rapid imagery. It is also useful in the surveillance of wildlife sanctuaries.

PLATE 2.30.1 X8 Fixed-wing UAV.
Source: Service Staff, Service Division, Feiyu Technology, Schenzen, China.

Useful References, Websites, and YouTube Addresses

http://www.feiyu-tech.com/contact/ (accessed Mar 28, 2017).
https://uavsysteminternational.com/product/x8-long-range-drone/ (accessed Mar 28, 2017).
http://www.feiyu-tech.com/about/ (accessed Mar 28, 2017).
https://www.youtube.com/watch?v=hFRsiL7BmnE (accessed Mar 28, 2017).
https://www.youtube.com/watch?v=HGCQiGiQVFw (accessed Mar 28, 2017).
https://www.youtube.com/watch?v=cB1AuKhYDpg (accessed Mar 28, 2017).

2.3 FIXED-WINGED UNMANNED AERIAL VEHICLE SYSTEMS IN AGRICULTURE: LIGHT-WEIGHT UAVS (5–50 kg)

There are several companies that produce fixed-winged UAVs that are categorized as "lightweight UAV." The lightweight UAVs are also of small size and made of carbon fiber and plastic composites. Their weight ranges from 5 to 50 kg. They are powered by electric batteries. Otherwise, some of them have IC engines supplied with petrol/diesel. They carry sensors. These agricultural UAVs have a slightly extended endurance. They are usually launched using catapult. The landing could be either via parachute, landing gear, or a few models that may land on belly through skidding. Light-weight UAVs are utilized to conduct aerial survey of land, entire farm, and water resources. They offer useful data to farmers about crop canopy growth, disease/pest status, water requirements (irrigation), and grain maturity.

Name/Model: Accipiter

Company and address: Carbon-Based Technology Inc., Central Science Park/3f, No 30, Keya Road, Daya District, Taichung City, 42881, Taiwan; phone + 886 4 2565 8558; e-mail: info@uaver.com; website: http://www.uaver.com

Accipiter is a light-weight UAV. This UAV's payload consists of cameras useful for aerial imagery. It must be launched using catapult, but it is autonomous during flight, aerial survey, and imaging events, as well as during landing. The cameras offer a set of very useful digital data. It includes unprocessed and processed imagery that is relayed to the GCS iPads. This UAV along with a series of other small UAVs are destined to throng the agricultural zones of Taiwan and adjoining countries. This UAV is easy to pack and transport to any location. It is not a costly UAV to purchase for a crop production company.

TECHNICAL SPECIFICATIONS

Material: It is made of advanced carbon composites. It has most intricate weave pattern that gives strength and rigidity to airframe.

Size: length: 120 cm; width (wingspan): 305 cm (19 ft); height: 65 cm; weight: 13–15 kg (MTOW 15 kg)

Propellers: It has one propeller placed at the rear end of fuselage. A 1500-W brushless motor is connected to propeller.

Payload: It is 2.5–3 kg. It includes high-definition HD video and still cameras, optical laser, radar, multispectral, and hyperspectral sensors.

Launching information: It is launched using a catapult or bungee. It has auto-launch and navigation software such as Pixhawk. Landing is autonomous, but it can be controlled manually, if needed. It also has option of parachute landing.

Altitude: The ceiling is 4000 m or 13,100 ft a.g.l.

Speed: It ranges from 64 to 120 km h^{-1} (35–70 knots)

Wind speed tolerance: It tolerates wind at Beufort scale 6. It can withstand rain force that is classified as drizzle and not beyond.

Temperature tolerance: –20 to +60°C

Power source: High-capacity lithium polymer batteries

Electric batteries: 28,000 mA h for power and 8000 mA h for avionics.

Fuel: No requirement for petroleum fuel.

Endurance: 2.5–3 h per flight

Area covered: It covers about 30 km^2 per flight

Remote controllers: The ground control equipment includes radio that operates at 1900 bps, half duplex with 10 channels. The radiocontrol zone extends for 30 km radius with antenna. The GCS invariably has a back-up battery set, to keep continuous contact with UAV. The GCS also has a series of iPads or PCs to collect/receive data from UAV's CPU. It has computer software for image processing, stitching orthomosaics, storing digital data for retrieval, and analysis using different programs. The usual image-processing software such as Pix4Dmapper, PhotoScan, and Agisoft are also included.

Photographic accessories: Cameras: General package includes Nikon D818 35-mm digital SLR camera or equivalent of other brands. Other cameras include multispectral, hyperspectral, NIR, IR, and red-edge cameras to photograph natural vegetation and crops in detail. The resolution of imagery relayed is 35 MP.

Computer programs: The orthomosaics received at the ground station are processed using PIx4D Mapper or Agisoft Photoscan or similar custom-made software. It adopts real time kinematics (RTK)/PPK techniques for navigation and labeling video/still images.

Spraying area: It is not a spraying UAV.

Volume: Not applicable

Agricultural uses: Major applications are in aerial survey and imagery of natural vegetation and agricultural cropping zones. This UAV is useful in scouting crops for canopy size, leaf area, leaf chlorophyll content, and crop-N status. It helps in calculating fertilizer-N requirement, using chlorophyll/N status data. It helps to assess crop water stress index. It is also used to collect spectral signatures of crops, weeds, healthy, and disease-afflicted crops. This UAV is most useful, when digital data obtained by its cameras are fed to precision equipment and variable-rate applicators.

Nonagricultural uses: This UAV has been utilized in scouting and identifying disaster-affected zones such as fire, earthquake, erupted volcanoes floods, etc. It is used to monitor large industrial installations, buildings, dams, electrical installations, oil pipelines, highway roads, and vehicles. It can be fitted with a device, to track vehicle in farm and in the highways.

Useful References, Websites, and YouTube Addresses

http://www.uaver.com/product-Accipiter-Accipiter.html (accessed Dec 20, 2016).
https://www.youtube.com/watch?v=Qaiy9ac4Iaw (accessed Dec 20, 2016).
https://www.youtube.com/watch?v=C5sEXzHPYK0 (accessed Dec 20, 2016).

Name/Models: Aerohawk

Company and address: RP Flight Systems Inc., Wimberley, TX, USA; phone: +1 512 665 9990, +1 408 832 1195; e-mail: info@rpflightsystems.com, sales@rpflightsystems.com; website: http://www.RPFlightSystems.com

RP Flight systems Inc. is head quartered at Wimberley, Texas, USA. This UAV company was initiated in 2001. Its focus is on developing designs and mass-producing fixed-winged UAVs. These UAVs are aimed at helping in surveillance of public events, policing city suburbs, tracking city vehicles, and monitoring buildings; these UAVs are also apt for monitoring agriculture fields and farm. They can provide crucial spectral data about crop status to farmers.

Aerohawk is small, fixed-winged UAV (Plate 2.31.1). It carries a full complement of sensors that offer high-resolution images of crop fields, city locations, buildings, and other installations. It needs a small runway to get air borne and land using a landing gear.

TECHNICAL SPECIFICATIONS

Material: This UAV is made of a composite containing carbon fiber, ruggedized plastic, and foam.

Size: length: 110 cm; width (wingspan):180 cm; height: 55 cm; weight: 10.2

Propellers: It has one propeller made of two plastic blades. The propeller is placed at the rear end of the short fuselage.

Payload: It accommodates 800–1200 g payload made of cameras and UAV's CPU.

Launching information: Aerohawk is an autolaunch UAV. It has landing gear made of three wheels with rugged rubber. It needs a runway of at least 100 m to obtain lift off and get airborne.

Altitude: It operates at 100–400 m a.g.l. during agricultural crop survey and digital collection.

Speed: It transits at 45–60 km h^{-1} during aerial imagery and loiter mode.

Wind speed tolerance: 20 km h^{-1} disturbance

Temperature tolerance: −20 to +50°C

Power source: Electric batteries

Electric batteries: 24,000 mA h; two cells

Fuel: No need for petrol or diesel

Endurance: 45 min per flight

Remote controllers: The GCS radio telelink and iPads operate and keep contact with the UAV, for up to 30 km radius. The GCS iPads have override facility to modify previously programed flight path, in fixing way points and camera shutter activity. The GCS iPad uses a "Pixhawk" software, to control the UAV's path. The UAV returns to point of launch in times of emergency or erroneous commands.

Area covered: 300 ha per flight

Photographic accessories: Cameras: The payload area on the UAV allows a gimbal with cameras such as visual (R, G, and B), Sony A5100, multispectral and hyper-spectral cameras, NIR, IR, and red-edge bandwidth camera.

Computer programs: The images derived by UAV's cameras are processed using software such as Pix4D Mapper or Agisoft's Photoscan or Trimble's Inphos or any other custom-made software. The image-processing software is available both at the UAV's CPU and at GCS iPad.

Spraying area: This is not a sprayer UAV.

Volume: Not applicable

Agricultural uses: Major agriculture-related functions of this model are scouting for seed germination, gaps in crop's stand establishment, estimation and mapping of leaf chlorophyll and crop's N status, mapping variations in crop-N content and deciding fertilizer-N supply dosages, estimation of canopy temperature, that is, crop's water stress index, etc.; the aerial imagery and spectral data maps are used to detect crop disease and pest-attacked zones and map them. Natural vegetation monitoring is also possible.

Nonagricultural uses: They are surveillance of industrial sites, buildings, city traffic and vehicles, oil and gas pipelines, electric installations, rail roads, rivers, dams, etc.

PLATE 2.31.1 Aerohawk by RP Flight Systems.
Source: Dr. Gene Robinson, President, RP Flight Systems, Wimberley, Texas, USA.

Useful References, Websites, and YouTube Addresses

http://www.asiatechdrones.com/ (accessed Sept 12, 2017).

http://www.unmannedsystemstechnology.com/company/rpflight-systems/ (accessed Sept 12, 2017).

https://www.tractica.com/resources/company-profiles/rp-flight-systems/ (accessed Sept 12, 2017).

Name/Models: AeroMapper 300

Company and address: AeroMao Inc. 33-7030 Copenhagen Rd., Mississauga, ON, Canada; phone: +1 647 928 4747; e-mail: info@aeromao.com; website: http://www.aeromao.com/home

Aeromao Inc. manufactures the "AeroMapper" series of UAVs. These UAVs are meant for acquiring image useful for mapping and surveying applications. The AeroMappers are currently used in more than 45 countries by many clients. The clientele includes government institutions, private companies, universities, and research organizations involved in a wide variety of applications and tasks. These UAVs are said to be efficient and accurate in performing tasks. Therefore, these UAV models outcompete other brands, it seems. AeroMao Inc. also imparts specialized training. The training covers aspects of UAV/GCS maintenance, flight control, imagery, and utilization of data efficiently.

The AeroMapper 300 offers unparalleled multisensor capabilities and 1.5 h flight endurance in its standard configuration. It is equipped with a 24-MP camera (Plates 2.32.1 and 2.32.2). Among its unique features is the capability of carrying several sensors simultaneously, in different combinations. It includes a 20-MP camera with one extra flight battery for a total of 2 h flight time. As usual, it features fully autonomous flight with several other flight modes available. It has facility for parachute-aided recovery or belly landing. The "AeroMapper 300" complies with all design standards stipulated for UAVs, by the government agency—Transport Canada.

TECHNICAL SPECIFICATIONS

Material: It is built using carbon fiber with kevlar reinforcements in key areas.

Size: length: 183 cm; width (wingspan): 300 cm; height: 45 cm; weight: 5.35 kg

Propellers: It has one propeller placed at the nose of the fuselage. It has two blades made of toughened plastic.

Payload: 650 g. It includes series of cameras for visual, NIR, IR, and thermal imagery. No gimbal is required. This is because of great stability, minimal roll, and pitch oscillations even in gusty windy conditions.

Launching information: This UAV has automatic take-off mode. The flight path and navigation are also under autopilot mode. Autopilot mode utilizes Pixhawk software developed by 3D Robotics Inc. Landing is done after two or three passes of the aircraft in the same location, but at a height of 40 m a.g.l. Parachute should be released, at least at 40 m a.g.l. The UAV is recovered, using parachute that pops with a single switch. It has strong kevlar linings. The parachute is very easy to pack into the UAV. It takes only 2 min to pack. Belly landing is also possible.

Altitude: It operates at 50–400 m a.g.l. for aerial survey of agricultural regions. However, the ceiling is 4500 m a.g.l.

Speed: The cruise speed is 60 km h^{-1}. Maximum speed attainable is 90 km h^{-1}.

Wind speed tolerance: 25 km h^{-1} for parachute downing and 45 km h^{-1} for general flight conditions. The UAV operates normally even in small drizzle.

Temperature tolerance: −20 to +60°C.

Power source: Electric batteries

Electric batteries: 28 mA h high capacity with four cells.

Fuel: No petroleum fuel required.

Endurance: 90 min per flight. It can be extended by opting for better battery power.

Remote controllers: The radiolink for commanding the UAV's flight is of 50 km radius. However, to receive aerial data, it is 20 km radius. The UAV/GCS uses a 915 or 868 MHz frequency contact. The UAV has a fail-safe mode. It automatically returns home, to the spot where it was launched, within 50 m^2 area. Flight modes can be manual, stabilized, return to home, FBW, and auto (see Plate 2.32.2).

Area covered: AeroMapper 300 covers 4.3 km^2 area when flown at 85 m a.g.l. and offers images of 1.6 cm pixel^{-1} resolution. It covers 13 km^2 at 250 m a.g.l. and offers images of 4.8 cm pixel^{-1}. It covers 32 km^2 at 600 m a.g.l. and offers images of 11.7 cm pixel^{-1}. The area is calculated considering a 30° overlap.

Photographic accessories: Cameras: Cameras are installed in panoramic orientation for optimal area coverage and with maximum side overlap. In general, cameras include Parrot Sequoia (R, G, and B) for obtaining NDVI data and orthomosaics, in a single flight. It has FLIR Vue and RGB Sony 24 MP for thermal imagery of crops and other features. FLir obtains thermal imagery while simultaneously Sony picks RGB images. This UAV also comes with options, such as single camera, for example, Sony α A 7r and NIR Sony A 5100 for chlorophyll estimation in the canopy. Additional cameras include mica-sense red-edge camera and hyperspectral imaging camera.

Computer programs: The UAV system including ground control equipment adopts software such as Pix4D Mapper, for stitching orthomosaics and processing images. Agisoft Photoscan offers high-resolution, accurate images that are tagged with GPS coordinates. The resolution is 5 cm for aerial images.

Spraying area: This is not a sprayer UAV.

Volume: Not applicable

Agricultural uses: The AeroMapper 300 is ideal for large-sized projects involving applications like 2D orthomosaics, 3D terrain models, and precision agriculture.

Nonagricultural uses: It includes surveillance of geological features, mines, mining activity and ore movement, vehicle tracking, monitoring electric lines, oil and gas pipeline, and water resources.

PLATE 2.32.1 AeroMapper 300.
Source: Mr. Mauricio Ortiz Sales & Support Aeromao Inc. 33-7030 Copenhagen Rd. Mississauga, Ontario. L5N-2P3, Canada.

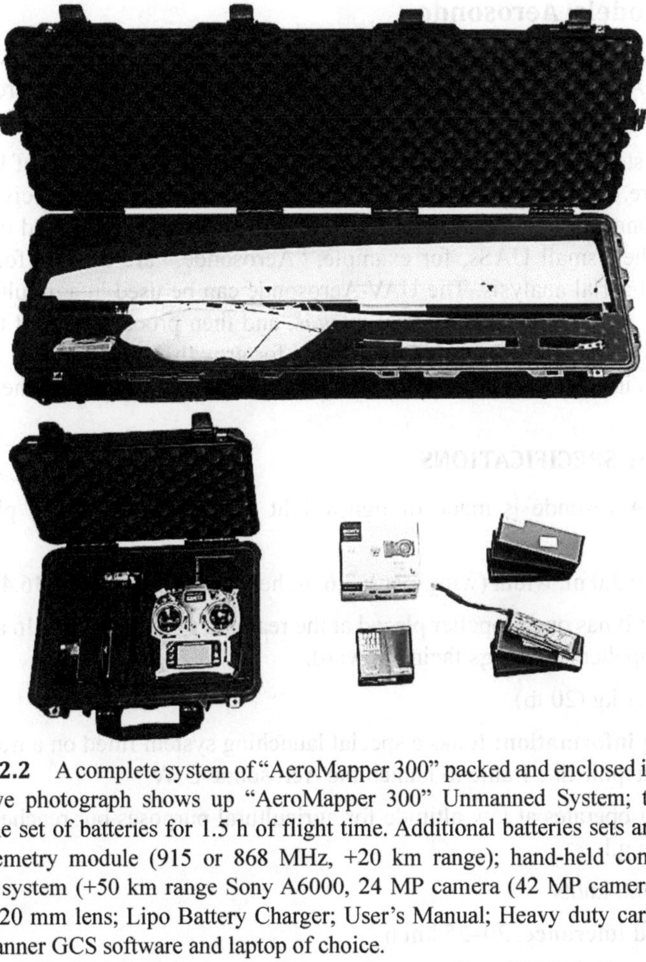

PLATE 2.32.2 A complete system of "AeroMapper 300" packed and enclosed in a suitcase.
Note: Above photograph shows up "AeroMapper 300" Unmanned System; the package includes one set of batteries for 1.5 h of flight time. Additional batteries sets are available; ground telemetry module (915 or 868 MHz, +20 km range); hand-held controller with long-range system (+50 km range Sony A6000, 24 MP camera (42 MP camera optionally available); 20 mm lens; Lipo Battery Charger; User's Manual; Heavy duty carrying cases; Mission Planner GCS software and laptop of choice.
Source: Mr. Mauricio Ortiz Sales & Support Aeromao Inc. 33-7030 Copenhagen Rd. Mississauga, Ontario, Canada.

Useful References, Websites, and YouTube Addresses

https://www.unmannedsystemssource.com/shop/vehicles/aeromapper-300/ (accessed July 27, 2017).

https://www.thedroneproshop.com/products/aeromao-aeromapper-300-pix4dmapper-pro (accessed July 27, 2017).

http://www.aeromao.com/aeromapper_300/ (accessed July 27, 2017).

https://www.youtube.com/watch?v=SLZN4XtSHdk (accessed July 27, 2017).

https://www.youtube.com/watch?v=UDB55k7cgvI (accessed July 27, 2017).

https://www.youtube.com/watch?v=SL0lAa5Juak (accessed July 27, 2017).

https://www.youtube.com/watch?v=SLZN4XtSHdk (accessed July 27, 2017).

Name/Models: Aerosonde

Company and address: Textron Systems Inc. 124 Industrial lane, Hunt Valley, MD 21030, USA; phone: +1 800 655 2615; e-mail: civilcommercial@textronsystems.com; website: http://www.tronsystems.com

Textron Systems Inc. is a multifaceted company regarding the types of UAVs they manufacture. They cater to wide range of human activities related to civil, agriculture, environment and meteorological analysis, industrial security, and importantly military. Their small UASs, for example, "Aerosonde," are utilized for civil and agricultural aerial analysis. The UAV-Aerosonde can be used in agricultural farms worldwide, to obtain accurate aerial images, and then process them at the ground station. Agricultural crops could be analyzed for growth, NDVI, crop-N status, and water stress index, using appropriate sensors in the payload region of the UAV.

TECHNICAL SPECIFICATIONS

Material: Aerosonde is made of lightweight and toughened foam, plastic, and carbon fiber.

Size: length: 3.0 m; width (wingspan): 3.6 m; height: 35 cm; weight: 36.4 kg (80 lb)

Propellers: It has one propeller placed at the rear end of the fuselage. In addition, it has two propellers on wings facing upward.

Payload: 9.1 kg (20 lb)

Launching information: It has a special launching system fitted on a trailer. It has a sliding catapult-like frame to launch the Aerosonde UAV.

Altitude: It operates at low altitude for agricultural purposes but reaches altitudes of 4500 m a.g.l.

Speed: 50–60 knots

Wind speed tolerance: 20–25 km h^{-1}

Temperature tolerance: −10 to +60°C

Power source: The Aerosonde (Mark 4.7) uses single cylinder, air-cooled, direct-injected, spark-ignited Lycoming EL-005 engine, for power and thrust.

Electric batteries: Consumes power up to 200 W.

Fuel: Jet fuel.

Endurance: 12 h per flight

Remote controllers: The remote-control station has video facility, to monitor the UAV's flight path, speed, and imaging sprees. Portable GCSs are also available. The remote-control station is usually equipped with interface box, miniground data terminal, ruggedized work stations, and iPads. Remote station can control flights of a few different Aerosonde UAVs, simultaneously, that is, a swarm of Aerosonde UAVs.

Area covered: It transits without interruption or refueling for 140 km (75 knots). It can cover 200–500 ac per flight.

Photographic accessories: Cameras: It supports a set of visual cameras, that is, R, G, and B, multispectral camera, NIR, and IR camera for thermal imagery. The imagery is well integrated for processing, at the GCS's PCs and iPads.

Computer programs: The EO systems are well connected with ground processors. The output from UAV is usually orthomosaics. They can be archived or processed immediately, by the ground-station computers (a touchscreen iPad). Computer software such as Pix4D Mapper is used. The imagery and videos are obtained and relayed, all through the day and night and under all weather conditions.

Spraying area: This is not a sprayer UAV.

Volume: Not applicable

Agricultural Uses: It includes aerial survey of land, topography, soil, and cropping systems. Crop scouting is done for NDVI, canopy characteristics, leaf chlorophyll (NIR sensors), water stress index (IR thermal imagery), disease/pest attack, drought, and flooded regions. Natural vegetation monitoring is also possible.

Nonagricultural uses: It includes geospatial, multispectral and thermal imagery of ground conditions, industrial installations, military installations, pipelines, rail roads, highway traffic, and city public events. Surveillance of mines and ore transport vehicles is also conducted.

Useful References, Websites, and YouTube Addresses

http://www.textronsystems.com/what-we-do/unmanned-systems/aerosonde-commercial (accessed Nov 16, 2016).

https://www.youtube.com/watch?v=lipyL9j05xk (accessed Nov 16, 2016).

https://www.youtube.com/watch?v=XTLmZQIA1pU (accessed Nov 16, 2016).

https://www.youtube.com/watch?v=3-3ZE6d33Nk (accessed Nov 16, 2016).

Name/Models: Apoena 1000

Company and address: XMobots, Rua Santa Cruz, 979, Sao Carlos, SP Brazil; phone: +55 (16) 3413 0655; e-mail: Not available; Website: http://www.xmobots.com

Xmobots is a UAV company initiated in 2007, in the Sao Paulo region of Brazil. They produce UAVs. These UAVs have been developed in collaboration with the Universidade de Sao Paulo and EMBRAPA research scientists. This UAV, Apoena 1000, was first launched in 2008. Apoena is a low altitude, long endurance UAV (Year Book, 2012). Apoena can stay afloat autonomously for 8 h at a stretch, covering 60 km. It has been in use in Amazonia, to monitor rivers, floods, boats, and natural resources.

TECHNICAL SPECIFICATIONS

Material: It is made of carbon fiber, foam, and steel.

Size: length: 2.66 m (8 ft 9 in.): width (wingspan): 2.52 or 3.97 m (8–11 ft) depending on model; height: 1.13 m (3 ft 8 in.); weight: 32–35 kg (MTOW)

Propellers: It has a strong propeller (two blades). Propeller is attached at the nose of the fuselage.

Payload: 10–26 kg depending on model size.

Launching information: It needs a small runway, to launch itself. It has undercarriage of two wheels on which it transits until take-off. Launch, flight, and landing are on autopilot mode, using Pixhawk software. The landing gear is retractable. When in flight, it folds into fuselage.

Altitude: <3000 m (9843 ft a.s.l.).

Speed: The cruising speed 115 km h^{-1} (62 knots)

Wind speed tolerance: 20 km h^{-1}

Temperature tolerance: −10 to +60°C

Power source: A two stroke IC engine generating 5.5 hp power

Electric batteries: Not applicable

Fuel: Petrol or gasoline

Endurance: 8 h

Remote controllers: GCS keeps radiocontact with UAV for up to 1 km distance.

Area covered: It covers about 60 km per flight. About 50–200 ha of crop land can be surveillance per flight.

Photographic accessories: Cameras: It has the usual complement of visual (R, G, and B), multispectral, NIR, IR, and red-edge band with cameras. It also has Lidar.

Computer programs: The UAV-derived images are processed using software such as Pix4D or Agisoft Photoscan or any other similar but custom-made software.

Spraying area: This is not a sprayer UAV. Perhaps with appropriate attachments, it could be converted to perform it.

Volume: Not applicable

Agricultural uses: They include aerial survey of terrain, land resources, fields, soil type, and water resources. Crop scouting for seedling establishment, canopy stand, leaf chlorophyll (crop-N status), canopy temperature, water stress index, and disease/pest attacked locations are the main uses. Natural vegetation monitoring is also possible.

Nonagricultural uses: It includes surveillance of industrial sites, large buildings, oil pipelines, electric transmission lines, rail roads, city traffic, tracking vehicles, monitoring convoys, transport vehicles, etc.

Useful References, Websites, and YouTube Addresses

Year Book. Brazilian UAS Community—Country Reports. *UAS—The Global Perspective*, 9th ed.; Byenburgh and Co., 2012, pp 85/216. http://www.usv-info.com (accessed Nov 26, 2016).

http://www.ebay.com/itm/XMobots-Apoena-1000-Brazil-Unmanned-Aerial-UAV-Aircraft-Desktop-Wood-Model-Small-/152129914784 (accessed Dec 25, 2016).

http://revistapesquisa.fapesp.br/en/2013/10/23/the-flight-of-the-falcon/ (accessed Dec 25, 2016).

https://www.youtube.com/watch?v=3gsAk2YGO1w (accessed Dec 25, 2016).

https://www.youtube.com/watch?v=hZYOBbGClpw (accessed Dec 25, 2016).

https://www.youtube.com/watch?v=-prcpso0buo (accessed Dec 25, 2016).

Name/Models: BAT 4 UAV

Company and address: Martin UAV LLC, 15455 North Dallas Parkway, Suite 1075, Addison, TX 75001, USA; phone: +1 512 520 7170; e-mail:info@martinuav.com; website: http://martinuav.com/uav-products/bat-4/

Martin UAV LLC is a UAV company based at Addison, Texas, USA. It offers a range of options regarding UAV models such as V Bat, Bat D-50, and Bat 4. They are efficient, powered by fuel engine and possess power generators (dynamos). These UAVs are being utilized in a range of locations and situations. They are sold all over, in North America and other continents. The Martin UAV's Bat 4 is a small UAV that has mission capabilities, typically found only in larger UAVs. It can carry a wide range of research payloads up to 20 lb weight (2.33.1).

TECHNICAL SPECIFICATIONS

Material: It is made of carbon fiber, toughened plastic, evelar, and foam.

Size: Total assembled size is 13 × 8 × 3 ft; length: 8 ft; width (wingspan): 13 ft; height: 50 cm; weight: 48 kg (includes 9 kg fuel)

Propellers: It has one propeller at the rear end of fuselage.

Payload: Maximum payload 20 lb (9 kg)

Launching information: It has autonomous DGPS take-off system from a location. Launching needs 150 ft runway. The flight path is regulated by autonomous software-guided system. Landing also needs 150 m runway. It has autoland facility, but ground control-guided landing is also possible.

Altitude: The ceiling is 10,000 ft

Speed: 40–70 knots

Wind speed tolerance: 25 knots

Temperature tolerance: −10 to +60°C

Power source: Internal combustion engine that runs on heavy fuel

Electric batteries: Generator develops 150 W current to run UAV's electronic system.

Fuel: Gasoline + oil (40:1)

Endurance: 6–12 h

Remote controllers: GCS has iPads and PCs. It can regulate flight path and aerial imagery spree, up to 20 km radius. Mission and flight path is regulated and preprogramed, using "Piccolo Command Centre Software."

Area covered: This UAV covers a linear distance of 300 km. It is limited by fuel in the tank. It covers an area within 20 km radius from GCS which is limited by telemetric connections between UAV and GCS. UAV can be flown even with small drizzle in the atmosphere, without affecting telemetry.

Photographic accessories: Cameras: Gimbal that is stabilized digitally holds a set of EO/IR cameras. Generally, sensors include DSLR visual cameras, multispectral and hyperspectral cameras, NIR, IR, and red-edge bandwidth cameras.

Computer programs: Images are preprocessed at the UAV's CPU. Otherwise, the images are relayed as orthomosaics that are tailored, using GPS tags. Then, images are used by the GCS.

Spraying area: This is not a sprayer UAV.

Volume: Not applicable

Agricultural uses: Major functions are aerial mapping of land resources, 3D imagery of farms, crop scouting, procuring data such as NDVI, leaf chlorophyll index, crop-N status, crop water status, disease/pest attack, and flood/drought-affected locations. Imaging natural vegetation and monitoring floral changes are also done.

Nonagricultural uses: It includes surveillance of buildings, industrial installations, rail road, city traffic, and vehicle movement. Patrolling borders, oil pipelines, and wildlife sanctuaries are also possible. Surveillance and providing 3D images of mines, ore distribution, mining progress, ore movement, and dumping sites are other important functions.

PLATE 2.33.1 BAT 4 UAV.
Source: Dr. Blake Sawyer, R.O., and Phil Jones, Martin UAV LLC Addison, Texas, USA.

Useful References, Websites, and YouTube Addresses

http://martinuav.com/uav-products/bat-4/ (accessed June 7, 2017).
https://www.youtube.com/watch?v=PJ_iMtsvBpE (accessed June 7, 2017).
https://www.youtube.com/watch?v=RRPV5-qF3w4 (accessed June 7, 2017).

Name/Models: Brican TD 100

Company and address: Brican Flight Systems Inc., 54 Van Kirk Dr, 54 Van Kirk Dr, Brampton, ON L7A 1B1, Canada

At Brican Flight Systems Inc., they design, develop, and manufacture complete UASs. The TD 100 series of UAVs that they produce can be fully customized, to exact requirements of the farmer. It can be modified to fly in all types of climate. Its data acquisition ability depends on sensors and accessories loaded. The TD 100 series is capable of advanced data collection and reporting to support scientific research and development. Brican Inc. is also capable of managing entire missions on a contractual basis. Training and certification of UAV operators and support personnel are additional services, available through Brican.

The "TD100 UAV" platform has high wing and t-tail configuration. It provides a flexible and stable base platform, for a wide variety of missions. The aircraft provides consistent performance over a wide range of payload, power plants, fuel systems, operating environments, sensor and data recording equipment, and mission profiles. The TD100 can achieve superior performance over a wide range of operating conditions, mainly because it adopts state-of-the-art structural and aerodynamic design components. The TD100 UAVs have several advantages, including advanced structural and aerodynamic advantages. Aircraft design is focused on key criteria and critical flight systems such as weight, dimensions, range, payload, rapid mission, specific reengineering and manufacturing, modularity of interchangeable mounting systems for payload, sensors, and data collection systems. Both electric and Wankel engine versions show up little or no vibration. So, it helps to get quality data. The multifuel engine and flight time of more than 25 h extend range of survey and imagery, proportionately. Fuel efficiency allows for a much lighter carbon footprint.

In general, remote piloting is safer. There is no risk of pilot fatigue, even for long or at-night missions. Low altitude flights enable closer range and more accurate imaging capabilities. Missions can be carried out as low as 300 ft. Launch and recovery systems allow the TD100 to be used in remote areas lacking smooth landing areas. The UAV platform is a modular by design, which lends itself extremely well to interchangeable propulsion systems. Additional features are magnetometer surveys, weather surveying, and high-resolution video/imagery (IR, hyperspectral, etc.).

A complete package of TD100 UAV includes airborne platform, user-defined payload complement, payload integration, launcher and recovery system, ground control and monitoring facilities, user training program, integrated logistics support, plus the 20-lb payload provides the user a full spectrum of sensing and sampling technologies.

TECHNICAL SPECIFICATIONS

Material: The airframe is designed for optimal durability and extended lifespan, even in harsh environments. It is made of light aluminum alloy, carbon fiber, kevlar, and ruggedized plastic. The modular and easily reconfigurable base airframe lends itself well to quickly adapting or repurposing with minimal costs incurred or time consumed for redevelopment or reengineering.

Size: length: 200 cm (79 in.); width (wingspan): 500 cm (196 in.); height: 50 cm (15 in.); weight: 24.5 kg (55 lb)

Propellers: It has one propeller at the tip of the nose. It has three plastic blades.

Payload: The TD100 UAS represents a new genre in flexible, affordable, and remotely operated airborne surveillance systems. In the under 55-lb gross take-off weight (Class 2) class, the TD100 allows 9 kg (20 lb) of free payload capacity. The payload includes all required power source, communications, and control functions necessary, to support and sustain. The cameras include visual, NIR, IR, red-edge, and Lidar sensor. They are all georeferenced and gyrostabilized for vibration-free photography.

Launching information: It is launched using a catapult. It has autolaunch, flight, and landing mode. It uses computer software such as Pixhawk, for autonomous navigation. Landing is accomplished using airbags that absorb shocks to airframe.

Altitude: It ranges from 100 to 400 m a.g.l. for close-up photography of natural vegetation, agricultural zones, and highway patrolling. However, the ceiling is 5500 m a.g.l.

Speed: The cruise speed is 65 km h^{-1}, maximum speed is 100 km h^{-1} (88 knots), and loiter speed is 33 knots.

Wind speed tolerance: 25 km h^{-1}

Temperature tolerance: −20 to +60°C

Power source: Interchangeable between IC engine and electric batteries

Electric batteries: 28 A h high-capacity lithium polymer batteries, four cells.

Fuel: Gasoline + oil (40:1)

Endurance: The TD100 also comes with a range of power plant options. It provides the user, the ability to select between 3 h of ultraquiet, zero-emission performance, all the way up to 30 h of extended operation.

Remote controllers: The remote-control station is "Stanag 4586"

Area covered: This UAV covers a linear distance of 200 nautical miles or equivalent in surface area, as per flight mission. Flight missions can be preprogramed or modified midway, using telemetric connectivity. Radiolink for commands and data download extends up 200 km radius.

Photographic accessories: It includes a wide range of high-resolution conventional cameras, a similar range of IR sensors, integrated multispectral cameras, precision geo-positioning and geo-mapping options for all sensors, integrated weather data collection, magnetometer, and associated geological sensors.

Computer programs: The aerial images can be processed in situ on the UAV or at the ground-station computers (iPads). The software adopted is usually the Pix4D Mapper or Agisoft Photoscan.

Spraying area: This is not used for spraying liquid or granular chemicals. But it could be used, if appropriate tanks and nozzles are fitted.

Volume: Data not available

Agricultural uses: Aerial survey and mapping of land, soil, and water resources, and cropping systems are major uses; monitoring farm operations, vehicle movement, and data collection. During the season, crop scouting for canopy growth, leaf area, leaf-N, fertilizer needs (chlorophyll content), water stress index (through thermal imagery), grain formation and maturity, etc. is possible. The aerial imagery and spectral data can allow us to identify disease/pest attacked zones and weed distribution farms.

Nonagricultural uses: The UAV's utility in nonfarming sector can be grouped as environmental monitoring and management, natural resource surveying, surveillance of mines and mining activity, monitoring transport vehicles, watching national borders, search/rescue and associated emergency management functions, and a few others.

Useful References, Websites, and YouTube Addresses

http://bricanflightsystems.com/services/ (accessed Dec 16, 2016).
http://bricanflightsystems.com/ (accessed Dec 21, 2017).

pdf.aeroexpo.online › Brican Flight Systems (accessed Dec 21, 2017).
https://www.youtube.com/watch?v=Yl7hQXaC9nI (accessed Dec 21, 2017).
https://www.youtube.com/watch?v=uWx3KnCMTEA (accessed Dec 21, 2017).
https://www.youtube.com/watch?v=ipk0wiv4qJg (accessed Dec 21, 2017).

Name/Models: BRV-X

Company and address: BRV UAV and Flight systems, Mogi das Cruzes, State of Sao Paulo, Brazil.

BRV flight systems have developed fixed-winged UAVs for military, civil, and agrarian purposes. This UAV company is operating since 2009 in Brazil. But it caters to many other nations in Latin America and elsewhere. BRV-X and BRV-02 are important flat-wing UAVs produced by the company. BRV-X has been tested and utilized by farm companies and military clients, in many countries (Year Book, 2012). BRV flight systems also provide engineering services and develop prototypes of custom-made UAVs to suit, say, aerial photography of crops, spraying pesticides, application of fertilizers (liquid fertilizers), etc. (Rangel, 2016).

TECHNICAL SPECIFICATIONS

Material: This small UAV is made of a composite derived from aluminum, carbon fiber, plastic, and toughened foam.

Size: length: 75 cm; width (wingspan): 90 cm; height: 35 cm; weight: 3.8 kg

Propellers: It has one propeller placed at the rear end of the fuselage. Propeller has two toughened plastic blades.

Payload: 800 g

Launching information: It has autolaunch, navigation, and landing software such as “Pixhawk Autopilot.” It returns to point of launch in case of emergency.

Altitude: 50–150 m for agricultural survey. However, the ceiling is 3200 m

Speed: 40 mi h^{-1}

Wind speed tolerance: 20 km h^{-1} disturbance

Temperature tolerance: −20 to +55°C

Power source: Electric batteries

Electric batteries: 14,000 mA h.

Fuel: No requirement for petrol or diesel

Endurance: 40 min per flight

Remote controllers: The UAV functions based on directions from software in the ground station. It also has autopilot mode. Software such as Pixhawk Autopilot software helps its navigation, in setting up preprogramed flight paths, and in regulating photographic schedules.

Area covered: It covers about 50–200 ac ground surface per flight.

Photographic accessories: Cameras: The sensors can be programed, using software to automatically collect pictures during aerial survey and photogrammetry. Cameras can also be regulated manually, using ground control (remote controller), to take images only where required. The sensors are placed in a stabilized gimbal. The gimbal avoids excessive vibrations and shocks to sensors. This is to help in obtaining sharp images. The UAV is usually fitted with high-resolution visual cameras (R, G, and B), multispectral camera, NIR, and IR range cameras for thermal imagery and survey of crops and installations.

Computer programs: The GCS adopts custom-made software to process the aerial images relayed, by the UAV. More commonly, software such as Pix4D Mapper or Agisoft Photoscan is used.

Spraying area: This is not a sprayer UAV.

Agricultural uses: Major utilities of this UAV are in aerial survey of land and water resources, crop scouting to assess canopy growth rates, to accrue phenomics data, detect disease/pest attack, and assess water status. Natural vegetation monitoring is also possible (Rangel, 2016).

Nonagricultural uses: Surveillance of industries, buildings, roads, rail roads, city traffic, tracking vehicles, etc. are important functions of this UAV.

Useful References, Websites, and YouTube Addresses

Rangel, R. K. Development of Pest's Biological Control Tool Using VTOL UAV Systems Brvant/BRV UAV & Flight Systems, Brazil. In: *30th International Congress of the Aeronautical Sciences, Track No 09-UAV-operations*; September, 2016, DCC, Daejeon, Korea; pp 1–7. http://icas.dglr.de/icas2016/index.html (accessed Nov 28, 2016).

Year Book. Brazilian UAS Community—Country Reports. *UAS—The Global Perspective*, 9th ed.; Byenburgh and Co., 2012, pp 85/216. http://www.usv-info.com (accessed Nov 26, 2016).

http://www.brvant.com.br/ (accessed Dec 1, 2016).

http://allaboutdrones.spruz.com/link-directory.htm?b=&tagged=BRV+Xn (accessed Dec 12, 2016).

Name/Models: Carolo P 360

Company and address: Institute of Aerospace Systems, Fakultät für Maschinenbau Technological University Braunschweig, Schleinitzstraße 20, D-38106 Braunschweig, Federal State of Brandenburg, Germany; phone: +49 531 391 4040; Tel./Fax: +49 531 391 4044; e-mail: info-fmbtu-braunschweig

The Aerospace Department (Mechanical Engineering) at TU of Braunschweig conducts research and develops models of small UAV (UAVs to suit different functions). Carolo P360 is among the UAV models that suit agricultural survey, rapid visual, and NIR imagery. This UAV is also well suited to conduct atmospheric survey for climatic parameters to detect dust particles, pollution, etc. (Altstadler et al., 2015; Wehren et al., 2016).

TECHNICAL SPECIFICATIONS

Chassis material: It is made of foam, wood, and carbon-ruggedized chassis. This is to withstand rough handling and soil surface during landing (Plate 2.34.1).

Size: length: 1.2 m; width (wingspan): 3.6 m; height: 0.6 m; weight: 22.5 kg (includes battery set)

Propellers: It has one propeller with two plastic blades. Propeller is located at the rear end of the fuselage and facing backward.

Payload: 2.5 kg (includes cameras and CPU)

Launching information: The UAV has rugged body. It is fixed with two wheels mounted on spring gears. It allows UAV to reach higher ground speed during take-off; also, landing is made smoother. Landing gears could be closed (folded) or opened, by the GCS.

Altitude: It is normally used at 400–1000 m a.s.l. for agriculture, but it can reach 3500 m a.s.l.

Speed: 35–45 km h^{-1}

Wind speed tolerance: 45–60 km h^{-1}

Temperature tolerance: −20–60°C

Power source: Electric batteries

Electric batteries: Lithium polymer batteries (each 10 cells and 10 Ah) for the electric motor and 2 lithium polymer batteries (3.55 Ah) for the autopilot system, CPU, and sensor operation.

Fuel: No requirement for petroleum fuel.

Endurance: 40 min

Remote controllers: The GCS consists of "MAVDesk Software." It enables operator to plan and control flight missions for the UAV. The GCS computers are endowed with software to control flight altitude, speed, imaging sprees and timing, etc. GCS computers also receive and store digital data and images, transmitted by UAV's CPU.

Area covered: 50–150 ha per flight

Photographic accessories: Cameras: Multispectral (Tetracam Inc. Chatsworth, CA, USA) and thermal sensors. Camera shutters are usually programed, to open in every 2 s, and cruise speed is regulated at 25 m s^{-1}. Images received are stored on a 2 GB computer memory card.

Computer programs: Digital images are processed and converted, using "Pixel-Wrench2" (Tetracam Inc. Chatsworth, CA, USA) software.

Spraying area: Not a spraying UAV.

Volume: Not applicable

Agricultural uses: It includes crop scouting for seedling establishment, canopy development, and grain formation in crop fields. It is used to accrue data about NDVI, LAI, leaf chlorophyll, and plant-N status. It has been used to monitor carbon cycle in farming regions. It has been used to study species diversity and biomass accumulation in mixed pastures and natural vegetation.

Nonagricultural uses: Surveillance of industries, buildings, oil and gas pipelines, electric installations, rail roads, city traffic, and public events are important uses.

PLATE 2.34.1 ALADINA or Carolo P 360 fixed-winged UAV.
Source: Institute for Aerospace, Technological University Braunschweig, Federal State of Brandenburg, Germany. http://www.drohnen.de/971/forschung-drohne-aladina-carolo-p360-misst-und-untersucht-feinstaub/

Useful References, Websites, and YouTube Addresses

Altstadler, B.; Platis, A.; Wehner, B.; Schultz, A.; Wildmann, N.; Hermann, M.; Kachner, K.; Baars, B.; Bange, J. and Lampert, A. ALADINA—An Unmanned Research Aircraft for Observing Vertical and Horizontal Distributions of Ultrafine Particles within the Atmospheric Boundary Layer. *Atmos. Measure. Tech.* **2015,** *8*, 1627–1639. DOI:10.5194/amt-8-1627-2015

Wehren, M.; Raunker, P. and Sommer, M. UAB-Based Estimation of Carbon Exports from Heterogeneous Soil landscapes—A Case Study from the CarboZalf Experimental Area. *Sensors* **2016,** *16*, 255–276. DOI:10.3390/s16020255.

https://www.tu-braunschweig.de/fmb/papierfliegerweltrekord (accessed Feb 7, 2017).

https://www.tu-braunschweig.de/fmb/studieninteressierte/studiengaenge/lur/index.html (accessed Dec 16, 2016).

http://www.mdpi.com/journal/sensors (accessed Feb 7, 2017).

https://www.youtube.com/watch?v=5caUxjXdAPI (accessed Feb 7, 2017).

https://www.youtube.com/watch?v=5caUxjXdAPI (accessed Feb 7, 2017).

https://www.youtube.com/watch?v=Kxn2QGHXlLo (accessed Feb 7, 2017).

Name/Models: EOS Mini-UAS

Company and address: Threod Systems, Kaare tec 3, 74010 Vimisi, Estonia, Europe; phone: +372 512 1154; e-mail: info@threod.com, sales@threod.com; website: http://www.threod.com

Threod Systems is a high-technology company located in Latvia. It specializes in UASs that are useful, in collection of aerial visual and spectral data about earth's surface and crop land. It also caters to military requirements of continuous surveillance. They have designed a range of fixed-winged small UAVs and multicopters, gimbals, autopilot software, sensors, and processing software (Plate 2.35.1). They produce a range of necessary items for adoption of UAV technology. Therefore, their UAV systems could be adopted more easily. Important UAV models are named "Theia Operational UAS," "Stream Tactical UAS," "EOS-Mini UAS," "K-4 multirotor UAS," and "K-4 Le Rotor UAS."

EOS-Mini UAS is a small UAV. It is launched either using catapult or by parachute. It delivers clear images of the surface features including a range of agricultural aspects. Regarding military, it is used to monitor platoon movement, monitor military installations, and international borders. Major advantages attributed to EOS mini UAS are over 2 h of endurance, low noise and acoustic signatures, excellent aerodynamics, and fully stabilized gimbal and photographic equipment. It offers excellent video depiction of surface features and sharp aerial maps of crops (Plate 2.35.1).

TECHNICAL SPECIFICATIONS

Material: It is made of foam, plastic, and carbon fiber ruggedized to withstand rough weather and handling.

Size: length: 160 cm; width (wingspan): 350 cm; height: 50 cm; weight: 6.9 kg

Propellers: A single propeller at the front tip of fuselage. It has two plastic blades.

Payload: 1.5 kg. It includes EO cameras, with 30× zoom, IR camera, geolocation, geo-tracking, and video-tracking devise to check farm, military, and commercial vehicles and also high-definition onboard recording of surface features and electronic storage. Optional payload includes single-axis photo-camera gimbal for aerial mapping (less than 2 cm/pixel).

Launching information: It has autolaunching and take-off facility and automatic navigation using autopilot software. It has loiter mode, scanning mode, and emergency scouting surveillance modes of flight. It also has autolanding mode.

Altitude: It flies at 50–400 m for aerial mapping of crops and other land features. It has a ceiling of 3500 m.

Speed: 50–100 km h^{-1}

Wind speed tolerance: 15 km h^{-1}

Temperature tolerance: −20 to +60°C

Power source: Electric engine

Electric batteries: 14,000 mA h; two cells

Fuel: No requirement for petroleum fuel.

Endurance: 2 h

Remote controllers: Ground control has iPad for control of UAV's movement and preprograming. It also comes with a joystick that has touchscreen to guide the UAV and conduct imagery. Ground control is equipped with software and iPad for conveying emergency and contingency routes for UAV's path. The UAV is kept in digital contact and telemetric control for up to 17 km radius from the ground control iPads. Ground control (iPad) can also receive encrypted data that could be processed later. It uses GPS, INS, and GLONASS, for locating the features and images.

Area covered: 50–250 ha per flight

Photographic accessories: Cameras: The sensors include visual cameras such as Sony Rx 100, multispectral mica-sense camera, NIR, IR, and red-edge cameras.

Computer programs: The images received could be preprocessed at the CPU, on the UAV itself, using custom-made software. It also transmits orthoimages for processing at ground station. Generally, software such as Agisoft Photoscan or Pix4D Mapper is used, to stich the images.

Spraying area: This is not a sprayer UAV.

Volume: Not applicable

Agricultural uses: Agricultural uses include crop scouting for growth, canopy characteristics, NDVI data collection, leaf chlorophyll estimation, crop-N status determination, and crop water stress index determination. This UAV provides spectral data about disease/insect pest attack.

Nonagricultural uses: It includes surveillance of military installations, vehicle movement, damage assessment in battle fields, offering reconnaissance data to ground control, etc. (Plate 2.35.1). It is used for commercial purposes such as building surveillance and protection, monitoring rail roads, highway traffic, mines, and mining activity.

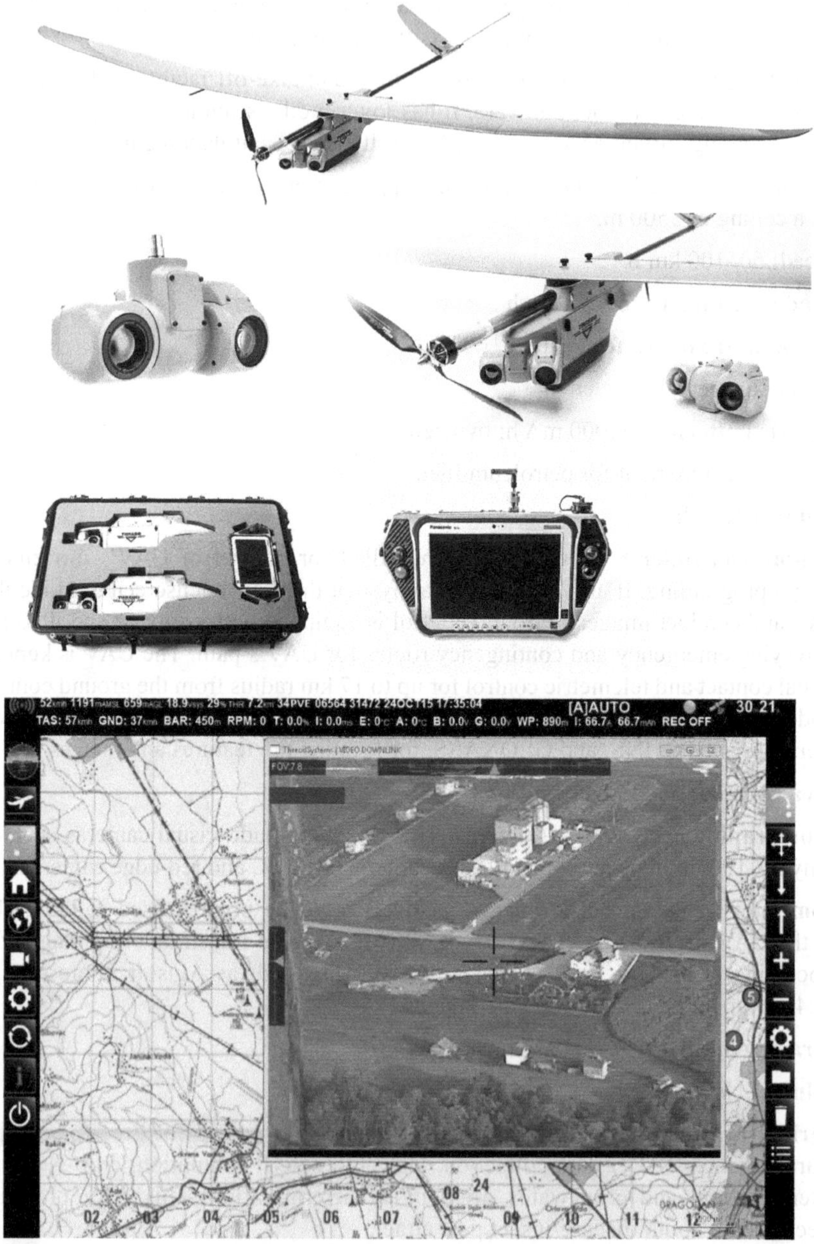

PLATE 2.35.1 Top: A EOS-Mini UAS; upper middle: gimbal and sensors; lower middle: ground control receiver and computer screen; bottom: aerial map and high-resolution image of a farm, fields, and installations.
Source: Mr. Siim Juss, Threod Systems Inc., Vimisi, Latvia, Europe.

Useful References, Websites, and YouTube Addresses

http://www.threod.com/products/uas-eos/description (accessed May 17, 2017).
http://www.threod.com/products/uas-eos/description (accessed May 17, 2017).
https://www.youtube.com/watch?v=dyLHgvh7XCA (accessed May 17, 2017).
https://www.youtube.com/watch?v=MVUBSp05dgA (accessed May 17, 2017).
https://www.youtube.com/watch?v=9xEgPR9AnmY (accessed May 17, 2017).
https://www.youtube.com/watch?v=jE9Kn_Q3eiM (accessed May 17, 2017).
https://www.youtube.com/watch?v=dyLHgvh7XCA (accessed May 17, 2017).
https://www.youtube.com/watch?v=9xEgPR9AnmY (accessed May 17, 2017).
https://www.youtube.com/watch?v=jE9Kn_Q3eiM (accessed May 17, 2017).

Name/Models: Greyhawk

Company and address: RP Flight Systems Inc., Wimberley, TX, USA; phone: +1 512 665 9990, +1 408 832 11195; e-mail: info@rpflightsystems.com, sales@rpflightsystems.com; website: http://www.RPFlightSystems.com

RP Flight Systems is a UAV company started in 2005, at Wimberley, Texas, USA. It specializes in small UAVs with a role in aerial surveillance and mapping ground resources. They produce a series of small UAVs such as "Aerohawk," "Greyhawk," "Vigilant," and "Spectra." A few of these are yet to be finally tested and released for regular commercial use. Recent spurt in use of small UAVs in agriculture could make this aerial robot more popular in USA and other regions.

Greyhawk is a highly streamlined small aerial robot. It is capable of aerial imagery. "Greyhawk" can be assembled in 10 min. There is no need for tools to fix the fuselage, wings, and other parts. UAV's components and parts are all easily joined (Plate 2.36.1).

TECHNICAL SPECIFICATIONS

Material: This UAV is built using carbon fiber, fiber glass, and foam composite. Carbon fiber is used throughout the body, to give rigidity.

Size: length: 160 cm; width (wingspan): 230 cm; height: 45 cm; weight: 10–12 kg

Propellers: It has one propeller at the rear end of the fuselage. The propeller is made of two plastic blades

Payload: 2.8 kg

Launching information: It is launched using a catapult. It can also be launched using bungee.

Altitude: It flies 50–150 m for agricultural and land survey. It can reach up to 3000 m a.g.l.

Speed: 45–60 km h^{-1}

Wind speed tolerance: 15 km h^{-1}

Temperature tolerance: −20 to +60°C

Power source: Electric batteries

Electric batteries: Lithium–polymer batteries, three cells, 14,000 mA h.

Fuel: Not an IC engine-fitted UAV. So, there is no requirement for petroleum fuel.

Endurance: 6–10 h

Remote controllers: The GCS has PCs or iPad supplied with software for preprogramed flight paths and those with ability for modification of flight path, midway. "Pixhawk" software allows autolaunch, autopilot, and autolanding.

Area covered: It covers 50–150 ha per flight.

Photographic accessories: Cameras: This UAV can carry usual complement of sensors such as visual cameras (R, G, and B), multispectral and hyperspectral cameras, NIR, IR, and red-edge cameras.

Computer programs: UAV's imagery is processed, using commonly available software such as Pix4D Mapper or Agisoft.

Spraying area: This is not a sprayer UAV.

Volume: Not applicable

Agricultural uses: Aerial mapping of agricultural zones is a primary function. It is specifically used to study land topography, soil type, water resources, and cropping systems. It could be used regularly for crop scouting, particularly for growth, NDVI data, leaf chlorophyll, and crop-N status. Natural vegetation monitoring is another important function of this UAV.

Nonagricultural uses: Surveillance of military installations, industries and their activity, pipelines, electric lines, mines, transport vehicles, city traffic, and public events.

PLATE 2.36.1 Grey Hawk.
Source: Dr. Gene Robinson, RP Flight Systems, Wimberley, Texas, USA.

Useful References, Websites, and YouTube Addresses

http://www.asiatechdrones.com/_p/prd1/4541470671/product/greyhawk (accessed Mar 12, 2017).
https://player.vimeo.com/video/196220480 (accessed Mar 12, 2017).
https://www.youtube.com/watch?v=ki-cz6BRZFY (accessed Mar 12, 2017).
https://www.youtube.com/watch?v=CTLZ-DmclFc (accessed Mar 12, 2017).

Name/Models: INDELA Berkut

Company and address: KB Indela Ltd., 11, Petra Glebski Street, Minsk 220104, Belarus; phone: +375 17 3638510; mobile +375 29 3951112; e-mail: mail@indela-group.com; website: http://www.indelauav.com

INDELA is a UAV company that offers both fixed-winged and copter versions. They also produce a series of other hardware and software, required for operation of their UAVs. This UAV company was started in 1995, basically, to produce military grade UAVs, plus to help a few other aspects such as aerial imagery, mapping, and general surveillance of skies and ground.

This UAV, "INDELA BERKUT," has turbo jet engine. It is capable of high speeds required for military reconnaissance and target seeking, in the air. It offers excellent service during detection of missiles and fighter jets. However, it can also be utilized for monitoring natural vegetation, wildlife parks, and studying ecological aspects of vegetation and crops. It has a role in agricultural crop surveillance.

TECHNICAL SPECIFICATIONS

Material: It is made of a composite of ruggedized foam, plastic, and carbon fiber.

Size: length: 238 cm; width (wingspan): 280 cm; height: 30 cm; weight: 34 kg (fuel tank may hold up to 8.2 l)

Propellers: It is a turbo jet engine.

Payload: Payload consists of visual, IR, and red-edge cameras. Thermal imagers and Lidar are also present. It can hold 6 kg payload by weight.

Launching information: It is an autolaunch UAV. It moves swiftly on the runway. This UAV lands using parachute. The speed at which parachute should be popped is 110 km h^{-1}.

Altitude: It flies 50–3500 m a.g.l.

Speed: Flight speed is 100–400 km h^{-1}. The cruise speed usually is 150 km h^{-1}. Take-off speed is 90 km h^{-1}

Wind speed tolerance: 30 knots

Temperature tolerance: −25 to +45°C

Power source: Turbo jet engine supplied with gasoline fuel.

Electric batteries: 5 mA h to run electronic aspects such as CPU, etc.

Fuel: Gasoline + oil (40:1). The turbo engine uses 180 mL fuel/min

Endurance: 30 min per flight

Remote controllers: Flight path is autocontrolled, using software such as Pixhawk or custom-made software. The flight navigation is autocontrolled but can be modified, by the ground control center. Ground control iPads/computers keep control for 100 km distance.

Area covered: It covers 60 km linear distance or 150–300 ha per flight.

Photographic accessories: Cameras: The cameras include visual, multispectral, hyperspectral, NIR, IR, and red-edge wave bands. They are primarily meant for military reconnaissance, target identification, and surveillance. The UAV's sensors can be effectively used, for a range of other functions, such as photographing agricultural cropping zones, land resources, water bodies, crop growth, fertilizer needs, etc.

Computer programs: The images from UAV could be either processed at the CPU on the UAV itself or by the ground control center.

Spraying area: Not a spraying UAV.

Volume: Not applicable

Agricultural uses: It includes aerial survey of natural vegetation and crops. The UAV is used to detect canopy status, crop nutritional status (via leaf chlorophyll), water stress index using thermal imagery, etc. Detection and mapping of pest/ disease-affected zones in the crop field are other uses.

Nonagricultural uses: This UAV has been designed and developed to serve military purposes such as tracking military jets, missiles, cruise missiles, intercept jets, and in general, reconnaissance of aerial targets in the sky. The UAV's imagery and mapping abilities can be used, during surveillance of infrastructure, buildings, rivers, dams, oil pipelines, electric installations, etc. It can be used for surveillance of mines, ore movement, mine dumping spots, etc.

Useful References, Websites, and YouTube Addresses

https://en.wikipedia.org/wiki/KB_INDELA (accessed Jan 20, 2017).
https://www.youtube.com/watch?v=HnP_OrUfVZo (accessed Feb 27, 2017).
https://www.youtube.com/watch?v=I5vcLkzVhNw (accessed Feb 27, 2017).

Name/Models: INDELA Strela

Company and address: Independent Development Laboratory Ltd. KB Indela Ltd., 11, Petra Glebski Street, Minsk 220104, Belarus; phone: +375 17 3638510; mobile +375 29 3951112; e-mail: mail@indelagroup.com; website: http://www.indelauav.com

"KB INDELA" is a dynamically developing company. It was founded in 1996. It has an extensive experience in the market of unmanned aircraft. They specialize in development and production of UAVs and multipurpose aircraft systems for various purposes.

INDELA STRELA is a small rapid take-off and swift unmanned aerial vehicle. It has turbo-jet engine fuelled by gasoline. It is useful during military reconnaissance of fighter jets, cruise missiles, enemy UAVs, etc. It can also be used, to monitor big buildings, oil rigs, industries, harbors, oil pipelines and storage facilities, rail roads, highway vehicles, etc. The above uses are in addition to agricultural uses.

TECHNICAL SPECIFICATIONS

Material: The UAV is composed of kevlar, carbon fiber, and ruggedized foam composite. It has landing gear made of steel and rubberized wheels.

Size: length: 238 cm; width (wingspan): 280 cm; height: 30 cm; weight: MTOW is 33 kg. It includes a fuel tank that holds 7.2 L petrol.

Propellers: It is a turbojet.

Payload: 4.5 kg

Launching information: It has autolaunching mode that needs a small runway (300 m). The take-off speed should be 90 km h^{-1}. Its navigation is autonomous and guided by computer software. It lands using a parachute. Maximum speed is 60 m s^{-1} and minimum speed is 15 m s^{-1} for parachute opening. The descent speed is 5 m s^{-1}. Minimum height for parachute release is 1000 m a.g.l. Landing speed is 115 km h^{-1}, if landing gear is used.

Altitude: This UAV generally operates at 50–400 m a.g.l., but the ceiling is 3500 m a.g.l.

Speed: It ranges from 100 to 300 km h^{-1}. Maximum speed is 280 km h^{-1} with open landing gear. General cruise speed is 150 km h^{-1}.

Wind speed tolerance: 30 knots

Temperature tolerance: −25 to +45°C

Power source: Turbo jet engine. It consumes 129–230 mL fuel min^{-1}

Electric batteries: 5–8 mA h to run the onboard computers.

Fuel: Gasoline

Endurance: 20–30 min

Remote controllers: This UAV is controlled from a GCS that has radars, telemetric instruments, PCs, and iPad connected to UAV's CPU. Flight control is automatic. Navigation is done using custom-made software. The UAV keeps contact with GCS for 60 km radius.

Area covered: 400 km per flight or 200 ha area.

Photographic accessories: Cameras: This UAV is equipped with radars, Lidar, high-resolution visual cameras, multispectral cameras, NIR, IR, and mica-sense red-edge cameras.

Computer programs to process orthomosaic: The UAV's imagery can be processed using software such as Pix4D Mapper, Enso Mosaic or Mince, etc.

Spraying area: This is not a sprayer UAV.

Volume: Not applicable

Agricultural uses: It has a series of uses in monitoring and surveillance of natural vegetation and crop fields. It has been used in ecological studies. This UAV is useful in crop scouting for growth, nutritional deficiencies, fertilizer needs, water stress index, disease, pests, and weed infestation.

Nonagricultural uses: This is a UAV used predominantly for military reconnaissance, target detection, identifying cruise missiles, fighter jets, battle tanks, and armored vehicles. It is used for surveillance of large industrial installations, buildings, rail roads, highway vehicles, mines, mining activity, ore movement, etc. Vehicles could also be tracked on highways and crop fields (e.g., tractors, harvest combines), using this UAV.

Useful References, Websites, and YouTube Addresses

http://www.indelauav.com/eng_lang/products.html (accessed Jan 10, 2017).
https://en.wikipedia.org/wiki/KB_INDELA (accessed Jan 30, 2017).
http://archive.li/Ff2pm (accessed Jan 30, 2017).

Name/Models: INDELA 6M

Company and address: KB Indela Ltd., 11, Petra Glebski Street, Minsk 220104, Belarus; phone: +375 17 3638510, mobile +375 29 3951112; e-mail: mail@indela-group.com; website: http://www.indelauav.com

"KB INDELA" is a dynamically developing company. It was founded in 1996. It has an extensive experience in the market of unmanned aircrafts. The company specializes in development and production of UAVs. They also produce aircraft systems for various purposes.

INDELA 6M is a more recent model of UAV offered by the KB INDELA Ltd. It is a robotic reconnaissance aircraft. It has several other uses in aerial surveying, mapping land resources, monitoring crop growth, survey for water resources, etc. It is fitted with an IC engine that runs on gasoline with relatively longer endurance. Modifications to suit agricultural operations are a clear possibility. It could be used in analyzing crop nutrient status, fertilizer needs, water status water needs, etc. Detecting weeds, diseases, and pests are also possible, using spectral signatures.

TECHNICAL SPECIFICATIONS

Material: It is made of a composite of aluminum, plastic, carbon fiber, and foam.

Size: length: 270 cm; width (wingspan): 601 cm; wing load: 235 g dm^{-2}; height: 33 cm; weight: take-off weight is 35 kg fuel weight equivalent of 7 L.

Propellers: It has a strong plastic propeller with two blades. Propellers are powered by IC engine.

Payload: Payload includes a series of sensors such as visual bandwidth cameras, multi-spectral, hyperspectral, NIR, IR, and mica-sense red-edge cameras. It also has Lidar facility and radar to detect fighters, cruise missiles, and other UAVs. However, these same cameras can offer excellent images and spectral data of agricultural crop land.

Launching information: It has automatic launching gear with three wheels. The UAV navigates autonomously based on preprogramed instructions, using software such as Pixhawk. Landing is achieved using landing gear, but, in times of emergency, parachute can be opened and safe landing could be made. Parachute is to be opened at least 100 m a.g.l. (Barometric height is <2000 m a.g.l.).

Altitude: General operations are conducted at 50–3500 m a.g.l.

Speed: Horizontal flight speed is 65–140 km h^{-1}, cruise speed is 85 km h^{-1}, landing speed is 35 km h^{-1}; and contact velocity is 60 km h^{-1}

Wind speed tolerance: 30 knots

Temperature tolerance: −25 to +45°C

Power source: internal combustion engine—two strokes, two cylinder engines (opposite). About 56 cm^3 is the volume of cylinders. Engine generates 4.0 kW power. It consumes 680 mL h^{-1} petrol.

Electric batteries: It has 5 mA h lithium–polymer batteries to support computers onboard.

Fuel: Gasoline + oil

Endurance: 5 h

Remote controllers: Flight path is controlled both via autonomous preprogramed instructions or dual, using GCS commands relayed to the UAV's computers. This UAV accomplishes a series of flight tasks and procedures autonomously. It picks aerial imagery and it keeps contact with GCS using both radio telemetry and internet. It sends videography to GCS. It waits for telemetric commands from the GCS. The communication rage is 50 km for video information and 100 km for telemetric contact and instructions.

Area covered: This UAV covers 440 km linear distance or 250–500 ha per flight.

Photographic accessories: Cameras: This UAV has gyro-stabilized gimbal that holds sensors in vibration-free condition. So, it offers haze/vibration-free imagery. It has INDELA OGD@—HIR system. The UAV carries a series of cameras for visual imagery, NIR, IR and thermal imagery, Lidar, etc.

Computer programs: The UAV's CPU has facility for immediate processing of images that are picked by sensors. It is also processed at the GCS, using computer software such as Agisoft or Pix 4D Mapper, using iPads.

Spraying area: This is not a spraying UAV, but it should be possible to convert it to spraying UAV, by attaching tanks and nozzles.

Volume: Not applicable

Agricultural uses: This UAV, with its array of cameras, can be effectively used, to procure aerial images and digital data about agricultural zones, mainly land resources, soil, water resources, crops, etc. It is also useful in crop scouting for growth, canopy characters, leaf/crop-N status, disease and pest attack, weed infestation, etc. Crop yield forecasting is also a possibility.

Nonagricultural uses: This UAV is used in monitoring of territories, ground features, and monitoring of power lines. It is used to prevent or control fire. It offers video information about city traffic. It is useful in prospecting ores and minerals. Surveillance of mines, mining activity, ore transport, and dumping zones are other functions.

Useful References, Websites, and YouTube Addresses

http://en.avia.pro/blog/indela-6m (accessed Oct 20, 2017).
http://www.indelauav.com/eng_lang/products.html (accessed Oct 20, 2017).
https://en.wikipedia.org/wiki/KB_INDELA (accessed Oct 20, 2017).
http://archive.li/Ff2pm (accessed Oct 20, 2017).
http://ur.avia.pro/blog/indela-6m (accessed Oct 20, 2017).
https://www.youtube.com/watch?v=gMZL3o4rrBE (accessed Oct 20, 2017).

Name/Models: INDELA 9M

Company and address: KB Indela Ltd., 11, Petra Glebski Street, Minsk 220104, Belarus; phone: +375 17 3638510, mobile +375 29 3951112; e-mail: mail@indela-group.com; website: http://www.indelauav.com

"KB INDELA" is an isodynamically developing company. It was founded in 1996. This company has experience in the market of unmanned aircraft. They specialize in development and production of UAVs and multipurpose aircraft systems for various purposes.

INDELA 9M is a slightly heavier, but still a small, UAV meant to accomplish a wide range of functions in military, commercial, and agricultural arena. It has been developed as a robotic airborne reconnaissance vehicle, if military option is considered. However, it serves to monitor and regulate various civilian and agricultural activities, through its aerial imagery and digital information sent to GCSs. This UAV stores aerial imagery in a black box. This UAV is used in Western European and Baltic region.

TECHNICAL SPECIFICATIONS

Material: This UAV is made of a composite of ruggedized foam, aluminum, plastic, and carbon fiber. It has rubber wheels in the landing gear.

Size: length: 340 cm; width (wingspan): 910 cm; height: 45 cm; weight: MTOW 45 kg

Propellers: It has one propeller. It is attached at the rear end of the fuselage. It is fixed at a slightly elevated location, using a stub at the end of fuselage. Propeller is powered either by an IC engine or electric batteries depending on the fixtures.

Payload: Maximum payload is 8 kg, excluding fuel and airframe weight. Payload usually includes gimbal with sensors, Lidar, radar, etc.

Launching information: INDELA 9M is launched using a runway (300 m length). It has software for autonomous launching and navigation during flight. Landing using gear with wheels is the general procedure. However, in times of emergency, like aircraft, there is also a parachute-landing facility.

Altitude: The operational height is 50 m, but ceiling is 3500 m a.g.l. (Barometric height feasible is <2000 m a.g.l.).

Speed: Horizontal speed is 55–140 km h^{-1}, cruise speed is 90 km h^{-1}, take-off speed is 51 km h^{-1}, and landing speed using the gear is 60 km h^{-1}

Wind speed tolerance: 30 knots

Temperature tolerance: −25 to +45°C

Power source: Electric engine with a power generation of 12 kW and an operating voltage of 45.8 V. This UAV comes with an option for IC engine (instead of electric motors). The IC engine-fitted is a 2-stroke 2-cylinder. It has a cylinder capacity of 97 cm^3. The IC engine generates 7.6 kW.

Electric batteries: 5–8 mA h for onboard computers and telemetric functions.

Fuel: IC engine uses gasoline + oil at a rate of 2.5 L h^{-1}

Endurance: 1.5 h if fitted with electric engine; 6 h if fitted with IC engine

Remote controllers: The GCS has series of monitors, telemetric connection equipment, and iPads, for sending command via internet. This UAV is both automatic and a dual control vehicle. The communication range is 25 km with video signals and internet connectivity. The telemetric connections operate up to 64 km radius. However, it operates up to 100 km, if telemetric connections are in a UAV model with IC engine.

Area covered: Linear distance covered by the UAV is 90 km per flight with electric engine. Area covered is 360 km^2 per flight with IC engine.

Photographic accessories: Cameras: This UAV has been fitted with a series of cameras such as high-resolution visual bandwidth cameras (R, G, and B), hyperspectral and multispectral cameras, NIR, IR, and red-edge cameras. INDELA 9M has fully integrated, gyro-stabilized gimbal. The gimbal that holds these cameras helps in offering vibration-free and haze-free images.

Computer programs: The raw imagery from UAV's sensors is processed using custom developed INDELA software or Agisoft Photoscan or Pix4D Mapper. The UAV can also process the images immediately, using software in the CPU on UAV itself and relay the images.

Spraying area: This is not a sprayer UAV, but it could be fitted with tanks to hold pesticides and fertilizers (liquid) and nozzles to spray.

Volume: Not applicable

Agricultural uses: This UAV, INDELA 9M could be utilized effectively, to conduct aerial survey of agricultural zones. The survey could be focused to provide images of land resources, topography, soil-type distribution, cropping pattern, crop growth characteristics, etc. Crop scouting for canopy traits, leaf-N, crop's fertilizer needs (fertilizer-N), water stress index maps, disease and pest attack maps, and drought/flood-affected zones is a clear possibility. Natural vegetation could be mapped and monitored, using this UAV. It is used to get NDVI values, study vegetational changes, and species diversity, if suitable spectral signature data bank is available.

Nonagricultural uses: INDELA 9M is a versatile UAV. It has series of useful functions in aerial surveillance, mapping, digital data collection of military installations, buildings, oil pipelines, highways and vehicular traffic, public events, mines, mining activity, ore movement and dumping, etc.

Useful References, Websites, and YouTube Addresses

http://www.indelauav.com/eng_lang/product_indela9.html (accessed Oct 12, 2017).
http://en.avia.pro/blog/indela-9 (accessed Oct 12, 2017).
https://www.youtube.com/watch?v=uCIDYHZ5QGE (accessed Oct 12, 2017).

Name/Models: Penguin C UAS

Company and address: UAV Factory Ltd. 1, Jaunbridagi, Marupe, Latvia LV-2167; phone: +371 29191590; e-mail: support@uavfactory.com; website: http://www.uavfactory.com/contacts

UAV Factory USA LLC. 750 NW Charbonneau street, Bend, Oregon, USA. Phone No + 1 (914) 591-3296

This UAV Factory was established in 2009 at Marupe, Latvia. Soon after its founding, the UAV factory was producing unmanned aircraft, pneumatic catapults, onboard generator systems, and portable GCSs. UAV Factory Ltd. is one of the leading developers of fixed-wing composite airframes, subsystems, and accessories, for small fixed-wing unmanned aircraft industry. UAVs based on their airframes and subsystems are delivered to over 47 countries. They manufacture UAVs in class of up to 25 kg (55 lb).

Penguin C is a long-endurance and long-distance UAV. It can cover large regions of agricultural land and other vast features on the surface of earth. It is equipped with appropriately matching sensors, to produce accurate imagery of land. The Penguin could be assembled and launched within few minutes. It comes with a portable launching and GCS (Plate 2.37.1). So far, Penguin C has been tried and adopted for surveillance, commercial, and agricultural purposes, in over 43 countries.

TECHNICAL SPECIFICATIONS

Material: It is made of a composite of carbon fiber, plastic, and steel.

Size: length: 1.9 m; width (wingspan): 3.3 m (10.8 ft); height: 45 cm; weight: MTOW 22.5 kg (50 lb). The entire UAVS system, including the launcher, can be transported in a minivan weighs about 265 kg (585 lb).

Engine: Penguin C has a fuel inject engine with electronic controls and remote contact. It exceeds an endurance of 20 h. Record timing is 54 h endurance during flight. The engine operates at wide range of temperature from freezing to very hot conditions, since it is supplied with suitable cooling fluid. The engine comes with a swappable propulsion module. The engine is supplied with a silent muffler (silencer) that reduces loud acoustic signatures.

Propellers: This UAV has one propeller with two blades. The propeller is placed at the rear end of the fuselage and facing backward.

Payload: Day and night gyro-stabilized gimbal holds cameras. The UAV allows a payload of 3.2 kg.

Launching information: Penguin C is launched using an autonomous catapult. The catapult is portable and lasts for over 10,000 cycles of launching. The catapult system is portable pneumatic with 6000 J launch energy. The UAV is generally on autopilot mode during flight. Landing is conducted using parachute. It pops out from the UAV itself. This avoids cumbersome landing nets and landing wires. Parachute landing reduces the risk of damage, to UAV and payload cameras.

Altitude: 4500 m a.g.l.

Speed: 32 m s^{-1} or 62 knots or 70 km h^{-1}

Wind speed tolerance: 20–25 km h^{-1}

Temperature tolerance: −25 to +60°C. It has anti-icing features to operate in Arctic zone.

Power source: Fuel engine: a 28-cm^3 fuel inject IC engine.

Electric batteries: 100 W generator system. There is also emergency battery power source. It offers 24 V power supply.

Fuel: 98 Octane and lubricant oil mix

Endurance: 20 h

Remote controllers: The GCS has user friendly, software, and portable systems with PCs, touchscreen iPad, and mobile systems. Flight control is done, using "Piccolo cloud cap technology." The ground control keeps contact with UAV for 100 km radius (60 miles radius). The imagery and commands can be transmitted at a rate of 12 Mbps. The tracking antenna and control systems are ruggedized, even to withstand military usage (Plate 2. 37.1.).

Area covered: It covers 500–1000 ha per flight above crop land

Photographic accessories: Cameras: Penguin C is equipped with advanced Epsilon series gimbal. The gimbal allows highly versatile tactical imagery. It offers advanced observation, inspection, and surveillance capabilities, during day and night. The cameras include EO sensor—Sony EV7500, HD 720p; EO sensor—Sony EV 7300; IR sensor; EO Sensor Sony H11. It has onboard videotracking and recording cameras. Onboard video-enhancing facility and electronic tracking devices help to track military or agricultural vehicles. The UAV is attached with videographing products such as "Epsilon Tractor tracking device," "Epsilon Moving target indictor," and long-range tacking device. The sensors that track another UAV are also fitted into the UAV. This UAV also comes with ability, to interface with another UAV and its cameras in the payload.

Computer programs: Image processing is done using software in the CPU of the UAV. The GCS computers adopt Pix4D Mapper or Agisoft or similar software, to process the orthomosaics rapidly. All the imagery is GPS or GNSS tagged for accuracy.

Spraying area: This is not a sprayer UAV.

Volume: Not applicable

Agricultural uses: This UAV with its long-range and endurance traits can be used, to surveillance crop land. The sensors are highly sophisticated. They allow surveillance and imagery of crops for NDVI, canopy characteristics, leaf chlorophyll, leaf-N, crop-N status, soil moisture, crop water stress irrigation, etc. The UAV's imagery could be utilized, to decide fertilizer-N inputs and irrigation schedules. The UAV's imagery could be used to study disease/pest progress and take remedial measures. This UAV is used for general crop scouting and natural vegetation monitoring. In the field, UAV can be used to monitor and track all farm vehicles. We can monitor their movement and rate of operations such as plowing, seeding, pesticide application, etc.

Nonagricultural uses: Penguin C is apt for military surveillance, particularly for tracking vehicle and to keep watch on the happenings in the military barracks. Its long endurance allows continuous surveillance. This UAV is also used to surveillance and track commercial vehicles on the highways, or in a city. Monitoring pipelines, rail roads, mining activity are other uses of the UAV.

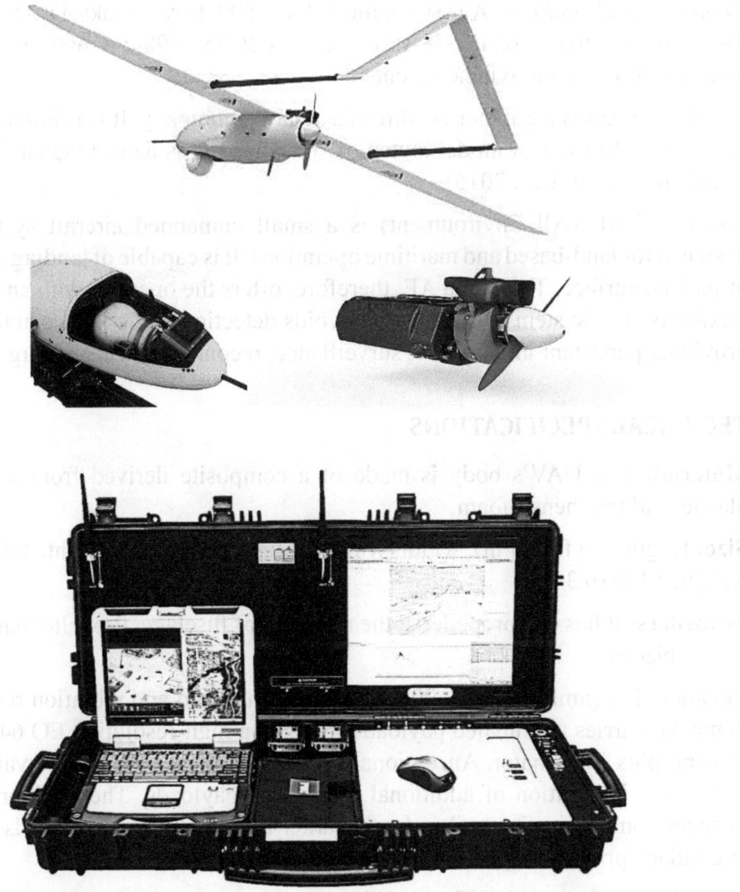

PLATE 2.37.1 Top: An overview of the UAV—"Penguin C"; middle left: nose/cockpit is with gimbal, CPU, and ground control interface software; middle right: the propeller with two blades. Bottom: A ground-control kit for "Penguin-C" UAV.

Source: Mr. Alexanders Manajenkovs and Mr. Konstantin, UAV Factory Ltd. Jaunbridagi, Marupe, Latvia.

Useful References, Websites, and YouTube Addresses

http://www.uavfactory.com/contacts (accessed Aug 12, 2017).
http://www.uavfactory.com/product/74 (accessed Aug 12, 2017).
http://www.uavfactory.com/page/gallery/video/all (accessed Aug 12, 2017).
https://www.youtube.com/watch?v=gOb4xJMUL8U (accessed Aug 12, 2017).
https://www.youtube.com/watch?v=gMjP1NKQkfE (accessed Aug 12, 2017).
https://www.youtube.com/watch?v=_RR_YoDflq8&feature=youtu.be (accessed Aug 16, 2017).
https://www.youtube.com/watch?v=HB1wxdnbc3Y (accessed Aug. 18, 2017).
https://www.youtube.com/watch?v=_RR_YoDflq8&feature=youtu.be (accessed Aug 18, 2017).

Name/Models: PUMA AE RQ-20B

Company and address: AeroVironment Inc., 800 Royal Oaks Drive, Suite 210, Monrovia, CA 91016 6347, USA; phone: +1 626 357 9983; e-mail: ir@avinc.com; website: https://www.avinc.com/contact

AeroVironment is a Southern California aviation company. It was initiated some 40 years ago. Most recent model touted by the company is named "Quantix (VTOL)" (AeroVironement Inc., 2016).

The Puma AE (All Environment) is a small unmanned aircraft system (UAS) designed for land-based and maritime operations. It is capable of landing on the water or on land surface. The Puma AE, therefore, offers the operator with an operational flexibility. The system is quiet. So, it avoids detection and operates autonomously, providing persistent intelligence, surveillance, reconnaissance, and targeting data.

TECHNICAL SPECIFICATIONS

Material: The UAV's body is made of a composite derived from carbon fiber, plastic, and toughened foam.

Size: length: 4.6 ft (1.4 m); width (wingspan): 9.2 ft (2.8 m); height: 1.5 ft (45 cm); weight: 14 lb (6.3 kg)

Propellers: It has one propeller at the nose end of fuselage. Propeller has two tough plastic blades

Payload: The gimbal tilts 0°–90° angle. It is stabilized and vibration resistant. The Puma AE carries a gimbaled payload including a high-resolution EO 640 × 480 IR camera, plus illuminator. An optional underwing "Transit Bay" is available. This is for easy integration of additional third-party payloads. The additional payload includes communications relay, laser marker to meet the diverse needs of military or civilian applications.

Launching information: The UAV has autolaunch facility. It is launched by hand. The UAV is held at shoulder height for launch. It could also be launched using a railed catapult. UAV is held on autopilot mode, using the software "Pixhawk." The Puma AE is durable due to a reinforced fuselage construction. It is portable. It requires no auxiliary equipment for launch or recovery operations.

Altitude: 500 ft (152 m) a.g.l.

Speed: 45 km h^{-1}

Wind speed tolerance: 15–20 km h^{-1}

Temperature tolerance: −20 to +60°C

Power source: Electric batteries

Electric batteries: Lithium polymer batteries 14,000 mA h.

Fuel: Not an IC engine-supported UAV. So, it has no requirement for petroleum fuel.

Endurance: 3+ h with an LE battery

Remote controllers: The GCS keeps contact with UAV for up to 15 km radius. The ground station can alter flight path based on requirements. AeroVironment's common GCS allows the operator to control the aircraft manually. We may also program it for GPS-based autonomous navigation using operator-designated waypoints. It has enhanced precision navigation system with secondary GPS. It provides greater positional accuracy and reliability of the system (AeroVironment Inc., 2016). The UAV comes with an AeroVironment custom-made decision support system (DSS). It has facility for rapid and intelligent data processing.

Area covered: It covers 50–200 ha in farms, crop land, and open space.

Photographic accessories: Cameras: The electrooptical component is usually a set of visual bandwidth cameras (red, green, and blue). The sensors also include NIR and IR bandwidth imagers, also red-edge cameras. It also carries Lidar for rapid assessment of forest growth, etc. The "Manits 145" gimbal fitted to Puma is compact. It allows light-weight cameras of wide bandwidth range.

Computer programs: The images are processed using the Pix4D Mapper and Agisoft Photoscan software. The processed images could be relayed to ground-station iPad/PCs. Alternatively, orthomosaics can be transmitted to ground-station computers, for further processing and analysis.

Spraying area: Not a spraying UAV.

Volume: Not applicable

Agricultural uses: Puma could be utilized in farms worldwide for a range of activities and crop analysis. Natural vegetation monitoring is also possible.

Nonagricultural uses: This UAV is well adapted to surveillance military installations, vehicle movement, and activity on ground. It is used in aerial mapping of geophysical features of earth, in geology and mining sites. It could be used extensively in monitoring oil pipelines (e.g., in Alaska), despite unfavorable weather conditions. It is useful in monitoring highway traffic, vehicle movement in cities, etc. Puma can also be used, to image and relay information about disasters, floods, droughts, soil-erosion-affected spots, etc. It can be used to watch public buildings, events, and city traffic.

Useful References, Websites, and YouTube Addresses

AeroVironment. *About Us—AeroVironement*; 2016; pp 1–5; AV-AeroVironment Inc.: California, USA. http://www.avinc.com/about (accessed Dec 3, 2016).

https//www.avinc.com/uas/small_uas/puma (accessed Dec 14, 2016).

https://www.avinc.com/uas/small_uas/puma (accessed Dec 14, 2016).

https://www.avinc.com/uas/view/puma (accessed Dec. 14, 2016).

http://www.army-technology.com/projects/puma-unmanned-aircraft-system-us/ (accessed Dec 14, 2016).

https://www.youtube.com/watch?v=uz4NLHUdmJE (accessed Jan 23, 2017).

https://www.youtube.com/watch?v=e27xjlEnVmg (accessed Jan 23, 2017).

https://www.youtube.com/watch?v=8RMB9WMdG6g (accessed Jan 23, 2017).

Name/Models: ScanEagle

Company and address: Boeings INSITU, 118, East Columbia River Way, Bingen, WA 98605, USA; Tel.: +1 509 493 5600; fax: +1 509 493 5601; e-mail: not available; https://insitu.com/information-delivery/service

INSITU is a subsidiary of Boeing Company, Bingen, Washington, USA. It produces the "ScanEagle" that is predominantly a military UAV. However, it is also a civilian surveillance and reconnaissance UAV. ScanEagle was designed and developed since 1994 but released in February 2002, by the Boeing Company (INSITU, 2016). ScanEagle was first deployed in 2004 by the US Marine Corps and US Navy. There have been several upgradations in terms of its engine, acoustic characteristics, and utility (Boeing International, 2014). Recent models have been tailored to suit even agricultural roles, such as aerial mapping of crops, measuring NDVI, crop water stress index via thermal imagery, estimating crop loss, and forecasting yield. ScanEagle X200 is a civilian variant of the original military grade UAV. It has been used to detect weed population in crop fields. It has been adopted in aerial mapping of crops and in detecting variations in crop productivity (soil fertility). This helps in prescribing fertilizers, water, and even pesticides based on precision farming principle. Commercial imagings of natural resources, city traffic, public functions, disaster zones, mines, pipelines, etc. are other functions possible with ScanEagle X200 or ScanEagle 2.

TECHNICAL SPECIFICATIONS

Material: It is made of metal alloy, plastic, and foam.

Size: Length: 5.1 ft (1.55 m); width (wingspan): 10.2 ft (3.11) height: 3.2 ft; weight: empty structure weight is 30.1–40 lb (14–18 kg), maximum weight at take-off is 48.5 lb (or 22 kg)

Propellers: It has one propeller on the fuselage. Propeller is placed at the center of wings and facing backward. It is powered by an IC engine of 1.5 hp.

Payload: Maximum payload is 7.5 lb (or 3.4 kg)

Launching information: This UAV is relatively heavy. It is launched using a "Mark 4 launcher." The launcher has an extraordinary design compatible with a range of heavy UAVs. Mark 4 launcher operates in all weather conditions and on rugged terrain. "ScanEagle" is a versatile UAV. It withstands a wide range of temperatures, from −18 to 50°C. The "Mark 4 launchers" come with an onboard diesel generator and air compressor. The total weight of Mark 4 launcher is 4200 lb (1950 kg) and it is pulled by a trailer. The "Mark 4 launcher" is usually kept mounted on a trailer.

Altitude: It reaches height of 19,500 ft or 5945 m a.s.l. The ScanEagle UAV is recovered, using a "Sky Hook recovery system." This allows versatile launch and recovery system.

Speed: Maximum horizontal speed is 80 knots (or 148 mi h^{-1} or 92 mi h^{-1}); cruise speed is 50–60 knots

Wind speed tolerance: Tolerates strong winds of 60 km h^{-1}

Temperature tolerance: −18 to +50°C

Power source: ScanEagle is attached with 1× orbital 3 W 2-stroke engine of 1.5 hp. This is used to drive the propeller.

Electric batteries: Not an electric engine-supported UAV, but onboard electric power is provided by 60 W batteries.

Fuel: UAV is powered by an IC engine that operates on gasoline.

Endurance: 24 h

Remote controllers: The ground control has a series of computers and iPads, to regulate, guide, and recover the UAV. The GCS keeps in touch with UAV via radio-link for over 2000 km from the GCS.

Area covered: It depends on fuel, endurance and intended flight path for either military or agricultural tasks.

Photographic accessories: Cameras: Sensors are placed in a stabilized turret that avoids vibrations. Electrooptic imager-EO 900 (two imagers) and mid-wave infrared (MWIR) dual imager are available. EO900 offer high-definition telescopic images with great precision and GPS coordinates. It is videolinked. It allows digital data and images to be transmitted, in encrypted form. The UAV is constantly linked to ground-station computers for instantaneous transmission of images. Imagery is done both during day and night. Quality thermal imagery could be done during night time too.

Computer programs: For agricultural and natural resource monitoring purposes, the UAV-derived images are processed immediately, using Pix4D Mapper or Agisoft Photoscan or similar software.

Spraying area: Usually, this is not a spraying UAV. It is flown at higher altitudes for surveillance and sharp imagery.

Volume: Not applicable

Agricultural uses: This is basically a military UAV. However, it can now be adopted for use in large-scale agricultural surveillance and studies. This is a highly pertinent UAV for rapid assessment of large agricultural belts. ScanEagle is used to study natural resources relevant to agriculture. It is used photograph cropping systems on a large scale, river projects, disease/pest spread, drought and flood effects on a wider scale, etc. Images drawn from higher altitude are essential for policymaking and planning crop-production strategies, on a larger scale, say a county. Scouting large expanses and natural vegetation monitoring is possible. "ScanEagle X200" has been used in farms, adopting precision agricultural techniques (INSITU, 2016; Koski et al., 2009).

Nonagricultural uses: It is an important military surveillance, spying, and combat UAV. ScanEagle has been used during first response to disasters, military combat, border security exercises, and asset protection. It is also used during fire-fighting, that is, locating fire in industrial location assets. ScanEagle is also used to monitor wildlife and historical monuments. Surveillance of mines, geological sites, and transport vehicles is also done using ScanEagle UAV.

Useful References, Websites, and YouTube Addresses

Boeing International. *ScanEagle Unmanned Aerial Vehicle*; Boeing Company: Washington, USA, pp 1–2.

INSITU. *ScanEagle Unmanned Aircraft Systems*; INSITU Pacific, 2016; pp 1–3. http://insitupacific.com.au/agriculture/ (accessed Nov 30, 2016).

Koski, W. R.; Abrll, P.; Yazvenko, B. A Review and Inventory of Unmanned Aerial Systems for Detection and Monitoring of Key Biological Resources and Physical Parameters Affecting Marine Life during Offshore Exploration and Production Activities. In *A Review and Inventory of Unmanned Aerial Systems*; LGL Ld., Ed.; E and P Sound and Marinelife Program, 2009; pp 1–32. http://www.soundmrinelife.org (accessed June 17, 2017).

https://insitu.com/information-delivery/hardware (accessed Nov 3, 2016).

https://insitu.com/information-delivery/unmanned-systems/scaneagle (accessed Nov 3, 2016).

http://theuavdigest.com/tag/scaneagle/

http://insitupacific.com.au/agriculture/

https://www.youtube.com/watch?v=6wSQSDi__-Q (accessed Nov 3, 2016).

https://www.youtube.com/watch?v=6wSQSDi__-Q (accessed Nov 30, 2016).

Name/Models: Sky Robot Fx 20

Company and address: Robot Aviation, Brandbu, Grinakerlinna 115N-2760 Brandbu, Norway; Phone: +47 994 30 582; e-mail: info@robotaviation.com; website: http://www.robotaviation.com/applications

Robot Aviation AS is a Scandinavian UAV company. It produces a range of fixed-wing and copter UAVs. This company caters to applications such as military surveillance, general inspection and security of large industrial installations, buildings, etc. Some models of UAVs are more suitable for monitoring natural vegetation, agricultural zones, and irrigation channels. Robot Aviation As offers a completely integrated UAV system. It develops and manufactures custom-made software, suitable cameras/sensors, and processing units (iPads).

Sky Robot Fx20 is a smaller, flat, fixed-winged UAV model produced by Robot Aviation As, Norway. It is supported by an electrical engine that generates 3.0 kW power. It has a range of applications in surveillance of agricultural farms, farm buildings, and crop fields. It could be effectively used to obtain aerial images of land resources. It can also be used to track and notify farm vehicle movement in the fields. Sky Robot Fx 20 is cost-effective. It is a high-performance UAV, particularly for farmers and other civilian users.

TECHNICAL SPECIFICATIONS

Material: The airframe is made of ruggedized foam, plastic, and carbon fiber

Size: length: 0.95 m; width (wingspan): 3.0 m height; weight: 6.5 kg

Propellers: It has one propeller at the end of the fuselage. Propeller has two plastic blades. It is connected to lithium polymer battery-powered motors.

Payload: 3.0 kg

Launching information: It takes off on a catapult launcher or bungee catapult. It has autonomous navigation and flight facility. It operates using Pixhawk software. It avoids obstacles automatically while on flight. It lands at the launch point, in case of emergencies. Landing is via parachute or belly landing.

Altitude: It operates at 50–150 m for agricultural crop surveys. However, the ceiling is 5000 m a.g.l.

Speed: 45–60 km h^{-1}

Wind speed tolerance: 25 knots

Temperature tolerance: −25 to +45°C

Power source: Electrical engine that generates 3 kW using lithium polymer batteries

Electric batteries: Lithium polymer batteries are of 14,000 mA h; two cells.

Fuel: Not an IC engine-fitted UAV. So, there is no requirement for petroleum fuel.

Endurance: 3 h

Remote controllers: This UAV keeps contact with GCS, using 3G and 4G communication modes. The UAV's flight mission is restricted to 200 km radius. The UAV keeps in contact with GCS, using telemetry and internet, for up to 200 km radius.

Area covered: It covers 300–500 ha of aerial imagery per flight

Photographic accessories: Cameras: It carries a usual complement of cameras such as high-resolution visual cameras, multispectral, hyperspectral, IR, red-edge, and Lidar. The gimbal that holds the sensors is gyro-stabilized. Therefore, we get vibration free high-resolution accurate images with GPS/GLONASS tags.

Computer programs to process orthomosaic: The images received from the UAV are processed using the software such as Pix4D Mapper or Agisoft's Photoscan. Similar other software are also compatible for use.

Spraying area: This is not a spraying UAV.

Volume: Not applicable

Agricultural uses: Sky Robot Fx 20 is relatively a small light-weight UAV. It fits well to agricultural farm operations, conditions, and requirements. This UAV with its sensor provides excellent aerial imagery about natural resources, land, soil type, cropping system, and crop density. Field scouting is done for seed germination, seedling establishment, plant stand, canopy growth, leaf area, leaf-N, crop-N status, crop water stress, drought-affected regions, floods and water stagnation in fields, soil erosion, etc. Surveillance of disease and pest attack and weed infestation can be performed regularly, using this UAV.

Nonagricultural uses: It includes aerial surveillance of military installations, vehicles, and tracking them electronically. Surveillance of large industries, buildings, highway patrolling, monitoring vehicle movement and tracking them, monitoring oil pipelines, electric power lines/installations, natural and international borders, etc. are other functions of this UAV.

Useful References, Websites, and YouTube Addresses

http://www.robotaviation.com/ (accessed Mar 12, 2017).

http://skyskopes.com/skyskopes-and-robot-aviation-partner-to-collect-long-endurance-uas-test-data-2/ (accessed Mar 24, 2017).

Name/Models: Spy Owl 200

Company and address: UAS Europe AB, Torvingegatan 13, 58278 Linkoping, Sweden; e-mail: info@uas-europe.com, sales@uas-europe.com; phone +46 (0) 13 560 222 40; website: http://uas-europe.com, http://www.uas-europe.se/index.php

UAS Europe AB is a UAV company located in Scandinavia. They have been serving farmers through supply of UAVs, since 2010. They are well spread out in their operations. It includes nations such as USA, Germany, Spain, United Arab Emirates, Uruguay, Lithuania, and Bulgaria. They offer the UAV "Spy Owl 200" plus a range of accessories. It includes custom-made gimbals, multispectral cameras, and IR cameras suited for agricultural operations. They also offer best suited remote-control equipment, datalink and transfer, software for autopilot, image processing, and agronomic analysis. They specialize in accurate aerial mapping of topography and crops. Fertilizer and water application calculations are also done, using appropriate software. There are several optional items, such as long duration propulsion batteries, laser altimeter for autolanding, optical zoom for sensors, video recording of ground surface events, and laptops to conduct flights.

TECHNICAL SPECIFICATIONS

Material: This UAV's airframe is made of toughened plastic and graphite. It includes an electric engine placed in the fuselage area.

Size: length: 1.53; width (wingspan): 2.01 m; height: 35 cm; weight: 6.7 kg or 8.5 kg versions

Propellers: It has one propeller on the fuselage, at the center of wingspan and facing backward.

Payload: It is 2.9 kg for hand-launched models and 4.7 kg for catapulted models.

Launching/Landing information: The hand-launched version weighs 6.7 kg. The catapult launched version weighs 8.75 kg. Landing is done typically on belly. The fuselage is ruggedized to withstand rough landing. Autolanding software is also available. The UAV operates using autopilot software known as "Easy pilot 3.0."

Altitude: Typical operational altitude is 75–1500 m, for this UAV. It is fitted with laser altitude meter and linked to data center.

Speed: The cruise speed is 16–22 m s^{-1} (i.e., 58–79 km h^{-1} or 31–43 knots); maximum speed attained is 144 km h^{-1} (i.e., 78 knots)

Wind speed tolerance: It tolerates 45 km h^{-1}

Temperature tolerance: −20 to +50°C

Power source: Electric batteries

Electric batteries: Lithium polymer batteries—24,000 mA h.

Fuel: Not an IC engine-fitted UAV. So, there is requirement for petroleum.

Endurance: It is 3 h per flight. However, it can be extended, using long endurance propulsion batteries.

Remote controllers: Radiocontrol is long range and works up to 50 km from ground station (or remote controller).

Area covered: The aerial scan area is 30 km^2

Photographic accessories: Cameras: The UAV carries combined R, G, B, and multispectral cameras; EO/IR Gimbal.

Computer programs: Agricultural version of the "Spy Owl" UAV has a GCS with license for a "mapping software." It has R, G, B, and multispectral sensor; Pix4D Mapper software and ADC micromultispectral camera. The UAV also comes with a video recorder for instantaneous recording of ground images. The package also includes a video transmitter that operates up to 16 km distance from the UAV. An analog video receiver and a video receiver box are also available. The agricultural version comes with GCS mapping software, such as Pix4D Mapper or Ag Software.

Spraying area: This is not a sprayer UAV.

Volume: Not applicable

Agricultural uses: Major uses in agriculture are surveillance of farm installations, crop fields, irrigation facilities, farm vehicles, and progress of different agronomic procedures, such as plowing, seeding, weeding, etc. Estimation of NDVI, crop water stress index, detection and estimation of xanthophyll pigmentation in crops, leaf chlorophyll and plant N status estimation, and topographical studies of land and crop fields are other uses.

Nonagricultural uses: Surveillance of industrial installations, city traffic, public events, pipelines, rail roads, etc. Monitoring mines and mining activity are other functions.

Useful References, Websites, and YouTube Addresses

http://www.uas-europe.se/ (accessed June 24, 2017).

http://www.uas-europe.se/index.php/mobile-products/mobile-spyowl-200 (accessed June 24, 2017).

http://www.uas-europe.se/index.php/mobile-products/mobile-spyowl-200 (accessed June 24, 2017).

https://www.suasnews.com/2014/10/spy-owl-200-surveillance/ (accessed June 24, 2017).

http://drone-rss.com/2014/10/spy-owl-200-surveillance/ (accessed June 24, 2017).

http://www.dronele.ro/tag/spy-owl-200/ (accessed June 24, 2017).

Name/Models: STC Orlan-3M multipurpose UAV

Company and address: Special Technology Centre, Russia

STC Orlan-3M is a modernized version of previous model. It is developed by "Special Technology Centre" situated in Russia. It is primarily a UAV meant to obtain aerial imagery with great details. It is also used to get panoramic view of ground features. It is an autonomous aerial robot. It is also used to make videoimages and thermal images of including crop fields. The creators have applied the principle of modular construction. It allows the rapid replacement of payload without any problems.

TECHNICAL SPECIFICATIONS

Material: STC Orlan is made of light avionics aluminum, carbon fiber, and plastic composite.

Size: length: 1.5 m; width (wingspan): 2.1 m; height: 0.40 m; weight: 7.2 kg

Propellers: It has one propeller made of plastic. Propeller is mounted at the front end of fuselage.

Payload: It includes visual range cameras, videocameras, and thermographic sensors.

Launching information: It is an autolaunch UAV. Its navigation and landing are also autonomous. The navigation is controlled, using a custom-made computer software or something like "Pixhawk." This UAV is recovered, using a parachute.

Altitude: Its operating height is 300 m. However, ceiling is 7250 m a.g.l.

Speed: Operating speed is 100 km h^{-1}, maximum cruising speed is 150 km

Wind speed tolerance: 20 km h^{-1}

Temperature tolerance: −10 to +35°C

Power source: Internal combustion engine (11.3 hp) based on gasoline

Electric batteries: They are used to support computers and CPU unit.

Fuel: Gasoline

Endurance: 3 h

Remote controllers: The GCS has mobile, portable laptop, and software to control UAV's navigation, fix its aerial imaging programs, and process images. The telelink with UAV operates within 50 km radius. The GCS can alter preprogramed flight path and frequency of camera shutter operation. The ground control instrumentation provided with the UAV can guide and control functions of 3–4 UAVs, simultaneously.

Area covered: 150–200 km in linear distance or up to 200 ha per flight

Photographic accessories: Cameras: It has a set of visual bandwidth and high-resolution multispectral and thermal (IR) cameras.

Computer programs: The photographic images are processed using custom-made software. Usually, software such as Pix4D Mapper or Agisoft Photoscan or Trimble's Inphos is utilized.

Spraying area: This is not a spraying UAV.

Volume: Not applicable

Agricultural uses: It includes aerial survey of land, soils, and water resources. Crop scouting is done to get data on phenomics and vigor. It is also used to detect disease/pest affliction. It helps to detect plus map floods, drought, or soil erosion. Natural vegetation monitoring is done, to get NDVI data and plant species diversity.

Nonagricultural uses: This UAV has been used to surveillance geological features, mines, mining activity, ore-dumping sites, and movement of ore-loaded vehicle. It is also used, to monitor industrial sites, buildings, public places, rivers, dams, etc.

Useful References, Websites, and YouTube Addresses

http://en.avia.pro/blog/stc-orlan-3m-mnogocelevoy-bpla (accessed June 27, 2017).

http://en.avia.pro/blog/stc-orlan-3m-mnogocelevoy-bpla (accessed June 27, 2017).

https://informnapalm.org/en/orlan-drones-the-sea-eagles-of-st-petersburg/ (accessed June 27, 2017).

https://www.youtube.com/watch?v=OZ-0auK9oIg (accessed June 27, 2017).

https://www.youtube.com/watch?v=XYh3wfU8TOc (accessed June 27, 2017).

https://www.youtube.com/watch?v=Pbfl3AyAdaE (accessed June 27, 2017).

Name/Models: Stream Tactical UAS

Company and address: Threod Systems, Kaare tec 3, 74010 Vimisi, Estonia, Europe; Phone: + 372 512 1154; e-mail: info@thread.com, sales@thread.com; website: http://www.threod.com

Threod Systems is a high-technology company located in Latvia. It specializes in UASs. They produce UAVs that are useful in collection of visual and multispectral data about earth's surface and crop land. It also caters to military requirements of continuous surveillance. They have designed a range of fixed-winged small UAVs and multicopters, gimbals, autopilot software, sensors, and image-processing software. Since they produce a range of necessary items for adoption of UAV technology, their systems could be adopted more easily. Important UAV models are named "Theia Operational UAS," "Stream Tactical UAS," "EOS-Mini UAS," "K-4 multirotor UAS," and "K-4 Le Rotor UAS."

Stream tactical UAS is robust and versatile aerial UAV. This model can be readied in 30 min and deployed for flight and surveillance (Plate 2.38.1). It launches using a catapult, navigates with autopilot software, and has an endurance of 8 h of uninterrupted flight. It is useful in military surveillance. It is also adopted for commercial monitoring and agricultural data collection.

TECHNICAL SPECIFICATIONS

Material: UAV's body is made of ruggedized foam, plastic, and carbon fiber

Size: length: 220 cm fuselage; width (wingspan): 350 cm; height: 45 cm; weight: 12 kg including payload

Propellers: It has one propeller placed at the nose tip of the fuselage. Propeller has two tough plastic blades.

Payload: This UAV has multiple payload bay that holds CPU, cameras, and other accessories, for ground control contact. It can carry 6.0 kg payload.

Launching information: It has an autolaunch system. It launches using catapult. It lands using landing gear made of three ruggedized wheels.

Altitude: It has a ceiling of 4500 m a.g.l., but operational ceiling is 500–2000 m.

Speed: 80–130 km h^{-1}

Wind speed tolerance: 25 km h^{-1}

Temperature tolerance: −20 to +60°C

Power source: It derives power from 50 cm^3 IC engines and 12/24 V 350 W generator.

Electric batteries: It has a dyno-generator of 12 V for propeller and CPU unit functioning.

Fuel: It utilizes gasoline

Endurance: 8 h flight time

Remote controllers: The UAV has autolaunch and navigation mode, using software such as Pixhawk. The flight path can be preprogramed. Autonavigation takes care of route between two use points. However, software allows real-time modification of flight path, based on commands from GCS computers. It has both loiter and scan-pattern modes of flight. Emergency landing, warning of obstacles, and autolanding are also possible. It returns to point of launch, in times of emergency. It is compatible with GPS, INS, and GLONASS based on tagging of location and navigation.

Area covered: It transits for 150–200 km distance per flight or covers 250 ha land surface depending on flight path.

Photographic accessories: Cameras: It is provided with triple camera gimbal. The gimbal and cameras are gyro-stabilized. They provide excellent GPS tagged geo-stabilized images. Sensors include EO sensor 30× optical zoom, 720 p HD; IR sensor-1 640 × 480, 25 mm lens; and IR sensor-2 640 × 480 mm, 60 mm lens.

Computer programs: UAV-derived data and imagery can be transmitted, using encryption. Datalink lasts for 100–120 km distance from ground control computers. The imagery is processed, using the common software such as Pix4D Mapper or Agisoft.

Spraying area: This is not a sprayer UAV.
Volume: Not applicable

Agricultural uses: It includes crop scouting for growth (NDVI), phenomics, crop-N status, disease/pest attack, floods, or drought affliction.

Nonagricultural uses: It includes surveillance of industrial sites, buildings, oil and gas pipelines, rail roads, city traffic, public events, mines and mining activity, etc.

PLATE 2.38.1 Stream Tactical UAS.
Source: Mr. Siim Juss, Threod Systems Inc. Vimsi, Estonia.

Useful References, Websites, and YouTube Addresses

http://www.threod.com/products/uas-stream/stream-desription (accessed Apr 20, 2017).
https://www.youtube.com/watch?v=KWK-nBNmVRg (accessed Apr 21, 2017).

Name/Models: Super BAT DA-50 UAV

Company and address: Martin UAV LLC, 15455 North Dallas Parkway, Suite 1075, Addison, TX 75001, USA; phone: +1 512 520 7170; e-mail: info@martinuav.com; website: http://martinuav.com/uav-products/super-bat-da-50/

Martin UAV LLC aims to provide the most cost-effective access to aerial information. They specialize in providing robust and versatile UASs, mainly to support the most demanding data collection needs. Super Bat DA-50 model has wide-ranging applications in agriculture and civilian settings. The improved Super Bat DA-50, now, features increased fuel capacity, and an onboard generator for 10 h flight duration, and it comes with a more powerful catapult launcher (Plate 2.39.1). This UAV has an extracamera attached to wings. It is capable of tracking vehicles. Improved landing gear means greater shock absorption for rougher fields.

TECHNICAL SPECIFICATIONS

Material: Kevlan, aluminum, and carbon fiber composite form most part of the body. The nose is made of carbon for easy retraction.

Size: Total assembled size is 8.5 × 5.3 × 2.25 ft; length: 5.3 ft; width (wingspan): 8.5 ft (2.6 m); height: 2.2 ft (65 cm); weight: 47 lb (21.3 kg)

Propellers: It has one propeller at the rear end of fuselage. It has two plastic blades. Propulsion is via heavy fuel engine.

Payload: Maximum allowed payload is 5.0 kg. It usually consists of a gimbal with 4–5 different types of sensors, onboard dynamo, and CPU.

Launching information: It has launching mode through "Piccolo software." Launching is done using bungee catapult. It lands on belly by skidding. Parachute-aided recovery of UAV is also an option included.

Altitude: Maximum ceiling is 10,000 ft

Speed: It transits at a speed of 40–70 knots

Wind speed tolerance: 25 knots. This UAV operates even in mild drizzle

Temperature tolerance: −20 to +60°C

Power source: 50 cm^3 IC engines and an in-built dynamo that generates 50 W power

Electric batteries: 75 W generators

Fuel: Gasoline–oil mix (40:1)

Endurance: 10 h.

Remote controllers: The GCS computers and telemetry keep contact with UAV for up to 10–16 km radius

Area covered: It covers 600 km linear distance that is limited by fuel. Its flight path is restricted to 20 km radius due to limitations with telemetric link.

Photographic accessories: Cameras: Sensors include cloud TASE 150/200 long range (90 km). It includes gimbal with EO sensors, IR cameras (NIR, IR, and red-edge bandwidths). It has an additional 20 MP wing mounted camera with automatic triggering facility. The UAV is also attached with video downlink 900 MHz spread with spectrum-2 modem. The gimbal that accommodates cameras is well stabilized to avoid vibrations and haziness to images. It also has facility for ×20 zoom. The UAV also comes with ability for tracking vehicles and pin-pointing the geolocation of the moving objects. It is useful during detection and guidance of agricultural vehicles in fields.

Computer programs: Images are processed by the software on the UAV and relayed instantaneously as stills or videos. Alternatively, images received as orthomosaics can be processed and analyzed, using software such as Pix4D Mapper or Agisoft Photoscan, at the GCS.

Spraying area: This is not a spraying UAV.

Volume: Not applicable

Agricultural uses: It includes aerial mapping of terrain, soil type, and cropping systems. They are done, accurately, using GPS tagging. UAV's images also provide details like NDVI, leaf area, canopy growth characteristics, and crop-N status (leaf chlorophyll). Other functions are crop scouting, for seedling gaps, for disease/pest attack, soil erosion, and drought-affected spots. Yield forecasting using NDVI and grain maturity data is also possible. Natural vegetation monitoring could be done using Super BAT UAV.

Nonagricultural uses: This UAV model has wide ranging capabilities. In addition to agricultural uses, it is applied in urban policing, city traffic monitoring, monitoring buildings, industrial installations, rail road, etc. It is used to monitor mining sites, movement of vehicles with ores, and ore-dumping yards. It has wide ranging uses in military surveillance, target identification, location, and even destruction.

PLATE 2.39.1 Super BAT DA-50 UAV.
Source: Dr. Blake Sawyer, R.O. and Phil Jones, Martin UAV LLC, Addison, Texas, USA.

Useful References, Websites, and YouTube Addresses

http://martinuav.com/uav-products/super-bat-da-50/ (accessed July 10, 2017).

https://www.youtube.com/watch?v=PJ_iMtsvBpE (accessed July 10, 2017).

https://www.youtube.com/watch?v=HX0EsUb2ljc (accessed July 10, 2017).

https://www.youtube.com/watch?v=gwcLK-bSZ8k (accessed Aug 1, 2017).

https://www.youtube.com/watch?v=HX0EsUb2ljc (accessed Aug 1, 2017).

Name/Models: THEIA Operational UAS

Company and address: Threod Inc., Kaare tec 3, 74010 Vimisi, Estonia, Europe; phone: +372 512 1154; e-mail: info@thread.com, sales@threod.com; website: http://www.threod.com

"Theia Operational UAS" is a relatively long endurance UAV. It allows heavy payload. This is in addition to the usual cameras for data collection and aerial mapping. The UAV can be assembled and launched in 50–60 min. It is suitable for Lidar missions. It has low acoustic signature. It is suitable for commercial, agricultural, and military data collection and surveillance (see Plate 2.40.1).

TECHNICAL SPECIFICATIONS

Material: It is made of carbon fiber, plastic, and foam composite.

Size: length: 320; width (wingspan): 420 cm; height: 60 cm; weight: 50 kg dry weight, up to 25 kg payload

Propellers: It has one propeller placed at the front tip of the fuselage. Propeller has two blades.

Payload: It accommodates up to 25 kg. The payload includes a series of EO/IR cameras and gyro-stabilized gimbal.

Launching information: It has automatic launching and landing facility, plus necessary software (Pixhawk). It has a landing gear made of three wheels. It needs strip of 150 m runway to land.

Altitude: The ceiling is 5000 m, but operational altitude is 50–400 m for agricultural purposes.

Speed: 90–150 km h^{-1}

Wind speed tolerance: 45 kg km h^{-1}

Temperature tolerance: −20 to +60°C

Power source: 4 strokes 110 cm^3 fuel injected engine; 12/24 V 650 W generator

Electric batteries: Lithium–polymer batteries 5000 mA h.

Fuel: Gasoline + oil (40:1)

Endurance: It is 24 h, depending on payload configuration

Remote controllers: This UAV could be controlled from launch, to flight path, and until landing. The flight path could be predetermined, using computer software, or it could be modified midway through the flight. It has loitered mode. It can be programed to take flight to scan the entire land surface as envisaged. Predetermined landing is possible. It is connected to GPS, INS, and GLONASS for navigation and depiction of images accurately.

Area covered: 50–150 ha or 400 km in linear length per flight

Photographic accessories: Cameras: EO sensors include wide-angle cameras and IR sensor with SWIR camera. Second IR sensor is MWIT camera. This UAV model has an optional Lidar range finder, IR pointer, and laser designator. The gimbal and sensors are gyro-stabilized. It also has target tracking facility. It helps in tracking military trucks, commercial vehicles, field tractors in a crop field, planters, combine harvesters, etc.

Computer programs: It adopts Pixhawk for determining navigation and flight path. It has in-built CPU with Pix4D Mapper or Agisoft software. Images relayed to GCS too could be processed, using similar software on iPad/PCs. Images are all tagged with GPS or GLONASSS coordinates for accuracy.

Spraying area: This is not a sprayer UAV.

Volume: Not applicable

Agricultural uses: It includes aerial mapping of land and water resources. Theia UAS is used to scout crops for a range of characteristics. They are detecting seed germination, seedling emergence, finding gaps in plant stand, canopy characteristics, leaf chlorophyll, crop-N status, and crop water stress index (using IR cameras). The UAV-derived data help in scheduling irrigation based on crop's water status. The UAV's imagery could also be used to detect disease/pest attack, if any. The UAV offers disease/pest attack maps. Therefore, it helps in spraying chemicals, accurately, only at required spots. Aerial maps showing drought, or flood or soil-erosion-affected spots, can be obtained. This UAV could also be used to monitor natural vegetation.

Nonagricultural uses: It is used to surveillance military installations, finding targets, and relaying information about troop and vehicle movement, in battle fields. It is used to surveillance commercial buildings. Other capabilities include surveillance of mines, oil pipelines, rail roads, transport vehicles, etc.

PLATE 2.40.1 Top: Theia Operation UAS—A UAV. Bottom: A UAV readied to be launched, using a catapult.
Source: Mr. Siim Juss, Threod Inc., Viimisi, Estonia, Europe.

Useful References, Websites, and YouTube Addresses

http://www.threod.com/products/uas-theia/theia-description (accessed Apr 10, 2017).
https://www.youtube.com/watch?v=ONoIrPglWQU (accessed Aug 1, 2017).

Name/Models: UAV S 360

Company and address: Hanseatic Aviation solutions, Hermann Kohl Str. 7, 28199, Bremen, Germany; phone: +49 (0) 421 59 67 942 0; e-mail: info@hanseatic-avs.de

Hanseatic Aviation is a German UAV producing company. They specialize in UAVs and accessories.

TECHNICAL SPECIFICATIONS

Material: This UAV is made of foam, carbon, and plastic composite. It is ruggedized to withstand harsh weather and landing impact (Plate 2.41.1).

Size: length: 2.7; width (wingspan): 3.6 m; height: 0.7; weight: 13.2

Propellers: A single two blade propellers is placed at the center of wings and just above the fuselage.

Payload: It is >5 kg but <25 kg. Payload bay is separated from engines. Impact of payload on engine is avoided. Payload volume is 15 l.

Launching information: UAV S360 has two wheels to move on terrain. It needs a short runway to gain speed and attain take-off. A catapult can also be used to put the UAV afloat. UAV lands automatically at the point of launch using autolanding gears.

Altitude: It is used at <400 m for agricultural purposes, but operates up to <4500 m a.s.l.

Speed: 80–170 mi h^{-1}

Wind speed tolerance: Withstands 45–60 km h^{-1} squally winds

Temperature tolerance: 0 to +55°C

Power source: Internal combustion engine

Electric batteries: Electric batteries support only CPU electronics.

Fuel: Gasoline

Endurance: 4 h per flight

Remote controllers: Remote control operates through long-range radiotelemetry connections. UAV can be controlled for several kilometers from the ground station. Flight mission can be preprogramed. It has autopilot mode and autolanding. The ground control has a ruggedized, all weather box. It protects software and instrumentation. The screen is 15-in square. The ground control operates on batteries and works for 5 h at a stretch. Total weight of GCS computers and related accessories is 9 kg.

Area covered: 350–500 ha per flight

Photographic accessories: Cameras: The camera chamber has visual cameras, a full HD videographer with up to 80 MP. It has a Lidar pod. An array of IR cameras could be placed in the payload based on purposes. A NIR camera is also available for photography. Specifically, this UAV is endowed with geomagnetic sensors for geomagnetic studies. It has a powerful CPU. It instantaneously processes and sends images to GCS or for storage and retrieval later.

Computer programs: The images are processed using the Agisoft's Photoscan or Pix4D Mapper or any other custom-made software.

Spraying area: This is not a spryer UAV.

Agricultural uses: It includes aerial survey and photography of crop belts and individual farmland. Crop scouting, monitoring crop growth, collecting NDVI values, disease/insect pest damage detection, spraying plant protection chemicals, drought and soil erosion survey, impact of floods on cop land, monitoring irrigation channels, and reservoirs. Natural vegetation monitoring is also possible.

Nonagricultural uses: It includes military and civil site inspection, security and land enforcement, search and rescue operations, surveillance of oil pipelines, electric installations, mines, transport vehicles, etc.

PLATE 2.41.1 Hanseatic's S360 fixed flat-winged UAV.
Source: Michael Schmidt, Managing Director, Hanseatic Aviation Solutions Inc., Germany.

Useful References, Websites, and YouTube Addresses

http://www.hanseatic-avs.de (accessed May 12, 2017).
http://www.hanseatic-avs.de/payloads.html (accessed May 12, 2017).
http://www.hanseatic-avs.de/s360.html (accessed May 12, 2017).
http://www.hanseatic-avs.de/ (accessed May 12, 2017).

Name/Models: UAV Strix

Company and address: Aerodreams, Buenos Aires, Argentina; Tel.: +5411 4393 3466; e-mail: Not available; http://www.aerodreams-uav.com

TECHNICAL SPECIFICATIONS

Material: This UAV is made of a composite material containing foam, metal alloy, rubber, and carbon fiber.

Size: length: 1.8 m; width (wingspan): 3.6 m; height: 2 m; weight: 22 kg

Propellers: It has one propeller larger with two blade propellers. It is placed at the rear end tip of fuselage, facing backward.

Payload: 5–8 kg depending on the purpose.

Launching information: This UAV has undercarriage with a pair of wheels at the front area of fuselage and two wheels at the middle portion near the wings. Launching needs a stretch of runway. This UAV can also be launched, using pneumatic catapult. Autolanding on a landing strip or using remote controller is possible. Parachute landing is also available in the accessories (Garcia and Becker, 2016).

Altitude: It operates at 400 m altitude for agricultural survey and imagery, but it can reach up to 4000 m a.s.l.

Speed: 160–180 km h^{-1}

Wind speed tolerance: 45 km h^{-1}

Temperature tolerance: −17 to +55°C

Power source: Electric motor of 10 hp

Electric batteries: 24,000 mA h for electronics.

Fuel: Not an IC engine-fitted UAV. No requirement for petroleum fuel.

Endurance: 15 h per flight

Remote controllers: The GCS is usually situated in a mobile van. It has several GCS computers and radiocontact equipment. UAV can be controlled, guided, and imagery could be well regulated, using ground computers. Radiocontrols work for up to 12 km from ground station.

Area covered: 500–800 ha of agricultural land surveyed per flight

Photographic accessories: Cameras include EO sensors, that is, a set of visual band cameras (R, G, and B) and NIR, and IR for thermal imagery.

Computer programs: The UAV has a CPU that processes images of atmosphere, clouds, haze, and agricultural land. It has satellite guidance and can be flown as a preprogramed autonomous UAV. Images are tagged using GPS coordinates. Images are sent to GCS, instantaneously. The computer software used for agricultural purposes is Pix4D, Airsoft, or any other custom-made.

Spraying area: It should be possible to fit tanks that hold plant protection chemicals or fertilizer formulation.

Agricultural uses: It includes aerial survey of land, soil, water, and natural vegetation. The UAV helps in scouting crop fields, monitoring farm activity and farm vehicles, irrigation equipment, and combine harvesters.

Nonagricultural uses: This UAV model is used to surveillance coast lines, riverine zone, mines, military installations, industrial area and transport vehicles, etc. (Garcia and Becker, 2016).

Useful References, Websites, and YouTube Addresses

Garcia, E. G.; Becker, J. UAV Stability Derivatives Estimation for Hardware-in-the Loop Simulation of Piccolo Autopilot by Qualitative Flight Testing; Aerodreams: Buenos Aires, Argentina, 2016; pp 1–9. http://www.aeroderams-uav.com/docs/aeroduav.pdf (accessed Nov 8, 2016).

http://www.aerodreams-uav.com/es-uav-strix.html (accessed Feb 11, 2017).

http://www.aerodreams-uav.com/es-sistema-ads.html (accessed Feb 11, 2017).

https://www.youtube.com/watch?v=PpMaKT-M9es (accessed Feb 11, 2017).

Name/Models: Venturer UAV

Company and address: Stratus Aeronautics Inc., #123, 3191 Thunderbird Crescent, Burnaby, British Columbia, BC, Canada V5A 3G1; phone: +1 604 444 9238; e-mail: Not available; website: http://stratus-aero.com/

Stratus Aeronautics is a UAV technology company located in Western Canada. They aim at development of the Venturer UAV and several types of applications that need aerial imagery. Stratus acquired Universal Wing Geophysics in 2012. In future, Stratus UAV could be serving farmers and other clients in a big way in the Canadian Plains, USA, and other areas. They have a team with backgrounds in engineering, computer programing, and geoscience. Stratus Aeronautics has placed a range of payloads into the UAVs. They are useful, during imagery and assessment of ground surface. They have a collection of processing software procedures. Stratus can use a variety of software to create 3D maps and other forms of processed data.

"Venturer UAV" is a small UAV. It has the ability for rapid assembly and take-off. In due course, it should be popular among crop producers, farm scientists, natural science, and geological experts. In addition to visual, NIR, IR, and Lidar, this UAV has ability for magnetometer aerial survey.

TECHNICAL SPECIFICATIONS

Material: It has carbon fiber airframe. Plastic and foam composite make up the other material of the UAV.

Size: length: 3 m long fuselage; width (wingspan): 4 m wingspan; height: 50 cm; weight: 13.8 kg

Propellers: It has one propeller. It is placed at the tip of nose on the fuselage.

Payload: It includes a series of sensors EO/IR and Lidar. It holds 2.2 kg load.

Launching information: It launches using a ruggedized launch and landing gear. The tires are fattened to withstand rough runway. A minimum of 100 m runway is required to get airborne.

Altitude: It operates at low altitudes of 50–400 m a.g.l. for aerial survey and crop imagery. Ceiling is 3000 m a.g.l.

Speed: 45–60 km h^{-1}

Wind speed tolerance: 20 km h^{-1}

Temperature tolerance: −20 to +55°C

Power source: An IC engine; 2 stroke fuel engines

Electric batteries: 8000 mA h for propellers, during lift-off.

Fuel: Gasoline. About 1.5 l of petrol is consumed per hour of flight.

Endurance: 10 h

Remote controllers: A 900-mHz radio and iridium keep contact between ground station and UAV. Telemetric contact is limited to 25 km. Flights can be preprogramed or modified midway, using ground support computers/iPads.

Area covered: 60 km linear distance per flight or 50–200 ha, if used for aerial survey and surveillance.

Photographic accessories: Cameras: Visual cameras include red, green, and blue bandwidths, multispectral, and hyperspectral cameras for aerial survey and sharp imagery, NIR for plant chlorophyll estimation, IR, and red edge for thermal imagery and to ascertain crop water status. Lidar cameras are also available.

Computer programs: Images are processed using Pix4D Mapper, Agisoft, or similar software at the ground station.

Spraying area: This is not a sprayer UAV.

Volume: Not applicable

Agricultural uses: Venture UAV has been custom-made, to derive aerial maps of ground surface, agricultural land, crops, and other features such as drains, etc. It could be used for crop scouting regularly. It adopts visual bandwidth sensors. It can offer data about NDVI of the surface vegetation and crops. It also provides data on crop growth trends, leaf-N, crop-N status, crop water stress index (using IR cameras), etc. Venture UAV could be adopted for natural vegetation monitoring.

Nonagricultural uses: This UAV is currently used for magnetometer and Lidar-based survey of geophysical conditions, geological features, and natural resources. It is used to monitor mines, their functioning, mine vehicle movement, and ore dumps. It is also used in military reconnaissance. It is useful to surveillance installations, such as industries, buildings, dams, railroads, pipelines, and city traffic.

Useful References, Websites, and YouTube Addresses

https://www.guavas.info/drones/Stratus%20Aeronautics%20Inc-Venturer%20UAV-224 (accessed Feb 10, 2018).

http://stratusaeronautics.com/?page_id=53 (accessed Mar 3, 2018).

Name/Models: Vigilant C 2

Company and address: RP Flight Systems Inc., Wimberley, Texas, USA; phone: +1 512 665 9990, +1 408 832 1195; e-mail: info@rpflightsystems.com, sales@rpflightsystems.com, Website: http://www.RPFlightSystems.com

RP Flight systems Inc. is head quartered at Wimberley in Texas, USA. This UAV company was initiated in 2001. Its focus is on developing designs and mass producing fixed-winged UAVs. These UAVs are aimed at helping in surveillance of public events, policing city suburbs, tracking city vehicles, and monitoring buildings; these UAVs are also apt for monitoring agriculture fields and farm. They can provide crucial spectral data about crop's status to farmers. Vigilant C2 is a light-weight fixed-winged drone. It is predominantly an aerial survey UAV. It has a range of uses in agricultural regions.

TECHNICAL SPECIFICATIONS

Material: It is made of carbon fiber, plastic, and foam composite (Plate 2.42.1).

Size: length: fuselage length is 120 cm; width (wingspan): 260 cm; height: 35 cm; weight: flying weight is 5.5–10 kg depending on payload

Propellers: It has one propeller at the rear end of the fuselage. Propeller has two plastic blades.

Payload: 4.0 kg. It includes sophisticated stabilized gimbal that holds a series of visual, NIR, and IR sensors.

Launching information: It needs catapult to launch. Catapults supplied are amenable for several different configurations of UAV. The UAV gains >2000 ft min^{-1}. Landing can be achieved via skid landing on belly or using a parachute in the mid-air (Plate 2.42.1).

Altitude: It is flown at 50–150 m altitude for ground survey and crop analysis.

Speed: The cruise speed is 50–200 km h^{-1}; dash speed is 100 km h^{-1}

Wind speed tolerance: 15 km h^{-1}

Temperature tolerance: −20 to +60° C

Power source: Electric batteries 100 amp

Electric batteries: 5500 mA h two cells litho-polymer.

Fuel: Not an IC engine-fitted UAV.

Endurance: 60 min per flight

Remote controllers: Ground control is connected to UAV via telemetry for 12 km radius. The flight path could be preprogramed or modified midway during flight, using GCS iPads/PCs.

Area covered: It covers a linear length of 60–100 km or 50–200 ha area.

Photographic accessories: Cameras: This UAV carries a gimbal with a series of visual, NIR, IR, and red-edge bandwidth cameras, for aerial (visual) and thermal photography. It also carries a Lidar pod.

Computer program: The UAV's CPU includes software for image analysis and transmission, to ground iPads. The imagery could be processed at the ground station and utilized, during crop analysis, and development of fertilizer and pesticide prescriptions.

Spraying area: This is not a sprayer UAV.

Volume: Not applicable

Agricultural uses: This UAV could be utilized to obtain aerial imagery of farm, its installations, crop fields, and crop growth status. Crop scouting for biomass accumulation, leaf chlorophyll content, and nitrogen status is possible. The UAV scouts for disease, insect pests, drought, floods, and soil erosion. Monitoring natural vegetation is also possible.

Nonagricultural uses: It includes surveillance of military installations, oil pipelines, electric power lines, city traffic, transport vehicles, etc.

PLATE 2.42.1 Top: Vigilant C2 UAV; bottom: a UAV and catapult assembly. The UAV is ready for launch.
Source: Dr. Gene Robinson, President, RP Flight Systems Inc., Wimberley, Texas, USA.

Useful References, Websites, and YouTube Addresses

http://www.asiatechdrones.com/ (accessed May 25, 2017).

http://www.asiatechdrones.com/_p/prd1/2762557611/product/vigilant-c1---bare-airframe (accessed May 25, 2017).

http://www.asiatechdrones.com/airframes (accessed May 25, 2017).

http://www.asiatechdrones.com/_p/prd1/1961646675/product/vigilant-e--kit (accessed May 25, 2017).

Name/Models: VYOM 01 and 02 UAVs

Company and address: TechnoSys Embedded Systems (P) Ltd., SCO 66, Sector 20 C, Chandigarh, India; phone: +91 172 5085909; +91 172 4671082; e-mail: info@technosysind.com, sales@technosysind.com, support@technosysind.com; website: http://www.technosysind.com

These UAVs are produced by TechnoSys Ltd. for farmers in Northern Plains of India. This company produces a few versions of VYOM fixed-winged UAVs. These fixed-winged UAVs produced are available in ready-to-use condition. It is also assembled rapidly. The payload is relatively more than other flat-winged UAVs (Plate 2.43.1). They are priced low to help the farmers and companies to adopt UAV technology.

TECHNICAL SPECIFICATIONS

Material: This UAV supposedly has a sturdy frame made of carbon fiber, foam, and steel.

Size: length: 3 m; width (wingspan): 6 m; height: 60 cm including wheels; weight: 13.8 kg

Propellers: It has one propeller with two blades placed at the rear end of the fuselage, facing backward.

Payload: 85 kg

Launching information: VYOM is launched using a runway. UAV has undercarriage with two wheels.

Altitude: It reaches<400 m altitude for agricultural surveillance. However, it can reach 5000 m, if needed.

Speed: 60 km h^{-1}

Wind speed tolerance: 25–30 km h^{-1}

Temperature tolerance: −20 to +60° C

Power source: Electric batteries and IC engine (hybrid). UAV is fitted with FUTUBA T10 Engine-3W210

Electric batteries: Lithium polymer batteries 8 mA h for propellers.

Fuel: Gasoline for IC engine

Endurance: 8 h

Remote controllers: GCS has ruggedized computers and radiocommunication module. The remote controller has video connection to UAV. Flight path can be altered midway through the flight, if needed. The remote control iPad or PC could be connected to other computers. Therefore, the data collected could be shared or transmitted to specific processing units. The datalinker has an amplifier that is operative within 100 km radius. The UAV flight is programed, using autopilot software known as DZY-02.

Area covered: 300–500 ac per flight of 8 h

Photographic accessories: Cameras: The UAV has a gimbal. It is stabilized and has antivibration device. The cameras on the UAV are interfaced with remote control center and its computers. The UAV carries a usual complement of visual range cameras operative at R, G, and B wavelength bands (e.g., Sony α7 series), plus Sony NIR, IR, and red-edge cameras. The UAV is also fitted with a videocamera (CG160).

Computer programs: The UAV is fitted with videographer camera-CG160. Videos could be instantaneously received, viewed, and edited at the ground-station computers.

Spraying area: This is not a sprayer UAV.

Volume: Not applicable

Agricultural uses: Crop scouting and natural vegetation monitoring.

Nonagricultural uses: This UAV is used in day and night observation of public events, city traffic, industrial units, etc. Aerial inspection of infrastructure, buildings, dams, rail road, and damage assessment after a disaster, such as earthquake, are possible. Surveillance of mines, ore transport vehicles, and dumping yards is a clear possibility.

PLATE 2.43.1 VYOM 01.
Source: Dr. Dhruv Arora, TechnoSys Embedded Systems (P) Ltd., Chandigarh, India.

Useful References, Websites, and YouTube Addresses

www.technosysind.com/product-details.php?prod=vyom_uav (accessed Nov 11, 2016).
https://www.youtube.com/watch?v=qY6bLSq_i7M (accessed Nov 13, 2016).
https://www.youtube.com/watch?v=9t9_2NmWQuE (accessed Nov 13, 2016).
https://www.youtube.com/watch?v=AUPuSmF9-QU (accessed Nov 13, 2016).

Name/Models: Wave Sight

Company and address: Volt Robotics Inc., Chesterfield, Missouri, USA; e-mail: Info@voltaerialrobotics.com

www.voltaerialrobotics.com; phone: +1636 410 0191

Volt Robotics is an early entrant into agriculture UAV technology. Since 2005, they have researched, developed, and mass produced a range of fixed-winged and copter UAVs. They have also developed several different software applications meant for aerial photography and processing of images, derived by UAV cameras. They have also developed accessories for sampling air and atmospheric dust and pollution causing agents.

Their main fixed-winged UAV in operation currently is "WaveSight." It is gaining in popularity among farmers. It is known as "Pros3 Wave sight." The wave sight is a small, light-weight UAV adopted most effectively, to conduct aerial imagery of agricultural crops. It has an advanced electrical system and cameras for spectral analysis. "Wave Sight" is a low-cost UAV. It is affordable to both small holder farmers and large grain producing companies. This UAV has several types of applications in other aspects such as industry, civil administration, city traffic policing, mines, geology, etc.

TECHNICAL SPECIFICATIONS

Material: It is made of reinforced plastic which is toughened to withstand harsh weather conditions.

Size: length: 1.36 m; width (wingspan): 2.3 m height; weight: 9 kg (19.95 lb)

Propellers: It has one propeller attached to the rear end of the fuselage.

Payload: 5–8 kg

Launching information: It is launched using a catapult. It is an autolaunch, navigation, and landing UAV. The UAV's flight path is designed and controlled using Pixhawk software. The landing is either on belly or it can be done using a parachute. Parachute has been opened to least 40 m above the ground level.

Altitude: 50–400 m a.g.l. for agricultural operations. However, ceiling is 3200 m a.g.l.

Speed: It ranges from 39 to 75 km h^{-1}. The cruise speed is 72 km h^{-1}

Wind speed tolerance: 45 km h^{-1}

Temperature tolerance: −20 to +60°C

Power source: Electric batteries

Electric batteries: 24,000 mA h, 2 cells

Fuel: No need for petrol or diesel.

Endurance: 60–120 min per flight

Remote controllers: The GCS telemetry and iPads keep contact with UAV's CPU for up to a radius of 30 km radius. The GCS iPads have override facility. So, we can modify the flight path and/or photography schedule.

Area covered: Up to 12 km^2 or 3000 ac per flight. It flies 300 km per flight.

Photographic accessories: It has a stabilized gimbal that accommodates a few different cameras. It has visual cameras of 20-MP resolution and a 20-MP NIR camera. Payload also has 20–50× optical zoom lenses. The UAV's CPU stores GPS-tagged images.

Computer programs: The autopilot is regulated using Pixhawk software. The aerial images are processed using Pix4D Mapper or Agisoft's Photoscan. Custom-made software is also adopted.

Spraying area: Not applicable (not a spraying UAV)

Agricultural uses: The agricultural applications of this UAV, "WaveSight," include aerial imagery of land resources, water bodies, rivers, and irrigation channels. The crop scouting is done to estimate NDVI, collect data on leaf area, leaf chlorophyll (crop's N status), canopy temperature (water stress index), and grain maturity. The UAV also scouts fields for crop diseases, pests, drought, and flood-affected areas and maps them. The digital data collected by the UAV's sensors are effectively utilized, during variable rate applications.

Nonagricultural uses: WaveSight can be used to surveillance international borders, military installations, industries, buildings, oil and gas pipelines, electrical installations, mines, mining activity, ore movement, city traffic and public events, tracking vehicles, etc.

Useful References, Websites, and YouTube Addresses

http://www. voltaerialrobotics.com (accessed Sept 20, 2015).
http://voltaerialrobotics.com/wavesight/ (accessed June 15, 2015).
http://www.voltaerialrobotics.com/wavesight/ (accessed June 15, 2015).
https://www.youtube.com/watch?v=nwu5d8JamTg (accessed June 15, 2015).
https://www.youtube.com/watch?v=AgPo8QPJzW4 (accessed June 15, 2015).

Name/Models: Wingo

Company and address: UAVision Aeronautics, Parque Empresarial de Torres Vedras, (Paul), Lotte A1, Armazem 1, 2560-383 Torres-Vedras, Portugal; phone: +351 261 311 552; e-mail: info@uavision.com; website: http://www.uavision.com/wingo

UAVision Aeronautics Ltd. is a UAV company initiated in 2014. It is located at Torres-Vedras in Portugal. It offers a few useful fixed-winged UAVs and couple of quad-copters. All of them may have big role to play, in the commercial and agricultural fields.

Wingo is a versatile UAV. It is capable of medium range transit from the ground control (or launch point). It has relatively larger volume of space for gimbal and sensors, in the cockpit/nose of the fuselage. It comes with an interchangeable nose, if we wish to enhance payload volume. It has wide range of applications in surveillance, border control, tracking military vehicles, etc. It has applications in commercial and agricultural realms.

TECHNICAL SPECIFICATIONS

Material: This UAV is made of a composite derived from carbon fiber, plastic, and ruggedized foam.

Size: length: 75 cm; width (wingspan): 125 cm; height: 35 cm; weight: 6.5 kg

Propellers: It has one propeller placed at the rear end of the fuselage. The propeller has two plastic blades.

Payload: 5 kg that includes sensors, military tracking devises if used for surveillance of trucks and convoys; visual, NIR, IR, and red-edge sensors, if used for agricultural crop analysis

Launching information: It has wheels and landing gear. It needs a small runway of 300 m to take-off. It is also launched, using a catapult. It has autolaunch and navigation software (Pixhawk).

Altitude: It is 50–400 m a.g.l. for aerial survey and crop's imagery. The ceiling is 4000 m a.g.l.

Speed: The cruise speed is 70 km h^{-1}. Maximum speed attainable by the UAV is 140 km h^{-1}. Stall speed is 50 km h^{-1}.

Wind speed tolerance: 20 km h^{-1}

Temperature tolerance: −20 to +55°C

Power source: Electric batteries (lithium–polymer)

Electric batteries: 14,000 mA h lithium polymer batteries, to run the motors connected to propellers.

Fuel: Not an IC engine-fitted UAV. No requirement for petrol/diesel

Endurance: 10 h per flight

Remote controllers: Remote control station has telemetric and internet connectivity. It lasts for up to 50 km radius, from GCS. iPads at GCS can preprogram the UAV's flight path and photographic schedule. Flight path can be altered midway, if needed, using ground computers. The UAV has failed safe mode that allows it to reach the spot of launch quickly.

Area covered: It covers a linear distance of 250 km per flight.

Photographic accessories: Cameras: It has an autostabilized gimbal that accommodates a set of 5–6 cameras. These are meant for aerial survey and high-resolution imagery of land, crops, and commercial infrastructure. The gimbal houses visual range sensors (R, G, and B), thermal cameras that operate at NIR, IR, and red-edge wave bands. A synthetic aperture radar, to detect other UAVs, airplanes, and ground features, is available. It has hyperspectral cameras, to provide very sharp images of crops and their canopies. It also has a separate adaptive tracking system for detailed surveillance of farm vehicles.

Computer programs: The aerial imagery is processed on-board, using software in the CPU, for example, using Pix4D Mapper or Agisoft's Photoscan. The images can be swiftly downloaded, at the GCS, using Internet and iPads.

Spraying area: This is not a sprayer UAV.

Volume: Not applicable

Agricultural uses: Major utility of this UAV is in obtaining aerial imagery of land resources, cropping systems, individual crops, etc. The visual and thermal bandwidth cameras provide a range of data related to NDVI, biomass accumulation, leaf area, leaf-N, crop-N status, and fertilizer-N requirement. The thermal cameras are used, to get detailed knowledge, about variations in soil moisture and crop water stress index. The aerial imagery also provides insights into disease, pest, drought, or flood. Natural vegetation monitoring is a clear possibility.

Nonagricultural uses: Wingo has been utilized for military surveillance, particularly tracking vehicles, convoys, and different types of infrastructures. It has been used to monitor happenings in and around the buildings, rail roads, highways, public places, etc. This UAV is apt to surveillance geological sites, mines, ore transport vehicles, and mine dumping zones.

Useful References, Websites, and YouTube Addresses

https://www.uavision.com/ (accessed Dec 15, 2017).
https://www.uavision.com/fixed-wing-uav-wingo/ (accessed Dec 15, 2017).
https://www.uavision.com/services/ (accessed Dec 15, 2017).
https://www.uavision.com/fixed-wing-uav-wingo/ (accessed Dec 15, 2017).

Name/Models: Wingo-S

Company and address: UAVision Aeronautics, Parque Empresarial de Torres Vedras, (Paul), Lote A1, Armazem 1, 2560-383 Torres-Vedras, Portugal; phone: +351 261 311 552; e-mail: info@uavision.com; Website: http://www.uavision.com/wingo

UAVision Aeronautics Ltd. is a UAV company initiated in 2014. It is located at Armezem, in Portugal. It offers a few useful fixed-winged UAVs and couple of quad-copters. All of them may have a big role to play, in the commercial and agricultural fields.

WINGO-S is a small UAV. It is versatile regarding its flight path, aerial imagery, and data capture. WINGO-S has been developed for flight missions needing agility and high-quality images. WINGO-S is adaptable and suitable to applications from tactical surveillance to search and rescue or even forest and agriculture monitoring.

TECHNICAL SPECIFICATIONS

Material: This UAV is made of a composite of kevlar, foam, plastic, and carbon fiber.

Size: length: 1.8 m; width (wingspan): 2.3 m; height: 0.3 m; weight: 10 kg MTOW

Propellers: It has one propeller placed at the center of the wing span, and above fuselage.

Payload: It allows 3 kg. Payload includes cameras and radar, plus CPU with useful software.

Launching information: Landing is done using a catapult or bungee catapult. Landing or recovery can be accomplished using net, parachute, or runway. Runway strip should be at least 300 × 100 m in size. Parachute should be popped, at least above 100 m from ground level.

Altitude: It is generally operated at 50–400 m a.g.l. However, ceiling is 2000 m a.g.l.

Speed: The UAV's cruise speed is 45–60 km h^{-1}, but the maximum speed attained is 100 km h^{-1}.

Wind speed tolerance: 20 km h^{-1}

Temperature tolerance: −20 to +60°C

Power source: Electric batteries

Electric batteries: Lithium polymer batteries. It generates 14,000 mA h.

Fuel: Not an IC engine-fitted UAV. No need for petrol/diesel

Endurance: 2 h per flight

Remote controllers: The GCS telemetric connectivity and internet contact extend for up to 40 km. Autopilot mode uses APM and PIX4. The flight path is either preprogramed or done manually. The GCS can relay commands, to change path midway during flight.

Area covered: It covers 200–400 ha per flight or 60–80 km linear distance.

Photographic accessories: Cameras: It has a gyro-stabilized gimbal. The gimbal supports a few different cameras. The sensors include visual, hyper and multispectral, NIR, IR, and red-edge bandwidths. It also holds radar and vehicle tracking sensor. In general, accessories include visible cameras, thermal cameras, multispectral cameras, adaptive vision tracking system, Geiger radiation sensor, multiple gas sensors, RFID Scanner, Lidar System, and Genlock System.

Computer programs: The imagery derived by UAV's cameras is processed, using CPU. The images are instantly relayed to ground-station iPad. The orthoimages could also be sent directly, to GCS and processed, using software such as "Pix4D Mapper" or "Agisoft" or any custom-made software that suits the UAV/GCS instrumentation.

Spraying area: This is not sprayer UAV.

Volume: Not applicable

Agricultural uses: They are aerial survey of land, soil and water resources, crop scouting, and natural vegetation monitoring.

Nonagricultural uses: It includes surveillance of military installations, industries, buildings, mines, mining activity, and ore-dumping yards. It is also used to monitor and keep vigil on oil and gas pipelines, electric lines and installations, rivers, dams, city traffic and transport vehicles, etc.

Useful References, Websites, and YouTube Addresses

https://www.uavision.com/fixed-wing-uav-wingo-s/ (accessed Dec 15, 2017).
https://www.uavision.com/products/ (accessed Dec 15, 2017).
https://www.uavision.com/fixed-wing-uav-wingo-s/ (accessed Dec 15, 2017).

2.4 FIXED-WINGED UNMANNED AERIAL VEHICLE SYSTEMS IN AGRICULTURE: MEDIUM (50–200 kg), HEAVY (200–2000 kg), AND SUPERHEAVY (>2 t) WEIGHT UAVS

At present, there are fewer agricultural UAV models falling under the classes medium (50–200 kg), heavy (200–2000 kg), and super heavy weight (>2 t) UAVs. They are relatively larger and hold petrol/diesel enough to support long-distance coverage. They are often flown at high altitudes and employed for surveillance of large areas. They are used in conducting vegetation surveys of districts, counties, etc. They are also called medium altitude long endurance or high-altitude long endurance UAV. It depends on the engines and fuel they carry with them. Heavy and super heavy weight UAVs transit long distances. They are used to conduct aerial survey and vegetation mapping of very large agricultural zones and cropping belts.

Name/Models: Arcturus UAV JUMP (VTOL)

Company and address: Arcturus UAV Inc., PO Box 3011, 539 Martin Avenue, Rohnert Park, CA 94928, USA; phone: +1 707 206 9372 e-mail: Not available; website: www.arcturus-uav.com

Arcturus UAV is an American company specializing in the production of small UAVs. It produces UAVs suitable for variety of purposes, including agriculture.

They offer VTOL flat-winged UAVs useful in aerial monitoring of farms, farm installations, and crops. There are at least three models produced by this company, namely, T-20, Jump-15, and Jump-20. The UAVs fitted with visual and IR sensors offer excellent data about crops, sensors offer data about crops, their growth pattern, leaf-N status, fertilizer-N requirements, crop water status, irrigation needs, etc. These UAVs could also be used, to detect drought effects or flooded regions, in a farm, at the earliest. Farm scouting by Arcturus UAVs helps in reducing labor needs.

TECHNICAL SPECIFICATIONS

Material: It is made of toughened foam, plastic, and carbon fiber composite.

Size: length: 25 cm; width (wingspan): 47 cm; height: 20 cm; weight: 90 kg

Propellers: There are four propellers on wings for VTOL. Plus, one propeller at the front tip of the fuselage is for forward thrust.

Engine: 190 cm^3, four stroke

Payload: 4100 cubic in.; 60 lb

Launching information: It has autopilot software (e.g., Pixhawk), for VTOL, flight, and landing. It does not require a special launching facility, such as catapult, etc.

Altitude: <15,000 ft a.g.l.

Speed: 72 knots

Wind speed tolerance: 35–45 km h^{-1}

Temperature tolerance: −10 to +60°C

Power source: Electric batteries for propellers during VTOL and MOGAS for IC engine

Electric batteries: Lithium polymer batteries 14,000 mA h.

Fuel: Gas

Endurance: 9–16 h per flight

Remote controllers: The GCS comes fully equipped with tracking antennas, data terminals, PCs, iPad, and shelter. UAV images can be collected via internet, Ethernet, etc. Telemetric tracking is also possible. The flight path is regulated, using software such as Pixhawk

Area covered: 50 ha/h

Photographic accessories: Cameras: The UAV is fitted with series of very useful cameras. The cameras usually fitted are the visual range (R, G, and B), multispectral image.

Computer programs: Images can be processed in situ in the CPU on the UAV or relayed as orthoimages to ground station. The postflight image processing is done, using Pix4D Mapper or Agisoft Photoscan.

Spraying area: This is not a spraying UAV.

Volume: Not applicable

Agricultural uses: This UAV is used to scout crop fields for seed germination, seedling establishment, and plant population, to detect gaps in crop stand, measure leaf area index, leaf chlorophyll (via NIR sensor), crop-N status, disease/pest attacked zones, drought or flood-affected zones, etc. (Koski et al., 2009).

Nonagricultural uses: It is useful in military intelligence, surveillance, and reconnaissance. It is used to monitor city traffic, public events, rail roads, oil/gas pipelines, dams, mines, and mining activity.

Useful References, Websites, and YouTube Addresses

Koski, W. R.; Abrll, P.; Yazvenko, B. A Review and Inventory of Unmanned Aerial Systems for Detection and Monitoring of Key Biological Resources and Physical Parameters Affecting Marine Life during Offshore Exploration and Production Activities. In *A Review and Inventory of Unmanned Aerial Systems*; LGL Ld., Ed.; E and P Sound and Marinelife Program, 2009; pp 1–32. http://www.soundmrinelife.org (accessed June 17, 2017).

http://arcturus-uav.com/ (accessed June 4, 2017).

http://arcturus-uav.com/product/JUMP-20 (accessed June 4, 2017).

https://www.youtube.com/watch?v=omkay1cJmXI (accessed June 4, 2017).

https://www.youtube.com/watch?v=omkay1cJmXI (accessed June 4, 2017).

Name/Models: Gamma UAV System

Company and address: AVASyS LLC is a Russian UAV company located at Krasnoyarsk, Siberia, Russia. The company is operated by the Siberian Federal University (Institute for Geology and Geo-technology) with the support of finance, from Prognoz Group LLC. AVASyS LLC aims at manufacturing of high number of autopilot UAVs. They aim at developing UAV systems that suit different purposes, such as mining, geology, road construction, inspection of buildings, and importantly natural vegetation and agricultural crop production (AVASyS LLC, 2016).

"Gamma UAV system" is suited for variety of aerial survey and assessment. It depends on the payload of cameras and other accessories. It has range of cameras that help in the analysis of crop and in forecasting crop yield (Plate 2.44.1). The laser scanner and magnetometer fittings are specifically helpful in geophysical studies.

TECHNICAL SPECIFICATIONS

Material: This UAV is made of toughened carbon fiber, plastic, and foam. It also has steel plates.

Size: length: 150 cm; width (wingspan): 400 cm; height; weight: MTOW is 50 kg

Propellers: It has one propeller at the rear end of the fuselage of the UAV. It has two blades.

Payload: 15 kg

Launching information: It has automatic take-off facility and software. It is equipped with autonomous flight control and preprogramed flight path. It lands autonomously, using landing gear.

Altitude: 100–3000 m a.g.l.

Speed: Reaches up to 120 km h^{-1}

Wind speed tolerance: 15 km h^{-1}

Temperature tolerance: −35 to +45°C. It withstands Russian cold winter

Power source: Internal combustion engine

Electric batteries: Lithium polymer batteries 8000 mA h

Fuel: Gasoline

Endurance: 12 h per flight

Remote controllers: It has autopilot mode. It has self-correcting software during flight. In case of wrong commands or obstacle, it avoids them. In emergency, it lands at the spot it was launched. It has software that checks full preflight using diagnostics.

Area covered: 300 ha per flight

Photographic accessories: Cameras: The payload often includes a full set of visual cameras (red, green, and blue wavelength), multispectral cameras, NIR, IR, and red edge for thermal imagery. Scanning laser radar, gamma ray spectroscope, and quantum magnetometer are other components of imaging and survey payload.

Computer programs: The images are instantaneously relayed to ground-station PCs and iPads, for processing. Images are also processed at the CPU, on the UAV itself, and final images are transmitted to iPads/PCs. Image processing is done using Agisoft 3D, Pix4D Mapper, or Photomod software.

Spraying area: This is not sprayer UAV.

Volume: Not applicable

Agricultural uses: It includes aerially photography and videography of land, topography, and cropping systems; crop scouting for NDVI, crop water stress index (via NIR and IR imagery), canopy traits such as leaf area, leaf chlorophyll, and crop-N status. Natural vegetation monitoring is also possible. Laser scanning of forests for height, biomass, and forest stand estimation could be done.

Nonagricultural uses: It includes aerial geophysical studies, magnetometer prospecting of earth's surface, video surveillance of mines, industries, rail roads, and buildings. Surveillance of mines and mining activity, monitoring and instructing transport vehicles, etc. are possible. This UAV could be specifically adopted, for magnetometer prospecting and geoelectric surveying (AVASyS, 2016).

PLATE 2.44.1 Gamma UAV system.
Note: Top, middle, and bottom left show different views of the UAV. Bottom right: Suitcase with entire UAV platform and accessories.
Source: Mrs. Lyudmila Piskunovs, AVASyS, Krasnoyarsk, Siberia, Russia.

Useful References, Websites, and YouTube Addresses

AVASyS LLC. Delta-M UAV System. *Autonomous Aerospace Systems-Geoservice. Krasnoyarsk, Siberia, Russia*; 2016; pp 1–8. http://uav-siberia.com/en/content/delta-m-uav-system?gclid=CNKdmZ3-0NACFdi (accessed Dec 1, 2016).

http://uav-siberia.com/en/content/gamma-uav-system pp 1–3 (accessed Dec 1, 2016).

http://vestnik-glonass.ru/en/news/implementation/russian-market-of-unmanned-aerial-vehicles-is-incremented-by-domestic-developments/ (accessed June 17, 2017).

Name/Models: Panther (and Minipanther)

Company and address: Israel Aerospace Industries, Telaviv, Israel.

Israel Aerospace Industries (IAI) is a government corporation. It conducts research on UAVs. It develops UAVS useful for civilians, military personnel, and agriculturists in their country and elsewhere. It was founded in 1953 by Mr. Shimon Peres.

IAI has developed its new revolutionary tilt-rotor UAV for tactical missions. It is named the "Panther." It was exhibited at the Latrun Conference in Israel, on October 5–7, 2010. This titled wing UAV technology combines, the efficiency and flight capabilities of both, the normal flat-winged airplane, and a helicopter-like VTOL (runway-free) plus hovering ability. Incidentally, these are traits that are most sought and useful to farming situations. Farmers/Specialists may make close observation of the crop, using the hovering abilities. At the same time, it has ability to cover large acreage of crop fields. It conducts aerial survey and photography, generally, possible with flat wing UAVs. This UAV was developed for use in the "Defence Department of Israel." Minipanther is a good example of tilted-winged UAV technology.

The portable minipanther system includes two planes and command control unit (GCS). It is easily carried on a backpack. A single operator can control both the UAV units. They are entirely automatic. "Minipanther" is very easy and quick to assemble. The larger version of "Panther" comes with a ground-station unit. It needs a mid-sized truck to carry it, to the place of installation.

TECHNICAL SPECIFICATIONS

Material: It is made of ruggedized plastic, carbon fiber, and foam

Size: length: 120 cm; width (wingspan): 240 cm; height: 45 cm; weight: 65 kg (minipanther 12 kg only)

Propellers: This UAV has tilted rotors. The propellers are attached to brushless motors. It adopts silent or very low acoustic noise motors for propulsion. This character is essential, if this UAV is used during defense missions.

Payload: Payload includes plug-in optronic cameras. They are meant for sharp aerial photography and detection of thermal changes.

Launching information: It is launched as a VTOL UAV. It does not need a runway. It has auto take-off and navigation, supported by custom-made software or Pixhawk. It lands automatically. In case of emergency, it returns to launch site.

Altitude: It operates at 50–300 or 400 m a.g.l., if adopted for aerial survey and scrutiny of crops. However, the ceiling is 3500 m a.g.l.

Speed: 60 km h^{-1}

Wind speed tolerance: 20 knots (Minipanther)

Temperature tolerance: −20 to +60°C

Power source: Electric batteries for minipanther

Electric batteries: 28 A h, two cell batteries.

Fuel: Not an IC engine-fitted UAV. No need for petrol or diesel.

Endurance: Minipanther loiters for 2 h per flight

Remote controllers: Its flight mission can be preprogramed using Pixhawk Autopilot software on an iPad or PCs at GCS. It is monitored and contacted using telemetry that operates for up to 50–60 km radius. The flight mission is often predetermined. However, it could be altered midway, using GCS telemetric commands.

Area covered: 60–120 km or 200–300 ha of crop land (minipanther).

Photographic accessories: Cameras: Cameras are held tightly without disturbance in a gyro-stabilized gimbal. The UAV has cameras such as visual (R, G, and B; Sony A 5100), multispectral, NIR, and IR (thermal). It has a Lidar pod. The images could be 2D or 3D.

Computer programs: The aerial images are processed instantaneously, by the software available at UAV's CPU. The orthoimages can be stitched and sharpened, at the ground station (iPad), using software such as Pix4D Mapper or Agisoft Photoscan.

Spraying area: This is not a sprayer UAV, but it could be modified using pesticide tanks and hovering mode.

Volume: No data available

Agricultural uses: Minipanther with its "tilted-wing technology" is best suited to serve farmers and companies with need, for aerial imagery of crops. The imagery could be used to assess seed germination percentage, seedling establishment, crop stand, and canopy traits. The UAV's imagery can provide details on leaf area, leaf-chlorophyll (NIR imagery), crop-N status, need for nutrients particularly N, and micronutrients. It is also used get canopy and air temperature data (i.e., crop water stress index) via thermal imagery. It is used to map flood or drought-affected area, soil pH disturbances, etc.

Nonagricultural uses: This is a UAV primarily developed to serve military/defense departments. In particular, it is used to survey, reconnaissance, and gather intelligence in battle fields. It is used to monitor international border regions. However, we must note that its versatility, as a swift flat-winged air plane and a hovering helicopter-like UAV, has massive advantages, particularly in agriculture and commercial fields, such as surveillance of large farms, industries, buildings, mines and mining activity, city traffic, etc.

Useful References, Websites, and YouTube Addresses

http://www.iai.co.il/32981-41360-en/MediaRoom_News.aspx (accessed Jan 10, 2018).

http://www.militaryfactory.com/imageviewer/ac/pic-detail.asp?aircraft_id=1093&sCurrentPic=pic1 (accessed Jan 10, 2018).

http://www.iai.co.il/2013/18892-en/BusinessAreas_UnmannedAirSystems.aspx (accessed Jan 10, 2018).
http://www.iai.co.il/Shared/UserControls/Print/PopUp.aspx?lang=en&docid=41360 (accessed Jan10, 2018).
http://www.iai.co.il/2013/36944-41636-en/BusinessAreas_UnmannedAirSystems.aspx (accessed Jan 10, 2018).
https://www.youtube.com/watch?v=IjXFnK8TtKc (accessed Jan 10, 2018).
https://www.youtube.com/watch?v=pIYEb-JrWTI (accessed Jan 10, 2018).
https://www.youtube.com/watch?v=QJcyx2qoWh8 (accessed Jan 10, 2018).

Name/Models: Sky Robot RX 450

Company and address: Robot Aviation, Brandbu, Grinakerlinna 115N-2760 Brandbu, Norway, Phone: +47 994 30 582; e-mail: info@robotaviation.com; Website: http://www.robotaviation.com/applications

"Robot Aviation AS" is a Scandinavian UAV company. It produces a range of fixed-flat wing and copter UAVs. They cater to applications such as military surveillance, general inspection and security of large industrial installations, buildings, etc. Some models of UAVs are more suitable for monitoring natural vegetation, agricultural zones, and irrigation channels. "Robot Aviation AS" offers a completely integrated UAV system. The UAV system has custom-made software, suitable cameras/sensors, and processing units (iPads).

"Sky Robot RX 450" is a swift launching UAV. It has a slightly larger wingspan. It is a relatively heavier UAV and runs on gasoline supported IC engine. It is used for several functions in farm or natural vegetation zone. The UAV offers detailed knowledge about crops, their growth pattern, need for nutrients/fertilizers, pesticides, weedicides, etc.

TECHNICAL SPECIFICATIONS

Material: It is made of light-weight composite of foam, aluminum, carbon fiber, and plastic. The material withstands long flight and endurance.

Size: length: 3.5 m; width (wingspan): 7.0 m; height: 1.5 m; weight: MTOW 180 kg, empty weight 80 kg

Propellers: It has one propeller at the front portion of fuselage (nose). The propeller has three blades.

Payload: External payload is 10 kg.

Launching information: This UAV launches automatically, using wheels for movement on runway. Usually, a 1000-m runway is needed. It has autonomous navigation based on preflight computer-aided instructions. However, the GCS can modify flight path, if necessary, using telemetry. The landing is usually done on hard ground of 100–1000 m, using a landing gear with four wheels.

Altitude: It operates at 1000–400 m a.g.l., but the ceiling is 8000 m a.g.l.

Speed: 60–90 km h^{-1}

Wind speed tolerance: 25–40 km h^{-1}

Temperature tolerance: −25 to +45°C

Power source: IC engine; a 2-stroke 2-cylinder engine

Electric batteries: 8 mA h lithium polymer batteries are utilized to support CPU and other electronic functions

Fuel: Gasoline, heavy petroleum fuel

Endurance: 30 h per flight

Remote controllers: This UAV has a long endurance. It is capable of long-distance flight and larger area coverage. The radius of operation that is telemetric contact with ground control lasts up to 1650 km.

Area covered: 1000–5000 ha per flight is a clear possibility

Photographic accessories: Cameras: photographic equipment includes gimbal that is fully gyro-stabilized. It holds several sensors that operate at visual, NIR, IR, and red-edge wavelength bands. It also has radar and Lidar facilities. The aerial imagery can be guided from GCS, using telemetric connections. Otherwise, it could be preprogramed using software.

Computer programs: The aerial imagery sent could be processed using Pix4D Mapper or Agisoft Photoscan or similar custom-made software. The images derived are of high resolution, very sharp, and tagged with GPS coordinates. This UAV is also compatible with GLONASS.

Spraying area: This is not a sprayer UAV.

Volume: Not applicable

Agricultural uses: It includes aerial survey, mapping, and data collection such as NDVI, green index, crop maturity status, and cropping system changes. This UAV with its ability for long-range flight and endurance can be effectively utilized, to study large agroecosystems, crops in large patches, sometimes covering the entire small nation. Crop scouting for disease and pest attack and weed infestation in an agroecosystem could be done. Consequently, remedial measures could be channeled quickly. This UAV suits to assess and monitor changes in natural vegetation. The data could be utilized by policymakers for macroplanning.

Nonagricultural uses: Sky Robot 450 suits well to surveillance military installations and vehicles. Monitoring is done from high altitudes. Therefore, the UAVs can track vehicles for longer distances. This UAV is useful in natural resource monitoring, mapping the various geographic features, soil types, water bodies, dams, rail roads, highway traffic, etc. This UAV can also be utilized, to monitor mines and mining activity, and ore movement. Large-scale disasters such as fire, droughts, floods, and destructions of infrastructure could be aerially surveyed and identified. The aerial data could be provided rather quickly for rehabilitation.

Useful References, Websites, and YouTube Addresses

http://www.robotaviation.com/ (accessed Dec 13, 2017).

2.5 FIXED-WINGED UNMANNED AERIAL VEHICLES HAVING WINGS INTEGRATED OR BLENDED WITH FUSELAGE

Fixed-winged UAVs with integrated fuselage (body) and wings are perhaps most popular with farmers in Europe and North America (e.g., eBee, UX5, Bramor, Delta, etc.). They are small, yet highly versatile regarding portability, flight path, speed, and aerial imagery. Their aerodynamics is said to enhance energy efficiency of electric batteries. These models are usually microweight UAVs and are launched by hand. They are easily packed in suitcase or backpack and transported.

Name/Models: ATLAS sUAS

Company and address: C-ASTRAL Aerospace Ltd. Gregorciceva ulica 20, 5270 Ajdovscina, Slovenia; phone: +386 (0) 40121119; e-mail: info@c-astral.com; website: http://www.c-astral.com

C-ASTRAL Aerospace is a fixed-wing UAV producer. They offer a few different models of flat-winged UAVs. These UAVs are capable of a variety of tasks related to aerial mapping, surveillance, and monitoring (tracking). They also offer custom-made software that suits their UAVs. Such software is relevant for guiding flight paths. Also, processing images received at the UAV's CPU or at the ground station. Their most recent models are named "Bramor ppX," "Bramor gEO UAV," "Bramor C4EYE," "Long Endurance sUAS," and "ATLAS sUAS." A few of the UAV models are yet to be released for public.

The ATLAS (The Advanced Light Acquisition System) is an ideal flat wing UAV for agriculture, forest, and land scape management (Plate 2.45.1). It is a light-weight UAV. It is capable of aerial survey of crop lands, swiftly, with a flight endurance of 1 h. It has a sleek design with antenna and GPS integrated to airframe.

TECHNICAL SPECIFICATIONS

Material: This UAV is made of foam (honeycomb), plastic, and carbon fiber. The material is ruggedized to withstand rough weather and landing.

Size: length: 82 cm (fuselage 61 cm); width (wingspan): 155 cm; height: 30 cm; weight: 2.3 kg

Propellers: It has one propeller placed at the front tip of the fuselage. Propeller has two strong plastic blades.

Payload: It has multimodular payload that includes gimbal. The gimbal holds 2–3 sensors. Payload is 1.2 kg

Launching information: We can assemble the UAV quickly. It is a hand-launched UAV

Altitude: It operates at 50–150 m during aerial survey of agricultural zones, crops, and ground features. However, ceiling is 3500 m

Speed: 45–60 km h^{-1}

Wind speed tolerance: 33 knots

Temperature tolerance: −25 to +45°C

Power source: Electric batteries

Electric batteries: 14,000 mA h

Fuel: Not an IC engine fitted UAV. No need for petrol/diesel

Endurance: 1 h per flight

Remote controllers: The GCS has iPads that are connected to UAV via Internet and telemetry. They keep contact for about 15 km radius from ground station. The UAV has autopilot mode, using Pixhawk software. The flight path of UAV can be modified midway, using ground iPads.

Area covered: 150–200 ha per flight

Photographic accessories: Cameras: A full complement of sensors includes visual cameras (e.g., Sony 7 A), Mica-sense multispectral camera, NIR, IR, and red-edge camera.

Computer programs: UAV-derived images are processed, using "PixD data processing software" bundle. The CPU on the UAV that processes and stores images and digital data is rainproof.

Spraying area: This is not a sprayer UAV.

Volume: Not applicable

Agricultural uses: The ATLAS UAV is supposedly most suited for aerial survey of agricultural zones, woods, forests, landscape, environ-ecological monitoring, monitoring flood, and drought-affected regions. Aerial imagery can also be used to depict maladies such as soil erosion, salt-affected regions, etc. Crop scouting for general growth, canopy traits, nutrient deficiencies, disease/pest attack, grain maturity, and forecast grain/biomass yield is also possible.

Nonagricultural uses: They are surveillance of Industrial units, buildings, vehicle traffic, tracking vehicles, mines, mining activity, oil pipelines, electrical installations, riverine belts, dams, etc. It has ability to collect and analyze atmospheric samples, for particulate and gaseous contamination. It is apt for getting data about climate change effects.

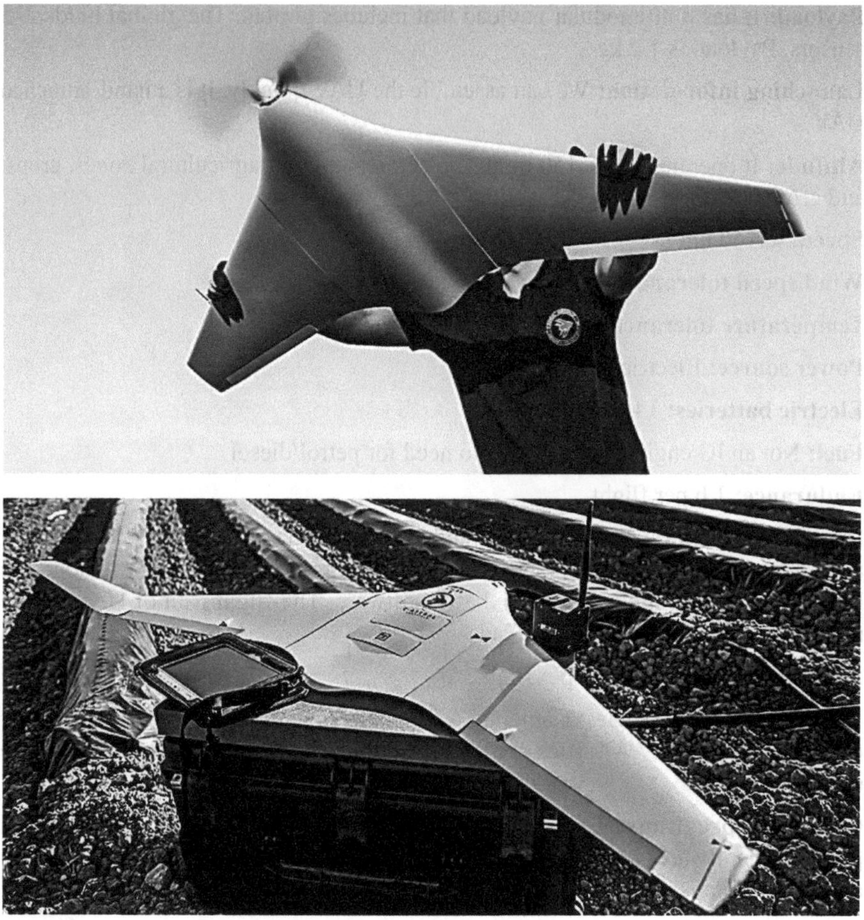

PLATE 2.45.1 ATLAS sUAS.
Source: Mr. Astrid Svara, C-Astral Ltd. Adjovscina, Slovenia.
Note: The wing and fuselage body are closely integrated.

Useful References, Websites, and YouTube Addresses

http://www.unmannedsystemstechnology.com/2015/05/c-astral-unveils-new-hand-launch-atlas-suas/ (accessed July 20, 2017).

http://www.unmannedsystemstechnology.com/2015/05/c-astral-unveils-new-hand-launch-atlas-suas/ (accessed July 13, 2017).

http://www.unmannedsystemstechnology.com/company/c-astral-aerospace-ltd/ (accessed July 13, 2017).

https://www.youtube.com/watch?v=aRNk2QQx8-g (accessed July 13, 2013).

Name/Models: Bramor C4EYE UAS

Company and address: C-ASTRAL Aerospace Ltd. Gregorciceva ulica 20, 5270 Ajdovscina, Slovenia; Phone: +386 (0) 40121119; e-mail: info@c-astral.com; website: http://www.c-astral.com

Bramor C4EYE is a UAV with blended fuselage (body) and wings. This has been done with a view to give smallest possible weight to the UAV and offer advanced aerodynamics, electric propulsion, and completely autonomous operation in the air. This UAS is useful to both commercial (including agricultural) and military operations (Plate 2.46.1).

TECHNICAL SPECIFICATIONS

Material: The airframe of the UAV is made of kevlar, carbon, and vectran composite. It provides high survivability and no radar signature.

Size: length: 96 cm, central module, length: 67 cm; width (wingspan): 230 cm; height: 35 cm; weight: take-off weight is 4.5 kg

Propellers: It has a propeller at the rear end of fuselage of the blended airframe. The propeller has two blades.

Payload: It includes gyro-stabilized gimbal. It holds sensors such as EO, IR, multispectral, hyperspectral, 10-MP videographer, thermal IR, and red-edge cameras.

Launching information: It takes-off using a catapult. It has autopilot that depends on custom-made software, such as "Pixhawk" or "C-ASTRAL Pilot 3C." Landing is accomplished, using a parachute that opens in the mid-air based on instructions, from ground control iPads or as the program directs.

Altitude: It operates at 40–5000 m depending on the purpose

Speed: Its cruise speed 16 m s^{-1}. Maximum horizontal speed is 30 m s^{-1}.

Wind speed tolerance: 30 knots

Temperature tolerance: −25 to +45°C

Power source: Lithium–polymer batteries. It has a brushless motor connected to battery.

Electric batteries: 14,000 mA h, two cell.

Fuel: Not an IC engine-fitted UAV. No need for petrol or diesel.

Endurance: 3 h

Remote controllers: The GCS contact extends for an operational range of 40 km. The ground command has an 868 or 900 MHz radiolink with the UAV, during its flight. The ground controls telemetric commands and videographic relay operates, for up to 40 km distance. This UAV has 4 Hz GPS/GLONSS antenna for navigation.

Area covered: It covers 150 km or 5000 km^2 per flight

Photographic accessories: Cameras: Sensors are in a gyro-stabilized gimbal (EYE-X EO/IR/LI Gimbal). Sensors include visual camera (Sony 7 A), mica-sense multispectral camera, NIR, IR, and red-edge cameras. The videographer offers 10-MP videoimagery. It has a thermal IR 640 × 480 LWIR and laser pointer.

Computer programs: The UAV-derived imagery is processed using software located in the CPU or at ground-station IPad. Software such as C-ASTRAL custom-made processors or Pix4D Mapper is utilized.

Spraying area: Bramor C4EYE is not a sprayer UAV.

Volume: Not applicable

Agricultural uses: It includes aerial survey, videography and monitoring of land resources, soils, and cropping systems. The UAV's sensors offer data about NDVI, leaf, and crop-N status, and also crop's water stress index. Crop scouting for disease, insect attack, floods or drought-affected zones, grain maturity status, etc. is other functions of the UAV. Natural vegetation monitoring is also done using this UAV.

Nonagricultural uses: They are aerial survey of forests and civilian zones in case of fires, and aerial surveillance of infrastructure such as industrial installations, buildings, bridges, oil pipelines, rail road, etc. This UAV could be used to track city vehicles, agricultural tractors, farm vehicles, etc. Surveillance of mines, tracking highway transport vehicles using electric tracker, etc. is possible. This UAV also has been so useful in search and rescue missions in disaster zones, military battle fields, mountains and difficult terrain, etc.

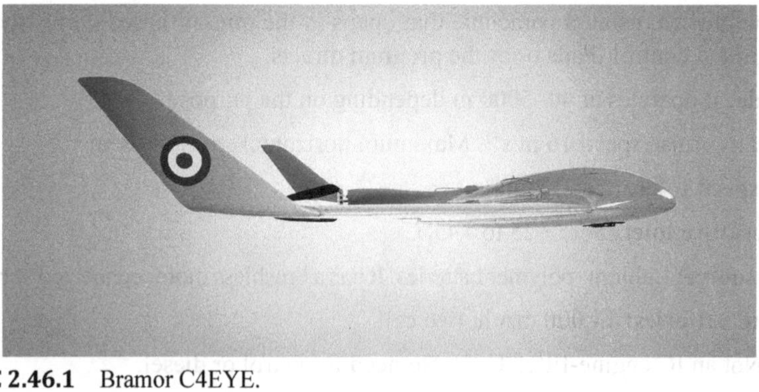

PLATE 2.46.1 Bramor C4EYE.
Source: Ms. Astrid Svara, C-Astral Aerospace Ltd., Ajdovscina, Slovenia.

Useful References, Websites, and YouTube Addresses

http://www.c-astral.com/en/products/bramor-c4eye/ (accessed July 14, 2017).
https://www.youtube.com/watch?v=w3S6zSG5GRk (accessed July 14, 2017).
https://www.youtube.com/watch?v=9pefhTuoNN4 (accessed July 14, 2017).

Name/Models: Bramor gEO UAV

Company and address: C-ASTRAL Aerospace Ltd. Gregorciceva ulica 20, 5270 Ajdovscina, Slovenia; phone: +386 (0) 40121119; e-mail: info@c-astral.com; website: http://www.c-astral.com

This UAV, "Bramor gEO" was first designed and developed, for testing in the year 2006. Most recent updating of instrumentation, software, and hardware was done in 2014. Bramor gEO UAV can be carried on shoulders and launched, from anywhere in a field, using catapult (Plate 2.47.1). It has typical applications in obtaining 2D and 3D aerial imaging and surveying of earth's surface features and conditions. The resolution of the imagery is excellent and down to 1 cm accuracy. Bramor gEO has been utilized in surveillance of infrastructure, industries, military installations, pipelines, etc.

TECHNICAL SPECIFICATIONS

Material: The body of the UAV is made of kevlar, veraclan, carbon fiber and plastic composite. It has wings blended with fuselage. This is supposed to enhance fuel efficiency. It also avoids detection by radars, when the UAV is in flight, spying on military and other activities.

Size: length: 90 cm, 67 cm at the fuselage zone; width (wingspan):230 cm; height: 15 cm; weight: 4 kg

Propellers: It has one propeller at the rear end of the fuselage portion. It has two blades. Propeller is shielded. Propulsion is achieved by drawing power from electric batteries.

Payload: 1.0 kg

Launching information: It is launched using a catapult. It is kept in autopilot mode during navigation and landing, by using "Pixhawk" software. The UAV returns to point of launch, in times of emergency or errors in guidance commands. The UAV usually lands within 15 m accuracy. Piloting the UAV requires only one ground crew. The crew can be trained in a matter of 15 h of UAV flight.

Altitude: The ceiling is 5000 m a.g.l.

Speed: 21 m s^{-1}

Wind speed tolerance: 15 km h^{-1}

Temperature tolerance: −25 to +50°C

Power source: Electric batteries

Electric batteries: 14,000 mA h

Fuel: Not an IC engine-fitted UAV. No need for petrol or diesel.

Endurance: 1 h per flight

Remote controllers: The GCS connections with UAV are based on telemetry and internet. The connectivity functions for 40 km radius. The UAV flight path is usually, fully preprogramed, and fixed. Flight planning and revisions if any are done, using the software "Geopilot." It has collision avoidance software. It has autonomous landing, in case of emergency. The GCS that comes along with the UAV is "Panasonic CF-19." The GCS computers have control over sensors and photographing events. It also has real time videodownload link.

Area covered: The UAV covers up to 150 km in linear distance per flight or 200 km^2 in an agricultural field.

Photographic accessories: Cameras: Onboard scanning of earths features are done using sharp cameras. Cameras include 24 MP with 19 mm or 30 mm lens. It has sensors for NIR, IR, multispectral, and hyperspectral imagery. It also has laser spectrometer. Onboard sensors are compatible with GPS, GLONASS, barometers, and compass.

Computer programs: The computer software included in the UAV are meant for block adjustment, camera self-calibration (Pioneering, EnsoMosaic, Pix4D, Agisoft, Menci), flight planning (GeoPilot), photogrammetric software (Pioneering, EnsoMosaic, Pix4D, Agisoft, Menci), and point-cloud processing software. The software is used to process orthomosaics, and obtain 2D and 3D images of landscape, etc.

Spraying area: This is not a sprayer UAV.

Volume: Not applicable

Agricultural uses: Bramor gEO is utilized to accomplish a series of operations, relevant to land resource evaluation, soil type survey, monitoring agronomic procedures such as planting, detecting gaps and planting density, crop canopy features, crop growth and biomass accumulation trends. Crop scouting leaf chlorophyll, leaf-N, crop water stress index (IR imagery), disease, weeds, insect attack (spectral signatures), drought, flood, or soil erosion effects is also done. Natural vegetation monitoring is also possible. This UAV is also capable of tracking farm vehicles and their progress using autotraffic devices. This UAV helps farmers to guide and detect progress of farm work from a ground station.

Nonagricultural uses: There are a range of uses for this UAV in surveillance of military installations, battle fields, convoys, etc. It is sued to monitor large buildings, pipelines, electric lines, rivers, dams, vehicular traffic in highways and in city. Surveillance of mines and mining activity are few other functions.

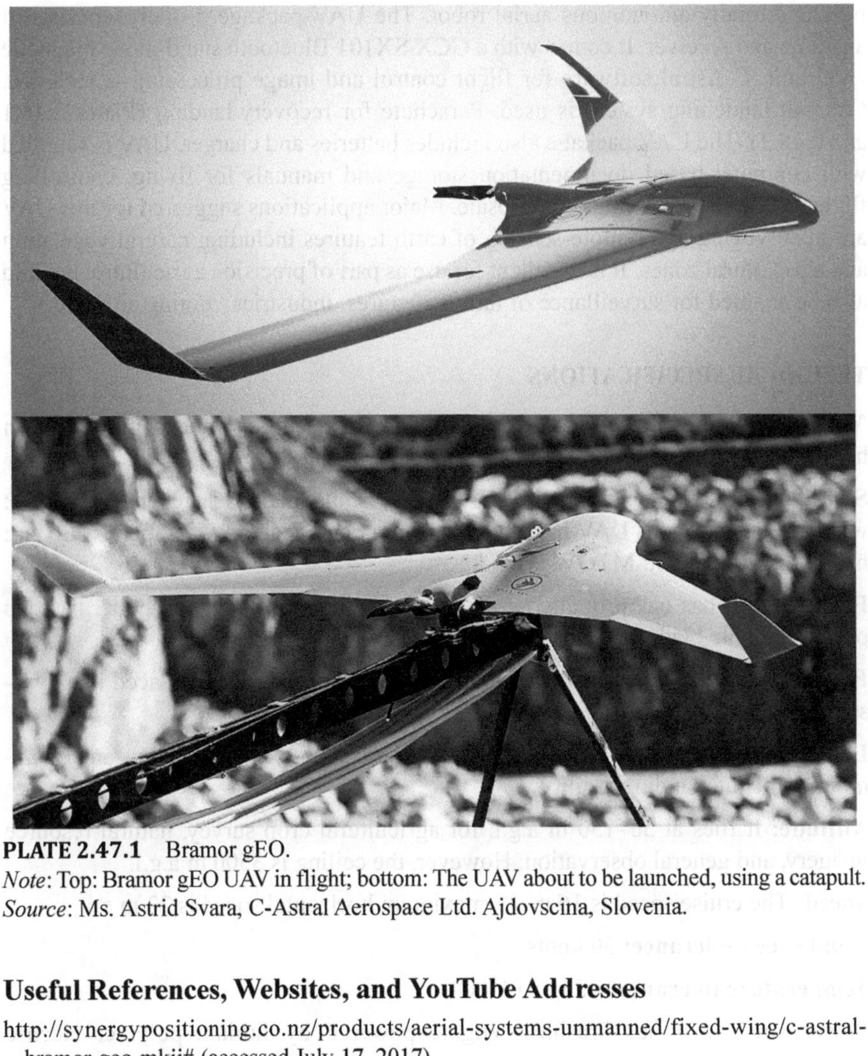

PLATE 2.47.1 Bramor gEO.
Note: Top: Bramor gEO UAV in flight; bottom: The UAV about to be launched, using a catapult.
Source: Ms. Astrid Svara, C-Astral Aerospace Ltd. Ajdovscina, Slovenia.

Useful References, Websites, and YouTube Addresses

http://synergypositioning.co.nz/products/aerial-systems-unmanned/fixed-wing/c-astral-bramor-geo-mkii# (accessed July 17, 2017).

http://geo-matching.com/products/id2510-bramor-geo-uav.html (accessed July 17, 2017).

https://www.youtube.com/watch?v=G0vwlN3RhgE (accessed July 17, 2017).

Name/Models: Bramor ppX

Company and address: C-ASTRAL Aerospace Ltd. Gregorciceva ulica 20, 5270 Ajdovscina, Slovenia; Phone: +386 (0) 40121119; e-mail: info@c-astral.com; website: http://www.c-astral.com

The "BRAMOR ppX" is the ultimate high productivity tool for global surveying and remote sensing. It suits for use in very remote and less remote areas. The Bramor ppX is a totally autonomous aerial robot. The UAV package, usually, consists of airframe and receiver. It comes with a GCX SX101 Bluetooth stand-alone magnetic GCS unit. C-Astral software for flight control and image processing is included. Catapult launching system is used. Parachute for recovery/landing (Plates 2.48.1 and 2.48.2). The UAV package also includes batteries and charger. UAV is supplied with computer-based documentation storage and manuals for flying, controlling flight path, and processing orthomosaic. Major applications suggested for this UAV are in surveying and remote-sensing of earth features including natural vegetation and agricultural zones. It is excellent for use as part of precision agriculture. It could also be adapted for surveillance of farm structures, industries, mining area, etc.

TECHNICAL SPECIFICATIONS

Material: Bramor ppX is made of kevlar, vectran, carbon fiber composite with honeycomb structure.

Size: length: 96 cm; central module length, that is, blended fuselage length is 67 cm; width (wingspan): this UAV has a blended wing body design. Wing span is 230 cm; height 22 cm; weight: MTOW is 4.7 kg

Propellers: It has one propeller at the rear end of fuselage. The propeller has two plastic foldable blades.

Payload: 2.2 kg. The payload includes a series of sensors EO/IR placed in a gyro-stabilized gimbal.

Launching information: This UAV is launched using a catapult. It has an autonomous parachute landing facility.

Altitude: It flies at 50–150 m a.g.l. for agricultural crop survey, natural resource imagery, and general observation. However, the ceiling is 3500 m a.g.l.

Speed: The cruise speed is 16 m s^{-1}, maximum horizontal speed is 23 m s^{-1}.

Wind speed tolerance: 30 knots

Temperature tolerance: −25 to +45°C

Power source: It has brushless engine powered by lithium–polymer electric batteries.

Electric batteries: 14,000 mA h lithium–polymer battery; two cells.

Fuel: Not an IC engine-fitted UAV. No need for petrol/diesel

Endurance: 3.5 h flight

Remote controllers: The GCS telemetry has 868 MHz radiofrequency connection with the UAV. The GCS iPads can control UAV's movement within a radius of up to 40 km. Flight path can be preprogramed by GCS iPad software such as "C-ASTRAL pilot software" or "Pixhawk." It comes with failsafe return mode. It leads the UAV to launching point, if errors occur in navigation. This UAV has software compatible with GPS, GLONASS, Beidou, and Galileo.

Area covered: It covers a linear distance of 150 km per flight or about 15 km^2 per flight

Photographic accessories: Cameras: It comes with usual complement of sensors such visual (R, G, and B, multispectral, hyperspectral, NIR, IR, and red-edge cameras.

Computer programs: The imagery provided by UAV's sensors can be processed at the CPU, on the UAV itself. Then, it can be relayed to ground control iPad touchscreens for analysis. As an alternative, ground control iPads can process orthomosaics received.

Spraying area: This is not a sprayer UAV.

Volume: Not applicable

Agricultural uses: "Bramor ppX" is useful during the conduct of "Precision Agriculture" procedures. Its sensors provide detailed maps of variations in crop-growth pattern, nutrient deficiencies, water stress index, and grain maturity maps. This helps in running planters, fertilizer application equipment and combine harvesters, appropriately, using the digital data provided by UAVs. The Bramor ppX UAV is also used to survey and map land surface, soil type, water resources, farm ponds, dams, etc. UAV's imagery also provides details about crop and, disease affliction and its variations, insect pest attacked spots, etc. This UAV is excellent for natural vegetation monitoring.

Nonagricultural uses: "Bramor ppX" conducts remote sensing both at low and mid-altitudes. It is used for surveillance of building, industrial infrastructure, mines, rail roads, city transport vehicles, etc.

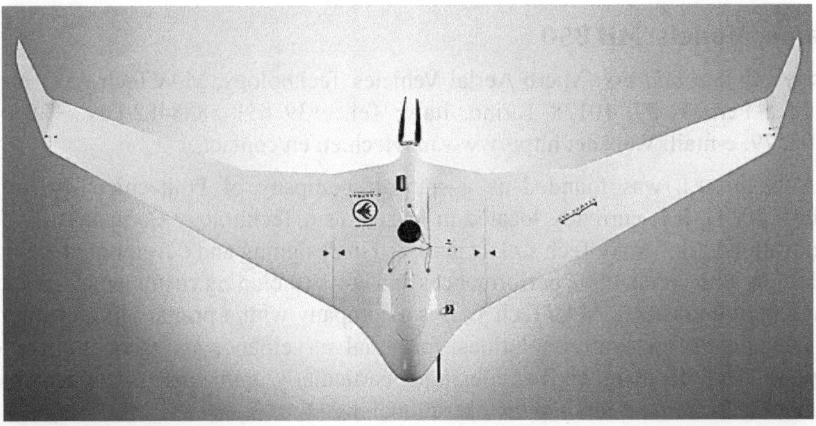

PLATE 2.48.1 Bramor ppX.
Source: Mr. Astrid Svara, C-Astral Aerospace Ltd., Adjovscina, Slovenia.

PLATE 2.48.2 An iPad ground control with C-ASTRAL Pilot C ³P Software. This UAV's computer is compatible with 3D survey, Agisoft photoscan, Pix4D Mapper, EnsoMosiac processor, and Menci software.
Source: Ms. Astrid Svara, C-Astral Aerospace Ltd., Ajdovscina, Slovenia.

Useful References, Websites, and YouTube Addresses

http://www.c-astral.com/en/products/bramor-ppx (accessed July 26, 2017).
https://www.youtube.com/watch?v=i_Pq0Sqn79U (accessed July 26, 2017).
https://www.youtube.com/watch?v=i_Pq0Sqn79U (accessed July 26, 2017).

Name/Models: MH 850

Company and address: Micro Aerial Vehicles Technology, MAVTech s.r.l., corso Galileo Ferraris 57, 10128 Torino, Italy; Tel.: +39 011 5808482/Fax: +39 011 5808579; e-mail: Website: http://www.mavtech.eu/en/contacts/.

MAVTech s.r.l. was founded as a spin-off company of Politecnico di Torino (2005–2014), It is currently located in Bozen, as a Technology Company of "TIS Innovation Park." MAVTech s.r.l. is involved in designing and developing technical solutions with competitive performances and costs (including customer support and training of end users). MAVTech s.r.l. is a company with a primary interest, in the development of innovative solutions, for aerial surveillance and tactical operation support. They focus on civil applications, particularly, transferring new aerospace technologies from research to the operational domain.

MH850 is a fixed-wing, tailless integrated wing-body aircraft. These configurations have a wider application, because of their payload capabilities and flight performances. They supposedly have better mission range and endurance. The design is compact with an adequate aerodynamic efficiency. UAV's structures are light, the stall is smooth, and the vehicle is spin resistant and stable, during flight (Plate 2.49.1).

TECHNICAL SPECIFICATIONS

Material: It is made of a composite including GF/EPP/ABS. The UAV is packed into suitcase. More frequently, it is packed in a tailor-made backpack. It is easily portable.

Size: length: 75 cm; width (wingspan): 85 cm; height: 35 cm; weight: 1.0 kg

Propellers: Propulsion is attained via brushless high-performance motor connected to propeller.

Payload: 200 g excluding batteries

Launching information: MH 850 is hand-launched. Otherwise, it could be done using catapult. Recovery is possible via autonomous or radiocontrolled landing on belly. Parachute-aided landing is also possible. The UAV is supplied with autopilot software for navigation and autolanding. A flight-simulator software is also available for use prior to programing the flight paths. A version of MH850 without autonomous navigation is also available. In that case, UAV must be guided, using "joy stick" or remote controller at the ground.

Altitude: It flies at 50–5000 m a.g.l. Typical operational altitude is 150 m a.g.l.

Speed: 7.5 + 20 m s^{-1}

Wind speed tolerance: It tolerates 20 km h^{-1} adverse wind disturbance.

Temperature tolerance: −20 to +60°C

Power source: Electric batteries

Electric batteries: 14 A h, two cell, lithium–polymer batteries.

Fuel: Not an IC engine-fitted UAV. No need for petrol or diesel.

Endurance: 60 min per flight

Remote controllers: The UAV is connected to GCS, via radiolink (2.4 GHz in flight override). MAVTech s.r.l has released a new software package ("JavaCube") that focuses on the specific needs and requirements of flight planning. It avoids collision and it includes grid waypoint planning for optimized photogrammetric surveys. The flight path of the UAV can be preprogramed using autopilot. It is also amenable for revision of flight path using commands from GCS computers (Guglieri and Ristorto, 2016).

Area covered: It covers 40–60 km linear distance per flight or 200 ha, if crop fields are surveyed.

Photographic accessories: Cameras: Standard visual, multispectral, NIR, IR and thermal cameras are attached to the UAV. Video camera for aerial survey comes as a default with the UAV.

Computer programs: The aerial images are processed, using real time data-link, to the GCS computer (iPad). A software such as Pix4D Mapper or Agisoft Phtotscan is utilized. It provides images with GPS tags (coordinates). The UAV can transmit digital data, for up to 10 km radius from the GCS radius.

Spraying area: This is not a sprayer UAV.

Volume: Not applicable

Agricultural uses: Crop scouting is a major purpose of this UAV. It is applied to survey crop fields for variety of aspects such as terrain mapping; land and forestry research; crop health and disease monitoring; vegetation identification; invasive species identification and analysis; vigor mapping; and frost mitigation.

Nonagricultural uses: They are surveillance of industries, buildings, highways, rail roads, mines, and tracking transport vehicles, etc. are important functions.

PLATE 2.49.1 Left: A MH 850 fixed-winged UAV. Right: A backpack, with a full complement of UAV and accessories.
Note: MH850 is UAV with wings integrated (or amalgamated) with fuselage of the body.
Source: Dr. Fulvia Quagliotti, President & CEO, and Mr. Gianluca Ristorto, Micro Aerial Vehicles Technology srl., Torino, Italy.

Useful References, Websites, and YouTube Addresses

Guglieri, G.; Ristorto, G. In *Safety Assessment for Light Remotely Piloted Aircraft Systems*. In International Conference on Air Transport; 2016; pp 1–7
http://www.mavtech.eu/en/company/ (accessed Sept 4, 2017).
http://www.mavtech.eu/en/applications/precision-farming/ (accessed Sept 4, 2017).
http://www.mavtech.eu/it/prodotti/drone-mh850/ (accessed Sept 4, 2017).
http://www.mavtech.eu/ (accessed Sept 4, 2017).
http://www.mavtech.eu/it/contacts/ (accessed Sept 4, 2017).
https://www.youtube.com/watch?v=exQlTpR2X68 (accessed Sept 4, 2017).

2.6 FIXED-WINGED UNMANNED AERIAL VEHICLES WITH VERTICAL TAKE-OFF AND LANDING (HYBRIDS)

These UAVs are known as "hybrids" because they possess propellers that allow vertical lift, just like a copter UAV. At the same time, they achieve forward thrust and higher speeds using propellers place horizontally. They do not require a runway because of VTOL facility. Many farmers may prefer to possess a VTOL fixed-winged UAV that seems easy to handle and versatile during operation. The fixed-winged VTOL UAVs may be small and light-weight or even very large and heavy.

Name/Models: ALTI Transition

Company and address: ALTI UAS Inc. 1, Waunhout Avenue, Knysna, Western Cap, South Africa; phone: + 27 44 382 7051; e-mail: info@altiuas.com, sales@altiuas.com; website: http://www.altiuas.com

This UAV company, ALTI UAS, was initiated in 2010 at Western Cape, South Africa. The ALTI Transition UAV and a few other models (e.g., SteadiDrone) were developed by this company.

ALTI transition UAV is a small multipurpose UAV. It has a great role in agricultural settings. It is supposedly efficient in terms of cost and economics of operation in crop fields. It offers reliable and accurate data, to farmers through its sophisticated gimbal and an array of payload cameras, particularly, EO visual and IR sensors (Plate 2.50.1).

TECHNICAL SPECIFICATIONS

Material: ALTI transition is made of carbon composite body and propellers. Wings and tail boom are easily detachable, without any tools.

Size: length: 250 cm; width (wingspan): 276 cm; height: 49 cm; weight: MTOW 15 kg

Propellers: It has one propeller placed at the rear end of the fuselage. It has 3 blades of 18 in. length. Propellers are attached to brushless electric motors. For thrust, propellers are connected to 80 A motor. It offers reliable thrust for UAV to get airborne and move at stipulated speeds. For vertical lift-off, there are four propellers fixed to wings. They are located on either side of fuselage.

Payload: Maximum payload 1 kg (mainly sensors EO/IR)

Launching information: The UAV can be deployed rapidly. Beginning from removal of parts from the suitcase till take-off, it takes only 10 min, to assemble UAV. Airframe is assembled without use of any tools (Plate 2.50.1).

Altitude: It operates at <400 ft for agricultural surveillance and aerial photography. However, the ceiling is 10,000 ft

Speed: 45–60 km h^{-1}

Wind speed tolerance: 25 km h^{-1}

Temperature tolerance: −20 to +60°C

Power source: Electric batteries. Electric fuel for take-off lasts for <10 min. It is a quick take-off UAV.

Electric batteries: 8–14,000 mA h lithium–polymer batteries.

Fuel: Not an IC engine-supported UAV. No need for petrol or diesel.

Endurance: 6 h per flight

Remote controllers: The GCS computer decides flight path, data collection timing and type, and videography. The GCS has weather-proof computers, iPads and telemetry options. Telemetric connections are limited to 6 km from GCS.

Area covered: It covers a distance 300 km linear distance or 50–200 ha of farmland, based on flight path.

Photographic accessories: Cameras: The UAV has a payload of visual bandwidth cameras (R, G, and B), multispectral and hyperspectral cameras, NIR, IR, and red-edge cameras.

Computer programs: Aerial images are processed, using Pix4D Mapper or Agisoft Photoscan

Spraying area: This is not a spraying UAV.

Volume: Not applicable

Agricultural uses: "ALTI Transition" is a small UAV capable of a very wide range of applications in agricultural regions. Its sensors offer high-resolution 2D and 3D images/video of terrain, land topography, soil type and cropping systems. It is an excellent UAV to scout crops for growth, biomass accumulation, NDVI, leaf area and canopy traits, and forecast grain yield. It is useful in detecting diseases/pest, drought affected, or flooded zones in large crop fields. Natural vegetation monitoring could be done with good accuracy using this UAV.

Nonagricultural uses: They are surveillance of industries, mines, transport vehicles, public events, city traffic and buildings are possible.

PLATE 2.50.1 ALTI Transition—A fixed-wing VTOL UAV.
Source: Dr. Ian Freemantle, ALTI IIAS, Western Cape, South Africa.

Useful References, Websites, and YouTube Addresses

http://www.altiuas.com (accessed Nov 27, 2017).

http://www.altiuas.com/?gclid=CNGi-qyZ5NACFZAEaAodnrUCfg (accessed Nov 27, 2017).

https://www.youtube.com/watch?v=L1Pptod6MZc (accessed Nov 27, 2017).

https://www.youtube.com/watch?v=mVovWJA3Fmg (accessed Nov 27, 2017).

Name/Model: F2 VTOL fixed-winged UAV

Company and address: GTF aviation Technology (Beijing) Co. Ltd., International Negotiate Garden Building Room 614, Shunyi District, 101300 Beijing, Beijing District, China (Mainland); phone: +86 1823 1133680; e-mail: info@GTF.com; Website: https://gtfdrone.en.alibaba.com/company_profile.html?spm=a2700.8304367.0.0.C5iZSV.

GTF Aviation Technology operates its manufacturing unit in China. It hires about 51–100 personnel, to develop, design and build the fixed-winged UAVs. The company has invested over US$ 2.5 million to manufacture the F2 model of the fixed-winged VTOL UAV. They have sold and popularized this UAV in different continents.

The "F2 model" is a vertical take-off fixed-winged UAV. It could be fitted with a series of sophisticated sensors for aerial survey, surveillance of ground activity, and imaging of crop fields in greater detail. It is relatively a smaller UAV. It can be packed in a suitcase for portability. It can be assembled rapidly anywhere in a crop field or urban location. The cost of the UAV ranges from US$ 15,000 to 45,000 per unit (February 2017 price level).

TECHNICAL SPECIFICATIONS

Material: It is made of carbon fiber and aviation aluminum, plus plastic.

Size: length: 65 cm; width (wingspan): 130 cm; height: 30 cm; weight: 7 kg including 4 kg payload

Propellers: It has four propellers. The propellers are foldable to face upward, while on lift-off or in horizontal position, during forward thrust and propulsion.

Payload: 4 kg. It includes a gimbal, series of cameras, and CPU.

Launching information: It is a VTOL fixed-winged UAV, that is, a "hybrid." It does not need a runway space. It is not hand launched. It is an autolaunch, autonavigation, and landing UAV. The flight path is usually predetermined, using software such as "Pixhawk Autopilot."

Altitude: It reaches 300–400 m above the crop surface during aerial imagery. It can also hover at low altitudes, if close-up imagery is to be conducted. The ceiling is 3000 m above the crop's canopy.

Speed: 45 km h^{-1}, during aerial survey and 60 km h^{-1}, during loiter mode.

Wind speed tolerance: 20 km h^{-1} disturbance

Temperature tolerance: −10 to +60°C

Power source: Electric batteries

Electric batteries: 28,000 mA h.

Fuel: Not an IC engine-fitted UAV. It does not require gasoline

Endurance: 1.2–4 h, depending upon flight pattern and payload.

Remote controllers: The GCS is an iPad with software for determining flight path, waypoints, and camera shutter operation speed and frequency. The iPad software allows us to alter, flight path, and shutter frequency midway through the predetermined flight path. It has override facility. The telelink keeps intact with UAV up to 15 km radius. The UAV can be operated manually, using a remote controller (joystick type).

Area covered: 50–250 ha/day of 6 h

Photographic accessories: Cameras: The UAV is equipped with gyro-stabilized gimbal. The gimbal holds cameras such as visual (R, G, and B), NIR, IR, and red edge. It also has Lidar sensor. It can be fixed with vehicle tracking device.

Computer programs: The UAV's imagery is processed, using Pix4D Mapper, Agisoft Photoscan, or Trimble's Inphos.

Spraying area: This is not a spraying UAV.

Volume: Not applicable

Agricultural uses: This UAV is used most often to conduct aerial survey. It is used prepare digital map and map of land resources, soil types, crop fields, and crop growth status. Crop scouting to measure canopy growth rate, leaf chlorophyll status (crop-N status), and water stress index (thermal imagery) could be done regularly. This UAV is also useful in detecting and mapping the spatial variations of disease/pest or weed attacks in the field. It allows farmers to pin-point and adapt remedial measures accurately. This way, it reduces use of harmful chemicals in the crop fields. The aerial imagery and digital data of crop is analyzed. Then, fertilizers are applied accurately, based on directions from computer DSSs. Natural vegetation monitoring to obtain NDVI values and to determine plant species diversity is a clear possibility.

Nonagricultural uses: They are surveillance of geological sites, mines and mining activity, industrial sites, buildings, highways, and roads and their traffic, electric lines, oil pipelines, public events, etc.

Useful References, Websites, and YouTube Addresses

http://gtfdrone.en.alibaba.com (accessed Jan 12, 2017).
https://gtfdrone.en.alibaba.com/company_profile.html (accessed Jan 12, 2017).
https://gtfdrone.en.alibaba.com/search/product?SearchText=f-2 (accessed Jan 12, 2017).

Name/Models: Quantix

Company and address: AeroVironement Inc. 800 Royal Oaks Drive, Suite 210, Monrovia, CA 91016 6347, USA; phone +1 626 357 9983; e-mail: pr@avinc.com; Website: https://www.avinc.com/contact

AeroVironment, a Southern California technology company, is set to release, a new commercial UAV soon. It is named "Quantix." It will have a variety of commercial applications. It is designed to assist businesses that focus on hybrid UAV in agriculture, energy, and transportation. AeroVironment employs several hundred workers at its Simi Valley facility. This company is touting the UAV as an easy-to-use and a largely automated tool. "Quantix" will be useful for farming business.

Quantix is an innovative fixed-winged VTOL. It is autonomous in flight and can be preprogramed to connect to different waypoints and conduct aerial photography. It is supposed to be immensely useful to farmers who intend to adopt precision farming techniques; also, those who adopt spectral imagery to analyze crops and observe fields. Quantix™ will give users the same air superiority, trusted certainty, and security. AeroVironment is a leading UAV supplier to the US Department of Defense. Quantix launches a new era of remote sensing for aerial inspections, mapping, and actionable insights. It combines the advantages of vertical lift-off and horizontal flight for seamless operations and maximum coverage. Quantix easily collects high-resolution imagery quickly and accurately to identify issues before they become costly problems (Aerovironment Inc., 2017; Grassi and Schrimpf, 2017).

TECHNICAL SPECIFICATIONS

Material: It is made of ruggedized foam, plastic, and carbon fiber.

Size: length: 75 cm; width (wingspan): 120 cm; height: 35 cm; weight: 4.2 kg

Propellers: It is a fixed-winged multirotor "hybrid" drone. It has four propellers. Each one is made of two plastic blades.

Payload: 1.4 kg

Launching information: Quantix has fully automatic take-off, navigation, and landing. It uses Pixhawk or a similar software to fix flight path. The flight path and landing are decided using touchscreen on iPad. It features an automatic "return home" and emergency landing software. It also has software that detects ground-based obstacles while flying low over the crops and corrects the flight path automatically.

Altitude: It operates at 50–150 m a.g.l. during imagery of crop fields. However, it can reach an altitude of over 1200 m, if needed.

Speed: It cruises at 40 km h^{-1}

Wind speed tolerance: It tolerates 20 km h^{-1} disturbance in the air.

Temperature tolerance: −20 to +55°C

Power source: Electric batteries

Electric batteries: 24,000 mA h.

Fuel: No need for petrol or diesel.

Endurance: 45 min per flight

Remote controllers: The GCS radiolink with UAV operates for up to 15 km radius. The GCS has computers with override facility and planning for flight and photographic schedule.

Area covered: It covers an area of 400 ac per flight of 45 min, at 400 ft a.g.l.

Photographic accessories: Cameras: Quantix carries a full set of cameras such as visual (R, G, and B), multispectral, hyperspectral, NIR, IR, and red edge. It also carries a Lidar pod. The "UAVs" CPU and DSSs automatically stores the images. The DSS usually stores aerial images for use by farmers. The aerial maps are offered quickly. They could be analyzed at the GCS and prescriptions could be prepared.

Computer programs: The GCS and UAV's CPU computers are supplied with software such as Pix4D Mapper or Agisoft's Photoscan. The GCS also has software for fertilizer-N and pesticide prescripts at variable rates. Therefore, the UAV can be useful during precision farming.

Spraying area: This is not a sprayer UAV.

Volume: Not applicable

Agricultural uses: Quantix is known to monitor and scout the crops effortlessly and quickly. It detects locations with nutrient deficiencies, using NDVI and leaf chlorophyll data. It offers spectral signatures of healthy and disease/pest-affected patches of crop fields. The map so prepared can be used in the variable-rate applicators, during precision farming. The aerial maps are also used to detect drought, floods, and soil-erosion-affected locations in a large field. Natural vegetation monitoring is also possible. It has also been used to monitor the movement of arm vehicles in crop fields and watch the progress of agronomic procedures (Grassi and Schrimpf, 2017.

Nonagricultural uses: Surveillance of mines mining activity if it is an open pit mine, ore movement, etc. are conducted by Quantix. It is used to monitor oil and gas pipelines, mainly, to detect leaks, spills, and pilferage. It has been used to surveillance international borders, coast lines, and ice melts. It is used to monitor industrial sites and buildings. It keeps a watch on city's vehicular traffic.

Useful References, Websites, and YouTube Addresses

Aerovironment Inc. *Quantix*; 2017; pp 1–2. http://www.avinc.com (accessed June 12, 2017).

https://www.avinc.com/resources/av-in-the-news/view/aerovironment-to-launch-latest-drone-next-spring (accessed Mar 23, 2017).

Grassi, M. J.; Schrimpf, P. *Top 10 Most Intriguing Technologies in Agriculture*; 2017; pp 3–4. http://www.precisionag.com/systems-management/top-10-most-intriguing-technologies-in-agriculture/ (accessed June 18, 2017).

http://www.avinc.com/contact/uas_general_inquiry (accessed Mar 23, 2017).

https://www.avinc.com/media_center (accessed Mar 23, 2017).

https://www.youtube.com/watch?v=AD_D591Q2WA (accessed June 11, 2017).

https://www.youtube.com/watch?v=jEISqo1xEkM (accessed June 11, 2017).

https://www.youtube.com/watch?v=h4sxC17HpNA (accessed June 11, 2017).

https://www.youtube.com/watch?v=RwBoEh6Ns2I (accessed June 11, 2017).

https://www.avdroneanalytics.com/ (accessed Jan 3, 2018).

Name/Model: Terra hawk VTOL Fixed-Winged UAV

Company and address: Phoenix-Aerial Systems, 3384 Robertson Pl suite 102, Los Angles, California, USA; phone: +1 323 577 3366; e-mail: info@phoenix-aerial.com

Phoenix-Aerial Systems Inc. is a UAV company situated in Southwest USA. It caters to variety of UAV-related services. Many of them are useful to farmers and civilian population in USA and elsewhere in other continents. It produces sleek, light-weight UAVs capable of rapid deployment. These UAVs are useful to surveillance agricultural farms and crop fields. This UAV is also used to surveillance industrial installations, city traffic, general policing of public events and monitoring and transport of ores, etc.

TECHNICAL SPECIFICATIONS

Material: These light-weight UAVs are made of composites of carbon fiber, foam, and plastic material. The UAV body is ruggedized to withstand environmental vagaries and rough landing on belly (Plate 2.51.1).

Size: length: 1.8 m; width (wingspan): 3 m; height: 30 cm; weight: 830 g

Propellers: It has one propeller at the front tip of the fuselage. The propeller has two plastic blades. There are four propellers attached to wings.

Payload: 2 kg

Launching information: This UAV has four propellers. Two of them are attached, each to left and right side of wing span. They aid in VTOL also in transit through the air. It has autolaunch, flight, and landing facility (Plate 2.51.1).

Altitude: Recommended altitude for scanning crops is 20–60 m above the crop's canopy.

Speed: 90 km h^{-1} (maximum attainable speed is 112 km h^{-1})

Wind speed tolerance: 20 km h^{-1}

Temperature tolerance: Withstands −10 to +60°C, rain and snow.

Power source: Electric batteries and gasoline. It has gasoline-powered engine for forward thrust and flight.

Electric batteries: Lithium–polymer batteries 8000 mA h; it consume 8 W power.

Fuel: Gas

Endurance: 4–6 h per flight

Remote controllers: The GCS is linked via both radio and satellite modes. Flight path could be regulated, using Pixhawk software. It has fully autonomous flight capability in autopilot mode.

Area covered: It covers 150–500 ha per flight of 4 h

Photographic accessories: Cameras: It offers excellent accuracy of images, since, it has Sensor STIM IMU. It has high-resolution cameras for photogrammetry. Set of cameras include Sony A 6000, Sony α7 series and NIR. Sensors on the UAV relay the processed images on to the mobile app or ground-station iPads and PCs.

Computer programs: The images relayed to ground station or iPads are processed using Pix4D Mapper or Agisoft Photoscan. Postflight processing software, such as "NovAtel's Inertial Explorer" is also used.

Spraying area: This is not a sprayer UAV.

Volume: Not applicable

Agricultural uses: Aerial mapping of agricultural terrain, land topography, crops, and cropping systems, irrigation channels, etc. are conducted. Crop scouting, phenotyping, and monitoring growth parameters are done efficiently. UAV's cameras are also useful in finding crop's water stress index, using IR wave bandwidth. It is used to image areas affected with drought, flood or stagnated regions. UAV cameras can also pick spots affected by disease and insect pest attack. They can use spectral signatures of crops and weeds. Then, detect weed infestation in fields. Natural vegetation monitoring is a clear possibility.

Nonagricultural uses: This is a sleek and noiseless UAV. It is useful in policing city traffic, monitoring public events, industrial installations, buildings, dams, rivers, vehicle movement in the highways, rail road, etc.

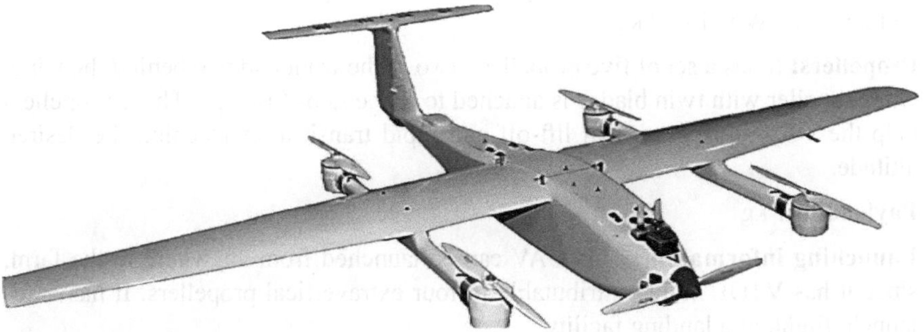

PLATE 2.51.1 Terrahawk (VTOL).
Source: Phoenix-Aerial Systems, Los Angles, California, USA. http://www.phoenix-aerial.com/products/arial-lidar-systems/terrahawk-v/.
Note: Terrahawk is a VTOL machine. It needs no runoff. It is also fitted with Lidar system.

Useful References, Websites, and YouTube Addresses

http://www.phoenix-aerial.com/products/arial-lidar-systems/terrahawk-v/ (accessed Sept 13, 2017).

http://old.phoenix-aerial.com/products/ (accessed Sept 13, 2017).

https://www.youtube.com/watch?v=tqgH9ANYaVU (accessed Sept 13, 2017).

https://www.youtube.com/watch?v=Ag7cRRTag9U (accessed Sept 13, 2017).

https://www.youtube.com/watch?v=Ag7cRRTag9U (accessed Sept 13, 2017).

Name/Models: Thresher-03 VTOL Hybrid UAV

Company and address: TechnoSys Embedded Systems, SCO 66, 1st floor, Sector 20 C, Chandigarh 160020, India; phone: +91 1725085909; +91 1724671082; e-mail: info@technosysind.com; sales@technosysind.com; support@technosysind.com Website: http://www.technosysind.com

Technosys Embedded Systems' UAV production operations were started in 2003. This UAV company caters to farmers and farming companies, in the Northern Indian

Plains. It offers a couple of useful UAV models with ability, for rapid aerial surveys, regular monitoring of crops, measurement of growth (NDVI), assessment of water status of crops, and estimation of leaf chlorophyll and nitrogen content. It helps in deciding fertilizer supply schedules to crops. This UAV, "Thresher-03" is a vertical take-off model, but, it is a "hybrid" with fixed-wings (Plate 2.52.1). It is gaining in popularity with farmers. This UAV model is also useful, to conduct several other nonagricultural functions.

TECHNICAL SPECIFICATIONS

Material: The UAV's body is built using a composite of carbon fiber, plastic, and ruggedized foam. The core of the UAV has a "honeycomb type" structure for rigidity.

Size: length: 161 cm; width (wingspan): 350 cm; wing area 70 dm^2; height: 30 mm; weight: MTOW is 14.40 kg

Propellers: It has a set of five propellers, two in the front and two behind the wing. One propeller with twin blades is attached to rear end of fuselage. These propellers help the UAV, during vertical lift-off and rapid transit after reaching the desired altitude.

Payload: 8.1 kg

Launching information: This UAV can be launched from anywhere in the farm, since it has VTOL ability attributable to four extravertical propellers. It has auto-launch, flight, and landing facility.

Altitude: For agricultural-related crop survey and imagery, this UAV operates at 400 m a.g.l. However, it reaches 4300 m a.g.l., if required.

Speed: It has a cruise speed of 90 km h^{-1}. Maximum speed attained is 110 km h^{-1}.

Wind speed tolerance: It tolerates 30–45 km h^{-1} disturbance/turbulence.

Temperature tolerance: −17 to +55°C

Power source: Electric batteries and fuel (petrol/gasoline)

Electric batteries: 8 A h lithium–polymer batteries for take-off

Fuel: Fuel tank holds 2.5 L petrol for rapid flight.

Endurance: 240 min per flight

Remote controllers: The GCS has radiolink that operates for up to 5 km radius. The take-off, flight path and autolanding can be managed, using computers at the ground station, iPads or a radiolinked remote controller. Flight path can be edited, midway, through the preprogramed flight path. Ground control computers could be used, to program cameras for exposures and videographing the land and cropped areas.

Area covered: The UAV covers 75–100 km in linear distance or 300 ha per flight.

Photographic accessories: Cameras: This UAV has a usual complement of sensors. They are visual (R, G, and B), multispectral, hyperspectral, NIR, IR, and red-edge cameras. It also has a Lidar pod.

Computer programs: The UAV-derived images could be processed in situ at the CPU and then transmitted to GCS iPad. However, orthoimages could be sent to GCS computers for processing, using software such as Pix4D Mapper or Agrisoft photoscan.

Spraying area: This is not a sprayer UAV.

Volume: Not applicable

Agricultural uses: They are aerial survey of farmland, its topographical features, soil type, water resources, natural vegetation, and cropping systems. Monitoring seed germination, seedling stand, and identifying gaps accurately. Crop scouting to collect and map NDVI values, leaf area index, and leaf chlorophyll contents. Detecting drought and flood-affected zones in the farms is possible using Thresher 03 UAV. A few other uses are monitoring progress of agronomic procedures, movement of tractors, autonomous vehicles, combines, etc. Detection of weeds, using their spectral signatures is possible. Aerial survey of forest stand and imaging log movement is done.

Nonagricultural uses: There is long list of nonagricultural functions that this UAV (Thresher-03) performs. They are fire detection and alerting, police services, power line inspection, media and filming of public events, surveillance of industrial installations, surveillance of mines, traffic and transport vehicles, etc.

PLATE 2.52.1 Thresher-03 VTOL Hybrid UAV.
Source: Mr. Dhruv Arora, TechnoSys Embedded Systems, Chandigarh, India.

Useful References, Websites, and YouTube Addresses

http://www.technosysind.com/octocopter.php (accessed Nov 10, 2016).
https://www.youtube.com/watch?v=qyJnFLZTiFM
https://www.youtube.com/watch?v=fNmQpkPnO_w (accessed June 17, 2017).
https://www.youtube.com/watch?v=qyJnFLZTiFM (accessed Apr 1, 2017).
https://www.youtube.com/watch?v=CZ41uG3EE30 (accessed June 17, 2017).

Name/Models: WINGTRA S125H

Company and address: Wingtra One, c/o ETH WEH H ETH Zurich (Wyss Zurich Project), Winberg Str., 35, 8092, Zurich, Switzerland; phone: e-mail: hello@wingtra.com; Website: https://wingtra.com/

Wingtra UAV company helps people to collect information using aerial robots. Since its entry into market in early 2017, the WingtraOne has become the preferred tool for aerial data collection in industries like, surveying, agriculture, forestry, mining, and infrastructure inspection. Wingtra is a spin-off from the Autonomous Systems Lab at ETH, Zurich. Wingtra has been supported as one of the Wyss Zurich projects since early 2016 (Plate 2.53.1).

The WingtraOne aerial robot takes off like a helicopter but covers large areas like a fixed-winged airplane (Plate 2.53.1; Lilian, 2016). WingtraOne drastically simplifies how aerial data is collected. Wingtra aims to have a positive impact for the human condition on a large scale. They like to achieve this through UAV technology and robots.

While conventional quadcopters are limited in range and speed, fixed-wing airplanes, need catapults and runways. Wingtra's hybrid UAV combines the advantages related to both aspects. It is as agile as multicopters and has the superior flight capability of a fixed-wing aircraft, regarding range, and speed. It opens new possibilities in improving wildlife protection, agriculture, parcel delivery, and many other applications.

TECHNICAL SPECIFICATIONS

Material: It is made of carbon fiber, foam, and toughened plastic (Plate 2.53.2).

Size: Length: 105; width (wingspan): 125 cm; height 18 cm; weight: 3.6 kg (maximum permissible take-off weight is 4.4 kg)

Propellers: Two propellers with blades. They are placed one each on right and left section of the wing

Payload: 800 g

Launching information: The UAV takes off from any surface vertically. It has VTOL system. Although it is a flat-winged UAV, it does not need the usual catapult or longer stretches of runway or even hand launching. It is a kind of hybrid between copter and fixed flat-winged versions of UAVs.

Altitude: UAV operates at 400 m a.g.l., during agricultural operations, but it can reach 4200 m a.g.l.

Speed: 55–60 km h^{-1}

Wind speed tolerance: 30 km h^{-1} during landing and 45 km h^{-1} during the cruise

Temperature tolerance: −20 to +55°C

Power source: Electric batteries

Electric batteries: Lithium polymer batteries.

Fuel: Not a UAV with internal combustion (IC) engines

Endurance: 4 h per flight

Remote controllers: The UAV keeps radiocontact for up to 4 km radius from the ground station. The radiocontact can reach up to 8 km if the atmosphere is clear. UAV's flight mission planning is done, using a software known as "Wingtra Pilot." Take-off, flight path, and landing could be fully automatic and preprogramed. This UAV is autonomous and has GPS connectivity for navigation.

Area covered: The UAV covers 310 ha at 141 m altitude; otherwise, 766 ha at 463 m attitude. Regarding image resolution and area coverage, it covers 220 ha at 2.1 cm pixel resolution and 544 ac at 0.8 cm pixel resolution per flight.

Photographic accessories: Cameras: UAV is fitted with R, G, and B (Sony QX1 with 20 mm lens) and multispectral cameras for imaging land and crop fields. Thermal imagery helps in detecting crop water status. Special lens of 15 mm can be fitted. Payload may also include Sony RX! RII with 35 mm lens. Micasense Rededge 5.5 mm lens is also included in the gimbal.

Computer programs: Usual image-processing software such as Agisoft's Photoscan or Pix4D Mapper are used to stich the pictures.

Spraying area: This is not a sprayer UAV.

Volume: Not applicable

Agricultural uses: Crop scouting for growth and phenomics data, monitoring natural vegetation, detection of drought, floods, and disease/pest attacked zones in the crop fields are major aspects.

Nonagricultural uses: Surveillance of mines, industrial installations, transport vehicles, railroads, pipelines, etc. are the main functions.

PLATE 2.53.1 Wingtra UAV.
Source: Dr. Sebastian Verling, Marketing Manager, Wingtra One, Zurich, Switzerland.
Note: Wingtra is a fixed-winged UAV with VTOL facility. It is a "Hybrid" between copter and fixed-winged UAV types. It is useful to agriculture.

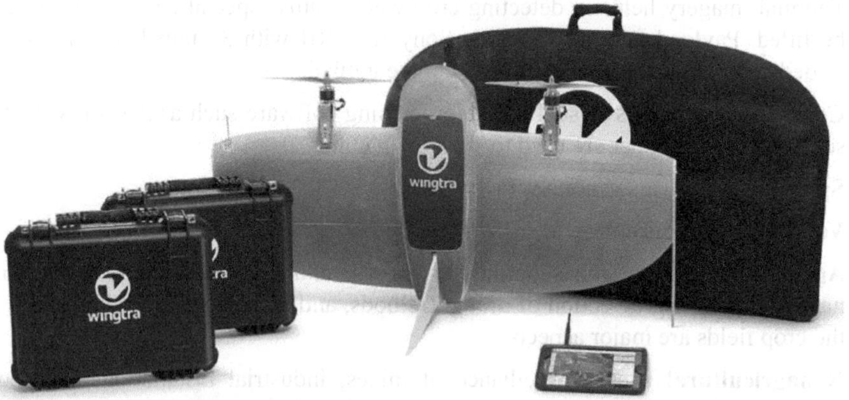

PLATE 2.53.2 Wingtra UAV, GCS iPad suitcase/baggage for portability.
Source: Dr. Sebastian Verling, Marketing Manager, Wingtra One, Zurich, Switzerland.

Useful References, Websites, and YouTube Addresses

Lilian, B. *Wingtra Start Up Develops Fixed Wing VTOL Drone. Unmanned Aerial*; 2016; pp 1–5. http://unmanned-arial.com/tech-start-up-develops-fied-wing-vol-drone/ (accessed Oct 1, 2016).

https://wingtra.com/news/ (accessed Dec 1, 2017).

https://www.youtube.com/watch?v=2xctOQexsHs&feature=youtu.be (accessed Dec 15, 2017).
https://www.youtube.com/watch?v=luQg25bEOEc (accessed Dec 15, 2017).
https://www.youtube.com/watch?v=12I_DiAQxao (accessed Dec 15, 2017).
https://www.youtube.com/watch?v=3HEx9Nbb8nc (accessed Dec 15, 2017).
https://www.youtube.com/watch?v=HgnTLdXTMrA (accessed Dec 15, 2017).

2.7 MISCELLANEOUS TYPES OF FIXED-WINGED UNMANNED AERIAL VEHICLES

A few companies offer models that do not comply with rigid classifications based on weight, structure of platform, propellers, payloads, size, etc. Here, in this volume, they have been grouped as "miscellaneous types of UAVs." They have utility in farming sector. Few of them offer special facilities not available with routine fixed-winged or copter UAV model.

Name/Models: AT-10 (Acuity Tilt winged VTOL UAV)

Company and address: The Acuity Technologies Inc. 3475 Edison Way Building Menro Park, CA 94025, USA; phone: +1 650 369 6783; e-mail: Not available; website: http://www.acuitytx.com

Acuity Technologies Inc. situated in Menro Park, California, USA is an UAV (drone) and aircraft producing company. It was founded in 1992. Acuity Technologies Inc. develops, manufactures, and supports components and systems for unmanned vehicles, robotics, and automation. Acuity concentrates on aerodynamics and vehicle design and control system development (Clark, 2016). AT-10, T-3 Owl, P-3 Orion are aerial robots being produced by this company. They also produce several accessories required for operation of UAVs. For example, they produce integrated UAV control systems and software for managing UAV flight and aerial imagery schedule. They produce SRI-500 which is a scanning imaging range finder and software for 3D mapping of terrain. AT-10 is a tilted-winged VTOL UAV. It could be utilized for aerial imaging of earth's surface features. The tilted rotor and tilted-winged UAVs have their known advantages, regarding vertical take-off, and higher efficiency, particularly, during flight and performance in the air. They do not require catapult or landing gears. It is useful in monitoring natural vegetation patterns, crop fields, and in analyzing crop growth phenomics, using digital spectral data. It is also useful in monitoring the activity of farm vehicles.

TECHNICAL SPECIFICATIONS

Material: This UAV is made of a composite of evelar, carbon fiber, and plastic.

Size: length: 8 ft; width (wingspan): 9.3 ft, wing area is 9.5 sq. ft; height; weight: maximum gross weight is 120 lb. maximum weight permissible during VTOL is 100 lb.

Propellers: There are two tilted propellers on tilted wing. The propellers have three ruggedized plastic blades.

Payload: The payload is 10 × 10 in. in area. Plus, it has a camera turret space in the nose. The UAV can carry a payload of 30 lb, during flight.

Launching information: It is a vertical take-off UAV. So, does not need a runway strip for launching. This UAV has autonomous flight path, if preprogramed. It can receive guidance from GCS and transit, using GPS, DGPS, and optical beacon.

Altitude: It is operated at 50–400 m a.g.l. during general aerial survey and mapping. The ceiling is 3500 m a.g.l.

Speed: 83 knots; maximum endurance loiter speed is 58 knots; and stall speed is 53 knots

Wind speed tolerance: 25 km h^{-1}

Temperature tolerance: −20 to +60°C

Power source: Electric batteries

Electric batteries: 28 A h, two cell lithium–polymer batteries.

Fuel: It needs heavy fuel for power generating system/for propulsion.

Endurance: 12 h per flight

Remote controllers: The UAV is connected to GCS via telemetric connectivity. The telemetric signals are operative for 50 km from the launch point. The GCS implements STANAG 4586 standard for autonomous systems.

Area covered: It is an excellent UAV for patrolling from height of 12–15,000 ft a.g.l. Therefore, it covers large areas during survey and surveillance trips.

Photographic accessories: Cameras: This UAV has larger payload space. It accommodates a series of cameras such as visual (R, G, and B), NIR, IR, mica-sense red-edge camera, and few other thermal imagery sensors.

Computer programs: The GCS utilizes Pix4D Mapper or Agisoft's Photoscan software, to process the images received.

Spraying area: This is not a sprayer UAV.

Volume: Not applicable

Agricultural uses: Crop scouting and natural vegetation monitoring.

Nonagricultural uses: surveillance of mines, transport vehicles, etc.

Useful References, Websites, and YouTube Addresses

Clark, R. *AT-10 Electric/HF Hybrid VTOL UAS*; 2016; pp 1–6. http://www.acuitytx.com (accessed Jan 11, 2017).

Stone, B.; Crandall, Z. *Tilt—Rotor Drone*; Electrical Engineering Department, California Polytechnic State University, Sn Luis Obispo, California, 2016, pp 1–23.

http://www.acuitytx.com/pdf/AT-10%20Brochure.pdf (accessed Mar 25, 2017).
https://www.youtube.com/watch?v=VbcITyB8MKU (accessed Mar 25, 2017).
https://www.youtube.com/watch?v=gIlQsdV9v08 (accessed Mar 25, 2017).

Name/Models: SPANA

Company Address: LAPCAD Engineering Inc., 8305 Vickers Street, Suite 202, San Diego, CA 92153 0454, USA; Tel: +1 858467 1947, +1 619 840 4403; e-mail: info@lapcad.com

LAPCAD Engineering was founded in 1988 and incorporated in the State of California in 1997. Its mission is to provide the best possible structural analysis and structural design capabilities to the aerospace, marine, mechanical, and medical sectors. LAPCAD is an Aeronautics company located at San Diego, California. It offers a variety of services to the US Government, the Department of Defense, and the Commercial and Public Sectors. They include consulting, design, fabrication, and software. Supported programs include UAVs for Teledyne Ryan, the Global Hawk unmanned aircraft for Northrup Grumman, the Vantage single engine business jet for Visionaire, the twin engine Safire business jet for Safire Aircraft, the landing gears for the Roton space craft for Rotary Rocket, life extension of the C-2 aircraft for the US Navy, helicopter doors for M7 Aerospace, 727 and 747 cargo conversions for Wagner Aeronautical, UAVs for AeroMech Engineering and A320 cargo door for Pacavi.

They have developed several UAVs and unmanned air vehicles. The UAV named "SPANA" has a Bridged Flying Wing. The maiden flight of this Proof of Concept (POC) vehicle took place April 12, 2005. The UAV has a flying wing, characterized by the aerodynamic bridge that supports the engine. It is powered by an 8 HP two-stroke twin cylinder Fuji/IMVAC gasoline engine. Wing span is approximately 12 feet.

TECHNICAL SPECIFICATIONS

Material: It is made of EPP, carbon fiber, toughened plastic, and foam.

Size: length: 160 cm; width (wingspan): 360 cm; height: 55 cm; weight: 18.5 kg

Propellers: One propeller fixed at the rear end of the fuselage.

Payload: 4.2 kg

Launching information: It adopts autolaunch software, for example, Pixhawk. The UAV needs a runway to become airborne. It is autonomous during light. The flight path is usually preprogramed using "Pixhawk Autopilot" or a similar custom-made software. It has landing gear made of three wheels. It and lands on strip of 1000 m length.

Altitude: Its operational height is 400 m a.g.l. but the ceiling 3200 m a.g.l.

Speed: 80 to 120 km h^{-1}

Wind speed tolerance: 20–35 km h^{-1} disturbance

Temperature tolerance: −20 to +55°C

Power source: Electric batteries. It is also attached with 8 hp 2 stroke, 2-cylinder Fuji/IMVAC engine for thrust.

Electric batteries: 14 mA h to support electronic circuits and CPU on the UAV.

Fuel: Gasoline plus lubricant.

Endurance: 3 h per flight

Remote controllers: The GCS computers (iPads) keep contact with UAV, using radio telelink for up to 150 km distance. The UAV is fail-safe. It returns to point of launch in case of emergency or errors in commands or program.

Area covered: 300–800 ha per flight of 6 h

Photographic accessories: Cameras: It carries a visual (R, G, and B), multispectral, hyperresolution, NIR, IR, and red-edge cameras. It also has a Lidar pod.

Computer programs: The images and digital data are relayed to GCS computers. They are processed using Pix4D Mapper or Agisoft's Photoscan.

Spraying area: This is not a spraying UAV.

Volume: Not applicable

Agricultural uses: SPANA is used to scout crop fields for seed germination, gaps in seedling establishment, weeds, etc. It is used to collect data on crop growth, phenomics, leaf chlorophyll (crop's N status), calculate fertilizer-N requirements, measure canopy temperature (thermal sensors) to decide irrigation needs, and find grain maturity status. Natural vegetation monitoring is also possible.

Nonagricultural uses: It includes surveillance of geological sites, mines, mining activity and ore movement, buildings, industrial sites, oil and gas pipelines, electric lines, city traffic and tracking vehicles, etc.

Useful References, Websites, and YouTube Addresses

http://lapcad.com/ (accessed June 5, 2017).
http://lapcad.com/?page_id=6 (accessed June 5, 2017).

Name/Models: V BAT

Company and address: Martin UAV LLC, 15455 North Dallas Parkway, Suite 1075, Addison, TX 75001, USA; phone: +1 512 520 7170; e-mail:info@martinuav.com; website: http://martinuav.com/uav-products/v-bat

Martin UAV LLC aims to provide the most cost-effective access to aerial information. They specialize in providing robust and versatile UASs. Their recent UAV model is the first-of-its-kind, vertical-take-off-and-land (VTOL) V-BAT UAS. It combines the ability to hover with the efficiency of high-speed horizontal flight. It is a wholly unique UAS capability. It bridges critical mission gap with a cost-effective solution and a near-zero footprint. The V-Bat is a long endurance VTOL design. It is ideal for shipboard and confined area operation. A ducted fan design maximizes operational safety by eliminating exposed rotors. It has a combination of VTOL operational convenience with the safety of a shrouded fan and fixed-wing duration in a small UAV system. It could revolutionize the availability of UAV operations from confined areas. Therefore, it bridges a critical mission gap at the tactical level, where it is needed the most (Plate 2.54.1; Koski et al., 2009).

TECHNICAL SPECIFICATIONS

Material: It is made of plastic, carbon fiber and foam composite.

Size: length: 8 ft (2.4 m); width (wingspan): 9 ft (2.74 m); height: 2 ft; weight: 76 lb

Propellers: It has a propeller placed at the bottom in a collar. Propeller is connected to 170 cm^3 two-stroke engine.

Payload: 5 (2.3 kg)

Launching information: This VTOL UAV launches autonomously within a space of 6 × 6 m. It is also amenable for catapult launching. It has autonomous flight software (Pixhawk). Landing is also autonomous and soft.

Altitude: It has a ceiling of 4200 m a.g.l.

Speed: Ranges from 0 to 90 km h^{-1}

Wind speed tolerance: Withstands disturbance up to 25 knots

Temperature tolerance: −20 to +60°C

Power source: 170 cm^3 IC engine that operates on fuel oil

Electric batteries: lithium–polymer batteries 8000 mA h.

Fuel: Heavy fuel (JP-8) or Gasoline oil mix (40:1)

Endurance: 8 h at 45 knots

Remote controllers: The GCS iPads/PCs have telemetric contact, with the UAV for up to 30 km.

Area covered: It transits for over 350 km linear distance and covers over 400 ha per flight.

Photographic accessories: Cameras: Sensors include SAR, SWIR, target marker, SIGNIT, hyperspectral camera, NIR, and IR cameras for thermal imagery. It also has mapping payload Hi-Res Digital SLR. Gimbal is stabilized for vibrations and distractions.

Computer programs: UAV's imagery is processed at the CPU itself and relayed to iPad for analysis.

Spraying area: Not a spraying UAV.

Volume: Not applicable

Agricultural uses: V-Bat has wide ranging applications in agricultural regions. Aerial mapping of land, its topography, and soil type are done using sensors. This UAV is excellent for crop scouting. It is used to monitor growth and get NDVI data, phenomics data, and grain maturity. It helps to monitor disease/pest. It is used to map-affected zones, droughts, floods and erosion-affected spots in fields. Natural vegetation monitoring is another possibility (Plate 2.54.1).

Nonagricultural uses: V-Bat UAV has been recommended for use in ship board operations, antipiracy surveillance, urban operations such as policing traffic, public events, large buildings, etc. This UAV has been employed as border petrol between nations. V-Bat could be used for surveillance of mines, ore-dumping yards, and ore movement.

PLATE 2.54.1 V Bat VTOl UAV.
Source: Dr. Blake Sawyer, coordinator, Martin UAV LLC, 15455 North Dallas Parkway, Suite 1075, Addison, TX 75001, USA.

Useful References, Websites, and YouTube Addresses

Koski, W. R.; Abrll, P.; Yazvenko, B. A Review and Inventory of Unmanned Aerial Systems for Detection and Monitoring of Key Biological Resources and Physical Parameters Affecting Marine Life during Offshore Exploration and Production Activities. In *A Review and Inventory of Unmanned Aerial Systems*; LGL Ld., Ed.; E and P Sound and Marinelife Program, 2009; pp 1–32. http://www.soundmrinelife.org (accessed June 17, 2017).

http://martinuav.com/uav-products/v-bat/ (accessed Feb 12, 2017).

http://martinuav.com/about/publictions-whitepapers/ (accessed Feb 12, 2017).

https://www.youtube.com/watch?v=lpRxkMEN3pY (accessed Feb 12, 2017).

https://www.youtube.com/watch?v=PJ_iMtsvBpE (accessed Feb 12, 2017).

https://www.youtube.com/watch?v=PJ_iMtsvBpE (accessed Feb 12, 2017).

KEYWORDS

- **fixed-winged UAVs**
- **vertical take-off and landing hybrids**
- **technical specifications**
- **engine**
- **fuel**
- **speed**
- **endurance**

GENERAL REFERENCES

Anderson, C. Agricultural Drones: Drones in the North American California. *MIT Technol. Rev.* **2014,** 1–4. http://www.technologyreview.com/featuredstorey/52649/agricultural-drones/ (accessed Aug 2, 2014).

Arjomandi, M. *Classification of Unmanned Aerial Vehicles*; The University of Adelaide: Adelaide, Australia, 2013. http://www.personal.mecheng.adelaide.edu/mazair/arjoomani/aeronautical%20engineering%20projects/2006/group9.pdf (accessed Sept 1, 2015).

ECA Group-Lima. Types of Drones-Explore the Different Models of UABs. *Circuit Today* 2017, 1–8 http://www.circuitstody.com/types-of-drone/ (accessed June 14, 2017).

Greenwood, F. Drones Are on the Horizon: New Frontier in Agricultural Innovation. *ICT Update* **2017,** 82, 1–4.

http://ictupdate.cta.Int (accessed Apr 1, 2017).

Keane, J. F. and Carr, S. S. A Brief History of Early Unmanned Aircraft. *Johns Hopkins Appl. Tech. Dig.* **2013,** *32*, 558–593.

Krishna, K. R. *Push Button Agriculture: Robotics, Drones and Satellite-Guided Soil and Crop Management*; Apple Academic Press Inc.: Waretown, NJ, 2016; pp 131–260.

Krishna, K. R. *Agricultural Drones: A Peaceful Pursuit*; Apple Academic Press Inc.: Waretown, NJ, 2018; p 475.

National Geographic. *Drones and the Future of Agriculture. Natl. Geogr.* 2016. https://www.youtube.com/watch?v=v3YcZtlVrls (accessed Nov 15, 2016).

Newcome, L. R. *Unmanned Aviation: A brief history of Unmanned Aerial Vehicles*; American Institute of Astronautics Inc.: Reston, Virginia, 2004; pp 1–29.

Nicole, P. *Sustainable Technology—Drone Use in Agriculture*; 2015; pp 1–7. https://wiki.usask.ca/display/~pdp177/Sustainable+Technology+Drone+Use+griculture (accessed July 6, 2015).

Tetrault, C. *A Short History of Unmanned Aerial Vehicles (UAVS)*; Dragon Fly Innovations Inc., 2014; pp 1–2. http://texasagrilife extension.edu/ (accessed July 24, 2014).

CHAPTER 3

Helicopter and Multirotor Unmanned Aerial Vehicle Systems Utilized in Agriculture

ABSTRACT

Unmanned aerial vehicle systems (UAVS) with rotors are also known as Copters. They have ability for vertical take-off and landing, hovering at a spot above crop's canopy and rapid transit. Such UAVS with rotors are being increasingly preferred, by several agencies. Autonomous copters are sought by farmers with small holdings as well as farm companies with large land holdings. The helicopters and multi-copters are utilized in agrarian regions for aerial photography, surveillance, assessment of crops' growth (i.e., to estimate NDVI and leaf chlorophyll), detection of pests, diseases, weeds and disasters such as soil erosion, flooding, or drought. Of course, UAV models with added applications are being manufactured in several countries. Several new models of helicopters, quad-copters, hexa-copters, and even octo-copters are being introduced into farms. Such small autonomous copters with ability for spectral analysis of crops, spraying fertilizers, and/or plant protection chemicals are gaining in popularity, in major agrarian belts. In all, technical specifications for about 100 copter UAV models are listed and discussed in this chapter. Knowledge about the manufacturer, UAVs approximate cost, its technical details such as size, number of rotors, payload, operational and ceiling altitude, power source, engines, propellers, endurance, fuel efficiency, remote control equipment, photographic accessories, computer programs to process aerial imagery and the array of agricultural and non-agricultural uses possible is essential. Some of these are mentioned in this chapter.

Helicopter UAVs are among the common autonomous aerial robots in agrarian regions. Several models of helicopter UAVs are currently in use in North America, Europe, Asia and Fareast. There are at least over 100 models of helicopters being manufactured and sold to farmers. The specifications of such small helicopters used above crops differ based on models. They usually try to add accessories based on farmer's preferences and purpose. A standard helicopter drone consists of series of sensors. Sensors that operate at visual band (red, green, and blue), near infra-red, red-edge, and multi-spectral band width and LIDAR. These are common attachments. A few models also possess air samplers and probes for atmospheric gases. In this chapter, brief information about manufacturers, plus specifications, agricultural, and non-agricultural uses of about 50 different models of helicopters are provided. At present, helicopters with accessories for spraying granules/fluid inputs into crop fields are popular. For example, in the United States, Europe, Japan, China, and other nations of Far East, several models (e.g., RMAX, AG-RHCD-80-15, Hercules, MJ-Agras, and AG-6A) are being utilized, to apply fertilizer top dress and pesticides/fungicides/herbicides. This is in addition to aerial survey and spectral analysis features. Helicopter sprayer drones differ in their endurance. They also differ in their specifications regarding liquid/granule tanks, spray bar size, spray width (swath), amount of agricultural chemicals sprayed, and land area covered. Farmers have a wide choice while they opt to procure a helicopter UAV. They say, helicopter sprayers have high fuel efficiency and acceptable levels of endurance while conducting agricultural operations.

Multi-rotor (VTOL) copters are getting popular in agrarian regions (e.g., DJI's MJ- AGRAS, Mattrice-600, HSE's Hercules, MD4-1000, MD-3000, AgFalcon, and AgStar X-8). UAV manufacturers have been flooding the market with wide range of multi-rotor drones with different sizes, payload capacity, sensors, and spray equipment. Farmers who wish to opt for a copter capable of both aerial imagery and spraying agricultural chemicals may have to select most appropriate model. Multi-rotors are capable of hovering and flight at very low altitude above the crop's canopy. They are relatively stable in the air. They possess gimbals that are vibration proof. Hence, they offer excellent sharp and close-up images of crops. Farmers can easily detect insects, disease, or any other maladies that occur on crops, rather quickly and easily. The copter drones with gas/liquid samplers are also available. They are used in assessing pollution levels of water bodies and atmosphere. Information about a few special types of copter drone models are listed in this chapter (e.g., UAV 180-120, Sprite). In all, details regarding manufacturer and specifications of 50 sprayer agricultural multi-copters are provided.

3.1 INTRODUCTION

Unmanned aerial vehicle systems (UAVs) with vertical take-off, hovering, and spraying ability have special significance to agrarian regions. Farmers and farming companies with large or those with small holdings, both, may utilize these UAVs with great advantage. No doubt, copter UAVs have a range of applications in surveillance of industries, large public buildings, railways, electric lines, oil pipelines, etc. In the present context, we are exclusively concerned with agricultural applications. In the farm sector, rotor UAVs are engaged in aerial photography of terrain, soil types, crops, leaf chlorophyll status (i.e., crop's N status), water stress index, pest and disease attacks, etc. They are also used in detection and first reaction to flood, drought, or soil erosion-related disasters in large farms. In this chapter, about 100 models of UAVs (helicopters and multicopters) have been listed along with their specifications and range of applications in agricultural and nonagricultural sectors. Several more models of copter UAVs are being introduced into the market each month, which provides greater pool of UAV models for farmers to choose.

3.2 UNMANNED AERIAL VEHICLES WITH ROTORS (COPTERS): TECHNICAL SPECIFICATIONS AND APPLICATIONS

The emphasis in this section is on the UAVs with rotors and "vertical take-off and landing" (VTOL) ability. Agricultural sector is now exposed to copter UAV models of wide range of capabilities. For example, there are innumerable helicopter (single rotor) models impinged into farm land. These helicopter models have been adopted to collect data about crop growth, its vigor, water status, pest, and diseases occurrence if any. They are also being utilized to spray plant protection chemicals. Next, there are copter UAVs with four rotors, that is, quadcopters, six rotors, that is, hexacopter, then copters with eight rotors, that is, octocopter. Farmers are indeed exposed to several models and their options to choose are enormous. They could consider specifications of each model, their cost, and economic advantages and suitability to their crop and farm per se. In the following paragraphs, several copter models, their specifications, and applications are listed. In all cases, website and YouTube addresses have been provided. In case, one wishes to see the performance and operation of specific UAV models they could refer to websites/YouTube.

3.2.1 HELICOPTER UAVS

Name/Models: AT-30 (Helicopter UAV)

Company and address: Mr. Pietro Amati, Advanced UAV Technology Ltd., 28 Red Post Hill, London SE24 9JQ, UK; phone: +44 20 7501 9345; e-mail: info@auavt.com; website: http://www.auavt.com

Advanced UAV Technology Ltd. is a UAV company in Europe. It was founded by Mr. Pietro Amati in 2010. Advanced UAV technology aircraft are all capable of fully autonomous flight. All functions are carried out through preprogramed instructions sent by an onboard computer, not by a pilot. The onboard computer interprets all commands and flies the aircraft, allowing the operator to concentrate on the mission. These helicopter UAVs could be deployed quickly. They require no airfield, runway, or launch/recovery equipment. The ground control station (GCS) is Windows software based, and operators will find its interface to be extremely user friendly. All advanced UAV technology products can be customized to meet the mission requirements. Many cameras, sensors, and other payloads can be integrated. The possible missions that could be accomplished are virtually unlimited. New applications are being constantly added. Advanced UAV technology offers seven VTOL UAV models. The range covers payloads from 1.5 to 350 kg. The mission endurance ranges from 30 min to over 16 h. The seven models range from the man-portable AT-10 machine to the 350-kg dry weight AT-1000.

AT-30 is a light-weight, medium-endurance VTOL UAV (Plate 3.1.1). Some of its features include autonomous waypoint navigation; real-time mission control from any computer connected to the internet; autonomous landing on moving platforms such as vehicles and ships; and ability to operate in a GPS-denied environment.

TECHNICAL SPECIFICATIONS

Material: This UAV is made of steel, carbon fiber, and ruggedized plastic.

Size: length: 163.8 cm; width (rotor): 35.5 cm, main rotor blade diameter 198.1 cm, tail rotor blades diameter: 33.7 cm; height: 62.2 cm; weight: 7 kg without payload

Propellers: It has two propellers; one main and larger propeller above the cockpit area and a tail propeller.

Payload: 5 kg. It includes cameras and tank for granules/liquid formations, if used for spraying.

Launching information: AT-30 is an auto-take-off and landing UAV. It has facility for fully autonomous flight with unlimited waypoints with in flight. It has fail-safe "return to home" software.

Altitude: This helicopter UAV can hover at a very low altitude over the crop canopy. Its normal altitude for aerial survey activity is 300 m above ground level, but ceiling could be over 1500 m.

Speed: Maximum speed 80 km h^{-1}

Wind speed tolerance: It withstands wind caused disturbance of up to 20 knots

Temperature tolerance: −10 to 45°C

Power source: Internal combustion engine (2.4 hp, 2-stroke gasoline engine). It is powered by gasoline fuel

Electric batteries: 8 A h to support electronic accessories

Fuel: 1 L fuel (gasoline) capacity

Endurance: Up to 2.5 h

Remote controllers: This UAV keeps in touch with GCS through telemetry. It has long-range telemetry up to 30 km or through 3G network or satellite link. It has flight controller software, if it semiautonomous and is being guided by ground station iPad or joystick. It also has fully autonomous preprogramed flight-path mode.

Area covered: It can cover about 150 km in linear distance per flight. If preprogramed, flight path can cover 200–300 ha per flight.

Photographic accessories: Cameras: It has still and videocameras that operate at visual bandwidth. Several other sensor packages are optional.

Computer programs: The digital data and images can be processed using Pix4D Mapper or Agisoft's Photoscan.

Spraying area: This UAV can be fitted with liquid-holding tank for spraying pesticides/fungicides in farms, if needed.

Volume: Data not available

Agricultural uses: This helicopter UAV is applied to accomplish a range of tasks related to crop production, such as survey and aerial mapping of land, water, and crop resources; monitoring seed sowing and germination activity, mapping plant stand and gaps, measuring normalized difference vegetation index (NDVI) of crops, leaf chlorophyll and crop's N status. Indirectly, this UAV can provide data useful to decide fertilizer-N dosages. It can also measure crop water stress index. So, it helps in deciding the need for irrigation. This UAV can be used to locate and map disease/pest-attacked zones in a crop field. Flood or drought-affected regions can also be marked.

Nonagricultural uses: Surveillance of large industries, oil pipelines, rail roads and highway patrolling, monitoring mines, and mining activity.

PLATE 3.1.1 AT-30—a helicopter UAV.
Source: Mrs. Kristy Hall, Advanced UAV Technology, Wellesley House, Duke of Wellington Avenue, Royal Arsenal, London SE18 6SS, UK.

Useful References, Websites, and YouTube Addresses

http://www.pegasusexecutive.com/uncategorized/757/ (accessed Oct 3, 2017).
http://auavt.com.existsite.com/ (accessed Oct 3, 2017).
http://www.zoominfo.com/c/Advanced-UAV-Technology-Limited/347665240 (accessed Oct 3, 2017).

Name/Models: AT-35

Company and address: Pietro Amati, Advanced UAV Technology Ltd., 28 Red Post Hill, London SE24 9JQ, UK; phone: +44 20 7501 9345; e-mail: info@auavt.com; website: http://www.auavt.com

As listed above, Advanced UAV Technology Ltd. is a UAV company. Regarding UAVS produced by them, it is said, all functions are carried out through preprogramed instructions sent by an onboard computer. The onboard computer interprets all commands and directs the UAV. They are VTOL models, so require no airfield, runway, or launch/recovery equipment. AT-35 is an advanced helicopter UAV released by the company.

TECHNICAL SPECIFICATIONS

Material: It is made of steel, carbon fiber, and plastic.

Size: length: 1670 mm; width: 510 mm; height: 690 mm; weight: 9 kg without payload

Propellers: Main rotor's diameter is 210 cm. It has a small propeller at the tail.

Payload: Maximum allowed is 1 kg. Payload includes gyro-stabilized gimbal with cameras, plus tank that stores pesticides/liquid fertilizer formulation.

Launching information: It has software for autolaunch, autonomous fight and navigation, and autolanding. The flight path can be predetermined and waypoints to cover during flight is almost unlimited (Plate 3.2.1).

Altitude: This UAV, AT-35, flies very close to crop canopy/ground surface. It operates at 50 m height for aerial survey and spectral analysis of crops. Maximum altitude, that is, ceiling is 1500 m above sea level.

Speed: Maximum speed is 55 knots. It can also hover over the crop without much movement (stand-still mode). Hovering at low altitude is useful in obtaining close-up shots.

Wind speed tolerance: 20–25 knots wind disturbance

Temperature tolerance: −10 to 60°C

Power source: A two-stroke internal combustion engine of 2.4 hp generates power for propulsion.

Electric batteries: 14 A h to supports electronics and CPU.

Fuel: Gasoline (lubricant oil)

Endurance: 3 h per flight

Remote controllers: AT-35 is capable of fully autonomous flight. It has auto-take-off and landing software (Plate 3.2.1). The software allows unlimited waypoints with in flight fail-safe "return to home" mode. The GCS and UAV keep contact via long-range telemetry up to 30 km or through 3G network or satellite link. Standard telemetry is only 1.5 km. Optional telemetry is 3–30 km.

Area covered: The UAV is capable of quickly covering large area during flight, pick aerial images, and relay it to GCS computers. Flight distance covered along with imagery is 50–60 km in linear distance or 50–200 ha per flight.

Photographic accessories: Cameras: This helicopter UAV carries in its gyro-stabilized gimbal a range of cameras. They are visual (daylight), near infrared (NIR) and infrared (IR) sensors. It also carries a Lidar sensor. The UAV has vehicle-tracking accessory.

Computer programs: The aerial images that UAV pick and relay to GCS are usually processed, using Agisoft's Photoscan or Pix4D Mapper software. The digital data that UAV picks could be stored in chips and utilized in variable-rate applicators during precision farming.

Spraying area: AT-35 UAV could be fitted with plastic tanks on either side of fuselage to hold plant protection chemicals or liquid fertilizer formulation.

Volume: Data not available

Agricultural uses: They include survey of land resources, crop scouting, and natural vegetation monitoring.

Nonagricultural uses: They include surveillance of geological sites, mines, mining activity, movement of ores, etc. It is used to monitor buildings, city vehicular traffic, public events, road transport, rail roads, oil pipelines, international borders, etc.

PLATE 3.2.1 AT-35
Source: Mrs. Kristy Hall, Advanced UAV Technology, Wellesley House, Duke of Wellington Avenue, Royal Arsenal, London SE18 6SS, UK.
Note: AT-35 is a light-weight, medium-endurance VTOL helicopter UAV.

Useful References, Websites, and YouTube Addresses

http://www.pegasusexecutive.com/uncategorized/757/ (accessed Oct 5, 2017).
http://auavt.com.existsite.com/ (accessed Oct 5, 2017).
http://www.zoominfo.com/c/Advanced-UAV-Technology-Limited/347665240 (accessed Oct 5, 2017).

Name/Models: AT-50

Company and address: Mr. Pietro Amati, Advanced UAV Technology Ltd., 28 Red Post Hill, London, SE24 9JQ, UK; phone: +44 20 7501 9345; e-mail: info@auavt.com; website: http://www.auavt.com

Advanced UAV Technology Ltd. is a UAV-manufacturing company situated in United Kingdom. As stated earlier, they offer several models of helicopter UAVs.

They are apt for use in agricultural, industrial, and military settings. Their model "AT-50" is one among series of AT helicopters. It is a small UAV capable of aerial survey, scouting and collecting important data about crops.

TECHNICAL SPECIFICATIONS

Material: This helicopter UAV (i.e., AT-50) is made of steel, carbon fiber, ruggedized foam, and plastic.

Size: length: 180 cm; width (wingspan): 76 cm; height: 80 cm; weight: 10 kg dry weight (without payload).

Propellers: It has two propellers, one large propeller at the top of cockpit area and second one at the tail that offers stability.

Payload: Maximum payload is 20 kg. It includes cameras for aerial imagery, a plastic tank to hold pesticide or liquid fertilizer.

Launching information: It is vertical take-off autolaunch UAV. It has software for autonomous flight and navigation (e.g., Pixhawk). It lands automatically. It could also be guided carefully using remote controller to land anywhere. It needs 5 × 5 m open space for launch or landing (Plate 3.3.1).

Altitude: It hovers just above the crop field and canopy. It operates normally at altitudes 10–50 m above crops/urban location. However, its ceiling is 1500 m above sea level.

Speed: 75 km h^{-1}

Wind speed tolerance: 20 knots

Temperature tolerance: −10 to 45°C

Power source: A 2 stroke gasoline fuel internal combustion (IC) engine

Electric batteries: 8–14 A h; two-cell batteries to support electronic parts and CPU computers in the UAV

Fuel: Gasoline

Endurance: Up to 3 h per flight

Remote controllers: The GCS is connected to UAV through telemetry that operates for up to 4–5 km range. The safety controller has 2.4 GHz, 800 m LOS. Standard telemetry in the open crop fields or an urban location is restricted to 1.5 km radius from GCS iPad.

Area covered: 200 ha per flight

Photographic accessories: Cameras: The helicopter has a gyro-stabilized gimbal. It holds usual complement of sensors. They are visual cameras, multispectral, NIR, and IR cameras, pus, Lidar, and videocamera. The videograph can be relayed to GCS instantaneously.

Computer programs: The UAV's CPU or GCS iPad utilizes software such as Agisoft's Photoscan, Pix4D Mapper, and any other custom-developed software.

For example, for identifying pest/disease-affected regions, a software to recognize spectral signature of health and disease-afflicted crops could be adopted.

Spraying area: This UAV could be used to spray chemicals/water.

Volume: No data available yet

Agricultural uses: They are aerial survey of land resources, crop scouting, and natural vegetation monitoring.

Nonagricultural uses: It includes surveillance of mines, transport vehicles, etc.

PLATE 3.3.1 AT-50—a helicopter UAV.
Source: Mrs. Kristy Hall, Advanced UAV Technology, Wellesley House, Duke of Wellington Avenue, Royal Arsenal, London SE18 6SS, UK
Note: AT-50 is a light-weight, medium-endurance VTOL UAV. It has several applications during crop production.

Useful References, Websites, and YouTube Addresses

http://www.pegasusexecutive.com/uncategorized/757/ (accessed Oct 15, 2017).
http://auavt.com.existsite.com/ (accessed Oct 15, 2017).
http://www.zoominfo.com/c/Advanced-UAV-Technology-Limited/347665240 (accessed Oct 15, 2017).

Name/Models: Chi-7 Helicopter UAV

Company and address: Aerodreams, Buenos Aires, Argentina. Tel./Fax: +54 11 4393 3466; e-mail: NA; website: http://www.aerodreams-uav.com

Aerodreams is a company that offers a range of UAV helicopters useful in military, civil, commercial, and agricultural realms. Aerodreams CHI-7 is a multipurpose UAV. It has been developed based on a previous model named "Cesare CH-6," which is again a helicopter UAV produced by the same company. The helicopter UAV CH-7 has found extensive use in military reconnaissance, civil aerial monitoring, commercial aerial imagery, survey of natural vegetation, and agricultural cropping zones. It has capacity for both low and high-altitude aerial imagery. Hence, it covers large areas at a time. CH-7 is a popular UAV among Argentinean UAV service companies. CH-7 is also used by agricultural aerial survey agencies. It has been effectively used for survey of terrain and in obtaining 3D imagery since 2014. It includes a range of electrooptical (EO) sensors to gather useful data about terrain.

TECHNICAL SPECIFICATIONS

Material: Steel, carbon fiber, and plastic

Size: length: 7.15 m; main rotor diameter: 6 m; height: 2.1 m; weight: 450 kg

Propellers: One main large propeller at the top of the cockpit area.

Payload: <230 kg

Launching information: It is an autolaunch, autonomous navigation, and autolanding UAV. Its flight path can be preprogramed using Pixhawk software.

Altitude: The ceiling is 4500 m

Speed: The cruising speed is 130 mi h^{-1}, and maximum speed is 180 km h^{-1}

Wind speed tolerance: 25 km h^{-1} wind turbulence

Temperature tolerance: −10 to 55°C

Power source: An internal combustion two-stroke, four cylinders, piston engine. It generates 100 hp.

Electric batteries: 14 A h; four-cell batteries for electronics and CPU

Fuel: Gasoline

Endurance: 8–12 h depending on payload

Remote controllers: The flight path, launch, and landing locations can be preprogramed, using Pixhawk software. This UAV is totally autonomous. If needed, it can be guided from ground control to change course of flight and waypoints.

Area covered: It can cover 1100 km linear distance per flight without refueling.

Photographic accessories: Cameras: A gyro-stabilized gimbal has series of 4–5 sensors and tracking device for ground vehicles. The cameras include daylight (visual) cameras (R, G, and B), NIR, IR, red-edge, and thermal bandwidth cameras. It has Lidar sensor.

Computer programs: This UAV comes with a custom-made photo-processing software. Usual software such as Pix4D Mapper and Agisoft's Photoscan is utilized to process ortho-images relayed by the UAV. The UAV's CPU can store a large amount of digital data and imagery collected during a flight.

Spraying area: CH-7 can be used to spray crops with pesticides, weedicides, and liquid fertilizer. We can do so by attaching plastic tanks to hold the chemicals and fixing appropriate spray nozzles. The digital data about variability in disease/pest attack or nutrient deficiencies is collected by the UAV. Later, it can be utilized in the variable-rate nozzles, if precision faming is in vogue.

Volume: Data not available

Agricultural uses: Crop scouting to detect disease or pest attack from mid-altitudes of 400 m is possible. It is accomplished with good efficiency by such helicopter UAVs. At low altitude, CH-7 is utilized to study the terrain, its features, soil type and its variations, cropping systems, etc. Natural vegetation monitoring and collection of data on NDVI and species diversity could be done efficiently. It is done by collecting spectral signatures of plants and crops. This UAV can be used to map the areas affected by drought, floods, or erosion in large areas of farm.

Nonagricultural uses: CH-7 is used during aerial reconnaissance and military target identification. It is specially adapted for surveillance borders and military installations from high altitudes. At altitudes of over 300 m above ground level, it can be used to conduct regular surveillance of highway traffic, oil pipelines, electric lines, rail roads, mines, and mining activity. It can also be used to collect atmospheric samples for detection of greenhouse gas contents and pollutants.

Useful References, Websites, and YouTube Addresses

http://www.aerodreams-uav.com/es-chi7-helicoptero-dual.html (accessed Nov 5, 2017).

http://www.aerodreams-uav.com/es-servicios.html (accessed Nov 5, 2017).

http://en.avia.pro/blog/aerodreams-chi-7-tehnicheskie-harakteristiki-foto (accessed Nov 5, 2017).

https://sites.google.com/site/stingrayslistofrotorcraft/aerodreams-chi-7 (accessed Nov 5, 2017).

http://en.avia.pro/blog/aeroderams-chi-7-technicheskie-harakteristiki-foto (accessed Nov 10, 2017).

https://ru.wikipedia.org/wiki/AeroDreams_Chi-7 (accessed Nov 10, 2017).

http://www.aerodreams-uav.com/es-chi7-helicoptero-dual.html (accessed Nov 10, 2017).

Name/Models: HEF 30

Company and address: High Eye, Nieuwe Rijksweg 66d, 4128 BN Lexmond, The Netherlands; phone: +31 (0) 34 7342631; e-mail: NA; website: http://www.higheye.nl/company/

High Eye was founded about 30 years ago, by Mr. Jan Verhaegen. He designed and built model helicopters, which he used for the filming of movies and commercials. Some of these helicopters were sold to third parties over the years. In 2003, Jan

Verhaegen decided to focus on aviation services as well as the designing, building and selling of unmanned helicopters (copter UAVs). In 2008–2010, several new models were designed. The HEF 26, HEF 30, HEF 80R, and HEF 150 are examples. In July 2013, Mr. Joost de Ruiter joined the company as CEO and majority shareholder.

The HEF 30 is a helicopter UAV. It has been designed to be a flexible, multi-sensor platform. This allows customers to perfectly adapt the system to their mission requirements. A wide variety of options are available to further expand the functionality of the HEF 30. There are many ways to expand the capabilities of the HEF 30 system. They are operating multiple sensors at once, increasing operational range, improving flexibility, and even integrating into air traffic control environments. They say that HEF 30 can do it all. The payload bay of the HEF 30 is located underneath the nose. It offers clear space without obstructions to mount large or odd-sized payloads. The built-in converters and IP-based datalink system allow the live transmission of many different types of payload data.

TECHNICAL SPECIFICATIONS

Material: HEF UAV is made of a composite of ruggedized foam, toughened plastic steel, and carbon fiber.

Size: length: 105 cm; rotor width: 65 cm; height; 75 cm; weight: 14 kg

Propellers: It has one larger propeller above fuselage and a smaller one at the end of tail piece.

Payload: 2.8 kg. The payload bay of the HEF 30 is located underneath the nose.

Launching information: This is a VTOL helicopter with autolaunch, navigation, and landing software. The flight path can be preprogramed to pass over at least 200–300 waypoints per flight.

Altitude: The regular operating altitude during surveillance and aerial imagery of crops is 5–50 m above the crop canopy. It can be used at 300–400 m above ground level, particularly for surveillance and monitoring of commercial installations, public events, etc. However, the ceiling is 1500 m above ground level.

Speed: 45 km h^{-1}

Wind speed tolerance: 20–25 knots wind disturbance and turbulence

Temperature tolerance: −10 to 60°C

Power source: Two-stroke IC engine fueled by gasoline

Electric batteries: 14 A h for supporting electronic parts, gimbals, and CPU

Fuel: Gasoline plus lubricant oil

Endurance: 3 h per flight

Remote controllers: The ground station equipment includes iPad, telemetry systems, antenna, and radiotracking devise and software. They are as follows:

Antenna-High Eye B.V. has designed a relatively simple to use tracking antenna system. The tracking antenna is a multifunctional Pan-Tilt Unit made by Flir®. It is directly controlled by the GCS using live autopilot data. The range of the antenna depends on the frequency, bandwidth, and amount of data an operator would like to send. In general, the range is increased beyond 50 km radius from the GCS.

Radiotracking: The RT-600 is an advanced radiolocation tracker, supplied by Rhotheta®. It can be used to search and track ELTs, PLBs, EPIRBs, MSLDs, medical beacons, covert transmitters, buoys, and similar equipment. The antenna is sensitive to a wide range of frequencies, like 118–470 MHz, A3E, F3E, A3X, F1D, and G2D. The tracker on the HEF 30 is positioned at the nose. It creates an unobstructed search window. The internal receiver and processer allow simple connection to the datalink system.

Area covered: It covers 50–250 ha crop/ground surface.

Photographic accessories: Cameras: This UAV carries a full complement of sensors such as visual (daylight), NIR, IR, red-edge, Lidar, and radiotracking devices.

Gimbals: This UAV could be fitted with gimbals with series of different sensors. They are as follows:

EPSILON 135: The Epsilon 135 is a compact single-sensor payload supplied by UAV Factory.

CM100: The CM100 is a lightweight, multisensor capable gimbal supplied by UAV Vision. It is designed to be very rugged. Both the video data and the control of the gimbal are communicated via IP/H.264.

OTUS-U135: The OTUS-U135 is a well-proven, multisensor capable gimbal. It is supplied by DST control. The Lidar Pod is a plug-and-play off-the-shelf Lidar system, specifically developed for UAVs and supplied by Routescene®.

Computer programs: The aerial images transmitted to GCS can be processed using the usual software such as Agisoft's Photoscan or Pix4D Mapper.

Spraying area: This UAV could be fitted with pesticide/liquid fertilizer holding tanks on the fuselage.

Volume: Data not available

Agricultural uses: It includes survey of land resources, crop scouting for growth, phenomics, disease/pest attack and drought effects, natural vegetation monitoring is also conducted.

Nonagricultural uses: They are surveillance of Industrial installations, buildings, rail roads, city traffic, public events, oil and gas pipelines, etc.

Useful References, Websites, and YouTube Addresses

http://www.higheye.nl/company/ (accessed Nov 20, 2017).
http://www.higheye.nl/hef-32/ (accessed Nov 20, 2017).

http://www.skeyebv.com/systems/hef-30/ (accessed Nov 20, 2017).
https://www.youtube.com/watch?v=gOYJEBoJNGM (accessed Oct 30, 2017).
https://www.youtube.com/watch?v=5SNx-yoBYkw (accessed Oct 30, 2017).
https://www.youtube.com/channel/UCE_hHTmGRYj3iQ8i0iLjCKg (accessed Oct 30, 2017).

Name/Models: RH2 "Stern"

Company and address: Delft Dynamics B.V., Molengraafsingle 10, 2629 JD Delft, The Netherlands; phone: +31 15 7111009; e-mail: info@delftdynamics.nl; website: http://www.delftdynamics.nl/index.php/en/products/rh4-en

Delft Dynamics is an UAV-producing company. It offers a few different models. Several of which are light in weight. They are efficient to use in the crop fields and economically feasible for farmers. These UAVs are currently operative in the European plains region.

RH2 "Stern" is a robotic helicopter attached with a series of cameras/sensors for aerial surveys. It is easily operated by farmers and civilians with little training. It comes with a joystick, if flight path and photographic events need to be altered. It has good hover capacity over buildings or crops while taking images. It offers images with GPS tags for greater accuracy. This helicopter UAV comes with custom-made ground station accessories and software (Plate 3.4.1).

TECHNICAL SPECIFICATIONS

Material: It is made of aluminum-based alloy, carbon fiber, and plastic.

Size: length: 100 cm; width: 30 cm; height: 50 cm; rotor length: 180 cm; weight: 3.2 kg dry weight (i.e., without payload)

Propellers: One main propeller with two plastic blades above fuselage. The propeller is connected to brushless motor for energy. A second propeller is small and placed at the tip of tail for stability and guidance.

Payload: 2.5 kg. It includes gyro-stabilized gimbal, cameras and pesticide tank, if opted.

Launching information: It is an autolaunch, autonomous flight, navigation and autolanding UAV. It is programed using software such as Pixhawk or one that is custom made.

Altitude: it hovers at 0 altitude over crop or installations. Its operational altitude is 50–300 m above ground level. However, the ceiling is 1000 m above ground level.

Speed: Maximum speed is 70 km h^{-1}. Operational speed is 35–50 km h^{-1}.

Wind speed tolerance: It tolerates wind disturbance of 25 km h^{-1}.

Temperature tolerance: −10 to 45°C.

Power source: Gasoline engine and electric batteries

Electric batteries: 8–14 A h for supporting electronic circuits and CPU

Fuel: Gasoline

Endurance: 15 min per flight

Remote controllers: The helicopter UAV—"RH Stern" comes with a custom-made ground station equipment. The accessories are light, portable, and highly efficient in controlling and commandeering the UAV's flight path. The UAV is easily controlled using a joystick, if it is not preprogramed (Plate 3.4.2).

Area covered: 50–250 ha h^{-1} depending on purpose

Photographic accessories: Cameras: It is provided with a full complement of cameras such as visual (R, G, and B), multispectral, hyperspectral, NIR, IR, and red-edge bandwidth cameras.

Computer programs: The UAV system is provided with custom-made software for processing digital data and storing it. Such data could to be converted into aerial images. We can also use regular software such as Pix4D Mapper or Agisoft's Photoscan.

Spraying area: It could be attached with a pesticide tank to store dusting material and be connected to nozzles for spraying.

Volume: Data not available

Agricultural uses: This is a multipurpose UAV meant for use in defense, civilian locations, and in agricultural farms. Agricultural surveillance and crop scouting is a major activity. It is used to detect disease/pest-attacked, drought, and flood-affected locations in a farm. It can detect soil erosion affected spots, etc. This UAV is also used to develop maps depicting variations in NDVI, leaf chlorophyll content (NIR), thermal indices (IR imagery), and crop/grain maturity trends. It can monitor the farm activity and vehicle movement, using tracking device. Natural vegetation monitoring could be conducted regularly.

Nonagricultural uses: They are surveillance of mines, mining activity, ore movement, and dumping zones; monitoring industrial zones, buildings, rail roads, highways, and vehicle movement, etc. It is also used to monitor activity of military camps and vehicles.

PLATE 3.4.1 RH2 "Stern" Helikopter (left: an UAV over the airport; middle: UAV showing gimbal and cameras; right: UAV in the air).
Source: http://www.delftdynamics.nl/index.php/en/products; Delft Dynamics, Delft, Netherlands.

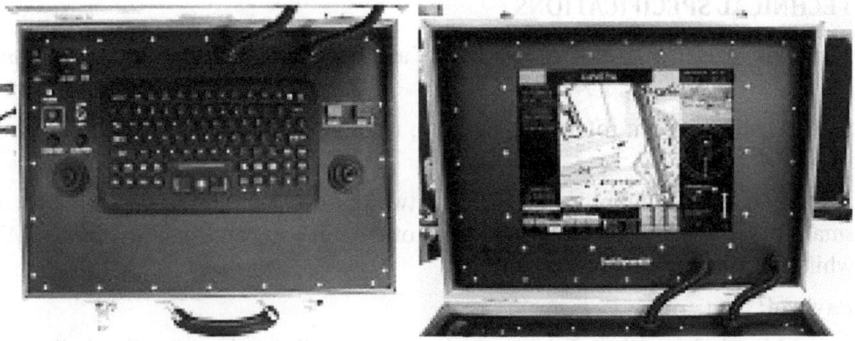

PLATE 3.4.2 Ground station with computer, screen, and software to program and control helicopter's flight path and imaging speed. It is also equipped to process the data and imagery received.
Source: http://www.delftdynamics.nl/index.php/en/products; Delft Dynamics, Delft, Netherlands.
Note: The custom-made ground station is small. The GCS apparatus fits into a suitcase. It is portable and has a computer, keypad, and screen along with requisite software. The software included helps to control the UAV and to obtain digital data. The ground station also receives videoimages for storage and analysis, at a later time.

Useful References, Websites, and YouTube Addresses

http://www.delftdynamics.nl/index.php/nl/producten/robothelikopter (accessed Oct 30, 2017).
http://www.delftdynamics.nl/index.php/en/products (accessed Oct 30, 2017).
http://www.delftdynamics.nl/index.php/en/video-en (accessed Oct 30, 2017).
https://www.youtube.com/channel/UCE_hHTmGRYj3iQ8i0iLjCKg (accessed Oct 30, 2017).

Name/Models: RH3 "Swift"

Company and address: Delft Dynamics B.C., Molengraafsingle 10, 2629 JD Delft, The Netherlands; phone: +31 15 7111009; e-mail: info@delftdynamics.nl; website: http://www.delftdynamics.nl/index.php/nl/producten/rh3

RH3 "Swift" is a robotic helicopter with autonomous flight. It has geo-optimized cameras for aerial photography and digital data capture. In addition to data relay, this helicopter UAV is also effectively used to collect air samples from atmosphere. Such samples are used to study climate change effects. This model was first tested and approved for use in 2016. It has been offered certification as unmanned rotorcraft system by the Dutch Aviation Agency (see Plate 3.5.1).

TECHNICAL SPECIFICATIONS

Material: This helicopter UAV is made of aluminum alloy, rugged carbon fiber and plastic

Size: length: 100 cm; rotor width: 30 cm; rotor diameter: 180 cm; height: 50 cm; weight: 4.2 kg

Propellers: One main propeller made of two plastic blades. A second propeller is small. It is placed at the tip of the tail. It offers direction and stability to the UAV while in flight.

Payload: 5 kg

Launching information: It is an autolaunch, autopilot and autolanding helicopter UAV. It just needs 5 × 5 space clearance to launch or land at a spot. It returns to the place of start, in times of emergency. Autonavigation and flight path is decided, using Pixhawk software.

Altitude: It flies close to crop canopy while obtaining close-up images. Also, during spraying, if used for applying pesticides. The operational height is 50–300 m above the crop canopy. However, the ceiling is 1000 m above ground level.

Speed: Maximum speed is 70 mi h^{-1}. Operating speed is 35–45 km h^{-1}

Wind speed tolerance: 15–20 km disturbing wind

Temperature tolerance: −10 to 40°C

Power source: Electric batteries

Electric batteries: 16,000 mA h

Fuel: Not an IC engine-fitted UAV

Endurance: 60 min per flight

Remote controllers: The UAV system comes with a custom-made GCS, a system known as "Delft Dynamics GCS." The telemetric connection is operative for up to 15 km radius. The GCS iPad allows us to modify the flight path midway through a preprogramed flight path.

Area covered: It covers 50–250 ha per flight, based on purpose for which the UAV is used.

Photographic accessories: Cameras: It has a gyro-stabilized gimbal with a series of five cameras. The cameras are visual (Sony 5100), multispectral, hyperspectral for close-up shots, NIR, IR, and red-edge cameras for detection of heat signatures and crop water stress index.

Computer programs: The aerial imagery is usually processed, using custom-made software or those like Agisoft or Pix4D Mapper.

Spraying area: This UAV can be fitted with a small tank to carry dusting material if needed.

Volume: Data not available

Agricultural uses: It includes detailed aerial imagery of farm land. Crop scouting to detect areas with gaps in seedlings, seedling/crop stand, crop nutritional status and its variations, water stress index variations, identifying disease/pest-attacked areas, and mapping them are other uses of natural vegetation monitoring and obtaining NDVI values.

Nonagricultural uses: They are surveillance of mines, mining activity, ore dump yards, industrial units, building, highways, rail roads, and tracking transport vehicles. It is also used to detect and obtain information from disaster areas.

PLATE 3.5.1 Swift-Helicopter UAV.
Source: http://www.delftdynamics.nl/index.php/nl/producten/rh3; Delft Dynamics, Delft, Netherlands.

Useful References, Websites, and YouTube Addresses

http://www.delftdynamics.nl/index.php/nl/producten/rh3 (accessed Sept 2, 2017).
http://www.delftdynamics.nl/index.php/en/video-en (accessed Sept 2, 2017).
http://www.delftdynamics.nl/index.php/en/ (accessed Sept 2, 2017).

Name/Models: Scout B 1

Company and address: Aeroscout Gmbh., Hochschule Luzern, Technik und Architektur, Technikumstrasse 21, Building III, Room D311, 6048 Horw, Switzerland; e-mail: info@aeroscout.ch; phone: +41 41 349 33 85; fax: +41 41 349 36 37; website: www.aeroscout.ch, http://www.aeroscout.ch/index.php/en/

Aeroscout GMBH is located at Horw in Switzerland. Aeroscout Gmbh. offers complete UAV solution. It is based on industrial unmanned helicopters. It offers customized UAV solutions for various applications. The helicopter UAVs produced by Aeroscout are currently on duty for several geometrics and surveillance. In addition, Aeroscout provides experienced flight crew. In cooperation with clients, Aeroscout develops new customized UAV solutions for specific usage.

The Scout B1-100 UAV helicopter developed by Aeroscout Gmbh. is an INS/GPS controlled unmanned helicopter UAV (Plate 3.6.1). This heavy lift UAV helicopter provides a unique payload capacity of 18 kg (40 lb). It includes digital cameras, video camcorders, laser scanners, IR, or hyperspectral camera equipment.

TECHNICAL SPECIFICATIONS

Material: It is made of aluminum alloy, carbon fiber, fiber glass, and plastic.

Size: length: 0.58 m; width (wingspan): 0.35 m; height: 0.44 m; weight: 19 kg

Propellers: One large propeller at the top of the cockpit area. It has four large plastic/carbon fiber blades. Propulsion is derived from a brushless motor, energized through batteries.

Payload: Payloads are successfully integrated on Scout B1-100 UAV helicopter. In addition, the task of payload integration includes electrical and software interfaces. On board data storage is possible. Generally, broadband data transmission to the GCS and data visualization is possible, during the flight. The payload system with gimbal and cameras are integrated. This allows well-stabilized and vibration-free photographic images.

Launching information: It has autolaunch mode. It is a vertical take-off and land helicopter.

Altitude: Flight altitude for aerial imagery of crops is 100 m above ground level.

Speed: 20–45 km h^{-1} (6 m s^{-1})

Wind speed tolerance: 20 knots

Temperature tolerance: This UAV tolerates 0 to 40°C in the open fields.

Power source: Power for propulsion is supplied at 110–220 V

Electric batteries: 28 A h high capacity lithium–polymer batteries, four cells

Fuel: No need for petrol or diesel.

Endurance: 2 h

Remote controllers: The Aeroscout GCS provides all capabilities to operate the "Scout B1-100" UAV helicopter. This includes mission planning, mission control, display of UAV status information, and UAV performance monitoring. The Aeroscout GCS features include easy transportability required for outdoor operations, a single source power supply (110–220 V), two sunlight readable displays, one touch screen for main UAV commands selection, the computer display or video display mode, a water-resistant keyboard, and two intelligent joysticks (Plate 3.6.3). The GCS also provides stabilized voltages for field operation, for example, for system startup or for the GCS datalink modules. The iPad operates on windows 10 software. It has facility for editing videos. The datalink module is 2 × RS232.

The GCS has two joysticks to regulate UAVs' flight path. It allows both preprogramed flight path as well as manually controlled flight path. It allows 200 flight points to be selected for the UAV to fly over. The GCS computers and radiolinks are all ruggedized. The computers can be operated and software utilized to rearrange or analyze flight path imagery right in the open crop fields.

Area covered: The area covered per flight is 1.0 km^2. It allows 200 points on flight path to be covered per flight.

Photographic accessories: Cameras: The AisaKESTREL 10 is especially designed for UAVs. The AisaKESTREL 10 fits perfectly on the Scout B1-100 UAV helicopter, which can carry a maximum of 18 kg. It leaves enough free payload capacity to integrate a high-grade dual-GPS antenna INS/GPS navigation system as well as a broadband datalink, batteries, and an advanced vibration damping system. This combination of components and UAV leads to great usability combined with an outstanding data quality (Plate 3.6.2).

Computer programs: Computer software such as Agisoft's Photoscan or Pix4D Mapper is found both in the CPU of UAV and GCS. The software is used to process ortho-images, stitch them and transmit to computers that analyze the digital data and images.

Spraying area: Not a spraying UAV. However, it can be modified to hold pesticide or plant protection chemicals in liquid or granular forms and spray them onto crop fields.

Volume: Data not available

Agricultural uses: It includes surveying land resources, crop scouting for variations in growth, also collecting phenomics and grain maturity data. It is used to detect locations affected by diseases, pests, drought or floods in a crop field (Plate 3.6.2). Natural vegetation monitoring is also possible (Plate 3.6.2; Aeroscout, 2018).

Nonagricultural uses: They are surveillance of buildings, industries, mines, ore dumps, city traffic, transport vehicles, oil and gas pipelines, electric lines, and fences/borders.

PLATE 3.6.1 Scout B1 UAV helicopter UAV.
Source: Dr. Christoph Eck., Aeroscout Gmbh., Horw, Switzerland.

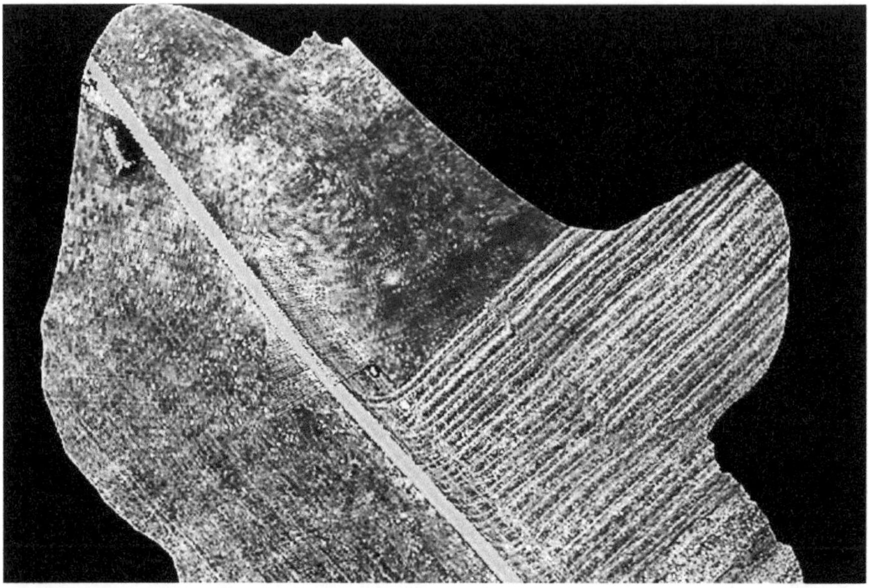

PLATE 3.6.2 An aerial image of crop fields obtained using cameras on Scout B1.
Source: Dr. Christoph Eck., CEO, Hochdorf, Switzerland.

PLATE 3.6.3 Scout B1 scanning a natural forest vegetation.
Source: Dr. Christoph Eck., CEO, Hochdorf, Switzerland.

Useful References, Websites, and YouTube Addresses

Aeroscout. *High Resolution Laser Ranging of Earth's Forests and Topography*; Aeroscout, Unmanned Aerial Technology, Newsletter March 2018, 2018; pp 1–4. http://www.aeroscout.com (accessed Mar 13, 2018).

http://www.aeroscout.ch/index.php/en/hyperspectral-imaging#sigProId1fdd3935e9 (accessed Mar 10, 2018).

http://www.aeroscout.ch/index.php/en/hyperspectral-imaging#sigProIdb4b7c06312inShare0 (accessed Mar 10, 2018).

http://www.aeroscout.ch/index.php/en/scout-uav-helicopter/ground-control-station#sigProIde572d58d8c (accessed Mar 10, 2018).

https://www.youtube.com/watch?v=Ddz2C6oUNo4 (accessed Mar 10, 2018).

https://www.youtube.com/watch?v=iTuPBpVDLVc (accessed Oct 12, 2017).

https://www.youtube.com/watch?v=x0_eKHlEeTI (accessed Oct 12, 2017).

Name/Models: Sniper

Company and address: Alpha Unmanned Systems, Calle Perales 9, San Sebastián de los Reyes (28703). Madrid, Spain; phone: NA; e-mail: info@AlphaUnmanned-Systems.com; websites: http://www.alphaunmannedsystems.com/products/sniper, http://www.alphaunmannedsystems.com/

Alpha unmanned systems is situated at Madrid, Spain. It has more than 25 years of combined experience, in the field of UAVs design and aircraft control systems. Their team of experts offers a range of UAVs and related services. This company designs, tests, and manufactures versatile high-performance UAV helicopter platform.

Alpha Unmanned Systems offers complete solutions to many types of clients in the UAV sector, including a completely integrated GCS, communications, control software, and all accessories tailored to suit the client's requirements. We distribute our products directly to end-user customers and indirectly through a growing network of UAV service providers. They also offer specialized services such as training UAV pilots, offering aerial imagery to clients, etc.

"SNIPER" combines three important aspects of any rotary wing UAV of this type, namely, (a) VTOL, (b) hovering, and (c) long endurance, due to its gasoline powered engine. SNIPER allows considerable payload capacity, enabling the use of high-quality EO/IR cameras and other sensors (Plate 3.7.1).

TECHNICAL SPECIFICATIONS

Material: The air frame is made of carbon fiber and ruggedized foam.

Size: length: 1.7 m; rotor diameter: 1.8 m; height; weight: 14 kg

Propellers: One propeller is placed at the top of the cockpit zone. A second one is small and is found at the tip of the tail.

Payload: Maximum payload is 2 kg. It includes gyro-stabilized gimbal that accommodates visual, IR, and thermal cameras. The sensors have GPS connectivity.

Launching information: It is a vertical take-off and launch type UAV.

Altitude: The ceiling is 3000 m. But it can be flown at low altitude of 50–100 m above the ground (crop) level.

Speed: Cruise speed is 55 km h^{-1}

Wind speed tolerance: 20 knots

Temperature tolerance: −20 to 60°C

Power source: Electric batteries and two stroke gasoline piston-engine

Electric Batteries: 28 A h, two cells, lithium–polymer batteries

Fuel: Gasoline + oil

Endurance: 2 h

Remote controllers: The UAV takes off automatically. It also has autonomous navigation facility. The flight path can be preprogramed using Pixhawk software. The landing is also autonomous, but it can be guided. This UAV model has facilities, such as autoflight plan execution (waypoints), fly-to-hover, auto-return-to-base (RTB) in case of communication failure, manual override. Day vision, low-light vision, night vision, or IR is also possible (Plate 3.7.1).

Area covered: It covers 80–120 km linear distance per flight. It can survey up to 50–100 ha of crop field per flight

Photographic accessories: Cameras: It includes sensors such as Sony 5100, Sony 7, mica-sense red-edge, and thermal sensor. It has Onboard 3D Digital Elevation Model for accurate flight control and camera pointing.

Computer programs: The aerial images are processed using Pix4D Mapper or Agisoft's Photoscan or any custom-made orthomosaic processor.

Spraying area: This UAV can be utilized for accurate spray of plant protection chemicals, pesticides, fungicides, liquid and granular fertilizer, etc.

Volume: Data not available

Agricultural uses: They are aerial mapping and surveillance of land resources, crop fields, and water resources. Crop scouting to detect diseases, pests, drought-attacked zones, flooded locations, and other damages to crop. Monitoring of seed germination, seedling emergence, crop stand establishment, grain maturity, etc. Identification of nutritional deficiencies (leaf chlorophyll, leaf-N). Crop monitoring for growth and canopy status. Natural vegetation monitoring could be conducted routinely (Plate 3.7.1).

Nonagricultural uses: It includes fire assessment, environmental inspection, border patrol, day/night traffic monitoring, inspection of commercial buildings, etc.; surveillance of mines and mining activity. This copter UAV has a range of applications in the realm of military. It includes intelligence gathering, silent surveillance, target acquisition, combat, etc.

PLATE 3.7.1 Left: SNIPER UAV in flight; right: SNIPER UAV resting on ground.
Source: http://www.alphaunmannedsystems.com/products; Ms. Neomi Tarancon; Alpha Unmanned Systems, Madrid, Spain.

Useful References, Websites, and YouTube Addresses

http://www.alphaunmannedsystems.com/contact-us (accessed Mar 10, 2017).
http://www.alphaunmannedsystems.com/products (accessed Mar 10, 2017).
https://www.youtube.com/watch?v=EFW-zWCK97c (accessed Mar 10, 2017).
https://www.youtube.com/watch?v=6wKk6bVv8-k (accessed Mar 10, 2017).
https://www.youtube.com/watch?v=jLl4H9lp9IU (accessed Mar 10, 2017).
https://www.youtube.com/watch?v=_8I3T9sbIhk (accessed Mar 10, 2017).

Name/Models: SW-4 "SOLO"

Company and address: Leonardo-Finmeccanica Helicopters, Augusta Westland, Via Giovanni Gust, 520-21017 Cascina Costa di Samarate, Italy; Registered Head Office: Piazza Monte Grappa no. 4, 00195 Rome, Italy; Tel.: +39 06 324731, +39 0331 229111; e-mail: NA; website: leonardocompany.com

"Leonordo Finmeccanica Helicopters" is a well-known and established helicopter manufacturing company, in Europe. It offers a range of models that suit military and civilian requirements. Helicopters are used in state defense, civilian aerial survey, including agricultural belts, and their natural resources. Their models such as "SOLO" have been utilized efficiently to evaluate natural resource, crop production trends, forest plantations, etc. The UAV helicopters are used by many nations in Europe, Africa, and Asia.

SW-4 "SOLO" is a rotary unmanned air system. But it is optionally a piloted helicopter. It is a relatively heavy flying machine at 180 kg unit^{-1}, particularly if we consider the agricultural and civilian tasks for which it is utilized. The SW-4 "SOLO" has been developed by PZL-Swidnik of Poland and Cascina Costa, Italy. The "SOLO" performs a series of tasks autonomously based on predetermined flight path and computer-based decisions and instructions. It includes aspects related to civilian and agricultural aspects. Its unusually long flight endurance and extended ability for aerial imagery allows it to be adopted to survey very large cropping belts. It is used to study large patches of cropping systems, flood, drought-affected zones, etc. It could be an excellent gadget in the hands of state agricultural agencies, which make decisions for larger crop belts. It is relatively a costly UAV.

TECHNICAL SPECIFICATIONS

Material: It is made of light avionics aluminum and tough carbon fiber.

Size: length: 907 cm; width: 228 cm; rotor diameter: 900 cm; height: 314 cm; weight: maximum take-off weight (MTOW) 1800 kg

Propellers: It has two propellers made of aluminum and carbon fiber. Main propeller is 900 cm in diameter. There is a smaller propeller at the rear end of fuselage and facing sideways for navigation.

Payload: 470 kg

Launching information: It is an autolaunch helicopter. It has autonavigation and shut-off and landing software. In case of emergency, the UAV returns to launch spot. It can also be recovered using GCS computers.

Altitude: This helicopter is flown at 20–300 m above the ground/crop canopy for routine aerial surveys.

Speed: Best operational speed is 80–150 km h^{-1}; maximum cruising speed 206 km h^{-1}

Wind speed tolerance: 45 km h^{-1} disturbance

Temperature tolerance: −20 to 60°C

Power source: Single internal combustion engine

Electric batteries: Two stroke IC engines

Fuel: Heavy fuel (jp 5/8). It holds 450 L of fuel in the tank.

Endurance: 6 h at a stretch per take-off

Remote controllers: This helicopter could be manually piloted during aerial survey, reconnaissance, or spraying sorties. It could be controlled, using ground station computers and telemetric link. The SOLO helicopter UAV has automatic flight mission software. It also takes commands from GCS.

Area covered: It can cover 940 km linear distance per flight. It could be used survey large areas such as a county/district to study natural resources and crop production trends.

Photographic accessories: Cameras: A very wide range of sensors could be attached to the helicopter UAV. It includes visual (R, G, and B), NIR, IR, red edge, Lidar, vehicle-tracking devices, etc.

Computer programs: The aerial images can be processed right at the helicopter's CPU or at the GCS computers. Pix4D Mapper or Agisoft software could be used to stitch and develop sharp images.

Spraying area: "SOLO" is predominantly a survey/reconnaissance UAV used in military and civilian locations. It could be fitted with sprayer equipment, if needed.

Volume: No data available

Agricultural uses: It includes agricultural ground survey and mapping of crop production trends. Surveillance of agro-forestry belts, crop belts, and forest plantations. Natural vegetation monitoring to obtain NDVI data and maps.

Nonagricultural uses: "SOLO" has major role in military and security services. It includes maritime and battlefield intelligence, surveillance, target acquisition, and reconnaissance (ISTAR); coastline and border monitoring; homeland security/law enforcement, disaster and battle damage assessment, logistics, cargo supply, and environmental monitoring. It can be effectively utilized to transport passengers and luggage.

Useful References, Websites, and YouTube Addresses

http://www.leonardocompany.com (accessed Sept 30, 2017).

http://www.leanardocompany.com/en/product-services/elicotteri_helicopters/unmanned-erial-systems (accessed Sept 12, 2017).

Name/Models: Vapor 35

Company and address: Pulse Aerospace Inc., 450 N, Iowa Street, Lawrence, KS 66044, USA; phone: +1 785 380 7209; e-mail: contact@pulseaero.com; website: http://www.puleaero.com

Pulse Aerospace is a leading UAV company. It was initiated in 2010. It has offered two helicopter Unmanned Aerial Systems (UASs) called "Vapor 35" and "Vapor 55." The vapor has an advanced autopilot mode and it is Federal Aviation Agency (FAA)-compliant. The vapor models are used for commercial, agriculture, research, and security applications. Pulse Aerospace's Vapor has slightly longer endurance, up to 1 h per flight. It offers enhanced flight stability over any other electric UAV helicopter on the market. The company at Lawrence, Kansas, also provides training on maintenance and engineering of vapor UAV models. It also helps farmers with simulations and knowledge about various analytical procedures use of data collected by the sensors.

Vapor 35 is a sleek and effective UAS. It comes with a series of sensors that includes visual, multispectral, NIR, IR, and Lidar. It has an endurance of 45-min flight. The UAV comes with accessories such as GCS. The GCS includes laptop, iPad, telelink, computer software to control flight path, waypoints, and aerial imaging (Plate 3.8.1). The UAV offers excellent digital data, for use on ground vehicles in a farm, particularly to control variable-rate applicators during precision farming.

TECHNICAL SPECIFICATIONS

Material: It is made of carbon fiber, light avionics aluminum, and plastic. It has landing gear made of aluminum bar.

Size: length: 120 cm; width: 45 cm; height: 60 cm; weight: 30 kg with all payloads; 16 kg without batteries and camera payload.

Propellers: Main propeller is 53 cm in length. Small rear propeller with two plastic blades is of 25 cm length.

Payload: Maximum payload allowed 5 lb or 2.2 kg

Launching information: This UAV can be operated manually or used entirely as autolaunch, navigation (using Pixhawk software), and landing vehicle.

Altitude: It is 0–5 m during close-up aerial imagery of crop canopy and for surveillance of disease/pests. Normally, it is flown at 100–300 m above crop canopy.

Speed: 25–35 km h^{-1}

Wind speed tolerance: 20 km h^{-1} wind disturbance

Temperature tolerance: −20 to 55°C

Power source: Electric batteries

Electric batteries: 16,000 mA h, two cells.

Fuel: Not an IC engine-fitted UAV. It does not require gasoline

Endurance: 60 min with entire permissible payload

Remote controllers: The ground station is a remote controller (joystick) or laptop that keeps contact with UAV's CPU for up to 5 km radius. The GCS computer has override facility. It can alter flight path, waypoints to touch, camera shutter functioning, and imaging frequency, if required.

Area covered: 50–300 ac per flight of 6 h

Photographic accessories: Cameras: The helicopter with gyro-stabilized gimbal holds a series of cameras such as visual, high-resolution multispectral, NIR, IR, red-edge bandwidth, Lidar sensor, and vehicle-tracking device.

Computer programs: The aerial images are processed using Pix4D Mapper or Agisoft's Photoscan or Trimble's Inphos software.

Spraying area: Not a spraying UAV. However, could be used to spray small volume, if attached with a fluid or granule holding tank.

Volume: It depends on the payload possible and volume of tanks.

Agricultural uses: Major use of this helicopter UAV is in high-resolution aerial imagery and data collection for agricultural ground vehicles with precision technique facility. For example, digital data obtained could be used for variable-rate pesticide/fungicide application. Vapor 35 is used extensively during crop scouting to determine canopy growth rate, NDVI data and maps, and leaf chlorophyll (crop's N status). It is used to measure canopy and ambient temperature (thermal imagery) to obtain maps depicting variations in water stress index. Natural vegetation monitoring for NDVI, water status, and species diversity is a clear possibility.

Nonagricultural uses: It includes surveillance of geological sites, mines, ore dumping yards, industrial sites, buildings, oil pipelines, electric lines, river flow, lakes and irrigation channels, monitoring highways, transport vehicles, etc.

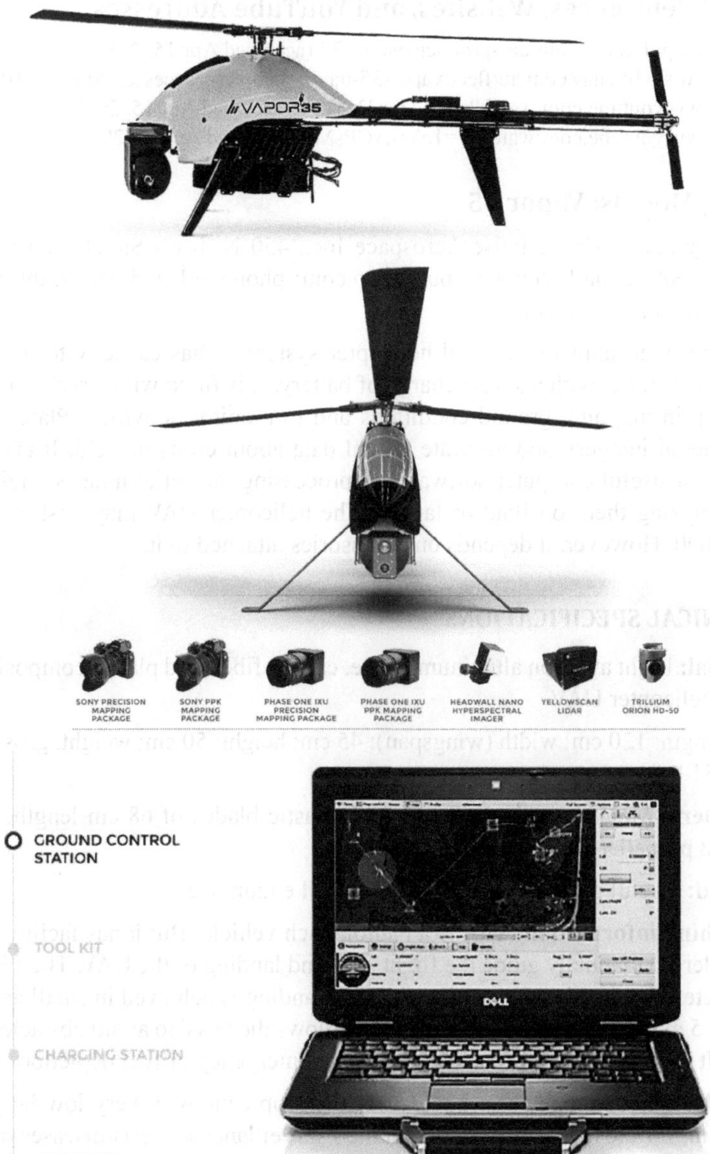

PLATE 3.8.1 Vapor 35—a helicopter UAV and ground station laptop with all necessary software.

Source: Donald Effren Pulse Aerospace Inc., Lawrence, KS, USA; http://www.pulseaero.com/uas-products/vapor-35.

Note: The Pulse Aerospace VAPOR 35 base package includes all common safety software, and those for determining flight path and landing.

Useful References, Websites, and YouTube Addresses

http://www.pulseaero.com/uas-products/vapor-35 (accessed Apr 15, 2017).
http://www.skylineuav.com.au/fleet/vapor-35-uav-helicopter/ (accessed Apr 15, 2017).
https://www.youtube.com/watch?v=ESHgb1NY-3Y (accessed Apr 15, 2017).
https://www.youtube.com/watch?v=1eYr-ACPsM8 (accessed Apr 15, 2017).

Name/Models: Vapor 55

Company and address: Pulse Aerospace Inc., 450 N, Iowa Street, Lawrence, KS 66044, USA; e-mail: contact@pulseaero.com; phone: +1 785 380 7209; website: http://www.pulseaero.com

Vapor 55 is an unmanned aerial helicopter system. It has capacity to be in flight for a longer time, with a single charge of battery. It is fitted with a series of sensors that help in mapping ground conditions and to monitor activities (Plate 3.9.1). It offers aerial imagery and accurate digital data about crops in field. It also comes with a few useful computer software for processing the aerial images, digital data, and analyzing them on iPad or laptop. The helicopter UAV may cost about US$ 25–30,000. However, it depends on accessories attached to it.

TECHNICAL SPECIFICATIONS

Material: Light aviation aluminum frame, carbon fiber, and plastic composite make up the helicopter UAV

Size: length: 120 cm; width (wingspan): 45 cm; height: 50 cm; weight: gross weight 55 lb (31 kg)

Propellers: Main propeller is larger with plastic blades of 68 cm length. Smaller rear end propeller is 35 cm in length

Payload: Useful payload 34 lb, 11 lb with full endurance

Launching information: This is an autolaunch vehicle. But it has facility (remote controller) to manually guide the flight path and landing of the UAV. The flight path is predetermined, using Pixhawk software. Landing is achieved in small areas with just 5 × 5 m clearance space. The software allows the UAV to avoid obstacles during flight. It returns to point of launch, in case of emergency or fuel depletion.

Altitude: The helicopter can hover over the crop canopy at very low height, that is, 1–5 m, for close-up images and detailed surveillance of pests/diseases on crops.

Operational height for aerial imaging is 5–300 m above the crop canopy. Ceiling is 1000 m above the ground level.

Speed: 45 km h^{-1} during aerial imagery. The UAV can even hover at a spot with error of just 1.0 m, during close-up imagery.

Wind speed tolerance: 20 km h^{-1} wind disturbance

Temperature tolerance: −20 to 55°C

Power source: Electric batteries

Electric batteries: 16,000 mA h, two cells.

Fuel: Not an IC engine-fitted UAV, does not require gasoline

Endurance: 60 min per flight; only 45 min with full payload.

Remote controllers: The GCS computers (laptop) and/or remote controller (joystick) keeps telemetric link and controls flight and cameras (shutters) for up to 1 km radius.

Area covered: 50–300 ac day^{-1} of 6–8 h of flight

Photographic accessories: Cameras: Vapor 55 has a gimbal that is gyro-stabilized. It holds a series of sensors. They include visual (R, G, and B), high-resolution multi-spectral, NIR, IR, and red-edge cameras. In addition, it has a 3D videocamera, Lidar sensor, and vehicle-tracking device. A package of sensors on Vapor 55 includes Sony Precision Mapping package, Headwall Nano hyperspectral mapping package, Yellow scan Lidar, Trillium Orion HD-50, and Phoenix aerial ranger Lidar.

Computer programs: The digital aerial images are processed, using computer software such as Pix4D Mapper, Agisoft's Photoscan, Trimble's Inphos, or similar custom-made software.

Spraying area: It could be used to spray plant protection chemicals, if suitably modified to hold a tank and is attached with spray nozzles. The digital data showing variations in disease/pest attack could be used to arrive at computer-decided dosage and regulate spray nozzles.

Volume: Data not available

Agricultural uses: Aerial survey and crop scouting for obtaining maps of NDVI, canopy growth, leaf chlorophyll (crop's N), and water (via thermal imagery) status are the major uses. It is also effectively used to detect and map the spatial variations of disease/pest attack on crops. This allows farmers to apply remedies only at spots affected by the maladies. Natural vegetation monitoring to obtain NDVI values, monitor biomass accumulation trends, and study species diversity is also possible. Spectral signatures of crop species could be utilized to identify the plant species.

Nonagricultural uses: It includes surveillance of natural resources, geological sites, mines, mining activity, and ore dumps; industrial sites, buildings and public spaces, highways, rail roads, oil pipelines and electric lines, tracking of transport vehicles, etc. In addition, Vapor 55 has a hold for packages/luggage with drop mechanism. This helps in rapid transport of small goods.

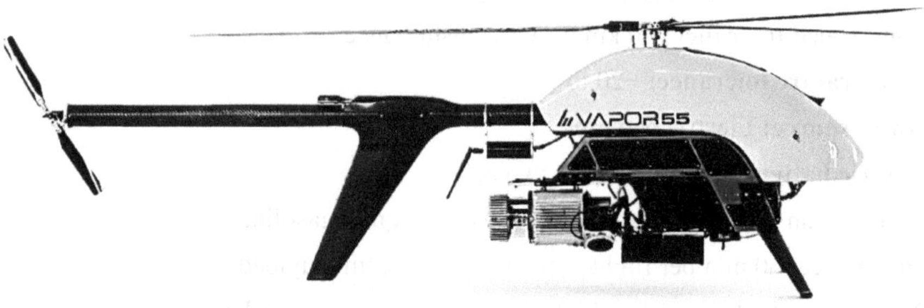

PLATE 3.9.1 Vapor 55.
Source: Mr. Donald Effren, Pulse Aerospace Inc., Lawrence, KS, USA; http://www.pulseaero.com/uas-products/vapor-55.

Useful References, Websites, and YouTube Addresses

http://www.pulseaero.com/uas-products/vapor-55 (accessed Apr 7, 2017).
http://www.skylineuav.com.au/fleet/vapor-55-uav-helicopter/ (accessed Apr 7, 2017).
http://www.pulseaero.com/ (accessed Apr 7, 2017).
https://www.youtube.com/watch?v=O4rFIrkLRAk (accessed Apr 7, 2017).
https://www.youtube.com/watch?v=J-bWDoRdszA (accessed Apr 7, 2017).

3.2.2 MULTIROTOR COPTER UAV

Multicopter UAVs have recently gained greater importance among farming regions. They have specific advantages. For example, they are VTOL aerial robots. They are highly flexible regarding altitude of flight over the crop canopy. Their ability to hover at stand-still condition over crop's canopy allows us close-up survey, detection, and imagery of crops. We can focus a single plant or even a branch/twigs, if required. They are amenable for attaching pesticide/liquid fertilizer tank, as part of regular payload. Therefore, they are highly useful during aerial spray of plant protection chemicals and fluid fertilizers (foliar application). Many of the models are cheaper. They are usually attached with a full complement of sensors such as visual (R, G, and B), NIR, IR, hyperspectral, and Lidar. Multicopters may range from being a quadcopter to hexacopter, octocopter, or even those with 12 propellers. Quadcopters are among most frequently adopted versions and are popular in many agrarian regions (Inspire 1, Phantom, MD-3000). Figure 3.1 is a diagrammatic representation of a quadcopter (e.g., Phantom).

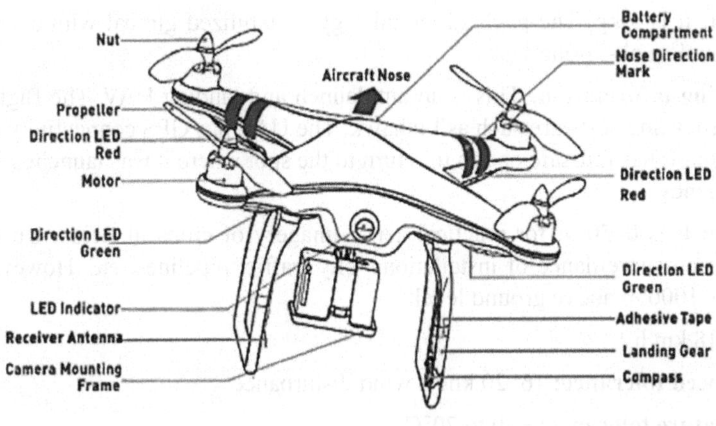

FIGURE 3.1 A small quadcopter with all its parts labeled.
Source: Mr. Adam Najberg, DJI, Shenzhen, China (Mainland).
Note: The labeling of parts of UAV is self-explanatory.

3.2.2.1 QUADCOPTERS

Name/Models: Altura Zenith

Company and address: Aerialtronics, Wassenaarseweg 75, 1e Mientlaan, 2223 LA, Katwijk, The Hague, Netherlands; phone: +31 (0) 70 322 32 24; e-mail: request@aerialtronics.com, sales@aerialtronics.com, pr@aerialtronics.com; website: https://www.aerialtronics.com/

Aerialtronics situated at Katwijk in Netherlands is a company that strives to produce high-utility UAVs. Their motto is enabling a multitude of professionals to easily survey, inspect, and analyze data in real time. Aerialtronics is an end-to-end solution developer. It utilizes UAVs, artificial intelligence, and digital data to provide services and insights. They try to create a perfect synergy between aerial applications and everyday business operations.

Altura Zenith is quadcopter. It is a versatile platform equipped with the latest high-performance and high-capacity battery. The Altura Zenith can fly for up to 40 min. It has been tested against TÜV industry standards to ensure a safe operation even in adverse weather and high electromagnetic fields (Plate 3.10.1).

TECHNICAL SPECIFICATIONS

Material: It is made of ruggedized carbon fiber, foam, and plastic.

Size: length: 60 cm; width (wingspan): 60 cm; height: 47 cm; weight: 9.65 kg

Propellers: This is a quadcopter. It has four propellers of 16 in. They help in vertical take-off and launch. They also aid in horizontal propulsion.

Payload: It is 3 kg. The payload includes gyro-stabilized gimbal with a series of sensors and an all-in-one sensor.

Launching information: This is an autolaunch and landing UAV. The flight path is decided, using software such as Pixhawk. The UAV has GPS connectivity during navigation. It has fail-safe option to return to the spot where it was launched in case of emergency.

Altitude: It is 0–50 m for practical aerial imagery of crops in fields. Up to 300 m, while on surveillance of installations, city traffic, pipelines, etc. However, the ceiling is 1000 m above ground level.

Speed: 18 km h^{-1}

Wind speed tolerance: 16–20 knots wind disturbance

Temperature tolerance: −10 to 70°C

Power source: Electric batteries

Electric batteries: 28 A h, two cells, 22.2 V

Fuel: Not an IC engine-fitted UAV

Endurance: 35–40 min

Remote controllers: The UAV retains connectivity with ground control (GCS) iPad or joystick for up to 5 km radius. The aerial datalink works until 5 km distance.

Area covered: It covers 50–250 ha per flight.

Photographic accessories: Cameras: It has a gimbal with several sensors, namely, visual, multispectral, hyperspectral, NIR, IR, and RedEdge. It is fitted with all-in-one sensor (Plate 3.10.2).

Computer programs: The orthomosaics depicting aerial images of crop fields, irrigation channels, farm vehicles, etc. are processed, using Pix4D Mapper or Agisoft's Photoscan software. This UAV relays orthomosaics to ground station computers for analysis.

Spraying area: Not a spraying UAV in the general course.

Volume: Not applicable

Agricultural uses: It includes aerial imagery of agricultural fields (2D and 3D). Crop scouting for general health and nutritional status, disease, pest attack if any, and drought- or flood-affected zones. Natural vegetation monitoring, obtaining NDVI data, and recording species diversity are possible.

Nonagricultural uses: There are several nonagricultural applications, wherein, this UAV is utilized. Its known applications are in firefighting, wind turbine inspection, phone mast inspection, high-voltage electric line inspection, search and rescue trips, traffic monitoring, gas pipeline monitoring, aerial photography of city events, and road traffic.

PLATE 3.10.1 Altura Zenith.
Source: Mr. Yohan Brochard, Marketing and Graphic Designer, Aerialtronics, The Hague, Netherlands.

PLATE 3.10.2 Dupla Vista—all-in-one sensor.
Source: Yohan Brochard, Aerialtronics, The Hague, Netherlands.

Useful References, Websites, and YouTube Addresses

https://www.aerialtronics.com/products/altura-zenith (accessed July 10, 2017).
https://www.youtube.com/watch?v=RKcFA_bVMXw (accessed July 10, 2017).
https://www.aerialtronics.com/#zenith (accessed July 10, 2017).
https://www.youtube.com/watch?v=H1MkXiFaw04 (accessed June 12, 2017).
https://www.youtube.com/watch?v=doC71R1wp-k (accessed June 12, 2017).
https://www.youtube.com/watch?v=JgqAKKtAjl4 (June 12, 2017).

Name/Models: Humming Bird

Company and address: Ascending Technologies Gmbh.; Konrad-Zuse-Bogen 4, 82152 Krailling, Germany; Tel.: +49 89 89556079 0; fax: +49 89 89556079 19; e-mail: team@asctec.de; website: www.asctec.de

Ascending Technologies Gmbh. is an UAV company located in Bavaria, Germany. The company was initiated in 2005. It produces a few different models of copter UAV. It includes the quadcopter, namely, "Humming Bird." The company serves with UAVs for range of purposes relevant to civilian administration and agricultural crop production

According to the UAV company, the quadcopter model "Humming Bird" has been designed for agricultural and civilian situations. It is robust materially and it has simple structure; so, it can be repaired quickly and easily by the owners themselves. It has high flight standards without risks or errors. Several different kinds of flight experiments and data collection have been conducted using Humming Bird (Plate 3.11.1). The company also states that it offers high precision with least aberrations. At the same time, it offers high flexibility in algorithms needed to conduct flights. It has several safety features. It is an excellent flying UAV over the crop fields. It is light and quick to operate. Humming Bird has been used in agricultural experimental stations. The UAV is priced moderately at US$ 12–15,000 per unit.

TECHNICAL SPECIFICATIONS

Material: Humming Bird is made of robust but light avionics aluminum. The frame is lightweight but tough. It has other components such as carbon fiber fuselage and plastic propellers (Plate 3.11.1).

Size: length: 54 cm; width (rotor to rotor): 54 cm; height: 85 cm; weight: MTOW is 2.7 kg

Propellers: It has four rotors made of tough plastic blades. Propellers are connected to brushless motors for propulsion.

Payload: 200 g

Launching information: It is an autolaunch UAV. Its navigation and flight path is usually predetermined, using computer software such as Pixhawk. It has direct vertical landing mode. We can either predetermine the spot or carefully recover it, using remote controller (joystick).

Altitude: Operational height is 1000 m above ground level for aerial imagery and reconnaissance flights. It can hover at low heights over the crops for close-up photography of canopy. However, ceiling is 4500 m above sea level.

Speed: 15 m s^{-1} during aerial survey. Climb rate is 5 m s^{-1}.

Wind speed tolerance: 20 km h^{-1}

Temperature tolerance: −20 to 55°C

Power source: Electric batteries

Electric batteries: 24,000 mA h

Fuel: This is not an IC engine-fitted UAV, no need for gasoline fuel.

Endurance: 20 per take-off

Remote controllers: The ground station computer is an iPad or laptop. It is used to keep track and fix the flight path of the UAV. Plus, it is used to decide aerial photo frequency. The UAV can also be manually controlled, using a joystick (remote controller). The UAV's flight path can be altered midway, if needed. The ground station computer is linked with UAV, using wireless telemetric link (2.4 GHz Xbee link, 10–63 mW, Wi-Fi). This has UAV GPS connectivity. Its flight is guided accurately, using GPS coordinates.

Area covered: It covers 50–100 ac of aerial survey/photography per flight of 20 min

Photographic accessories: Cameras: This UAV has a gimbal that is gyro-stabilized. It hosts a series of cameras such as visual (R, G, and B), high-resolution multispectral, NIR, IR, and red-edge bandwidth specifications. It has Lidar sensor.

Computer programs: The aerial imagery received via UAV's CPU is processed using computer software such as Pix4D Mapper, Agisoft's Photoscan or Trimble's Inphos, or a similar custom-made software.

Spraying area: This is a small UAV for any sort of liquid or pesticide payload in a tank.

Volume: Not applicable

Agricultural uses: Humming Bird is useful in obtaining aerial images of crops at different spectral bandwidth. Analysis of such spectral data allows the farmers to fix fertilizer and irrigation needs, also to decide pesticide, fungicide, and weedicide sprays. This UAV is used to get data on NDVI variations in a crop field, canopy growth rate, leaf chlorophyll, canopy temperature, and disease/pest attack. Natural vegetation monitoring to obtain NDVI data and identifying plant species diversity is a clear possibility.

Nonagricultural uses: It includes surveillance of industrial sites, mines, buildings, highways, rail roads, vehicles, oil pipelines, electric lines, international borders, etc.

PLATE 3.11.1 Humming Bird.
Source: Mr. Mathias Beldzik and Prof Guido Morganthal, Ascending Technologies Gmbh., Kreilling, Germany.

Useful References, Websites, and YouTube Addresses

http://www.asctec.de/en/uav-uas-UAVs-rpas-roav/asctec-hummingbird/ (accessed June 1, 2017).

http://www.asctec.de/en/uav-UAV/ascending-technologies/asctec-research-line/asctec-hummingbird/ (accessed June 1, 2017).

http://wiki.asctec.de/display/AR/UUser+Manual (accessed June 1, 2017).

http://wiki.asctec.de/display/AR/AscTec+Hummingbird/ (accessed June 1, 2017).

https://www.youtube.com/watch?v=CuuUXqU6eS8 (accessed June 1, 2017).

https://www.youtube.com/watch?v=dZLqk2-kERU (accessed June 1, 2017).

https://www.youtube.com/watch?v=5HjPlDPyjfE (accessed June 1, 2017).

Name/Models: INSPIRE 1

Company and address: SZ DJI Baiwang Technology Company Ltd., 14th Floor, West Wing, Skyworth Semiconductor Design Building, No. 18 Gaoxin South 4th Ave., Nanshan District, Shenzhen 518057, China; phone: +86 (0)755 26656677.

DJI of Shenzhen, China, produces a wide range of copters. Its models are currently popular worldwide. They are handy, often small, and versatile. They have many great applications in civil and agricultural sectors (DJI, 2016).

"INSPIRE 1" is a versatile quadcopter with several applications. It has a range of 5–6 cameras that offer aerial imagery. They allow us to conduct live surveillance, using telemetric connection with ground computers/mobile or iPad. The UAV is packaged in a case. It is portable. It can be assembled in 10 min and launched, from anywhere in field, since it is a VTOL UAV. Inspire UAV costs US$ 2000 at present (February 2017) rates. This model has applications in civilian and agricultural tasks (Plate 3.12.1).

TECHNICAL SPECIFICATIONS

Material: This UAV is made of a composite of carbon fiber, toughened foam, and plastic.

Size: length: 55 cm; width (rotor to rotor): 75 cm; height 18 cm; weight: 6.27 lb (2.85 kg) including propellers and battery. MTOW 7.71 lb (3.5 kg)

Propellers: It is a quadcopter. It has four propellers situated above landing gear risers. The propellers are attached to brushless motors energized by electric batteries.

Payload: 800–1200 g

Launching information: This is a small quadcopter with VTOL ability. The auto-launch, navigation, and landing are made possible, using software such as Pixhawk.

Altitude: It is 0–5 m for close-up shots of crops/ground surface. Operating height in an agricultural farm is 5–50 m above crop canopy. However, the ceiling is 500 m above crop canopy.

Speed: It cruises at 79 km h^{-1} (49 mi h^{-1}).

Wind speed tolerance: It withstands 10 m s^{-1} disturbing wind

Temperature tolerance: −10 to 40°C

Power source: Electric batteries with custom made DJI batteries (A14-100P1A)

Electric batteries: 6000 mA h 2S lithium–polymer; flight battery TB48. Net weight of batteries is 670 g

Fuel: Not an IC engine-fitted UAC. No need for petrol or diesel.

Endurance: 18 min

Remote control: The operating frequency for telemetric-link with GCS is 922.7–927.7 MHz. The GCS is equipped with mobile (DJI GO model) and tablet (iPad). Operating distance and control is 5 km radius (3.1 miles).

Area covered: It covers 50 ha per flight of 1 h.

Photographic accessories: Cameras: The UAV carries a series of cameras custom-made by DJI, such as Zen muse X3, Zen muse X5, Zen muse X5R, Zen muse XT, and Zen muse Z3. The X3 model FC350 is a 12.7-MP camera. It picks video images. It has 20-mm lens. The imagery is done from 3 m above the crop canopy, up to a limit of 50 m for agricultural aspects. The video images are 12 mega pixels. The cameras are placed in a 3-axis gimbal with ability to rotate 360°.

Computer programs: The aerial images are processed using Pix4D Mapper or Agisoft's Photoscan or similar custom-made software. The processed image is usually 720p HD image. The HD images can be viewed on a mobile or iPad and analyzed. Later, computers with various agriculture crop production-related software could be adopted. Finally, it allows us to calculate fertilizer supply rates or pesticide spray rates and timing can be fixed appropriately.

Spraying area: Not a spraying UAV.

Volume: Not applicable

Agricultural uses: They are aerial imagery and crop scouting. This UAV (Inspire) is extensively used to scout for the crop production zones. The aerial imagery is used to judge seedling emergence and crop stand, find gaps in seedling establishment, trace variations in nutrient assimilation and status, measure water stress indices, map disease/pest-attacked zones, and weed distribution. It could be used to collect data about crop water status, then decide on irrigation timing and quantity. Natural vegetation monitoring for NDVI and to detect species diversity is also a possibility.

Nonagricultural uses: It includes surveillance of industrial locations, buildings, highways, rail roads, mines and mining activity, electric lines, oil pipelines, etc.

PLATE 3.12.1 INSPIRE 1.
Source: Adam Najberg, DJI Inc., Shenzen, China.
Note: The DJI's "INSPIRE 1" shown above is integrated with Zen muse XT camera for thermal imagery, that is, to assess canopy and air temperature.

Useful References, Websites, and YouTube Addresses

DJI. Above the World. *UAS Mag.* **2016,** 238. http://www.uasmagazine.com/articles/1591/dji-explains-new-book-of-UAV-captured-images (accessed Apr 20, 2017).

http://enterprise.dji.com/agriculture?gclid=CIvJ59PZvtACFcIQaAod5FwCkw (accessed May 7, 2017).

http://store.dji.com/product/inspire-1-v2 (accessed May 7, 2017).
https://www.youtube.com/watch?v=ZnJcZfsVLAQ (accessed May 7, 2017).
https://www.youtube.com/watch?v=tyERIL8YMXI (accessed May 7, 2017).
https://www.youtube.com/watch?v=Vv8hzWaZ0eY (accessed May 9, 2017).
https://www.youtube.com/watch?v=Leul115VcmI (accessed May 9, 2017).

Name/Models: KX 4 Multirotor Copter

Company and address: Threod Systems Inc. Kaare tec 3, 74010 Vimsi, Estonia; phone: +372 512 1154; e-mail: info@threod.com, sales@threod.com; website: http://www.treod.com/products

Threod Systems Inc. is a UAV producing company located in Estonia (Europe). It offers a few different models to suit the purposes in military, civil, and agricultural aspects. Threod Systems Inc. is a company started in 2010. It offers variety of services related to UAV usage, such as aerial photography and surveillance of natural resources.

K4 multirotor UAV is a light-weight machine. It is easily transportable in a backpack. It could be swiftly assembled and made ready to fly within 10 min or less (Plate 3.13.1). The UAV is fitted with gyro-stabilized gimbal and EO and IR cameras. It also carries a videocamera for instantaneous relay of action movies. It has been used in military and civilian locations. It has important applications in agricultural landscape.

TECHNICAL SPECIFICATIONS

Material: It is made of a composite containing carbon fiber, ruggedized foam, and plastic. The landing gear is made of light alloy.

Size: length: 85 cm; width (rotor to rotor): 85 cm; height: 45 cm; weight: MTOW is 6.9 kg. It includes a payload capacity of 1.5 kg

Propellers: It is a quadcopter. It has four propellers made of two plastic blades each.

Payload: 1.6 kg. It includes a series of sensors for aerial imagery.

Launching information: It is an auto-take-off (VTOL), autopilot, and autolanding UAV. The flight way, loiter points, and contingency routes are usually predetermined. Yet, real-time modifications to the flight path and photographic events can be altered, using GCS mobile/iPad.

Altitude: It flies at 100–200 m above ground level.

Speed: 6–12 m s^{-1}

Wind speed tolerance: 20 km h^{-1} disturbance

Temperature tolerance: −10 to 50°C

Power source: Lithium–polymer electric batteries

Electric batteries: 16,000 mA h, two cells, lithium–polymer batteries

Fuel: Not an IC engine-fitted UAV. No need for petrol or diesel

Endurance: It is 30–40 min

Remote controllers: The GCS has a mobile or iPad. The UAV's flight path, speed, and photographic events could be controlled, using GCS iPad. The controller operates for up to 4–5 km radius. The remote video terminal is compliant with STANAG 4609. The telemetric signals are protected, using encrypted (AES-256) messages.

Area covered: It covers 50 ha per flight

Photographic accessories: Cameras: This UAV is attached with EO sensors: usually 30× optical zoom 720p HD; IR sensor: TAU 640-480, 35 mm or 60 mm. it also carries range finder and tracking device for vehicles. The images are gyro-stabilized and geo-reference (tagged with GPS coordinates).

Computer programs: The aerial imagery is processed at the UAV's CPU or at GCS (IPad) using software such as Pix4D Mapper or Agisoft's Photoscan.

Spraying area: This UAV model is not used to spray chemicals, but modifications could be done.

Volume: Not applicable

Agricultural uses: Crop scouting for canopy characters such as NDVI, leaf area, leaf-N (chlorophyll) status, canopy temperature (water stress index), etc. is possible. Natural vegetation monitoring and detection of plant species diversity are also possible. This UAV is also used to detect and map areas with drought, flood, or erosion damage.

Nonagricultural uses: It includes surveillance of mines and mining activity, industries, buildings, oil pipeline, electric lines, dams, rivers, lakes, etc. Transport vehicles could be tracked using tracking device on the UAV.

PLATE 3.13.1 KX-4 Multirotor.
Source: Mr. Siim Juss, Project Manager of Sales and Marketing, Threod Systems Inc. Vimsi, Estonia.

Useful References, Websites, and YouTube Addresses

http://www.threod.com/products/kx-4-multirotor-description/ (accessed June 14, 2017).
https://www.youtube.com/watch?v=mIuSU774HHA (accessed June 14, 2017).
https://www.youtube.com/watch?v=JVXka49xYIc (accessed June 14, 2017).

Name/Models: KX-4 LE Multirotor

Company and address: Threod Systems Inc. Kaare tec 3, 74010 Vimsi, Estonia; phone: +372 512 1154; e-mail: info@threod.com, sales@threod.com; website: http://www.treod.com/products

As stated earlier, Threod Systems Inc. is a UAV producing company located in Estonia. This company manufactures a few different models to suit the purposes in military, civil, and agricultural aspects. KX-4LE is a relatively longer endurance UAV compared to other models in the same series that are produced by Threod Systems Inc. of Estonia (3.14.1). It is also a heavy lift UAV with multiple payload options. It has been preferred in several rescue missions. It is used regularly in conducting aerial imagery.

TECHNICAL SPECIFICATIONS

Material: It is made of carbon fiber, kevlar, plastic, and rugged foam. It has landing gear made of light alloy.

Size: length: 78 cm; width (rotor to rotor): 78 cm; height: 50 cm; weight: MTOW is 16.7 kg

Propellers: It is a quadcopter with propellers. Each propeller has two plastic blades.

Payload: It carries a payload of 6.1 g. It includes series of cameras in a gimbal, a CPU, radio-tracking device, and laser range finder.

Launching information: KX4-LE has automatic take-off and landing software. It has automatic navigation software such as Pixhawk. The number of waypoints and photography by sensors is predetermined, but they can be modified, using GCS override facility. The GCS mobile controller can fix/modify loiter mode, change waypoints, and flight path. Usually, emergency flight path and photographic events are decided. Warning displays are available, in case of error in flight path. The UAV's sensors and computer decision software are made compatible with GPS, INS, and GLONASS navigation.

Altitude: The optional height above the crop canopy or ground surface is 100 m. Maximum altitude is 500 m above ground level.

Speed: 8–14 m s^{-1}

Wind speed tolerance: 25 km h^{-1} wind disturbance

Temperature tolerance: −10 to 55°C

Power source: Lithium–polymer electric batteries

Electric batteries: 28,000 mA h

Fuel: Not an IC engine-fitted UAV

Endurance: It is 70 min without payload, about 55 min with 1 kg payload and 45 min with 2.5 kg payload.

Remote controllers: The GCS has a mobile controller/phone and or iPad. A fixed PC with monitor screen may also be used to control the UAV's flight pattern and photographic survey. The GCS has a single digital datalink that is integrated with the hand-held mobile/iPad. The datalink operates for up to 4–5 km radius. The data transmission is made secure by encryption using AES-256.

Area covered: It covers 50–250 ha depending on flight path and intensity of aerial survey and mapping.

Photographic accessories: Cameras: The aerial imaging payload includes gimbal with gyro-stability. The EO sensors include cameras with 30× optical zoom, 720p HD. IR sensor is TAU2 640 × 480 with 25 mm lens and a second IR sensor with TAU2 640 × 480, 60 mm lens is also available. It has an optional laser range finder, IR illuminator, and laser-target designator. This UAV also has radiotracking sensor to pick vehicles on the highways. All the images produced by these cameras are geo-reference for greater accuracy.

Computer programs: The aerial images are processed instantaneously at the UAV's CPU. It could be stored in a chip. They are used in farm vehicles with variable-rate applicator, in case it deals with crop growth variability and sprays. The data can also be transmitted in an encrypted state via GCS mobile/iPad and processed. The computer software such as Pix4D Mapper or Agisoft's Photoscan is used to process the images.

Spraying area: Not used for spraying pesticide. However, it could be modified to apply fertilizers and chemicals.

Volume: Not applicable

Agricultural uses: They are aerial survey, imagery, and crop scouting. It is done mainly to study canopy characters, leaf area, leaf-N (chlorophyll) status, canopy, and air temperature (i.e., to measure water stress index). The aerial imagery could also be used to detect and map disease and pest attack or weed infestation. Knowledge about spectral signatures of crop species and weeds is required. We can collect spectral signatures of diseased/healthy canopy or weeds and compare spectral reflectance patterns. Natural vegetation monitoring to obtain NDVI data, prepare maps, and detect species diversity is possible.

Nonagricultural uses: It includes aerial survey and regular live surveillance of geologic sites, mines and mining activity, industrial establishments, large buildings, public places, nature reserves and sanctuaries, city traffic and vehicle movement, etc. It is also used to keep watch and ward on oil pipelines, electric lines, and borders between states.

PLATE 3.14.1 KX-4 Le Multicopter.
Source: Mr. Siim Juss, Project Manager of Marketing and Sales, Threod Systems Inc. Vimsi, Estonia.

Useful References, Websites, and YouTube Addresses

http://www.threod.com/products/kx-4-le/description (accessed June 10, 2017).
https://www.youtube.com/watch?v=JVXka49xYIc (accessed June 10, 2017).

Name/Models: Matrice 100

Company and address: SZ DJI Baiwang Technology Company Ltd., 14th Floor, West Wing, Skyworth Semiconductor Design Building, No. 18 Gaoxin South 4th Ave., Nanshan District, Shenzhen 518057, China; phone: +86 (0)755 26656677

Major application of this UAV is aerial survey of land, collecting 3D images showing topographic details, soil type, and cropping systems. Matrice 100 is used for aerial photography, particularly when greater details are needed, such as close-up shots or 3D images, also when high-resolution 3D videos need to be downloaded immediately.

Matrice 100 was released for public in July 2015, after a series of trials. This UAV is among popular models sold worldwide. It is used extensively in civilian and agricultural situations. Matrice 100 with most accessories for aerial survey and spectral analysis of crops costs US$ 3300 (February 2017 price levels). It is supposedly a relatively expensive quadcopter (Plate 3.15.1). The UAV selling outlets are available in all continents.

TECHNICAL SPECIFICATIONS

Material: The platform and body of Matrice 100 is made of carbon fiber, plastic, and foam. It has landing gear made of light alloy or hardened plastic (Plate 3.15.1).

Size: length: 65 cm; width (rotor to rotor): 65 cm (25.6 in.); height: 35 cm; weight: 2.4 kg (5.2 lb)

Propellers: It is a quadcopter. Each of the four propellers has two blades made of plastic.

Payload: 1.2 kg. It includes a gimbal and series of cameras.

Launching information: It is an autolaunch, autopilot, and landing UAV. The flight pattern can be determined using computer software. It keeps to autonomous flight for 40 min and lands at stipulated point. We can select waypoints. It returns to a location close to ground control (iPad) in times of emergency or expected collision or errors in software and flight planning.

Altitude: The ceiling is 1500 m above ground level.

Speed: The cruise speed can reach up to 55 km h^{-1}, ascent speed 5 m s^{-1}, descent speed 4 m s^{-1}

Wind speed tolerance: 20 km h^{-1} disturbing winds

Temperature tolerance: −10 to 40°C

Power source: Lithium–polymer electric batteries with a charger (26.3 V).

Electric batteries: 4000–6000 mA h, lithium–polymer 2S, 22.2 V power. Battery weight is 600 g.

Fuel: Not an IC engine-fitted UAV

Endurance: 40 min per take-off

Remote controllers: The ground station has iPad (Android 4.1.2) or tablet or a mobile (DJI GO) to control flight path and camera shutters. The UAV flight path can be predetermined, using "Pixhawk" or similar software. It is GPS compatible.

Area covered: It covers 50–250 ha per flight (iPad or mobile)

Photographic accessories: Cameras: Matrice 100 has a gyro-stabilized gimbal. It has cameras such as visual, multispectral, hyperspectral, NIR, IR, and red-edge cameras. The DJI company offers custom-made Zenmuse X5 and series of cameras. It also comes with Zenmuse XT, a thermal camera capable judging crop's water needs and soil surface moisture status (Plate 3.15.2).

Computer programs: The aerial images are processed right at the UAV's CPU or at the GCS computers (iPad) using software such as Pix4D Mapper or Agisoft's Photoscan.

Spraying area: Not a spraying UAV.

Volume: Not applicable

Agricultural uses: Aerial survey of land, collecting 3D images showing topographic details, soil type, and cropping systems is a major application of this UAV. Matrice 100 is used for aerial photography, particularly for those needing greater details, such as close-up shots or when high-resolution 3D videos are required. Matrice 100 is useful in scouting crops right from seedling emergence, detection of gaps in crop stand, uneven growth patterns, nutritional deficiencies/disorders, their spatial variations, etc. Aerial data about leaf area, leaf-chlorophyll, and crop-N status help in deciding fertilizer supply schedules (DJI, 2016). The thermal cameras help in detecting canopy and ambient temperature, that is, water stress index. Such data help in deciding timing and quantity of irrigation. This UAV helps in avoiding farmer's drudgery in fields, particularly when detecting disease/pest-attacked patches or flood/drought-affected spots. The aerial imageries clearly depict the zones that need remediation. Aerial detection is done in a matter of minutes. Natural vegetation monitoring for NDVI data, canopy reflectance, and plant species diversity is also possible.

Nonagricultural uses: It includes surveillance and aerial imagery of geologic sites, mines and mining activity, industrial zones, large buildings, harbors, rail roads, highways, etc. If the UAV is fitted with tracking device, then vehicles in military camps and highways can be tracked, using connection with GCS. Large dams, rivers, channels, and their water status can be monitored all through the day and night.

PLATE 3.15.1 Matrice 100.
Source: Mr. Adam Najberg, Sales and Marketing division, Shenzen, China (Mainland).

PLATE 3.15.2 Zen muse XT Thermal Camera.
Source: Mr. Adam Najberg, Sales and Marketing division, Shenzhen, China (Mainland).
Note: Zen muse XT is an integrated thermal imaging solution for use with the Inspire 1 or the Matrice Series UAVs that can monitor plant health or track irrigation systems. Modular design allows users to integrate third-party sensors or DJI's own Zen muse XT thermal camera.

Useful References, Websites, and YouTube Addresses

DJI. Above the World. *UAS Mag.* **2016,** 238. http://www.uasmagazine.com/articles/1591/dji-explains-new-book-of-UAV-captured-images (accessed Apr 20, 2017).

http://enterprise.dji.com/agriculture?gclid=CIvJ59PZvtACFcIQaAod5FwCkw (accessed May 25, 2017).

https://developer.dji.com/products/ (accessed May 25, 2017).

http://UAVs.spaceout.com/1/219/DJI-Matrice-100 (accessed May 25, 2017).

Http://enterprise.dji.com/agriculture (accessed May 25, 2017).

https://www.youtube.com/watch?v=0aNX6sDpA4g (accessed May 25, 2017).

https://www.youtube.com/watch?v=_eL8rmjV870 (accessed May 25, 2017).

https://www.youtube.com/watch?v=1ro99KyvAoA (accessed May 25, 2017).

Name/Models: Mavic Series by DJI

Company and address: SZ DJI Baiwang Technology Company Ltd., 14th Floor, West Wing, Skyworth Semiconductor Design Building, No. 18 Gaoxin South 4th Ave., Nanshan District, Shenzhen 518057, China; phone: +86 (0)755 26656677

The UAV company, DJI Technology Ltd. was founded in 2006. It is now among the foremost UAV suppliers in the world. It is situated in the "Silicon Valley" of China, that is, Shenzhen area. They have UAV outlets and servicing centers in many locations within United States of America, Germany, Netherlands, Japan, Hong Kong, and Beijing in China. They churn out a wide range of copter UAVs meant for a range of activities (DJI, 2016).

Mavic series of UAVs is again a small, light-weight quadcopter series. They are low-cost aerial robotic machines (Plate 3.16.1). They are used primarily in civilian and agricultural locations. They provide excellent aerial images, videos of happenings on ground, and in crop fields. The cameras offer useful data that can be analyzed using computer programs, to arrive at most appropriate fertilizer, irrigation, and/or pesticide/disease control chemicals. These UAVs also offer useful data to forecast crop yield.

TECHNICAL SPECIFICATIONS

Material: It is made of a composite of carbon fiber, ruggedized rubber and foam, and plastic.

Size: length: 75 cm; width (wingspan): 90 cm; height: 45 cm; weight: 3.4 kg

Propellers: It has four foldable propellers made of two plastic blades each. The propulsion is through connection with brushless motors.

Payload: 0.8–1.2 kg

Launching information: This is an autolaunch, navigation, and landing quadcopter. It can detect obstacles to predetermined flight path, if any. They make appropriate corrections. In the event of emergency, it returns to point of launch.

Altitude: The operating altitude is 5–50 m during crop surveillance. The general operating height is 300–400 m above ground level. The ceiling is 5000 m above ground level.

Speed: Mavic UAV flies at 14–40 km h^{-1}. The ascent speed is 6.6 ft s^{-1} (2 m s^{-1}) and descent speed is 3.3 ft s^{-1} (1 m s^{-1}). The UAV can also hover over a location without much movement while picking close-up shots of crops. This is useful, particularly to identify disease, pest species, or nutritional disorders if any.

Wind speed tolerance: It tolerates 20–25 km h^{-1} disturbance.

Temperature tolerance: The UAV functions normally at temperature range from −10 to 40°C. General operating temperature is 5–30°C.

Power source: Lithium–polymer electric batteries

Electric batteries: 3830 mA h, 11.4 V, lithium–polymer 35. It gives out 43.6 W energy. The batteries weigh about 0.5 lb (0.250 kg).

Fuel: Not an IC engine-fitted UAV. No need for petrol or diesel

Endurance: 21 min per take-off

Remote controllers: The Wi-Fi at the GCS operates at 2.4 G/SG. Mavic is compatible with both GPS and GLONASS. All the still images and videoframes are tagged with GPS coordinates for accuracy.

Area covered: It covers a linear distance of 13 km per take-off. It covers 50–250 ha of aerial imagery per day of 8 h.

Photographic accessories: Cameras: The UAV has an axis gimbal. It has a series of visual; NIR, IR, and red-edge cameras for aerial imagery and spectral analysis of crops and other objects (Plates 3.16.2–3.16.4). The cameras offer single shots at a burst rate of 3/5/7 frames. The image size is 4000 × 3000 mm. It has autoexposure. The videocamera has a bit rate of 60 Mps. It offers MP4 video or JPEG images. The still and video images could be received and downloaded using an iPhone or iPad or mobile devices or android tables.

Computer programs: The UAV carries a set of microSD with a maximum capacity of 64 GB for storage of still images and videos. The GCS has mobile app (DJI GO), using which UAV flight and camera shutters could be controlled. Android is also used at the GCS. The images are processed using Pix4D Mapper or Agisoft's Photoscan.

Spraying area: Mavic UAV is not a spraying UAV. It has very small payload capacity.

Volume: Not applicable

Agricultural uses: Mavic quadcopter is apt for aerial survey, imagery, and crop scouting. It could be utilized to study the growth pattern of crop seedlings and canopy. The cameras can provide data about NDVI, leaf area, leaf chlorophyll, leaf-N (crop-N) status and variations, canopy and ambient temperature (i.e., water stress index), etc. The data derived can help in channeling fertilizers and irrigation more accurately to exact locations and at accurate rates. The UAV's imagery could also be used to detect and trace variations in disease/pest-attacked regions in a crop field. Natural vegetation monitoring, determination of NDVI, and plant species diversity are possible (DJI, 2016).

Nonagricultural uses: Mavic is used in surveillance of industrial sites, buildings, electric lines, oil and gas pipelines, rail roads, city traffic, and vehicle traffic.

PLATE 3.16.1 Mavic Pro-A quadcopter.
Source: Mr. Adam Najberg, Sales and Marketing division, Shenzen, China (Mainland).

PLATE 3.16.2 Zenmuse X5S.
Source: Mr. Adam Najberg, Sales and Marketing division, Shenzen, China (Mainland).
DJI Technology Ltd., Shenzhen, China.
Note: Zenmuse X5 packs a powerful sensor that is capable of recording 4K videos at up to 30 fps and capturing still images at 16 MP. Zenmuse is equipped with an uprated micro 4/3 sensor. The Zenmuse X5S has a dynamic range of 12.8 stops and much improved signal-to-noise ratio and color sensitivity than the X5R. It supports up to eight standard M4/3 lenses (including zoom lenses). The focal lengths range from 9 to 45 mm (equivalent to 18–90 mm on a 35-mm camera). So, it allows more creative flexibility. The new CineCore 2.0 image processing system on the Inspire 2 makes the Zenmuse X5S capable of capturing 5.2K 30 fps CinemaDNG video and Apple ProRes video, as well as 4K 60 fps using H.264 and 4K 30 fps using H.265 (both at 100 Mbps). Continuous DNG burst shooting at 20 fps with 20.8 MP images is also supported. The Zen muse X5S was designed to keep pace with the rigors of high-end professional aerial imaging.

PLATE 3.16.3 Zen muse X4S.
Source: Mr. Adam Najberg, DJI Technology Ltd., Shenzhen, China.
Note: The Zen muse X4S is a camera featuring a 20-MP 1-in. sensor and a maximum ISO of 12,800. It uses a DJI-designed compact lens with low dispersion. It has low distortion 24-mm equivalent prime lens. Combined with CineCore 2.0, the Inspire 2's powerful image-processing system, it can record 4K/60 H.264 and 4K/30 H.265 videos at a 100-Mbps bit rate. It can make the difficult balance between agility and image quality.

PLATE 3.16.4 Zen muse X5R.
Source: Mr. Adam Najberg, DJI Technology Ltd., Shenzhen, China.
Note: The Zenmuse X5R is the world's first micro-four-thirds aerial camera capable of recording 4K videos.

Useful References, Websites, and YouTube Addresses

DJI. Above the World. *UAS Mag.* **2016,** 238. http://www.uasmagazine.com/articles/1591/dji-explains-new-book-of-UAV-captured-images (accessed Apr 20, 2017).

http://www.dji.com/mavic/info (July 25, 2017).

http://enterprise.dji.com/agriculture?gelid=CIvJ59PZvtACFclQaAod5FwCkw/ (accessed July 25, 2017).

https://www.youtube.com/watch?v=p1d_ptE6yrc (accessed July 25, 2017).

https://www.youtube.com/watch?v=hyZr74YaZs8 (accessed July 25, 2017).

https://www.youtube.com/watch?v=semSRD-mQ9U (accessed July 25, 2017).

https://www.youtube.com/watch?v=ia5o97OY4Rw (accessed July 25, 2017).

Name/Models: MD4-200

Company and address: MicroDrones Gmbh., Gutenbergstrasse 86, 57078 Siegen, Germany; phone: +49 271 7700380; fax: +49 271 77003811; websites: info@microdrones.com, www.microdrones.com

MicroDroness, 625 Bomber Drive, Rome, NY 13441, USA; phone: +1 866 874 3566; websites: info@microdrones.com, www.microdrone.com

MicroDroness Gmbh. is a quadcopter production industry. It was initiated in 2006 in Germany. MicroDrones is a UAV company that specializes in development and production of very useful agricultural and civilian UAVs. They produce powerful UAVs for professional pursuits, such as city surveillance, oil pipeline monitoring, etc. Regarding their use in agricultural crop production, their UAVs are called "Flying Farmers." This company offers a series of 4–5 models that suit farmer's requirement. These quadcopters are useful during precision farming. MD series copters have also found use in vineyards of Europe.

MD4-200 is a small UAV. It is touted as an agile model suitable for rapid aerial imagery, at low latitudes. It is very easy and quick to deploy from any spot with small clearance. It is a rain and dust-resistant UAV. It is supposedly a reliable professional solution to land natural resource surveyors. It maps about 35 ha of ground surface per flight. It has light-weight cameras of 24 MP. It is not a spraying UAV. The MD4 200 UAV costs US$ 2000 (February 2017 price level) (Plate 3.17.1).

TECHNICAL SPECIFICATIONS

Material: It is made of a composite of carbon fiber, rugged foam, and plastic. It has landing gear made of light metal alloy.

Size: length: 103 cm; width (rotor to rotor): 103 cm; height: 23 cm; weight: MTOW is 1.1 kg

Propellers: It is a quadcopter. It has four rotors with two plastic blades each.

Payload: Maximum payload is 250 g. It includes visual and multispectral camera.

Launching information: It is an autolaunch UAV. It is autonomous in flight and its navigation is guided by Pixhawk software or similar custom-made software.

Altitude: It operates at 0–50 m above crop canopy during close-up aerial photography. Operating height is 100–300 m above ground level. The ceiling is 1000 m above ground level.

Speed: 5–8 m s^{-1}

Wind speed tolerance: 15 km h^{-1} disturbance

Temperature tolerance: −10 to 50°C

Power source: Electric batteries

Electric batteries: 2.300 mA h, lithium–polymer 4S, battery power is 14.8 V.

Fuel: No need for petrol/diesel

Endurance: 30 min per flight

Remote controllers: The UAV keeps telemetric contact with ground station for up to 1000 m height and 5 km radius. The flight controller frequency is 2.4 GHz.

Area covered: 50–100 ha per flight

Photographic accessories: Cameras: It has a gyro-stabilized gimbal. The gimbal holds a visual (Sony 5100) and multispectral camera for rapid aerial imagery. It usually carries a GoPro action camera.

Computer programs: The ortho-images are processed using Pix4D Mapper or Agisoft's Photoscan or similar custom-made software.

Spraying area: It is not a spraying UAV.

Volume: Not a spraying UAV

Agricultural uses: It has extensive application while hovering and flying over crop fields. They include obtaining NDVI data and maps, measuring crop water stress index, leaf chlorophyll (crop's N) status, and general phenomics data. Natural vegetation monitoring is also possible.

Nonagricultural uses: MD4-200 has several applications such as aerial photography and surveillance of industries, civil buildings, monitoring mines and mining activity, oil pipelines, rail roads, highways and tracking vehicles, etc. It has been used in disaster zones, to get on the spot information. It helps in avoiding poaching in natural reserves (see Plate 3.17.1).

PLATE 3.17.1 MD4-200—a small, light-weight quadcopter.
Source: Ms. Miriam Baumer, Marketing Division, MicroDornes Gmbh., Seigen, Germany.

Useful References, Websites, and YouTube Addresses

http://www.apspecialists.com.au/microdrone-md4-200/ (accessed May 1, 2017).
http://UAVs.specout.com/l/129/Microdrones-MD4-200 (accessed May 1, 2017).
https://www.youtube.com/watch?v=gyNdqetEfLQ (accessed May 1, 2017).
https://www.youtube.com/watch?v=vG4YzZ4lJJs (accessed May 1, 2017).
https://www.youtube.com/watch?v=iuwOwP0evHM (accessed May 1, 2017).
https://www.youtube.com/watch?v=_EFZQSog610 (accessed May 1, 2017).

Name/Models: MD4-1000

Company and address: MicroDrones Gmbh., Gutenbergstrasse 86, 57078 Siegen, Germany; phone: +49 271 7700380; fax: +49 271 770038 11; websites: info@ microdrones.com, www.microdrones.com

MicroDrones, 625 Bomber Drive, Rome, NY 13441, USA; phone: +1 866 874 3566; websites: info@microdrones.com, www.microdrones.com

MicroDrones Gmbh. is a copter manufacturing unit situated at Seigen in Germany. It was founded in 2006. MicroDrones Gmbh. specializes in manufacture of very useful agricultural and civilian UAVs. MD4-1000 is a medium-sized quadcopter. It is robust, powerful, stable in flight, and dependable. It has a versatile platform. MD4-1000 is supposedly a perfect flying tool/vehicle in the hands of European farmers. It is capable of complex tasks in agricultural fields and other locations. It has a relatively longer endurance. Hence, it can accomplish tasks in single flight or at best a few launches. The MD4-1000 UAV comes with its own custom-made and ready-to-use software for flight path control, aerial imagery, and processing the images (e.g., mapper 1000) (Plate 3.18.1).

TECHNICAL SPECIFICATIONS

Material: This quadcopter is made of carbon fiber, rigid foam, and plastic. It can withstand wet conditions to a certain extent. It can fly even during a mild drizzle in the atmosphere.

Size: length: 103 cm; width (rotor to rotor): 103 cm; height: 50 cm; weight: MTOW is 6 kg. Vehicle dry weight is 2650 g.

Propellers: It is quad-copter. It has four propellers, each made of two plastic blades. The UAV emits sound at 71.0 dBa@3 m

Payload: Maximum payload is 1200 g (1.2 kg or 2.8 lb).

Launching information: It is a VTOL UAV. It has autolaunch, navigation, and landing facility. Its navigation is predetermined, using Pixhawk or similar custom-made software. The UAV package comes with options in software such as "mdWaypoint" and "md Landing assistant." These computer programs can be used to command, track, and land the UAV as required.

Altitude: It is 0–50 m above crop's canopy during close-up aerial survey and imagery. Its normal operational altitude is 300 m above crop canopy. Its ceiling is 2000–4500 m above ground level depending on rotor.

Speed: 12 m s^{-1}

Wind speed tolerance: It tolerates 20–25 km h^{-1} wind disturbance. Steady pictures are got up to 6 m s^{-1} wind disturbance but maximum permissible is 12 m s^{-1}.

Temperature tolerance: −10 to 55°C; humidity: it withstands 90% relative humidity

Power source: Lithium–polymer electric batteries

Electric batteries: Smart batteries and charging for maximum flight endurance. It has 6S2P lithium–polymer, 22.2 V, 13,000 mA h.

Fuel: Not an IC engine-fitted UAV. No need for petrol or diesel.

Endurance: For a small and light-weight UAV, it has exceptional endurance of 90 min per flight.

Remote controllers: This UAV comes with custom-made computer (iPad) and software programs to monitor, command, direct, and keep control of the UAV. The telelink allows control of UAV, for up to 5 km radius. It allows rapid transfer of digital data and processed imagery to GCS iPad.

Area covered: 50–220 ha per 90 min single flight.

Photographic accessories: Cameras: The photographic equipment found in the gyro-stabilized gimbal includes Sony A 6300 visual camera (24 MP), multispectral and hyperspectral camera, NIR, IR, and red-edge cameras. Current options for payload cameras include Sony NEX-7, Sony HDR-CX-740VE, Olympus E-P3, Panasonic HDC SD 909, daylight camera, FLIR Tau 640, Tetracam ADC lite, or GIG Microtector.

Computer programs: The ortho-images are stitched and processed to great accuracy and resolution, using computer software such as Pix4D Mapper or Agisoft's Photoscan or custom-made software. There is optional software that is helpful in precision farming practices. They help in reading digital data about variations in crop growth and disease/pest attack or nutritional deficiency. Such data can be relayed instantaneously to ground vehicles fitted with variable-rate methods. The images are all tagged with GPS coordinates. It has GPS waypoint and myUAV software.

Spraying area: This is not a spraying UAV, but it could be modified if needed.

Volume: Not applicable

Agricultural uses: MD-4 1000 is a UAV model apt for use in agricultural regions. It is used to scout crops for seedling and crop stand establishment to detect gaps if any and mark those using GPS coordinates or map them. It is utilized to analyze crops through multispectral cameras. It is used to obtain data about crop canopy, crop growth rate, leaf area, leaf chlorophyll, crop-N status, crop's canopy temperature and air temperature (i.e., to estimate crop's water stress index, weed infestation, etc. (Penia et al., 2015). Natural vegetation monitoring to obtain NDVI values, chlorophyll data, and species diversity is possible. Detection of species diversity is possible, but it is dependent on availability of spectral signatures of different crops/plants and weeds in the data bank so that computers can match them (Penia et al., 2015).

Nonagricultural uses: Major uses are in aerial imagery and surveying of terrain. It is also used for inspection and surveillance of industrial installations, construction of buildings, mines and mining activity, oil pipelines, electric lines, city traffic and transport vehicles, etc. This model is also used in disaster, fire, and flood/drought detection services.

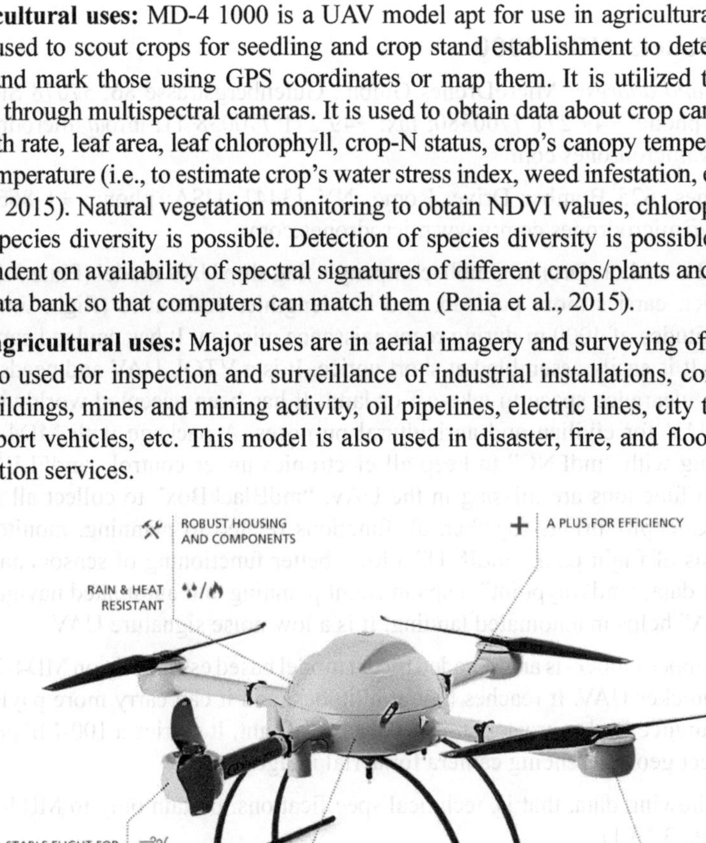

PLATE 3.18.1 MD4-1000.
Source: Ms. Mirjam Baumer, Marketing Manager-Europe, MicroDrones Gmbh., Siegen, Germany.

Useful References, Websites, and YouTube Addresses

Penia, J. M.; Torres-Sanchez, J.; Serrano-Perez, A.; de Castro, A. I.; Lopez-Granados, F. Quantifying Efficacy and Limits of Unmanned (UAV) Technology for Weed Seedling Detection as Affected by Sensor Resolution. *Sensors* **2015,** *15*, 5609–5626. DOI:10.3390/s150305609 (accessed Sept 23, 2017).

https://www.youtube.com/watch?v=-CLIpBpLV90 (accessed July 2, 2017).

https://www.youtube.com/watch?v=viRQhw6AgZE (accessed July 2, 2017).

https://www.youtube.com/watch?v=-CLIpBpLV90 (accessed July 2, 2017).

https://www.youtube.com/watch?v=ZQsMtqW-Lwg (accessed July 2, 2017).

https://www.youtube.com/watch?v=PcEWhbdY3zI (accessed July 2, 2017).

Name/Models: MD4-3000

Company and address: MicroDrones Gmbh., Gutenbergstrasse 86, 57078 Siegen, Germany; phone: +49 271 7700380; fax: +49 271 770038 11; info@microdrones.com; www.microdrones.com

MicroDrones, 625 Bomber Drive, Rome, NY 13441, USA; phone: +1 866 874 3566; info@microdrones.com;www.microdrones.com

MD4-3000 is a MicroDrones Gmbh. company's flag ship UAV model (Plate 3.19.1). It is heavier, carries more payload, and has lengthier endurance (flight time). It reaches altitudes of 4000 m during reconnaissance mission. It has payload capacity of 3–5 kg. It is easily assembled at short notice. It is a VTOL UAV and needs only 10 × 10 m clearance space to take-off or land. It has been accepted worldwide as a useful UAV for civilian and agricultural purposes. A package with MD4-3000 comes along with "mdFNC" to keep all electronics under control, "mdFLEX" to see that no functions are missing in the UAV; "mdBlackBox" to collect all flight data; "mdCockpit" brings together, all functions related to planning, monitoring, and analysis of flight path; "mdIMU" allows better functioning of sensors and the analysis of data; "mdWaypoint" helps in flight planning and automated navigation; and "mdLA" helps in automated landing. It is a low noise signature UAV.

"mdMapper3000G" is an upgraded recent model based essentially on MD4-3000. This is a quicker UAV. It reaches higher altitudes, and it can carry more payloads. Flight endurance is also greater at 45–90 min per flight. It carries a 100-MP performance direct geo-referencing camera for aerial imagery.

The following data, that is, technical specifications, pertain only to MD4-3000 model (Plate 3.19.1).

TECHNICAL SPECIFICATIONS

Material: It is made of a composite derived from tough carbon fiber, ruggedized foam, and plastic. It has landing gear made of light aluminum-based alloy.

Size: length: 188 cm; width (rotor to rotor): 205 cm; height: 36 cm; weight: maximum permissible take-off weight is 15 kg

Propellers: This is a quadcopter. It has four propellers made of two plastic blades each.

Payload: Maximum allowed is 4.0 kg, but recommended load is 3.0 kg

Launching information: It is VTOL vehicle. It has autolaunch, navigation, and landing facility.

Altitude: The ceiling is 4000 m above sea level.

Speed: 20 m s^{-1}

Wind speed tolerance: It withstands wind disturbance of 6 m s^{-1}.

Temperature tolerance: −10 to 50°C

Power source: Electric batteries

Electric batteries: 10S1P lithium–polymer, 37.0 V, 21,000 mA h

Fuel: Not an IC engine-fitted UAV

Endurance: 45–90 min depending on batteries and payload

Remote controllers: Telemetry and GCS computer links operate for up to 50 km radius.

Area covered: About 50 ha h^{-1}, but it depends on flight altitude, speed, and resolution needed.

Photographic accessories: Cameras: the CPU is housed in a closed carbon box for safety. The cameras are situated in the gimbal.

Computer programs: The aerial imagery is processed using software such as Pix4D Mapper or Agisoft's Photoscan or custom-made software depending on the purpose.

Spraying area: It has not been touted or used as spraying UAV. It is a high-altitude rapid imaging UAV that covers large areas of agricultural crop production.

Volume: Not applicable

Agricultural uses: It is an apt UAV model for aerial survey and mapping of crop fields. It is used to conduct crop scouting aimed at measuring crop's growth rate, phenomics, grain maturity, water stress index, and nutritional status. Natural vegetation monitoring is also possible.

Nonagricultural uses: MD4-3000 is used for surveillance of industries, mines, oil and gas pipelines, rail roads, city traffic, tracking vehicles delivery of small parcels, etc.

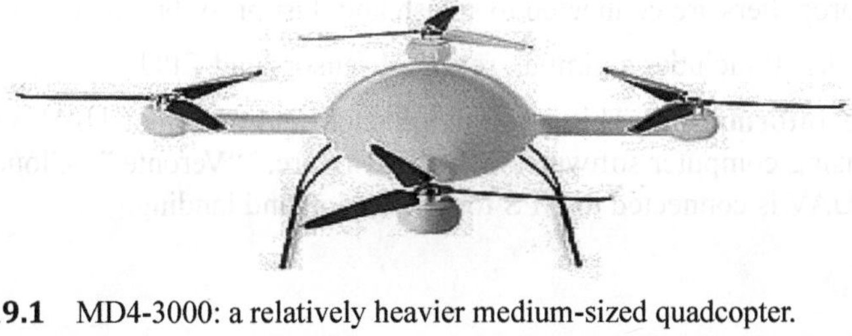

PLATE 3.19.1 MD4-3000: a relatively heavier medium-sized quadcopter.
Source: Ms. Mirjam Baumer, Marketing Manager-Europe, MicroDrones Gmbh., Siegen, Germany.

Useful References, Websites, and YouTube Addresses

http://www.uasvision.com/2013/03/28/microdrones-unveils-the-new-md4-3000/ (accessed May 30, 2017).

https://www.microdrones.com/en/mdaircraft/md4-3000/ (accessed May 30, 2017).

https://www.guavas.info/drones/microdrones%20GmbH-MD4-3000-279 (accessed May 30, 2017).

https://www.youtube.com/watch?v=L6BbXZTZCtc (accessed July 9, 2017).

https://www.youtube.com/watch?v=Y4jtguSF0n4 (accessed July 9, 2017).

Name/Models: Nemesis 44

Company and address: Allied Drones, 3480 W. Warmer Ave., Santa Ana, CA 92704, USA; phone: NA; e-mail: NA; website: http://alliedrones.com; http://www.ttfly.com/com/allieddrones/

Allied Drones is a company that produces UAVs. They are in operation, since 15 years. The company is situated in California, USA. The small UAVs produced by Allied Drones Inc. are suitable to conduct a range of functions in the realms of aerial surveillance, spectral analysis of ground features, traffic control, policing, agricultural, and other civilian aspects. They also produce accessories such as camera gimbal adapter kits, accessory mounting plates, antivibration isolators, and custom one-off projects. They adopt 3D printing technology to arrive at most useful UAV models.

Nemesis 44 is a quadcopter. It is simple in design and operation. It is also priced moderately low so that even farmers with small holdings could purchase it. The UAV's airframe and all parts are packed in a small suitcase and transported to any location in the farm or urban areas. It can be assembled without need for any tools and started in 2 min (Plate 3.20.1). It has a longer flight endurance and is powered by electric batteries. It is an excellent UAV for aerial photography of crop fields.

TECHNICAL SPECIFICATIONS

Material: UAV's air frame is made of aerospace grade carbon fiber and aluminum. It is attached with plastic or carbon fiber propellers.

Size: length: 90 cm; width (rotor to rotor): 90 cm; height: 35 cm; weight: 3.2 kg

Propellers: It is a quadcopter. It has four rotors made of tough plastic or carbon fiber. The propellers are connected to splash and dust proof brushless motors.

Payload: 2 kg. It includes a gimbal, set of all sensors and CPU.

Launching information: This is an autolaunch and autopilot UAV. Its flight is regulated using computer software such as "Airware," "Veronte," "Cloud Cap," or 3DR. The UAV is connected to GPS for navigation and landing.

Altitude: The UAV can hover over the crop canopy at 2–5 m altitude and obtain high-resolution images. The operating height during aerial survey and photography is 5–50 m above ground level. The ceiling is 500 m above ground level.

Speed: It cruises at 45 km h^{-1} with payload

Wind speed tolerance: 15–20 km h^{-1} disturbance

Temperature tolerance: −10 to 50°C

Power source: Lithium–polymer electric batteries

Electric batteries: 16,000 mA h, two cells, lithium–polymer

Fuel: Not an IC engine-fitted UAV. No need for petrol or diesel

Endurance: 2 h per flight

Remote controllers: Remote control has both manual and GPS connected flight control options. The GCS keeps telelink with UAV using 110–250 VAC 50–60 Hz ground power. The GCS has a PC or mobile or a regular joystick (remote controller) to control the UAV's flight and aerial photography. The camera shutters are controlled remotely.

Area covered: It covers 50–250 ha per flight.

Photographic accessories: UAV carries a full complement of sensors such as visual (R, G, and B), high-resolution multispectral camera, NIR, IR, and red-edge cameras. It also has video-photographic camera.

Computer programs: The images are processed using Pix4D Mapper or Agisoft's Photoscan software.

Spraying area: This is not a spraying UAV.

Volume: Not applicable

Agricultural uses: It could be used extensively for aerial photography of land resources, soil types, crops, and farm vehicles. It aids crop scouting to detect variation in canopy growth, NDVI values, leaf chlorophyll data (crop's N status), and crop's water stress index. It helps in deciding supply of fertilizer-N and irrigation, accurately. Natural vegetation monitoring for NDVI values and detect plant species diversity is also possible.

Nonagricultural uses: It includes surveillance of geological sites, mines and mining activity, industrial regions, oil pipelines, electric lines, river flow, lakes, irrigation channels, etc.

PLATE 3.20.1 Nemesis HV 44 quadcopter.
Source: Allied Drones, Santa Ana, CA, USA; https://allieddrones.com/portfolio-item/hv44-nemesis/.

Useful References, Websites, and YouTube Addresses

https://allieddrones.com/portfolio-item/hv44-nemesis/ (accessed July 11, 2017).
http://allieddrones.com/ (accessed July 11, 2017).
http://www.ttfly.com/com/allieddrones/ (accessed July 11, 2017).
https://productz.com/en/allied-drones-hl88-nemesis-UAV (accessed July 11, 2017).

Name/Models: Patriot 100 (quadcopter)

Company and address: Homeland Surveillance and Electronics Inc. (HSE LLC.), IL, USA

Home land Surveillance and Electronics Inc. is a UAV-manufacturing and UAV-related service provider. It manufactures several models of UAVs. They comprise both fixed-wing and rotor copters. These UAV models were results of over 10 years of research and development (R&D) by the company. The HSE was started in 2009 by Mr. Dave Sanders and his entourage. They have industrial facility in Denver, Colorado, USA, that produces UAVs.

"Patriot 100" UAV is an expertly designed UAV. It is capable of aerial surveillance of agricultural crop fields, city installations, and public events. HSE's Patriot has cutting-edge avionics and camera options. This UAV has been in operation with military groups for over two decades. It has a rugged body, made of carbon fiber, and ruggedized foam (Plate 3.21.1). Patriot costs less than US$ 8000 for a complete system.

TECHNICAL SPECIFICATIONS

Material: Patriot quadcopter is made of kevlar, carbon fiber, ruggedized plastic, and foam.

Size: length: 85 cm; width (rotor tip to rotor tip): 80 cm; height: 45 cm; weight: MTOW is 3.5 kg

Propellers: It is a quadcopter. It has four rotors supported by brushless motors.

Payload: 1 lb 8 oz (680 g). It includes mostly gyro-stabilized gimbal plus cameras.

Launching information: This is a VTOL multicopter. It has autolaunch, navigation (predetermined), and autolanding software. It needs 5 × 5 m clearance in vegetated zones or urban location to take-off or land accurately.

Altitude: It flies at 50–3650 m above ground level.

Speed: Maximum flight speed is 35 km h^{-1} (15 m s^{-1}).

Wind speed tolerance: It tolerates a wind disturbance of 22 km h^{-1} (11 m s^{-1}).

Temperature tolerance: −10 to 50°C

Power source: Lithium–polymer electric batteries. This UAV consumes about 30 W (0.4 hp).

Electric batteries: 28 A h, four cells.

Fuel: Not an IC engine-fitted UAV. No need for petrol or diesel.

Endurance: 40 min per flight.

Remote controllers: The GCS computers (iPad) keep contact with the UAV for up to 1.5 mi (2.4 km in a farm). Telemetric connections allow GCS to regulate UAV's flight path, speed, and landing. In case of emergency, patriot quadcopter returns to point of launch swiftly.

Area covered: It covers 50–60 km linear distance per flight. It covers about 200 ha ground with aerial imagery per flight.

Photographic accessories: Cameras: The quadcopter carries a full complement of sensors. They include visual (R, G, and B; e.g., Sony alpha 7), multispectral, and hyperspectral cameras, NIR, IR, and red-edge bandwidth cameras and thermal sensors for estimating water stress index.

Computer programs: The aerial images can be processed right at the CPU of the UAV and relayed to GCS iPad. The images can also be relayed to GCS iPad for processing, using Pix4D Mapper or Agisoft's Photoscan. The digital data collected and stored in computers/chips can be directly used in the ground vehicles for spraying and inoculating with fertilizers.

Spraying area: Not a spraying UAV

Volume: Not applicable

Agricultural uses: It includes crop scouting for detecting seed germination trends, seedling establishment, and crop stand to identify spots with gaps in rows. The aerial images can depict locations with weed infestation and disease/pest attack. This will help in concentrating sprays only at spots affected by the malady. UAV can also be used to surveillance and estimate drought, flood or soil erosion spots, and their effects. This helps in appropriately devising the remedial measures. Natural vegetation monitoring, study of plant species diversity, and spectral signatures could be accomplished with accuracy.

Nonagricultural uses: Surveillance of military camps, borders, mines, mining activity, industrial installations, buildings, highways, rail roads, tracking transport vehicles, dams, reservoirs, rivers, etc. are important uses of this UAV.

PLATE 3.21.1 Patriot 100—a light-weighted quadcopter.
Source: Mr. Terry Sanders, Homeland Surveillance and Electronics Inc. HSE LLC., Denver, CO, USA.

Useful References, Websites, and YouTube Addresses

http://www.hse-uav.com/patriot_100_quadcopter_uav_UAV.htm (accessed May 30, 2017).
http://www.hse-uav.com/products.htm (accessed May 30, 2017).
https://www.youtube.com/watch?v=h5BJmtipi4Y (accessed May 30, 2017).
https://www.youtube.com/watch?v=VrMHAUhRfCk (accessed May 30, 2017).

Name/Models: PD4-AW

Company and address: CEO, Mr. Masakazu Kono; ProDRONE—Head Office: 16F Bantane Sakae Bldg. 2-4 Shinsakae-machi, Naka-ku, Nagoya-shi, Aichi 460-0004, Japan; Tel.: +81 52 950 1278; fax: +81 52 950 1277; e-mail: info@prodrone.jp; Tokyo Office: 1F Hirakawacho-Daiichiseimei Bldg. 1-2-10 Hirakawa-cho, Chiyoda-ku, Tokyo 102-0093, Japan; Tel.: +81 3 5212 5132; fax: +81 3 5212 5102

ProDRONE company is situated in Japan. It was incorporated in the year 2015. It is involved, in developing models and producing UAVs useful in variety of tasks such as aerial photography, etc. It concentrates on commercial use of UAV system and consulting, UAV operator training and UAV maintenance. ProDRONE company offers a series of multicopters useful in agriculture, civilian, and military aspects, for example, PD4-AW, PD6B, PD6E2000, PD6B-AW, etc.

ProDRONE 4-AW is a relatively small quadcopter. Yet, it is highly flexible in usage. It has a wide range of applications in agriculture, civilian, and military aspects. It is used mainly as surveillance UAV for crops and detecting the fertility variations. The aim is to make fertilizer supply accurate and efficient. The quadcopter has replaceable parts, in case they are overused. It has amphibious ability of taking-off and landing on ground or water surface. It is easily transportable. The entire UAV is foldable and can be packaged in a small suitcase (Plate 3.22.1).

TECHNICAL SPECIFICATIONS

Material: This UAV, PD4-AW is made of a composite of kevlar, ruggedized foam, carbon fiber, and plastic propellers. It has waterproof body. It is amphibious. It can take-off from or land on water surface.

Size: Length (rotor to rotor) 86 cm; width (wingspan) 86 cm; height: 34.5 cm; weight: dry weight 5.5 kg

Propellers: Four propellers each of 43.2 cm diameter

Payload: Maximum payload is 3.5 kg. It includes gyro-stabilized gimbal with sensors, and a small tank for dusting, if opted for it.

Launching information: It has autolaunch and landing software, such as Pixhawk. It returns to the place of take-off, in times of emergency, or commands being erroneous or fuel depletion. It has landing gear and skid landing facility to land safely. The skids may have to be replaced after a certain period of usage.

Altitude: It operates at 0–50 m for agricultural crop's surveillance and aerial imagery. General operation height is 300 m. However, ceiling is 1000 m above ground level.

Speed: Maximum speed attainable is 60 km h^{-1}. Operational speed is 8 m s^{-1}.

Wind speed tolerance: 25 km h^{-1} disturbance

Temperature tolerance: −10 to 45°C

Power source: Lithium–polymer electric batteries

Electric batteries: 24,000 mA h, four cells.

Fuel: Not an IC engine-fitted UAV

Endurance: 10–40 per flight depending on payload and fuel reserve

Remote controllers: The UAV keeps contact with GCS using radiolinks for up to a radius of 5 km. Flight path could be altered midway, using GCS computers.

Area covered: It covers about 50 km linear distance or 50–250 ha, if flight path involves aerial survey of crop fields or a location.

Photographic accessories: Cameras: It has a small tri-axis gimbal that allows cameras to get focused at any angle. It carries a set of cameras such as visual (Sony 5100), multispectral, hyperspectral for close-up, and red-edge bandwidth.

Computer programs: The digital data can be starred in the CPU of the UAV or processed using Pix4D Mapper or Agisoft's Photoscan software. Images could be instantaneously transmitted to ground control iPad for processing.

Spraying area: It could be used to spray small areas by flying it repeatedly with a small tank that holds pesticide.

Volume: Data not available

Agricultural uses: Crop scouting could be a major application of this UAV, along with collecting data about NDVI, leaf chlorophyll, crop-N status, and crop water stress index (thermal imagery). The hovering function allows the UAV to pick close-up images, also to detect disease/pest-attacked locations, in a crop field. The UAV also detects areas affected by floods, drought, or rampant soil erosion.

Nonagricultural uses: It includes surveillance of industries, mines, mining activity, oil pipelines, electric lines, rail roads, highways and vehicle movement, natural resources such as rivers, tanks, channels, etc.

PLATE 3.22.1 ProDRONE PD4-AW.
Source: Ms. Aya Kawkami, ProDRONE Co. Ltd., Aichi, Nagoya, Japan.

Useful References, Websites, and YouTube Addresses

https://www.prodrone.jp/en/products/pd4-aw/ (accessed May 23, 2017).
https://www.prodrone.jp/en/products/ (May 23, 2017).
https://www.youtube.com/watch?v=IiThAiAeAfk (accessed May 23, 2017).
https://www.youtube.com/watch?v=HoV0QLDcs4c (accessed July 1, 2017).
https://www.youtube.com/watch?v=T6kaU2sgPqo (accessed July 1, 2017).

Name/Models: Pelican

Company and address: Ascending technologies Gmbh.; Konrad-Zuse-Bogen 4, 82152 Kreilling, Germany; Tel.: +49 89 895560790; fax: +49 89 89556079 19; e-mail: team@asctec.de; website: www.asctec.de

Ascending Technologies Gmbh. is a UAV company located in Bavaria, Germany. It produces a few different models of copter UAVs. It includes the quadcopter namely "Humming Bird." The company serves with UAVs for range of purposes. The UAVs are relevant to civilian administration and agricultural crop production.

Pelican is a small, light-weight quadcopter. It is a tower-like model with fuselage, instruments, and accessories arranged one above the other. It can be attached with a series of sensors capable of visual, IR, and red-edge capabilities. It carries a Lidar scanner. The UAV is easily packed and transported to any location. It can be launched quickly, say in 10 min. It could cost US$ 10–15,000 per unit plus accessories (February 2017 price levels) (see Plate 3.23.1).

TECHNICAL SPECIFICATIONS

Material: It is made of light avionics aluminum frame, carbon fiber, and toughened plastic. This quadcopter can be easily packed in a suitcase of the size 70 × 70 × 50 cm. It can be assembled quickly, using tools. AscTec Pelican can be stored, in fully assembled condition in a suitcase, and launched immediately.

Size: length: 65 cm; width (rotor to rotor): 65 cm; height: 19 cm; weight: MTOW is 1.65 kg

Propellers: It has four propellers attached to four brushless motors (160 W) for propulsion. The propellers have two plastic blades of 10 in. length.

Payload: 650 g

Launching information: It is an autolaunch quadcopter. It can be controlled, using remote controller or could be preprogramed, using appropriate computer software

Altitude: It can be used in crop fields by flying it at low heights over the crop's canopy. This UAV is flown at low heights of 1–5 m above crop's canopy, for close-up shots and detailed analysis of crop stand. UAV's operational altitude is 1000 m above ground level. However, the ceiling is 4500 m above sea level.

Speed: 16 m s^{-1}

Wind speed tolerance: 8 m s^{-1}

Temperature tolerance: −5 to 35°C

Power source: Lithium–polymer electric batteries

Electric batteries: 6250 mA h, two cells

Fuel: This is a UAV fitted with IC engine. It does not require gasoline or diesel as fuel.

Endurance: 16 min. Maximum flight time is 30 min without payload.

Remote controllers: The GCS is usually an iPad or mobile or remote controller (joystick). The quadcopter can be flown, both, using predetermined flight path or it could be altered midway during the flight. The waypoints, aerial images, and their frequency can be modified by the ground computer. The GCS computer overrides predetermined flight path and waypoints.

Area covered: It covers 5–50 ha per flight of 15 min

Photographic accessories: Cameras: The cameras include visual (R, G, and B), high-resolution multispectral, NIR, IR, and a laser scanner. There are a few variants of laser scanners that could be opted. The recommended laser scanner is Hokuye UST-20LX.

Spraying area: Not a spraying UAV

Volume: Not applicable

Agricultural uses: It is used to conduct aerial survey and photography of natural features relevant to land and soil resources, crop fields, and grain maturity status. Pelican could be used for crop scouting aimed at noting NDVI values and its spatial variations, canopy growth rate, leaf chlorophyll (crop's N status), canopy and ambient temperature (thermal imagery), water stress index, diseases, pests and weed infestation in fields, etc. Natural vegetation monitoring could also be conducted, using a Pelican quadcopter.

Nonagricultural uses: They are surveillance of geological sites, mines, mining activity, ore dumps, industrial sites, buildings, rail roads, oil pipelines, electric lines, dams, irrigation channels, highways, and transport vehicles.

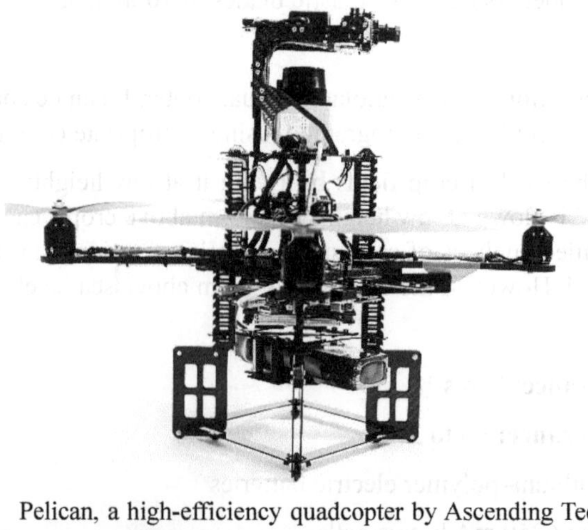

PLATE 3.23.1 Pelican, a high-efficiency quadcopter by Ascending Technologies Gmbh., Kreilling, Germany.
Source: Dr. Matthais Beldzik, and Prof. Guido Morganthal, Ascending Technologies Gmbh., Kreilling, Germany.

Useful References, Websites, and YouTube Addresses

http://www.asctec.de/uav-uas-drohnen-flugsystem/asctec-pelican/ (accessed June 17, 2017).
http://wiki.asctec.de/display/AR/AscTec+Pelican/ (accessed June 17, 2017).
https://www.youtube.com/watch?v=jDzI9mNi9Pw (accessed June 17, 2017).
https://www.youtube.com/watch?v=FT-NqI-Cocw (accessed June 17, 2017).
https://www.youtube.com/watch?v=VIpPC6IPVGw (accessed June 17, 2017).

Name/Models: Phantom 3

Company and address: SZ DJI Baiwang Technology Company Ltd., 14th Floor, West Wing, Skyworth Semiconductor Design Building, No. 18 Gaoxin South 4th Ave., Nanshan District, Shenzhen 518057, China; phone: +86 (0)755 26656677

Phantom 3 is among the most popular quadcopters produced and sold worldwide. It is a light-weight autonomous aircraft. It has a streamlined body made of carbon fiber. It has wide range of sensors attached for aerial visual and thermal imagery of ground features. Phantom 3 has advanced and modified version to suit different purposes. For example, "Phantom 3 Professional" is a UAV with better electronics and it serves multiple purposes. "Phantom 3 Advanced" takes to sky rapidly. It is a VTOL. It has a fully integrated electronic and intelligent system. There is also "Phantom 3 Standard" which is safe to fly. It is easy to operate in general urban conditions and crop fields (Plate 3.24.1; DJI, 2016). It is used to obtain aerial imagery and identify disease, weeds (Lottes et al., 2017), drought, and flood damages to crop fields.

TECHNICAL SPECIFICATIONS

Material: It is made of lightweight but toughened carbon fiber and plastic

Size: length: 32 cm; width (rotor to rotor): 32 cm; height: 25 cm; weight: 1.22 kg without payload.

Propellers: There are four propellers made of plastic. They are 35 cm in length.

Payload: It includes batteries, CPU, and gimbal with sensors.

Launching information: It is a VTOL autolaunch quadcopter. Its navigation is regulated, using Pixhawk software. This UAV has GPS connectivity. It can be landed, using manual or predetermined determined, using computer software. It has autoreturn facility. It returns to point of launch in case of emergency or as per telelink instructions.

Altitude: The operating height while surveying is 300–500 m above ground level. It can reach close to crop canopy at 0–5 m above crop and hover for close-up shots. Its ceiling is 19,850 ft above ground level.

Speed: Maximum ascent speed is 5 m s^{-1}, maximum descent speed is 3 m s^{-1}, and maximum flight speed is 16 m^{-1}. Hover accuracy is 1.5 m.

Wind speed tolerance: It tolerates 15–20 km h^{-1} disturbing wind

Temperature tolerance: 0–40°C

Power source: Lithium electric batteries

Electric batteries: 16,000 mA h, voltage is 17.4 V

Fuel: It has no requirement for gasoline. It is not an IC engine-fitted UAV.

Endurance: 25 min per flight

Remote controllers: The GCS has a laptop or mobile (DJI GO) to regulate flight path and shutter speed, during aerial survey and imagery (Plate 3.24.2). The ground station telelink operates at 2.4 GHz ISM. Other recommended device to regulate the UAV is iPhone, Androids such as Samsung Tabs 705c, etc.

Area covered: It covers 25–30 km linear distance or 15 ac per flight.

Photographic accessories: The cameras attached to this UAV depend on the purpose. CMOS ½3.5 sensor with FOV 20 mm lens is recommended. ISO range for video is 100–3200 and for still photos is 100–1600. The electronic shutter speed ranges from 1 to 1/8000 s. Image size is usually 4000 × 3000 mm. Videorecording is possible. It can be instantaneously sent to ground station, if necessary. HD pictures of 1280 × 720 P are possible (Plates 3.24.3–3.24.5).

Computer programs: The imagery is processed, using several different types of software. Most commonly used are Pix4D Mapper, Agisoft's Photoscan, and Trimble's Inphos.

Spraying area: This is not a spraying UAV

Agricultural uses: It is used for aerial survey of natural resources such as land, soil type, water availability, etc., that are relevant to crop production. It is used in crop scouting to obtain NDVI data, canopy growth rate, leaf chlorophyll (crop's N) status, and water stress index (via thermal imagery). It is also used to survey, identify, and map disease/pest-attacked zones, flood/drought-affected areas, pH anomalies and its effect on crop growth, etc. Natural vegetation monitoring and detection of plant species diversity are clear possibilities.

Nonagricultural uses: It includes surveillance of geo-physical features, mines, mining activity, mine vehicle movement, etc. It utilized to monitor international borders, oil pipelines, dams, irrigation channels, highways, rail roads, and vehicle movement.

PLATE 3.24.1 Top: Phantom 3 Professional; middle: Phantom 3 Advanced; bottom: Phantom 3 Standard.
Source: Mr. Adam Najberg, Sales and Marketing division, Shenzen, China (Mainland).

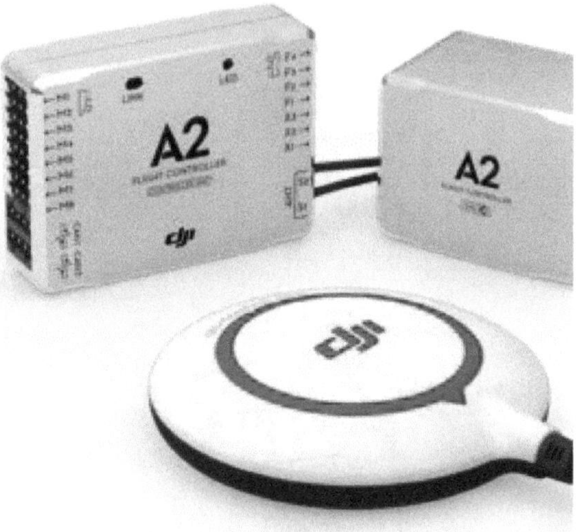

PLATE 3.24.2 DJI multirotor stabilization and flight controller.
Source: Mr. Adam Najberg, Sales and Marketing division, Shenzhen, China (Mainland).

PLATE 3.24.3 Zenmuse X5 mirrorless, compact microcamera made for aerial photography.
Source: Mr. Adam Najberg, Sales and Marketing division, Shenzhen, China (Mainland).

PLATE 3.24.4 Zenmuse XT.
Source: Mr. Adam Najberg, Sales and Marketing division, Shenzen, China (Mainland).
Note: Gimbal and image transmission technology, thermal imager (FLIR).

PLATE 3.24.5 DJI Zenmuse Z3.
Source: Mr. Adam Najberg, Sales and Marketing division, Shenzhen, China (Mainland).
Note: It has 7× zoom made of 3.5× optical and 2× digital zoom.

Useful References, Websites, and YouTube Addresses

DJI. Above the World. *UAS Mag.* **2016,** 238. http://www.uasmagazine.com/articles/1591/dji-explains-new-book-of-UAV-captured-images (accessed Apr 20, 2017).

Lottes, P.; Raghav, K.; Pfefer, J.; Siegwart, R.; Stachniss, C. *Drone-Based Crop and Weed Classifications for Smart Farming*; 2017; pp 1–22. http://www.ipb.uni.bonn.de/wp-content/papercite-data/pdf/lottes17icra.pdf (accessed Sept 20, 2017).

https://developer.dji.com/products/ (accessed Aug 10, 2017).

https://www.youtube.com/watch?v=YJmFHrCwVVY (accessed Aug 10, 2017).

https://www.youtube.com/watch?v=9kJPT2F_bPU (accessed Aug 10, 2017).

https://www.youtube.com/watch?v=AivufAt8FcI (accessed Aug 10, 2017).

https://www.youtube.com/watch?v=S3divEXOFr0

https://www.youtube.com/watch?v=ZvMpElv31DQ

https://www.youtube.com/watch?v=42sYHpX1uQc

https://www.youtube.com/watch?v=tnLdOsi2kuc

Name/Models: Phantom 4 Series

Company and address: DJI SZ DJI Baiwang Technology Company Ltd., 14th Floor, West Wing, Skyworth Semiconductor Design Building, No. 18 Gaoxin South 4th Ave., Nanshan District, Shenzhen 518057, China; phone: +86 (0)755 26656677

Phantom 4 series are among most popular multicopter UAVs. They are commonly utilized in the urban and agricultural zones, depending on the purpose. It is not a very costly UAV at US$ 7–10,000 for a unit, with all accessories. Phantom 4 is small quadcopter with facility for autolaunch, navigation, descent, and aerial photography. Its flight path is usually predetermined and camera shutter is regulated, using computer decision support or manually (Plate 3.25.1; DJI, 2016).

TECHNICAL SPECIFICATIONS

Material: Light-weight carbon fiber and plastic.

Size: length: 35 cm; width (rotor to rotor): 35 cm; height: 25 cm; weight: 1.380 g without payload.

Propellers: It is a quadcopter with propellers made of two plastic blades each.

Payload: Gyro-sized gimbal with cameras for visual, IR, and red-edge aerial imagery; CPU

Launching information: It is a VTOL autolaunch aerial robot. Its navigation can be predetermined using Pixhawk software. This UAV can be recovered, using remote controller manually or using predetermined location. It has vertical descent.

Altitude: It is 0–10 m above ground in general locations, but 5–50 m above crop's canopy, during aerial imagery. It has a ceiling of 6000 m above ground level.

Speed: 25 km h^{-1} when in free loiter. Maximum speed while in ascent is 6 m s^{-1}. During descent, the speed is 4 m s^{-1}. However, accuracy is 0.1–1.5 m.

Wind speed tolerance: Resists 10 m s^{-1} or 15–20 km h^{-1} wind disturbance

Temperature tolerance: 5 to 40°C

Power source: Electric batteries

Electric batteries: 6000 mA h, lithium–polymer battery. Weight of batteries 462 g. Operating current voltage is 1.2 A/7.4 V.

Fuel: It does not require gasoline. It is not fitted with IC engine.

Endurance: 28 min

Remote controllers: The operating frequency of the remote controller is 2.4 GHz. It keeps contact with UAV for up to 5 km radius. The GCS is a laptop or mobile (DJI GO). Androids such as Samsung 705c or Samsung 6 are also used. The UAV's flight is regulated, using flight controllers such as A2 or A3.

Area covered: It covers 25–30 ac of aerial survey or imagery per flight.

Photographic accessories: The UAV has a gyro-stabilized gimbal (Plate 3.25.1). It has been fitted with wide range of cameras such as Zenmuse X5, Zenmuse XT, and Zenmuse Z3. The UAV comes with CMOS 3.5 with 12.4 effective pixels, ISO range is 100–3200 for video and 100–1600 for still photos. The electronic shutter speed is 8–1/8000 s. The image size is 400 × 300 mm. Images are swiftly transmitted as JPEG, MP4.

Computer programs: Still and video photography output could be an USB or transmitted directly to ground station laptop. It is usually processed, using Pix4D Mapper or Agisoft's Photoscan or Trimble's Inphos or similar custom-made software.

Spraying area: This is not a spraying UAV.

Volume: Not applicable

Agricultural uses: DJI's Phantom is a general-purpose UAV. It is capable of sharp and accurate aerial survey. It has been used to obtain NDVI data, canopy growth rate, nutrient distribution in fields, leaf chlorophyll content, and crop's N status variation (Kokila et al., 2017). Its thermal cameras help in detecting variations in crop's water stress index. Therefore, it helps in deciding quantity and timing of irrigation. Phantom 4 is excellent in relaying images depicting disease/pest-attacked zones, in a crop field. It also shows effect of drought/flood, if any, on the crop. Natural vegetation monitoring to get NDVI data and information about plant species diversity is possible. Spectral signatures of plant species are used while detecting diversity.

Nonagricultural uses: It has a wide range of uses in urban, industrial, and public locations. It is used to survey and map geological features, mining locations, ore dumping sites, etc. This UAV is useful in monitoring industrial installations, buildings, roads, rail roads, vehicle traffic, oil pipelines, international borders, etc.

PLATE 3.25.1 Top: Phantom 4; bottom: Phantom 4K.
Source: Mr. Adam Najberg, Sales and Marketing division, Shenzen, China (Mainland).

Useful References, Websites, and YouTube Addresses

DJI. Above the World. *UAS Mag.* **2016,** 238. http://www.uasmagazine.com/articles/1591/dji-explains-new-book-of-UAV-captured-images (accessed Apr 20, 2017).

Kokila, M.; Karthi, J.; Mdhuvski, E.; Sathya, S.; Vignesh, B. Estimation of Chlorophyll Content in Maize Leaf: A Review. In *International Conference on Emerging Trends in Engineering, Science and Sustainable Technology*, 2017; pp 73–78.

http://enterprise.dji.com/agriculture?gclid=CIvJ59PZvtACFcIQaAod5FwCkw (accessed Apr 23, 2017).

https://www.youtube.com/watch?v=QAlDtUy5kB8 (accessed Apr 23, 2017).

https://www.youtube.com/watch?v=0kMyeyf6KeA (accessed Apr 23, 2017).

https://www.youtube.com/watch?v=zdbdX7thMW4 (accessed Apr 23, 2017).

https://www.youtube.com/watch?v=TSJgOeSrVdA (accessed Apr 23, 2017).
https://www.youtube.com/watch?v=2tNEhWmuSnk (accessed Apr 23, 2017).

Name/Models: Q4L Multicopter

Company and address: MAVTech s.r.l., corso Galileo Ferraris 57, 10128 Torino, Italy; Tel.: +39 011 5808482; fax: +39 011 5808579; e-mail: mavtech@mavtech.eu; website: http://www.mavtech.eu/

MAVTech s.r.l. is an Italian avionics company. It specializes in manufacture of small UAVs and UAV-related products. "Q4L" is a small quadcopter. It is meant for aerial photography of ground surface and crop status. It is easy to transport, since it fits into a small suitcase or backpack. It is accompanied with remote controller and flight path software, payload cameras, and orthomosaic processing software. It can be assembled and launched in <5 min, from anywhere in crop field or an urban location. This UAV "Q4L" is affordable for farmers with 200–500 ac of crop land (Plate 3.26.1). It is currently being used in European farm belt. The UAV is utilized in the following aspects of agricultural and natural resources, namely, terrain mapping, vegetation identification and metrics, invasive species identification and analysis (invasive plants), land and forestry research, vigor mapping and frost mitigation, crop health and disease monitoring, agricultural insecticide spraying, and fertilizer dispensing.

TECHNICAL SPECIFICATIONS

Material: It is made of a composite (GF-ABS).

Size: length: 98 cm; width (rotor to rotor): 98 cm; height: 30 cm; weight: 2.0 kg

Propellers: It has four propellers made of two blades each.

Payload: 300 g

Launching information: It is an autolaunch UAV. Its navigation is predetermined using computer software such as Pixhawk. It has both manual and autolanding capabilities. Landing or recovery could be conducted, using remote controller. The UAV is VTOL and needs only 5 × 5 m clearance space to launch or land.

Altitude: Operational height during aerial photography of crops and land resources is 70 m above ground surface.

Speed: 10 m s^{-1}

Wind speed tolerance: 20 km h^{-1} disturbance

Temperature tolerance: −10 to 55°C

Power source: Electric batteries.

Electric batteries: 16,000 mA h, two cells lithium–polymer.

Fuel: Not an IC engine-fitted UAV. No need for petrol or diesel

Endurance: 20 + 25 min per flight.

Remote controllers: The remote controller (joystick or iPad or mobile) has 250 m + 1000 m operational control over the UAV's flight path and camera shutter.

Area covered: 50–100 ha of aerial photography or surveillance

Photographic accessories: Cameras: It has a full complement of sensors such as visual (R, G, and B), high-resolution multispectral, NIR, IR, and red-edge bandwidth cameras.

Computer programs: Post flight images are processed, using Pix4D Mapper or Agisoft's or Trimble Inphos software.

Spraying area: This is not a spraying UAV. Its payload is too small for a pesticide tank with 5–10 l fluid to be accommodated.

Volume: Not applicable

Agricultural uses: This UAV's major applications are in aerial surveillance and close-up imagery of land resources, soil type, and crop status. Crop scouting for canopy growth rate, mapping variations in NDVI, leaf chlorophyll (crop's N), and canopy/ambient temperature (water stress index) is done regularly. Aerial surveys help in early detection and mapping of diseases, pests, or weeds. Therefore, it allows farmers to take accurate and timely remedial measures.

Nonagricultural uses: It includes surveillance of industries, buildings, river flow, dams, lakes, irrigation channels, oil pipelines, electric lines, highways, and vehicle movement.

PLATE 3.26.1 Q4L—A quadcopter.
Source: Dr. Fulvia Quagliotti, President and CEO, and Mr. Gianluca Ristorto, Micro Aerial Vehicles Technology s.r.l., Torino, Italy.

Useful References, Websites, and YouTube Addresses

http://www.mavtech.eu/ (accessed Jan 20, 2017).
http://www.mavtech.eu/en/products/q4l-UAV/ (accessed Jan 20, 2017).
http://pdf.aeroexpo.online/pdf/mavtech/q4l/181321-4114.html (accessed May 12, 2017).
http://www.mavtech.eu/en/applications/precision-farming/ (accessed May 12, 2017).
http://www.mavtech.eu/en/applications/aerial-mapping/ (accessed May 12, 2017).

Name/Models: Q4P Multicopter

Company and address: MAVTech s.r.l. corso Galileo Ferraris 57, 10128 Torino, Italy; **Tel.:** +39 011 5808482; fax: +39 011 5808579; e-mail: mavtech@mavtech.eu; website: http://www.mavtech.eu/

Q4P multicopter is another quadcopter UAV offered in a series of small UAVs, by MAVTech s.r.l. of Italy. It has specific advantages during aerial photography of terrain and agricultural fields.

TECHNICAL SPECIFICATIONS

Material: The airframe and other parts of UAV are built, using a composite (GF-ABS)

Size: length: 980 cm; width (rotor to rotor): 980 cm; height: 300 cm; weight: 1.8 kg (<2.0 kg)

Propellers: It has four rotors made of two plastic blades each. The rotors are attached to brushless motors for propulsion.

Payload: <300 g (it includes a series of sensors)

Launching information: It is an autolaunch VTOL quadcopter

Altitude: The operational height for the UAV is 5–70 m above crop's canopy. However, the ceiling altitude is 1000 m above ground level. It is an auto-take-off, autopilot, and landing UAV. The UAV is recovered using a remote controller. UAV operator must guide it to land carefully on landing gear.

Speed: 10 m s^{-1} during aerial photography and slightly higher at 12–14 m s^{-1} during loiter. It can also hover with higher accuracy of 1 m at spot just above the crop's canopy. This mode is helpful in close-up aerial photography.

Wind speed tolerance: 20 km h^{-1} wind disturbance. It also tolerates rain to a certain extent.

Temperature tolerance: −10 to 55°C

Power source: Electric batteries

Electric batteries: 14,000 mA h.

Fuel: Not an IC engine-fitted UAV

Endurance: 20 + 25 min per flight

Remote controllers: The GCS computers [iPad or PC or remote controller (i.e., joystick) keeps contact with UAV for up to 300–500 m]. The telemetric contact operates using 2.4 Hz for overriding predetermined programs in flight.

Area covered: It covers 15–20 ha h^{-1}.

Photographic accessories: Cameras: It has a complement of visual, high-resolution multispectral, NIR, IR, and red-edge bandwidth cameras. It also has facility for video-imagery.

Computer programs: The postflight processing of images is done, using software such as Agisoft's Photoscan or Pix4D Mapper or Trimble Inphos, etc.

Spraying area: This is not a spraying quadcopter. Its payload is too small to carry pesticides.

Volume: Not an IC engine-fitted UAV

Agricultural uses: It is an apt UAV model for rapid aerial imagery of land resources, soil types, and crop fields, prior to tillage and planting. It is useful in scouting crops for detecting variations in canopy growth pattern, NDVI values, leaf chlorophyll, and canopy temperature (water stress index). It is also highly useful in detecting spatial and temporal variations of disease, pest, or weed. Natural vegetation monitoring for NDVI and plant diversity is a clear possibility.

Nonagricultural uses: They are surveillance of industrial sites, buildings, river flow, lakes, irrigation channels, oil pipelines, electric lines, highways and transport vehicles, etc.

Useful References, Websites, and YouTube Addresses

www.mavtech.eu/site/assets/files/1158/q4p-rotor.pdf (accessed May 30, 2017).
http://www.mavtech.eu/ (accessed May 30, 2017).
http://www.mavtech.eu/en/applications/precision-farming/ (accessed May 30, 2017).
http://www.mavtech.eu/en/applications/aerial-mapping/ (accessed May 30, 2017).

Name/Models: RH 4 Spyder

Company and address: Delft Dynamics B.V. Mollengraaftsingel 10, 2629 JD Delft, The Netherlands; phone: +31 15 7111009; e-mail: info@delftdynamics.nl; website: http://www.delftdynamics.nl/index.php/en/

The Delft Dynamics is a UAV company located in Netherlands. It was instituted 10 years ago. It offers a series of small multicopters such as "RH 2 Stern" and "RH3 Swift." These copters can be applied to accomplish a variety of tasks, related to commercial and agricultural activities. Besides, designing and building robot helicopters such as the "RH2 Stern," Delft Dynamics B.V. also carries out engineering assignments. This company makes use of years of experience in the fields of hardware and software integration, real-time simulation, and control design. They help in increasing the applicability of UASs. Delft Dynamics B.V. carries out research projects on a regular basis with research institutes and other end users. In these projects, the extensive experience and knowledge regarding unmanned helicopters as well as the availability of adaptable test platforms of Delft Dynamics prove very useful.

The multicopter "RH 4 Spider" is a compact and light UAV. It has a series of 4–5 cameras fitted to its fuselage. It includes a videocamera. The UAV is made of robust composite so that it withstands vagaries of weather and handling shocks.

TECHNICAL SPECIFICATIONS

Material: It is made of steel and composite material that has ruggedized foam, carbon fiber, and plastic.

Size: length 28 cm; width 35 cm; height: 45 cm; weight: 2 kg

Propellers: "RH 4 Spyder" is a quadcopter. It has four propellers attached vertically. Each propeller has four blades made of toughened plastic. Blades are 15 cm in length and 3 cm in breadth. There are four motors, each connected to a rotor for propulsion.

Payload: It includes a combination of visual (daylight), NIR, IR cameras, and Lidar sensor.

Launching information: It is autolaunched. It is VTOL UAV. It has autonomous flight and landing facility.

Altitude: 50–100 m above ground/crop canopy level. It can reach higher ceiling at 1500 m AGL.

Speed: It is 20–40 km h^{-1}

Wind speed tolerance: 10–20 km h^{-1}

Temperature tolerance: −10 to 40°C

Power source: Electric batteries

Electric batteries: 14 A h, two cells

Fuel: Not an IC engine-fitted UAV

Endurance: 15–30 min per flight

Remote controllers: The UAVs flight path can be preprogramed, using Pixhawk software. It could also be controlled, using a remote controller (joystick). About 100 waypoints can be fixed for the UAV to transit and capture aerial images. The UAV can be controlled using a ground station.

Area covered: It covers 20–30 km h^{-1} or 50 ha per flight

Photographic accessories: Cameras: In addition to the usual component of visual (Sony 5100) and multispectral cameras, the UAV is fitted with videocamera. The videocamera transmits images of ground/crop instantaneously. It has 10× zoom and is connected to GCS via computers/telemetry. All images are tagged with GPS coordinates for accuracy.

Computer programs: The aerial images are processed, using software such as Agisoft's Photoscan or Pix4D Mapper. The data captured by cameras could be stored in the computer chips, at the CPU of the UAV and/or at the GCS iPad.

Spraying area: This is not a spraying UAV.

Volume: Data not available

Agricultural uses: This model is used in aerial surveys of terrain, land resources, soil types, crops, and cropping systems in vogue in an area. Crop scouting could be done for obtaining NDVI values and mapping them to measure chlorophyll index, crop's N status, water stress index, and to monitor grain maturity. Natural vegetation monitoring is also possible.

Nonagricultural uses: It includes surveillance of mines, industrial sites, buildings, oil and gas pipelines, railroads, city vehicular traffic, tracking individual vehicles, etc.

Useful References, Websites, and YouTube Addresses

http://www.delftdynamics.nl/index.php/en/contact-en (accessed May 1, 2017).
http://www.delftdynamics.nl/index.php/en/products (accessed May 1, 2017).
www.geo-informatie.nl/.../Noorbergen_GLOSSY_9May2014.pdf (accessed May 1, 2017).

Name/Models: SkyRanger

Company and address: Aeryon Inc. **Aeryon Labs Inc.,** 575 Kumpf Drive, Waterloo, ON, Canada N2V 1K3; **Head Office**: +1 519 489 6726; **fax**: +1 519 489 6726x310 (technical customer support), +1 519 489 6726x320 (sales), +1 519 489 6726x360 (media); website: http://www.aeryon.com/aeryon-skyranger

Aeryon Labs Inc. is a provider of **small** UAS. This company's UAVs are popular in different regions of North America and Europe. **Aeryon Labs'** copters have great potential in civil, agricultural, and military aspects. Aeryon Labs was founded in 2007. It is situated in Waterloo, Ontario, Canada. The company offers training in launching, flight control, and data acquisition via sensors. However, piloting and obtaining maps of crop fields is so easy that no prior training seems needed to use the joystick or laptop. The UAV is operated, using touch screen laptop. It offers excellent aerial images and digital data to the operator.

Aeryon Labs Quadcopter, namely, "SkyRanger," is a popular autonomous, aerial UAV. It has a range of applications in agriculture, civilian surveillance, and military reconnaissance. The quadcopter package consists of wide range of accessories, such as sensors to surveillance, obtain aerial imagery, and thermal data of crops/ambient atmosphere. It has real-time advanced video processing facility. Hence, aerial images are seen instantaneously, at the GCS iPad. It allows rapid alterations in flight path and imagery. It offers quick analysis of data. The "SkyRanger" of Aeryon Labs Inc. costs about US$ 12–16,000 (February 2017 price level) (Plate 3.27.1).

TECHNICAL SPECIFICATIONS

Material: The quadcopter is made of a composite of carbon fiber, ruggedized foam, and plastic. Light-weight aviation aluminum is utilized for landing gear.

Size: length: 50 cm; width (rotor to rotor): 102 cm; height: 24 cm; weight: 5.3 lb (2.4 kg)

Propellers: It has four propellers. They are fitted to brushless motors for propulsion. The two blades are made of tough plastic.

Payload: 2.0 kg. It includes a series of sensors, CPU accessories, and batteries with charger.

Launching information: SkyRanger is a vertical take-off and launching UAV (VTOL). The UAV has autopilot and autolanding facility. It needs 5 × 5 m clearance space to land anywhere on the ground. It is waterproof and can be flown during a small drizzle.

Altitude: The operational altitude during aerial imagery and civilian surveillance is 400 m above ground level. Its ceiling is 15,000 m above the ground level.

Speed: 50–60 km h^{-1} depending on flight path and imaging frequency.

Wind speed tolerance: 40 km h^{-1} of sustained wind disturbance or up to 90 km h^{-1} wind that occurs in gusts.

Temperature tolerance: −22 to 50°C

Power source: Electric batteries

Electric batteries: 28,000 mA h, two cells. The battery charger device is also included in the package.

Fuel: This UAV is not fitted with a gasoline-supported IC engine.

Endurance: Up to 50 min per flight

Remote controllers: The GCS keeps telelinks and flight control ability, for up to 5 km radius. The radiolink operates at 900 MHz, 2.4 GHz. The GCS includes a joystick (remote controller) for manual operations. The iPad/laptop is utilized for programing fight path and frequency of photography.

Area covered: The UAV covers an area of 50–300 ha aerial imagery per flight.

Photographic accessories: Cameras: This quadcopter carries a gyro-stabilized gimbal. Internally stabilized gimbals keep the camera focused, on its target. It has a few cameras for aerial imagery, thermal photography, and tracking vehicles in the farm or highways. They are Aeryon SR EO/Li MII, Aeryon SR IR EO/IR (for thermal imaging), Aeryon SR 3SHD (high definition multispectral camera), Lidar, etc.

Computer programs: The UAV package comes with several image processing and data analysis software. They are utilized in crop fields for specific purposes, such as application of fertilizer and pesticides. The software includes (a) vector video-processing platform and software, (b) advanced video analysis programing, (c) real-time video enlargement software, and (d) Aeryon map development software. The software package also includes "Aeryon Scout" to study the ground surface features and crop fields.

Spraying area: This is not used commonly to spray plant protection chemicals or for dusting powders

Volume: No data available

Agricultural uses: This quadcopter is useful in aerial imagery of land and soil resources, crop fields, and farm activity, including tracking farm vehicles and monitoring work progress. Crop scouting is done to obtain data regarding NDVI, biomass accumulation trends and spatial variations, leaf-chlorophyll (crop's N), thermal imagery to detect canopy water status, and need for irrigation quantum (Common, 2014). Aerial images and spectral data can also be used shrewdly to study the variations in disease/pest-attacked zones in the crop fields. Similarly, we can assess and map the flood or drought-affected zones. Natural vegetation monitoring is possible to get NDVI values and study the plant species diversity, using spectral signatures of various species.

Nonagricultural uses: They are surveillance of geological sites, mines, mining activity, ore dump sites, industrial locations, buildings, oil pipelines, electric lines, highways, and vehicle movement. We can track vehicles if a GPS-connected tracking device is fitted.

PLATE 3.27.1 SkyRanger by Aeryon Inc.
Source: Mr. Brad Young, Aeryon labs Inc., Waterloo, Ontario, Canada.

Useful References, Websites, and YouTube Addresses

Common, D. *UAVs Go Commercial, Take on Tasks from Industry to Farming*; 2014; pp 1–4. http://www.cbc.c/news/technology/UAVs-go-commercial-take-on-tasks-from-industry-to-farming (accessed June 10, 2015).

https://www.aeryon.com/aeryon-skyranger (accessed June 14, 2017).

https://www.aeryon.com/aeryon-skyranger/features (accessed June 14, 2017).

https://www.youtube.com/watch?v=YVtNi0CWGg8 (accessed June 14, 2017).

https://www.youtube.com/watch?v=7kQ196B13Go (accessed June 14, 2017).

https://www.youtube.com/watch?v=pD5YMnqy_dY (accessed June 14, 2017).

https://www.youtube.com/watch?v=vYi5-zHWCs8 (accessed June 14, 2017).

Name/Models: UX 401

Company and address: UAVision Aeronautics, Armazem, Portugal. UAVision Lda Casais da Arriota, 26 Bonabal, 2565-835 Ventosa, Portugal; e-mail: info@uavision.com; website: www.uavision.com; phone: +351 261 311 552

UAVision is UAV-manufacturing company located at Amerzem, Portugal. They produce models such as Spyro and Wingo. They also supply sensors for UAVs. The applications of their UAVs are in civil and military inspection, crowd monitoring, terrain scouting, monitoring oil and gas pipelines, and detection of flood or fire-related disasters. It has a role during adoption of precision agricultural methods.

UX 401 is a versatile quadcopter. This UAV has been designed to achieve maximum stability and vibration-free operation. It has automatic take-off and landing system. UX-401 provides assisted-piloted mode. Anyone with minimum training could operate the quadcopter, in a natural simple and secure way. Some of the features quoted by the UAV company are as follows.

- Highly stable autohovering
- Automatic take-off and landing
- Autostabilized payload (pan and tilt gimbal)
- Autonomous navigation—waypoint-based mission (up to 20)
- Bi-directional data-link (range up to 1 km)
- Analog or digital encoded video link

TECHNICAL SPECIFICATIONS

Material: The UAVs frame is made of carbon fiber and ruggedized plastic.

Size: length: 65 cm; width (wingspan): 65; height: 45 cm; weight: 5 kg MTOW

Propellers: It has four propellers. Each propeller has two plastic blades.

Payload: 2 kg. It includes gimbal, cameras, and CPU.

Launching information: It has autopilot, navigation, and landing software. It avoids obstacles.

Altitude: 50–100 m above crop canopy level if it is used for surveying crop canopy. Generally, it is flown at 300–400 m above ground level. The ceiling altitude is 1000 m above ground level.

Speed: 25–40 km h^{-1}

Wind speed tolerance: It withstands 25 km h^{-1} wind disturbance.

Temperature tolerance: −20 to 50°C

Power source: Electric batteries

Electric batteries: 14,000 or 28,000 mA h, two cells.

Fuel: No need for petroleum/gasoline fuel.

Endurance: 25 min

Remote controllers: The UAV keeps contact with the GCS for up to 1.0 km radius. The GCS iPad operates on Pixhawk software. It has override facility and can alter UAV's flight path, if necessary. It has high-speed communication with GCS via internet.

Area covered: 50,250 ha of aerial survey and imagery

Photographic accessories: Cameras: It has the usual complement of cameras such as visual rage (R, G, and B), NIR, IR, red edge and Lidar. The cameras are placed in a stabilized gimbal with 360° flexibility.

Computer programs: The UAV flight is controlled using "Pixhawk" or similar software. The images are processed using Pix4D Mapper or Agisoft's Photoscan or any other similar custom-made software.

Spraying area: This not a spraying UAV, perhaps could be modified to spray small quantities in localized areas.

Agricultural uses: Aerial survey and mapping of ground features and crop land. Crop scouting is done to observe seed germination, canopy growth and crop stand, collect data such as NDVI, water stress index, and chlorophyll content (i.e., crop's N status). Natural vegetation monitoring is also possible.

Nonagricultural uses: They are surveillance of industrial sites, buildings, railroads, highway, and city traffic, tracking vehicles and convoys, and transport of small parcels and hazardous material.

Useful References, Websites, and YouTube Addresses

https://www.uavision.com/quadcopter-uav-spyro/ (accessed Feb 24, 2017).

https://www.uavision.com/quadcopter-uav-spyro/ (accessed Feb 24, 2017).

https://www.uavision.com/products/ (accessed Feb 24, 2017).

https://www.uavision.com/applications-for-multi-rotor-and-fixed-wing-uavuas/ (accessed Feb 24, 2017).

https://www.uavision.com/quadcopter-uav-spyro/ (accessed Feb 24, 2017).

https://www.uavision.com/products/ (accessed Feb 24, 2017).

https://www.uavision.com/applications-for-multi-rotor-and-fixed-wing-uavuas/ (accessed Feb 25, 2017).
https://www.uavision.com/products/ (accessed Feb 25, 2017).
https://www.uavision.com/applications-for-multi-rotor-and-fixed-wing-uavuas/ (accessed Feb 25, 2017).
http://en.avia.pro/blog/quadcopter-ux-401-tehnicheskie-harakteristiki-foto (accessed Feb 25, 2017).
http://bcbin.com/product/sq6-quadcopter-ux-401/ (accessed Feb 25, 2017).
https://www.youtube.com/watch?v=Uakk-PjZBkk (accessed Feb 25, 2017).
https://www.youtube.com/watch?v=OY8uNEXkjec (accessed Feb 25, 2017).
https://www.youtube.com/watch?v=EiO86Iy2fhc (accessed Feb 25, 2017).
https://www.youtube.com/watch?v=Kj0IsXbuflo (accessed Feb 25, 2017).
https://www.youtube.com/watch?v=_PNugTbFRog (accessed Feb 25, 2017).

Name/Models: UX SPYRO

Company and address: UAVision Aeronautics, Amerzem, Portugal, UAVision Lda Casais da Arriota, 26 Bacabal 2565-835 Ventosa, Portugal; phone: +351 261 311 552; e-mail: info@uavision.com; website: www.uavision.com

UX SPYRO is supposedly a new copter UAV released to market. It has several commercial and agricultural users. It is a VTOL copter. It supposedly offers new dimension in aerodynamics, flight stability and control, imagery, and payload options. It is a noiseless UAV. It withstands disturbance by wind a bit better than other models. It has ultralight carbon gyro-stabilized gimbal. The gimbal has visible and thermal IR cameras that can be easily installed on SPYRO.

TECHNICAL SPECIFICATIONS

Material: It is made of carbon fiber and plastic.

Size: length: 80 cm; width (wingspan): 80 cm; height: 80 cm; weight: MTOW is 12 kg

Propellers: It has four propellers. Propellers are fitted with two plastic blades. Propellers are connected to brushless motors for propulsion.

Payload: 2–3 kg. It is made of cameras, gimbal, and CPU.

Launching information: It is an autolaunch UAV. Navigation can be fixed using software such as Pixhawk.

Altitude: It is 350 m above ground level

Speed: The cruising flight speed is 30 km h^{-1}; maximum flight speed is 40 km h^{-1}; maximum flight distance is 3.5 km.

Wind speed tolerance: It tolerates 20 km h^{-1} wind disturbance and turbulence

Temperature tolerance: −20 to 50°C

Power source: Electric batteries

Electric batteries: Litho-polymer batteries.

Fuel: No need for petroleum fuel

Endurance: 35 min

Remote controllers: The GCS has an iPad with necessary software for determining and preprograming UAV's flight path (e.g., Pixhawk). It also has override facility to change the flight path, waypoints, and photographic events. The GCS iPad keeps in touch with UAV for up to 2 km radius.

Area covered: It covers 50–250 ha h^{-1}

Photographic accessories: Cameras: It has a visual, multispectral, NIR, IR, and red-edge camera plus a Lidar pod.

Computer programs: The images relayed by the UAV are processed using Pix4D Mapper or Agisoft's Photoscan or similar custom-made software.

Spraying area: Not a spraying UAV

Agricultural uses: Aerial survey and mapping of land resources and crops. Crop scouting is possible to monitor crop stand and growth rate, collect NDVI data, leaf chlorophyll (i.e., crop's N status), canopy and air temperature (i.e., crop's water stress index), grain maturity, etc. Natural vegetation monitoring is also possible.

Nonagricultural uses: Surveillance of mines, ore movement, industrial installations, buildings, railways, city traffic, etc., tracking vehicles and convoys, and transport of parcels and hazardous chemicals, etc.

Useful References, Websites, and YouTube Addresses

https://www.uavision.com/quadcopter-uav-spyro/ (accessed Jan 27, 2017).

https://www.uavision.com/products/ (accessed Jan 27, 2017).

https://www.uavision.com/applications-for-multi-rotor-and-fixed-wing-uavuas/ (accessed Jan 27, 2017).

https://www.uavision.com/applications-for-multi-rotor-and-fixed-wing-uavuas/ (accessed Jan 27, 2017).

https://cineaerials.com/my-product/quadcopter-ux-spyro/ (accessed Jan 27, 2017).

https://www.youtube.com/watch?v=Uakk-PjZBkk (accessed Jan 27, 2017).

3.2.2.2 HEXACOPTERS

Name/Models: AgFalcon

Company and address: Falcon UAV Australia, Dr. Phil Lyoons, 2/10 Thea Avenue, Baldwyn North, VIC 3104, Australia; mobile: +61 0416029896; e-mail: phil@falconuav.com.au; website: www.falconuav.com.au

The AgFalcon UAV system is designed to instantly provide an eye-in-the-sky image of any type of "plant matter." A closer inspection of any paddock section or live stock is possible because of its ability to hover. Using the NIR camera, the AgFalcon will autoland whenever low battery power is detected. It uses "Return Home function."

AgFalcon system includes durable hexacopter copter flight controller, flight planning software, autopilot firmware and software, lithium–polymer battery packs and smart fast charger, NDVI-modified camera, optional video kit, flight instructions manual, and spare parts kit (Plate 3.28.1). On-site training is also available.

TECHNICAL SPECIFICATIONS

Material: This UAV is made from advanced carbon fiber composites.

Size: length: 90 cm; width (rotor to rotor): 90 cm; height: 45 cm; weight: 7.2 kg:

Propellers: It has six propellers. Clip-on/clip-off carbon fiber props give quiet robust flight.

Payload: 5–10 kg

Launching information: This is an autolaunch VTOL UAV. It has facility for preprograming flight path using software such as "Pixhawk." The UAV returns to point of launch, in case of emergency.

Altitude: It flies at 50–100 m for spraying agricultural fields. However, the ceiling is 500 m above ground level.

Speed: 15–35 km h^{-1}

Wind speed tolerance: 20 km h^{-1}

Temperature tolerance: −20 to 55°C

Power source: Lithium–polymer electric batteries

Electric batteries: 24,000 mA h, two cells.

Fuel: No need for petrol/diesel

Endurance: 30 min per flight

Remote controllers: The GCS has radio telelink and iPad. They keep in touch with UAV for up to 5 km radius. The GCS computers have override facility and can modify the UAV's flight path, if required.

Area covered: This is a precision agriculture UAV system designed and built for Australia. The AgFalcon hexacopter is a tough agricultural tool, capable of scanning between 50 and 120 ac of crop, orchard, or vineyard, in 25 min of flight.

Photographic accessories: Cameras: It has spectral camera, mounted underneath, taking images in the NIR spectrum; these can be used to create NDVI maps to pinpoint problem areas, or to show those plants that are very healthy and do not need any chemical treatment. The AgFalcon is ideally suited to operate on smaller areas, those that are surrounded by trees, hills and can benefit from this UAV's ability to hover.

With onboard high definition videocamera option, the "AgFalcon" can transmit live images back to the laptop. Using laptop, we can set a flight path to observe farm infrastructure, gates, water holes, fence lines, etc. With a flight time of up to half an hour in video mode, the AgFalcon can be an "eye-in-the-sky."

Computer programs: The GCS computers process the ortho-images received from UAV's CPU using software such as Pix4D Mapper or Agisoft's Photoscan.

Spraying area: It can spray 50–100 ac, if used for spraying.

Agricultural uses: They are crop scouting for detecting seedling establishment and canopy development. This UAV is used to collect data such as NDVI, leaf chlorophyll, and water stress index of crops. It is also used to monitor farm vehicles and animals. Natural vegetation monitoring is also possible.

Nonagricultural uses: It includes surveillance of mines, industrial installations, buildings, public places, events, railways, city traffic, etc.

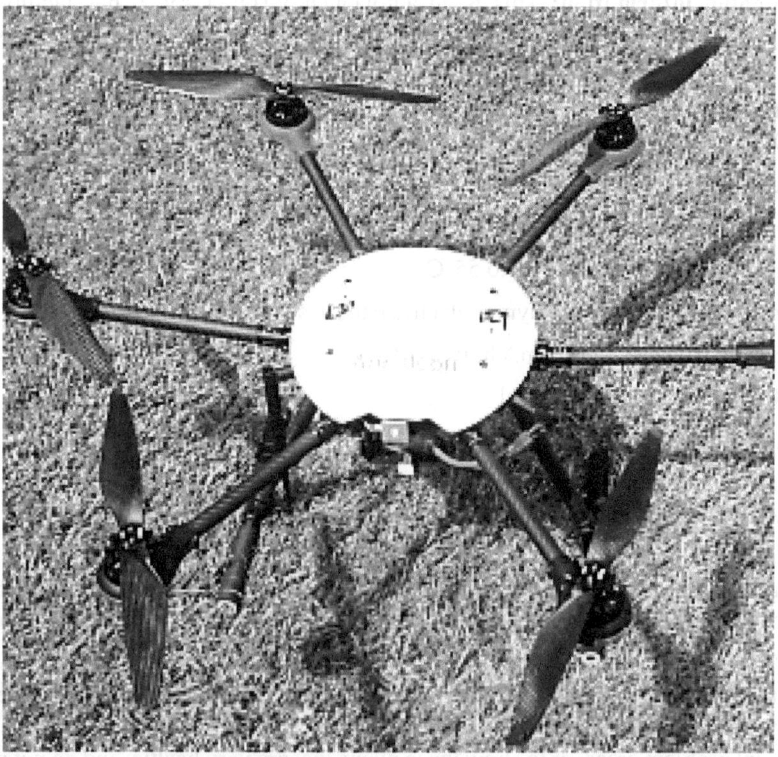

PLATE 3.28.1 AgFalcon—a hexacopter.
Source: http://www.falconuav.com.au/; Falcon UAV, Baldwyn North, Victoria, Australia.

Useful References, Websites, and YouTube Addresses

http://www.falconuav.com.au/ (accessed July 3, 2017).
http://www.falconuav.com.au/uav-products/agfalcon-quad-copter (accessed July 3, 2017).
http://www.falconuav.com.au/uav-agriculture-videos (accessed July 3, 2017).

Name/Models: ATTA Copter

Company and address: LAPCAD Engineering, Inc., 8305 Vickers Street, Suite 202, San Diego, CA 92111, USA; Tel.: +1 858 467 1947; websites: info@lapcad.com, http://lapcad.com/wordpress/

LAPCAD Engineering Inc. offers a variety of services to the US Government, the Department of Defense, and the Commercial and Public Sectors, including consulting, design, fabrication, and software. They have developed several UAVs. In addition to providing support for the aerospace sector, they also provide structural analysis of complex building steel structures, for the benefit of the civil engineering community.

ATTA Cargo HEXA Copter is a VTOL UAV. The rotors are connected to electric motors. Each motor delivers 2 hp, for a total of 12 hp. Max thrust estimated at 64–80 lb.

TECHNICAL SPECIFICATIONS

Material: It is made of carbon fiber, light aluminum, and toughened plastic.

Size: length: 40 cm; width (rotor to rotor): 40 cm; height: 40 cm; weight: 8.0 kg

Propellers: It has six propellers, each one attached with two plastic blades. The propellers are attached to brushless motor for propulsion.

Payload: 3–8 kg

Launching information: This is an autolaunch, navigation, and VTOL UAV.

Altitude: 50–300 m above ground level. However, the ceiling is 1000 m above ground level.

Speed: 15–35 km h^{-1}

Wind speed tolerance: 20 km h^{-1} wind disturbance

Temperature tolerance: −20 to 65°C

Power source:

Electric batteries: Electric batteries.

Fuel: No need for petroleum fuel

Endurance: 35 min

Remote controllers: The GCS has an iPad or mobile to control the flight path of the UAV. The radiolink operates for up to 15 km radius. The UAV returns to point of launch in times of emergency.

Area covered: 5–10 km distance

Photographic accessories: It has a full complement of photographic equipment such as visual, NIR, IR, and red-edge cameras.

Computer programs: The GCS iPad adopts software such as Pix4D Mapper or Agisoft's Photoscan to process the orthomosaics sent by the UAV's CPU.

Spraying area: Generally, it is not a spraying UAV but can be modified by adding a tank and sprayer bar.

Agricultural uses: They are aerial survey of land and soil resources, cropping systems, irrigation channels, etc. It is used to conduct crop scouting aimed at recording data such as seed germination, seedling establishment, canopy growth, leaf chlorophyll (i.e., crop's N status), canopy and air temperature (i.e., crop's water stress index), grain maturity, etc. Natural vegetation mapping is also possible.

Nonagricultural uses: Surveillance of mines, transport vehicles, etc.

Useful References, Websites, and YouTube Addresses

http://lapcad.com/?page_id=6 (accessed June 1, 2017).
http://lapcad.com/wordpress/ (accessed June 1, 2017).
https://www.youtube.com/watch?v=72vj69Hluwk (accessed June 1, 2017).

Name/Models: Firefly

Company and address: Ascending technologies Gmbh.; Konrad-Zuse-Bogen 4, 82152 Krailling, Germany; Tel.: +49 89 89556079 0; fax: +49 89 89556079 19; e-mail: team@asctec.de; website: www.asctec.de

Ascending technologies Gmbh. is a German avionics company. This UAV company designs, develops, tests, and makes large-scale production of small aerial UAVs. It was initiated in 2005. It offers a range of copter UAVs capable of surveillance, aerial imagery, and spraying formulations. These UAV models such as Firefly, Falcon, or Humming Bird are traceable in different regions of the world. They conduct tasks such as aerial surveillance of civilian installations, dams, highways, farm structures, and crop fields. Soon, they could be serving global agriculture in a perceptible way.

AscTec's "Firefly" is a lightweight micro-UAV. It is a hexacopter with multiple capabilities in farm, and in civilian locations (Plate 3.29.1). Firefly package comes with a range of computing capacities that help farmers and others to process the digital data and images. Then, analyze them using different software. It leads them to more accurate and useful decisions, in applying inputs to farms, also in taking appropriate measures in the civilian activities. It has high-performance onboard computer in the general CPU, in the fuselage. It has several safety measures added.

They help to achieve smooth flight path and aerial imagery. Sensor outputs are checked immediately. It has a series of sensors to obtain aerial images and thermal data, from both indoor and outdoor crop field locations. This hexacopter may cost US$ 7–10,000. The full package consists of platform, sensor accessories, and GCS instruments.

TECHNICAL SPECIFICATIONS

Material: AscTec's Firefly is made of a composite of carbon fiber, rugged foam, plastic, and light aviation aluminum. It can be packed and transported easily in a suitcase. It can be launched in just few minutes.

Size: length: 60.5 cm; width (rotor to rotor): 66.5 cm; height: 16.5 cm; weight: MTOW is 1.6 kg including payload.

Propellers: It has six propellers. Each propeller is made of two plastic blades. The rotors are connected to brushless motors for propulsion.

Payload: 600 g made of gimbal, sensors, and CPU.

Launching information: It is an autolaunch aerial robot. Its navigation is predetermined, using software such as Pixhawk or similar custom-made computer software. This UAV has computer-guided flight vibration stabilization, gyro-stabilization, and flight path correction facility.

Altitude: The hexacopter hovers at just 1.0 m above ground or crop canopy. Its operational height is 50–300 m above ground level. The ceiling altitude is 4500 m above sea level.

Speed: 15 m s^{-1} without payload; it is 8 m s^{-1} with payload.

Wind speed tolerance: 20 km h^{-1} disturbance

Temperature tolerance: −20 to 55°C

Power source: Electric batteries

Electric batteries: 24,000 mA h, two cells lithium–polymer

Endurance: 12–14 min per flight

Remote controllers: The hexacopter, "Firefly," has an onboard CPU, with third-generation Intel Core processor. The ground station has telemetric connection, using 2.4 GHz, XBee link and Wi-Fi on board. The UAV can be utilized in autopilot mode or using manual mode with a remote controller (joystick). It has several autocorrection facilities in-built in the software, particularly to avoid obstacles, make flight corrections, smooth turning, correcting errors in programing, etc. The UAV lands at the point of launch in case of emergency, fuel shortage, or errors in programing. This UAV has GPS connectivity throughout flight and the images are all tagged with GPS coordinates. A special feature of this hexacopter is that the GCS and UAV's CPU has facility for swarming. It has facility for "inter-UAV communication" during rapid aerial survey and imagery.

Area covered: It covers 50–250 ac day^{-1}

Photographic accessories: Cameras: Hexacopter hosts a gimbal known as "AscTec's master mind" + "Kamerahalter." The gimbal holds sensors for 2D and 3D imagery, high-resolution videographic camera, laser scanner, laser detection and ranging (Lidar) device, and traffic vehicle-tracking device.

Computer programs: All important sensor output, that is, the orthomosaics received from UAV's computers/sensors are processed, using custom-made software, at the ground station computers (iPad).

Spraying area: This is not a spraying copter. Its payload area and permissible weight are too small.

Volume: Not applicable

Agricultural uses: This is a micro-lightweight agricultural UAV (Plate 3.29.1). It is launched rapidly to obtain aerial images and digital about NDVI, biomass distribution, canopy growth rate, leaf chlorophyll, crop's N status, canopy temperature (thermal imagery), and water stress index of the crop. Crop scouting is done to detect and map diseases, pests, and weed flora in a crop field. Natural vegetation monitoring for obtaining NDVI values and detecting plant species diversity using specific spectral signatures of canopy is possible.

Nonagricultural uses: It includes surveillance of natural features on ground, geological sites, mining zones, ore dumps and ore movement, oil pipelines, electric lines, dams, buildings, rail roads, highways and tracking vehicle movement, etc.

PLATE 3.29.1 Firefly.
Source: Mr. Matthias Beldzik and Prof Guido Morganthal, Ascending Technologies Gmbh., Kreilling, Germany.

Useful References, Websites, and YouTube Addresses

http://www.ascetec.de/en/asctec-research-uav/ (accessed Sept 7, 2017).
http://wiki.asctec.de/display/AR/AscTec+Firefly (accessed Sept 7, 2017).
https://www.youtube.com/user/AscTecVideos (accessed Sept 7, 2017).
https://www.youtube.com/playlist?list=PLSl4B94nu1UwKXGqqJFepiA6HhSAX1VT1 (accessed Sept 7, 2017).
https://www.youtube.com/watch?v=NQB86hxVPZY&list=PLSl4B94nu1UwKXGqqJFepiA6HhSAX1VT1&index=15 (accessed Sept 7, 2017).

Name/Models: Hercules HL6, Hercules HL10, and Hercules HL20

Company and address: Homeland Surveillance and Electronics (HSE) LLC., Denver, CO, USA; phone: +1 309 361 7656; e-mail: tsanders@hse-uav.com; website: http://www.hse-uav.com/

Homeland Surveillance and Electronics Inc. (HSE) is a UAV and UAV-related service provider. It manufactures several models of UAVs. They comprise both fixed-wing and rotor copters. These UAV models, it seems, were results of over 10 years of R&D by the company. The HSE was started in 2009 by Mr. Dave Sanders and his entourage. They have industrial facility in Denver, Colorado, USA, that produces UAVs.

"Hercules" heavy lift UAVs are of great use during farming and other types of activity. There are three main models considered here. They are Hercules heavy lift UAV—HL6, HL10, and HL 20. However, the data pertains to only "HL20′ multirotor UAV." HL 20′ is a long endurance six rotor UAVs. It is capable of aerial photography, including close-up shots. It carries over 20 kg payload. It could also be spray material if adopted for use during plant protection procedures (see Plate 3.30.1).

TECHNICAL SPECIFICATIONS

Material: HL20 is made of steel, high-strength carbon fiber, foldable arms, and propellers made of plastic.

Size: length: 120 cm; width (rotor to rotor): 120 cm; height: 35–60 cm based on the model; weight: empty weight of HL 6 is 8 kg, HL10 is 8 kg, and HL 20 is 14 kg.

Propellers: Hercules UAV has 6–8 propellers based on Hercules model HL6, HL10, or HL20. Propellers are 18.0 cm long. They are made of plastic in case of HL 20.

Payload: It is 5–10 kg depending on the model.

Launching information: The multicopter has autolaunch software. Navigation and landing are controlled by using Pixhawk software. The copter can be launched from anywhere in the field. It is a VTOL UAV. It just needs 16 × 16 ft clearance on the ground for launch or landing.

Altitude: It hovers or flies at very low heights over the crop canopy during hyperspectral imagery and pesticide application. Pesticide drift is avoided effectively because it flies close to the crop canopy. Operating altitude ceiling is 1200 m above ground level.

Speed: Maximum cruise speed is 60 km h^{-1}.

Wind speed tolerance: Cross winds of 25 km h^{-1} are easily tolerated, by the UAV HL 20.

Temperature tolerance: −10 to 55°C

Power source: Electric batteries

Electric batteries: 16,000 mA h, lithium–polymer Batteries. Maximum power is 1200 W.

Fuel: Not an IC engine-fitted UAV. No need for petrol or diesel

Endurance: 30 (HL6)–60 min (HL20) per flight, based on model

Remote controllers: The GCS communication is conducted through 2.4 GHz radiocontrol. It has a single antenna—92 dbm. Radio frequency bandwidth is 1.25/2.5 MHz or 6/7/8 MHz. Ground control iPad receives video-images. The data transmission is protected, using encryption.

Area covered: It (HL20) covers about 200 ha per flight. Flight range is 1.83 miles per flight.

Photographic accessories: Cameras: The UAV carries visual, multispectral, hyperspectral, video, NIR, IR, and red-edge cameras.

Computer programs: Pix4D Mapper or Agisoft's Photoscan is utilized to process raw imagery received from the UAV.

Spraying area: Spraying accessories could be attached, if needed.

Agricultural uses: They are aerial survey of agricultural terrain, land resources, soil types, water resources, crops, and cropping systems. This UAV conducts crop scouting for seedling and canopy establishment, weed infestation, diseases, and pests. Its sensors collect data about leaf area and leaf chlorophyll, that is, crop's N status, canopy, and air temperature, that is, crop's water stress index and grain maturity. Natural vegetation monitoring and identifying species diversity using spectral signatures are also possible. Data banks of spectral signatures of different crop species are required.

Nonagricultural uses: It includes surveillance of geological sites, water bodies, mines, industries, buildings, railways, city traffic, vehicle tracking, etc.

PLATE 3.30.1 Hercules UAV (Heavy lift UAV).
Note: Top and bottom left: Copter UAV in flight. Bottom right: UAV monitoring electric line.
Source: Mr. Terry Sanders, VP Marketing and Innovations HSE LLC., Denver, CO, USA.

Useful References, Websites, and YouTube Addresses

http://www.hse-uav.com/ (accessed Apr 13, 2017).
http://www.hse-uav.com/hercules_heavy_lift_uav.htm (accessed Apr 13, 2017).
https://www.youtube.com/watch?v=8JddjmDA9AE (accessed Apr 13, 2017).
http://www.uavcropdustersprayers.com/ (accessed Apr 13, 2017).

Name/Models: Matrice 600

Company and address: SZ DJI Baiwang Technology Company Ltd., 14th Floor, West Wing, Skyworth Semiconductor Design Building, No. 18 Gaoxin South 4th Ave., Nanshan District, Shenzhen 518057, China; phone: +86 (0)755 26656677; e-mail: tsanders@hse-uav.com; website: https://www.dji.com/products

DJI of Shenzhen, China produces a wide range of copters. Its models are currently among the popular UAVs worldwide. They are handy, often small and versatile, with great applications in civil and agricultural sectors.

Matrice 600 is a hexacopter. It is usually held in ready-to-use condition (Plate 3.31.1). The flight is controlled by a custom-made remote controller or a GCS iPad. It is more frequently used to obtain aerial images, maps and spectral data about land resources, crop fields, installations, etc. It is supposedly a powerful UAV. It integrates with several other autonomous vehicles on the ground. Therefore, it is preferred by agronomists. It is also used for recreation.

TECHNICAL SPECIFICATIONS

Material: It is made of light metal alloy, carbon fiber, ruggedized foam, and plastic. It has foldable platform and propellers for easy portability.

Size: length: 168 cm; width (rotor to rotor): 100 cm; height: 75 cm; weight: 9.6 kg. Matrice 600 is a relatively heavier UAV, when compared with similar models produced by DJI Ltd., Shenzhen, China. But overall, this UAV model is classified as medium weight UAV.

Propellers: It is a hexacopter. It has six rotors with two plastic blades each.

Payload: 6.0 kg

Launching information: It is an autolaunch, autopilot, and landing UAV.

Altitude: The ceiling is 2500 m (8200 ft)

Speed: Maximum speed is 18 m s^{-1}. Ascent speed is 5 m s^{-1} and descent speed is 3 m s^{-1}.

Wind speed tolerance: 20 km h^{-1} disturbance

Temperature tolerance: −10 to 40°C

Power source: Electric batteries 22.2 V. Battery charger 26.3 V

Electric batteries: Lithium–polymer 6S.

Fuel: Not an IC engine-fitted UAV

Endurance: 35 min per take-off

Remote controllers: The UAV is GPS compatible and takes commands from GCS computers accurately. It has facility such as altitude hold, position hold, autonomous flight and navigation, automatic landing, collision avoidance, and return-to-launch-spot, in times of emergency. Remote controller has a mobile (DJI GO). The mobile is used to work out flight path and control the UAV's movement and camera shutter. The GCS also has Android 4.1 for flight control, image recovery, its storage, and analysis. Matrice 600 can also be powered by A3 flight controller. It can be upgraded with D-RTK GNSS system to attain cm level accuracy. This accuracy is useful, while hovering over crop canopy and detecting, and disease/pest attack zones. The GCS telemetric connections operate, for up to 15 km radius.

Area covered: It is 50–250 ha per flight

Photographic accessories: Cameras: The DJI's Matrice 600, like other models, is attached with custom-made cameras, for example, Zenmuse X5 for visual imagery, videos, and close-up shots. It has Zenmuse X5S for thermal imagery and detection of moisture on the soil surface and crop's canopy. The visual and IR cameras offer views, images, and videos of crops canopy, leaf, and panicles. The spectral differences in crops species, disease affected, and healthy plants are utilized, advantageously.

Computer programs: The aerial images are processed, using software such as Pix4D Mapper or Agisoft's Photoscan.

Spraying area: This is a heavy lift UAV. Therefore, platform can be fitted with pesticide/fertilizer spray tanks and nozzles and spray equipment. The computer software must decipher digital data and then supply liquid (pesticide) at variable rates.

Volume: Data not available

Agricultural uses: It has several applications in the realm of agricultural crop production and maintenance of farm installations and vehicles. It is used for crop scouting. The crop scouting is done to obtain data for plant physiological traits, such as growth rate, NDVI, and leaf-chlorophyll (leaf/crop-N status). It is used to record canopy and ambient temperature, which help to determine crop's water stress index and need for irrigation. It is also used to obtain close-up shots of crop canopy to detect and identify crop disease/pest attack. It is used to survey and map drought or flood-affected regions in a farm. Natural vegetation monitoring for NDVI and recording plant species diversity in a region is also possible. This is a UAV with capability for aerial spray of plant protection chemicals and fertilizers.

Nonagricultural uses: It includes surveillance of industrial sites, buildings, highways and city traffic, tracking vehicles, monitoring mines and mining activity, electric lines, oil pipelines, etc.

PLATE 3.31.1 Matrice 600 and gimbal with a camera.
Source: Mr. Adam Najberg, DJI Inc., Shenzhen, China.

Useful References, Websites, and YouTube Addresses

http://www.dji.com/matrice600 (accessed June 20, 2017).
https://www.dji.com/matrice600-pro (accessed June 20, 2017).
https://www.youtube.com/watch?v=MZ2vCfQ4aJs (accessed June 20, 2017).
https://www.youtube.com/watch?v=LKbBratqPNs (accessed June 20, 2017).
https://www.youtube.com/watch?v=jayVMmvLxOA (accessed June 20, 2017).
https://www.youtube.com/watch?v=o2Ko1rtJrrk (accessed June 20, 2017).
https://www.youtube.com/watch?v=3aaxcQbLq54 (accessed June 20, 2017).

Name/Models: Vulcan UAV/Raven, Black Widow

Company and address: Vulcan UAV Ltd. Ground floor Building 11, Vantage Point Business Village, Mitcheldean, Gloucestershire GL 17 OSZ, UK; phone: +44 (0) 1989 555 025; e-mail: info@vulcanuav.com; website: http://vulcanuav.com/

Vulcan UAV Ltd. is a UAV-manufacturing company. They have offered different models of UAV for the past 17 years. They also produce UAV and its accessories to suit a situation and offer custom-made machines. Important UAVs offered by this company are "Harrier Industrial," "Black widow," "Raven," and "Airlift." Ranger is new model yet to hit the market. Vulcan multirotor UAVs have wide range of applications. Following is a list: photogrammetry, crop spraying, thermal imaging, structural imaging, building surveillance.

According to the company, Raven is a heavy lift workhorse (Plate 3.32.1). It is a highly reliable and well-proven platform. It is ideal for heavy and expensive payloads from movie cameras to Lidar. The Raven includes a well-proven anti-vibration system for payloads of 10 kg or more, a 400A power distribution board, DayBright LED Navigation lights with controller, and servoless retracts.

TECHNICAL SPECIFICATIONS

Material: It is made of carbon fiber, 6061 high-grade light aluminum, and plastic. The composite is ruggedized to withstand harsh weather and landing impact.

Size: length: 120 cm; width (rotor to rotor): 35 cm; height: 55 cm; weight: 8.0 kg

Propellers: It has six propellers. Each propeller attached with two blades made of toughened plastic. Each rotor is attached to brushless electric motor.

Payload: It is a heavy lift UAV. It ordinarily carries 10 kg payload. It includes heavy cameras and parcels.

Launching information: It is an autolaunch, navigation, and landing UAV. It is a VTOL and needs very small area of 10 × 10 m area to lift off with luggage.

Altitude: It is used at 100–300 m above ground. However, ceiling is 1000 m above ground level.

Speed: 10–35 km h^{-1}

Wind speed tolerance: 25 km h^{-1} wind disturbance

Temperature tolerance: −20 to 50°C

Power source: Electric batteries

Electric batteries: 24,000 mA h, two cells

Fuel: No need for petroleum fuel

Endurance: 45 min

Remote controllers: The GCS has an iPad to program the UAV's flight path. It adopts a software such as "Pixhawk." The GCS computer has facility to modify the flight path, waypoints, and photographic events, if needed.

Area covered: It covers 15–20 km linear distance or 250 ha.

Photographic accessories: Cameras: The vibration resistant gimbal includes a range of cameras. The UAV carries a heavy camera and few others such as visual, NIR, IR, red-edge, and Lidar sensors.

Computer programs: The images relayed by the UAV CPU are processed using Pix4D Mapper or Agisoft's Photoscan.

Spraying area: Not a spraying UAV.

Agricultural uses: They are aerial survey and photography of crop belts and individual farm land, crop scouting for variations in growth, collecting NDVI values, detecting disease/insect pest damage, spraying plant protection chemicals if carrying a pesticide tank, and sprayer. Natural vegetation monitoring is also possible.

Nonagricultural uses: It includes military and civil site inspection, security and land enforcement, search and rescue operations, surveillance of mines, industries, buildings, highways, and railroads. It is very useful in transporting heavier payload of 10–30 kg weight.

PLATE 3.32.1 Raven.
Source: Dr. Alex Hardy, Director, Vulcan UAV Ltd., Gloucestershire, Great Britain.

Useful References, Websites, and YouTube Addresses

http://vulcanuav.com/aircraft/ (accessed July 31, 2017).
http://vulcanuav.com/#1#applications (accessed July 31, 2017).
http://vulcanuav.com/applications/#crop (accessed July 31, 2017).
https://www.youtube.com/watch?v=hSq2Pb8dZ54 (accessed July 31, 2017).
https://www.youtube.com/watch?v=TncgB1dcM7k (accessed July 31, 2017).
https://www.youtube.com/watch?v=54JsqWysk6M (accessed July 31, 2017).

Name/Models: Y6 Multirotor Copter UAV

Company and address: Aerovision Unmanned Aerial Solutions Pty. Ltd., 43 Sea Cottage Drive, Noordkoek, Cape Town 7979, South Africa; Tel.: 0825641809, +27 825641809 (international); e-mail: info@aerovision-sa.com; website: https://www.aerovision-sa.com/

Aerovision is a UAV company that offers several UAV models such as AltiMapper, Y6, etc. They are small UAVs. They are versatile. They carry wide range of facilities related to aerial surveillance and assessment. The UAVs produced by Aerovision Inc. are used in mining projects, residential project management, monitoring fire damage to forests and reporting, and golf estate management. These UAVs are of great utility during crop production activity. They are used in assessing crop health, disease/pest incidence, nutritional status, assessment of crop's N status and water requirements, etc.

"Y6" is a small hexacopter produced by Aerovision Inc. It is a VTOL with crescent-shaped propeller blades. It is moderately priced hexacopter affordable to many farmers, industrial units, etc. It is a small UAV that can be packed in a suitcase and transported. It can be assembled and launched in few minutes (Plate 3.33.1).

TECHNICAL SPECIFICATIONS

Material: It is made of carbon fiber, ruggedized foam, and plastic.

Size: length: 90 cm; width (rotor to rotor): 90 cm; height: 45 cm; weight: 7.2 kg

Propellers: It has propellers. They are fitted two plastic propellers each. Propellers are connected to brushless electric motors for propulsion.

Payload: 2 kg

Launching information: It is an autolaunch, navigation, and landing UAV. It has "Pixhawk" software to control flight path.

Altitude: It is 50–100 m above ground for crop scouting.

Speed: 35 km h^{-1}

Wind speed tolerance: 20 km h^{-1} wind disturbance can be tolerated

Temperature tolerance: −20 to 50°C

Power source: Electric batteries

Electric batteries: 28 mA h

Fuel: No need for petroleum fuel.

Endurance: 20 min

Remote controllers: The GCS has an iPad or mobile to control the UAV's activities. The iPad has Pixhawk software. It has override facilities, in case flight path is to be modified.

Area covered: 50–200 ha per flight

Photographic accessories: Cameras: It has a set of visual, NIR, IR, red-edge, and Lidar sensor.

Computer programs: The images relayed by the UAV are processed using software such as Pix4D Mapper or Agisoft's Photoscan.

Spraying area: Not a spraying UAV

Agricultural uses: It includes aerial survey and mapping of natural resources, terrain, soil types, and cropping systems. The UAV is used to scout crop fields for seedling and canopy growth rate, leaf chlorophyll (i.e., crop's N status) canopy and air temperature (i.e., water stress index), grain maturity, etc. Natural vegetation monitoring is also possible.

Nonagricultural uses: It includes surveillance of mines, industrial sites, buildings, railways, city vehicular traffic, coastal activities, etc. It can also be used to pick parcels and deliver. It can safely transport hazardous goods such as chemicals, explosives, etc.

PLATE 3.33.1 Y6 copter UAV.
Source: Dr. Ian Freemantle, Aerovision, Noerdhoek, South Africa.

Useful References, Websites, and YouTube Addresses

https://www.aerovision-sa.com/ (accessed Aug 27, 2017).
https://www.aerovision-sa.com/portfolio (accessed Aug 27, 2017).
http://diyUAVs.com/forum/topics/y6-and-hexa-equal-lift (accessed Aug 27, 2017).
https://www.youtube.com/watch?v=PhD-Zu7LQ3o (accessed Aug 27, 2017).
https://www.youtube.com/watch?v=a1DtbhbxfGc (accessed Aug. 27, 2017).
https://www.youtube.com/watch?v=Oz-CiJBvp2Q (accessed Aug 27, 2017).
https://www.youtube.com/watch?v=PU5ZCCGDNTk (accessed Aug 27, 2017).

3.2.2.3 OCTOCOPTERS

Name/Models: Falcon 8

Company and address: Ascending Technologies Gmbh.; Konrad-Zuse-Bogen 4, 82152 Krailling, Germany; Tel.: +49 89 89556079 0; fax: +49 89 89556079 19; e-mail: team@asctec.de; website: www.asctec.de

Ascending Technologies Gmbh. situated in Kreilling, Germany is a UAV-related company. It was founded in 2005. It offers a few different types of copters. They are useful in accomplishing tasks such as aerial surveillance and crop analysis in farms. Falcon 8 is among popular model sold in different continents. They are being regularly used to monitor rivers, dams, irrigation channels, water resources like lakes, and impact on natural vegetation.

"Falcon 8" model produced by the Ascending Technologies Gmbh. is a v-type octocopter (Plate 3.34.1). Its production began in 2009. It has been tested and used widely in Germany and many other nations. It is used to surveillance public installations, agricultural farms, and crop production fields. It has three flight modes, namely, GPS guide mode, altitude mode, and manual mode. We can switch the modes, if necessary. It has computer memory to record and retrieve entire flight and aerial photographic data, at any time. The payloads that include cameras are quickly inter-changeable, even within a flight program. The sensor data can be verified and analyzed immediately by the GCS computer. The UAV with all the accessories may cost US$ 7–12,000 (February 2017 price level).

TECHNICAL SPECIFICATIONS

Material: It is made of a composite having light avionics aluminum, carbon fiber, rugged foam, and toughened plastic. The entire UAV is usually packed in a suitcase or a rucksack. It includes batteries, battery charger, and a remote controller (joystick or mobile).

Size: length: 77 cm; width (rotor to rotor): 82 cm; height: 12.5; weight: MTOW is 2.3 kg

Propellers: It has eight propellers. They are made of two plastic blades each. The rotors are attached to brushless motors for propulsion. These are low noise motors. They do not leave sound signatures.

Payload: 0.8 kg

Launching information: The UAV has autolaunch, navigation (e.g., Pixhawk software), and landing facility. It has flight documentation and flight data recorder. It records each flight path, waypoints selected and commands, precisely. In case of emergency or errors in programing, the UAV has fail-safe mechanism. It returns to point of launch automatically.

Altitude: Operational height is 5 m above ground or crop canopy to 300 m above canopy. Ceiling is 1000 m above ground level.

Speed: 16 m s^{-1}

Wind speed tolerance: 20 km h^{-1} disturbance

Temperature tolerance: −5 to 35°C

Power source: Electric batteries

Electric batteries: 6250 mA h, lithium–polymer batteries.

Fuel: This is not an IC engine-fitted UAV, so does not require gasoline/diesel

Endurance: 12–22 min based on payload and flight plan

Remote controllers: The remote-control station (iPad/mobile) receives data via radiolink (2 × 2.4 GHz, 10–63 mW). The flight control is done, using a customized software. The UAV can be controlled about flight path, speed, and aerial photography. It is done using different modes, such as autopilot, GPS mode, and Hohen mode.

Area covered: It covers 50–250 ha of aerial survey/photography per take-off when endurance is 20 min.

Photographic accessories: Cameras: The set of digital cameras fitted to Falcon 8 includes Sony Alpha 7R (36 MP); Sony Alpha 6000 full high-definition visual camera (24 MP); Panasonic Lumix DMC TZ71 12 MP zoom camera; Panasonic IR camera, FLIR TAU 2 640+; Infra-rotkamera FLIR TAU 640 × 512 Pixel plus AscTec IR Rohdatenlogger; and Camcorder for RGB videography (Sony Camcorder HDR-PJ810E).

Computer programs: It has camera halter (gimbal) that is gyro-stabilized. It allows 360° focusing of points of interest. It hosts imaging and detection cameras such as "GeoExpert" to study ground installations, "InspectionPro" to photograph and inspect the activities, "GeoExpert" to image natural resources, geographical features, land surface, and soil characteristics and crops. "VideoExpert" is used to videograph the ground features and activities in public places. The cameras have automatic mode based on predetermined shutter activity. The camera shutters could also be controlled manually, using GCS telelink.

Spraying area: It is a small copter without fluid tank as payload. It is not a spraying UAV.

Volume: Not applicable

Agricultural uses: Falcon 8 is useful to conduct aerial surveys of land resources, soil types, crops, and water resources. Such aerial surveys can help in planning development of crop fields, demarcating them based on soil types, in laying irrigation channels and pipelines. Time tables for plowing, planting, interculture, irrigation, and fertilizer application are all done, using aerial images, got using UAVs. Falcon 8 has been used to keep vigil on crop fields, pastures, and cattle. It has been utilized to observe crop stand, seedling emergence, and gaps. Crop scouting to obtain data about variations in NDVI values, canopy growth rate, leaf chlorophyll status (crop's N content), canopy, and ambient temperature (thermal imagery) to detect crop's water status and decide on irrigation schedule, etc., has also been accomplished. It is also useful in detecting disease, pest, or weed infestation in crop fields. Natural vegetation monitoring to obtain data about biomass index, NDVI, and species diversity is a clear possibility (see Plate 3.34.1; Krishna, 2018; Von Bueren et al., 2015).

Nonagricultural uses: Surveillance of natural geographic features such as volcanoes, glaciers, rivers, storms, floods and their impact, geological features, mines, mining activity, ore dumps and their transport, oil pipelines, electric lines, rail roads, roadways and vehicular traffic, public places, buildings and events, etc.

PLATE 3.34.1 Falcon 8—an octocopter.
Source: Mr. Mathias Beldzik and Prof Guido Morganthal, Ascending Technologies Gmbh., Kreilling, Germany.

Useful References, Websites, and YouTube Addresses

Krishna, K. R. *Agricultural UAVs: A Peaceful Pursuit*; Apple Academic Press Inc., Waretown, NJ, 2018; p 425.

Von Bueren, S. K.; Burkart, A.; Hueni, A.; Rascher, U.; Touhy, M. P.; Yule I. J. Deploying Four Optical UAV-Based Sensors over Grassland: Challenges and Limitations. *Biogeosciences* **2015,** *12*, 163–175. DOI:10.5194/bg-12-163-2015 (accessed Sept. 26, 2017).

http://www.asctec.de/en/uav-uas-UAVs-rpas-roav/asctec-falcon-8/ (accessed June 19, 2017).

https://www.youtube.com/user/AscTecVideos (accessed June 19, 2017).

https://www.youtube.com/watch?v=wptZQp4xUrc (accessed June 19, 2017).

https://www.youtube.com/watch?v=VtbAf8CFXPQ (accessed June 19, 2017).

https://www.youtube.com/watch?v=4ebc-seMRR8 (accessed June 19, 2017).
https://www.youtube.com/watch?v=tawkrn_OpIE (accessed June 19, 2017).
https://www.youtube.com/watch?v=djIgBPspudk&list=PLSl4B94nu1UyvT9_pOgT9ygBDncR38sRc&index=2 (accessed June 19, 2017).

Name/Models: Nemesis 88

Company and address: Allied UAVs, 3480 W. Warmer Ave., Santa Ana, CA 92704, USA; e-mail: NA; website: http://alliedUAVs.com; http://www.ttfly.com/com/alliedUAVs/

Allied UAVs is a company that produces UAVs. They are in operation, since 15 years. The company is situated in California, USA. The small UAVs produced by Allied UAVs Inc. are suitable to conduct a range of functions in the realms of aerial surveillance, spectral analysis of ground features, traffic control, policing, agricultural, and other civilian aspects. They also produce accessories such as camera gimbal adapter kits, accessory mounting plates, antivibration isolators, etc. They adopt 3D printing technology to arrive at the most useful UAV models.

At US$ 2000, Nemesis 88 is termed an affordable Octocopter UAV. It is used more frequently for aerial photography and surveillance of civilian and agricultural regions. It has wide-ranging applications in farm surveillance and spectral analysis of crops. Its applications are based on the type of accessories attached to it. Some of the advantages attributed to Nemesis 88 are as follows: It is usually kept in ready-to-fly mode, it is GPS compatible, allows complicated flight plan and alterations midway, if needed; it has automatic VTOL and returns to home in times of errors in flight plan (Plate 3.35.1).

TECHNICAL SPECIFICATIONS

Material: The airframe and body are made of carbon fiber, aluminum alloy, and plastic.

Size: length: 135 cm; width (rotor to rotor): 135 cm; height: 45 cm; weight: 8.2 kg

Propellers: Nemesis 88 is an octocopter; so, it has eight rotors. Each rotor is made of two plastic blades. Propellers are 30–32 cm in length. Propellers are connected to 8 × 1000 W brushless DC motors for power.

Payload: 8.2 kg (or 18 lb).

Launching information: It is a VTOL UAV and needs only 5 × 5 m clearance space to take-off or land. The autopilot system used is called micropilot MP 2128 multi-cloud cap Picallo SL or Airware osFlexPilot. It allows larger number of waypoints and good payload control.

Altitude: It hovers above the crop at 2–3 m above the canopy. It has good hover accuracy of 1.0–1.5 m. Its operational height for close-up aerial survey of land and crop field is 5–50 m above crop canopy. Its ceiling is 1000 m above ground surface.

Speed: 45 km h^{-1}

Wind speed tolerance: 15 km h^{-1} disturbance

Temperature tolerance: −10 to 55°C

Power source: Electric batteries

Electric batteries: 16,000 mA h, two cells, lithium–polymer, 22.2 V.

Fuel: Not an IC engine-fitted UAV

Endurance: It is 30-min flight time. But it can be enhanced to 30 + 22 min, if payload is decreased.

Remote controllers: The GCS remote controller keeps radiolink with UAV, for up to 1 km radius. Telelink uses 990 MHz or 2.4 Hz datalink. The flight path and aerial imagery, that is, camera shutters, can be regulated using the controller. The GCS has an iPad of mobile controller.

Area covered: It covers 15 km in linear distance. In crop fields, it may cover up to 50 ha per flight depending on altitude, flight pattern, and intensity of aerial survey work.

Photographic accessories: Cameras: It includes a series of visual, high-resolution multispectral, NIR, IR, and red-edge bandwidth cameras. These are used for aerial imagery, detection of moisture status of crop, etc.

Computer programs: The aerial images transmitted as orthophotos are stitched and processed using software, such as Pix4D Mapper or Agisoft's Photoscan or Trimble Inphos. The software offers sharp and clear images.

Spraying area: This octocopter has not been used to spray crop fields but could be adapted, if suitable tank and sprayers are attached.

Volume: Not applicable

Agricultural uses: Major utilities of this UAV are aerial photography and spectral analysis of crops. It is also utilized to assess NDVI, leaf chlorophyll and crop's N status, and canopy temperature (crop's water stress index). It offers close-up imagery to get an idea about disease and pest attack, if any, on the canopy. It is also used to determine panicle and grain maturity, via close-up aerial survey of the entire field.

Nonagricultural uses: It includes surveillance of geologic sites, natural resources, mines, and mining activity. It also used to monitor oil pipelines, electric lines, dams, bridges, rivers, coast line, and city traffic.

PLATE 3.35.1 Nemesis 88—an octocopter.
Source: Allied UAVs Inc. Santa Ana, CA, USA; http://UAVs.specout.com/1/134/Allied-UAVs-HL88-Nemesis.

Useful References, Websites, and YouTube Addresses

http://UAVs.specout.com/1/134/Allied-UAVs-HL88-Nemesis (accessed Apr 28, 2017).
http://UAVlife.cm/cms/product (accessed Apr 28, 2017).

3.2.3 SPRAYER UAVS FOR AGRICULTURAL CROP PRODUCTION

The UAVs have been touted to be useful in variety of ways in farm land, industries, public services, and of course in military. They have been successful in many of these aspects. A few of them involved a good deal of modifications to UAV vehicles and attachments. In the present context, we are concerned with various ways that UAVs have been adopted during crop production. There are several applications listed for UAVs in agriculture. Among them, crop sprayer UAVs are getting ever popular. At present, they seem to be successful. Sprayer UAVs are a popular already in China, Japan, and other far-eastern nations.

To spray pesticide using UAV technology, a few preparatory steps need to be standardized. They are, first, deciding the UAV model and matching its sprayer with the requirements. We should develop a sprayer device or fit one that suits best with standard number of nozzles. We should evaluate droplet atomization methods. We can adopt centrifugal type or hydraulic atomization. The size of droplet can be regulated. We should standardize droplet size and fix it for a specific crop and its stage of growth. We have to note the droplet transport and sedimentation process. We should take note of meteorological conditions and its impact on pesticide drift. Otherwise, it may affect uniformity of spray and deposition of pesticide droplets. The size of droplet, quantity of pesticide discharged, and number of sprayings required should be considered. As a precaution, we ought to possess an operating manual and details of spraying system, prior to spraying the crop. We should select most appropriate UAV model and its spray equipment prior to actual spray. Therefore, previous data and software that evaluate the current requirements should be useful.

3.2.3.1 HELICOPTER SPRAYER UAVS

Name/Models: AG-RHCD-01 Helicopter

Company and address: Homeland Surveillance and Electronics Ltd., HSE LLC., IL, USA

Homeland Surveillance and Electronics LLC. is a UAV-manufacturing company. It offers both single rotor helicopter and multirotor UAVs capable of spraying farm chemicals, efficiently and rapidly. This company is situated in the state of Illinois and several others. This company is known to produce a wide range of crop duster UAVs. They also specialize in application of any kind of liquid and granular formulation on any terrain, hilly, or plains, and whatever crop species that is being grown.

AG-RHCD-01 is an efficient sprayer helicopter UAV (Plate 3.36.1). It operates in all kinds of complex terrain conditions such as (hilly, plains, or undulated topography). It is used over farms, cropland, natural shrub or forest land, etc. It is a versatile helicopter. It reaches different heights to conduct aerial imagery and to spray pesticides with great accuracy. It is best suited for precision farming operations that need variable-rate application. It economizes and reduces usage of pesticides/fungicides. Hence, it is environmental-friendly. Remote operation avoids farmer's contact with hazardous chemicals.

TECHNICAL SPECIFICATIONS

Material: AG-RHCD-01 helicopter is made of carbon fiber, steel, ruggedized plastic, and foam.

Size: length: 180 cm; width (rotor to rotor): 55 cm; height: 70 cm; weight: dry weight without payload: 20.5 kg. MTOW: 30 kg

Propellers: Two. The main rotor is 196 cm and tail rotor is 40 cm.

Payload: Maximum payload is 8 kg. It is composed of cameras, tanks (2), and pesticide formulation.

Launching information: AG-RHCD-01 is an autolaunch helicopter. It is a VTOL UAV. It has software for autonomous flight, navigation, obstacle avoidance, and autolanding. In case of emergency, the UAV returns to point of take-off.

Altitude: It is advised to fly the helicopter at 1–3 m above the ground or crop canopy.

Speed: 3–6 m s^{-1} during spray. It can be increased to 20 km h^{-1}, during aerial imagery.

Wind speed tolerance: 20 knots of disturbing winds

Temperature tolerance: −10 to 60°C

Power source: A Kamatsu two stroke IC engine supports the rotors and spray equipment.

Electric batteries: 14 A h, lithium–polymer batteries to support electronics and CPU.

Fuel: Gasoline plus lubricant oil.

Endurance: 12–15 min per flight during spraying mode.

Remote controllers: The GCS is an iPad or PCs in a cabin. The UAV keeps in contact with GCS iPad for 12–km radius.

Area covered: 50 ac per flight of 12 min.

Photographic accessories: Cameras: The helicopter can be fitted with a full complement of sensors, such as visual (R, G, and B), multispectral, hyperspectral, NIR, IR, and red edge. The gimbal may also allow to make oblique shots of crops, that is, to know the crop's height. It offers excellent close-up aerial images of crop canopy, leaves, stem, and twigs prior to spraying chemicals.

Computer programs: The aerial images are processed, using the usual software such as Agisoft's Photoscan or Pix4D Mapper. The spray locations and rates could be marked, using color codes.

Spraying area: It has four nozzles placed within 2.2 m length. The helicopter covers about 4–5.5 m s^{-1}, while in spraying mode. The spray flow rate is 0.8–1.2 l min^{-1} (four nozzles spray the fluid uniformly.

Volume: 8 kg liquid/granular formulation capacity.

Agricultural uses: The primary functions of the helicopter are aerial imagery, particularly close-up shots of crops and pest/disease affected zones. It is an excellent sprayer UAV with four nozzles that distributes pesticide uniformly onto the crop. This UAV suits for regular use during precision farming. The nozzles could be guided using computer software for variable-rate release of plant protection chemicals. The sensors on the copter can be utilized to study canopy development, get data on NDVI, seedling/plant stand, and gaps if any, leaf area, leaf-chlorophyll/crop-N status to decide fertilizer-N supply (fluid fertilizer). The thermal camera can offer data on variability of crop's water stress index. So, water requirements of crops could be estimated. The imagery can also help us in forecasting yield. Natural vegetation monitoring and study of weed flora and its intensity of infestation are other possibilities with this helicopter UAV.

Nonagricultural uses: Surveillance of industrial installations, buildings, highway routes, dams, water bodies, rivers, electric lines, and oil pipelines are possible, using this helicopter. It offers excellent close-up shots of infrastructure, pipes, and roads.

PLATE 3.36.1 AG-RHCD-01 (Helicopter UAV).
Source: Mr. Terry Sanders, HSE LLC., IL, USA.

Useful References, Websites, and YouTube Addresses

http://www.uavcropdustersprayers.com/agriculture_helicopter_uav_helicopter_spray_orrh_001.htm (accessed May 5, 2017).

http://www.hse-uav.com/contact_us.htm (accessed Apr 28, 2017).

http://www.hse-uav.com/ (accessed Apr 28, 2017).

http://www.uavcropdustersprayers.com/ (accessed Apr 28, 2017).

http://www.agricultureuavs.com/ (accessed Apr 28, 2017).

Name/Models: AG-RHCD-80-1 (Helicopter Sprayer)

Company and address: Homeland Surveillance and Electronics (HSE) LLC., Denver, CO, USA

As stated earlier, Homeland Surveillance and Electronics LLC. is a UAV company. It specializes in production of multirotor UAVs capable of spraying farm chemicals, efficiently and rapidly. This company is situated in the state of Illinois and several others. HSE LLC. is known to produce a wide range of copter and multicopter crop duster UAVs. They specialize in application of any kind of liquid and granular formulation on any terrain, hilly or plains, and whatever crop species that is being grown.

AG-RHCD-80-1 is an agricultural UAV. The UAV system has two parts. One is the copter platform and the other is the ground station accessories, and computers and their software (Plate 3.37.1). It is a helicopter with great versatility while surveying, imaging, or spraying crops with liquid/granular formulation. It is a multipurpose UAV of great utility to farmers and farming companies. RHCD-80-1 is suitable for operation in crop fields, woods, dry terrain/arid land. It is among most perfect UAVs for rapid and efficient crop dusting.

TECHNICAL SPECIFICATIONS

Material: It is made of a composite of carbon fiber, plastic, and steel.

Size: length: 292 cm; width (wingspan): 60 cm; height: 75 cm; weight: 12 kg dry weight

Propellers: One main propeller of 810–950 cm. A smaller propeller is attached at the rear end of tail. It is meant for stability and direction of flight.

Payload: 5–10 kg

Launching information: This is an autolaunch VTOL helicopter UAV. It needs 10 × 10 m clearance, to take-off. It has autonavigation and landing software. Manual guidance is also possible.

Altitude: It ranges from zero (hovering just above crop canopy) to 1000 m above canopy. While spraying, it is kept at 2–3 m above crop canopy.

Speed: It flies at 0–15 m s^{-1}

Wind speed tolerance: 20 knots wind disturbance

Temperature tolerance: −15 to 60°C

Power source: Komatsu two stroke engine

Electric batteries: 14 A h, two cells to support electronics and CPU

Fuel: Gasoline plus lubricant oil

Endurance: 15 min per flight

Remote controllers: This UAV keeps contact with GCS for up to 15 km distance.

Area covered: 8–10 ha h^{-1} depending on crop and spray intensity prescribed and number of waypoints selected.

Photographic accessories: Cameras: This UAV carries a gimbal, with full complement of sensors, such as visual (R, G, and B), multispectral, NIR, and IR cameras. The cameras can be switched on at waypoints using software or could be handled, manually.

Computer programs: The aerial imagery is processed right at the CPU of the UAV. Otherwise, it could be relayed to GCS iPad for processing. Usually, software such as Pix4D Mor Agisoft's Photoscan is used to process the orthomosaics.

Spraying area: It covers 8 ha h^{-1}; spraying speed is 0–6 m s^{-1} (adjustable).

Volume: Volume of tank is 10 L. Nozzle spray rate is 500–1000 mL of fluid pesticide or water min^{-1}. However, it is adjustable depending on the number of nozzles and pesticide dosage recommended. This UAV is suitable to release 4.5–7.5 1 of formulation per ha.

Agricultural uses: AG-RHCD-80-1 is an excellent crop scouting and dusting helicopter UAV (Plate 3.37.1). It is used to conduct aerial survey of farms, collect useful imagery of crop fields, and identify locations with pest/disease affliction. Aerial maps and digital data about pest/disease spread are utilized, while deciding spray quantity and locations. The digital data are usually supplied to variable-rate applicators on the ground. Digital data could be used by the UAV's CPU to regulate pesticide release by the nozzles. This UAV is also useful in obtaining aerial images that depict crop health and nutrient status. Such data help in supplying fertilizer (liquid formulation/granules) at variable rates, if precision techniques are adopted. Natural vegetation monitoring is also conducted using the helicopter UAV.

Nonagricultural uses: Surveillance of industries, mines, highways, transport vehicles, rail roads, electric lines, dams, reservoirs, and waterways are possible using AG-RHCD-80-1 helicopter UAV.

PLATE 3.37.1 AG-RHCD-80-1 (helicopter sprayer UAV).
Source: Mr. Terry Sanders, HSE LLC., IL, USA.

Useful References, Websites, and YouTube Addresses

http://www.agricultureuavs.com/ (accessed June 1, 2017).
http://www.uavcropdustersprayers.com/agriculture_helicopter_uav_crop_duster_or_80_1.htm (accessed June 1, 2017).

Name/Models: AG-RHCD-80-15

Company and address: Mr. Terry Sanders, HSE LLC., Denver, CO, USA

Homeland Surveillance and Electronics LLC. specializes in production of multirotor UAVs capable of spraying farm chemicals, efficiently, and rapidly.

AG-RHCD-80-15 is a low-cost crop spraying helicopter (Plate 3.38.1). It is easy to operate. It offers high-efficiency crop spraying, covering about 1300 m^2 area min^{-1}. The payload tank holds over 10 kg liquid or granular formulation. It leads to rapid coverage of a large area of crop field affected by pest/disease. Fields needing fertilizer are supplied with it rapidly. Reports suggest that, so far, results of aerial spray of chemicals using AG-RHCD-80-15 are excellent (Plate 3.38.1), particularly in terms of uniformity of spray. It can be used during precision farming, if appropriate nozzles and digital data are used. Most importantly, this UAV is environmental-friendly. It avoids exposure of farmers to hazardous pesticides and fungicides.

TECHNICAL SPECIFICATIONS

Material: It is made of steel, carbon fiber, fiber glass, ruggedized plastic, and foam.

Size: length: 220 cm; rotor width: 240 cm; height: 180 cm; weight: MTOW 36 kg

Propellers: Main propeller is supported by brushless motors. It is situated above the fuselage. A small propeller is found at the tail.

Payload: 15 kg. It includes sensors and pesticide-filled tank.

Launching information: This UAV can be started using a remote-control switch. It has autolaunch, navigation, and landing software. It needs at least 10 × 10 m space to lift off. In times of emergency, the UAV returns to point of launch, using safety mode in the software.

Altitude: It ranges from zero (hovering) to 3000 m above ground level.

Speed: 7 m s^{-1} during spray. It has 3 m spraying width (swath).

Wind speed tolerance: 20 knots disturbing wind

Temperature tolerance: −10 to 60°C

Power source: It has a Komatsu 80-cm^3 double cylinder IC engine and electric batteries.

Electric batteries: 14 A h, two cells.

Fuel: Gasoline (plus two stroke engine oil). It consumes 6 L h^{-1}

Endurance: The flight endurance during spray is 8–10 min per take-off.

Remote controllers: The UAV is kept in contact with ground control iPad for up to 15 km distance. The flight path and spray are usually predetermined, using Pixhawk software. However, iPads can be utilized to alter the path midway during a spray schedule.

Area covered: This UAV covers about 8–15 ha of area h^{-1}, depending on spray intensity.

Photographic accessories: Cameras: This UAV has full complement of visual (Sony 5100), multispectral, NIR, IR, and red-edge cameras.

Computer programs: The aerial imagery is processed using Pix4D Mapper or Agisoft's Photoscan or custom-made software.

Spraying area: 8 ha h^{-1}; spray width: 3–6 m; flight speed during spray: 0–6 m s^{-1}; droplet size: 300–500 μm

Volume: Pesticide box size is 545 × 560 × 85 mm

Agricultural uses: Main function is crop scouting to detect pest and disease-attacked zones. Spraying pesticides, fungicides, and fertilizer are other important utilities of this UAV.

Nonagricultural uses: It includes surveillance of installations, farms, rail roads, city traffic, carrying parcels, etc.

PLATE 3.38.1 AG-RHCD-80-15 (Helicopter sprayer UAV).
Source: Mr. Terry Sanders, HSE LLC., Denver, CO, USA.

Useful References, Websites, and YouTube Addresses

http://www.agricultureuavs.com/ (accessed July 28, 2017).
http://www.uavcropdustersprayers.com/agriculture_uav_crop_duster_8015.htm (accessed Aug 10, 2017).
http://www.uavcropdustersprayers.com/agriculture_uav_crop_duster_8015.htm (accessed Aug 10, 2017).

Name/Models: Eagle Brother (model 3WD-TY-17L)

Company and address: Brother UAV Innovation Co. Ltd., China Mainland, 1&11/F, HTM Sci. Park, No. 3 Ganli Rd., Buji Str., Longgang Dist., Guangdong, Shenzhen 518112, P.R. China; website: https://ebuav.en.alibaba.com/

This helicopter UAV is meant predominantly to help farmers, in spraying pesticides and liquid fertilizers to crop fields. It has been utilized by farmers growing crops such as paddy, sugarcane, corn, orchard fruit trees, and pastures. This helicopter UAV can be effectively utilized during precision farming. Computer software and decision-support software are necessary to control nozzles and apply pesticides, at variable rates. The digital data obtained previously show pest-attacked zones. The helicopter has a main propeller and tail propeller. The UAV costs about US$ 12–25,000 per unit (June 2017 price level).

TECHNICAL SPECIFICATIONS

Material: This helicopter UAV is made of fiber glass shell with light but good insulation. The material is resistant to heat, corrosion, dust, and sand.

Size: length: 198.5 cm; width (rotor): 45.5 cm; height: 62 cm; weight: 10.5 g without payload, MTOW 31.5 kg; packed size: 210 cm × 50 cm × 65 cm.

Propellers: It has one larger propeller made of toughened plastic and carbon fiber. One small propeller is placed at the rear end of the helicopter.

Payload: 17 L pesticide in a plastic tank; battery weight 3.8 kg

Launching information: It is an autolaunch autonomous navigation vehicle. Its flight path, navigation, and spraying maps can be prepared, using software such as "Pixhawk." It is recovered in couple of minutes.

Altitude: Operating height is 5–200 m above crop fields. Altitude ceiling is 1000 m above ground level.

Speed: 25–45 km h^{-1}

Wind speed tolerance: Resists 20 km h^{-1} km h^{-1}

Temperature tolerance: −10 to 50°C

Power source: Electric batteries

Electric batteries: 24,000 mA h, two cells

Fuel: Not an IC engine supported UAV. Gasoline is not required.

Endurance: 15–25 min per flight.

Remote controllers: The GCS has laptop and mobile that are used to control the helicopter manually or via a preprogramed flight path. The radiolink operates for up to 5 km radius.

Area covered: It covers 26–34 ac sprayed h^{-1}.

Photographic accessories: This is primarily a sprayer agricultural UAV. It could be fixed with necessary sensors such as visual, NIR, IR, and red edge. These are sensors are required only if aerial survey is also envisaged.

Computer programs: The CPU in the UAV could be included with software necessary for image processing. Images could then be transmitted to GCS laptop.

Spraying area: It sprays 30 ac of crop fields h^{-1}. In a single flight, it covers about 17 ac.

Agricultural uses: It is used predominantly to spray crops with plant protection chemicals and liquid fertilizer formulations.

Nonagricultural uses: It could be used to transport hazardous chemicals and luggage packets.

Useful References, Websites, and YouTube Addresses

http://www.ebuav.com (accessed Mar 20, 2017).

http://111uav.com/ (accessed Mar 20, 2017).

http://www.alibaba.com/product-detail/3WD-TY-17-Smart-Agriculture-Farming (accessed Mar 20, 2017).

https://www.youtube.com/watch?v=wG0PXQJidNg (accessed Mar 20, 2017).

https://www.youtube.com/watch?v=ZMUZNxvUwIo (accessed Mar 20, 2017).

https://www.youtube.com/watch?v=qvhV3vPbISE (accessed Aug 18, 2017).

https://www.youtube.com/watch?v=IBARi8CVVz8 (accessed Aug 18, 2017).

https://www.youtube.com/watch?v=4J94RpU91us (accessed Mar 20, 2017).

Name/Models: Hercules 30

Company and address: Homeland Surveillance and Electronics, HSE LLC., Denver, CO, USA

"Hercules 30" is an automatic helicopter UAV (Plate 3.39.1). It has facility for aerial imagery and to spray chemicals. It comes with a ground controller telemetric instrumentation and iPad for control of flight path and speed of UAV. Flight path and spray schedules can also be preprogramed by adopting user-friendly software. It flies closely above crops, canopy, and causes ruffles and disturbance in leaves and sprays pesticides. This gives good distribution of pesticides on the crop's canopy. "Hercules 50" is yet another improvised model offered by HSE LLC. It is an excellent sprayer helicopter UAV (Plate 3.39.2).

TECHNICAL SPECIFICATIONS

Material: Steel-based alloy, carbon fiber, kevlar, and plastic composite.

Size: length: 91 cm; width (rotor diameter): 330 cm; height: 117 cm; weight: 30 kg

Propellers: Main and tail propellers are powered by a motor "Hirht-F33 ES TBO."

Payload: 12–15 kg. It includes cameras for visual and thermal imagery. Payload includes and pesticide tank depending on the purpose.

Launching information: It is launched using remote controller. It has autolaunch, navigation, and autolanding software, namely, Pixhawk. The helicopter UAV returns to point of launch in times of emergency.

Altitude: It flies close to crop canopy while taking close-up images or during spraying. The recommended altitude is 5–30 m, above ground level. However, the ceiling is 1000 m above ground level.

Speed: 10–20 km h^{-1} working speed

Wind speed tolerance: 20 knots (25 km h^{-1}) disturbance

Temperature tolerance: −10 to 55°C

Power source: internal combustion engine. The engine generates 28 hp.

Electric batteries: 14 A h to support electronics and CPU.

Fuel: Gas

Endurance: 12 min per flight

Remote controllers: Helicopter is kept in contact with GCS iPad, for up to 30 km radius. The GCS computer or telemetric instruments can alter the flight path of the UAV midway, if needed. Flight path is usually preprogramed, using Pixhawk software.

Area covered: It covers 15 km linear distance per flight, about 50 ha aerial survey per flight.

Photographic accessories: Cameras: The sensors are placed in a gyro-stabilized gimbal. The series includes visual (Sony 5100), multispectral, hyperspectral, NIR, and IR bandwidth cameras and Lidar sensor. The gimbal allows the UAV to pick oblique shots of crop fields and natural vegetation stands.

Computer programs: The aerial images are processed, using Pix4D Mapper or Agisoft's Photoscan.

Spraying area: The spraying area covered per flight may reach about 50 ha.

Volume: This UAV empties about 8–10 L pesticide per flight of 12 min, over the crop. The UAV flies very close to crop canopy to release the plant protection chemicals. It avoids drifting and contamination of neighboring zones with hazardous chemicals. The UAV has a 3.0-m long spraying bar (pipe). It has several nozzles that could be controlled, based on digital data. It is useful for variable-rate application of chemicals, if precision techniques are adopted.

Agricultural uses: This UAV is utilized in crop scouting. It is used to detect and collect data about seed germination trends, gaps, seedling, and crop stand variations. Detection and mapping of disease/pest-attacked zones, drought, flooded, or soil erosion zones are other uses. Natural vegetation monitoring and obtaining data about NDVI are possible.

Nonagricultural uses: It includes surveillance of industrial zones, mines and mining activity, tracking transport vehicles, etc. Monitoring oil pipelines, electric lines, dams, water bodies, rivers, and canals are also done using this UAV.

PLATE 3.39.1 Hercules 30 (Helicopter Sprayer UAV).
Source: Mr. Terry Sanders, HSE LLC., IL, USA.

PLATE 3.39.2 Hercules 50 (Helicopter Sprayer UAV).
Source: Mr. Terry Sanders, HSE LLC., IL, USA.

Useful References, Websites, and YouTube Addresses

http://www.hse-uav.com/ (accessed Sept 20, 2017).
http://uavcropdustersprayers.com/hercules_uav.htm (accessed Sept 20, 2017).

Name/Models: SDO 50V2

Company and address: Swiss UAVs Operating AG, Bahnweg Nord 16, CH-9475 Sevelen, Switzerland; Tel.: +41 81 785 20 10; e-mail: info@swissUAVs.com; website: http://www.swissUAVs.com/

The Swiss UAV Operating AG is a young, high-tech company based in Sevelen, Switzerland. It specializes in the development and production of unmanned helicopters (VTOL UAVs) for civil purposes. It includes surveillance of crops and spraying plant protection chemicals. In addition to development of UAV models and their production, SDO offers excellent engineering and after-sale services. It advices on crop production to farmers. They also train UAV users.

The "SDO 50V2" is a helicopter UAV with VTOL. This UAV can be assembled and launched in 15 min. It just needs two persons conversant with the UAV model. It has unique design and features that allow superior payload, prolonged endurance in flight, stable flight path, and several safety features. It has custom-made cameras (Leica) and software. The SVO 50V2 comes along with GCS. The GCS includes iPad and telemetric accessories to command and control operation of the helicopter UAV. It can be used effectively as a spraying helicopter UAV. It is supplied with software that reads the digital data during precision farming. It has variable-rate spraying equipment (nozzles) to spray plant protection chemicals or fertilizers.

TECHNICAL SPECIFICATIONS

Material: The helicopter UAV is made of excellent rigid carbon fiber, alloy, ruggedized foam, and plastic. It has landing-gear made of tough alloy.

Size: length: 232 cm; width (at fuselage): 70 cm; height: 98 cm; weight: MTOW is 86 kg; empty weight is 36 kg

Propellers: It has double rotor system. It has two sets of propellers made of plastic. The rotor diameter is 2 × 280 cm.

Payload: Maximum payload is 50 kg

Launching information: This is VTOL helicopter is of medium weight. It needs 15 × 15 m clearance space for launching. It has autolaunch, navigation, and landing software. It lands on landing-gear made of steel or light alloy.

Altitude: Maximum ceiling is 4000 m above ground level. Operational height is 350 m above ground/crop surface. It flies at very low altitude of 0–50 m above crop canopy, while spraying plant protection chemicals. This is done to avoid drift of harmful chemicals.

Speed: Maximum speed is 15 m s^{-1} in flight.

Wind speed tolerance: 25 km h^{-1} wind disturbance

Temperature tolerance: −10 to 60°C

Power source: It is fitted with turbine jet engine for propulsion.

Electric batteries: It has lithium–polymer batteries 14,000 mA h for electronics and CPU functioning.

Fuel: The jet engine uses diesel jet fuel. Maximum fuel capacity is 13 L. Additional fuel tanks can be attached.

Endurance: 18 h

Remote controllers: The GCS is small. All the accessories and instruments are held in a compact aluminum suitcase. The GCS computer is an iPad integrated with the suitcase frame. Flight path can be preprogramed or modified midway, during fight. The telemetric contact works for up to 50 km distance from the GCS.

Area covered: It covers 400–500 ha depending on flight altitude, resolution of aerial photography required and intricate flight path.

Photographic accessories: Cameras: The helicopter UAV is fitted with special Leica RCD 30 camera in the payload area. The gimbal is gyro-stabilized and operates in dual-axis while imaging crop fields or ground feature.

Computer programs: The aerial images are processed, using Pix4D Mapping or Agisoft's Photoscan software.

Spraying area: This is a crop-spraying UAV. It comes with 2–3 tanks fitted under the fuselage. It can carry up to 30 L of pesticide or fungicide or weedicide or fertilizer formulations such as granules/powder or liquids. It can be fitted with variable-rate nozzles and software for reading digital data. The crop spraying facility on this UAV has been tested and used in undulated terrain within grape yards, vegetable fields, wide turf areas, etc.

Volume: Data not available

Agricultural uses: They are crop scouting to detect seedling establishment pattern, canopy growth, to estimate leaf area, NDVI, and crop water stress index (canopy and air temperature). Natural vegetation monitoring to collect NDVI data, canopy traits, and plant species diversity is possible.

Nonagricultural uses: Major nonfarm uses for this helicopter UAV are listed by the company. They are as follows: oil and gas pipeline inspections; powerline inspections; wind turbine inspections; photovoltaic system inspections; roadwork, ramp, bridge, and canal inspections; railroad infrastructure inspections; safety assessments; flood damage detection; erosion monitoring; etc.

Useful References, Websites, and YouTube Addresses

http://www.swissUAVs.com/overview.html (accessed Oct 12, 2017).
http://www.swissUAVs.com/applications.html (accessed Oct 12, 2017).
https://www.youtube.com/watch?v=3CKToI6Y6TE (accessed Oct 12, 2017).
https://www.youtube.com/watch?v=_hRIN8qCoOM (accessed Oct 12, 2017).
https://www.youtube.com/watch?v=azoxU_tYGkw (accessed Oct 12, 2017).
https://www.youtube.com/watch?v=N9H_Xk_qhCY (accessed Oct 12, 2017).

Name/Models: Terra 1

Company and address: Mr. Toru Tokushige, CEO; Terra UAV, 5-53-67, Jinguumae, Shibuya-Ku, Tokyo, Japan; Tel.: +81 03 6419 7193; e-mail: info.jp@terra-UAV.co.jp; website: http://www.terra-UAV.co.jp/en/comppany/message/

Terra UAV is an avionics company that manufactures agricultural helicopters. It is a company with specialists from motor and computer industries from Japan and Silicon Valley in USA. It offers a range of UAV models. They suit the farming stretches of Japan and other nations.

Terra 1 is an agricultural UAV. This is a pesticide and fertilizer formulation spraying small UAV. It is a helicopter with versatile abilities. It hovers over the crop for close-up shots or during spraying hazardous chemicals. The hovering helps in avoiding drift of chemicals in the air. It is useful in aerial imagery, land surveying, and monitoring crop growth and maturity. It is a light-weight helicopter. It can be transported, using pick-up van (see Plate 3.40.1).

TECHNICAL SPECIFICATIONS

Material: It is made of light-weight alloy, carbon fiber, ruggedized foam, and plastic. It has a landing-gear made of alloy to withstand rough landing, if any.

Size: length: 186 cm; width (fuselage): 75 cm; spraying equipment width: 167 cm; height: 85.5 cm; weight: 14.9 kg without payload of pesticide and fuel.

Propellers: Main rotor diameter is 195 cm and tail rotor is 30 cm wide.

Payload: It includes a series of cameras that are useful in aerial imagery. Sensors are useful in obtaining digital data plus the pesticide tank with its content. It holds 2 l gasoline fuel for the IC engine to operate. Pesticide tank holds 14 l of plant protection chemical in liquid or granular or dust form. It has a longer endurance, during spraying compared to other UAV models.

Launching information: It is a VTOL helicopter and needs only 5 × 5 m clearance to swiftly gain height. It is an autonomous helicopter. Its navigation and flight path can be predetermined, using computer software, such as Pixhawk.

Altitude: It can fly at 0–50 m above crop canopy, during spraying schedules. It takes a height of 300 m above crop fields, while obtaining aerial photography. The ceiling is 1000 m above ground level.

Speed: It is 45–60 km h^{-1} during aerial imagery. It is 15–25 km h^{-1} during intense spraying schedules. It depends on flight path and spray volume and nozzle rates.

Wind speed tolerance: 25 km h^{-1} wind disturbance

Temperature tolerance: −10 to 50°C

Power source: IC engine operated using gasoline

Electric batteries: 14,000 mA h for supporting electronics and CPU.

Fuel: Gasoline

Endurance: 3–4 h per take-off

Remote controllers: Terra 1 adopts the GCS equipment and programs manufactured by DJI. The autohovering and GPS tagging are also produced by DJI of China. The GCS has facility to control flight path and speed of the helicopter. It regulates the motors connected to rotor.

Area covered: It covers about 100–150 ha of crop field during spraying pesticide or fertilizer formulation. It moves swiftly and covers over 400 ha of land/crop canopy if used only for aerial photography.

Photographic accessories: Cameras: It has a gyro-stabilized gimbal that holds a series of cameras. They are visual (R, G, and B), multispectral, hyperspectral for close-up examination of crops for pest/disease and nutritional disorders, NIR for chlorophyll and crop-status estimation, IR and red-edge for detecting canopy and air temperature (i.e., crop's water stress index), and Lidar sensor to get crop height.

Computer programs: The aerial imagery is processed, using Pix4D Mapper or Agisoft's Photoscan.

Spraying area: The UAV covers about 300–400 ha of crop field per flight with an endurance 5–6 h.

Volume: Data not available

Agricultural uses: They are aerial imagery of terrain, land resources, soil types, cropping systems, irrigation, drainage channels, etc. Crop scouting to is possible detect spatial variation and map diseases, pests, nutrient deficiencies, and water status of crops, at various stages. It sprays plant protection chemicals and fertilizer formulation, using variable-rate nozzles, if precision techniques are practiced. It is used to monitor and track farm vehicles and their work schedules. Natural vegetation monitoring is done to collect NDVI data, water stress, if any. We can use the copter to detect and map species diversity, using spectral signatures of each plant species, etc.

Nonagricultural uses: It includes surveillance of civilian installations like urban and farm buildings, industrial regions, irrigation dams, channels, and their functioning. It can then be used in farm to track vehicle and inspect work progress by autonomous tractors, sprayers, and combine harvesters.

PLATE 3.40.1 Terra 1 series.
Source: Dr. Toru Tokushige, CEO and founder, Terra UAV, Shibuya-Ku, Tokyo, Japan.

Useful References, Websites, and YouTube Addresses

https://www.terra-UAV.net/en/software/utm/ (accessed July 2, 2017).
http://www.terra-UAV.co.jp/en/agriculture/ (accessed July 2, 2017).
https://www.terra-UAV.net/en/terra-UAV-sets-eyes-on-agriculture-introduces-companys-first-crop-spraying-uav/ (accessed July 2, 2017).
https://www.youtube.com/watch?v=dkarS7Ao9Ws (accessed July 2, 2017).

Name/Models: Yamaha's RMAX Helicopter UAV

Company and address: Yamaha Corporation, 10, Nakazawacho, Naka-Ku, Hamamatsu-Shi, Shizuoka, 430-8650, Japan; or the other address is Scott Noble, Business Development Manager, 489-493 Victoria Street, Wetherill Park, NSW 2164, Australia; phone: +61 2 9827 7500; e-mail: skyinfo@yamaha-motor.com.au

Yamaha's RMAX was developed in 1990s by upgrading the predecessor helicopter UAV called "Yamaha R-50." This helicopter UAV was developed in response to demand by the Japanese Agricultural Department, for a spraying UAV. It is said that flat-winged airplanes were not efficient, due to small farm holdings. So, a hovering type small UAV was opted. "R-MAX" autonomous helicopter has been tested and used extensively, in several agrarian regions of the world. As a commercial agricultural UAV, "R-MAX" has surpassed over 2 million hours of aerial photography and/or spraying.

Yamaha's RMAX agricultural UAV is among the most popular agricultural UAV (Plate 3.41.1). It is a surveying and spraying UAV. It offers great agility, accuracy, and efficiency, regarding applications of plant protection chemicals (Cornet, 2013; Frey, 2014). It has been cleared by FAA of United States of America for use in commercial agricultural farms. Agricultural uses include spraying, seeding, remote sensing, precision agriculture, frost mitigation, and variable-rate dispersal. In Japan, RMAX helicopters are primarily used for seeding, scouting, and spraying rice crop. Given the unique features of the RMAX (RMAX, 2015), the opportunities in other Asian nations, and in Australia, seem promising.

TECHNICAL SPECIFICATIONS

Material: It is made of light-weight aluminum alloy, tough carbon fiber, plastic, and foam.

Size: length: 3.63 m; width (wingspan):0.72; height: 1.08 m; weight: empty weight 64 kg (i.e., without payload). MTOW is 94 kg.

Payload: Maximum permitted payload is 28–31 kg of liquid, granular, or dust/powder formulation of pesticides, herbicides, or fertilizer formulation.

Propellers: Main propeller is of 3.11 m length and the small rear rotor is of 1.2 m length.

Launching information: This is an autolaunch vehicle. Its take-off can be manually controlled, using joystick or remote controller (iPad/mobile). Its navigation is controlled, using predetermined flight path software. Its avionics is controlled, using Yamaha's Attitude Control System.

Altitude: It hovers just above the crop canopy. Its operation altitude is 50–100 m above crop canopy, during spraying. Its ceiling is 1500 m above ground level.

Speed: 45–60 km h^{-1}

Wind speed tolerance: 20 km h^{-1} disturbance

Temperature tolerance: −20 to 55°C

Power source: Two stroke internal combustion engine

Electric batteries: 14,000 mA h to support CPU and other circuits.

Fuel: Gasoline plus lubricant oil

Endurance: 1 h

Remote controllers: The GCS has an iPad/mobile or remote-control joystick. This UAV can be a flight path which can be predetermined, using software such as Pixhawk, or could be manually controlled. The UAV is kept in contact with ground computer, using telelink that operates up to 15 km radius.

Area covered: It covers 50–250 ha day^{-1} of 6 h.

Photographic accessories: Cameras: It has a full complement of sensors, such as visual (R, G, and B), high-resolution multispectral cameras, NIR, IR, and red edge to get thermal images and estimate crop's water stress index, in other words to estimate irrigation requirements. Lidar sensor has also been used on RMAX.

Computer programs: The aerial images transmitted as ortho-images are processed, using Pix4D Mapper, or Agisoft's Photoscan or Trimble's Inphos.

Spraying area: It sprays pesticide on over 50–250 ha of crop land in a day.

Volume: The pesticide tank holds about 28–31 l of pesticide/liquid fertilizer formulation. The entire tank is emptied in a matter 35–40 min per flight. This UAV has been used in precision farming in cereal farms and plantations. It has been fitted with variable-rate spray nozzles. It adopts digital data obtained through aerial imagery to control spray nozzles.

Agricultural uses: It has been used most frequently to conduct aerial surveys, photography, and spraying chemicals. Crop scouting for canopy growth traits, leaf chlorophyll to decide fertilizer-N supply, and thermal images to decide irrigation needs have been done, using RMAX. Natural vegetation monitoring to get NDVI values to detect variation in biomass accumulation pattern and to study plant species diversity is possible. RMAX is used to detect and map disease, pest attack, and infestation in crop fields (Quackenbush, 2017). The digital data so collected are used to spray plant protection chemicals at variable rates. It is done using appropriate spray nozzles, meant for precision farming (Plate 3.41.1).

Nonagricultural uses: Yamaha's RMAX has been used extensively in military reconnaissance, patrolling, and launching attacks. This is now more frequently used in surveillance of industrial sites, buildings in urban zones, mines and mining activity, oil pipelines, dams, river flow, lakes and irrigation channels, international borders, highways and transport vehicles, etc.

PLATE 3.41.1 Yamaha's "RMAX" Helicopter UAV.
Source: Yamaha Corporation, Hamamatsu, Shizuoka, Japan; http://www.max.yamaha-motor.UAV .au/specifications.

Useful References, Websites, and YouTube Addresses

Cornett, R. *UAVs and Pesticide Spraying a Promising Partnership*; WesternFrmPress.com, 2013; pp 1–3 (accessed May 23, 2014).

Frey, T. *Future Uses for Flying Drones*; 2014, pp 1–7. http://www.futuristspeaker.com/2013/08/griculture-the-new-game-of-UAVs/ (accessed Oct 28, 2016).

Quackenbush, J. North Coast Vineyards See More UAV Use as Agriculture Market Soars. *North Bay Bus. J.* **2017,** 1–7. https://www.vineview.com/single-post/2017/01/17/VineView-Featured-in-North-Bay-Business-Journal-Flight-of-the-Vineyard-Skybots (accessed Sept 25, 2017).

RMAX. *RMAX Specifications*; Yamaha Motor Company: Japan, 2015; pp 1–4. http://www.max.yamaha-motor.UAV.au/specifications (accessed Sept 8, 2015).

https://www.youtube.com/watch?v=qUuOBC_OHv4 (accessed Sept 4, 2017).

https://www.youtube.com/watch?v=ydfPzqaNkuA (accessed Sept 4, 2017).

https://www.youtube.com/watch?v=aE0LYwDnx7M (accessed Sept 4, 2017).

https://www.youtube.com/watch?v=BekkmalOM-U (accessed Sept 4, 2017).

Name/Models: ZHNY 15 Agricultural Sprayer

Company and address: Hennan Wonderful Industrial Co. Ltd., No. 106, Floor 21st, Unit 1, Building No. 4, Zhidi Plaza, No. 1188, Zhongzhou Ave., Henan, Zhengzhou 450007, China (Mainland)

Henan Wonderful Industrial Co. Ltd. is a recently initiated UAV company. It has churned out helicopter UAVs since 2012. It offers a helicopter sprayer UAV useful to farmers. It offers a helicopter sprayer that holds relatively larger payload of 15 l pesticide. These helicopter UAVs cost US$ 15–50,000 per unit, but it is based on the accessories preferred by farming companies.

ZHNY is a heavier and larger helicopter UAV. It is used to spray a variety of farm chemicals to crops from a close range. It helps farmers to avoid contact with hazardous chemicals. It could be used to apply liquid or granular fertilizer formulation. It is used most commonly to dust the crops with insecticides. It has a good spray rate and spray width. Hence, it is highly efficient in performing the task quickly and in time. It saves farm labor significantly.

TECHNICAL SPECIFICATIONS

Material: It is made of carbon fiber, light aluminum alloy, and plastic.

Size: length: 210 cm; width (rotor to rotor): 36 cm; height: 72 cm; weight: MTOW: 36 kg

Propellers: One propeller with 98 cm length. It is made of two plastic blades. The rear end propeller is smaller at 22 cm length.

Payload: 15 kg. It includes pesticide tank, CPU, and sensors if attached.

Launching information: It is an autolaunch VTOL helicopter. It needs 10 × 10 m clearance to lift-off or land on landing gear. Landing spot can be predetermined or the UAV could be carefully controlled to land on the landing-gear. The landing gear is made of toughened but light aluminum bars. Navigation can be done using autopilot mode or manually controlled.

Altitude: The operational height during spraying is 1–3 m above the crop's canopy height. The ceiling is 3000 m above ground level.

Speed: Flight speed in normal flying mode (not spraying) is 4–8 m s^{-1}.

Wind speed tolerance: 20 km h^{-1}

Temperature tolerance: −10 to 50°C

Power source: Electric batteries for computers and telemetric connectivity. A 2-stroke, dual cylinder internal combustion engine for lift-off and forward thrust.

Electric batteries: 14,000 mA h.

Fuel: It utilizes gasoline with lubricant. The helicopter UAV consumes 6 l of gasoline h^{-1} of flight, while spraying.

Endurance: It is 25 min per flight

Remote controllers: The GCS has both manually operated remote control (joystick) and IPad or mobile for predetermined flight path. The UAV's flight path is determined using software such as Pixhawk. The telemetric link between GCS and UAV operates up to 3000 m radius.

Area covered: It covers 50–250 ac day^{-1} of 6–8 h flight

Photographic accessories: Cameras: This helicopter UAV is predominantly a sprayer UAV. However, if needed, farmers may attach sensors and procure digital data about crops, their growth, and disease/pest attack, if any.

Computer programs: This UAV is supplied with digital data in a chip. The digital data are utilized by the CPU and computer programs for variable-rate application of pesticide.

Spraying area: The spray width (spray bar with variable-rate nozzles) is 3–6 m. Sprayer capacity is 1500 m^2 min^{-1}; spray speed is 0–6 m s^{-1}

Volume: Dimensions of pesticide box is 54.5 × 35.8 × 8.5 cm. It holds a volume of pesticide up to 15.2 l. The sprayer unloads 1.2 l of plant protection chemical per minute in liquid form.

Agricultural uses: This helicopter UAV has wide range of uses in crop fields. It is used extensively to spray plant protection chemicals and nutrients. It can supply fertilizer granules or in liquid form as foliar spray. Foliar sprays are conducted rapidly in a matter of few minutes to cover 50–200 ha. Foliar sprays require very low quantities of fertilizer formulation when compared to soil application. During plant protection, the need for pesticides/fungicides is again reduced significantly, compared to traditional blanket sprays, because UAV utilizes digital maps and computer decision-support system to apply chemicals only at spots that are attacked.

Nonagricultural uses: ZHNY helicopter can be used to transport luggage or parcel up to 15 kg, from one location to another. It can be used to carry and unload chemicals, particularly hazardous chemicals at predetermined spots.

Useful References, Websites, and YouTube Addresses

https://zzwonderful.en.alibaba.com/?spm=a2700.8304367.0.0.67501fdb2k2Iwu (accessed Nov 20, 2017).

https://zzwonderful.en.alibaba.com/product/2017636988-800306424/ZHNY_15_agriculture_sprayer_remote_control_UAV_helicopter.html (accessed Nov 20, 2017).

3.2.3.2 MULTIROTOR SPRAYER UAVS

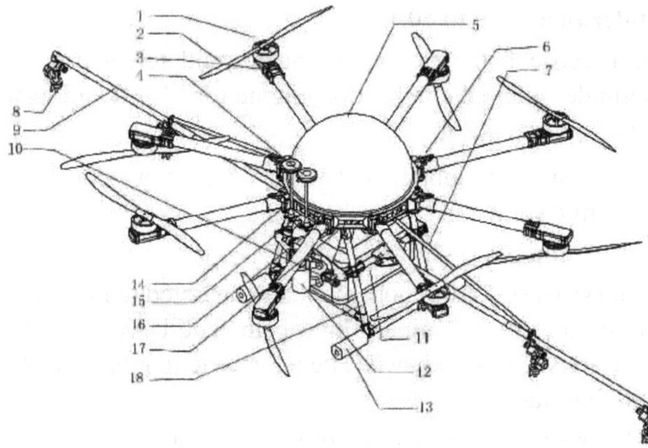

FIGURE 3.2 A typical multirotor sprayer UAV.
Note: The description of numbered parts in figure of the UAV is available here under the following:
1 = motor; 2 = propeller; 3 = motor mount; 4 = GPS antenna; 5 = dome; 6 = folding arms; 7 = support rod; 8 = nozzle; 9 = sprayer pipe; 10 = pump interface; 11 = tank; 12 = pump; 13 = landing gear; 14 = GPS holder; 15 = top plate; 16 = on board electronics; 17 = bottom plate; 18 = landing gear vertical support
Source: Mr. Terry Sanders, Home land Surveillance and Electronics (HSE) LLC., Denver, CO, USA.
Note: An agricultural UAS (UAV Sprayer) generally includes the following items:

Ready-to-fly aircraft/autopilot—one set; spraying equipment—one set; liquid carrying tank—one set; radio controller—one piece; Wi-Fi/data communication module—one piece; ground station tablet—optional; industrial carrying case—one piece; assembly tools—one set; warranty—one card and related information; and aerial imaging package—optional.

Name/Models: AG-6A (Sprayer Hexacopter)

Company and address: Homeland Surveillance and Electronics (HSE) LLC., Denver, CO, USA

Homeland Surveillance and Electronics LLC. is a UAV company that specializes in the production of multirotor UAVs capable of spraying farm chemicals, efficiently and quickly (Plate 3.42.1). A diagrammatic representation of sprayer hexacopter is made available in Figure 3.2.

HSE's AG-UAV sprayer-AG-6A is a hexacopter. It has versatile ability for low altitude surveillance of crops. This trait is used for detection of disease/pest and for spraying plant protection chemicals on crop's canopy. This UAV can hover over a spot and pick high-resolution images for confirmation of disease/pest. The pest species that has attacked the crop's canopy can be identified.

TECHNICAL SPECIFICATIONS

Material: It is made of carbon fiber, ruggedized foam, and plastic composite.

Size: length: 160 cm; width (wheel base): 145 cm; height: 65 cm; weight: empty weight 14.5 kg; MTOW: 20 kg including a payload of plant protection chemical

Propellers: It has six propellers. Each propeller has two plastic blades. The propellers are attached to brushless motors for vertical take-off and propulsion.

Payload: It ranges from 7 to 12 kg depending on fluid tank size.

Launching information: It is a VTOL vehicle. It needs only 5 × 5 m clearance to launch itself. It has autonomous mode for navigation, waypoints to cover and distance could predetermined, and landing is automatic.

Altitude: It flies very close to crop canopy. It can hover at a spot very close to a single plant, if needed. The general height for spraying is close to canopy, that is, at 5–10 m above crop field. The UAV can reach 400 m above ground level.

Speed: It ranges from zero (hovering) up to 25–40 km h^{-1} depending on purpose.

Wind speed tolerance: 20 km h^{-1}

Temperature tolerance: −10 to 45°C

Power source: Electric batteries

Electric batteries: 14–28 A h, two cells

Fuel: Not an IC engine-fitted UAV

Endurance: It is 15–30 min spraying time per flight depending on flight speed, spray volume, and wind speed.

Remote controllers: This UAV is autonomous about launch (VTOL), navigation, and landing. It has facility to return to launch site, if telemetric signals or spraying difficulties arise. In semiautonomous mode, spraying speed and path can be modified midway, using GCS iPad.

Area covered: The endurance and spray volume decide the area covered by the UAV. The short endurance allows only small areas up to 50 ha h^{-1}, to be sprayed.

Photographic accessories: Cameras: AG-6A is supplied with a full complement of cameras such as visual (Sony 5100), NIR, IR, mica-sense, and red-edge bandwidths.

Computer programs: The digital data and close-up imagery can be stored, in the UAV's CPU or GCS iPad. They can be processed, using Pix4D Mapper or Agisoft's Photoscan software. Then, images can be used to decide spray quantity and areas.

Spraying area: It covers 50 ha per flight.

Volume: AG-6A has tanks with 20 L capacity for distributing plant protection chemicals. The spray flow is 0.9–2.4 L min^{-1} (Plate 3.42.2).

Agricultural uses: AG-6A is used for close-up surveillance of crops. However, it is predominantly a spraying UAV. It is adopted in crop fields to survey and identify locations affected by diseases/pests or weeds. The digital data can be used by the UAV to spray the crops using variable-rate nozzles. Crop scouting to obtain data about NDVI, canopy characteristics, leaf-chlorophyll, and water stress index can also be accomplished by this hexacopter. Natural vegetation is monitored for biomass accumulation pattern and species diversity.

Nonagricultural uses: It includes surveillance and monitoring of industries, public buildings, city traffic, tracking vehicles, guarding pipelines, electric installations dams, reservoirs, etc.

PLATE 3.42.1 "AG-6A" HSE's hexacopter sprayer UAV.
Source: Mr. Terry Sanders, Homeland Surveillance and Electronics LLC. (HSE LLC.), Denver, CO, USA.

PLATE 3.42.2 AG-6A UAV in flight and spraying pesticide.
Source: Mr. Terry Sanders, HSE LLC., Denver, CO, USA.
Note: AG-6A can fly at very low altitudes above the crop. Therefore, it avoids wind disturbance and drifts that are otherwise common, while spraying pesticides. The above UAV has lengthier spray bar. Therefore, it allows 16 ft spray width (swath).

Useful References, Websites, and YouTube Addresses

http://www.uavcropdustersprayers.com/agriculture_octo_copter_uav_crop_duster_v8a.htm (accessed Aug 12, 2017).
https://www.youtube.com/watch?v=1bcfwJJAq3w (accessed Aug 12, 2017).
https://www.youtube.com/watch?v=v4OWVNDvpa4 (accessed Aug 12, 2017).

Name/Models: AG 10—a sprayer quadcopter

Company and address: GTF Aviation Technology Ltd., International Negotiate Garden No. 1 Building Room 614 Shunyi District, Beijing 101300, China (mainland); e-mail: gtfdroen@outlook.com; Tel.: +86 18231133680; website: http://em.gtfdronne.com/goods/goods_id=45&gclid=CPDe3P6QltlCFZMXaAodt

GTF Aviation Technological Company is situated in Beijing, China. It offers a series of quad and hexacopters capable of spraying agricultural chemicals. They cater to farmer's requirement by offering sprayer UAVs of wide range of costs, capacity to hold pesticides, spray rates, and efficiency. These UAVs are easy to pack and export. These models are made available in different countries of Asia, Europe, and Latin America.

"AG 10" is a sleek, easy to assemble, and uses quadcopter. It is meant mainly to spray the crops. It covers about 8–10 ac of crop land per each take-off and endurance of 12 min. It is highly efficient and accurate in releasing pesticides, using digital data. We can also adopt the UAV to conduct blanket sprays.

TECHNICAL SPECIFICATIONS

Material: Its airframe and body is dust proof, waterproof, and corrosion resistant. It is made of carbon fiber, rugged foam, and plastic.

Size: The dimension of this UAV, "in stretched out ready to fly state," is as follows: length: 116.8 cm; width (rotor to rotor): 116.8 cm; height: 49.7 cm. The dimension of this UAV "in folded state" is 51.2 × 512 × 49.7 cm. weight: MTOW is 23.0 kg; but without payload, the take-off weight is 9.8 kg.

Propellers: It has four propellers of 30.8 in. in size. They are made of toughened plastic blades.

Payload: Maximum payload is 10 kg granules/powder or 10 L if it is a liquid formulation.

Launching information: It has both manual and auto-take-off mode during launch, navigation, and landing. It is a VTOL vehicle. It needs just 5 × 5 m clearance space. UAV's navigation can be controlled using both manual or GPS connected remote controllers. Predetermined flight is possible using appropriate software, for example, Pixhawk.

Altitude: The operational height suggested is 2–3 m above the crop canopy. It is operated at 1–100 m above canopy, during loiter mode.

Speed: It is 0–10 m s^{-1} under GPS mode. The suggested speed for optimal spray efficiency is 3–5 m s^{-1}.

Wind speed tolerance: It tolerates 20 km h^{-1} (or 8–10 m s^{-1} disturbing wind)

Temperature tolerance: 5 to 50°C

Power source: Lithium–polymer electric batteries

Electric batteries: 16,000 mA h, lithium–polymer battery or 21,600 mA h, lithium–polymer battery

Fuel: Not an IC engine-fitted UAV.

Endurance: It is 9–12 min, depending on payload and spraying pattern and rate.

Remote controllers: This UAV could be operated manually, particularly using remote controller to decide its flight path, speed, and spray schedules. It can also be predetermined using software, such as Pixhawk. The UAV is kept in contact with GCS computer/remote controller for up to 1 km radius.

Area covered: It covers 20–100 ac day^{-1} of 8 h.

Photographic accessories: Cameras: This is basically a sprayer UAV. However, payload with sensors can be attached if necessary.

Computer programs: The UAV uses digital data and aerial photography of the field with crop. The areas damaged are marked. The UAV's CPU utilizes digital data to conduct aerial spraying at variable rates. The variable-rate spray nozzles are direct by computer data. Previously processed digital data are utilized.

Spraying area: It covers 8–10 ac h^{-1}; spray flight time is 12 min per take-off; spray width is 4–6 m depending on size of spray bar; and spray speed is 3–4 m s^{-1}

Volume: The payload tank holds 10 l of pesticide or liquid fertilizer formulation.

Agricultural uses: Main purpose of this UAV is in spraying crops with pesticide, fungicide, herbicide, or fertilizers. It discharges chemicals in the form of granules, dust (powder), or liquid. So far, it has been tested and utilized to spray crops such as wheat, soybeans, corn, cotton, potato, sunflower, pepper, tomato, and fruit trees.

Nonagricultural uses: Transport of goods as payload is a clear possibility whenever required.

Useful References, Websites, and YouTube Addresses

http://en.gtfUAV.com/single_page?id=28 (accessed Sept 1, 2017).
http://en.gtfUAV.com/aricle_cat?id=27 (accessed Sept 1, 2017).
http://en.gtfdrrone.com/single_page?id=25 (accessed Sept 1, 2017).

Name/Models: AG-MRCD-6 (Hexacopter Sprayer)

Company and address: Homeland Surveillance and Electronics—HSE LLC., CO, USA

Homeland Surveillance and Electronics LLC. is a UAV company. It specializes in the production of multirotor UAVs capable of spraying farm chemicals, efficiently and rapidly.

AG-MRCD-6 is a small easily foldable and portable hexacopter. It fits into a small suitcase for ease of portability. It is used to spray a range of liquid and granular formulations onto crop fields. It sprays at a fast pace of 2–3 ac min^{-1} of flight. AG-MRCD-6 flies at low height above the crop's canopy. Therefore, it avoids drifts of chemicals, during spraying (Plate 3.43.1; Bowman, 2015).

TECHNICAL SPECIFICATIONS

Material: The frame is made of steel, carbon fiber, and ruggedized plastic.

Size: length: 130 cm; width (wheel base): 110 cm; height; weight: dry weight 5.5 kg, and with payload, it is 12–13 kg

Propellers: This is a hexacopter. It has six propellers made of plastic blades. Each propeller is connected to a brushless motor for propulsion.

Payload: This UAV holds a payload of 6 kg per flight. The payload includes sensors and liquid tank that hold spray material.

Launching information: AG-MRCD-6 has autolaunch, navigation based on predetermined flight path, and autolanding software. The VTOL allows the UAV to be launched from small areas. It lifts-off from in field location anywhere with 5 × 5 m clearance.

Altitude: It ranges from 1.0 ft over the canopy to 50 m above ground level. However, ceiling is 500–1000 m AGL.

Speed: Flight speed ranges from 0 (hovering or stand still) to 10 m s^{-1}

Wind speed tolerance: This UAV withstands 20 km h^{-1} disturbance. The spray rate and uniformity depend largely on wind caused disturbance

Temperature tolerance: −15 to 40°C

Power source: Electric batteries

Electric batteries: 10,000 mA h, 60 C

Fuel: Not an IC engine-fitted UAV

Endurance: Flight endurance during spraying pesticides or other plant protection chemicals is 5–10 min only.

Remote controllers: It has flight controller that directs its speed, height above the crop's canopy, and speed-FUTABA 8FG.

Area covered: This UAV covers about 5–50 ac per flight of 15–20 min. It can utilize variable-rate nozzles to spray pesticides/fungicides. This UAV great advantage over manual application of pesticides, using knap sack sprayers.

Photographic accessories: Cameras: When not in spraying mode or even during spraying schedules, the UAV can simultaneously obtain digital data and aerial images of the crop. It can obtain close-up shots of crop's canopy, leaves, twigs, and branches. The clarity is high. We can identify the pest/disease attack on the plants easily. In general, the gyro-stabilized gimbal is equipped to hold a series of visual (daylight), NIR, IR, red-edge, and thermal sensors.

Computer programs: Aerial imagery is processed using the software, such as Pix4D Mapper, Agisoft's, or similar custom-made computer programs.

Spraying area: It covers about 1–2 ac min^{-1}; diameter of spray drops is 80–120 μm; spray width (swath) is 3–4 m; spray speed is 0–8 m s^{-1}. It is adjustable by manipulating the nozzles or when variable-rate nozzles are used. Spray height is 1–3 m to the top of the crop.

Volume: This UAV is equipped with a submersible pump with an output flow rate of 0.2–0.4 L min^{-1}, but it is adjustable. The volume of the sprayer fitted to this UAV AG-MRCD-6 is relatively small and it allows stability during spray schedules.

Agricultural uses: AG-MRCD-6 is predominantly a sprayer UAV (Platte 3.43.1). It is used to spray crops with pesticides, fungicides, herbicides, or even liquid fertilizer formulations, such as urea. It is equipped with a series of sensors. Therefore, surveillance of crop fields, monitoring seeding, crop stand, and crop canopy development is also possible. Data about leaf-chlorophyll, crop's N status, and water stress index can also be collected and stored in the CPU of UAV itself. Otherwise, the data are relayed to ground control iPad. The close-up video images allow farmers to study the pest and disease progression, if any, rather closely and accurately. Natural vegetation monitoring is another possibility (Bowman, 2015).

Nonagricultural uses: They are surveillance of industrial installations, buildings, farm structures, highways, rail roads and dams, and water bodies, and also observing work progress in farms and tracking farm vehicles.

PLATE 3.43.1 AG-MRCD-6: A multirotor precision sprayer UAV.
Source: Mr. Terry Sanders, HSE LLC., Denver, CO, USA.

Useful References, Websites, and YouTube Addresses

Bowman, L. *UAVs for Farmers and Ranchers*; Homeland Surveillance and Electronics LLC., 2015; p 103 http://www.griculturuavs.com/uav_UAVs_for_farmers_rancher.htm (accessed May 23, 2015).

http://www.uavcropdustersprayers.com/agriculture_uav_crop_duster_sprayer_agmrcd6.htm (accessed June 1, 2017).

http://www.uavcropdustersprayers.com/ (accessed June 1, 2017).

https://www.youtube.com/watch?v=F_BG6KWkbTg (accessed June 1, 2017).

https://www.youtube.com/watch?v=S0mjaoqSm_Y (accessed June 1, 2017).

Name/Models: Ag-MRCD-18 (Agricultural Sprayer: Octocopter)

Company and address: Homeland Surveillance and Electronic LLC. (HSE LLC.), Denver, CO, USA

"AG-MRCD-18" is a multirotor UAV. It is primarily meant for agricultural crop dusting. It has autolaunch and landing facility. It is a low noise fertilizer duster. It adopts brushless motors to energize the rotors. It is a VTOL copter that launches from anywhere in a crop field. AG-MRCD-18 can fly with greater stability due its relatively smaller size, superpower system, autonavigation, and stability control accessories. It is easily portable, since the center UAV is fitted and packed into a small suitcase and carried as a backpack.

TECHNICAL SPECIFICATIONS

Material: This octocopter is built using a composite of carbon fiber, steel, ruggedized plastic, and foam.

Size: length: 180 cm; width of airframe: 250 cm; height: 45 cm; weight: 16 kg including payload

Propellers: The UAV has eight main rotors attached to small motors for propulsion. They are energized by electric batteries. Rotor diameter is 43 cm each.

Payload: Maximum payload is 5 kg. The payload usually includes tank that holds spray liquid/granules.

Launching information: This is an autolaunch (VTOL), autonomous flight, navigation, and autolanding UAV. The UAV has a wireless remote start for autolaunch facility. This facility is helpful while handling hazardous chemicals or even fertilizer formulations.

Altitude: It flies close to crop canopy during spraying. It can reach 50 m above canopy for aerial imagery. However, ceiling could be as high as 100 m altitude or even more, while on surveillance of large areas of crop land or urban zones.

Speed: This is relatively a slow transit copter. Mainly because, it is predominately a fertilizer spray UAV that adopts precision variable-rate nozzles. This procedure reduces its speed.

Wind speed tolerance: It tolerates 20 km h^{-1} wind disturbance. Ag-MRCD 18 comes with extra-gyro-stabilization and wind disturbance tolerance. Therefore, uniform and accurate spraying is possible.

Temperature tolerance: −15 to 60°C

Power source: Electric batteries

Electric batteries: 28 A h, four cells

Fuel: Not an IC engine-fitted UAV

Endurance: 5 min per flight, during an intense spraying schedule. It is usually 10–15 min, during rapid spraying flight.

Remote controllers: The GCS iPad keeps contact with UAV's CPU for up to 5000 m radius. The UAV's flight path and spray intensity can be altered, midway, by the GCS computers.

Area covered: Aerial photography of 50–100 ac h^{-1} flight is possible and about 15 ac per flight, if the UAV is used for intense spraying.

Photographic accessories: Cameras: It has a full complement of cameras, such as visual (daylight), NIR, IR, and mica-sense red-edge bandwidth.

Computer programs: The aerial images transmitted are processed, using software such as Agisoft's Photoscan or Pix4D Mapper.

Spraying area: It covers 10–15 ac per flight; spraying altitude is close to canopy, that is, up to 2–4 m above crop canopy; spraying speed is adjustable between 0 and 4 m s^{-1}; spraying swath width is 3–5 m wide. AG-MRCD-18 has two spray nozzles with ability for 0.2–0.25 L spray dispensation min^{-1} of flight. Spray droplet size is 150–300 μm.

Volume: This UAV unloads (sprays) about 10–15 L of liquid fertilizers/granules per flight of 10 min.

Agricultural uses: This is mainly a spraying UAV. It is used to spray plant protection chemicals and fertilizer formulations (Plate 3.44.1). However, aerial survey and crop scouting are also possible using visual camera. Natural vegetation monitoring is feasible.

Nonagricultural uses: They are surveillance of industrial sites, buildings, dams, reservoirs, mines, rail roads, city traffic, and public events.

PLATE 3.44.1 AG-MRCD 18 (octocopter sprayer).
Source: Mr. Terry Sanders, Homeland Surveillance and Electronics-HSE LLC., Denver, CO, USA.

Useful References, Websites, and YouTube Addresses

Bowman, L. *UAVs for Farmers and Ranchers*; Homeland Surveillance and Electronics LLC., 2015; p 103 http://www.griculturuavs.com/uav_UAVs_for_farmers_rancher.htm (accessed May 23, 2015).

http://www.uavcropdustersprayers.com/agriculture_uav_crop_duster_sprayer_agmrcd18.htm (accessed Aug 4, 2017).

www.youtube.com/watch?v=JsyXIvEhrZI (accessed Aug 4, 2017).

Name/Models: AG-MRCD-24 (Octocopter Sprayer UAV)

Company and address: Homeland Surveillance and Electronics (HSE) LLC., IL, USA

The AG-MRCD-24 is an agricultural crop duster UAV. It can take-off and land vertically in a narrow space compared with fixed-wing UAV. It has low noise signature. This multirotor crop duster is used predominantly as a fertilizer sprayer. It adopts the ultraquiet brushless motor compared with traditional brush motors. This kind of motor is characterized with high energy conversion efficiency and low noise. AG-MRCD-24 is a relatively small farm UAV (Plate 3.45.1). It is easily packed into a backpack for portability. It has superpower system that helps it to resist wind disturbance.

TECHNICAL SPECIFICATIONS

Material: AG-MRCD-24 is an octocopter made of composite material that includes steel frame, carbon fiber, and ruggedized foam.

Size: length: 180 cm; airframe width: 320 cm; height: 45 cm; weight: 19 kg

Propellers: It has eight rotors. The rotors have two plastic blades each. They are energized by brushless motors. The diameter of rotors is 43 cm.

Payload: 10 kg. It includes cameras, CPU, and tank with pesticide/fertilizer formulation.

Launching information: This is an autolaunch, autonomous flight, and autolanding octocopter. It needs a small clearing of 5 × 5 m for VTOL. The octocopter can be started, using remote control. This is helpful while handling hazardous pesticides/fungicides.

Altitude: It ranges from zero (hovering) to 4 m above crop canopy, during intense spraying schedules. Its flight altitude is 50 m above ground level, for general flight mode. However, ceiling is 1500 m above sea level.

Speed: 25–45 km h^{-1}; Spraying speed is 3–4 m s^{-1}

Wind speed tolerance: 25 km h^{-1} wind disturbance.

Temperature tolerance: −10 to 60°C

Power source: Electric batteries

Electric batteries: 28 A h; four cells

Fuel: Not an IC engine-fitted UAV

Endurance: 5 min of spraying per flight

Remote controllers: The GCS computer (iPad) is utilized to guide the launch, flight path, waypoints, and landing spot of the UAV. The telemetric zone is small. Yet, the UAV keeps in contact with GCS, for up to 15 km radius. Under semiautonomous mode, the GCS iPad can alter flight path, spray schedule speed/volume, and total area.

Area covered: It covers a linear distance of 15–20 km or 30–50 ha land area per flight.

Photographic accessories: Cameras: This UAV carries a set of visual, NIR, IR, and red-edge thermal cameras, for high-resolution imagery, thermal imagery, and detection of crop growth variations.

Computer programs: The GCS computers and CPU in the UAV adopt software, such as Pix4D Mapper for processing ortho-images.

Spraying area: The vehicle covers a spray area of 30 ac min^{-1}.

Volume: The UAV has three nozzles to spray liquid or granules. Spray rate: The UAV sprays 0.2–0.25 L min^{-1} (double-spray nozzle) and 0.3–0.6 L min^{-1} (triple-spray nozzle). Droplet diameter is 300 μm. The usual spraying altitude is 2–4 m above crop canopy (Plate 3.45.1).

Agricultural uses: AG-MRCD-24 is predominantly a pesticide/fertilizer sprayer UAV. It is used to spray the crop. It adopts digital data and/or images of crop depicting the spread of pest attack or disease. The UAV adopts digital data and variable-rate applicators. It distributes fertilizers based on variations in the nutrient status of the soil/crop. Crop scouting for canopy growth, NDVI data, leaf chlorophyll, and water stress are major aspects for which the octocopter is utilized (Bowman, 2015).

Nonagricultural uses: Surveillance of large industrial installations, mines, mining activity, highway traffic, rail roads, and oil pipelines are possible, using this octocopter.

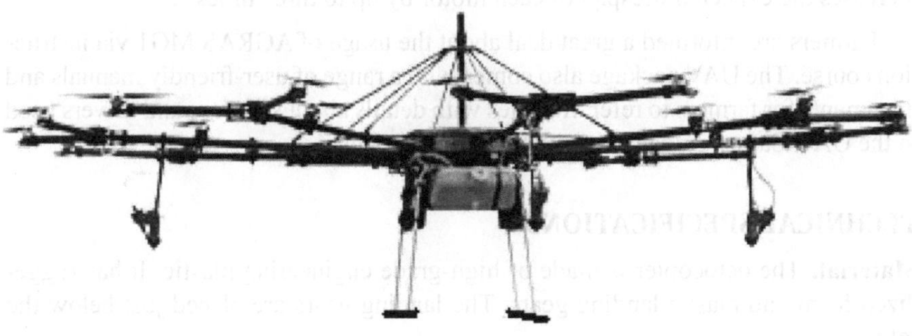

PLATE 3.45.1 AG-MRCD-24—an octocopter sprayer UAV.
Source: Mr. Terry Sanders, Homeland Surveillance and Electronics-HSE LLC., Denver, CO, USA.

Useful References, Websites, and YouTube Addresses

Bowman, L. *UAVs for Farmers and Ranchers*; Homeland Surveillance and Electronics LLC., 2015; p 103 http://www.griculturuavs.com/uav_UAVs_for_farmers_rancher.htm (accessed May 23, 2015).

http://uavcropdustersprayers.com/agriculture_uav_crop_duster_sprayer_agmrcd24.htm (accessed June 4, 2017).

Name/Models: AGRAS MG-1

Company and address: SZ DJI Baiwang Technology Company Ltd., 14th Floor, West Wing, Skyworth Semiconductor Design Building, No. 18 Gaoxin South 4th Ave., Nanshan District, Shenzhen 518057, China; phone: +86 (0)755 26656677

DJI of Shenzhen, China, produces a wide range of copters. Its models are currently the most popular worldwide. They are handy, often small, and versatile. They have many applications in civil and agricultural sectors.

AGRAS MG-1 is a specialty UAV product by DJI Ltd., China. It is meant to be adopted during precision farming, also during regular blanket sprays of plant protection chemicals and fertilizer. It has been adopted already in many agrarian regions worldwide, successfully. AGRAS MG-1 is an octocopter with VTOL capability. Hence, it is preferred in farms. It holds about 10 l by volume. It sprays rapidly in the air (Plate 3.46.1). So, it reduces costs on farm labor. It avoids contact with hazardous chemicals.

The equipment used in plant protection operations is susceptible to dust and corrosion. It can lead to high cost of maintenance and shortened lifespans. To counteract degradation, the MG-1 is designed with a sealed body and an efficient, integrated centrifugal cooling system. The combination of cooling and filtering increases the expected lifespan of each motor by up to three times.

Farmers are informed a great deal about the usage of AGRAS MG1 via instruction course. The UAV package also comes with a range of user-friendly manuals and documents for farmers to refer. It comes with details about software and drivers used in the UAV. Safety guidance manuals are also included.

TECHNICAL SPECIFICATIONS

Material: The octocopter is made of high-grade engineering plastic. It has ruggedized foam and plastic landing gears. The landing gears are placed just below the rotors.

Size: length: the frame arm length is 62.5 cm; width (rotor to rotor): 1471 cm; height: 48.2 cm (all dimensions in unfolded state). The dimensions reduce to 780 × 780 × 482, if arms are folded; weight: total weight without batteries is 8.888 kg. Standard take-off weight is 22.5 kg. MTOW is 24.5 kg.

Propellers: Propulsion is through motors connected directly to propellers. The size of the motor is 6.0 cm × 1.0 cm. It generates 130 rpm/V. Maximum thrust per motor is 5.1 kg. Maximum power generated is 780 W. The weight of the motor is 280 g each. The propellers are foldable. They are made of rugged plastic. They weigh 58 g each and are 21 × 7.0 in. in size.

Payload: 10 kg plant protection chemical, fuel batteries plus cameras, and CPU computer

Launching information: This octocopter is autolaunch, autopilot, and navigation and autolanding type. It has a lightweight landing gear made of aluminum alloy. Its flight can be predetermined, using Pixhawk or custom-made software. The flight path can be altered midway, using GCS mobile/iPad. The UAV has series of fail-safe and early warning systems. It returns to point of launch in times of emergency, depletion of spray material, or malfunctioning of nozzles or errors in reading digital data by precision farming software.

Altitude: This UAV can fly close to the crop canopy to obtain aerial pictures or to spray/dust chemicals, avoiding drift. The operating height during spraying is zero to 5–50 m. It can fly swiftly at 300 m above crop canopy. However, ceiling is 500 m above crop canopy.

Speed: The operating speed is 8 m s^{-1}. Maximum flying speed is 22 m s^{-1}.

Wind speed tolerance: 20 km h^{-1} disturbance

Temperature tolerance: Operating temperature is 0–40°C.

Power source: Lithium–polymer electric batteries

Electric batteries: It has a designated battery (MG 12,000 mA h). It consumes 6400 W power, while spraying. It takes 3250 W, if hovering at a spot for aerial imagery.

Fuel: Not an IC engine-fitted UAV

Endurance: 45 min per flight

Remote controllers: The custom-made remote controller model is GL628C. The operating frequency for telemetric link is 2.400–2.483 GHz. The signal transmission range is 1.0 km, in a farm location. Remote controller has a built-in battery of 6000 mA h, two lithium–polymer batteries. The batteries are charged, using DJI charger that has 9 W output. Remote controller with latest lightbridge two-transmission systems enables ultralow-latency controls. The remote is designed with water and dust protection. A special low-energy display panel gives real-time flight information. It lasts over extended periods of time on a single charge. Overall, joysticks and convenient function keys help in mastering the UAVs flight and functioning.

Area covered: It covers 7–10 ac of crop field for each 1 h of spraying flight. The rate can be modified to hasten spray and cover larger area, if needed.

Photographic accessories: Cameras: The UAV can be fitted with a series of cameras. For example, it can carry cameras such as visual (Sony 5100), multispectral cameras, NIR, IR, red-edge, and a Lidar sensor.

Computer programs: The aerial imagery is processed, using custom-made software or Pix4D Mapper or Agisoft's Photoscan.

Spraying area: AGRAS MG-1 carries 10 kg of liquid payloads, including pesticide and fertilizer (Plate 3.46.1). The combination of speed and power means that an area of 7–10 ac can be covered, in just 10 min of flight. It amounts to 40–60 times faster than manual spraying operations, by farm laborers. The praying system automatically adjusts its spray according to the flying speed so that spray is always uniform. Therefore, amount of pesticide or fertilizer is precisely regulated to avoid accumulation in soil or on the canopy. It avoids pollution. It economizes on pesticide quantity and reduces cost of production proportionately. The MG-1 automatically records its current coordinates. It also remembers its past coordinates as it makes its way, across the field, in case the operation is interrupted, for example, due to depleted battery or spraying liquid. Then, flight could be easily resumed from the last point, in its memory, after changing the battery or refilling its tank.

Volume: The liquid tank holds 10 l in liquid form or about 100 kg plant protection chemical in granular form.

Agricultural uses: It conducts aerial surveys. Crop scouting is done to collect data for a series of physiological and growth parameters, leaf chlorophyll content (crop-N status), etc. (Caturegli et al., 2016; DJI, 2016) Spraying plant protection chemicals is a major application of AGRAS MG 1 octocopter. Natural vegetation monitoring and detection of flooded or drought-affected zones are other functions possible, using AGRAS MG1 UAV.

Nonagricultural uses: Surveillance of industries, mines, highways, and oil pipelines are also possible, using the octocopter. However, the UAV is designed primarily to conduct aerial spraying of crop field.

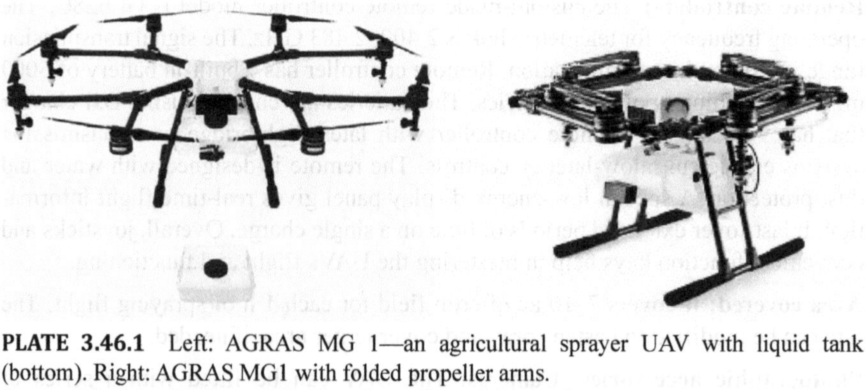

PLATE 3.46.1 Left: AGRAS MG 1—an agricultural sprayer UAV with liquid tank (bottom). Right: AGRAS MG1 with folded propeller arms.
Source: Mr. Adam Najberg, DJI, Shenzhen, China (Mainland).

Useful References, Websites, and YouTube Addresses

Caturegli, L.; Comeglia, M.; Gaetani, M.; Grosssi, N.; Magni, M.; Miglizzi, M.; Angelini, L.; Mazzoncini, M.; Silvestri, N.; Fontanelli, M.; Raffaelli, P.; Peruzzi, A.; Volterrani, M. Unmanned Aerial Vehicle to Estimate Nitrogen Status of Turfgrass. *PLoS One* **2016,** *11*, 1–9. https://doi.org/10.1371/journal.pone0158268 (accessed July 27, 2017).

DJI. Above the World. *UAS Mag.* **2016,** 238. http://www.uasmagazine.com/articles/1591/dji-explains-new-book-of-UAV-captured-images (accessed Apr. 20, 2017).

https://www.dji.com/mg-1 (accessed Sept 3, 2017).

http://www.dji.com/mg-1/info (accessed Sept 3, 2017).

http://enterprise.dji.com/agriculture (accessed Sept 3, 2017).

http://www.smallUAVsreview.com/2015/11/28/dji-agras-mg-1-octocopter-UAV/ (accessed Sept 3, 2017).

https://www.youtube.com/watch?v=P2YPG8PO9JU (accessed Sept. 3, 2017).

https://www.youtube.com/watch?v=ahrMJlYAiFk (accessed Sept 12, 2017).

https://www.youtube.com/watch?v=3swAXfkiWvk (accessed Sept 12, 2017).

Name/Models: AGStar X8, the Precision Agriculture UAV

Company and address: Rise Above Custom UAV Solutions; Unit 2, 40 Dunn Rd.; Smeaton Grange, NSW 2567, Australia; phone: +61 (02) 4647 3450; sales support—sales@riseabove.com.au; technical support—support@riseabove.com.au; website: http://www.riseabove.com.au/agstar-precision-ag-UAV

The AgStar X8 is touted by the company as a UAV meant for precision agriculture. It is a sprayer (seed spreader), feed distributer in a cattle ranch and a bird scarer in a cereal/legume field with ripe panicles, a versatile sprayer UAV indeed. A complete quadcopter package consists of thoroughly tested frame (body). It has a folding frame for easy portability, a folding landing gear made of light Al alloy, a set of aerial photography cameras, camera shutter controllers, and flash controllers. It also has UAV flight controllers for smooth and stable flight mode and throttle to change speeds (Plate 3.47.1). The UAV also comes with DJI Naza V2 flight controller, a Futuba flight radio for telemetric link, 2.4 Hz waypoint module, 125,000 mA h battery for power supply, a lithium–polymer charger and tester. Really, it includes a series of items that make the UAV complete. This versatile UAV along with accessories currently (February 2017 price level) costs about Australian $ 9000 per unit.

TECHNICAL SPECIFICATIONS

Material: The UAV's fuselage and frame are made of a composite of carbon fiber, rugged foam, and plastic. The landing gear is made of lightweight alloy.

Size: length 92 cm; width (rotor to rotor): 120 cm; height: 65 cm; weight: maximum weight with payload accessories is 10.0.kg plus seed/feed weight.

Propellers: The Quadcopter has four plastic propellers. Each propeller has two plastic blades.

Payload: It allows 4–6 kg payload that includes multiple sensors, a bird scarer, or water sampler or seed/feed spreader or sprayer. The seed sprayer or spreader/feed spreader payload includes Vulcan quick installation kit, automated seed/feed spreader (sprayer), and an additional bird scarer to ward-off bird pests while seeding. The AgStar Bird Scare payload includes Vulcan quick release plate, bird guard bird scarer system, and an additional speaker to create louder noise.

Launching information: The quadcopter is an autolaunch, autopilot, and autolanding vehicle. The flight path is decided using software such as Pixhawk. The landing gear made of alloy tubes allow smooth landing.

Altitude: It can hover over crop's canopy for long stretches of close-up photography. The operation altitude is 5–20 m above crop's canopy. The ceiling is 50–100 m above crop canopy.

Speed: 30 km h^{-1} during aerial photography, 20–25 km h^{-1} during seed spreading or bird scaring flights.

Wind speed tolerance: 15–20 km disturbing winds

Temperature tolerance: −10 to 50°C

Power source: lithium–polymer electric batteries

Electric batteries: It is recommended to use 10,000 mA h lithium–polymer battery, if winds are 20 km h^{-1}. If heavier payload is anticipated, a 12,000-mA h (maximum) lithium–polymer battery is preferred. It supports flight time of 30 min.

Fuel: Not an IC engine-fitted UAV

Endurance: It is 30 min with 12,000 mA h batteries. It is 25 min, if bird scarer mission is opted. If dual payload is opted, the expected flight time is only 15–10 min. Flight duration is also affected by the flight pattern and intensity of seed/feed spreading. In case of extended flight with higher payload, 2 × 12,000 mA h batteries are needed. The UAV comes with a single port battery charger system.

Remote controller: It has a Black Pearl monitor of 7 in. diameter, long-range antenna set, an intelligent on-screen display of telemetry, and data display. The GCS has Futuba 10J flight radio, an iPad system to decide flight pattern and waypoints. The telemetric with UAV is restricted to 1 km radius. The GCS has a Futaba T14SG Remote Controller + Charger.

Area covered: It covers 50–100 ac day^{-1} of 6 h.

Photographic accessories: Cameras: AgStar quadcopter is usually fitted with visual sensors (Go Pro Hero with H3/H4 stabilized brushless gimbal and a customized 20× zoom camera). The multispectral NDVI sensor has a mica-sense red-edge camera with stabilized brushless gimbal. It has MAP IR hard mounted sensor and thermal sensors to identify variations in canopy temperature and water status of crops. An advanced package with two-axis stabilized gimbal is attached. A big thermal two-axis stabilized gimbal can also be included. Most models are attached with both visual (GoPro Hero) and thermal sensors.

Computer programs: The aerial photographs, both stills and videos, are processed, using Pix4D Mapper or Agisoft Photoscan or a custom-made software.

Spraying area: The crop-spraying payload usually includes spray nozzles and precision spray software. They are available specifically for the octocopter. But it could be modified if needed and attached to a quadcopter.

Volume: Not data available for pesticide spray tanks.

Agricultural uses: Major applications of this UAV are in precision agriculture, particularly in obtaining NDVI data, deciphering crop's N requirement, water needs, and irrigation scheduling, determining crop grain maturity, etc. Water sampling from dams, rivers, and irrigation is possible. Bird scaring in areas affected by bird pests, particularly during grain maturity and harvest stages (Plates 3.47.2–3.47.4) is possible.

Nonagricultural uses: They are surveillance of buildings, industrial installations, mines, mine ore dumps, railways, city traffic, public events, dams, reservoirs, rivers, etc.

PLATE 3.47.1 A quadcopter sprayer UAV.
Source: Riseabove Inc. Australia; http://www.riseabove.com.au/agstar-precision-ag-UAV.
Note: This UAV is used to spray seed and microbial inoculants or biological control formulations on crop fields. It is also used as a bird scarer.

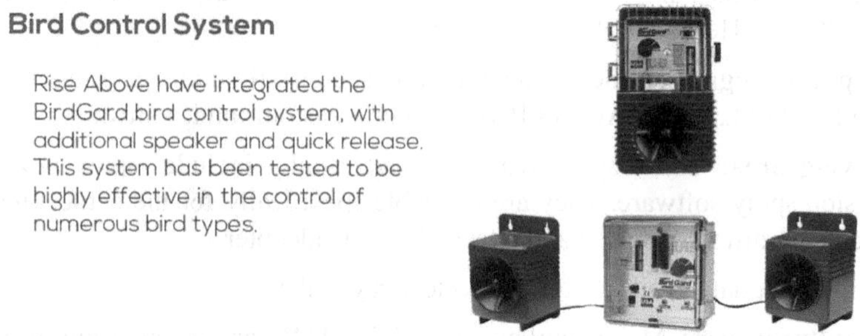

PLATE 3.47.2 A bird control system.
Source: Mr. Vick M. Sales and Marketing Division, Rise-Above Custom UAV Solutions Inc. Smeaton Grange, New South Wales, Australia; http://www.riseabove.com.au/agstar-precision-ag-UAV.

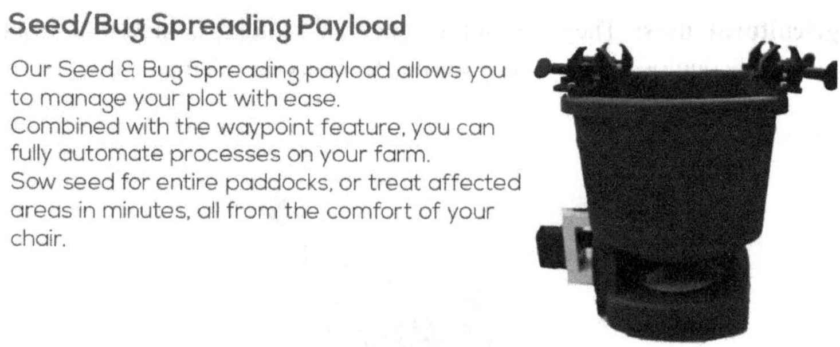

PLATE 3.47.3 A seed spreader.
Source: Mr. Vick M. Sales and Marketing Division, Rise-Above Custom UAV Solutions Inc. Smeaton Grange, New South Wales, Australia; http://www.riseabove.com.au/agstar-precision-ag-UAV.

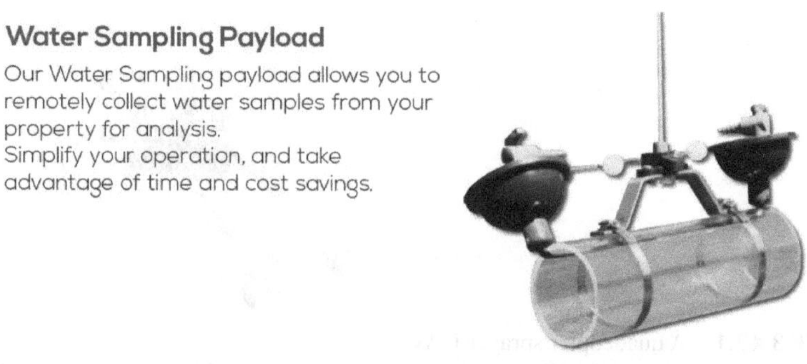

PLATE 3.47.4 A water-sampling equipment.
Source: Mr. Vick M. Sales and Marketing Division, Rise-Above Custom UAV Solutions Inc. Smeaton Grange, NSW, Australia; http://www.riseabove.com.au/agstar-precision-ag-UAV.

Useful References, Websites, and YouTube Addresses

http://www.riseabove.com.au/agstar-precision-ag-UAV (accessed May 17, 2017).
http://www.riseabove.com.au/agstar-go-pro-fpv-payload (accessed May 17, 2017).
http://www.riseabove.com.au/rise-above-seed-feed-pellet-spreader-dry-material (accessed May 17, 2017).

Name/Models: AG-UAV 8 A (Octocopter)

Company and address: Homeland Surveillance and Electronics (HSE) LLC., Denver, CO, USA

"AG-UAV Sprayer Model 8A" is among the most advanced octocopter sprayer UAV. It has immense value to farmers, particularly to those who aim to cover large areas of crops with pesticide sprays, in short time. It covers over 1000 ac in 30 h of flight. It equates almost 34 ac h^{-1} of flight. There are four submodels with slight variations to suit the purpose. This UAV may cost around US$ 25,000 and it comes with 1-year warranty from the company—HSE LLC.

TECHNICAL SPECIFICATIONS

Material: This octocopter is made of light aluminum and carbon fiber (Plate 3.48.1).

Size: Wheel base is 130 cm (115–145 cm); height: 65 cm; weight: 9.5 kg (plus payload 10 kg), MTOW is 30.5 kg

Propellers: There are eight rotors. Each rotor is supported by a brushless motor 380 rpm/V × 8. Propeller is 18.5–30-in. wingspan. Propellers are made of carbon fiber.

Launching: This is an autolaunch VTOL model. It can take-off from any point in field with 5 m × 5 clear space. It has software for fail-safe. The UAV returns to point of launch in case of emergency.

Payload: Payload includes pesticide tank, spray bar, and nozzles. This is in addition to CPU and gimbal.

Altitude: The octocopter can fly at very close heights above the crop's canopy or ground, during spraying. It flies at 10–50 m above crop's canopy, during spraying. The ceiling could be increased to 300–400 m above ground level, if needed, during aerial survey and mapping. Maximum launch altitude is 8000 ft above sea level.

Speed: The flight speed is controlled using Type 60A and radiofrequency of 400 Hz.

Wind speed tolerance: It works optimally at 20–40 km h^{-1} disturbing winds. It is water resistant and can withstand slight drizzle. Electronic components are sealed tightly.

Temperature tolerance: It is rated to work normally between −20 and 65°C.

Power source: Electric batteries

Electric batteries: This UAV is equipped with 16,000 mA h, 6S, 22.2 C, DC, lithium–polymer × four cells.

Fuel: Not an IC engine-fitted UAV.

Endurance: It is 10–20 min. But it depends on wind disturbance, flight speed, and volume of spray.

Remote controllers: The GCS includes a joystick, if the UAV's flight and spray rate are not preprogramed.

Area covered: During spraying, this UAV covers an area of 33 ac h^{-1} with a swath width of 16 ft.

Photographic accessories: Cameras: This UAV can be utilized for aerial imagery. It has the usual complement of visual, NIR, and IR cameras. It can also be utilized to obtain close-up shots of the crop, leaf, and canopy by using it in hovering mode. Aerial images showing pest/disease-affected areas could be mapped, using appropriate spectral signatures of healthy and affected areas.

Computer programs: The aerial images can be processed using Pix4D Mapper or Agisoft's Photoscan. The areas affected with pest/disease should be marked with prior knowledge about spectral signatures of healthy and disease-affected crop.

Sprayer and spraying area: Tank capacity is 10 kg; spray flow is 0.8–1.8 L min^{-1}. Sprayers fitted with high efficiency nozzles are fitted. They can be controlled via computer software during variable-rate spray.

Agricultural uses: Spraying crops with plant protection chemicals and fertilizer formulations are the functions of this UAV.

Nonagricultural uses: They are surveillance of industrial sites, buildings, mines, railways, city traffic, etc.

PLATE 3.48.1 AG-UAV 8A.
Source: Mr. Terry Sanders, Homeland Surveillance and Electronics-HSE LLC., CO, USA.

Useful References, Websites, and YouTube Addresses

http://www.uavcropdustersprayers.com/agriculture_octo_copter_uav_crop_duster_v8a.htm (accessed Jan 12, 2018).

http://www.hse-uav.com/ (accessed Jan 12, 2018).

https://www.youtube.com/watch?v=1bcfwJJAq3w (accessed Jan 12, 2018).

https://www.youtube.com/watch?v=juZm-LPy680https://www.youtube.com/watch?v=juZm-LPy680 (Jan. 12, 2018).

Name/Models: Cellularlink LSL 818—An Agricultural Sprayer UAV

Company and address: Guangdong Cellularlink Electronic Incorporated Company, Cellularlink Innovation Park, No. 5, Huangdongduan 2nd Rd., Golden, Industrial Area, Fenggang Town, Dongguan, Guangdong 523690, China; phone: +86 755 84712786; fax: +86 755 84712786; e-mail: amy@cellularlinkcn.com; website: http://www.globalsources.com/cellularlinkcn.co

Guangdong Cellularlink Electronic Incorporated produces multicopter UAVs. They are meant exclusively for spraying plant protection chemicals and fertilizer formulations to crops. At present, these spraying UAV models are spreading into farm belts of northeast and southern China. They are sold to farming stretches in Australasian nations, Far-East (Japan), and parts of Eastern Europe. The relatively lower cost of the UAV, versatility in flight and spraying, and commensurate returns on the use of UAVs are quoted as advantages. The adoption of UAV's reduces need for farm labor to a great extent, during spraying and applying fertilizer. The UAV model LSL 818 costs US$ 5000 per unit (February 2017 price level).

"Cellularlink LSL-818" is a quadcopter of simple design. It has a 2-m long spraying bar with nozzles. The nozzles could be directed to apply liquid at variable-rates based on digital data. It has a moderately larger tank for pesticide. It unloads pesticide at a rapid rate so that farmers can cover larger areas of fields per hour or day. The use of UAV reduces quantity of pesticide needed, because we can apply chemicals at only spots attacked by and deteriorated.

TECHNICAL SPECIFICATIONS

Material: It is made of light aluminum-based alloy, carbon fiber air frame, plastic, and rugged foam.

Size: length: 70 cm; width (wingspan): 70 cm; wheel base 130 cm; height: 70 cm; weight: MTOW is 32 kg (it includes two pieces of batteries).

Propellers: It is quadcopter with four propellers. Propellers are made of two plastic blades. Rotors are attached to brushless motors for propulsion. Motors have an rpm of 5330 rotations min^{-1}.

Payload: 10 kg of liquid/granular/dust formulation of plant protection chemicals/fertilizer. Maximum permissible payload is 15 kg.

Launching information: It is an autolaunch and navigation vehicle. Its flight path pattern is determined using software such as Pixhawk.

Altitude: The working height during pesticide spray is 2–5 m based on crop and formulation. The ceiling is 3200 m above crop's canopy.

Speed: The working speed is 3 m s^{-1}. It is a relatively slow UAV while spraying.

Wind speed tolerance: It tolerates wind speeds of 15–20 km h^{-1} during flight.

Temperature tolerance: −20 to 55°C

Power source: Electric batteries

Electric batteries: 16,000 mA h, two pieces

Fuel: Not an IC engine-fitted UAV

Endurance: 25 min per flight

Remote controllers: The UAV's flight pattern and height are usually predetermined. However, it can be altered and controlled, using joystick or iPad or PC (i.e., GCS). The UAV has GPS connectivity.

Area covered: It covers 50–250 ha day^{-1}, depending on intensity of spraying schedules and flight pattern.

Photographic accessories: Cameras: More commonly used for spraying. It could be attached with sensors, if needed.

Computer programs: The digital data are usually obtained prior to use of this UAV for spraying. The digital maps and pesticide/fertilizer recommendations are obtained on chip. This information is placed in the UAV's CPU. The UAV conducts variable-rate spray at high accuracy, using computer directions.

Spraying area: It covers an area of 33,350 m^2 h^{-1}. Spraying rate is 1–1.5 l min^{-1}. Spraying width is 3 m.

Volume: The pesticide tank holds 10 L formulation.

Agricultural uses: "LSL-818" is designed for spraying pesticides. It avoids contact of farm labor with hazardous chemicals. It covers large expanses of crop field with pesticide/fungicide/weedicide in a matter minutes/hour. It can also be used to apply fertilizer formulations as foliar fertilizer (liquid formulation).

Nonagricultural uses: It could be used to carry payload from one location to other. It can transport hazardous chemicals and explosives safely.

Useful References,W, and YouTube Addresses

http://www.cellularlinkcn.com/product/cellularlink-agriculture-spraying-UAV-with-10kgs-spray-tan.html (accessed Mar 8, 2017).

http://static.cellularlinkcn.com/images/productimages/2016/1130/agricultural-UAV-4-axis.jpg (accessed Mar 8, 2017).

http://www.globalsources.com/gsol/I/Agricultural-UAV/p/sm/1146180788.htm?tests (accessed Mar 8, 2017).

http://www.cellularlinkcn.com/ (Mar. 8, 2017).

Name/Models: Digital Eagle YM6160

Company and address: Jiangsu Digital Eagle Technology Co. Ltd., Room 201-208, Building A1, No. 999, Gaolang East Road, Wuxi 214000, Jiangsu, China; phone: NA; e-mail: NA; website: NA

Jiangsu Digital Eagle Technology Development Co. Ltd. is a high technology avionics company. It was initiated in 2013. It specializes in producing multicopter UAVs for use in agricultural fields. They churn out over 100 units of crop duster units per month. It has servicing centers in different parts of China. Its UAVs are utilized worldwide.

This is a professional agricultural UAV meant to spray a crop field with plant protection chemicals, quickly and efficiently. They are commonly called crop-dusting UAVs or flying dusters in China. They could be used both to conduct blanket sprays of entire farm quickly, also to spray the crop using variable-rate technology. The precision techniques (i.e., variable-rate methods) need CPU, software to read, and then command activity of spray nozzles. They should be fixed with variable-rate nozzles. It holds 5–15 l of pesticide formulation based on model (Plate 3.49.1). Digital Eagle's UAV is a hexacopter and is easily portable inside a custom made aluminum suitcase. This UAV with accessories costs but US$ 5000 per unit (March 2017 price level).

TECHNICAL SPECIFICATIONS

Material: It is made of carbon fiber and toughened plastic.

Size: length: 120 cm; width (rotor to rotor): 120 cm; height: 45 cm; weight: MTOW 21.9 kg

Propellers: Six rotors attached to brushless motors and energized by electric batteries

Payload: 10 kg carried in a plastic can or tank

Launching information: It is an autolaunch, autonomous navigation UAV. It adopts software such as Pixhawk to decide UAV's flight path.

Altitude: The stipulated altitude is 0.5–5 m above crop's canopy.

Speed: Working speed is 5–10 s^{-1}.

Wind speed tolerance: It resists 15 km h^{-1} wind speed, and to avoid drifting of pesticide, it should be used in low wind condition.

Temperature tolerance: −15 to 50°C

Power source: Electric batteries

Electric batteries: Two units of 12,000 mA h lithium–polymer batteries

Fuel: It does not need gasoline

Endurance: 10–15 min per flight

Remote control: The GCS has laptop, mobile, and telelink apparatus. The telelink operates only for 5000 m radius. The laptop has necessary software to predetermine flight path using Pixhawk software. It can also be used to override and alter the flight path spray speed midway during a flight program.

Area covered: It covers 25–50 ac per flight of 15 min.

Photographic accessories: Cameras could be fitted if needed

Computer programs: This UAV is not utilized much for aerial photography, so no emphasis on processing images.

Agricultural uses: This UAV is meant specifically to spray crop fields and natural vegetation with liquid/granular/dust (powdered) formulation. It can also be used to conduct foliar spray of liquid fertilizer formulation, for example, to spray 0.2% urea on crop canopy.

Nonagricultural uses: Digital Eagle can be used to carry small packets with useful luggage. It can be used effectively to transport hazardous chemicals such as acids, insecticides, etc.

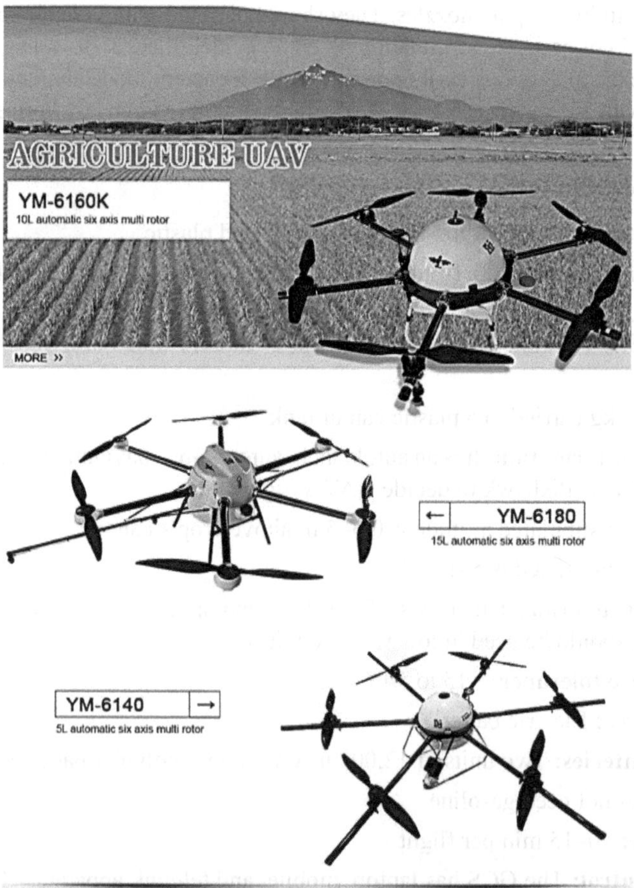

PLATE 3.49.1 Digital Eagle—agricultural sprayer UAV.
Source: Jiangsu Digital Eagle Technology Development Centre, Wuxi, Jiangsu, China Main land. https://uav.en.alibaba.com/?spm=a2700.8304367.0.0.zubs6E.
Note: These are three different models of Digital Eagle agricultural UAV with different pesticide payload.

Useful References, Websites, and YouTube Addresses

https://uav.en.alibaba.com/?spm=a2700.8304367.0.0.zubs6E (accessed Jan 12, 2017).
https://www.youtube.com/watch?v=cB_izfunnn4 (accessed Jan 12, 2017).
https://www.youtube.com/watch?v=VzQ5K6LlPuo (accessed Jan 12, 2017).
https://www.youtube.com/watch?v=cB_izfunnn4 (accessed Jan 12, 2017).
https://www.youtube.com/watch?v=OHPsUrL3YGg (accessed Jan 12, 2017).

Name/Models: FH-8Z-5 Crop Duster Sprayer

Company and address: Shandong China Coal and Mining Industries, No. 11, North of Kaiyuan Road, High-Tech Zone, Jining, Shandong 272000, China; phone: +86 537 2350961; http://chinacoalintl.en.made-in-china.com

FH-8Z-5 is a multirotor sprayer UAV. This UAV was manufactured in 2015. It has option for six or eight rotors. The pesticide payload tank holds 5 kg formulation. It has an excellent sprayer system, with atomizer nozzles and high-efficiency delivery system. Sprayer kit includes dispersion spray system and booster pump. It offers uniform spray. FH-8Z-5 creates turbulence that shakes the crop foliage. This occurs during spraying from low heights above the crop canopy. It covers about 200–300 ac of crop fields per day. This sprayer UAV can also be provided with software for fertilizer formulation application at variable rates. It is a low-cost UAV costing US$ 2000–4000 per unit and depending on accessories opted.

TECHNICAL SPECIFICATIONS

Material: Carbon fiber, plastic, and light avionics aluminum

Size: length: 115 cm; width (rotor to rotor): 115 cm; height: 50 cm; weight: 7 kg without payload, MTOW is 14 kg, pesticide load is 5 kg; folded size: 43.5 × 43.5 × 50 cm.

Propellers: It has six or eight toughened propellers made of carbon fiber. They are specifically meant for sprayer UAVs.

Payload: 5 kg liquid/granular formulation in plastic tank.

Launching information: It is an autolaunch aerial robot. Its navigation during loiter and spraying is autonomous. It is a VTOL UAV and can be recovered to land at even small locations, in the farm. This model can take-off or land in 1 min. The software such as Pixhawk allows the UAV to return to spot of launching in times of emergency.

Altitude: The operating height is 0–200 m above crop's canopy. However, ceiling is 1000 m above crop's canopy.

Speed: Zero (hovering) to 8 m s^{-1}. Flying height is 200 m above crop's canopy.

Wind speed tolerance: Resists 15–20 mi h^{-1} disturbing winds

Temperature tolerance: −15 to 40°C

Power source: Lithium–polymer electric batteries

Electric batteries: 22.2 V/16,000 mA h (plus two sets of chargers).

Fuel: It is not based on gasoline.

Endurance: It is 10–15 min per flight

Remote controllers: The ground control has nine channel radiolink with the sprayer UAV. The remote control is equipped with remote controller (joystick), a laptop with software programs to control the UAV's flight path and spray rate. The UAV keeps link with GCS for 3000 m radius during spraying.

Area covered: It covers over 200–300 aces sprayed with pesticide per day.

Photographic accessories: This is a sprayer UAV. It is not used commonly to obtain detailed aerial imagery. However, this UAV can be fixed with a full complement of visual, NIR, and IR cameras if needed. If utilized for precision farming, then this UAV takes command from computers. The computer decision-support is based on digital data obtained previously, by other UAV cameras.

Computer programs: This is a sprayer UAV and GCS may not require facility to obtain or process orthomosaics.

Spraying area: It covers 200–300 ac day^{-1} of 3–5 h. The spray system has dispersion device and six diffuse atomizer nozzles. Motor for spray pump operates at 1200 rpm. Spray width is 4 m. Spray height is 0.5–4 m above crop canopy. Spray flow is 0.25–0.75 L min^{-1}. Spray efficiency is 1–3 ac min^{-1}

Agricultural uses: Crop spraying with pesticides and/or liquid fertilizer formulations.

Nonagricultural uses: It could be used to transport small luggage, hazardous chemicals packets for delivery at different spots.

Useful References, Websites, and YouTube Addresses

http://www.alibaba.com/product-detail/FH-8Z-5-uav-helicopter-crop_60326161665... (accessed Feb 17, 2017).

http://chinacoalintl.en.made-in-china.com/product-group/FoRJKPGHsUrp/UAV-Sprayer-UAV-1.html (accessed Feb 17, 2017).

http://chinacoalintl.en.made-in-china.com/product/yKgQHItrumRY/China-Agriculture-UAV-Agriculture-UAV-Fh-8z-5-Details.html (accessed Feb 17, 2017).

http://chinacoalintl.en.made-in-china.com (accessed Feb 17, 2017).

Name/Models: FH-8Z-10 Crop Duster Sprayer

Company and address: Shandong China Coal and Mining Industries, No. 11, North of Kaiyuan Road, High-Tech Zone, Jining, Shandong 272000, China; phone: +86 537 2350961; website: http://chinacoalintl.en.made-in-china.com; https://www.chinacoalintl.com

FH-8Z-10 is a multirotor sprayer UAV. This UAV has been manufactured since 2015. It has option for six or eight rotors. The pesticide payload tank holds 10 kg formulation. It has an excellent sprayer system with atomizer nozzles and high efficiency delivery system. The sprayer kits help in dispensing pesticides from very low height above crop's canopy. It covers about 200–300 ac of crop fields per day. This UAV can also be provided with software for fertilizer formulation application at variable rates. It is a low-cost UAV at US$ 2000–4000 per unit and depending on accessories opted.

TECHNICAL SPECIFICATIONS

Material: It is made of carbon fiber, plastic, and light avionics aluminum.

Size: length: 135 cm; width (rotor to rotor): 135 cm; height: 60 cm; weight: 15 kg without payload, MTOW is 29 kg, pesticide load is 10 kg; folded size: 80 × 80 × 60 cm.

Propellers: Six or eight toughened plastic propellers of 23 in. and made of carbon fiber and meant specially for sprayer UAVs.

Payload: 10 kg liquid/granular formulation in plastic tank.

Launching information: It is an autolaunch aerial robot. Its navigation during loiter and spraying is autonomous. It is a VTOL UAV and can be recovered at any small location in the farm. This model can take-off or land in 1 min. The software such as Pixhawk allows the UAV to return to spot of launching in times of emergency.

Altitude: Operating height 0–200 m above crop's canopy. Ceiling is 1000 m above canopy.

Speed: It ranges from 0 (hovering) to 8 m s^{-1}. Flying height is 200 m above crop's canopy.

Wind speed tolerance: Resists 15–20 mi h^{-1}

Temperature tolerance: −15 to 40°C

Power source: Lithium–polymer electric batteries

Electric batteries: 22.2 V/16,000 mA h (plus two sets of chargers)

Fuel: It is not based on gasoline

Endurance: It is 10–15 min per flight

Remote controllers: The ground control has nine channel radiolink with the sprayer UAV. The remote control is equipped with remote controller (joystick), a laptop with software programs to control the UAV's flight path and spray rate. The UAV keeps link with GCS for 3000 m radius, during spraying.

Area covered: It sprays 100–200 ac of crop land with pesticide per day.

Photographic accessories: This is a sprayer UAV. It is not used commonly to obtain detailed aerial imagery. However, this UAV can be fixed with a full complement of visual, NIR, and IR cameras if needed. If utilized for precision farming, the UAV takes command from computers based on digital data obtained separately from other UAV cameras.

Computer programs: This is a sprayer UAV. So, GCS may not have facility to obtain or process orthomosaics.

Spraying area: It covers 200–300 ac day^{-1} of 3–5 h. The spray system has dispersion device, and six diffuse atomizer nozzles. Motor for spray pump operates at 1200 rpm. Spray width is 6 m. Spray height is 0.5–4 m above crop canopy. Spray flow is 0.25–0.75 L min^{-1}. Spray efficiency is 1–3-ac min^{-1}.

Agricultural uses: Spraying crops with pesticides and/or liquid fertilizer formulations are the major purpose.

Nonagricultural uses: It could be used to transport small luggage, hazardous chemicals, and small packets for delivery, at different spots.

Useful References, Websites, and YouTube Addresses

https://www.zmgkmachinery.com (accessed Mar 10, 2017).

https://m.zmgkmachinery.com (accessed Mar 10, 2017).

https://www.chinacoalintl.com (accessed Mar 10, 2017).

http://www.alibaba.com/product-detail/FH-8Z-5-uav-helicopter-crop_60326161665 (accessed Mar 10, 2017).

https://alibaba.com/product-detil/FH-8Z-10-Unmanned-aerial-crop_60311726../ (accessed Mar 10, 2017).

https://www.alibaba.com/product-detail/FH-8Z-10-UAV-Unmanned-aerial_60312420290.html (accessed Mar 10, 2017).

Name/Models: Hercules 20 (Sprayer UAV)

Company and address: DRONE Volt, 14, rue de la Perdrix, Lot 201 Paris Nord 2, Roissy CDG, 93420 Villepinte, France; phone: +33 180894444; e-mail: contact@UAVvolt.com; website: http://www.dronevolt.com/en/contact-us/

DRONE Volt is a drone-producing company located at Paris, France. It offers several UAV models with options, for aerial survey and aerial spraying of pesticides. It adopts digital data and precision techniques. The UAVs are being produced and offered in different countries such as Denmark, Canada, France (Roissy), Switzerland, Italy, and USA (Woodinville, WA).

"Hercules 20" is a multirotor copter with efficient sprayer system. The spray bar has 20 nozzles. It releases pesticides rapidly, uniformly and accurately. It dispenses about 3 L of liquid pesticide formulation per minute and covers about 1.8 ha per each 10 min of flight. It is said that "Hercules 20" sprayer UAV is 10 times faster and efficient, in terms of economic aspects compared to manual spraying methods that are used at present. It is a compact foldable UAV and can be packaged in a suitcase. The UAV can be assembled quickly, anywhere within the farm for launch (Plate 3.50.1).

TECHNICAL SPECIFICATIONS

Material: It is made of carbon fiber, ruggedized foam, and plastic.

Size: length: 120 cm; width (rotor to rotor): 180 cm; height: 45 cm; weight: 10.5 kg including spray tank and spray system. The MTOW is 30 kg.

Propellers: Four rotors with blades of size 18 cm × 6 cm. Propulsion is achieved by connecting propellers to motors (T-Motor by UAV VOLT U7V2).

Payload: 20 kg. It includes cameras and spray tank with fluid/granules. Payload tank holds 12 L of pesticide.

Launching information: It has autolaunch, navigation, and autolanding software, namely, "Pixhawk." The flight path is controlled using this software.

Altitude: It can hover just above the crop canopy. This UAV has hovering accuracy of ±5 m vertically and ±1 m horizontally. Generally, it is used at 1–3 m height above the crop canopy, during spraying. However, the ceiling is 1000 m above ground level.

Speed: The cruising speed is 50–60 km h^{-1}. Maximum speed attainable is 75 km h^{-1}.

Wind speed tolerance: 20 knots (25 km h^{-1}) disturbing wind

Temperature tolerance: −10 to 55°C

Power source: Electric lithium–polymer batteries

Electric batteries: 22 A h, four cells.

Fuel: Not an IC engine-fitted UAV.

Endurance: 12 min per flight

Remote controllers: The GCS and CPU at the UAV use software known as "UAV VOLT PILOT." The GCS telemetry and computers control the UAV for up to 10 km radius. Flight control is done by Pixhawk software.

Area covered: It covers 1.8 ha per 10 min when in spraying mode. Aerial surveying of over 50 ha can be accomplished in 10 min.

Photographic accessories: Cameras: It is fitted with full complement of sensors for aerial survey and digital data collection. It includes visual, multispectral, hyperspectral, NIR, and IR cameras. It is also equipped with videocamera. The close-up 2D and 3D shots are useful while diagnosing the pest/disease afflictions on the crop canopy.

Computer Programs to process orthomosaic: This UAV relays digital data and imagery to ground control computers (iPad), instantaneously. The orthomosaics are processed using Agisoft's Photoscan or Pix4D Mapper software. Other custom-made software may also be utilized at the GCS. All imageries (photos) are tagged with GPS coordinates to enhance accuracy.

Spraying area: The "Hercules 20" sprayer system is fully foldable and easy to pack. We can shift to any location. The sprayer system has flat nozzles fitted at an angle of 110°. Hercules sprayer distributes about 3 L pesticide formulation per minute. It covers approximately 1.8 ha per 10 min flight. The sprayer width (swath) is 3 m. The sprayer system is designed to be accurate for constant spraying schedules while precision techniques are adopted. The sprayer could be programed or it could also be controlled manually (Plate 3.50.1).

Volume: The integrated pesticide tank holds about 12 L of pesticide formulation or equivalent volume of granules.

Agricultural uses: Crop scouting for disease/pest attack, drought, floods, or soil erosion is easily done, using a low flying multirotor copter UAV such as Hercules 20. This UAV can be effectively used to monitor seed germination, seedling and crop stand, gaps in plant stand, etc. The crop's canopy growth, NDVI data, leaf chlorophyll data, and crop-N status can also be obtained, using the UAV. The imagery showing variations in crop nutrition (particularly N) and water stress index (thermal imagery) can be used to control variable-rate applicators. Natural vegetation monitoring is also possible (Plate 3.50.2).

Nonagricultural uses: Surveillance of mines, mining activity, industrial zones, buildings, highways, and city traffic tacking transport vehicles could be accomplished with good accuracy.

PLATE 3.50.1 A Hercules 20 multirotor UAV with sprayer.
Source: Ms. Céline Vergely, Corporate Communication Manager, Drone Volt, ROISSY CDG Paris Nord, Villepinte, France.
Note: This UAV is equipped with an efficient sprayer system. It has a wide (3-m long) sprayer bar and several nozzles. Multiple nozzles hasten spraying.

PLATE 3.50.2 Hercules 20 UAV flying over crop field.
Source: Ms. Céline Vergely, Corporate Communication Manager, Drone Volt, ROISSY CDG Paris Nord, Villepinte, France.

Useful References, Websites, and YouTube Addresses

http://www.UAVvolt.com/en/expert-solutions/hercules-20/ (accessed June 12, 2017).
http://www.UAVvolt.com/en/expert-solutions/hercules-20-spray/ (accessed June 12, 2017).
https://www.youtube.com/watch?v=AZbSKHAMl2k (accessed June 12, 2017).
https://www.youtube.com/watch?v=8JddjmDA9AE (accessed June 12, 2017).
https://www.youtube.com/watch?v=7taJ4iBsIZo (accessed May 31, 2017).
https://www.youtube.com/watch?v=v4OWVNDvpa4 (accessed May 31, 2017).

Name/Models: JMR-V 1000x4

Company and address: Shenzhen GC Electronics Co., Ltd., JunFeng Science Park, HuangMaBu Village, XiXiang Town, Bemoan District, Shenzhen 518100, China; e-mail: sales@4fpv.com; WeChat ID: 147924130; website: http://www.4fpv.com/fpv1

Shenzhen GC Electronics deals with design and manufacture of electronic gadgets, useful in agricultural farming. They also produce and supply specialized parts such as remote controllers, custom-made software, gimbals, nozzles and sprayer bars. This company offers a wide range of agricultural UAVs. They are all mainly spraying UAVs of different capacities and payloads (e.g., 5, 10, and 15 l pesticide tank). The UAVs are capable of different rates of spray speed. The UAVs are fitted with variable-rate nozzles for use in precision farming. Otherwise, they could be used to conduct blanket spray. It is said the sprayer UAV is primarily meant for farmers/ companies with medium to large sized farms (500–10,000 ha). Following are a few variations in capacity and size of UAV models: JMR models offered by this company:

(1) JMR-X1000 4-rotor 5L agricultural UAV;
(2) JMR-V1000 6-rotor 5L agricultural UAV;
(3) JMR-V1200 6-rotor 10L agricultural UAV;
(4) JMR-X1380 4-Rotor 10L agricultural UAV;
(5) JMR-01550 8-Rotor 10L agricultural UAV;
(6) JMR-V1650 6-rotor 20L agricultural UAV; and
(7) JMR-V2000 8-rotor 20L agricultural UAV.

JMR-V 1000 model is available both in quad- and hexacopter format. It is supposedly a highly economical and practically efficient spraying UAV. It has been designed and produced specially for farmers, who own about 100–500 ac. JMR-V 1000 is basically an agricultural UAV. It also conducts aerial survey and obtains spectral data via cameras. This UAV is efficient in spraying pesticides (Plate 3.51.1). It avoids farmer's contact with hazardous plant protection chemicals. It reduces use of farm labor. It seems, maintenance and repair of UAV parts is easy. User friendly manuals are supplied with the UAV. Some of the functions listed by the company for this UAV are stable mode, GPS mode, intelligent orientation, flight mode with a maximum of 128 waypoints for selection, click-and-fly mode, fail-safe mode, low-voltage protection and early warning facility, break point continue to spray, and flow-rate control.

TECHNICAL SPECIFICATIONS

Material: The UAV is made of a composite of light-weight, corrosion resistant alloy. The fuselage is made of tough carbon fiber. Propellers are made of toughened all weather tolerant plastic blades. The UAV is foldable, particularly the propeller arms and spray nozzles. The fuselage has space for CPU, cameras, and gimbal.

Size: length: 75 cm; width (rotor to rotor): 75 cm; height: 48 cm; weight: MTOW is 15 kg

Propellers: They are made of toughened plastic. Each propeller has two plastic blades. Propellers are connected to brushless motors for propulsion.

Payload: It is 5 l of pesticide in a plastic tank and camera accessories, if opted.

Launching information: It is an autolaunch, autopilot and landing UAV. Its flight and navigation is guided by software, such as "Pixhawk." The UAV adopts safe technology, such as return to point of launch in case of emergency. The radar keeps the UAV height above the canopies of crops like wheat, rice, or maize.

Altitude: Maximum altitude during spraying is 50 m above crop's canopy. This UAV is connected to GPS. This facility allows the UAV to be kept at constant height and accurate locations for spraying. While dusting, UAV's height can be adjusted to obtain maximum uniform spread of pesticide formulation on leaves.

Speed: It is 15–20 km h^{-1}, during spray, but 45 km h^{-1} in loiter mode.

Wind speed tolerance: 15–20 km h^{-1} disturbance

Temperature tolerance: 0 to 50°C

Power source: Lithium–polymer electric batteries

Electric batteries: 16,000 mA h, two cells

Fuel: Not an IC engine-fitted UAV

Endurance: Spraying time is 11–13 min per take-off. Hovering time is 16–20 min.

Remote controllers: The GCS iPad or remote controller regulates UAV for up to 900 m radius.

Area covered: It covers 50 ha h^{-1}

Photographic accessories: Cameras: It comes with a full complement of visual, NIR, and IR cameras. They can be used to obtain aerial imagery and to conduct surveillance of crop fields.

Computer programs: UAV-derived images are processed, using software such as Pix4D Mapper or Agisoft's Photoscan.

Spraying area: The UAV covers 166,750 m^2 day^{-1} of 6 h; spray speed is 3–8 m s^{-1}; spray width is 3–8 m; spray height is 1.2–4 m. The pesticide tank holds about 4.5 l liquid or granules. The flow rate of nozzle is 1.2–1.3 l s^{-1}. The flight-mode UAV is semiautonomous. It reaches the starting point; then, it waits hovering at a spot to receive commands for change of flight path. The autospray mode helps in completing the task efficiently.

Volume: The tank holds 5 or 10 l of pesticide (liquid or dust or granules) based on tank size and model).

Agricultural uses: JMR-V 1000 models are basically agricultural UAVs. They are of utility to farmers. They are capable of crop scouting for seedling emergence and crop stand, detection of nutritional deficiencies, and variations in crop's nutrient status in fields, detection of disease and pest attack, mapping the attacked zones, and also mapping flood and drought-affected zones. The UAV is usually supplied with digital data depicting crop growth variability. The sprays are conducted based on such data, during precision farming. Natural vegetation monitoring for NDVI values, spraying plant protection chemicals if needed and studying plant diversity using spectral signatures of canopy and leaves are also possible.

Nonagricultural uses: It includes surveillance of industrial installations, buildings, roads, rail roads, city traffic, mines and mining activity.

PLATE 3.51.1 JMR-V 1000 a hexacopter version for agricultural purposes. Top: normal spraying mode; bottom: folded mode.
Source: Mr. Jenny Wang, Schengen GC Electronics Ltd. XiXiang Town, Bemoan District, Shenzhen, China; http://www.4fpv.com/fpv1, http://www.4fpv.com/product/277174153http://www.4fpv.com/product/277174153.
Note: The hexacopter UAV is foldable. Folding helps in easy portability.

Useful References, Websites, and YouTube Addresses

http://www.4fpv.com/product/277174153 (accessed Apr 25, 2017).
http://www.4fpv.com/product/277174153 (accessed Apr 25, 2017).
http://www.4fpv.com/fpv1 (accessed May 4, 2017).
https://www.youtube.com/watch?v=9ouTPyAgqwg (accessed May 4, 2017).
https://www.youtube.com/watch?v=JXj-P42SzZo (accessed May 4, 2017).

Name/Models: JMR-V 1200 (10 kg model)

Company and address: Shenzhen GC electronics Co., Ltd.; JunFeng Science Park, Hanau Village, XiXiang Town, BaoAn District, Shenzhen 518100, China; Tel.: +86 0755 29037325; phone/WhatsApp: +86 15173192084; e-mail: sales@4fpv.com, r9(at)gcele.net; Skype: gcelectronic; WeChat: yyy147888; WeChat ID: 147924130; QQ: 1401079384

Shenzhen GC Electronic Co. Ltd. specializes in the production of multirotor UAVs and their accessories. The company's motto is providing customers with high-quality agricultural UAV products. They design carbon fiber frames for four, six, and eight rotor UAVs. They also produce the high-efficiency brushless power system.

Major emphasis of their UAV technology is in products for aerial photography, entertainment, agriculture, security, etc. They are aiming to make UAV's frame lighter with reasonable structural strength.

JMR-V 1200 is an agricultural UAV with six rotors. It is a foldable UAV. It is easy to carry and launch from anywhere in a crop field. It is primarily a sprayer UAV with 10 L capacity pesticide tanks. It has sprayer system that is computer controlled and totally autonomous regarding flight path, spraying speed, and rate. This UAV is meant for farms with 50–200 ha area. Farmers can purchase this UAV and accessories for US$ 3500 (February 2017 price level). This is relatively a light-weight UAV with capability for 10 kg liquid spray. When compared with traditional manual sprayer, it reduces labor needs markedly and avoids worker's contact with hazardous farm chemicals (Yin, 2016). Major advantages quoted for this UAV are flexibility, stability in flight, simple operation, safe and reliable machinery and computer software, and low maintenance costs. This UAV has been tested for performance and economic advantages repeatedly in different locations.

TECHNICAL SPECIFICATIONS

Material: It is made of carbon fiber, ruggedized plastic, and light-weight corrosion resistant alloy for landing gear.

Size: length: 108 cm; width (rotor to rotor): 120 cm; height: 49 cm; weight: MTOW 24 kg

Propellers: It is a hexacopter. It has six toughened, lightweight, anticorrosion-treated propellers. They are attached to brushless motors for propulsion.

Payload: 10 kg. It includes cameras for aerial imagery and spectral analysis. It excludes pesticide tanks.

Launching information: JMR V 1200 is an autolaunch VTOL copter UAV. It needs 5 × 5 m clearance to get airborne. It is an autopilot UAV. Its navigation is made totally autonomous, using "Pixhawk" or similar software for navigation.

Altitude: Maximum flight height is 35 m above crop canopy.

Speed: Hovering speed without any a payload or in spraying mode is 15–20 min. Speed during spray schedules is 3–6 m s^{-1}.

Wind speed tolerance: 0–50 km h^{-1}

Temperature tolerance: 0–50°C

Power source: Lithium–polymer electric batteries

Electric batteries: 16,000 mA h, two cells

Fuel: Not an IC engine-fitted UAV.

Endurance: It is 35–50 min per flight.

Remote controllers: The telemetric link with UAV has a limit of 1000 m radius. The UAV has GPS and compass module. This facility allows the UAV to attain and then stabilize at constant height. This aids in accurate spraying. The UAV has autopilot and autospray mode. Yet, GCS computers/mobile can alter flight/spray rates. The UAV has autocorrection facility. It adjusts spray rate, height, and quantity of pesticide to be unloaded.

Area covered: 50 ac per flight of 6 h.

Photographic accessories: Cameras: The UAV carries a series of visual, high-resolution multispectral, NIR, IR, and RedEdge cameras. They are held in a gyro-stabilized gimbal.

Computer programs: The aerial imagery could be processed using Pix4D Mapper or Agisoft's Photoscan. The images could be instantaneously processed, at UAV's CPU. The raw images (orthomosaics) could also be transmitted to GCS iPad for processing. Agricultural service companies usually take 36 h to collect images, process them, and analyze using fertilizer/pesticide recommendation software.

Spraying area: The pesticide spray equipment and atomizer are designed to provide good distribution of pesticide droplets on canopy/leaves. The spray equipment can also distribute dust/powders and granules. Spray area covered is 240,120 m^2 per 6 h. Spraying endurance is 12–16 min, spraying rate is 3–6 m s^{-1}, spraying width is 3–6 m, spraying height 1.2 m; and flow rate of nozzles is 1.4–1.6 l min^{-1}.

Volume: The UAV carries a 10-l capacity pesticide tank.

Agricultural uses: JMR-V 1200 is primarily an agricultural UAV. It is also useful in scouting for canopy growth rate, leaf-chlorophyll (crop-N), canopy and ambient temperature (crop water stress index). It helps in deciding irrigation rate and timing. The aerial imagery is also used to detect and map drought, flood, weeds, and soil erosion-affected spots. Natural vegetation monitoring for getting NDVI values and in determining plant species diversity using spectral signatures is also possible. It is also useful for spraying pesticide, herbicides, fungicides, and liquid/granular formulations of fertilizer-N.

Nonagricultural uses: It includes surveillance of mineral, mining activity, ore dumps, industrial zones, buildings, rail roads, highways, and vehicular traffic; for aerial imagery of useful rivers, dams, lakes, and tanks in crop production.

Useful References, Websites, and YouTube Addresses

Yin, D. *JMR-V1200, 10 kg Agricultural Crop Sprayer UAV for Farm, Spraying Pesticides UAV*; 2016. https://www.youtube.com/watch?v=r1haAxVvIpc (accessed Nov 15, 2016).

https://www.UAVvibes.com/forums/threads/jmr-v1200-agricultural-uav-UAV-for-spraying-for-sale.31389/ (accessed June 7, 2017).

http://www.4fpv.com/product/277301557 (accessed June 7, 2017).

https://www.youtube.com/watch?v=r1haAxVvIpc (accessed June 7, 2017).

https://www.youtube.com/watch?v=r1haAxVvIpc (accessed June 7, 2017).

https://www.youtube.com/watch?v=K3ooDsoy62Y (accessed June 7, 2017).

Name/Models: JT Sprayer 15-Octocopter Sprayer UAV

Company and address: Joyance Technology Ltd., 19/F, 2321 Beihai Rd., Weifang, Shandong 261061, China; Tel.: +86 5362791566/+86 5362797166/+86 5368648566; e-mail: sales@joyancetech.com (sales), info@joyancetech.com

Shandong Joyance Intelligence Technology Co., Ltd. is a UAV technology company. They have a professional R&D team involved in developing efficient agriculture sprayer UAVs. They have invested a capital of 60 million RMB. They churn out 100 units of agriculture UAVs monthly. The company's motto is to make the "UAVs that farmers love." The company is located at Weifang High-Tech Industrial Development Zone. About 10 experienced engineers ensure prompt delivery, and every UAV is tested.

JT Sprayer 15-608 is relatively a heavier multicopter UAV at 40 kg MTOW. It could be used extensively to spray plant protection chemicals and fertilizer formulations. It unloads 15 kg plant protection chemicals in under an hour. Aerial sprays are done using JT Sprayer 15 (Plate 3.52.1). It is said to save cost on farm laborers rather perceptibly. Use of JT Sprayer results in 40–60% reduction in quantity of pesticide and cost incurred on spraying. Reduced use of pesticide avoids contamination of soil and water. JT Sprayer is priced moderately. This is to help farmers who own just 50 ha of cropped land. The UAV package includes maintenance and repair kit plus a series of user friendly manuals.

TECHNICAL SPECIFICATIONS

Material: It is made of carbon fiber, light aluminum alloy, plastic, and rugged foam. The landing gear is made of alloy.

Size: length: 1.55 m; width (rotor to rotor): 2.65 m; height: 0.35 m; weight: 15 kg without payload of cameras or pesticide; MTOW is 40 kg. This UAV is foldable for easy portability. In folded state, the UAV's size is 0.95 m length, 0.7 m width, and

0.65 m height. It is packed in an aluminum suitcase in a folded state and carried to any spot for launch (see Plate 3.52.1).

Propellers: It is a 6-axis UAV, but it has eight rotors. The rotors are made of tough carbon fiber, which is corrosion resistant. The rotors are connected to brushless motors for propulsion.

Payload: The pesticide tank attached under the fuselage holds 15 L of formulation.

Launching information: This is an autolaunch, autopilot, and landing copter UAV. The flight path is predetermined, using "Pixhawk" or similar software. The UAV returns to point of launch, in case of emergency or errors in flight programing or obstacles to flight. It has facility to restart from the point where the previous flight was discontinued.

Altitude: It flies at an altitude of 0–200 m above ground level. The UAV can hover at very low height above the crop's canopy for close-up shot and intensive spray of plant protection chemicals. A low-flying sprayer UAV helps to avoid drift of hazardous chemicals in the atmosphere.

Speed: UAV's speed while on general flight is 0–12 m s^{-1}. Speed during spraying is 0–8 m s^{-1}.

Wind speed tolerance: It resists 10 m s^{-1} wind disturbance.

Temperature tolerance: −10 to 70°C

Power source: Electric batteries

Electric batteries: 6S 16,000 mA h, 15 C, two PCs. The batteries are charged using a synchronous balance charger.

Fuel: Not an IC engine-fitted UAV

Endurance: 10–15 min

Remote controllers: JT Sprayer 15-608 is usually controlled using a joystick or remote controller model Tutuba T8FG. This UAV's telelink covers only short distance of 1000 m radius. Flight control is done using "V & 1 software." The GCS also has telemetric link to contact or alter flight course midway through a flight program.

Area covered: It covers about 100–150 ac of crop field, while conducting aerial imagery or live surveillance/monitoring.

Photographic accessories: Cameras: The UAV carries a full complement of sensors for aerial survey, to collect still images and videography. It has gyro-stabilized three-axis gimbal that holds visual, high-resolution multispectral, NIR, IR, and RedEdge cameras.

Computer programs: The orthomosaics are stitched and processed, using software such as Pix4D Mapper or Agisoft's Photoscan. The images can be processed either at the UAV's CPU itself or relayed to GCS computer screen. It can then be processed at the GCS using iPad with appropriate software.

Spraying area: This UAV sprays 50–100 ac of crop land per day for 6–8 h. The spray efficiency is 2000–3500 $m^2 min^{-1}$. Spray width is 5–7 m. It has 6 nozzles that can be used, during variable-rate application or for blanket application of liquid formulation.

Volume: The UAV's pesticide tank holds 15 l of pesticide by volume.

Agricultural uses: "JT sprayer 15" is a versatile spraying UAV. It is used to spray plant protection chemicals or fertilizer formulations. It can be used to apply blanket prescriptions all through the crop fields or if the nozzles are connected to digital data; they could be used for variable-rate application (i.e., precision farming) (Plate 3.52.1; Joyance Tech., 2016). This UAV can also be used for aerial imagery of crop fields. Its ability to hover allows close-up and detailed picturization of crop canopy. Crop scouting for canopy size and growth rate, leaf-chlorophyll (crop-N) status, canopy, and ambient temperature (i.e., crop's water stress index) can also be done efficiently using this UAV. Natural vegetation monitoring is possible.

Nonagricultural uses: It uses surveillance of geologic sites, mines and mining activity, monitoring rivers, dams and irrigation channels for water flow, monitoring highways, and transport vehicles, etc.

PLATE 3.52.1 JT sprayer 15-608—a sprayer UAV.
Above: A UAV about to be launched; below: a close-up of rotors and spray nozzles.
Source: Mr. Jenny Wang, Shandong Joyance Technologies Ltd., Shandong, China.

Useful References, Websites, and YouTube Addresses

Joyance Tech. *Agricultural Sprayer UAV Operation Video*; 2016. https://www.youtube.com/watch?v=zHDIMMhBoJ0 (accessed Nov 15, 2016).

http://www.wecanie.com/html/support/videos/6-Axles-8-Rotors-15L-Sprayer-UAVs.html (accessed June 1, 2017).

http://www.wecanie.com/html/support/videos/JT-15-sprayer-UAV-15Kg-full-load.html (accessed June 1, 2017).

https://www.youtube.com/watch?v=SlYvvILJJGc (accessed June 1, 2017).

https://www.youtube.com/watch?v=8zHHaG-O9lo (accessed May 25, 2017).

https://www.youtube.com/watch?v=FHQHGD7Ro0g (accessed May 25, 2017).

https://www.youtube.com/watch?v=l6oOL2iE1Pk (accessed May 25, 2017).

Name/Models: Lieber LB-410 Agricultural Sprayer UAV

Company and address: Shenzhen Lieber Electronics Co., Ltd., B Building 6 Floor, Zhong Yu Green High-Tech Industrial Park, He Shui Kou, Gongming Town, Guangming New District, Shenzhen, China; phone +86 15989394601; +86 755 27170878; **fax:** +86 755 29860079; e-mail: lieber_rc@126.com; website: www.lieber-rc.com

Lieber Group Co. Ltd. was founded in 2004. It is composed of semiconductor division and model aircraft division. Lieber Electronics (Shenzhen) is the subsidiary company of Lieber Electronics in China founded in 2009. It is a high-tech company that produces UAVs and their accessories. They offer a wide range of models that could be used to spray crops with pesticides, herbicides, and fertilizers.

"Lieber's LB410" is primarily a sprayer UAV with capacity to hold plant protection chemicals ranging from 5 to 15 l. "LB410" is an autonomous quadcopter. It might cost US$ 12,900–15,200 based on the model and attachments requested. It is an efficient sprayer UAV and applies pesticide to 50–150 ac^{-1} of crop fields in 1 h.

TECHNICAL SPECIFICATIONS

Material: It is made of carbon fiber, plastic, and rugged foam

Size: length: 115.2 cm; width (rotor to rotor): 115.2 cm; height: 39 cm; weight: 8.2 kg empty weight (i.e., without payload); MTOW is 34 kg

Propellers: This is a quadcopter. It has four propellers made of toughened plastic blades. Propellers are attached to brushless motor for propulsion.

Payload: 10 L plastic tank with pesticide or fluid fertilizer

Launching information: It is a VTOL UAV. It is an autolaunch and autopilot vehicle. It has software that makes the UAV return to point of launch automatically in times of emergency.

Altitude: The operating altitude is 5–50 m above crop canopy.

Speed: 3–6 m s^{-1}

Wind speed tolerance: Resists 15–20 km h^{-1} disturbance. It is gyro-stabilized UAV.

Temperature tolerance: −15 to 50°C

Power source: Electric batteries

Electric batteries: Lithium–polymer batteries 24,000 mA h.

Fuel: No requirement for gasoline

Endurance: 10–20 min depending on payload

Remote controllers: It has manual mode that adopts a hand-held remote controller to guide the UAV above the crop fields and spray pesticides. The autonomous mode uses preprogramed flight path. The spray schedule is dictated by computer decision-support systems. Yet, the GCS (laptop) can alter the preprogramed flight path, if necessary.

Area covered: It covers 50–150 ac^{-1} of crop fields per hour.

Photographic accessories: Cameras could be attached, if aerial imagery is also envisaged.

Computer programs: This UAV is primarily a sprayer UAV. Aerial images and digital data are required only, if precision techniques are adopted. Blanket spraying does not need precise digital data.

Spraying area: It is 50–150 ac h^{-1}. Spray width (swath) is 3–4 m with four nozzles. Options for six nozzles are available.

Volume: Options for tanks with 5, 10 and 15 l payload are also available.

Agricultural uses: Crop dusting/spraying with pesticides is the major function of this UAV. This UAV could be used to obtain NDVI, crop biomass, and water status data, provided the sensors are fixed.

Nonagricultural uses: They are used for transport of small luggage and hazardous chemicals and objects. Fixing sensors could add to its utility in nonagricultural sector too.

Useful References, Websites, and YouTube Addresses

http://www.lieber-rc.com/en/ (accessed July 6, 2017).
https://www.youtube.com/watch?v=pkFHr24InGg (accessed July 6, 2017).

Name/Models: MMC F4-5-17-4

Company and address: Shenzhen Micro-Multi-Copter Aero Technology Co., Ltd., 17F Han's Laser Building, No. 9988, Shennan Ave., Nanshan District, Shenzhen 518052, Guangdong, China; phone +86 755 27836344, +86 755 23286506; fax: +86 755 27836344; mobile: +86 13926559135; e-mail: info@mmcuav.com; website: http://www.globalsources.com/mmcuav.co

Shenzhen MicroMultiCopter Aero Technology Co., Ltd. (MMC) is an avionics company founded in 2010. They produce multirotor UAVs that are useful in aerial photography, film and television, advertising, electric power-line stringing and inspection, aerial remote sensing, land monitoring, city planning, forest-fire fighting, traffic management, aerial reconnaissance, and disaster relief. This company is situated in Shenzhen, China (Mainland). It has more than 30 top technicians in UAV section. It has collaboration with Harbin Institute of Technology. Most of the customers dealing with UAVs come from the United States of America, United Kingdom, Germany, France, Australia, Brazil, and other countries worldwide. This UAV model MMC F4 costs US$ 5000–6000 per unit.

"MMC F4" is an agricultural UAV. It is a quadcopter meant specifically for spraying pesticides and fertilizer formulations, to crop fields. It is among the newest crop sprayer models released by this company. It can be handled efficiently, even by a newcomer farmer. Most of the operations are streamlined and are computer controlled. This UAV is easy packed into a suitcase. It can be transported to any location in a farm/field. It can be assembled in a matter of 5 min without need for tools.

TECHNICAL SPECIFICATIONS

Material: The UAV is made of carbon fiber, foam and plastic

Size: length: 200 cm; width (rotor to rotor): 200 cm; height: 40 cm; weight: 9 kg; MTOW is 22.5 kg

Propellers: It has four propellers made of tough plastic blades. The length of the rotor is 41 cm.

Payload: 10 l of liquid formulation in a plastic tank that is attached to fuselage.

Launching information: It is an autolaunch, autonomous navigation, and landing copter. It is a VTOL UAV. It just needs 5 × 5 clearance space to take-off or land.

Altitude: The ceiling is 4500 m above sea level. The operating height is determined by the crop species, pesticide/fungicide formulation, and disturbing wind speed/direction.

Speed: It is 35 km h^{-1} during loiter and without payload. Spraying speed depends on formulation quantity to be delivered and area to be covered.

Wind speed tolerance: 20 km h^{-1} wind disturbance

Temperature tolerance: −20 to 55°C

Power source: Lithium–polymer electric batteries

Electric batteries: 16,000 mA h, two cells

Fuel: Not an IC engine-fitted UAV

Endurance: 16 min with normal payload

Remote controllers: The GCS (iPad) or joystick (remote controller) keeps contact with the UAV for up to 2 km radius. The radio telemetric link can be used to alter or override predetermined flight path and aerial spray schedule, if needed.

Area covered: It covers 50–250 ha day^{-1} of 6 h of UAV flight.

Photographic accessories: Cameras: This UAV is exclusively a sprayer UAV. However, if needed, payload containing sensors could be attached.

Computer programs: This is basically a spraying UAV. The aerial images, if any, derived out of sensors are processed, using the common commercial software such as Pix4D Mapper.

Spraying area: It is 20 ac (8 ha h^{-1}); spraying height is 3 m above canopy. Spraying width is 3 m wide.

Volume: 10 l pesticide tank. It unloads 1–1.5 L pesticide min^{-1}.

Agricultural uses: MMC F4 is used predominantly to spray liquid/granular formulation or dusting powders on crops (MMC UAV, 2016). The UAV covers large areas of crop fields rapidly. It reduces cost of manual spraying. Farm labor needs are reduced enormously. It sprays accurately at only spots that are affected and need remediation. Therefore, pesticide or fertilizer formulation consumption gets perceptibly reduced.

Nonagricultural uses: It is limited to transport of hazardous goods, chemicals, and dumping in yards. It can be used to collect water samples from dams, rivers, and irrigation channels by fixing suitable payload (samplers).

Useful References, Websites, and YouTube Addresses

MMC UAV. *Precision Agricultural Spraying UAV*; 2016; pp 1–7. http://www.globalsources.com/mmcuav.co (accessed Nov 22, 2017).

https://www.youtube.com/watch?v=WQ7I34fJwRI (accessed Nov 15, 2016).

http://www.mmcuav.com/contact-us/

https://www.youtube.com/watch?v=WQ7I34fJwRI&list=PLjHr05tQzac2THKbqRB0ydrIKSCrCN2nr (accessed Sept 4, 2017).

https://www.youtube.com/watch?v=bMfNRuMZQb8 (accessed Sept 4, 2017).

Name/Models: MS-Z10 Agricultural UAV

Company and address: Henan Machinery Equipment Company Ltd. No. 1202, 12th Floor, Unit 1, Bldg. 24, Shangwu Neuhaus Road, Zhengdong New District, Zhengzhou, Henan, China; phone: +86 371 55619981; fax: +86 371 8702612; e-mail: http://www.msmining.com/, maiseng.en.alibaba.com; website: http://www.ms-machinery.net

Henan Machinery and Equipment Company Ltd. is a multifaceted company. It produces products ranging from rotary kilns, engines, etc. The UAV "MS-Z 10" is model produced since past 5 years by this company. MS-Z10 is a sprayer UAV. It is used to spray pesticides, herbicides, and fertilizer formulations. It is a multicopter with variable-rate spray nozzles. This multicopter may cost about US$ 5–10,000 based on accessories and modifications sought. It is used in the agricultural zones of entire China and Far-East. It is an efficient sprayer that dispenses about 1.5–2 l of plant protection chemical per minute during flight. There are three models based on the payload. They are UAVs with 5, 10, and 15 l capacity pesticide tanks. The UAV is usually packed in an aluminum suitcase in folded condition. It is then transported to locations in a farm.

TECHNICAL SPECIFICATIONS

Material: This hexacopter is made of fiber glass, light aluminum, and tough plastic.

Size: length: 118 cm; width (rotor to rotor): 118 cm; height: 52 cm; weight: 8.7 kg without battery and pesticide payload. MTOW is 20–24 kg; folded size of the UAV is 53 × 48 × 52 cm.

Propellers: Six propellers (22 in.) made of plastic blades and attached to brushless motors for propulsion.

Payload: Payload is usually 10 kg pesticide stored in a plastic tank, but maximum load could reach 12 kg, if opted.

Launching information: It is a VTOL autolaunch vehicle. Its navigation is controlled, using "Pixhawk" software.

Altitude: It flies at 0–5 m above crop's canopy. It loiters at a height of 50 m above crop canopy.

Speed: The UAV's working speed is 5–10 m s^{-1}. Spraying speed is 7 m s^{-1}.

Wind speed tolerance: Resists 15 km h^{-1} disturbance. It is preferable to use the UAV in low wind condition to avoid drift of hazardous plant protection chemicals.

Temperature tolerance: −15 to 45°C

Power source: Electric batteries

Electric batteries: 24,000 mA h.

Fuel: No gasoline requirement, not an IC engine-fitted UAV

Endurance: 12 min per take-off and without recharging

Remote controllers: The GCS has remote controller or laptop. Laptop has sufficient software to control UAV's flight path, spraying speed dictating launch, and landing location and timing.

Area covered: It covers 25–50 ac per flight for 12 min.

Photographic accessories: Cameras could be fitted, if necessary. This is basically a sprayer UAV used for blanket spray of chemicals.

Computer programs: Aerial images are obtained using sensors, then, it could be processed by the ground control laptop, using Pix4D Mapper or Trimble's Inphos software.

Spraying area: Spray width is 1.2 m, but it depends on spray bar's size. The UAV flies at 3–5 m above crop's canopy during spraying. The spray bar has four pressured nozzles for blanket sprays. The variable-rate nozzles are needed, if used in precision farming.

Volume: The hexacopter dispenses 1–12 l of pesticide in 12–15 min of flight endurance.

Agricultural uses: "MS-Z10" is used predominantly to spray the crop fields in China and other Far-East nations. Its variant model with different payload capacities is available in many other countries.

Nonagricultural uses: It could be used to carry hazardous chemicals in the payload tank. It can be used to transport parcels to different locations in a farm or urban location.

Useful References, Websites, and YouTube Addresses

https://www.alibaba.com/product-detail/10L-Professional-Unmanned-Aerial-Vehicle-UAV_60597002590.html (accessed June 25, 2017).

Name/Models: PD6-AW

Company and address: CEO, Mr. Masakazu Kono, ProDRONE—Head Office: 16F Bantane Sakae Bldg. 2-4 Shinsakae-machi, Naka-ku, Nagoya-shi, Aichi 460-0004, Japan; Tel.: +81 52 950 1278; fax: +81 52 950 1277; e-mail: info@proUAV.jp; Tokyo Office: 1F Hirakawacho-Daiichiseimei Bldg. 1-2-10 Hirakawa-cho, Chiyoda-ku, Tokyo 102-0093, Japan; Tel.: +81 3 5212 5132; fax: +81 3 5212 5102; website: https://www.prodrone.jp

ProDRONE is located in Japan. It was incorporated in the year 2015. It is involved mainly in developing modes and producing UAVs useful in variety of tasks such as aerial photography, etc. It concentrates on commercial use UAV system, training, and UAV maintenance. ProDRONE company offers a series of multicopters useful in agriculture, civilian, and military aspects, for example, PD4-AW, PD6B, PD6E2000, PD6B-AW, etc.

ProDRONE PD6-AW is a hexacopter with wide range of applications in agriculture, civilian, and military aspects (Plate 3.53.1). It is predominantly used to obtain aerial photography, surveillance installations, and spray crops with plant protection chemicals. It is a small copter that can be easily assembled or dismantled. It fits into a small suitcase. Therefore, it is easily portable.

TECHNICAL SPECIFICATIONS

Material: It is made of carbon fiber, alloy, ruggedized foam, and plastic. It has all weather and waterproof body.

Size: length: 5.5 kg; width (rotor to rotor): 90.0 cm; height: 30.0 cm; weight: 4.2 kg

Propellers: It is a hexacopter. It has six propellers with a diameter of 43.2 cm.

Payload: Maximum payload is 5 kg

Launching information: It is an autotake, navigation, and landing rotor UAV. It has fail-safe facility. It returns to point of launch or predetermined spot in times of emergency or if computer commands go erroneous.

Altitude: It ranges from zero, that is, hovering very close to crop canopy to 50 m above crop canopy, for aerial photography, particularly assessment of leaf chlorophyll and canopy temperature.

Speed: Maximum speed is 76 km h^{-1}; 10 m s^{-1} operating speed

Wind speed tolerance: 25 km h^{-1} crosswind

Temperature tolerance: −10 to 55°C

Power source: Lithium–polymer electric batteries

Electric batteries: 16,000 mA h, 4.4 V, 6000 mA h

Fuel: Not an IC engine-fitted UAV

Endurance: 10–50 min per flight

Remote controllers: The GCS computer (iPad) is connected to UAV's CPU for up to 5 km distance. The telemetric instructions can be used, to alter preprogramed flight path. Flight path is usually programed using "Pixhawk" software or any other custom-made program. The UAV's flight path, speed, and camera's shutters are controlled by GCS computers.

Area covered: It covers about 50–250 ha aerial photography per hour. Area covered by UAV reduces to 50–100 ha, if used as a sprayer.

Photographic accessories: Cameras: It comes with options such as tri-axis gimbal plus cameras, dual axis gimbal plus cameras, a large gimbal, etc. UAV carries visual, multispectral, hyperspectral, NIR, IR, and RedEdge cameras, and Lidar sensor.

Computer programs: The aerial images received at the GCS computer is processed, using Pix4D Mapper or Agisoft's Photoscan software. Images and digital data can be relayed quickly to GCS iPad. It helps in rapid analysis of crop's imagery. It also helps in developing recommendations for fertilizer/pesticide/weedicide application.

Spraying area: Aerial spraying depends on the size of the tank and volume of plant protection chemical/liquid fertilizer formulation held, number of nozzles, and their regulation also affects spraying speed. Usually, 50–100 ha of crop field is sprayed per hour.

Volume: Data not available

Agricultural uses: It includes crop scouting to assess seed germination, seedling establishment, and crop stand. Aerial photography to assess crop damage by disease or pest or weed infestation is possible. Aerial survey and photography to mark the areas affected by drought, floods, or soil erosion are done rapidly by UAV. Natural vegetation monitoring to collect data on NDVI and species diversity is feasible.

Nonagricultural uses: It includes surveillance of international borders, industrial sites, mines, mining activity, mine dumps and ore movement, public events, rail roads, highway traffic, rivers, lakes, channels, etc.

PLATE 3.53.1 ProDRONE hexacopter PD6-AW.
Source: Ms. Aya Kawakami, ProDRONE Ltd., Aichi, Nagoya, Japan.

Useful References, Websites, and YouTube Addresses

https://www.wearechampionmag.com/proUAV-pd6aw-UAV-water-resistant-and-all-weather-capable (accessed Oct 10, 2017).

http://www.roboticgizmos.com/proUAV-pd6-aw-weather-UAV-winch/http://www.roboticgizmos.com/proUAV-pd6-aw-weather-UAV-winch/ (accessed Oct 10, 2017).

https://www.proUAV.jp/en/products/pd6-aw/ (accessed Oct 10, 2017).

https://www.youtube.com/watch?v=9_D_7jGHYf0 (accessed Nov 4, 2017).

https://www.youtube.com/watch?v=gWaKffbCZAY (accessed Nov 4, 2017).

Name/Models: PD6B

Company and address: CEO, Mr. Masakazu Kono; ProDRONE—Head Office: 16F Bantane Sakae Bldg. 2-4 Shinsakae-machi, Naka-ku, Nagoya-shi, Aichi 460-0004, Japan; Tel.: +81 52 950 1278; fax: +81 52 950 1277; e-mail: info@proUAV.jp; Tokyo

Office: 1F Hirakawacho-Daiichiseimei Bldg. 1-2-10 Hirakawa-cho, Chiyoda-ku, Tokyo 102-0093, Japan; Tel.: +81 3 5212 5132; fax: +81 3 5212 5102; website: https://www.prodronejp

"ProDRONE PD6B" is a hexacopter (Plate 3.54.1). It is capable of aerial photography and spraying of chemicals on crop fields. It is capable of autonomous flight. It transmits images of high quality. The images are useful in judging crop growth. Spectral analysis and the data can be used, in prescribing pesticide, fungicide, and fertilizer spray dosages. It is a compact UAV that fits into a suitcase. It is portable and can be launched from anywhere in crop fields or city locations. It is a very handy UAV useful in surveillance of installations, rivers channels, etc.

TECHNICAL SPECIFICATIONS

Material: PD6B is made of carbon fiber, ruggedized foam, and plastic. It has steel frame. It is not a waterproof UAV. It cannot be used on water surface.

Size: length: 78 cm; width (rotor to rotor): 162 cm; height: 85 cm; weight: 11 kg without payload

Propellers: This is hexacopter. It has six propellers of size 68.6 cm diameter

Payload: Maximum payload is 30 kg. It includes all cameras inside a gyro-stabilized gimbal, a tank that holds pesticide/fungicide or liquid fertilizer formulation or even granules.

Launching information: It is a VTOL autolaunch copter UAV. Its navigation is programed using "Pixhawk" software. It has autolanding. It can land by skidding or careful autolanding using software. This UAV is nonwater resistant. It cannot be used for launching or landing on water surface.

Altitude: It flies at 0–50 m during aerial imagery of crop fields and ground conditions. Normal operational height during civilian surveillance is 300 m. It has a ceiling of 1000 m above ground level.

Speed: Maximum speed is 60 km h^{-1} without payload and spraying schedules. The speed is regulated to 30–45 km h^{-1}.

Wind speed tolerance: 25 km h^{-1} disturbance

Temperature tolerance: −10 to 55°C

Power source: Electric batteries

Electric batteries: Lithium–polymer batteries 44.4 V/16,000–20,000 mA h.

Fuel: Not an IC engine-fitted UAV

Endurance: 10–30 min per flight

Remote controllers: This UAV can also be connected using a cable. The cable powers the UAV with electrical energy. It also transmits video images with great accuracy to ground control computers. In case of interruption of radiosignal, the UAV could be controlled using the cable connection.

Area covered: It covers a linear distance of over 25–40 km per flight of 30 min. It covers about 50–100 ha spraying per hour.

Photographic accessories: Cameras: The aerial photography and thermal imagery is done using a series of sensors located inside a gimbal. It has options for different types of gimbals such a model that holds only one camera, or two cameras, one that holds all six cameras, Lidar sensor and vehicle-tracking device, etc.

computer programs to process orthomosaic. The aerial images (orthomosaics) are processed using software such as Pix4D Mapper or Agisoft's Photoscan.

Spraying area: 50–100 ha h^{-1}

Volume: Data not available

Agricultural uses: Aerial photography of land resources, topography, soil type, and water resources in a farm; crop scouting to obtain data on growth and NDVI. It is used to get data on crop canopy growth rate, leaf area, leaf chlorophyll, crop's water status and stress index values (thermal imagery). It is also used to survey and detect spots affected by disease, pests, drought, floods or soil erosion. Such data is directly introduced into variable-rate applicators on ground vehicles or UAV itself, for effective control measures. Natural vegetation monitoring, obtaining NDVI values and detection of species diversity is also done, using this UAV.

Nonagricultural uses: It includes surveillance of military camps, industrial installations, aerial supply, and dropping of food packets and goods in times of emergency, aerial inspection of rail roads, highways, pipelines and electric lines, tracking transport vehicles, etc.

PLATE 3.54.1 ProDRONE PD6B.
Source: Ms. Aya Kawakami, ProDRONE Ltd., Aichi, Nagoya, Japan.

Useful References, Websites, and YouTube Addresses

https://www.proUAV.jp/en/products/pd6b/ (accessed Oct 22, 2017).
https://www.proUAV.jp/en/products/pd6b/ (accessed Oct 22, 2017).
http://www.alamy.com/stock-photo-a-proUAV-pd6-b-with-an-attached-dslr-camera-on-display-during-the-83009347.html (accessed Oct 22, 2017).
https://www.youtube.com/watch?v=T6kaU2sgPqo (accessed Oct 22, 2017).
https://www.proUAV.jp/products/pd6b-aw/ (accessed Oct 22, 2017).

Name/Models: PD6B-AW

Company and address: CEO, Mr. Masakazu Kono, ProDRONE—Head Office: 16F Bantane Sakae Bldg. 2-4 Shinsakae-machi, Naka-ku, Nagoya-shi, Aichi 460-0004, Japan; Tel.: +81 52 950 1278; fax: +81 52 950 1277; e-mail: info@proUAV.jp; Tokyo Office: 1F Hirakawacho-Daiichiseimei Bldg. 1-2-10 Hirakawa-cho, Chiyoda-ku, Tokyo 102-0093, Japan; Tel.: +81 3 5212 5132; fax: +81 3 5212 5102; website: https://www.proUAV.jp

"ProDRONE PD6B-AW" is a relatively heavier UAV among the UAVs classified as light UAVs (5–50 kg). It weighs approximately 18.8 kg. It is a hexacopter with versatile ability to hover at low altitudes above ground or crop canopy. It has a wide range of applications in aerial photography, crop scouting, inspection of public installations, industries, mines, etc. It is portable using light vehicles. It is moderately priced so that small cooperatives and large company farms may own them. Its major role in agriculture is spraying plant protection chemicals and fertilizer formulations. The spraying is done based on commands by computer decision-support system (Plate 3.55.1).

TECHNICAL SPECIFICATIONS

Material: It is made of carbon fiber, ruggedized foam, and plastic.

Size: length: 122 cm; width (rotor to rotor): 162 cm; height: 80 cm; weight: 18.8 kg without payload

Propellers: This is a hexacopter. It has six propellers with 68.6 cm diameter.

Payload: 10–15 kg

Launching information: It is an autolaunch, autonomous flight and autolanding UAV. Its navigation is predetermined, using Pixhawk software or a similar custom-made software. It launches itself from within a clearance (space) of 5 × 5 m. It has fail-safe software, which directs the UAV to return to point of launch, particularly in case of malfunction of computer program, depletion of battery charge, or error in flight path.

Altitude: 0–50 m above crop for aerial photography. Operational altitude is 300 m above ground level. However, ceiling is 1000 m above ground level.

Speed: Maximum speed attainable is 60 km h^{-1}; Maximum operational speed is 10 m s^{-1}

Wind speed tolerance: 25 km h^{-1} disturbing wind

Temperature tolerance: −10 to 60°C

Power source: Electric batteries

Electric batteries: 44.4 V/16,000–20,000 mA h

Fuel: Not a gasoline requiring engine

Endurance: 10–30 min per flight

Remote controllers: The GCS computers (iPad) keep contact with UAV's CPU for up to 15 km radius. The flight path can be altered by the GCS computers midway through a preprogramed flight path and aerial photographic schedule. The UAV's cameras are started or shut based on ground commands relayed, using telemetric connections. The spray schedule, pattern, volume, and rate can all be altered midway, despite preprograming.

Area covered: 50 km linear distance. It covers 50–200 ha of land surface with aerial photography per flight of 1 h. If used for spraying, then area covered reduces based on spray schedule and quantity to be released.

Photographic accessories: Cameras: The UAV is fitted with a couple of different gimbal options, such as small tri-axis gimbal, dual axis servo gimbal, or tri-axis servo gimbal. Plus, it has large gimbal. The cameras attached depend on the purpose. The gimbals have a series of 5–6 cameras, such as visual (Sony 5100), multispectral, hyperspectral, NIR, IR, and RedEdge. It is also equipped with Lidar sensor that is used in detecting forest plantation height and volume.

Computer programs: The aerial images relayed are processed, using Pix4D Mapper or Agisoft's Photoscan software. Images can be processed right at the UAV's CPU or at GCS iPad. The GCS computers analyze the data and arrive at recommendations, for aerial spray of plant protection chemicals and fertilizers. The UAV could be fitted with variable-rate nozzles. They operate in response to spatial variability detected in the aerial photography. The chip with spatial variations (digital data) can be directly introduced to ground vehicles with variable rate sprayers, if precision techniques have in vogue.

Spraying area: 50–200 ha h^{-1}

Volume: Data not available

Agricultural uses: It is used crop scouting to detect variations, if any, in seedling establishment, canopy development, leaf area, leaf-chlorophyll content, crop-N status, crop's water stress index, grain maturity, and forecast of grain yield. It is used to detect disease and pest-attacked patches and map them digitally. It is also used to map weed affected locations, drought and flood-affected patches. Natural vegetation monitoring to obtain NDVI values, chlorophyll index, canopy temperature and air temperature, and study plant species diversity, using their spectral signatures is possible.

Nonagricultural uses: It includes surveillance of military camps, vehicles, and borders; mines and mining activity, industrial installations, buildings, oil pipelines, dams, rivers and lakes, etc. This UAV tracks public vehicles and their movement, if fitted with tracking sensor.

PLATE 3.55.1 ProDRONE PD6B-AW.
Source: Ms. Aya Kawakami, ProDRONE Ltd., Aichi, Nagoya, Japan.

Useful References, Websites, and YouTube Addresses

https://www.proUAV.jp/products/pd6b-aw/ (accessed Oct 19, 2017).
https://www.youtube.com/watch?v=kyWqbLxblbY (accessed Oct 19, 2017).
https://www.youtube.com/watch?v=8ARNmQckYhw (accessed Oct 19, 2017).
https://www.youtube.com/watch?v=T6kaU2sgPqo (accessed Oct 19, 2017).

Name/Models: PD6E2000-AW-High

Company and address: CEO, Mr. Masakazu Kono; ProDRONE—Head Office: 16F Bantane Sakae Bldg. 2-4 Shinsakae-machi, Naka-ku, Nagoya-shi, Aichi 460-0004, Japan; Tel.: +81 52 950 1278; fax: +81 52 950 1277; e-mail: info@proUAV.jp; Tokyo Office: 1F Hirakawacho-Daiichiseimei Bldg. 1-2-10 Hirakawa-cho, Chiyoda-ku, Tokyo 102-0093, Japan; Tel.: +81 3 5212 5132; fax: +81 3 5212 5102; website: https://www.proUAV.jp/products

"PD6E2000-AW-High" is a light-weight hexacopter. It has an endurance of just 10–50 min. Yet, it covers a reasonably large area of 50–100 ha h^{-1} with aerial imagery and/or spraying. It has a range of sensors that provide reasonable sharp images and accurate data about crop's performance and spatial variations. It offers data about disease/pest attack, weeds, and nutritional status. It can be fitted with variable-rate nozzles and software to apply chemicals for precision farming (Plate 3.56.1).

TECHNICAL SPECIFICATIONS

Material: PD6E2000-AW-high is made of carbon fiber, aluminum alloy, and plastic. It is all weather capable UAV.

Size: length: 80 cm; width (rotor to rotor): 126 cm; height: 30 cm; weight: dry weight 6 kg (without spray tank)

Propellers: This is a hexacopter. It has six propellers of 572 cm diameter each

Payload: 10 kg

Launching information: It is a VTOL autolaunch UAV. It needs just 5 × 5 m clearance to take-off from any location in crop field or urban or industrial zones. It comes with an option of different landing gears such as motor mount skids, center mount skids, retractable skids or detachable arms.

Altitude: It is 0–50 m for aerial photography and surveillance of crop fields.

Speed: 65 km h^{-1}; operating speed 10 m s^{-1}

Wind speed tolerance: 25 km h^{-1} wind disturbance

Temperature tolerance: −10 to 60°C

Power source: Lithium–polymer electric batteries

Electric batteries: 44.4 V/6000–16,000 mA h.

Fuel: Not fitted with gasoline engine

Endurance: 10–50 min per flight

Remote controllers: The hexacopter is kept in control, using telemetric link for up to 15 km radius from GCS (iPad). The GCS iPad can modify predetermined flight path, if needed, using datalink.

Area covered: It covers about 50–200 ha h^{-1} flight

Photographic accessories: Cameras: The UAV is fitted with tri-axis small gimbal, dual axis-servo gimbal, and a large gimbal. These gimbals have different cameras such a single visual or multispectral camera, a series of five cameras that include visual, multispectral, hyperspectral, NIR, IR, and RedEdge bandwidth cameras and Lidar sensor.

Computer programs: The aerial imagery is processed, using Pix4D Mapper or Agisoft's Photoscan or similar custom-developed software. The images are usually processed and transmitted. Images could also be sent to GCS iPad for processing. The images are tagged with GPS. It provides good overlap so that they are highly accurate, when stitched together.

Spraying area: This UAV could be fitted with tank to hold pesticide, fungicide, or herbicide. It is attached with appropriate variable-rate nozzles. It covers 50–100 ha of crop land per day for about 6–8 h. The UAV's imagery and digital data can also be used to detect spatial variations in nutrient status, nutrient-deficient areas, and calculate fertilizer requirement. Digital instructions to nozzles help in supplying fertilizer granules or liquid at variable-rates.

Volume: Data not available

Agricultural uses: It has a wide range of applications in agricultural sector. It is used to obtain aerial photographs of land resources with borders depicted accurately (GPS coordinates). Crop scouting for seedling establishment and crop stand, NDVI data, canopy characteristics, leaf area, leaf-N (chlorophyll), and water stress index (thermal imagery) is possible. This UAV is used to spray pesticide, fungicide, herbicide, or granule/liquid fertilizers to crops. Natural vegetation monitoring is also possible.

Nonagricultural uses: They are surveillance of military camps and vehicles, industrial installations, buildings, mines, mining activity, oil pipelines, electric lines, dams, rivers, lakes, and irrigation channels. It is also used to track transport vehicles using tacking sensors.

PLATE 3.56.1 PD6E2000-AW high.
Source: Ms. Aaya Kawakami, ProDRONE, Ltd., Aichi, Nagoya, Japan.

Useful References, Websites, and YouTube Addresses

https://www.proUAV.jp/products (accessed Sept 12, 2017).

https://www.wearechampionmag.com/proUAV-pd6e2000aw-UAV-water-resistant-and-all-weather-capable (accessed Aug 30, 2017).

https://www.youtube.com/watch?v=XE8U4W5yFCI (accessed Aug 30, 2017).

https://www.youtube.com/watch?v=BcaN4oTUwyA (accessed Aug 30, 2017).

Name/Models: Pulverizador Sprayer UAV (5 kg)

Company and address: Joyance Technology Ltd., 19/F, 2321 Beihai Rd., Weifang, Shandong 261061, China; Tel.: +86 536 2791566, +86 536 2797166, +86 536 8648566; e-mail: sales@joyancetech.com; info@joyancetech.com; website: http://www.joyance.tech/

Shandong Joyance Intelligence Technology Co., Ltd. is a UAV technology company. They have a professional R&D team involved in developing efficient agriculture sprayer UAVs. They have invested a capital of 60 million RMB. They produce 100 units of agriculture UAVs monthly. Pulverizador sprayer UAV is a special type of multirotor UAV produced by this company in China. The company's motto is to make the "UAVs that farmers love." It is located at Weifang High-tech Industrial Development Zone. There are 10 experienced engineers to ensure prompt delivery. Every UAV is tested and confirmed for its good working condition before shipment.

Pulverizador sprayer has three models. These models differ based on load of plant protection chemicals that they carry and spray onto crops. This model has 5 kg payload capacity (plastic tank) with 15–25 min. endurance. It is a hexacopter with six axles. It is supposedly an efficient sprayer UAV. It costs about US$ 2500–4000 (March 2017 price level) per unit with basic accessories of GCS.

TECHNICAL SPECIFICATIONS

Material: It is made of carbon fiber and avionics light aluminum.

Size: length: 120 cm; width (rotor to rotor): 120 cm; height: 30 cm; weight: 7.5 kg without is payload, pesticide load 5 kg, MTOW is 15 kg; folded size is 50 × 40 × 50 cm

Propellers: It has six propellers. They are made of tough plastic and placed on six axles. They are quick release propellers. Propellers are powered by brushless motors.

Payload: It has a plastic container that holds 5 kg liquid or granular or powdered formulation for spraying.

Launching information: This is an autolaunch and autopilot UAV. The UAV returns to point of launch in case of emergency.

Altitude: The operational height is 0–200 m above crop canopy. However, ceiling height is 1000 m above crop canopy.

Speed: It is 0–12 m s^{-1} flying speed; spraying speed is 8 m s^{-1}.

Wind speed tolerance: Wind resistance is 10 m s^{-1}. Flying down air flow is 4–15 m s^{-1}.

Temperature tolerance: −10 to 70°C

Power source: Electric lithium–polymer batteries

Electric batteries: 6S 12,000 mA h 10C 1 PC

Fuel: This does not have IC engine, no need for gasoline.

Endurance: It is 15–25 min depending on payload and flight path envisaged.

Remote controllers: The GCS has mobile, laptop (7 in. screen), and telelink apparatus. The remote controller uses "V8.3 application." It uses nine channel radio connections to control the UAV and alter its flight path, if necessary.

Area covered: Flight radius is 1000 m from remote controller.

Photographic accessories: This is primarily a sprayer UAV. Perhaps visual camera may be fitted, if necessary.

Computer programs: This UAV is not usually used to obtain, transmit, or process images of ground features.

Spraying area: 1000–1500 m^2 min^{-1}. Sprayer flow is 0.2–2.0 l min^{-1}. Spray width is >2 m but depends on the spray bar, if fitted.

Agricultural uses: This sprayer UAV is capable of dusting/spraying fluid to crops. It can also be used to apply liquid or granular formulation of fertilizers.

Nonagricultural uses: It can be used to carry small luggage, packets, or hazardous chemicals in plastic tanks.

Useful References, Websites, and YouTube Addresses

http://www.joyance.tech (accessed July 29, 2017).

http://www.joyancetech.com/html/support/knowledge/Joyance-UAV-Vs-Cheap-UAV-with-spraying-pole.html (accessed July 29, 2017).

https://www.youtube.com/watch?v=ot2HWvkbaEI (accessed July 29, 2017).

https://www.youtube.com/watch?v=Ljl4nfGL8dc (accessed July 29, 2017).

https://www.youtube.com/watch?v=ot2HWvkbaEI (accessed July 29, 2017).

https://www.youtube.com/watch?v=IRibZXRZ8iU (accessed July 29, 2017).

https://www.youtube.com/watch?v=q93PK1JjxhY (accessed July 29, 2017).

https://www.youtube.com/watch?v=fbc96qLhIq4 (accessed July 29, 2017).

Name/Models: Pulverizador Sprayer UAV (10 kg)

Company and address: Joyance Technology Ltd., 19/F, 2321 Beihai Rd., Weifang, Shandong 261061, China; Tel.: +86 536 2791566, +86 536 2797166, +86 536 8648566; e-mail: sales@joyancetech.com (sales), info@joyancetech.com; website: http://www.joyance.tech/

Shandong Joyance Intelligence Technology Co., Ltd. is a UAV technology company. They have a professional R&D team involved in developing efficient agriculture sprayer UAVs. Load does affect the size of several parts and features of drone models. It is a hexacopter with six axles. It is supposedly an efficient sprayer UAV and it costs about US$ 2500–4000 (March 2017 price level) per unit with basic accessories of GCS.

TECHNICAL SPECIFICATIONS

Material: It is made of carbon fiber and avionics light aluminum.

Size: length: 120 cm; width (rotor to rotor): 120 cm; height: 30 cm; weight: 11.0 kg without is payload, pesticide load 10 kg, MTOW is 25 kg; folded size is 70 × 50 × 55 cm

Propellers: It has six or eight propellers. They are made of tough plastic and placed on six axles. They are quick-release propellers. Propellers are powered by brushless motors.

Payload: It has a plastic container that holds 10 kg liquid or granular or powdered formulation for spraying.

Launching information: This is an autolaunch and autopilot UAV. The UAV returns to point of launch in case of emergency.

Altitude: The operational height is 0–200 m above crop's canopy. However, ceiling height is 1000 m above crop canopy.

Speed: It is 0–12 m s^{-1} flying speed; spraying speed is 8 m s^{-1}.

Wind speed tolerance: Wind resistance is 10 m s^{-1}. Flying down air flow is 4–15 m s^{-1}

Temperature tolerance: −10 to 70°C

Power source: Electric lithium–polymer batteries

Electric batteries: 6S 15,000 mA h, 10 C, 1 PC

Fuel: This does not have IC engine, no need for gasoline

Endurance: 15–25 min depending on payload and flight path decided.

Remote controllers: GCS has mobile, laptop (7 in. screen), and telelink apparatus. The remote controller uses "V8.8 application." It uses nine channel radio connections to control the UAV and alter the path, if necessary.

Area covered: Flight radius is 1000 m from remote controller.

Photographic accessories: This is primarily a sprayer UAV. Perhaps visual camera may be fitted if necessary.

Computer programs: Generally, this UAV is not used to obtain, transmit, or process images of ground features. It is a sprayer UAV.

Spraying area: It covers 1000–1500 m^2 min^{-1}. Sprayer flow is 0.2–2.0 l min^{-1}. Spray width is >2 m but depends on the spray bar, if fitted. Sprayer nozzles are 3–4 in number.

Agricultural uses: This is sprayer UAV is capable of dusting/spraying fluid to crops at various stages of growth. It can also be used to apply liquid or granular formulation of fertilizers.

Nonagricultural uses: It can be used to carry small luggage, packets, or hazardous chemicals in plastic tanks.

Useful References, Websites, and YouTube Addresses

http://www.joyance.tech (accessed Sept 16, 2017).
http://www.joyancetech.com/html/support/knowledge/Joyance-UAV-Vs-Cheap-UAV-with-spraying-pole.html (accessed Sept 16, 2017).
https://www.youtube.com/watch?v=ot2HWvkbaEI (accessed Sept 16, 2017).
https://www.youtube.com/watch?v=Ljl4nfGL8dc (accessed Sept 16, 2017).
https://www.youtube.com/watch?v=ot2HWvkbaEI (accessed Sept 16, 2017).
https://www.youtube.com/watch?v=IRibZXRZ8iU (accessed Sept 16, 2017).
https://www.youtube.com/watch?v=q93PK1JjxhY (accessed Sept 16, 2017).
https://www.youtube.com/watch?v=fbc96qLhIq4 (accessed Sept 16, 2017).

Name/Models: Pulverizador Sprayer UAV (15 kg)

Company and address: Joyance Technology Ltd., 19/F, 2321 Beihai Rd., Weifang, Shandong 261061, China; Tel.: +86 536 2791566, +86 536 2797166, +86 536 8648566; e-mail: sales@joyancetech.com (sales), info@joyancetech.com; website: http://www.joyance.tech/

Shandong Joyance Intelligence Technology Co., Ltd. is a UAV technology company. They have a professional R&D team involved in developing efficient agriculture sprayer UAVs.

Pulverizador sprayer has three models based on load of plant protection chemicals each carries and sprays onto crops. This model (Pulverizador Sprayer UAV-15) has 15 kg payload capacity (plastic tank) with 15–25 min endurance. It is a hexacopter with six axles. It is supposedly an efficient sprayer UAV and it costs about US$ 4500 to 5000 (March 2017 price level) per unit with basic accessories of GCS.

TECHNICAL SPECIFICATIONS

Material: It is made of carbon fiber and avionics light aluminum.

Size: length: 270 cm; width (rotor to rotor): 130 cm; height: 35 cm; weight: 13.0 kg without is payload, pesticide load 15 kg, MTOW is 30 kg; folded size is 120 × 50 × 55 cm

Propellers: It has 6–10 propellers. They are made of tough plastic and are placed on six axles. They are quick release propellers. Propellers are powered by brushless motors.

Payload: It has a plastic container that holds 15 kg liquid or granular or powdered formulation for spraying.

Launching information: This is an autolaunch and autopilot UAV. The UAV returns point of launch in case of emergency.

Altitude: Its operational height is 0–200 m above crop canopy. However, ceiling height is 1000 m above crop canopy.

Speed: It is 0–12 m s^{-1} flying speed; spraying speed is 8 m s^{-1}.

Wind speed tolerance: Wind resistance is 10 m s^{-1}. Flying down air flow is 4–15 m s^{-1}.

Temperature tolerance: −10 to 70°C

Power source: Electric lithium–polymer batteries

Electric batteries: 6S 15,000 mA h, 10 C, two PCs

Fuel: This does not have IC engine, no need for gasoline.

Endurance: 15–25 min, depending on payload and flight path decided

Remote controllers: GCS has mobile, laptop (7 in. screen) and telelink apparatus. The remote controller uses V8.8 application. It uses nine channel radioconnections to control the UAV and alter the flight path, if necessary.

Area covered: Flight radius is 1000 m from remote controller.

Photographic accessories: This is primarily a sprayer UAV. Perhaps visual camera may be fitted, if necessary.

Computer programs: This UAV is not used to obtain, transmit, or process images of ground features.

Spraying area: It covers 1000–1500 m^2 min^{-1}. Sprayer flow is 0.2–2.0 l min^{-1}. Spray width is >2 m but depends on the spray bar, if fitted. Sprayer nozzle number is 3–6.

Agricultural uses: This is sprayer UAV capable of dusting/spraying fluid to crops, at various stages of growth. It can also be used to apply liquid or granular formulation of fertilizers.

Nonagricultural uses: It can be used to carry small luggage, packets, or hazardous chemicals in plastic tanks.

Useful References, Websites, and YouTube Addresses

http://www.joyance.tech (accessed Sept 27, 2017).

http://www.joyancetech.com/html/support/knowledge/Joyance-UAV-Vs-Cheap-UAV-with-spraying-pole.html (accessed Sept. 27, 2017).

https://www.youtube.com/watch?v=ot2HWvkbaEI (accessed Sept 27, 2017).

https://www.youtube.com/watch?v=Ljl4nfGL8dc (accessed Sept 27, 2017).

https://www.youtube.com/watch?v=ot2HWvkbaEI (accessed Sept 27, 2017).

https://www.youtube.com/watch?v=IRibZXRZ8iU (accessed Sept 27, 2017).

https://www.youtube.com/watch?v=q93PK1JjxhY (accessed Sept 27, 2017).

https://www.youtube.com/watch?v=fbc96qLhIq4 (accessed Sept 27, 2017).

Name/Models: Spraying UAV (5 L Model)

Company and address: Spraying UAV, 406 Padre Palomino, Machala, El Oro 05202, Ecuador; phone: +593 982991481 (Ecuador), +1 805 308 0409 (USA); e-mail: info@sprayingUAV.com; website: http://sprayingUAV.com/index.html

Spraying UAV Inc. is an avionics company located in El Oro, Ecuador. It was started in 2010. Their aim is to supply different models of UAVs, particularly those relevant to agricultural farmers. This company offers a few models and options of different accessories. They deal with farmers involved in fields/plantation crops in Ecuador and other nations in the continent.

This UAV model belongs to light-weight category of UAVs. Yet, it holds optimum levels of pesticide/fertilizer fluid or granules to be efficient in the crop field. It is predominantly a sprayer UAV. It is of great utility to farmers while conducting plant protection. It comes in ready-to-fly condition with many fail-safe features. It is packaged in a small aluminum suitcase and is easy to transport across crop fields and launch from anywhere. It comes with a tool kit to assemble or dismantle and repair. At current price levels (2017), it costs about US$ 5300 unit. It is not very costly for a farm company with large acreage.

TECHNICAL SPECIFICATIONS

Material: The UAV's frame is built with tough carbon fiber. Aluminum adjustable legs are provided to alter height of the UAV, during take-off and landing or resting.

Size: length: 112 cm; width (rotor to rotor): 135 cm; height: 90 cm; weight: MTOW is 15 kg. It includes pesticide tank and CPU.

Propellers: It is a hexacopter. It has six motor-driven propellers made of toughened plastic. They are connected to brushless motor (NB5208-300KV) for power and propulsion. The propellers are kept folded, but they are quick-release type.

Payload: Plastic pesticide tank filled with 5 l of formulation.

Launching information: It is an auto-take-off, navigation, and landing UAV. It has VTOL mode. It is controlled by flight controller system "V8.5 Pixhawk software."

Altitude: Flying height while on normal spray schedules is 0–200 m above crop's canopy. However, ceiling height is 1000 m above-ground level.

Speed: It is 0–12 m s^{-1}

Wind speed tolerance: Wind disturbance of 10 m s^{-1} is tolerated by this UAV

Temperature tolerance: −10 to 70°C

Power source: Electric batteries

Electric batteries: 6S 12,000 mA h, lithium–polymer batteries with charger; 16,000 mA h batteries are optional **Fuel:** Not an IC engine-fitted UAV

Endurance: It is 10–25 min, based on spray rate and intensity of spray.

Remote controllers: The remote controller has nine channels (telemetric link) to contact and control UAV's flight path.

Area covered: It covers 50–100 ha per flight depending on spray rate and flight path restrictions.

Photographic accessories: Cameras: Usually, it carries a complement of visual and multispectral cameras. They are fixed to a gyro-stabilized gimbal.

Computer programs: This UAV is meant for spraying. Yet, the images obtained and transmitted to GCS can be stored or processed using Pix4D Mapper or Agisoft's Photoscan.

Spraying area: It covers 50 ha per flight; spraying speed is 8 m s^{-1}; spraying width (swath) is >4/4 nozzles; spray flow is 0.2–2 l min^{-1}; spraying time is 10–25 min per flight.

Volume: It discharges 0.2–2 l fluid min^{-1}.

Agricultural uses: It is basically a sprayer UAV. This UAV could be utilized for aerial survey and spectral analysis of crop growth, nutritional status, water stress index, and grain maturity. Natural vegetation monitoring is also possible.

Nonagricultural uses: It includes surveillance of industrial installations, mines, ore dumps, railways, city traffic, rivers, dams, irrigation projects, etc.

Useful References, Websites, and YouTube Addresses

http://sprayingUAV.com/ (accessed June 24, 2017).

http://sprayingUAV.com/models.html (accessed June 24, 2017).

http://www.ecuador.com/blog/ecuador-launches-locally-made-UAV (accessed June 24, 2017).

https://www.youtube.com/watch?v=hEpaD6sbsoo (accessed June 24, 2017).

Name/Models: Spraying UAV (10 L Model)

Company and address: Spraying UAV, 406 Padre Palomino, Machala, El Oro 05202, Ecuador; phone: +593 98299148 (Ecuador), +1 8053080409 (USA); e-mail: info@sprayingUAV.com; websites: http://sprayingUAV.com/, http://sprayingUAV.com/index.html

The Sprayer UAV Inc. is a UAV company. It is in El Oro, Ecuador. It began producing and testing copter UAVs in 2010. It offers a few different models of multirotor UAVs capable of spraying pesticides, to crop fields. They cater to farms of different sizes and economic viability.

Spraying UAV (10 L model) is usually kept in "ready-to-fly" condition. It has autonomous navigation, autolaunch and landing gears and appropriate software. The airframe is made of carbon fiber, and propellers are plastic. It has an umbrella folding frame. This UAV is at present adopted to survey the natural vegetation, forest, and crops. It is being used in Central American and South American farming regions. This UAV is small and foldable. It is packed into an aluminum case and transported easily to anywhere in the crop field. This UAV costs US$ 8400 per unit (2017 price line).

TECHNICAL SPECIFICATIONS

Material: This UAV is made of carbon fiber frame and plastic propellers. It has foldable aluminum joints to adjust height.

Size: length: 120 cm; width (rotor to rotor): 280 cm; height: 65 cm; weight: 11 kg; MTOW is 25 kg.

Propellers: This is an octocopter. It has eight propellers attached to brushless motors (NB5020-200 KV) for energizing the propeller.

Payload: 12 kg. It includes gyro-stabilized gimbal, a series of five cameras and pesticide tank with 10 L of fluid/granules/powder.

Launching information: It is an autolaunch VTOL UAV. It needs 5×5 m^2 space to lift-off. It returns to point of launch, in case of emergency.

Altitude: The ceiling is 1000 m above ground level. Altitude adopted during aerial imagery and spraying is 0–200 m above ground level.

Speed: It is 0–12 m s^{-1}

Wind speed tolerance: It withstands wind disturbance of 10 m s^{-1}

Temperature tolerance: −10 to 70°C

Power source: Electric batteries

Electric batteries: Two units of battery 6S 16,000 mA h, plus a charger, and adapter.

Fuel: Not an IC engine supported UAV

Endurance: Spraying time is 10–25 min

Remote controllers: The remote controller has nine radio channels to communicate and send commands to the UAV. The telemetric connection operates in 15 km radius zone. The GCS computers can override the preprogramed flight path and alter it, using software such as "Pixhawk" or "Flight control V8.8."

Area covered: It covers 50–200 ha of aerial photography and/or spraying per hour.

Photographic accessories: Cameras: This UAV carries a gyro-stabilized gimbal that holds visual (R, G, and B), multispectral, hyperspectral, NIR, IP, and red-edge cameras. Image transmission accessories are attached.

Computer programs: Aerial imagery relayed to ground computers (iPads) is processed, using Pix4D Mapper or Agisoft's Photoscan.

Spraying area: It covers 50 ha per flight. Spraying width (swath) is >2.5 m. It uses four nozzles. Spray flow is 0.2–2.0 l min^{-1}.

Volume: The tank fitted under the UAV holds 10 L of pesticide or fertilizer granules. The entire volume is released in 12–15 min of flight over the crop field.

Agricultural uses: This UAV is primarily a multirotor sprayer UAV. It is used for crop scouting and imaging disease/pest-attacked zones or those showing nutrient deficiencies. The digital data is used to conduct variable-rate spraying. The cameras on the UAV also pick NIR and IR images. Such images are useful during determination of crop's water stress index and scheduling irrigation. Estimation of chlorophyll, crop-N status, and deciding fertilizer-N spray schedules, using appropriate computer software is another important function of this UAV.

Nonagricultural uses: It includes surveillance of mines, mining activity, ore dumps, buildings, industries, river flow, dams and irrigation channels, oil pipelines, electric lines, and public events. Tracking farm and traffic vehicles are other functions of this UAV.

Useful References, Websites, and YouTube Addresses

http://sprayingUAV.com/ (accessed June 28, 2017).

http://sprayingUAV.com/models.html (accessed June 28, 2017).

http://www.ecuador.com/blog/ecuador-launches-locally-made-UAV (accessed June 28, 2017).

https://www.youtube.com/watch?v=hEpaD6sbsoo (accessed June 28, 2017).

Name/Models: Spraying UAV (15 L Model)

Company and address: Spraying UAV, 406 Padre Palomino, Machala, El Oro 05202, Ecuador; phone: +593 982991481 (Ecuador), +1 8053080409 (USA); e-mail: info@sprayingUAV.com; website: http://sprayingUAV.com/index.html

Spraying UAV Inc. is an agricultural UAV company situated at El Oro in Ecuador. It offers a few different models of multirotor agricultural UAVs. They are meant mainly to spray chemicals such as pesticides, fungicides, and herbicides onto crops. These UAVs are also equipped with sensors for aerial surveillance and imagery.

"Spraying UAV—model 15 kg" is among the biggest in the series of multirotor UAVs produced by the company. It is made of aviation aluminum, carbon fiber, and plastic. It is a hexacopter with larger pesticide tank. It is an efficient and rapid release UAV. It uses six nozzles to unload pesticide formulation rapidly, say in 12–15 min per flight. The hexacopter has facility for aerial imagery and processing, using apt software such as Pix4D Mapper. The 15 L capacity UAV costs US$ 10,400 per unit (2017 price levels).

TECHNICAL SPECIFICATIONS

Material: This UAV is made of carbon fiber, light aviation aluminum, and plastic.

Size: length: 120 cm; width (rotor to rotor): 180 cm; height: 65 cm weight: dry weight is 13 kg, MTOW is 30 kg

Propellers: This is an octocopter and has eight rotors. Propellers are made of toughened plastic. Propellers are connected to brushless motors (NB5020-200KV). It is fitted with Quick release propellers 3131.

Payload: 15 kg

Launching information: It is an autolaunch, navigation, and landing UAV with appropriate software. The flight path is decided, using "Pixhawk" software. It returns to point of launch in times of emergency.

Altitude: The ceiling is 1000 m. Normal flying height during spraying and aerial imagery of crops is 0–200 m above ground level.

Speed: Flying speed is 0–12 m s^{-1}

Wind speed tolerance: 10 m s^{-1} disturbance

Temperature tolerance: −10 to 70°C

Power source: Electric batteries

Electric batteries: 16,000 mA h lithium–polymer batteries, two cells

Fuel: No need for gasoline

Remote controller: It has nine channels for telemetric command and flight guidance. It is operative for up to 15 km radius.

Area covered: It covers 50–200 ha h^{-1}.

Photographic accessories: Cameras: It has a full complement of sensors that includes visual (R, G, and B), multispectral, hyperspectral for close-up shots, NIR for chlorophyll estimation, IR, and RedEdge for water stress index determination.

Computer programs: Aerial images are processed, using Pix4D Mapper or Agisoft's Photoscan software.

Spraying area: It covers 50 ha per flight; spraying speed is 0–8 m s^{-1}; spraying width (swath) is >3 m, using six nozzles; spray flow of pesticide formulation is 0.2–2 l min^{-1}.

Volume: Pesticide tank holds 15 kg formulation and releases it all in 15 min.

Agricultural uses: It is used for crop scouting. The scouting is done to detect seed germination trends and gaps in rows, seedling, and crop stand establishment, disease/pest-attacked zones, if any, and to map them. The digital data about variations in disease/pest attack and nutrient deficiency spots are relayed. Such data are used in variable-rate applicators of ground vehicles or at the nozzles in the UAV itself. This UAV is effectively used to obtain aerial maps of crops. They are useful to recommend fertilizer and water application schedules. It is also used to estimate/forecast grain yield.

Nonagricultural uses: They are surveillance of mines, mining activity, ore dumps, industrial zones, buildings, city public events, rail roads, highways, tracking transport vehicles, etc.

Useful References, Websites, and YouTube Addresses

http://sprayingUAV.com/ (accessed Oct 30, 2017).
http://sprayingUAV.com/models.html (accessed Oct 30, 2017).
http://www.ecuador.com/blog/ecuador-launches-locally-made-UAV (accessed Oct 30, 2017).
https://www.youtube.com/watch?v=hEpaD6sbsoo (accessed Oct 30, 2017).

Name/Models: SZM-X4-1200 Agricultural Sprayer UAV

Company and address: Shan Zhuang Machinery Co., Ltd., China (Mainland) No. 1, Yintai Street, Qingzhou Economic Development Zone 262500, Shandong, Shandong, China (Mainland); website: https://cszmc.en.alibaba.com/?spm=a2700.8304367.0.0.e83844a5mvHDI

SZM-X4-1200 is a multirotor crop duster UAV produced and utilized in China (mainland). It is made of light-weight carbon fiber. It withstands vagaries of agricultural fields and its environment. It is a low-flyer UAV with ability to hover while spraying. It holds 10 kg pesticide or other plant protection chemicals. The UAV has autonavigation facility. It is an efficient UAV sprayer. It covers about 1–2 ac min^{-1}. It can effectively replace farm labor during crop dusting. It costs about US$ 7500–10000 per unit, along with accessories such as battery chargers, nozzles, etc.

TECHNICAL SPECIFICATIONS

Material: This UAV is made of light-weight carbon fiber and toughened plastic.

Size: length: 38 cm; width (rotor to rotor): 38 cm; height: 45 cm; weight: 12.2 kg without payload

Propellers: It has four or six propellers depending on option. Propeller size is 30.8 cm.

Payload: 10 kg fluid/granular pesticide in a plastic tank.

Launching information: It is an autolaunch, autonomous aerial robot. Its navigation is controlled by using "Pixhawk" software. It lands in 1–2 min. It is a VTOL UAV and needs only small area to lift-off.

Altitude: Its operating height is 0–5 above crop's canopy.

Speed: Flying speed is 0–10 m s^{-1}; flying range is 2 km, and flying height is 1.5 m above crop's canopy

Wind speed tolerance: It resists 15–20 km h^{-1} disturbance.

Temperature tolerance: −15 to 40°C

Power source: Electric batteries

Electric batteries: 24,000 mA h

Fuel: It does not require gasoline

Endurance: It is 10–18 min per flight

Remote controllers: This UAV keeps contact with ground control (remote controller or joystick) for up to 2 km radius. The ground control laptop could be used to override and modify flight path and spray rates, if required.

Area covered: 1–2 ac min^{-1}

Photographic accessories: Cameras are not fitted, if it is meant exclusively to spray the pesticide, based on blanket recommendations.

Computer programs: Aerial images are not offered by this UAV. It is basically a sprayer drone.

Spraying area: Spraying width (swath) is 3–4 m depending number of nozzles used. Spray flow is 1.25 l min^{-1}.

Volume: 10 L per flight of 15 min; UAV's speed during spraying is 0–6 m s^{-1}

Agricultural uses: This is basically a sprayer UAV. It could be used to apply pesticides, fungicides, herbicides or liquid fertilizers as foliar dosages.

Nonagricultural uses: It could be used to carry hazardous luggage and chemicals from one location to another.

Useful References, Websites, and YouTube Addresses

https:www.alibaba.com/product-detail/10litre-pesticide-Sprayer-UAV-UAV-for_60/ (accessed Apr 20, 2017).

https://www.alibaba.com/product-detail/10liter-Pesticide-Sprayer-UAV-UAV-for_60482082090.html (accessed Apr 20, 2017).

https://www.alibaba.com/product-detail/6-Alxes-crops-plant-sprayer-UAV_60491711783.html (accessed Apr 20, 2017).

Name/Models: Trump Sprayer UAV (Model S4-1400)

Company and address: Shenzhen Trump Aero Technology Co., Ltd. Donglong Zing Building, Huarong Rd., Dalang, Bao'an District, Shenzhen 518000, Guangdong, China; phone: +86 755 86185833; e-mail: sales@trumpuav.com; websites: http://www.trumpuav.com/, http://www.trumpuav.com/LEDdisplay/

The Shenzhen Trump Aero Technology Co., Ltd. is located near the Shenzhen city, in China. It was initiated in 2011 with an aim to produce small UAV models, useful in agriculture. It offers a series of UAV models, to suit middle income farms (few 100 ha) to large farms (10,000 plus ha). They cater to requirements of aerial photography, aerial spraying of pesticides and fertilizer formulations. They are useful in rapid scouting of crop fields. In addition, these UAV models are used in other activities, such as surveillance of natural vegetation, water resources, rivers, lakes, etc.

The Trump Sprayer UAV is a high-tech product manufactured by the above Chinese UAV company. This UAV called "Trump Sprayer UAV" is used extensively in aerial photography, film, and television. It has a major role to play in agricultural crop survey, aerial surveillance, and spectral analysis of ground features, including crops. Most important function is crop dusting/spraying with pesticides. It reduces cost of production immensely compared to traditional system of spraying. It unloads 10 l of pesticides in a matter of minutes (Plate 3.57.1). The UAV costs US$ 5000 (February 2017 price level). It has been exported to different nations from China.

TECHNICAL SPECIFICATIONS

Material: The UAV's frame and fuselage are made of carbon fiber and light aluminum alloy. Frame and fuselage are fire and dust resistant. This UAV can be used even during rain (waterproof).

Size: length: 90 cm; width (rotor to rotor): 90 cm; height: 55 cm; weight: 9.0 kg dry weight without payload; MTOW is 28 kg

Propellers: It has four propellers. They are made of two plastic blades each. The length of propeller is 41 cm.

Payload: Maximum payload is 12 kg.

Launching information: It is an autolaunch UAV. It is a VTOL UAV. The UAV is totally autonomous during flight. Its flight path is predetermined, using "Pixhawk" software.

Altitude: The ceiling is 1000 m above crop canopy and 4500 m above sea level. Operating height while spraying pesticide is 5–20 m above crop canopy. It has a hover accuracy of ±1.5 m, while making close-up shots or during spraying.

Speed: It is 45 km h^{-1}, while on aerial scouting. It is 20 km h^{-1}, while spraying pesticide.

Wind speed tolerance: 28 km h^{-1} wind disturbance

Temperature tolerance: −20 to 50°C

Power source: Electric batteries

Electric batteries: 10,000 mA h lithium–polymer.

Fuel: Not an IC engine-fitted UAV

Endurance: It is 40 min per flight

Remote controllers: The GCS computer (iPad) keeps contact with UAV for up to 5000 m radius. The UAV has fail-safe mode. It reaches the point of launch, in case of emergency.

Area covered: It transits a linear distance of 15 m or covers 50–100 ha in aerial survey and photography

Photographic accessories: Cameras: It has a full complement of cameras such as visual (R, G, and B), high-resolution multispectral, NIR, IR, RedEdge and Lidar sensor.

Computer programs: The aerial images are processed, using computer software, such as Pix4D Mapper or Agisoft's Photoscan.

Spraying area: The UAV covers an area of 50–200 ha of crop field per day.

Volume: The plastic pesticide tank holds about 10 l of liquid or granular formulation.

Agricultural uses: Major utilities of this "Trump sprayer UAV" is in agricultural crop spraying. It sprays rapidly above the crop canopy. It avoids contact of farmers with hazardous chemicals (Plate 3.57.1). It does the task at low cost, low labor requirement and reduces usage of pesticides. Crop scouting for leaf area, leaf-N (crop's N status), canopy and ambient temperature (crop's water stress index), panicle, and grain maturity are other uses. Natural vegetation monitoring for NDVI and plant diversity is also possible.

Nonagricultural uses: It includes surveillance and aerial photography of geologic sites, land resources, soils, mines, mining activity, electric lines, oil pipelines, highways and vehicles, etc.; aerial reconnaissance of disaster zones and relief movement.

PLATE 3.57.1 Trump UAV.
Source: Mr. Allen, J. Sales and Marketing Division, Dalang, Bao'an District, Shenzhen 518000, Guangdong, China, Shenzhen Trump Aero Technology Co., Ltd., Shenzhen, China (Mainland).

Useful References, Websites, and YouTube Addresses

http://www.trumpuav.com/Product/ (accessed Sept 1, 2017).
http://www.trumpuav.com/application/AgriculturalSpray/ (accessed Sept 1, 2017).

Name/Models: Unzerbrechlich X406

Company and address: Shandong Xin Mei Mining Equipment Group Co., Ltd.; The South of Yanghuli Village Commercial Street, Changgou Town 27210, Shandong, Rencheng District, China (Mainland); phone: +86 537 3288771; e-mail: NA; website: http://xinmeimachinery.en.alibaba.com

Unzerbrechlich is a 6-kg sprayer agricultural UAV. It is a hexacopter. This UAV was developed and sold in 2017. There are four models with different payload and dusting efficiency. This UAV has gyro-stabilization and provides stable flight pattern. It can be operated using remote controller or laptop. Hence, it avoids contact of farmers with pesticides. It is known for easy operation and low maintenance cost. It is a small UAV. Therefore, it can be transported and launched, even, from small areas in a crop field.

TECHNICAL SPECIFICATIONS

Material: It is made of aviation carbon fiber and light aluminum

Size: length: 143 cm; width (rotor to rotor): 143 cm; height: 590 cm; weight: machine wt. 5 kg, MTOW is 12 kg; folded size: 67 cm × 38 × 49 cm.

Propellers: Four propellers made of tough plastic. They are connected to brushless motors for propulsion.

Payload: 6 kg plant protection chemicals, plus CPU.

Launching information: It is an autolaunch vehicle. Flight path planning using computer software helps in autonomous navigation. It has automatic return system in case of emergency.

Altitude: It is 5 m above crop canopy to 50 m above crop canopy. Ceiling is 1000 m.

Speed: It ranges from 1 to 10 m s^{-1}

Wind speed tolerance: 20 km h^{-1} disturbance. This UAV is gyro-stabilized to provide stable flight path.

Temperature tolerance: −5 to 50°C

Power source: Electric batteries

Electric batteries: 24,000 mA h, lithium–polymer

Fuel: Not an IC engine-fitted UAV. It does not need gasoline.

Endurance: 18 min per flight

Remote controllers: The remote controller has a laptop with all appropriate computer programs. The UAV can also be controlled, using mobile. The UAVs flight path is usually predetermined, using software such as Pix4D Mapper or Trimble's Inphos.

Area covered: Flight radius is small at 2000 m. Area covered is about <35 ha.

Photographic accessories: Cameras: This is primarily a sprayer UAV. The aerial imagery accessories may be fitted, if needed.

Computer programs: This is a sprayer UAV. Image processing software in the GCS could be used, if it has visual camera. It is also capable of sending ortho-images for processing.

Spraying area: It covers about 66 ac day^{-1}; spray flow is 1–2 l pesticide min^{-1}; spray width (swath) is 3–5 m, depending on the spray bar. It has four spray nozzles. The spray control system is WFT08/WFT0911.

Volume: 6 kg pesticide per flight

Agricultural uses: This is a crop duster or sprayer UAV. It is useful in spreading plant protection chemicals or liquid fertilizer formulation.

Nonagricultural uses: It can be used to carry hazardous chemicals. Otherwise, any luggage of small size (volume) and within 6–8 kg weight is transportable.

Useful References, Websites, and YouTube Addresses

https://xinmeimachinery.en.alibaba.com/product/60608644477-804401285/X820_Manufacture_Of_UAV_UAV_Crop_Sprayer.html (accessed July 2, 2017).

https://www.alibaba.com/product-detail/Unzerbrechlich-6kg-UAV-agriculture-sprayer (accessed July 2, 2017).

Name/Models: Unzerbrechlich X810

Company and address: Shandong Xin Mei Mining Equipment Group Co., Ltd.; The South of Yanghuli Village Commercial Street, Changgou Town, 27210 Shandong, Rencheng District, China (Mainland); phone: +86 537 3288771; website: http://xinmeimachinery.en.alibaba.com

Unzerbrechlich X810 is a 10.0-kg sprayer agricultural UAV. It is a hexacopter. There are four models with different payload and dusting efficiency. This UAV was developed and sold in 2017. Unzerbrechlich X810 is a slightly larger version of sprayer UAV compared to Unzerbrechlich. UAV has gyro-stabilization and provides stable flight pattern. It can be operate using remote controller or laptop. Hence, it avoids contact of farmers with pesticides. It is known for easy operation and low maintenance cost. It is a small UAV. Therefore, can be transported and launched even from small areas in a crop field.

TECHNICAL SPECIFICATIONS

Material: Aviation carbon fiber and light aluminum

Size: length: 249 cm; width (rotor to rotor): 164 cm; height: 84.5 cm; weight: machine wt. 10 kg, MTOW is 23 kg; folded size: 70 × 60 × 74 cm

Propellers: There is option of models with 4, 6, or 8 propellers. Propellers are made of tough plastic. They are connected to brushless motors for propulsion.

Payload: It is 10 kg plant protection chemicals plus CPU

Launching information: It is an autolaunch vehicle. Flight path planning using computer software helps in autonomous navigation. It has automatic return system in case of emergency.

Altitude: It flies at 5 m above crop canopy to 50 m above crop canopy, during spraying on crops. However, ceiling is 1000 m.

Speed: It ranges from 1 to 10 m s^{-1}

Wind speed tolerance: 20 km h^{-1} disturbance. This UAV is gyro-stabilized to provide stable flight path.

Temperature tolerance: −5 to 50°C

Power source: Electric batteries

Electric batteries: 24,000 mA h, lithium–polymer

Fuel: Not an IC engine-fitted UAV. It does not need gasoline.

Endurance: It is 18 min per flight

Remote controllers: The remote controller has a laptop with all appropriate computer programs. The UAV can also be controlled using mobile. The UAVs flight path is usually predetermined, using software such as Pix4D Mapper or Trimble's Inphos.

Area covered: Area cover is about 1–100 ac day^{-1}.

Photographic accessories: Cameras: This is primarily a sprayer UAV. The aerial imagery accessories may be fitted, if needed.

Computer programs: This is a sprayer UAV. Image processing software in the GCS could be used, if it has visual camera and capable of sending ortho-images for processing.

Spraying area: It covers about 100 ac day^{-1}; spray flow is 1–2 l pesticide min^{-1}; spray width (swath) is 4–6 m, but, depending on the length of spray bar. The spray bar usually has four spray nozzles; spray control system is WFT08/WFT0911.

Volume: 6 kg per flight

Agricultural uses: This is primarily a crop duster or sprayer UAV. It carries plant protection chemicals or liquid fertilizer formulation.

Nonagricultural uses: It can be used to carry hazardous chemicals or any luggage of small size and within 6–8 kg weight.

Useful References, Websites, and YouTube Addresses

https://www.alibaba.com/product-detail/Unzerbrechlich-6kg-UAV-agriculture-sprayer (accessed Jan 2, 2018).

https://xinmeimachinery.en.alibaba.com/product/60608644477-804401285/X820_Manufacture_Of_UAV_UAV_Crop_Sprayer.html (accessed Jan 2, 2018).

Name/Models: Unzerbrechlich X815

Company and address: Shandong Xin Mei Mining Equipment Group Co., Ltd., The South of Yanghuli Village Commercial Street, Changgou Town, 27210 Shandong, Rencheng District, China (Mainland); phone: +86 537 3288771; website: http://xinmeimachinery.en.alibaba.com

Unzerbrechlich X815 is a 15.0-kg sprayer agricultural UAV. It is a hexacopter. As stated earlier, there are four models with different payload and dusting efficiency. This UAV was developed and sold in 2017.

TECHNICAL SPECIFICATIONS

Material: It is made of aviation carbon fiber and light aluminum

Size: length: 255 cm; width (rotor to rotor): 167 cm; height: 63.5 cm; weight: machine wt. 11 kg, MTOW is 28 kg; folded size: 75 × 50 × 70 cm.

Propellers: It has 6–8 propellers made of tough plastic. They are connected to brushless motors for propulsion.

Payload: 15 kg plant protection chemicals, plus CPU

Launching information: It is an autolaunch vehicle. Flight route can be fixed using computer software. It helps in autonomous navigation. It has automatic return system, in case of emergency.

Altitude: It flies at 5 m above crop canopy to 50 m above crop canopy. However, ceiling is 1000 m.

Speed: 1–10 m s^{-1}

Wind speed tolerance: 20 km h^{-1} disturbance. This UAV is gyro-stabilized to provide stable flight path.

Temperature tolerance: −5 to 50°C

Power source: Electric batteries

Electric batteries: 48,000 mA h lithium–polymer

Fuel: Not an IC engine-fitted UAV. It does not need gasoline.

Endurance: It is 18 min per flight.

Remote controllers: The remote controller has a laptop with all appropriate computer programs. The UAV can also be controlled, using mobile. The UAVs flight path is usually predetermined, using software such as Pix4D Mapper or Trimble's Inphos.

Area covered: Area cover is about 132 ac day^{-1}

Photographic accessories: Cameras: This is primarily a sprayer UAV. The aerial imagery accessories may be fitted, if needed.

Computer programs: This is a sprayer UAV. Image processing software in the GCS could be, if it has visual camera and it can send ortho-images, for processing.

Spraying area: It covers about 132 ac day^{-1}; spray flow is 1–2 l pesticide min^{-1}; spray width is 4–8 m depending on the spray bar. It has four spray nozzles; spray control system is WFT08/WFT0911.

Volume: 15 kg per flight

Agricultural uses: This is a crop duster or sprayer UAV. It carries plant protection chemicals or liquid fertilizer formulation.

Nonagricultural uses: It can be used to carry hazardous chemicals or any luggage of small size, and within 6–8 kg weight.

Useful References, Websites, and YouTube Addresses

https://www.alibaba.com/product-detail/Unzerbrechlich-6kg-UAV-agriculture-sprayer (accessed Jan 3, 2018).

https://xinmeimachinery.en.alibaba.com/product/60608644477-804401285/X820_Manufacture_Of_UAV_UAV_Crop_Sprayer.html (Jan 3, 2018).

https://www.alibaba.com/product-detail/Unzerbrechlich-6kg-UAV-agriculture-sprayer (accessed Jan 3, 2018).

Name/Models: Unzerbrechlich X820

Company and address: Shandong Xin Mei Mining Equipment Group Co., Ltd., The South of Yanghuli Village Commercial Street, Changgou Town 27210, Shandong, Rencheng District, China (Mainland); phone: +86 537 3288771; website: http://xinmeimachinery.en.alibaba.com

Unzerbrechlich X820 is again a hexacopter. It is a 20.0-kg sprayer agricultural UAV. It is the largest among the series of sprayer models produced by the company. It is a hexacopter. This UAV was developed and sold in 2017. There are four models with different payload and dusting efficiency.

TECHNICAL SPECIFICATIONS

Material: Aviation carbon fiber and light aluminum.

Size: length: 200 cm; width (rotor to rotor): 175 cm; height: 75 cm; weight: machine wt. 14 kg, MTOW is 34 kg: Folded size: 75 × 50 × 75 cm.

Propellers: It has 6–8 propellers made of tough plastic. They are connected to brushless motors for propulsion.

Payload: 20 kg plant protection chemicals, plus CPU

Launching information: It is an autolaunch vehicle. Planning the flight route using computer software helps in autonomous navigation. It has automatic return system, in case of emergency.

Altitude: It takes a height of 5–50 m above crop canopy, during spraying. However, ceiling is 1000 m.

Speed: It is 1–10 m s^{-1}

Wind speed tolerance: 20 km h^{-1} disturbance. This UAV is gyro-stabilized to provide stable flight path.

Temperature tolerance: −5 to 50°C

Power source: Electric batteries

Electric batteries: 48,000 mA h lithium–polymer

Fuel: Not an IC engine-fitted UAV. It does not need gasoline.

Endurance: It is 18 min per flight

Remote controllers: The remote controller has a laptop with all appropriate computer programs. The UAV can also be controlled, using mobile. The UAVs flight path is usually predetermined, using software such as Pix4D Mapper or Trimble's Inphos.

Area covered: Area covered is about 165 ac day^{-1} of 6–8 h.

Photographic accessories: Cameras: This is primarily a sprayer UAV. The aerial imagery accessories may be fitted if needed.

Computer programs: This is a sprayer UAV. Image processing software in the GCS could be used, if it has visual camera and capable of sending ortho-images for processing.

Spraying area: It covers about 66 ac day^{-1}; spray flow is 1–2 l pesticide min^{-1}; spray width is 6–10 m depending on the spray bar. It has four spray nozzles; spray control system is WFT08/WFT0911.

Volume: 15 kg per flight

Agricultural uses: This is a crop duster or sprayer UAV. It carries plant protection chemicals or liquid fertilizer formulation.

Nonagricultural uses: It can be used to carry hazardous chemicals or any luggage of small size and within 6–8 kg weight.

Useful References, Websites, and YouTube Addresses

http://xinmeimachinery.en.alibaba.com (accessed Jan 8, 2018).

https://www.alibaba.com/product-detail/Unzerbrechlich-6kg-UAV-agriculture-sprayer (accessed Jan 8, 2018).

https://xinmeimachinery.en.alibaba.com/product/60608644477-804401285/X820_Manufacture_Of_UAV_UAV_Crop_Sprayer.html (accessed Jan 8, 2018).

Name/Models: WJD Octocopter Sprayer

Company and address: Shenzhen GC Electronics Co., Ltd., B/5, Block E, Jue Feng Technology and Industrial Zone, LeZuJiag Xing Village, HuangMaBu XiXiang, Baoan District, Shenzhen, China; phone: NA; e-mail: NA; website: NA

Shenzhen GC Electronics Co. is a UAV producer. The UAV models are highly useful in application of pesticide and herbicide to vast stretches of crop fields in different parts of China. These aerial robots are now finding their way into agrarian zones of different countries.

The new sprayer UAV with 10 l capacity is known as "WJD agricultural helicopter." It is an efficient machine and dispenses pesticides into 0.6–1.0 ac min^{-1}. Hence, it replaces a large section of labor requirement during plant protection procedures.

TECHNICAL SPECIFICATIONS

Material: It is made of carbon fiber, plastic, and lightweight steel.

Size: length: 162 cm; width (rotor to rotor): 162 cm; height: 40 cm; weight: 3.7 kg without payload.

Propellers: Four pairs. They are made of toughened plastic.

Payload: 10 kg pesticide in a plastic container, CPU and gimbal with camera optional

Launching information: It is an autolaunch UAV. Its navigation is easily controlled, manually, using remote controller. Otherwise, it could be predetermined, using Pixhawk software. It lands on a landing gear, automatically. The GC laptop could still alter the flight path, midway, if required.

Altitude: The operating height is 0–3 m above crop canopy during spraying. The general loiter altitude is 300 m above crop field.

Speed: It is 8–12 m s^{-1}.

Wind speed tolerance: Resists 20 km h^{-1} wind disturbance

Temperature tolerance: −10 to 40°C

Power source: Lithium–polymer electric batteries

Electric batteries: 24,000 mA h.

Fuel: It does not require gasoline. It is not an IC engine-fitted UAV.

Endurance: It is 18 min, depending on battery and payload.

Remote controllers: The ground station has a laptop, mobile, and remote controller (joystick) to control flight path of the UAV. The laptop uses "Pixhawk" software to decide or alter the flight path.

Area covered: This UAV covers 0.6–1.0 ac min^{-1}.

Photographic accessories: Cameras are optional. This is basically a sprayer UAV.

Computer programs: If aerial images are picked using cameras, then they are processed, using Pix4D Mapper or Agisoft's Photoscan.

Spraying area: It sprays 0.6–1.0 ac min^{-1}. The spray width is 4 m. It has four nozzles that operate using pressure. Spray speed can be controlled, using remote controller. Spray speed is 1–1.2 l min^{-1}. Spray height is 40 cm.

Volume: It dispenses about 10 l of pesticide in 10–12 min

Agricultural uses: Spraying crops with pesticides/fungicides or fertilizer formulations are the primary function.

Nonagricultural uses: They are transport of hazardous chemicals, transport and delivery of small luggage/packets.

Useful References, Websites, and YouTube Addresses

https://gcele.en.alibaba.com/contactinfo.html?spm=a2700.8304367.0.0.38gxsr (accessed Jan 7, 2018).

http://www.gcele.net (accessed Jan 7, 2018).

http://www.aliexpress.com/store/602529 (accessed Jan 7, 2018).

Name/Models: X430-A10 Quadcopter for Agriculture

Company and address: Beijing Dagong Technology Co. Ltd., 1F, Founder Bldg., No. 9, 5th Street, Shangdi, Hidian District, Beijing 100085, P.R. China; phone: +86 010 82177816; e-mail: info@dagontech.com; website: http://www.dagongtech.com/en/

Beijing Dagong Technology Co., Ltd. is concerned with UAV R&D, since 2012. Researchers from this company collaborate with Tsinghua University or Beijing University of Aeronautics and Astronautics. Beijing Dagong Technology Co., Ltd. produces tethered UAV system, high-payload UAV, long-endurance UAV and agriculture UAV. They are utilized to surveillance, communication relay, TV broadcasting, agriculture, geological, and mining exploration.

"X430-A10" hexacopter is a light UAV. It is made of avionic aluminum. It offers excellent efficiency and highly accurate atomized spray of pesticides onto crops. This UAV comes with all its accessories and operating manual. It is easy to operate. The entire UAV is packed in small suitcase and transported to any location in crop fields. The software and CPU in the UAV allow predetermination of flight path, waypoints, and spray rate (Plate 3.58.1).

TECHNICAL SPECIFICATIONS

Material: It is made of light avionics aluminum and carbon fiber.

Size: length: 132 cm; width (rotor to rotor): 132 cm; height: 45 cm; weight: 8.2 kg without payload. MTOW is 23 kg.

Propellers: It has six rotors with two plastic blades each. The rotors are connected to brushless motors for propulsion.

Payload: 10 l of pesticide contained in a plastic tank plus the operating batteries

Launching information: It is a VTOL, autolaunch vehicle with predetermined navigation. It can land in areas with 5 × 5 m clearance.

Altitude: It reaches 0–5 m above crop canopy, while spraying. The operating height is usually 3–5 m above canopy. This avoids excessive drift of chemicals that are sprayed on the canopy, also adds accuracy to the spraying process. It ascends and descends, from 50 to 100 m height in a short time. It can also hover at a spot on the crop canopy for intensive spraying.

Speed: It is 3–8 m s^{-1}, while spraying and 8–12 m s^{-1} on free loiter mode.

Wind speed tolerance: It resists 15 km h^{-1} wind disturbance.

Temperature tolerance: −15 to 50°C

Power source: Electric batteries

Electric batteries: 24,000 mA h

Fuel: It does not require gasoline.

Endurance: It is 15 min per flight

Remote controllers: The GCS consists of laptop with software to regulate UAV's path and its spray schedule. It also has facility for cable attachment to control the UAV's flight above the crop. It can also be controlled by connecting it to a moving pick-up van that travels at <25 km h^{-1} speed. The GCS computers allow modification of flight path. The UAV returns to point of launch in case of emergency.

Area covered: 15–25 ac per flight of 12 min

Photographic accessories: Cameras could be fitted to gyro-stabilized gimbal, but usually not done because it is basically a sprayer UAV.

Computer programs: This UAV could be fitted with visual cameras. The digital data/orthomosaics could be processed, using appropriate software, such as Pix4D Mapper or Agisoft's Photoscan.

Spraying area: It is 15–25 ac h^{-1}. Spray width (swath) is 4 m, but it depends on the spray bar size. It has 6 nozzles that are pressurized for blanket application of fluid pesticides. If precision farming is adopted, the, variable-rate nozzles and computer-based decision-support are needed, along with digital data/map, showing areas that need pesticide spray.

Volume: 10 l of pesticide per flight 12–15 min spraying endurance.

Agricultural uses: This UAV is basically a pesticide/herbicide spraying vehicle. It is useful in crop fields and natural vegetation zones.

Nonagricultural uses: It is used to carry hazardous chemicals from one location to another without exposure to farm workers or public. It can carry small packets and deliver them at predetermined spots.

PLATE 3.58.1 Top: X430-A10 hexacopter with accessories; bottom: A X430-A10 UAV in action, spraying a paddy crop with pesticides.
Source: Beijing Dagong Technology Co. Ltd., Beijing, China; excerpted from http://www.dagongtech.com/en/plus/view.php?aid=89, http://www.dagongtech.com/en/plus/view.php?aid=81.
Note: The turbulence caused by the low, close-flying UAV causes the plant canopy to get disturbed and ruffled. This process helps in distributing pesticides more evenly right into the lower layers of the crop's canopy.

Useful References, Websites, and YouTube Addresses

http://www.xUAVs.com.cn/en/ (accessed Sept 9, 2017).
http://www.xUAVs.com.cn/en/plus/list.php?tid=1 (accessed Sept 9, 2017).
http://usuavs.com/agricultural-uav.html (accessed Sept 9, 2017).

Name/Models: YF-04D Agricultural UAV System

Company and address: Shenzhen Geruibang Technology Co. Ltd., Sokkia Science and Technology, Main Building, Baoan District, Shenzhen 518000, Guangdong, China; Tel.: +86 755 26648819; fax: +86 755 26648818; mobile: +86 18123732548; e-mail: not available; website: http://www.globalsources.com/geruibang.co

Shenzhen Geruibang Technology Ltd. is a Chinese UAV company. It specializes in production of agricultural UAVs. These UAV models are user friendly, easy to assemble, and dismantle. They can be operated by farmers with ease. They are being sold and distributed in many of the Asian and European farming regions.

"YF-04D" is a sprayer UAV. It belongs to light-weight class of UAVs. It is a recent introduction into crop land in China (Mainland). Farmers call it an intelligent sprayer to control disease/insect pests. It has been designed to help farmers who are new comers into UAV technology. The cost is around US$ 4000. It is affordable even for a farmer with medium-sized holding (20–30 ac). It is supposed to gain in acceptance in several other agricultural regions.

TECHNICAL SPECIFICATIONS

Material: The fuselage is made of all-weather resistant carbon fiber. It is fire proof, dust proof, and rain proof. Hence, the UAV can be flown even during slight drizzle.

Size: length: 75 cm; width (rotor to rotor): 90 cm; height; 45 cm; weight: It is classified as light-weight UAV (5–50 kg). MTOW is 20 kg

Propellers: It is quadcopter; so, it has four rotors with *crescent*-shaped plastic propellers. Propellers are attached to brushless motors for power.

Payload: 12 kg. It includes sensors, spray tank, and spray nozzle assembly.

Launching information: It is an auto-take-off aerial vehicle. Navigation is predetermined using software, such as "Pixhawk" or a similar custom-made software. It navigates autonomously based on fight path selected. It has automatic obstacle forewarning and avoiding facility in the air.

Altitude: This UAV hovers just above the crop canopy during pesticide spray, also, during high-resolution close-up imaging sprees.

Speed: It is 45 km h^{-1}, but it depends on payload and spraying intensity.

Wind speed tolerance: 20 km h^{-1} disturbance

Temperature tolerance: −10 to 50°C

Power source: Electric lithium–polymer batteries

Electric batteries: 16,000 mA h lithium–polymer batteries, two cell

Fuel: Not an IC engine-fitted UAV

Endurance: 25–30 min based on payload

Remote controllers: GCS has telemetric link for up to 1 km radius. The GCS is an iPad or mobile that regulates UAV's speed, alters flight path, cameras shutter timing, and frequency, during aerial imagery. The GCS (iPad) also regulates spray nozzles during plant protection exercise.

Area covered: It covers 50 ac of aerial imagery or scouting per hour. Otherwise, 20–30 ac of aerial spray h^{-1}.

Photographic accessories: Cameras: The quadcopter has a full complement of visual, high-resolution multispectral, NIR, IR, and red-edge cameras.

Computer programs: The aerial images of crop fields are processed using software such as Pix4D Mapper or Agisoft Photoscan.

Spraying area: It covers 20–30 ac h^{-1}. The spray width (swath) is 3.5 m. It is supposed to enhance work efficiency and reduce the need for longer flight time. The spray nozzles are usually tested for accuracy and uniformity, particularly while using them on precision farming fields. There are at least four different models of spray nozzles and controllers known to be compatible with this platform.

Volume: The payload tank holds 10 l of pesticide or fertilizer formulation.

Agricultural uses: "YF-04D" has been touted as a UAV with major applications in agriculture and forestry. It is used to scout for crop growth, NDVI data, leaf-chlorophyll (crop's N), canopy temperature (thermal imagery), water stress index, etc. It could be used regularly and frequently to spray the crop with pesticides, fungicides, bactericides, and weedicides. Natural vegetation monitoring, obtaining NDVI values, and studying plant diversity using spectral signatures is also a possibility.

Nonagricultural uses: It includes surveillance of industries, buildings, oil and gas pipelines, electric lines, highways, and vehicular movement. As first reaction vehicle, it can be used to get imagery of disaster zones, map flood or drought effects, and droughts.

Useful References, Websites, and YouTube Addresses

http://www.szghdz.com/products/show-htm-itemid-15556.html (accessed Oct 12, 2018).

http://geruibang.manufacturer.globalsources.com/si/6008850437255/pdtl/Agricultural-UAV/1143871686/Agricultural-UAV.htm (accessed Oct 12, 2018).

http://www.globalsources.com/gsol/I/Agricultural-UAV/p/sm/1143869956.htm#1143869956 (accessed Oct 12, 2018).

Name/Models: Z8P Agricultural Sprayer UAV

Company and address: Beijing Sage Town Technologies Company Ltd., Beijing, China

"Z8P agricultural UAV" is an octocopter with automatic VTOL. It is a light UAV made entirely of carbon fiber and plastic. Its payload includes a plastic tank that holds 10 l of pesticide. It is an efficient UAV and dispenses pesticide for over 30–40 ac h^{-1}. It is easy to operate, using a remote controller. Otherwise, its flight path could predetermined. This UAV is packed in a suitcase in folded condition and transported, anywhere in the field. Z8P sprayer UAV costs US$ 4000–7000 per unit.

TECHNICAL SPECIFICATIONS

Material: Light weight but toughened carbon fiber and plastic.

Size: length: 138 cm; width (rotor to rotor): 138 cm; height: 49 cm; weight: UAV weight without payload is 10.9 kg, MTOW is 25 kg.

Propellers: It has eight rotors made of plastic blades. They are connected to brushless motor for propulsion.

Payload: It includes 10 kg plastic container with pesticide filled and ready for spraying

Launching information: It is an autolaunch octocopter. Its navigation can be decided based on digital data and maps depicting pest/disease/weed-attacked zones in the crop fields. Generally, software such as "Pixhawk" is used to decide flight path and control UAV's flight pattern. It has autoreturn facility in case of emergency. It has autolanding facility and adopts VTOL method.

Altitude: It is used to spray the crop from 0 to 3 m height above canopy. In case of forest plantation, the UAV's altitude is kept at 5–10 m above tree canopy. Working altitude during loiter is up to 200 m above ground level.

Speed: The cruising speed under automatic mode is 6–10 m s^{-1}.

Wind speed tolerance: It resists 15 km h^{-1} disturbance. It is gyro-stabilized. It is preferable to spray in no or low wind conditions to avoid pesticide drift.

Temperature tolerance: −5 to 40°C

Power source: Electric batteries

Electric batteries: 22.2 V, 22,000 mA h.

Fuel: Gasoline is not required.

Endurance: 40 min in free loiter mode; 25–30 min in spraying mode

Remote controllers: Z8P is primarily a sprayer UAV. The ground station (remote controller or laptop) keeps in touch with the UAV for 5000 m distance during spray schedules. The GCS software can alter UAV's flight path and spray speed, if needed.

Area covered: It covers 5500 m^2 per flight or 30–40 ac h^{-1}

Photographic accessories: Cameras could be fixed, if opted

Computer programs: Aerial images are usually preprocessed and digital data are utilized to prepare flight path.

Spraying area: 5500 m^2 per flight; spraying height is 0.5–3 m above canopy; spray flow is 0.5–1 L min^{-1}, flying speed during spraying is 2–8 m s^{-1}, spraying width (swath) is 3–4 m, depending on spray bar and number of nozzles.

Volume: It holds about 10 L of pesticide in the tank and dispenses it in 15–20 min.

Agricultural uses: Z8P is primarily a sprayer octocopter used in agricultural fields.

Nonagricultural uses: It is used to carry and transport hazardous chemicals such as pesticides, inflammable, and explosive items.

Useful References, Websites, and YouTube Addresses

https://www.alibaba.com/product-detail/Z8P-10L-agriculture-use-Professional-Unmanned_60576890192.html

https://www.alibaba.com/showroom/uav-drone-agricultural-sprayer.html

3.3 MISCELLANEOUS AND SPECIAL TYPES OF UNMANNED AERIAL VEHICLES WITH ROTORS

Name/Models: ASIO-B

Company and address: Selex ES Inc. Rome, Italy; Tel.: +39 06 41 504651; e-mail: pressit@selex-es.com; website: www.selex-es.com

Selex ES is an innovative company that offers at least two different types of very small and versatile UAVs. They are ASIO-B and SPYBALL-B. These small UAVs could be used in military reconnaissance, commercial, and agricultural uses. Selex ES employs large workforce to produce and disseminate these UAVs in the United Kingdom, Italy, the United States, Germany, Turkey, Romania, Brazil, Saudi Arabia, and India.

ASIO-B is a small VTOL UAV (Plate 3.59.1). It has visual and IR cameras in the gimbal. It can be used for aerial surveillance of military camps, installations, and monitor vehicle movement. It can be used extensively in agricultural farms to scout the crops, pick aerial images, and transmit them to ground control for detailed analysis. This UAV is easily packed and carried anywhere in farms, using backpacks.

TECHNICAL SPECIFICATIONS

Material: It is made of tough carbon fiber and plastic. This UAV meets with military standards regarding ruggedness.

Size: length: 35 cm; width (diameter): 35 cm; height: 20 cm; weight: MTOW is 8.0 kg with payload.

Propellers: It has a single propeller with three blades. It has a fan-ducted system.

Payload: 1.2 kg

Launching information: It is an autolaunch vehicle. Its flight path can be predetermined. It has autonavigation facility based on custom-made software. It can land in a small location with clearance space 5 × 5 m.

Altitude: It reaches 5–300 m above ground level during the spray operation. However, ceiling is 1000 m above ground level.

Speed: It is 45 km h^{-1}

Wind speed tolerance: 20 km h^{-1} disturbance

Temperature tolerance: −6 to 45°C

Power source: Electric batteries

Electric batteries: 14,000 mA h, lithium–polymer

Fuel: It is not fitted with IC engine—no need for gasoline

Endurance: 30 min per flight

Remote controllers: The GCS has a remote controller for real-time routing of the UAV. The UAV could be flown using predetermined flight path and GPS guided navigation. The telemetric links operate for up to 8 km radius from the ground station computer (laptop).

Area covered: It covers 50 km per flight or 50–150 ha (aerial imagery)

Photographic accessories: Cameras: The gyro-stabilized gimbal has visual (R, G, and B), and IR (thermal) camera.

Computer programs: The aerial imagery could be relayed to GCS computers for processing using software, such as Pix4D Mapper, Agisoft's or Trimble's Inphos.

Spraying area: This is not a spraying UAV.

Volume: Not applicable

Agricultural uses: Aerial imagery and scouting of crops for NDVI, leaf chlorophyll, and canopy temperature; natural vegetation monitoring for NDVI and to detect plant diversity.

Nonagricultural uses: They are surveillance of military camps, vehicles and troops; geologic sites, mines and mining activity; industrial sites, buildings, roads, oil pipelines, electric lines, etc.

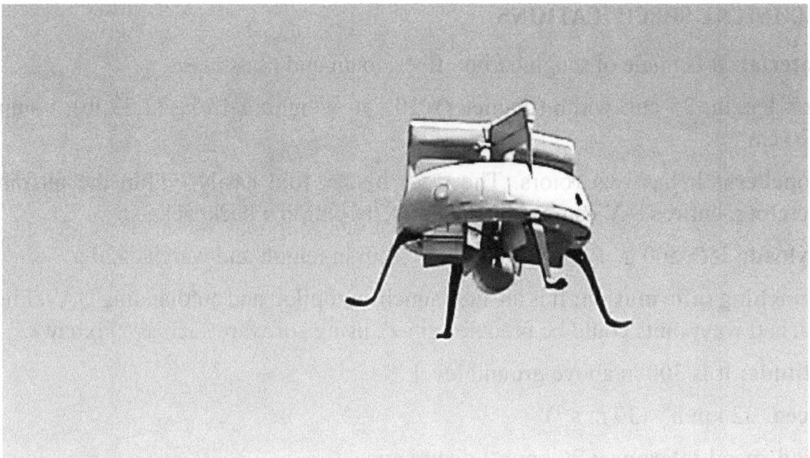

PLATE 3.59.1 ASIO-B.
Source: http://www.seles-es.com; Selex Es., Rome, Italy.

Useful References, Websites, and YouTube Addresses

http://www.seles-es.com (accessed Dec 10, 2017).
http://www.shephardmedia.com/news/uv-online/seles-es-uvs-ready-italian-army-details/ (accessed Dec 12, 2017).
http://www.indiandefencereview.com/news/finmeccanica-selex-ess-mini-and-micro-uavs-see-success-in-tests/ (accessed Dec 12, 2017).
http://www.leonardocompany.com/-/forza-nec-new-tests (accessed Dec 12, 2017).

Name/Models: Sprite

Company and address: Ascent Aero Systems Inc., New Town, CT, USA; e-mail: orders@ascentaerosystems.com; website: http://www.ascenterosystems.com

Ascent Aero Systems Inc. got started as a UAV producing unit in 2010. Ascent Aero Systems LLC. is situated at New Town, Connecticut. This company designs, manufactures, sells, and supports small UAVs commonly known as "UAVs." They have developed a UAV named "Sprite." It has a unique "coaxial" configuration that is more portable, more durable, and simpler to operate than typical multirotor designs. It is suited for professional, commercial, agricultural, and military applications. Ascent's revenue model is based on vehicle and accessory equipment sales, development contracts and recurring monthly revenue from turn-key systems. Their initial commercial product, "Sprite," was launched in May 2015. It has no protruding motor arms or delicate electronics or rigid propellers. Sprite is rugged with a polycarbonate body. It is small, lightweight, and compact. It could be transported anywhere in a backpack. It is easy to assemble, but it is often carried in ready to use condition (Plate 3.60.1). Payloads (sensors) can be easily fitted or replaced with just a half turn. The entire UAV comes with a warranty of 1 year.

TECHNICAL SPECIFICATIONS

Material: It is made of tough carbon fiber, foam and plastic.

Size: length: 35 cm; width (diameter): 10 cm; weight: 1.15 kg (2.53 lb); volume: 3520 cm^3

Propellers: It has two rotors. The rotor blades fold easily within the airframe. Therefore, entire UAV can be fitted compactly within a backpack.

Payload: 250–500 g. Typical gimbal is 15 cm in length and weighs 420 g.

Launching information: It is an autolaunch, autopilot, and autolanding UAV. Flight path and waypoints could be predetermined, using software such as "Pixhawk."

Altitude: It is 300 m above ground level

Speed: 32 km h^{-1} (30 m s^{-1})

Wind speed tolerance: 20 km h^{-1} disturbance

Temperature tolerance: −6 to 49°C

Power source: Lithium–polymer batteries

Electric batteries: 3700 mA h

Fuel: Not an IC engine-supported UAV

Endurance: It covers 25 min without payload; 12–15 min with payload

Remote controllers: This UAV operates in modes, namely conventional real-time controller (joystick) and fully autonomous GPS navigation. It has autopilot mode that covers all waypoints, survey, circling, and return to home in case of emergency or when the flight functions have been completed. Real time telemetry (915 MHz) is used to obtain data from the UAV and to alter the predetermined flight path.

Area covered: 50 ha per flight

Photographic accessories: Cameras: The sensors, namely, GoPro visual (R, G, and B), IR are fitted into a gyro-stabilized gimbal. High-definition videocamera could be fitted as regular payload.

Computer programs: The images collected by the UAV's sensors are processed, using software such as Pix4D Mapper or Agisoft's Photoscan

Spraying area: This is not a sprayer UAV.

Volume: Not applicable

Agricultural uses: They are aerial survey of land and water sources; crop scouting to obtain data on NDVI, leaf chlorophyll (crop's N status), canopy, and ambient temperature (water stress index). Sprite could also be used to detect disease, pest, drought, or flood effects; and natural vegetation monitoring for NDVI data and to study plant species diversity using spectral signatures.

Nonagricultural uses: They are resource exploration, industrial and infrastructure security, real estate surveillance, environmental engineering, insurance, public safety, monitoring roads, and vehicle traffic.

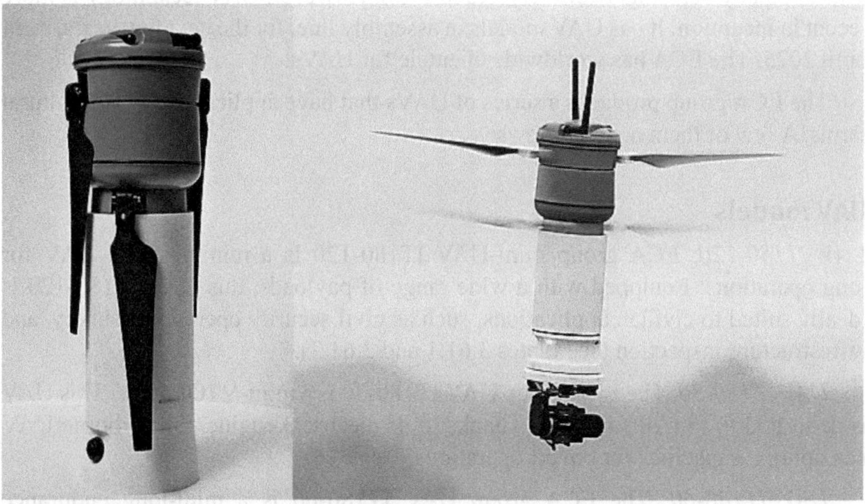

PLATE 3.60.1 Sprite—an upright rotocopter of great utility.
Source: Ascent Aero Systems Inc., New Town, CT, USA; excerpted from https://www.ascentaerosystems.com/.

Useful References, Websites, and YouTube Addresses

https://www.ascentaerosystems.com/ (accessed Sept 15, 2017).
https://www.ascentaerosystems.com/Applications (accessed Sept 15, 2017).
https://www.ascentaerosystems.com/pages/environment-conservation (accessed Sept 15, 2017).
https://www.ascentaerosystems.com/pages/about-us (accessed Sept 15, 2017).
https://www.youtube.com/watch?v=_l6CQRHIGyg (accessed Sept 15, 2017).
https://www.youtube.com/channel/UCCNbULm2mweUuu5ryeRbx1A (accessed Sept 15, 2017).

Name/Models: UAV 180-120, UAV 180-30, UAV 180-60, and UAV 180-999

Company and address: ECA Group, A GORGE Group Company, Z.I. Toulon Est 262, rue des frères Lumière 83130, La Garde Z.I. Toulon Est B.P. 242, 83 078 Toulon Cedex 9, France; ECA UAV—UAV services: operation, rental, maintenance; Tel.: +33 (0)4 94 08 90 00; fax: +33 (0)4 94 08 90 70; e-mail: ecagroup@ecagroup.com; website: https://www.ecagroup.com/en/about-us

The ECA group at La Garde in France is a company known for its expertise in production of a series of robots, UAVs, automated machines, simulation software, and industrial process equipment. ECA group offers a range of novel UAVs. They possess great utility in farms, crop fields, and in monitoring natural vegetation. This company began its R&D services in 1936. However, UAV technology is more recent in inception. It has UAV models in assembly line, for the next few years, right until 2025. The ECA has worldwide clientele for UAVs.

The ECA group produces a series of UAVs that have applications in agricultural farms. A few of them are as follows:

UAV models

UAV IT180-120: ECA group mini-UAV IT180-120 is a mini-gasoline UAV for long operations. Equipped with a wide range of payloads, this UAV IT180-120 is ideally suited to civilian applications, such as civil security operations, survey, and infrastructure inspection (see Plates 3.61.1 and 3.61.2).

UAV IT180-30: The ECA group UAV IT180-30 is a mini-VTOL UAV. This UAV is dedicated to ISTAR missions. Thanks to its electrical engine, this airborne UAV has optimal capacities for covert operations.

UAV IT180-60: The ECA group UAV IT180-60 is a mini-long endurance electrical VTOL UAV. It can carry multiple payloads and perform a wide range of missions, both for civilian and military needs. Thanks to its electrical engine and low radar cross-section, the UAV IT180-60 has optimal capacities for covert operations, such as ISTAR, and prevents noise pollution.

UAV IT180-999: ECA group supplies the UAV IT180-999, a tethered UAV solution for permanent operations. Powered by a ground power supply, the Airborne UAV IT180-999 can fly for an almost unlimited period and perform various types of missions, such as surveillance or radio relay.

TECHNICAL SPECIFICATIONS

The technical specifications listed below refer only to one model, that is, UAV 180-120 (UAV).

Material: It is made of high stability and robust material. It is made of carbon fiber, metal alloy, ruggedized foam, and plastic. This UAV model has been designed to withstand extremes of weather and natural conditions such as rain, snow, heat, high wind, all through the day and night.

Size: length: 35 cm (landing gear size); width (vertical rotor): 125 cm; height: 55 cm; weight: maximum weight 16 kg without payload

Propellers: It has two propellers placed one above the other. Each one has two blades made of plastic. The size of propellers is 80 cm in length and 15 cm in breadth. The two propellers are counter rotating type.

Payload: Maximum payload is 5 kg. It includes cameras inside a gyro-stabilized gimbal. It has dual payload slots, one at the bottom and other at the top.

Launching information: It is a VTOL UAV. It has landing-gear made of alloy. It is an autonomous UAV with predetermined flight path capability. It has autolanding software. In times of emergency it has fail-safe device; So, the UAV returns to its place of launch.

Altitude: The maximum altitude ceiling is 3000 cm. Operational height while hovering over crop fields and obtaining close-up images of crops/individual plants is 0–50 m above ground level.

Speed: 70 km h^{-1}

Wind speed tolerance: 25 km h^{-1} wind disturbance

Temperature tolerance: −10 to 40°C

Power source: gasoline and electric lithium–polymer batteries

Electric batteries: 14,000 mA h lithium–polymer batteries for CPU and electronics

Fuel: It consumes gasoline to run the motors that energize propellers. It has brushless motors connected to the counter-rotating propellers.

Endurance: It is 120 min per flight.

Remote controllers: The GCS has an iPad for relaying commands and modifying, any predetermined flight path or camera exposure schedules. The UAV keeps contact with GCS for up to 10 km range, using telemetric link.

Area covered: It covers about 150–200 km linear distance per flight of 2–2.5 h.

Photographic accessories: Cameras: It carries a series of cameras that includes visual (e.g., Sony 5100), hyperspectral for close-up imagery, multispectral to analyze crop growth and nutritional status, and videocamera to record ground events in farms. Farm activities such as vehicle movement, crop harvesting, and seed transport could be tracked. It has magnetometer to obtain maps of magnetic field variations, Lidar, CBRN sensors, anti-UAV kits, etc.

Computer programs: The images are processed rapidly by the UAV's CPU. Otherwise, it is done at the GCS, using iPad with Pix4D Mapper or Agisoft's software.

Spraying area: This not a spraying UAV, normally.

Volume: Not applicable

Agricultural uses: UAV IT180-120 has a range of applications in the agricultural farms. Crop scouting to estimate or detect crop stand variations, canopy development, leaf-area, leaf-chlorophyll (NIR imagery), crop's N status, crop's water stress index (thermal imagery), etc. The UAV's aerial photography and ability for spectral analysis can be used to detect disease/pest attacks on crops and map them. It is also used to detect spots affected by drought, floods, or soil erosion; natural vegetation monitoring to obtain NDVI values, canopy images, and species diversity.

Nonagricultural uses: It includes surveillance of military installations and tracking movement of military vehicles. Geophysical mapping for mineral distribution on ground surface, magnetic analysis and mapping its variations, mapping water resources, etc. It is used to surveillance mines, mining activity, ore-dumping yards, and ore movement in vehicles; monitoring oil pipelines, electric lines, highways, and rail roads.

Major customers for these UAVs (i.e., UAV IT 180 series) other than in the realm of agriculture are as follows: Aerospace, airborne monitoring and survey, airlines and MROs, aviation simulation, defense, nuclear industries, naval training and merchant navy companies, borehole irrigation agencies, land survey agencies, homeland security and public events monitoring agencies such as police, etc.

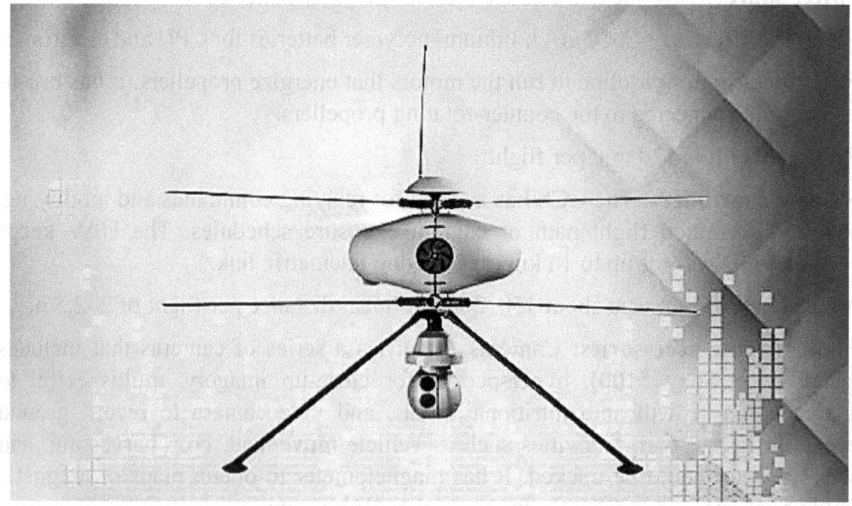

PLATE 3.61.1 UAV IT180-120.
Source: Ms. Meliha Boucher, Marketing Director, ECA group, La Garde, France.
Note: This is a highly versatile mini-gasoline engine supported UAV. It has wide range of applications in military, civilian, and agricultural activities.

PLATE 3.61.2 UAV 180-120 copter UAV on a grass patch and ready to take-off.
Source: Ms. Meliha Boucher, Marketing Director, ECA group, La Garde, France.

Useful References, Websites, and YouTube Addresses

https://www.youtube.com/watch?v=zHDIMMhBoJ0 (accessed Sept 18, 2017).

KEYWORDS

- **unmanned aerial vehicles**
- **surveillance**
- **vertical take-off**
- **endurance**
- **foliar fertilizer**
- **helicopter**
- **multi-rotor**
- **sprayer**

CHAPTER 4

Parachutes, Blimps, Balloons, and Kites as Unmanned Aerial Vehicles Useful in Agriculture

ABSTRACT

Farmers are experimenting with a wide range of unmanned aerial vehicle systems that offer them aerial imagery (spectral data) quickly with least drudgery and at low cost. In addition to the long lists of UAV aircrafts discussed in the previous two chapters, there are indeed a few more types of autonomous aerial vehicles (robots) that could be eventually utilized in farm world. They improve farmer's efficiency in data collection and its utilization. In this chapter, the primary focus is on UAVs such as semi- or fully autonomous parachutes (including parafoils), blimps, aerostats, helikites, and kites. The published literature and efforts to adopt the above UAVs are relatively feeble. However, certain field trials with parafoils, blimps, and tethered aerostats convince us that they can be applicable in farm land. They can be utilized to collect aerial imagery, spectral data about land resources (topography, fertility, vegetation, water resources, etc.), also, crops and their growth parameters and general status. Autonomous parafoils with payload of sensors offer aerial imagery of farms. Parafoils offer aerial imagery showing variations in soil types, fertility, and water resources. Parafoils could be piloted or they could be totally autonomous. They are slow in movement over crop's canopy. Therefore, parafoils offer haze-free sharper images. They can be flown low over crops. They possess relatively longer endurance. Parafoils can be flown repeatedly after re-fueling. Autonomous parafoils could be utilized to transport small quantities of agricultural cargo. Autonomous blimps and aerostats are gaining in popularity among farmers. Blimps offer continuous data about the crop and its growth pattern. They could be stationed above the field for entire crop season. Blimps are apt to data about phenomics of crops on a continuous basis, for the whole season. Blimps are larger aerial vehicles. They are usually filled with lighter-than-air gas helium. They are relatively costlier than many other types of UAVs. Helikites are hybrids prepared using a lighter-than-air helium balloon and kite. They are relatively more stable than an aerostat. They withstand wind currents better. Therefore, they offer sharp images of crops. A few models of parafoils (e.g., Hawkeye and SUSI-62), blimps (e.g., Anabaetic Aero and Solar Blimp), aerostats (e.g., AeroDrum's Compact Aerial Photography system and Allsopp Helikites), and kites are discussed in greater detail. In this chapter, information regarding manufacturers, specifications and agricultural uses of only a few UAVs such as parafoils, aerostats, and helikites are included. However, currently, there are several models of above types of UAVs being flooded into markets.

4.1 INTRODUCTION

Unmanned aerial vehicle (UAV) systems encompass a wide range of fully or semi-autonomous aerial vehicles. They include parachutes, blimps, and airships that are filled with lighter-than-air gas (e.g., helium), hot-air (thermal) ships, helium-filled balloons, and aerostats. They even include kites that could be controlled, using strings. The more important types of UAVs that are gaining ascendency in terms of popularity and usage among farmers and farming companies in North America, Europe, and Far East are the fixed-winged and multirotor aircrafts. However, parachutes, blimps, and aerostats have certain clear advantages that need to be utilized favorably, during agricultural crop production. For example, longer endurance, consistent surveillance of a location from above crop fields or urban setting, better economic efficiency, and high-resolution imagery possible with blimps, aerostats, and parachutes need emphasis.

Now, let us define and briefly explain a few of these UAVs. In some cases, the difference between UAV types is minimal. Yet, they are considered separately, as a class of UAVs.

4.1.1 PARACHUTES

A powered parachute consists of motor, propeller, wheels, a payload area, and parachute (parafilm). Powered parachutes may be guided, using remote control. Otherwise, their flight path could be predetermined, using computer software such as eMotion or Pixhawk. Parachutes move at a speed of 25–30 km/h above the crop's canopy. They may be made to fly low over the crop's canopy. Say, at a few meters (10–20 m) or at 150–500 m above ground level. The powered parachute uses a set of lithium polymer batteries or a two-stroke IC engine run on petrol/diesel, to energize the motor. They have a relatively longer endurance of 3–5 h per flight. They can float freely without requiring power for long durations. Parachutes are often included in the fuselage of fixed-winged or multirotor UAVs. This is to recover UAV aircrafts safely. On the other hand, there are UAVs (aircrafts) that fly and release parachutes into atmosphere, at different heights. Parachutes are affordable UAVs and are purchased at less than 5000 USD.

4.1.2 BLIMPS AND AIRSHIPS

Blimps are filled with a lighter-than-air gas, such as helium. They are powered with lithium batteries or a two-stroke IC engine. A blimp has no rigid internal or external frame work; hence, it collapses, if deflated. Blimps do carry a range of sensors for aerial photography. They are slow and can fly at low altitudes. Therefore, they offer high-resolution images of terrain, agricultural farms, and crops. Blimps are shaped oval and have fins to control the direction of navigation. An airship is a powered, free-flying aerostat that can be steered. Airships can be divided into rigid and nonrigid types. In fact, the nonrigid airships are often known as blimps. There are a few different types of lighter-than-air gases (lifting gas) utilized, they are hydrogen, coal gas, and helium. Hydrogen is the lightest gas used in balloons. Hydrogen was a common gas filled into balloons and airships during 1930s and 1940s. However, hydrogen is inflammable and is prone to disasters. Balloons could also be filled with "coal gas". This is a lifting gas. Coal gas is composed of mixture of methane and a few other lighter gases. Helium is both nonflammable and a lifting gas. Its lifting power is about 92% of that experienced, if hydrogen were to be used in the balloon. It is nontoxic. Most of the balloons, aerostats, and airships use helium. An alternative envisaged for future airships or aerostats is to create a tight vacuum inside the structure. This will help it to levitate into atmosphere. It then allows the UAV to float without the use of lifting gas or any heating.

4.1.3 AIRSHIPS

Airship could be rigid with a frame inside or they could be collapsible. They are similar in their functions regarding aerial photography of crop plants. Their utility in agriculture depends much on the set of sensors that they may carry underneath and the resolution of visual and thermal imagery.

4.1.4 THERMAL OR HOT-AIR AIRSHIPS

Thermal airships are said to cost much less. Usually, it is only about 5% of helium-filled ones. The running costs are also marginal. They do not require a hanger, no portable mooring mast, and no expensive ancillary equipment. Hot-air airships require only a few crew to manage flight path and derive aerial imagery. General maintenance of thermal airships costs much less than the helium version (Cameron Balloons, 2017).

4.1.5 HYBRID AIRSHIPS

Hybrid blimps are being manufactured by several aviation companies. They are as large as previous models/versions (e.g., Hindenburg). They are 280 ft. in length and carry a cargo of 20–22,000 kg. They fly at relatively slow speeds but stay afloat for long periods. They could be used to transport solid cargo packages, oil, and gas (Grothaus, 2016). A hybrid type uses both static buoyancy and dynamic airflow, to provide lift. The dynamic movement may be created, using either propulsive power as a "hybrid airship" or by tethering in the wind like a kite, as a "kytoon." Of course, hybrid airships could be utilized effectively to conduct aerial survey, regular surveillance of ground features and photograph them.

4.1.6 BALLOONS

A balloon is a kind of "unpowered aerostat." It has no means of propulsion. So, balloons must be tethered on a long cable. Otherwise, they should be allowed to drift freely with the wind. Therefore, balloons are stationery, unless they are steered using strings. Balloons are referred to as less intrusive and are not commonly used, to trespass into neighbor's plots. The operation of balloons filled with helium gas above crop fields may not need permission, from aviation departments or similar agencies. A tethered balloon is held down by one or more mooring lines or tethers. It has sufficient lift to hold the line tight. Its altitude is controlled by winching the line in or out. A tethered balloon does feel the wind. A round balloon is usually unstable and bobs about in strong winds. Therefore, the "kite balloon" was developed with an aerodynamic shape similar to a "nonrigid airship." Both kite balloons and nonrigid airships are sometimes called "blimps." Notable uses of tethered balloons include aerial observation. The uses of untethered balloons include espionage, collecting weather and air pollution related parameters.

Tethered balloon could accommodate a full set of visual, NIR, IR, multispectral, and LiDAR sensors on them, in the payload area. So, it could be effectively utilized during agricultural crop production. They offer high-resolution imagery useful to assess and map variations in growth (normalized difference vegetative indices, NDVI), crops N-status (leaf chlorophyll), water stress index, etc. They do relay set of images to GCS computers that can be quickly processed. Balloons could be easily hired or purchased permanently. They can be held safely, after systematic folding until next use. They are easily transportable. There are also instruction manuals for "do it yourself" low-cost tethered balloons with sensors. Balloons could be easily purchased by co-operatives and used for entire small villages, if individual farms are small and farmers are not economically well to do (Sudhoff, 2012; The McGraw-Hill Companies Inc. 2005).

4.1.7 AEROSTATS

Aerostats are relatively small. They are tethered lighter-than-air vehicles. Aerostats were initially called balloons. Powered aerostats are capable of horizontal flight. They are commonly called as dirigible balloons or simply dirigibles (dirigible—meaning steerable). These powered aerostats later came to be called "airships." They may carry a payload of 10–12 kg, that is, a full complement of cameras (sensors) for effective aerial survey, surveillance, and analysis of cropped fields. They are relatively stationery and offer high-resolution accurate images. Aerostats could be moved and deployed at various locations in a farm. Aerostats are useful in collecting digital data about variations in soil fertility and crop production, which is required during precision farming (Aeros, 2016). Aerostats can provide spectral data of crops affected by drought, floods, and disease/pests. They can be used to keep record of farm activities, progress of different agronomic procedures, also vehicle and animal movement in farms. Sooner, we may find almost all farms to have an aerostat floating slowly and in stand-still position. Farmers could be using them as sentinels-cum-data-collecting agents about their crops. They are relatively cheaper to own and operate compared to UAV aircrafts.

4.1.8 KITES

Kite is defined as tethered flying object made of foils or tough boards derived out of variety of materials. It flies using the aerodynamic lift to generate air flow over the lifting surface. A few types of kites are provided with lighter-than-air gases. So, they lift off and stay afloat even without wind or being towed. There are now indeed innumerable types of kites. They are each named based on material used, shape, size, design, purpose, region, or the profession for which it is used (Wikipedia, 2017; NASA, 2016; Robinson, 2003). In the present context, we are concerned with kite models used for agricultural purposes. These kites are mostly employed to obtain aerial photography of the agricultural terrain, natural vegetation, rivers, lakes, soil types, irrigation lines, cropping systems, drought/flood-affected regions, and crop nutrition. Essentially, agricultural kites carry on them, a full complement of sensors such as; visual (R, G, and B), multispectral, NIR, and IR sensors.

In the following pages, only few examples of each of the UAV types, listed above, have been discussed. The emphasis is on the models of UAVs utilized above agricultural fields. Particularly, those used to obtain aerial imagery (visual, NIR, or IR) of terrain, topography, land and soil resource, crops, and cropping systems. These UAV types are also used to monitor progress of agronomic procedures and guide ground vehicles.

4.2 PARACHUTES AS UAVS IN AGRICULTURE

The history of parachutes dates to 200 B.C. Evidence found in the historical archives of Peking, in China, and translated by the French monk, Vasson, indicates that, parachute-like devices were used as early as the 12th century A.D. These reports indicate that some form of parachute was used during circus-type stunts arranged to entertain guests, at the Chinese court ceremonies. The relation between the umbrella, which is known to have been invented by the Chinese and this early device appears obvious (Pat Works, 2014).

The first known written account of a parachute concept is found in Leonardo da Vinci's notebooks (c. 1495). The sketch he drew consisted of a cloth material pulled tightly over a rigid pyramidal structure. Although da Vinci never made the device, he is given credit for the concept of lowering man to the earth safely, using a maximum drag decelerator. Faust Vrancic is said to have constructed the first parachute in 1617. It was based on designs explained by Leonardo da Vinci. Often the credit for first practical use of parachute is attributed to Sebastian Lenormand. Paintings by Etienne Chevalier de Lorimier suggest that parachutes were in vogue by 1797. Andrew Garnerin is said to have recorded the first jump from a tower, using a parachute. He also developed a parachute without oscillations, using an air vent. Parachute harness and strapping was designed by Thomas Baldwin, in 1887. The method to fold the parachute and preserve it without damage was standardized by 1890. In 1920, parachute was used for the first time, to jump from an airplane. It avoided free fall and injuries. Parachute jumping became a regular sport in 1960s (Bellis, 2017; Kerman, 2012).

The development of modern parachutes deployed at high speeds and low/high altitudes, started in the 1930s. In Germany, Knacke and Madelung developed the ribbon parachute, for decelerating heavy high-speed payloads. After World War II, Knacke invented the ring slot parachute which is used for moderate subsonic speeds. This parachute is used primarily for cargo delivery and aircraft deceleration. The ring slot parachute is significantly cheaper to manufacture than the ribbon parachute. The ring sail parachute, developed by Ewing, is used to decelerate payloads at low to moderate subsonic speeds. Historically, the ring sail parachute was used in the final stages of the Mercury, Gemini, and Apollo space exploration projects (Kerman, 2012).

During past two decades, parachutes have served many types of purposes. Their potential for application in agriculture too seems high. They could serve the farmer and agricultural researcher/extension agent in a variety of ways. They include most simple tasks such as aerial survey, monitoring farm activities, preparing maps of cropping systems, providing data about crop growth (NDVI), drought/flood hit patches in farm, soil erosion, and pest/ disease-attacked zones.

Several models of parachutes have been adopted during crop production. Most recently, they have been evaluated in detail for their utility in assessing crops, their genotypes, influence of agronomic procedures and inputs, etc. (Hunt et al., 2015; Thamm, 2011; Meyer, 1985). There are indeed several commercial models of parachutes that can be directly adopted. Otherwise, they could be modified slightly to suit the agricultural purposes. In this book, the main emphasis is on the UAV aircrafts, that is, UAV systems (small robotic aircrafts). Hence, detailed specifications of only three most recently utilized parachute models have been listed.

Thamm (2011) and Thamm et al. (2013) suggest a few reasons that make UAVs such as parachutes more suitable for obtaining high-quality aerial photos, thermal images, and digital data about terrain and crops. As stated earlier, they argue that classical aircraft-aided aerial photos can be costly and prohibitive; particularly, if one needs them repeatedly and in detail. Large stretches cannot be easily picked. Airplane campaigns are generally high-cost options. Satellites offer only low-resolution images. Plus, the turn-around times are uncongenial. Hence, slow moving and low-flying parachutes (e.g., SUSI 62; Pixy; and Hawkeye) with different sensors as payloads are better alternatives. For example, Pixy motorized parachute fitted with Canon EOS 350D and Sony DSC F828 sensors were flown, to collect high-resolution images. Parachutes have been used to collect data regarding crop canopy characteristics, leaf chlorophyll (leaf-N), and canopy temperature. For example, wheat genotype grown at different seeding densities was assessed in experimental station, using a parachute—Pixy (Lelong et al., 2016). Parachutes have been utilized to conduct low-altitude aerial survey of archaeological sites. Their ability for consistent flight over the sites and large-scale photography makes them suitable, for aerial studies of archaeological sites (Hailey, 2005). Parachutes have also served well during safe recovery of spacecrafts. The spacecrafts have been made to descend slowly and safely (NASA STAFF, 2017). Parachutes (e.g., IRIS ultra-parachutes) are frequently used during the recovery of small fixed-winged and multicopter UAVs. Some of the common models of small UAVs that possess parachutes and adopt them during landing are AAI Textron, Delta-M, Quest UAV, Aeromao, etc. Parachutes can also launch UAV aircrafts into the atmosphere (e.g., Skycat Parachute launches 6–8 kg multicopters).

Now, let us consider specifications of a few parachute models that are in vogue at present, particularly, to serve agricultural purposes:

Name/Models: Hawkeye UAV Parachute

Company and address: Tetracam, Inc. Devonshire Street—Chatsworth, CA; Tel.: 818-288-4489; Fax: 818-718-7103; e-mail: info@tetracam.com; website: www.tetracam.com

Tetracam Inc. produces multispectral cameras and the software required to interpret the imagery. Tetracam serves an international base of customers ranging in size from individuals to multinational corporations with sensors (cameras) and UAV parachutes (Hawkeye). Tetracam is predominantly a sensor design and manufacture company. Tetracam imaging systems are in use in farmers' fields, in botanical reserves, forestry, environmental research, surveillance, and infrastructure inspection. Tetracam has manufacturing units in Hong Kong and Shenzhen, China. Tetracam Inc. is also engaged in manufacture of parachutes and its accessories.

Tetracam's "Hawkeye" is a parafoil that lifts the company's multispectral cameras into the air, to capture visible and near-infrared radiation (NIR) images objects. Growers fly these systems over crops to detect crop's problems (Plates 4.1.1–4.1.4). The Hawkeye is unique among UAVs, due to its comparative low cost, ease of operation, and its inherent safety, with its chute continually deployed (Tetracam, 2017; Huang et al. 2013).

The Hawkeye floats in case there is a malfunction in engine, propellers, or GCS computer commands. Hawkeye is a light weight, foldable, and easily portable model. According to manufacturers, Hawkeye is designed and developed to be docile and easy to use, even with little training. Farmers could be trained to fly a Hawkeye parachute in a matter of 5-h training. Hawkeye parachute UAV is also built to order, if special accessories are to be attached. It holds series of sensors and offers high-resolution images, to the GCS iPad. The Hawkeye parachute can be equipped with narrow band sensors. Then, it offers details of soil surface and crop plants (Hunt et al., 2015).

TECHNICAL SPECIFICATIONS

Material: The Hawkeye is made of bright green and orange parafoil, which is attached to a carbon–aluminum light-weight frame. The frame holds engine, sensors, and any other payload. The bright foil makes Hawkeye very conspicuous in the sky. The Hawkeye UAV has an airframe. It holds payload box, Goose autopilot and GPS, wiring harness, adjustable steering and gear system, a brushless motor with two propellers attached to it, and a 2.4 GHz receiver (Spektrum AR 7000). It has a series of four lithium polymer batteries.

Size: Length: vehicle's length is 35"; width: 120"; height: 32"; and weight: 4.5 kg (8 lb).

Propellers: It has one propeller made of four blades and attached to motor at the rear side.

Payload: It holds about 1.8 kg payload that includes sensors, CPU, and tool kit.

Launching information: It needs a runway to launch. Parachute can land itself based on preprogrammed flight path. In case of incorrect command or engine trouble-shoot, the parachute floats and lands slowly. The parachute has "return home feature," if radio control fails. The Hawkeye can be controlled manually, using Spektrum RC controls. The Hawkeye has two power chords that control altitude and speed. Then, roll servo that helps in turning the aircraft right or left, during navigation. It has an adjustable centre of gravity (CG) steering system.

Altitude: The ceiling is 1000–1200 m above ground level.

Speed: On an average, the parachute UAV moves in the sky at 10 knots h^{-1}. The forward speed of Hawkeye depends on parafoil wing and wind speed. It cannot be controlled like a conventional aircraft UAV.

Wind speed tolerance: 6 m s^{-1}

Temperature tolerance: −5°C to 45°C

Power source: Lithium polymer batteries (four numbers).

Electric batteries: It has 24 mAh, two-cell batteries to support electronic circuits and CPU.

Endurance: 10–30 min depending on the batteries.

Remote controllers: In the autopilot mode, ground control station (GCS) computers keep contact and send commands to parachute, for up to a radius of 3.5 km. This can be extended to 7.5 km, if required. The flight path is controlled by using "Goose Autopilot" software and other programs available with ground control station computers (iPad). The Hawkeye's GCS computers fix the imaging schedule, sensors' shutter functioning, and waypoints. The "Goose Autopilot" integrates with Tetracam multispectral sensors and captures the images.

Area covered: The area covered depends on the speed, altitude, swath of the sensors, and flight path design. Generally, it can cover about 30–50 ha per h and offer detailed images.

Photographic accessories: The Hawkeye airframe accommodates a series of sensors. The cameras function at R, G, and B wavelength band. It also has NIR, IR cameras, and LiDAR pod. The images are all tagged with GPS co-ordinates. Hawkeye could be provided with MCA narrow band sensors. Then, it offers highly sharp images of crops showing details of pest/disease, and other minute features. The spectral data it offers allows us, to distinguish healthy and disease/insect pest-attacked crops.

Computer programs: GCS has laptops that operate using Windows 8. The software includes "Pixhawk" for fixing flight path, or Goose Autopilot. It has Pix4D Mapper to develop images.

Agricultural uses: Hawkeye is mostly utilized to conduct aerial survey of large crop production companies/zones. It is utilized to monitor farm activities and send instantaneous reports to GCS. The sensors offer detailed images of crops (Plates 4.1.2–4.1.4). The spectral data is highly useful. The data collected includes NDVI that depicts crop growth pattern and canopy growth rate. The sensors offer leaf chlorophyll (crop's-N) and crop water stress index (CWSI) (Hunt et al., 2016).

Nonagricultural uses: Recording public events, surveillance of mines, transport vehicles, buildings, dams, lakes, riverine regions, etc.

PLATE 4.1.1 Hawkeye parachute's metallic frame and payload area.
Source: Dr. John Edling, Tetracam Inc., Chatsworth, California, USA.

PLATE 4.1.2 A Sugar beet field in the German plains imaged, using Hawkeye parachute carrying sensors at 450–750 nm.
Source: Mr. John Edling, Tetracam Inc., Chatsworth, California, USA.

PLATE 4.1.3 Citrus trees in an orchard imaged, using Hawkeye parachute with visual and multispectral sensors, in Florida, USA.
Source: Mr. John Edling, Tetracam Inc., Chatsworth, California, USA.
Note: The image depicts variations in NDVI across the orchard.

PLATE 4.1.4 Citrus orchard affected by frost damage photographed, using sensors on a Hawkeye parachute.

Source: Mr. John Edling, Tetracam Inc., Chatsworth, California, USA.

Note: Light-colored zones indicate citrus trees severely affected by frost and are dead. Darker patches show trees tolerant and healthy (in a colored picture, healthy trees in the orchard appear violet and dead ones are cyan and pale).

References, Websites and You tube Addresses

Huang, Y.; Thomson, S. J.; Clint Hoffman, W.; Lan, Y.; Fritz, B. K. Development and Prospect of Unmanned Aerial Vehicle Technologies for Agricultural Production Management. *Int. J. Agr. Biol. Eng.* **2013,** 6, 1–10.

Hunt, E. R.; Horneck, D. A.; Gadler, D. J.; Bruce, A. F.; Turner, R. W.; Spinelli, C. B.; Brungardt, J. J. Detection of Nitrogen Deficiency in Potatoes Using Small Unmanned Aircraft Systems, 2015. https://www.ars.usda.gov/research/publications/publication/?seqNo115=301000. pp 1–4 (accessed Oct 27, 2017)

Tetracam. Hawkeye: Autonomous Unmanned Aerial Imaging System, 2017. http://www.tetracm.com/ProductHawkeyewindow2.htm. pp 1–4 (accessed Nov 15, 2017).

http://www.tetracam.com/hawkeye_video.html (accessed Oct 12, 2017).

http://www.tetracam.com/ProductHawkeyewindow2.htm (accessed Oct 12, 2107).

https://www.youtube.com/watch?v=ERSAm7k6clI (accessed Nov 1, 2017).

https://www.youtube.com/watch?v=5qOJARBWJlc (accessed Nov 1, 2017).

https://www.youtube.com/watch?v=ERSAm7k6clI (accessed Nov 1, 2017).

https://www.youtube.com/watch?v=2-MqdmXug0M (accessed Nov 1, 2017).

https://www.youtube.com/watch?v=V1V6BwIMd0A (accessed Nov 1, 2017).

https://www.youtube.com/watch?v=04H1PU63O0E (accessed Nov 1, 2017).

https://www.youtube.com/watch?v=oIXdICdP2zI (accessed November 1, 2017).

https://www.youtube.com/watch?v=vtF3XqqbQMM (accessed November 1, 2017).

Name/Models: Pixy ABS le drone (Moto-glider)

Company and address: ABS Aero light, French Research Institute for Development (IRD), Agropolis Fondation, 1000, Avenue Agropolis, 34394 Montpellier, Cedex-5, France. Tel: +33 (0)4 67 04 75 74; Fax: +33 (0)4 67 04 75 43; e-mail: agropolis-fondation@agropolis.fr; website: www.ird.fr

Pixy ABS parachute is a light-weight UAV. It is made of plastic foil and aluminum pipes. It has an ability to gain height quickly, using a short runway. It has a propeller attached to brushless motor. The motor is powered by lithium polymer batteries. It is easily foldable. Therefore, it is easily portable to any place in the farm or an urban location. Pixy has a usual complement of sensors placed in the payload area. It offers a range of spectral data very useful to farmers. Such data help while deciding inputs and agronomic procedures. "Pixy" has been already tested in farms across many locations in the European plains (Lelong et al., 2016).

TECHNICAL SPECIFICATIONS:

Material: It is made of plastic foil and toughened ropes. The payload region is made of light aluminum tubes.

Size: Vehicle's length is 35"; width: 120"; height: 32"; weight: total weight is 6–8 kg, including payload of 2–3 kg.

Propellers: It has one propeller with three plastic blades. Propellers are attached to a brushless motor. Propeller helps in forward thrust.

Payload: It allows 2–3 kg payload that includes cameras, CPU, and lithium polymer batteries.

Launching information: It needs a runway of 10–15 m before it gains height into the atmosphere. It has "auto-return to home" facility. The Pixy parachute has four, light, but sturdy wheels. They withstand rigors of launching and landing.

Altitude: The ceiling is 1500 m above ground level. It can be flown low over the crops at 100–300 m, if close up view is required.

Speed: The parachute cruises at 15–20 kmph

Wind speed tolerance: It tolerates a wind speed disturbance of 5–6 m s^{-1}.

Temperature tolerance: −5°C to −45°C

Power source: Litho-polymer batteries

Electric batteries: 24 mAh Li-Po; two cells.

Fuel: Not applicable.

Endurance: 30–45 min, but depends on the batteries and payload.

Remote controllers: Pixy is relatively easy to guide. Its flight path, speed, and camera shutter activity are controlled, using remote controllers (Pudelko et al.,

2012). An iPad at GCS can design the flight path or can alter it, in case of emergencies. The Pixy has auto-return home facility, in case of erroneous flight commands or exhaustion of fuel.

Area covered: The area covered in the aerial photography depends on the speed of the parachute, altitude, swath of the gimbal/camera. Usually, 50 ha of cropland is covered in 15–20 min of flight, but, it depends on flight path and altitude.

Photographic accessories: Pixy parachute is usually fitted with a R, G, and B camera, multispectral camera, and IR/NIR sensors, for aerial photography of crop fields. An 8 Mpixels SONY DSCS F828 camera and/or Canon EOS 350D are the common aerial photography accessories (Lelong et al., 2016). The gimbal that holds the sensors allows semivertical and vertical axis imagery. The parachute is generally flown at 20–700 m above crop fields, to obtain images of crops/terrain. High-resolution images are best obtained at 10–50 m above the crop canopy. All the images are tagged with GPS co-ordinates.

Computer programs: The flight path can be designed, using "Pixhawk" or "eMotion" software. The images can be processed using Pix4D Mapper or similar software.

Agricultural uses: Aerial scouting of terrain, crop fields, and water resources are the most common functions in the agrarian regions. Pixy is used to draw digital data about crops' growth (NDVI) and leaf chlorophyll (crop's-N) status. The thermal imagery provides data on crop's water stress index. It can also be used to collect air samples, in the atmosphere above the crops. It can be used to assess gaseous quality of atmosphere above crop's canopy. Pixy can be used to monitor farm activity. Pixy has been utilized in agricultural experimental stations. It can be used to collect periodic data during experimental evaluation of crop species and genotypes, for grain yield performance (Pudelko et al., 2012).

Nonagricultural uses: Pixy parachute could be flown above installations, roads, and public places to monitor human/vehicle activity. It is used to surveillance mines and ore-transporting vehicles. It helps in providing aerial view of disaster prone or affected areas. Pixy can be used to surveillance natural resources such as water bodies, dams, rivers, etc.

References, Websites, and YouTube Addresses

Lelong, C. C. D.; Burger, P.; Jubelin, G.; Roux, B.; Labbe, S.; Baret, F. Assessment of Unmanned Aerial Vehicle Imagery for Quantitative Monitoring of Wheat Crop in Small Plots. Sensors **2016,** 8, 3557–3585.

Pudelko, R.; Stuckzynski, T.; Borzecka-Walker, M. The Suitability of an Unmanned Aerial Vehicle (UAV) for the Evaluation of Experimental Fields and Crops. Zemdirbyste Agr. **2012,** 99, 431–436.

https://en.ird.fr/the-ird; http://www.abs.aero/ (accessed Sept 22, 2017).

https://www.researchgate.net/figure/LAvion-Jaunes-powered-glider-left-and-Pixy-motorized-parachute-right_26547944_fig2 (accessed Sept 22, 2017).

https://www.youtube.com/watch?v=x39pkCHPtQI (accessed Sept 22, 2017).

Name/Model: SUSI 62 UAV

Company address: FU-Berlin, Department of Geography, Neustr 40, Linz am Rhein, Germany.

SUSI 62 is a parachute developed to withstand certain levels of vagaries in weather parameters. It has been designed to carry a payload. It can accommodate different kinds of objects, parcels, or photographic equipment. SUSI 62 has been used to conduct variety of tasks. In Germany, it has been used for recreation purposes in the city of Cologne, primarily, to carry visitors above the city and other locations (Thamm, 2011). In Africa, it has been utilized to photograph migrants and immigration routes of the tribes. SUSI 62 with its full complement of sensors such as R, G, B, IR, NIR, and LiDAR has been utilized, to survey agricultural land, crops, and water resource, in the German plains and other agrarian locations in Europe (Thamm, 2011; Thamm et al., 2013). SUSI 62 has provided excellent data about soil fertility variations, biomass accumulation patterns, vegetation, and its species diversity. Hence, it has been adopted to survey land resources and mark "management blocks" during routine crop management.

A list of the attributes of SUSI 62 (Plate 4.2) as listed by Thamm and Judex (2006) are as follows:

SUSI 62 is very robust, easy to operate, and easy to maintain. It has high safety features in case of failures mid-way in the air. It carries higher payloads. It has long flight endurance. It moves around at low speed and thus offers high-quality images. It has enough space for sensors. The entire kit of the SUSI 62 is portable and easy to transport.

SUSI 62 has certain advantages that make it more amenable, for use above crop fields. It is a safe UAV. It moves at low speed and offers clearer pictures. A short training period is required to fly a parachute UAV. It can be packed into a small case for transport. Several sensors can be operated simultaneously, from the UAV. The images are relatively sharp since it avoids haze.

A few disadvantages quoted are that, it needs a runway to start and land the SUSI 62 parachute UAV. The two-stroke petrol engine is noisy and leaves noise signature. Parachute is affected by wind. It drifts if wind speed is high or if there is air turbulence (Thamm, 2011; Thamm and Judex, 2006).

TECHNICAL SPECIFICATIONS:

Material and body: SUSI 62 has two main parts. They are the aluminum frame and foldable parachute. It has four robust wheels (under carriage) that help in rapid movement, on the ground surface. The wheels are made of carbon fiber. The wheels and frame withstand rough landings. All major parts of the SUSI 62 are easy to dismantle. It can be reassembled in minutes.

Size: SUSI 62 parachute is a "four liner". It has two holding points and two flaps. Flaps help in changing the course of parachute. A larger version is 6.0 m^3 and smaller type is 4.0 m^3 in volume.

Payload: Parachute allows a payload of 8 kg in all. Usually, payload has four to five sensors and a computer unit. They are located in the payload basket. The sensors can be operated using remote controller (GCS laptop) or they could be preprogrammed. The payload includes 2 L of petrol/diesel fuel and batteries for computer.

Launching information: Launching of the parachute needs some thrust and a runway of 10–50 m. Landing the parachute SUSI 62 can be preprogrammed or handled manually. In the autopilot mode, the parachute returns to the point of launch.

Altitude: Maximum attainable altitude is 3500 m above ground level.

Speed: Maximum flight speed of SUSI 62 ranges from 45 to 55 kmph.

Wind speed tolerance: The parachute withstands wind-caused distraction, only to a small extent. It is usually attributed to low speed of the UAV. It withstands up to 6 m s^{-1} wind speed.

Temperature tolerance: −5°C to −45°C.

Power source: The power source is a 62 cm^3 two-stroke engine (5.5 hp) made of aluminum alloy. The engine is robust enough to withstand shocks, while movement and landings. The engine has a remote-controlled (RC) starter. Therefore, it can be kept "on or off" during the flight, based on the requirement.

Electric batteries: 24 mAh, two cells. It has rechargeable batteries that support the activity of computers and data relay system.

Fuel: IC engine utilizes diesel or petrol. Petrol tank holds 2 L of fuel. It is situated in the payload area.

Endurance: Generally, parachutes have long endurance. A free-floating parachute does not require power from engine. SUSI 62 has been flown for 600 h at a stretch. The parachute can float freely for a long duration.

Remote controllers: The GCS includes remote controller (a laptop). It keeps contact with parachute for up to 6 km radius. Modifications to signal strength may allow GCS, to keep in touch with parachute UAV, for up to 40 km radius. In practice, above the agricultural areas, SUSI 62 parachute is remote controlled within a radius of 2.0–2.5 km from GCS. The RC computers can program the flight path, way points, start and off the engine, alter direction and speed of the parachute, and also the frequency of photographs and shutter activity of the sensors. The autopilot mode allows the parachute to trace the path and reach starting point, in case of emergency or upon completion of aerial survey. SUSI 62 is a stable parachute. The parachute can also be managed in the manual mode. It just requires 2 days of training for the farmer, to learn to control the flight path and sensors, in order to obtain photography. Due to low speed of the parachute, wind may cause deviation, to plight path by 15 m, particularly, if it is too windy.

Area covered: The area covered per hour depends on the flight path, altitude, speed of the parachute, and swath of the cameras. About 50–100 ha h^{-1} could be surveyed, using a SUSI 62 parachute.

Photographic accessories: SUSI 62 can be fitted with a series of different sensors. Optical sensors include mid-range SLR (e.g., Nikon 300 D or a sophisticated Canon D5 Mark 11). It also carries a thermal-range sensor (Hasselblad 60 megapixel). The IR sensor is used to capture surface temperature variation in agricultural regions, crop canopy temperature, and ambient air temperature. Sensors are held in a robust gimbal that resists shocks. Gimbal is stabilized. Therefore, sensors offer haze-free images/maps. All images are tagged with GPS co-ordinates.

Computer programs: The sensors' output in digital format can be processed. Ortho-images could be stitched together and developed into complete images and maps. Usually, Pix4D Mapper is used to process the images. Ground control station computers may receive digital data and process them into images.

Agricultural uses: The parachute UAV (SUSI 62) is used to conduct aerial survey of land, decipher land use pattern, detect soil type variation, cropping systems, water resources, etc. The sensors offer data about crop growth, canopy size, its reflectance, and temperature. They offer data about NDVI; leaf chlorophyll (crop's-N status); crop's water stress index; disease/pest attack; disasters due to flooding, droughts, and excessive soil erosion; etc. SUSI 62 is also used to monitor farm activity and animals in the pasture land. Aerial images from SUSI 62 have been effectively used, to mark management blocks during precision farming (Thamm, 2011).

Nonagricultural uses: SUSI 62 parachute can be left to float above buildings and other installations, to surveillance and monitor activities on ground. In game sanctuaries, it has been used to monitor animal activity. In agrarian regions, it has been used to visualize and get pictures of drought or disaster-affected areas. It has been used to monitor wood logging and movement above forest plantations. SUSI 62 has been used to collect thermal maps of terrain and crop fields. The data from IR sensors have been utilized in assessing crop water status. It helps in quantum and timing of irrigation.

PLATE 4.2 SUSI 62 UAV—An autonomous (pilotless) parachute.
Source: Dr. Hans Peter Thamm, Geo-Technic, Germany, Hp.thamm@gmail.com; thamm@geo-technic.de.

References, Websites, and YouTube Addresses

Thamm, H. P. SUSI 62 A Robust and Safe Parachute UAV with Long Flight Time and Good Payload. Int. Arch. Photogramm. Remote Sens. Spatial Inform. Sci. **2011,** 38: 19–24.

Thamm, H. P.; Judex, M. The Low-Cost Drone. An Interesting Tool for Process Monitoring in a High Spatial and Temporal Resolution. Int. Arch. Photogramm. Remote Sens. Spatial Inform. Sci. ISPs Commission 7th Mid-Term Symposium. Remote sensing: From Pixels to Process. Enchede: The Netherlands **2006,** 36, 140–144.

Thamm, H. P.; Menz, G.; Becker, M.; Kuria, D. N.; Misana, S.; Kohn, D. The Use of UAS for Assessing Agricultural Systems in AN Wetland in Tanzania, in the Wet-Season, for

Sustainable Agriculture and Providing Ground Truth for Terra Sar X Data. ISPRS Int. Arch. Photogramm. Remote Sens. Spatial Inform. Sci. **2013,** XL-1/w2, 401–406 DOI: 10.5194/isprsarchives-XL-1-W2-401-2013.
http://www.geo-technic.de/?q=en/node/15 (accessed Nov 10, 2017).
https://www.youtube.com/watch?v=6XwTj3teSuI (accessed Nov 10, 2017).
https://www.youtube.com/watch?v=qwQ_wZBlNG0 (accessed Nov 10, 2017).
https://www.youtube.com/watch?v=sEXn7h_2jYA (accessed Nov 10, 2017).
https://www.youtube.com/watch?v=O4CYCMK1Hxg&feature=youtu.be (accessed Nov 10, 2017).
https://www.youtube.com/watch?v=qwQ_wZBlNG0&feature=youtu.be (accessed Nov 10, 2017).
https://www.youtube.com/watch?v=y0SEzLv6Tgo (accessed Nov 10, 2017).
https://www.youtube.com/watch?v=Qg6xxRZWnI4 (accessed Nov 10, 2017).

4.3 BLIMPS IN AERIAL SURVEY OF AGRICULTURAL ZONES

Historically, by 1920s, airships were regularly used for trips across Atlantic ocean. A Zeppelin had navigated the planet in just 21 days (Jowit, 2010). A brief discussion about the etymology and history of blimps is available, in Chapter 1. Here, it suffices to define blimp as a kind of airborne vehicle that levitates due to the pressure of "lifting gas". It is usually filled with lighter-than-air gas, for example, helium. It is a nonrigid vehicle. Therefore, the blimp collapses, if it is deflated. This is unlike rigid airships or Zeppelins that possess a rigid inner frame made of metal. They are filled with helium-like lighter gas mixture (Gardner, 2015: Plate 4.3).

Blimps are frequently used in advertising, aerial photography, and surveillance missions. However, during recent few years, blimps are being actively explored for their utility, in surveillance and collecting detailed spectral data of crops grown in large farms. Blimps fitted with sensors could provide valuable spectral data about their crops, to farmers. However, blimp-derived photographs could be hazy due to wind interference, particularly, if the UAV functions in windy conditions (Khot et al., 2016; Sankaran, 2015).

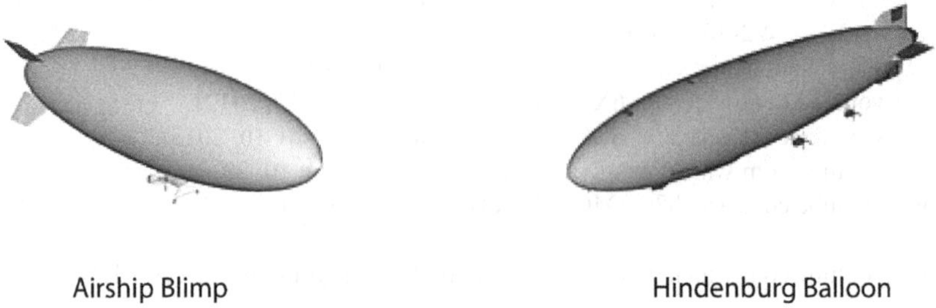

PLATE 4.3 A blimp with sensors.
Source: Dr. Gilles Dutruy http://www.anabatic.aero.

As stated previously in the first chapter, there are over 100 different uses of blimps, airships, and aerostats (Mothership Aeronautics, 2016; Airstar, 2017). Among the uses for blimps, most are about aerial survey and surveillance. They have been utilized to survey and monitor changes in natural resources such as rivers, water bodies, lakes, etc. (Vericat, 2008; Guglieri and Ristorto, 2016; Airstar, 2017). During recent years, however, there are innumerable uses identified for blimps/aerostats, in agriculture. Agricultural uses range from simple aerial survey of topography, land and soil, cropping systems, and water resources. Monitoring farm activities and happenings on the ground is an important use. The digital data can be procured at a fast pace. Such data can be stored efficiently or transmitted quickly to ground computers. Scouting the crops is possible at very low cost, if blimps located above the crop fields or experimental plots are utilized. Blimps have been utilized to monitor disease progression on crops. It is also used to monitor spraying activity in crop fields. Blimps have the ability to offer detailed digital data about variations in crop growth, soil fertility, and moisture content. Blimps are apt for use during precision farming. The digital data derived by blimps can be channeled into variable-rate applicators.

During past 3–5 years, there has been an intensive search for the most suitable UAVs, for use in routine experimental evaluation of crops/genotypes, in agricultural experimental stations. They are being tested for efficiency and ability to replace human skilled technicians, particularly, during collection of data. Blimps are among the most recently tested UAVs in many of the world's top agricultural research centers. For example, International Maize and Wheat Centre (CIMMYT) at Mexico has evaluated blimps, during phenotyping of wheat and maize genotypes. Blimps are also used to obtain a series data about physiological aspects of crop genotypes (CIMMYT, 2012).

Zhou et al. (2017) state that, recent trends in farm management involve the use of noninvasive remote sensors and aerial imaging systems, in solving many aspects that range from monitoring land preparation, seeding interculture, input supply right till harvest, etc. The use of RC blimps is still rudimentary. The RC blimps are being evaluated in many agricultural experimental stations (e.g., CIMMYT, Mexico). They are being utilized mainly to assess land and soil conditions prior to sowing, seeding, and crop stand management. The RC blimps that hover over the experimental plots collect valuable spectral data about crop growth (NDVI), greenness index, leaf chlorophyll index, disease/pest tolerance, grain yield levels, etc. Generally, a standstill RC blimp above the plots/crops may be used to collect data useful for agricultural researchers. Researchers can analyze such data using appropriate computer software, later. In other words, RC blimps could be of great utility during phenotyping of crops, particularly, in grain production farms or experimental plots. Further, it is believed that helium-filled blimps could serve as regular farm cargo carriers. They could be transporting fresh fruit, flowers, and other products.

We can compare the salient features of RC blimps with other UAVs such as fixed-winged ones, helicopters, and multirotor copters. Blimps have ability for vertical take-off. They have an excellent ability for hovering for hours, without much disturbance. Blimps can carry cargo. Their payload may also include sensors, computers, electric batteries, etc. The RC blimps possess ability to float, due to lighter-than-air helium gas in them. Therefore, they have long endurance compared to all other known UAVs. The RC blimps can easily keep moving for 60 min and more, depending on fuel and engine. It can also float above fields for days. The blimps could be controlled, using GCS computers. They can be commanded using radio control. Otherwise, they could be entirely programmed and kept totally autonomous. Blimps are very slow-moving UAVs above crop fields. Their speed is around 2.5 m s^{-1} compared to 22 m s^{-2} for a fixed-winged UAV aircraft, 11 m s^{-1} for multirotor UAV, and 20 m s^{-2} for a low-weight helicopter UAV (Shi, 2016).

4.3.1 USES OF BLIMPS IN AGRICULTURE

Like small fixed-winged or single/multirotor UAV aircrafts, blimps and aerostats are also gaining in acceptance in farm land. Blimps are sought by farming agencies, companies holding large acreage and agricultural experimental stations. To quote an example, a blimp operator in the United States has switched from regular aerial photography of public events, sports, football, etc., to monitoring crops and noting spectral data of crops. He has been engaged in aerial surveillance of large farms of 10,000–50,000 ha (Hendrix, 2014).

Let us consider an example involving the usage of blimps to oversee, monitor growth and collect crucial spectral data, about cotton crop grown in Georgia, USA (Associated Press, 2004). Here, the blimp (15 ft. long) is used to judge the cotton crop's reaction to drought. The blimp carries a series of visual, a multispectral and most importantly a set of NIR and IR cameras, underneath. The healthy plants appear white in the infrared (IR) images derived by the blimp. The drought-affected plants show up shades of red due to low water status. Clearly, a blimp-mounted IR camera provides a full view of the crop fields. The maps allow farmers to judge the extent and location of drought-affected patches. In case, a few skilled technicians are employed, such a view of the entire crop in-one-go is not possible. It takes a lot of walking and notes about crop's health and water status. Human fatigue-related errors too creep in. Further, data from blimps sensors about crops' water stress index helps in judging the irrigation requirements. Comparison with ground truth, data collected using soil sensors indicate high correlation. For farmers in Georgia, USA, it seems, timing of irrigation to cotton crop is crucial. Particularly, if one wants to harvest a high boll yield. In this regard, blimps could provide them with accurate data. Such data gives a scientific basis to fix the frequency and quantum of irrigation.

Within the farm sector, blimps are also used to transport different types of food stuffs. However, blimps have consistently faced tough competition with airplanes, for transport of perishables and dry food stuff. Prentice et al. (2017) state that, during early 20th century blimps and airplanes competed as vehicles for transport of large quantities of food grains and other eatables.

A careful analysis has suggested that, at present, highly perishable stuff such as seafood, meat, and flowers are mostly transported, using conventional airplanes. Marine routes, that is, sea transport on ships is utilized, to transport commodities such as grains, root crops, citrus, and bananas. Airships are utilized to transport agricultural commodities such as fresh fruits with better shelf life, such as pineapples, peaches, mangoes, papaya, several types of vegetables, etc. Clearly, blimps could be important transport vehicles for certain types of agricultural goods.

Name/Models: Anabatic Aero

Company and address: Anabatic Aero Chemin du Grand-Pré 2 1184 Luins, Switzerland; website: www.anabatic.aero ; e-mail: contact@anabatic.aero; telephone: N.A.

Anabatic Sarl is a company initiated in Switzerland, around 2000 A.D. Anabatic Aero designs and builds UAVs and blimps between 4 and 22 m length. They are used for advertising, cinema, scientific missions, agriculture, or any other specific project. The company offers two sets of models of blimps: a twin engine (two-stroke petrol engine) outdoor blimp of 6–9 m length and another of 9–14 m length. They also manufacture gimbals and other accessories for the blimps. These blimp models are easy to launch. The blimp is to be filled with helium gas for vertical liftoff. Blimps float slowly or hover at same location. Hence, they are utilized in urban areas for advertising. However, if adopted for agricultural uses, these blimps are guided to cover large areas of farm land from a high altitude (Plate 4.4).

TECHNICAL SPECIFICATIONS

Material: It is made of poly-urethane foil in double layer. It is attached with a light-weight aluminum payload area—a basket with gimbal, sensors, batteries, and central processing unit.

Size: length: 6 m (19.7 ft.); width: 1.7 m; height: 2 m; volume 9 m^3; weight: 14.5 kg.

Propellers: One at the rear end, to get thrust.

Payload: 1.5 kg.

Launching information: It is a vertical take-off and landing (VTOL) vehicle with auto-launch facility. The blimps' movements could be totally autonomous for up to 15 km distance. This is done using appropriate software and computer programming.

Altitude: 50–1500 m above ground level. Spectral measurements on crops are mostly done from 100 m above the crop canopy.

Speed: Maximum transit speed 30 kmph.

Wind speed tolerance: 6.2 m s^{-1}.

Temperature tolerance: −5°C to −45°C.

Power source: Electric batteries (lithium polymer).

Electric batteries: 2 × 2.50 W.

Fuel: Petrol

Endurance: 45–60 min. The blimp floats freely, if power gets exhausted.

Remote controllers: The GCS computers are connected to the blimp, for up to 15 km radius.

Area covered: The blimp is a relatively slow mover. The area covered and photographed depends on flight path, speed fixed, and altitude from which the ground details are photographed.

Photographic accessories: The payload area supports a gimbal, set of visual, multispectral, NIR, IR cameras, and a LiDAR sensor.

Computer programs: The images could be immediately relayed to ground computers (iPad) and processed in situ. The images are processed using Pix4 D Mapper or Agisoft PhotoScan.

Agricultural uses: Blimps are used to survey land resources, map the distribution of soil types, cropping pattern, irrigation lines, flood/drought-affected zones, extent of soil erosion, etc. These blimps could be utilized for aerial survey of crop fields, regular surveillance farms, and their installations. They can be used to collect data about crop growth (NDVI), leaf chlorophyll, and canopy temperature (water stress index). Blimps could be very useful, during monitoring of agronomic procedures. Surveillance of farm vehicles and grain storage facility are other uses.

Nonagricultural uses: They are used for commercial advertising, surveillance of public places, events, city traffic, buildings, installations, etc. Blimps could be used to identify and relay instantaneous images of disaster-prone area.

PLATE 4.4 A blimp above a crop field.
Source: Dr. Gilles Dutray, Anabatic Aero Sarl, Luins, Switzerland.
Note: The above blimp is on a surveillance mission.

References, Websites, and YouTube Addresses

http://www.anabatic.aero (accessed Dec 18, 2017).
https://plus.google.com/108359840123061818523 (accessed Dec 18, 2017).
https://www.youtube.com/watch?v=Opm7T97xEbo (accessed Dec 18, 2017).
https://www.youtube.com/watch?v=ni39eweIXJw (accessed Dec 18, 2017).
https://www.youtube.com/watch?v=KDDWGvZcWKQ (accessed Dec 18, 2017).
https://www.youtube.com/watch?v=2nB71Uz524s (accessed Dec 18, 2017).
https://www.youtube.com/watch?v=W6AjiL29UE0 (accessed Dec 18, 2017).
https://www.youtube.com/watch?v=FP9Tq4HdJBw (accessed Dec 13, 2017).
https://www.youtube.com/watch?v=wcNWxI4YG_c (accessed Dec 13, 2017).
https://www.youtube.com/watch?v=RHBpRQg_bpE (accessed Dec 13, 2017).
https://www.youtube.com/watch?v=W6AjiL29UE0 (accessed Dec 13, 2017).

Name and Model: Solar Blimp

Company and address: Aero Drum Ltd., Vojislava Ilica 99a, 11 000, Belgrade, Serbia, Europe; **Tel.:** +381 11 4071580; **Mobile:** +381 64 56 37 679; **e-mail:** rczeppelin@gmail.com; **Skype:** alexmijat

Solar blimps could find immense popularity among public enterprises, advertising agencies, and most importantly, farmers and farming companies with very large acreage (Plate 4.5). They could serve the large farms as sentinels-cum-data collectors. Since, they are autopowered via solar cells, they could be kept in the air for longer durations. They have recharging facility. Of course, solar blimp will avoid an extra expenditure for fuel. It seems **Aero Drum Ltd is actively working on the "Solar Blimp"** for several years. The company states that their goal is to have a blimp that is able to fly almost indefinitely. They have tested several designs of blimp with solar cells. Tubular blimps were tested in 2015 and 2016 (T-Blimp 1 and 2). Although promising, the main problems were that the contemporary building materials are still too heavy, to get successful working design (Aero Drum, 2017). Their research effort has led them to the development of a 22 m solar blimp. Technically, it is an autonomous UAV with power drawn from solar batteries. The highlight of this solar blimp (UAV) is that, it costs 21,000 USD. Not much indeed.

It takes about 3–5 months to produce a custom-designed solar blimp. Currently, major clients are in the transport department. Solar blimps are used mainly, to transport parcels and post. They may find larger clientele in the farm world, if they offer aerial images and digital data necessary for precision farming. The blimp needs helium gas only under inclement weather. These solar blimps have been designed to undergo minimum aerodynamic resistance and drag.

TECHNICAL SPECIFICATIONS

Material: Solar blimp is made of 150 μ poly-urethane foils. The payload cabin is made of aluminum strips welded, using "double-welding technology". Helium loss is restricted to less than 5% of original volume.

Size: length: 22 m; width: 2.3 m; height: 3 m; volume: 45 m^3, and weight: 14 kg.

Propellers: It has one propeller at the rear.

Payload: It accommodates useful payload of sensors, batteries, and CPU totally 5 kg.

Launching information: It is a VTOL and is launched by filling helium. It has autonavigation facility. It has two fins and ailerons for setting the direction of blimp's movement. The autopilot mode is optional.

Altitude: It floats at 100–300 m above crops canopy/terrain.

Speed: Top speed is 70–80 kmph. However, general cruise speed is 40 kmph.

Wind speed tolerance: The blimp withstands wind disturbance of 15 kmph, if in transit.

Temperature tolerance: −5°C to −45°C.

Power source: Solar cells/batteries.

Electric batteries: It has two sets of lithium polymer batteries.

Fuel: Solar batteries.

Endurance: 1–2 h, but could be extended depending on solar cells.

Remote controllers: The blimp can be controlled using a radio-link and GCS computers, for up to 5 km radius. It has GPS locater in case it drifts away.

Area covered: It depends on flight path programed, speed of the blimp fixed by GCS, and altitude from which the aerial survey and imagery is contemplated.

Photographic accessories: It has a full complement of sensors such as visual (R, G, and B) cameras, multispectral, NIR, IR, and LiDAR sensors. All images are tagged with GPS co-ordinates.

Computer programs: The ortho-images sent by the sensors are processed, using Pix4D Mapper.

Agricultural uses: The possible uses of solar blimps are in general surveillance of farms and crop fields, also, in aerial survey and crop imagery. They could be employed in collecting data about crop growth (NDVI), leaf chlorophyll (crops N-status), and thermal images (CWSI).

Nonagricultural uses: It includes surveillance of public places, advertising, collecting data on happenings around buildings, and monitoring city traffic.

PLATE 4.5 A solar blimp.
Source: Dr. Alexander Mijatovic—Aero Drum Inc. Belgrade, Serbia.

References, Websites, and YouTube Addresses

Aero Drum. RC Solar Blimps. Aero Drum Ltd.: Belgrade, Serbia, 2017. http://www.rc-zeppelin.com/solar-blimp.html Pp 1–8 (accessed Jan 12, 2018).

http://www.rc-zeppelin.com/aerial-photography-systems.html (accessed Jan 3, 2018).

http://www.rc-zeppelin.com/solar-blimp.html (accessed Dec 18, 2017).
https://www.youtube.com/watch?v=N8dXDgt0iCg (accessed Dec 18, 2017).
https://www.youtube.com/watch?v=QrjCAF5fl7M (accessed Dec 18, 2017).
https://www.youtube.com/watch?v=Stvsg1UOtYQ (accessed Jan 12, 2018).
https://www.youtube.com/watch?v=vhra329IbXA (accessed Jan 12, 2018).
https://www.youtube.com/watch?v=cpp-5QiIYa0 (accessed Jan 12, 2018).
https://www.youtube.com/watch?v=JEVmTdpXYsg (accessed Jan 12, 2018).

4.3.2 AEROSTATS HAVE A ROLE IN AERIAL SURVEY OF CROPS

Aerostats could be deployed almost above each small farm, because, they are low-cost, lighter-than-air, and easily deployable UAVs. Like UAV aircrafts, they can carry full complement of visual, multispectral, NIR, IR, and LiDAR pods. They do offer accurate and high-resolution aerial images and useful spectral data, about crops and terrain. Unlike UAV aircrafts, aerostats can be kept in the air for longer durations, above crops fields. Therefore, they consistently draw aerial images and offer farming agencies with regular data. Aerostats can be deployed rapidly. They are best suited for persistent surveillance of cropped fields.

Aerostats can be held at a precise location (GPS tagged) for a long duration. They can withstand wind disturbance better, because, they are tethered. Hence, they avoid large drifts in flight path. In terms of agricultural utility, aerostats are best suited for aerial photography of land resources, soils and crop plants. Crop growth analysis could be carried out for extended periods, using the continuous flow of data from aerostat's cameras. Several of the crop management decisions could be based on aerial imagery drawn via aerostats. Aerostats could be held in air for 14–16 days when they are filled with helium. Therefore, farmers may receive the data from aerostats consistently, for a stretch of 2 weeks or even more. Aerostats can be lowered and re-deployed. Recent reports suggest that, aerostats could be excellent farm gadgets, during precision farming. They offer valuable digital data to demarcate management blocks. Aerostats provide spectral images showing the variations in land resources, topography, soil type, crops, soil fertility, and productivity. It is believed that, among the various types of UAVs adopted by farmers/farming agencies in different continents, sizeable number of farmers in any region may opt for aerostats. Major reasons for selecting aerostats would be low cost of operation, very few repairs or maintenance, high-resolution accurate images, and continuous spree of images relayed by the sensors.

Aerostats are usually easily foldable and transportable, to any location in the farm. They need just a single technician or at best another helper. They are operated at a height of 100–150 m above the crop canopy. They can carry a payload of 30–35 kg that includes sensors and computer data processing units. Now, let us consider technical specifications of an Aerostat model.

Name/Model: Aerial Photographic System

MA-01-09E—**Compact Aerial Photography System**

MA-01-07 DS1/B—DSLR Aerial Photography System

Company/address: Aero Drum Ltd., Vojislava Ilica 99a, 11 000, Belgrade, Serbia, Europe; Tel.: +381 11 4071580; Mobile: +381 64 56 37 679; e-mail: rczeppelin@gmail.com; Skype: alexmijat

Aero Drum Ltd. produces "RC blimps," aerial photography systems and aerostats. They offer blimps and aerostats of variety of sizes, shapes, and photographic/sensor components. Blimps of 10–40 m size are also produced by the company. Aerostats of several sizes and attached with wide range of sensors (visual, NIR, IR, and FLIR) are offered. They are also custom-made, to suit an end-use (Plates 4.6.1–4.6.4). For example, to collect high-resolution spectral data about crops in an experimental station, etc. The UAVs are made of double layer of strong polyurethane foil. The payload area is built using light aluminum and double-welding technology. They usually carry high-definition video camera, if the blimp/aerostat is used, for surveillance and crop photography. All models are amenable for remote control. They could be preprogrammed regarding flight path, photography, and landing. Following are a few specifications of aerostat (Mijatovic, 2014).

TECHNICAL SPECIFICATIONS

Material: Blimp/aerial photographic system is made of double layer poly-urethane foils. It holds helium without leak. The payload area is made of aluminum strips welded together, to form a basket that holds gimbal and electric batteries.

Size: The blimps are ellipsoid with dimensions 2.8 × 2.8 × 1.9 m. The blimp's total volume is 8 m^3.

Payload: The maximum payload is 3 kg, excluding the weight of the UAV.

Maximum helium loss: Maximum helium loss is 0.3% of total volume per day.

Camera rig: The camera rig is made of 8 mm aluminum. It is "T-" shaped and holds gimbal and sensors.

Gimbal: The aerial photographic system has a gyro-stabilized gimbal. The tilted gyro-stabilized gimbal is optional. It allows oblique shots.

Electric power: The UAV has 3000 mAh 11.1 V lithium polymer batteries, for the camera, gimbal and 5.8 GHz video sender.

Endurance: The blimp stays in the air for 120–180 min. The blimp can be lowered to change the batteries and continue aerial survey.

Speed: Blimp's maximum speed is 2.5 3.0 m s^{-1}. It can be altered using GCS radio controls.

Remote controller: The blimp keeps contact with GCS computers for up to 3 km radius. The GCS has six radio channels, and a 2.4 GHz wireless receiver. The GCS has a wireless video of 5.8 GHz 1000 mW with eight channels. Video images are sent out for GCS computers. In addition, GCS has 20 cm color monitor.

Repair kit: The aerial photographic system is supplied with a repair tool kit.

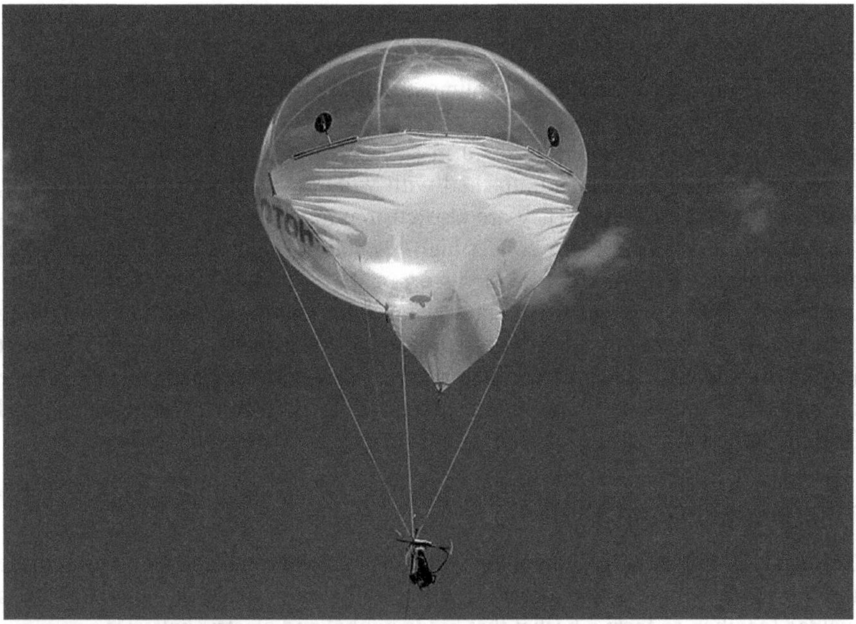

PLATE 4.6.1 Compact aerial photography system: MA-01-09E.
Source: Dr. Alexander Mijatovic—Aero Drum Ltd., Belgrade, Serbia.
Note: This "compact aerial photography system" is designed for cameras of all manufacturers up to 500 g. It is the simplest system to use. It can produce high-quality aerial photographs depending on the camera selected.

PLATE 4.6.2 DSLR aerial photography system: MA-01-07 DS1/B.
Source: Dr. Alexander Mijatovic—Aero Drum Ltd., Belgrade, Serbia.
Note: This system is designed for DSLR cameras like The Canon Rebel series, Nikon DSLR, Olympus, or other DSLR cameras. The camera rig has a mechanical gyro system that levels the camera with the horizon line, all the time.

PLATE 4.6.3 SLR aerial photography system—MA-01-07 DS1.
Source: Dr. Alexander Mijatovic—Aero Drum Ltd., Belgrade, Serbia.
Note: This system is suited for professional photographers. The lift and stability of blimp is designed for heavy cameras like Canon 5D Mark II, III, IV; Nikon D3X or similar. It has minimum 2–3 h of autonomy.

PLATE 4.6.4 3D Aerial Photography System.
Source: Dr Alexander Mijatovic—Aero Drum Ltd, Belgrade, Serbia.
Note: This aerostat is used to obtain 3D Aerial Photography. The camera rig has facility for adjusting camera's positions depending on the distance to the object or scenery to be photographed.

References, Websites, and YouTube Addresses

Mijatovic, A. Compact Aerial System: Specifications. Aero Drum Ltd. Technical Specifications, 2014. http://www.rc-zeppelin.com/compact-aerial-photography-systems.html pp 1–2 (accessed Nov 23, 2017).

http://www.rc-zeppelin.com/compact-aerial-photography-systems.html (accessed Nov 23, 2017).

http://www.rc-zeppelin.com/aerostat-surveillance.html (accessed Nov 23, 2017).

http://www.unmannedsystemstechnology.com/category/supplier-directory/platforms/lta-manufacturers/ (accessed Nov 23, 2017).

https://www.youtube.com/watch?v=nXUZIcwdLwY (accessed Jan 3, 2018).

https://www.youtube.com/watch?v=p0n-_MWE7vk (accessed Jan 3, 2018).

4.4 BALLOONS IN AERIAL PHOTOGRAPHY OF CROP FIELDS

Historically, balloons were first utilized as platforms for aerial photography in 1858. It was utilized to obtain pictures of Paris suburbs (France). Since, then, these balloons have found use in variety of situations. They are predominantly used to advertise products in a bold fashion, in the open air. They are used to surveillance public events, by attaching them with necessary visual cameras. The IR cameras attached to them may help in getting IR images of terrain, buildings, industries, crops, etc.

In practical urban and agricultural locations, we may trace several types of balloons, their models of different sizes and payload capacities. Balloon's ability of aerial imagery and relay of images may also differ. A few diagrammatic representations are made available in Plate 4.7.

Types of balloons that could be employed during aerial survey of land resources, agricultural fields, and crops—few examples:

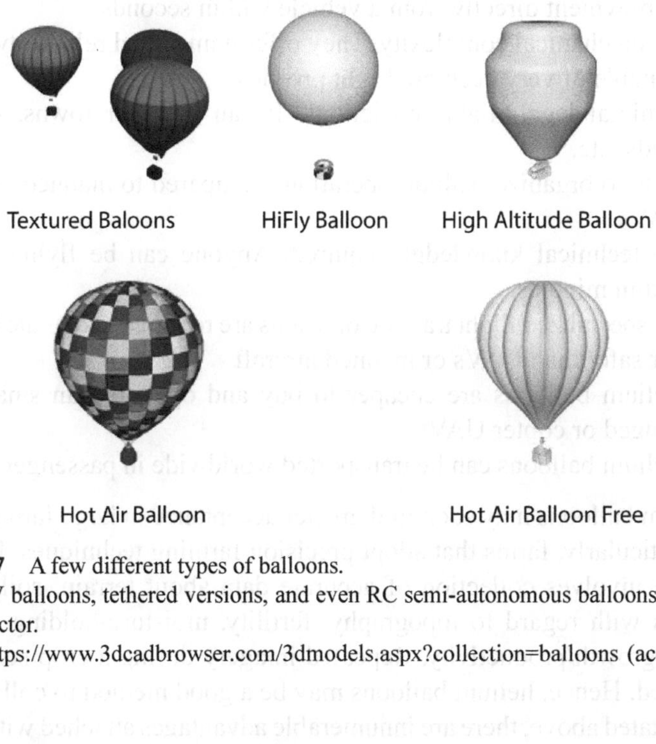

PLATE 4.7 A few different types of balloons.

Note: UAV balloons, tethered versions, and even RC semi-autonomous balloons are used in farming sector.

Source: https://www.3dcadbrowser.com/3dmodels.aspx?collection=balloons (accessed Nov 23, 2017).

The balloons (UAVs) carry a payload of gimbal with sensors, CPU, and batteries. They can be classified as "light" if the payload is about 4 kg, "medium" if the payload comprises of two packages of more than 4 kg, and as "heavy" if the balloon carries two or more packages of 6 kg weight. Normally, the density of packages is held to less than 13 g cm^2. The suspension rope used to hold the payload (cameras, packages, etc.) is usually tough. It withstands impact force of 230 N (Newtons) (The McGraw-Hill Inc., 2005). Many of the commercial models of balloons available in the market are provided with instruction manuals, for effective use above crop fields. Further, there are manuals to prepare custom-made balloons that costs less. They are effective in obtaining useful spectra data about crops (Sudhoff, 2012).

Some of the advantages attached with using helium balloons, during aerial survey and data collection (spectral) are listed as follows by Allsopp Helikites Ltd. (2017):

- Automatic flight in all weather conditions.
- Steady flying platform reduces camera shake.
- Suitable for every type and make of cameras.
- Deployment directly from a vehicle within seconds.
- No mechanical complexity. They offer unmatched reliability.
- Capable of very accurate flight profiles.
- Minimal legal flight restrictions. It can fly over towns, railways, roads, etc.
- Easy to organize balloon operations compared to manned aircraft or UAVs.
- No technical knowledge required. Anyone can be flying balloons within minutes.
- No specialized flight training or exams are required to operate balloons.
- Far safer than UAVs or manned aircraft.
- Helium balloons are cheaper to buy and operate than small fixed-winged or copter UAVs.
- Helium balloons can be transported worldwide in passenger luggage.

Helium balloons may soon find greater acceptance in large farms, world-wide, particularly, farms that adopt precision farming techniques. Precision technique involves collection of accurate data about terrain, soils and its variations with regard to topography, fertility, moisture-holding capacity, and crop growth/productivity. Repeated imagery during a crop season will be required. Hence, helium balloons may be a good method to collect aerial data. As stated above, there are innumerable advantages attached with helium balloons, when used in agricultural fields. It may not be long before almost each large farm may fly a helium balloon for long periods, above their crops. This is just to monitor crop growth and note the needs of crops with regard to soil nutrients, pesticides, herbicides, and water.

Let us consider an example, where in a “helium balloon” has been utilized in agricultural settings. It has been used to conduct aerial survey and offer useful imagery of crops and surroundings. Also, the spectral data has immediate utility in judging crop growth, leaf chlorophyll (i.e., crops N-status), and water stress index (i.e., to judge irrigation needs). This is a study conducted in Cyprus, in the Middle-east region (Agapiou et al., 2012; Themistocleous, 2014). Themistocleous (2014) has compared the efficiency, accuracy, and usefulness of different platforms. The low-altitude aerial system—a helium balloon was equipped with sensors. The sensors operate at visual band width (R, G, and B), VNIR, and at IR range. They used Sony NEX-7, Tetracam multispectral camera, and FLIR thermal sensor. They flew the helium balloon up to 200 m above the crop canopy, for aerial photography. It had a payload of 6 kg that included the sensors and CPU. They examined an archeological site of relevance to crop production tactics followed. The aerial imagery clearly showed the historical crop production sites. The spectral signatures of archaeological crop sites are useful to researchers. The archaeological remains of crops were identified, using their spectral signatures. The images showed that, historically, the agricultural site supported production of alfalfa. Results showed that, the NIR spectral data was more useful in distinguishing the "crop remains." The helium balloons could be held for a long duration in the atmosphere. So, detailed photography from almost a stand-still helium balloon was possible. However, Themistocleous (2014) states that, during aerial survey and photography, wind and breeze can disturb the helium balloon and cause haziness. The size of the helium balloon needs due consideration. Helium must be filled afresh at each location, if we re-start the balloon. The balloon must be inflated repeatedly. Helium tanks need to be carried along with the helium balloon inside the farm and to different locations.

In Japan, specialists dealing with Kushiro, a wetland natural reserve of great biodiversity have utilized balloons, to obtain aerial photography. They have aimed to study the heterogeneity of plants, their spatial and temporal changes, biomass accumulation trends, etc. These balloons (kite balloons) offer high-resolution images (15 cm pixel) and detailed maps of the marshy and dry areas, in the natural reserve (Miyamoto et al., 2004; Koizumi et al., 1986; Nagano, 1990). Balloons have also been used to study effects of bush fires on natural vegetations (Amoroso and Arrowsmith, 2012)

A few models of helium balloons and their salient features are listed. Again, since the focus of this book is on UAV aircrafts, only very few examples of balloons have been included.

Name: Helikite Balloons

Company address: Allsopp Helikites Ltd.—Unit 2, Fordingbridge Business Park, Fordingbridge, Hampshire, SP6 1BD, England (UK); Tel.: +44(0)1425-654967; Fax: +44(0)1425-839367; e-mail: info @ helikites.com

In the mid 1990's, Allsopp Helikites Ltd. pioneered the use of digital cameras to produce the world's first practical, low cost, "Airborne Geo-Information System." Nick Russell of "TerraDat" was the first surveyor to try Helikite photogrammetry, to measure the ground. He used aircraft measurement software that was accurate to 0.4 mm. Then, the engineering company "Bechtel Inc." purchased a system to measure railways. It can very accurately survey 10 km of rail track within 2 h, while the trains run normally underneath. This is a huge cost saving, as traditional survey methods take days and can be done only with the trains stopped. In New Zealand, NIWA started using Helikite balloons for power line surveys. Eventually, Helikite balloons became mainstream airborne GIS platforms. They are now used by numerous operators, worldwide (Allsopp Helikites Ltd. 2017).

Allsopp Helikites Ltd. is involved in producing balloons and aerostats. Their aerostats are utilized for aerial survey, aerial photography, monitoring natural resources, aerial advertising, developing videos of public events, monitoring traffic, etc. , (Plates 4.8.1–4.8.3).

Cameras: Helikites can lift almost every type of camera. The Helikite has a unique "Universal Camera Mount" capable of steadily holding all designs of cameras and pan/tilt gimbals. Larger Helikites can lift any GIS camera or sensor, up to 40 kg weight. Many cameras have wireless remote controls or internal intervalometers. Therefore, it is easy for them to take photos when at height. Most commonly used sensors on the Helikite balloons are as follows: Canon EOS 5DS (50 Mp); Nikon D800 (36 Mp); Nikon D5200 (24 Mp); Sony NEX-7 (24 Mp); Ricoh GR2 Digital (12 Mp); Pentax Optio W60 (12 Mp); GoPro (12 Mp); and many other brands. LiDAR: Small LiDARs designed for UAV use can be fitted on a Helikite. This is far easier, cheaper, and safer than using a UAV.

Magnetometers: Helikites lift large magnetometers and hover over areas, where humans cannot reach. They fly on minefields, marshes, water bodies, or other types of difficult or dangerous terrain.

Meteorology sensor: Tethersondes can be lifted on the Helikite. It helps to produce real-time wind speed profiles, wind direction, humidity, etc.

Radio-Rely and WiFi: Surveyors working in remote areas need communication for safety. Also, to send the data back to base. Helikites are superb at persistently lifting radio-relays and WiFi systems. The maximum range obtained, so far, from small Helikites is 60 miles. Large Helikites have increased radio range to 100 miles. The images obtained by the sensors on the Helikites are processed, using software such as: Leica tridicon (http://www.tridicon.de/software/tridicon-pointcloud/?L=1); "pix4Dmapper" (https://pix4d.com/products/); and "AGISOFT" (http://www.agisoft.com/). Open source software too could be adapted, to process aerial images (https://en.wikipedia.org/wiki/List_of_geographic_information_systems_software). The raster calculator tool of ArcMap 9.0 is used to overlay R, G, and B and NIR images. NDVI are estimated, using the relationship: NDVI = (NIR − RED)/(NIR + RED)

Agricultural uses of Helikite balloons: They include aerial survey of crop land, monitoring farm activities, that is, movement of tractors, combine harvesters, and grain transport vehicles in farms. Regular monitoring of irrigation pipes and leakage, if any, is also possible. Mapping weed-infected zones in crop land is possible. Heliokites are used in detection of flood- or drought-affected patches. Helikite balloons have been used to monitor ecological sites and map the changes. It has also helped in studying the botanical diversity of natural vegetation. Helikite balloons provide excellent digital data, maps, and botanical composition of pastures. They serve as excellent alternatives, to UAV aircrafts during aerial survey of crop fields (Allsopp Helikites Ltd., 2017).

Nonagricultural uses of Helikite balloons: They are used for oil and gas pipeline inspection, refinery inspection, solar energy farm inspection, rail road and road traffic inspection, surveillance of big buildings, bridges, monitoring flood- and erosion-affected locations, and offering aerial pictures. Surveillance of rivers, borders, and military convoy movements are other functions. These balloons are also used in monitoring mines, ore movement, and ore stockpiles. Helikite's balloons have acted as regular IR eyes in the sky, to monitor a wide range of activities. It includes measurement of temperature changes on the terrain. Helikite's balloons are also used in environmental management, for example, monitoring waste dumps. Collection of air samples from different levels above ground is another easily feasible function, using these balloons. Helikite balloons aid in collecting air samples at low altitudes over cattle dairies and crop fields. This is to assess methane, CO_2, and CO. For example, it has been successfully used in the dairy cattle and pasture-growing regions of New Zealand, to procure samples of air, for assessing methane and N-emissions.

PLATE 4.8.1 Allsopp Helikite balloon.
Source: Mrs. Sandy Allsopp, Allsopp Helikite Ltd., Fordingbridge, Hampshire, United Kingdom.
Note: The Allsopp Helikite is a combination of a helium balloon and a kite. It forms a single, aerodynamically sound tethered aerial vehicle. It exploits both wind and helium for its lift.

PLATE 4.8.2 Preparing Helikite for aerial survey of terrain and mapping natural vegetation.
Source: Mrs. Sandy Allsopp, Allsopp Helikite Ltd., Fordingbridge, Hampshire, United Kingdom.

PLATE 4.8.3 A near-infrared imagery of wheat fields in England.
Source: Mrs. Sandy Allsopp, Allsopp Helikite Ltd., Fordingbridge, Hampshire, United Kingdom.
Note: The IR imagery was done using Helikites balloons fitted with a visual and IR sensors.

References, Websites, and YouTube Addresses

Allsopp Helikites Ltd. GIS, Geomatics, Surveying and Inspection Helikites Balloons, 2017. http://www.allsopp.co.uk/index.php?mod=page&id_pag=63. pp 1–15 (accessed Nov 21, 2017).
https://www.easyaerialphotography.com/ (accessed Nov 17, 2017).

http://www.allsopp.co.uk/index.php?mod=page&id_pag=60 (accessed Nov 17, 2017).
http://www.allsopp.co.uk/index.php?mod=page&id_pag=37 (accessed Nov 17, 2017).
https://www.youtube.com/watch?v=20Pqdfa9Vd4 (accessed Dec 13, 2017).
https://www.youtube.com/watch?v=fb86AMwdPkM (accessed Dec 13, 2017).
https://www.youtube.com/watch?v=7S8BgoqUwWc (accessed Dec 13, 2017).
https://www.youtube.com/watch?v=lbjw9f8R47A (accessed Dec 13, 2017).

Name/Model: Helio Balloon

Company and address: Aerovision—Unmanned Aerial Solutions PTY Ltd. 43 Sea Cottage Drive, Noordhoek, Cape Town-7979, South Africa.

Helio balloon kites are useful during aerial surveillance and reconnaissance. They can be used as "relay stations," to boost radio signal's strength. They are portable and could be deployed rapidly. Helio balloons are ideal for difficult-to-reach areas with poor communications. Aerovision Pty Ltd. designs and builds helium balloons, to suit customer specification with payloads up to 30 kg or more. Helio balloons are capable of withstanding disturbing wind speeds of up to 35 knots (Plate 4.9).

Agricultural uses of Helio balloons include: regular monitoring and survey of grape vine yard. Helio balloons are used in reporting crop growth trends (NDVI), crops N-status (leaf-chlorophyll data), and water stress index. Also, large scale disease/pest attack, if any, in the vineyards could be detected. The Helio balloons are utilized, to monitor and collect data about crop growth from experimental plots. Helio balloons have been utilized for aerial survey and mapping of farm terrain, installations, etc. They have been utilized for regular monitoring of buildings, electrical installations, golf courses, mines and mining activity, oil prospecting, etc. (Johnson et al. 2014; Stombaugh et al. 2017). They have also been utilized to report about disaster zones, like fire-damaged area, flood- and erosion-affected areas, river-bank erosion and overflow, etc.

PLATE 4.9 A Helio balloon up in the sky picturizing farmland.
Source: Mr. Ian Freemantle, Aerovision Unmanned Aerial Solutions, Cape Town, South Africa.

References, Websites, and YouTube Addresses

Johnson, K.; Nissen, E.; Saripalli, S.; Arrowsmith, J. R.; McGarey, P.; Scharer, K.; Williams, P.; Blisniuk, K. Rapid Mapping of Ultrafine Fault Zone Topography with Structure from Motion. Geosphere **2014,** *10,* 969–986.

Stombaugh, T.; Smith, S.; Thamann, M. The Use of Unmanned Aircraft Systems in Agriculture, 2017; pp 1–6. http://www.uky.edu/bal/sites/UASFctsheetFinal.pdf (accessed Apr 25, 2017).

http://www.aerovision-sa.com/unmanned-aerial-solutions (accessed Aug 12, 2017).

https://www.aerovision-sa.com/unmanned-aerial-solutions (accessed Aug 12, 2017).

https://www.youtube.com/watch?v=e9vDCCxUu-0 (accessed Aug 12, 2017).

https://www.youtube.com/watch?v=N4Lq3NBOcGo (accessed Aug 12, 2017).

https://www.youtube.com/watch?v=Ryo0Ji9TX6o (accessed Aug 12, 2017).

https://www.youtube.com/watch?v=QsckKBOjrtY (accessed Aug 12, 2017).

4.5 KITES IN AERIAL IMAGERY OF AGRARIAN TERRAIN

A brief history of kite aerial photography is available in Chapter 1. Kite aerial photography, it seems, got started in 1880s. It experienced a kind of rejuvenated interest in late 1900s. Kite aerial photography kits, camera cradles, and sensors were miniaturized. So, kites found a new enthusiasm among scientists requiring aerial imagery of natural vegetation and crops (Conrad, 2017). Historically, it seems, it took about a decade to initiate kite aerial photography after the first direct photos became possible in 1839 (Mulders, 1987). So, kites have been tested for their ability to offer good aerial photos over a century ago.

Some points to note about "kite aerial technology (KAT)" are that kites are quick to fly. They offer images and digital data at a rapid pace. They are flexible regarding launch and recovery. They are perhaps cheapest among the platforms we know, so far. Any number of kites could be used repeatedly by the farmers. Storage and parking of kites, either small or larger versions are easy. They could be placed in any "farm house," safely. Kites could be applied to study growth, physiological parameters, nutrient, and water status of crops such as wheat, rice, maize, sorghum, lentil, cowpea, cotton, sunflower, etc., grown in different agrarian belts. At present, there seems to be a search for a platform that is versatile, cheap, and one that could be utilized repeatedly at short notice. Plus, the platform should be less cumbersome regarding technology. Kites seem to fit many of these requirements for crops grown in different continents. It is a matter of time, before agricultural kites adorn the skies, rather, too frequently and serve the farmer usefully. Research effort to test and improve kite models and adapt them, to different agrarian belts and crops is necessary.

Kites are among the earliest of the platforms used to place a camera and obtain aerial photographs. Usage of kites to obtain aerial images or digital data about the agrarian terrain, crops or water resources is yet to gain acceptance. It has not made a big dent in the procedures adopted to analyze the crop's growth status. Kites are relatively low-cost platforms. They are easily accessible to farmers. Perhaps, in future, kites may find popularity with resource-poor farmers and those with small holdings.

Abe et al. (2010) have opined that, as need for geographical data increases, farmers will rely increasingly on aerial data of their crops. Consequently, need for small-format aerial photography too increases. The small-format aerial photography requires a platform such as kites, parachutes, blimps, or balloons. Reports made about two decades ago suggest that, kites could potentially become useful gadgets for overhead surveillance of crops in agrarian regions. As such, aerial photography has been adopted since few decades. In recent years, kites have found use in gathering information about crops, cropping systems, yield, etc. Kites could also be effective in gathering agricultural intelligence (Gasser, 1993).

Now, let us consider a few examples, where in, kites have been utilized to produce good-quality, high-resolution clear images of landscapes, vegetation, crop fields, and other agricultural aspects. One of the reports presented at the International Kite Conference suggests that, aerial imagery of natural vegetation, agricultural landscapes, and individual cropped field is possible, it could become common (Kipemed-10, 2015; Benton, 2016; Digital Earth Watch, 2012). The kite aerial photography is usually conducted from vantage points, say, about 150 m above the soil or crop canopy. Resulting imagery offers a spatial resolution of 2–5 cm. Kites are well suited for farmers holding 5–10 ha of cropped land (Aber et al., 2009).

Soybean is a major legume crop in the Central Great Plains of North America. It is perhaps the most important rotation crop in the wheat- and maize-growing regions of the plains. Soybean does experience periodic droughts. Intermittent drought stress is common in the plains. Farmers need to collect data about canopy and ambient air temperature, to decipher crop's water stress index. Then, they can decide on the quantum and frequency of irrigation, based on water stress index values. Purcell (2014) has reported the use of kites, a set of three of them, to collect accurate temperature of the soybean's canopy and air above it. The three kites help in obtaining haze-free, high-resolution multispectral images and temperature maps of the crop. Kites are flown when it is cool and wind-free. Usually, temperature data is collected by the thermal sensors, that is, an IR band width camera located on the kite, from early in the morning (06.00 a.m.) till evening (17.30 p.m.). A certain amount of standardization of data collection time and its correlation with accuracy of water stress index values is needed. Kites are launched and then it is walked through the field while the camera shoots visual and thermal images. At a time, images could be relayed, for up to 30 min to GCS computers, that is, a laptop. The aerial photos of different genotypes of soybean are downloaded and analyzed. Soybean genotypes that tolerate drought (or dearth for water) and show clearly a delay in wilting are selected.

In Poland, KAT was practiced by Aber and Galazka (2000), to obtain visual imagery and map the terrain. They have travelled extensively around Warsaw and northward in Gdansk region and obtained detailed images of geographical features, water resources, and cropping patterns, using KAT. They flew different types of kites such as Sutton Flowforms, Delta Conyne, and Giant Rokkaku. Kites were transported across the terrain using a small car. The topographic map constructed after processing the ortho-images is quite simple to use. We have to just click at the spot on the computer screen, to find out the details of terrain, topography, soil type, and crops.

Kites aided aerial photography of difficult terrains such as mountainous ecosystems, their vegetation, and diversity of botanical species. In Norway, kites have been effectively used to obtain aerial photos of botanical species, algae, and other tundra vegetation. In such mountainous tundra, resolution of images is a problem. Also, since we cannot reach vantage spots. However, flying kite with visual cameras can help us to obtain good images. They have been used it to construct spatial and temporal changes in vegetation of the icy mountainous region in Norway (Wundram and Lofflier, 2014). Kites have also been utilized to conduct aerial survey of Estonian bogs (Aber et al., 2002)

In the western region of the United States, kites have been utilized to study the effect of bio-control agents on rapid spread of Saltcedar (Aber et al., 2005). Saltcedar is a tree/shrub that is growing rampantly fast. Diorhabda elongata has been used to see, if it reduces the spread of Saltcedar. Kite aerial photography was conducted to obtain high-resolution images of Saltcedar spread, in areas treated with beetle. Aerial images clearly showed defoliation due to beetles. It tallied excellently with the ground truth data collected by scouts. They say it is also possible to make analysis of number of beetles released, their intensity in a particular area, and its impact on saltcedar growth. So, kites could be of low-cost, quick to operate UAVs offering useful data, to botanists and geographists. Also, kites could be tried to assess the effects of variety of inputs, pesticides, fungicides, and herbicides on crops.

Kites, like other UAVs are versatile. Kites find uses in variety of terrain, weather patterns, and geographic locations. For instance, kites have been utilized in the dry Sahelian region of West Africa, by researchers of International Crops Research Institute for the Semi-Arid Tropics (ICRISAT, Niamey, Niger). Gerard (2016) reports that, use of aerial photography of Sahelian agricultural zones is not new. The fragile agroecosystems have been consistently studied, using aerial photography. Aerial photos taken from 100 to 300 m above crop/vegetation canopy have provided useful clues about environmental effects, on Sahelian crops. The crop species such as sorghum, pearl millet, and cowpeas have been photographed using kites. Kites are ideal for resource-poor farmers with small holdings. It provides aerial photos of workable quality and resolution in the dusty/windy conditions of Sahel, during crop season. Careful maneuvering of kite and control of camera's shutter can help in detecting disease/pest attack, if any, on the Sahelian crops such as groundnut.

KEYWORDS

- **parachutes**
- **blimps**
- **aerostats**
- **helikites**
- **kites**
- **aerial survey**
- **aerial photography**

GENERAL REFERENCES

Abe, J.; Marzoff, I.; Ries, J. B. Small-Format Aerial Photography, 2010. DOI: 10.1016/B978-0-444-53260-2.10008-0. pp 1–5 (accessed Apr 28, 2017).

Aber, J. S.; Aaviksoo, K.; Karofeld, E.; Aber, S. W. Patterns of Estonian Bogs Depicted in Colour Kite Aerial Photographs. *Suo* (Mires and Peat) 2002, 53, 1–15.

Aber, J. S.; Aber, S. W.; Buster, L.; Jensen, W. E.; Sleezer, R. L. Challenge of Infrared Kite Aerial Photography: A Digital Update. *Trans. Kansas Acad. Sci.* **2009,** *112,* 31–39

Aber, J. S.; Eberts, D.; Aber, S. W. Applications of the Kite Aerial Photography: Biocontrol of Salt Cedar (Tamrix) in the Western United States of America. *Trans. Kansas Acad. Sci.* **2005,** *108,* 63–66 -of-intelligence/kent-csi/vol11no4/html. pp 1–5 (accessed Apr 28, 2017).

Aber, J. S.; Galazka, D. Kite Aerial Photography of Poland, 2000. http://www.geospectra.net/kite/polnd/polnd.htm. pp 1–22 (accessed Apr 28, 2017).

Aeros. Aerostats in Agriculture. Aeros Corporation, 2016. http://aeroscraft.com/precision-agriculture/4584020239 (accessed Dec 24, 2017).

Agapiou, A.; Hadjimitsis, D. G.; Sarris, A.; Georgopoulos A.; Alexakis, D. D. Optimum Temporal and Spectral Window for Monitoring Crop Marks Over Archaeological Remains in the Mediterranean Region. *J. Archaeol. Sci.* 2012, 40, 1479–1492. DOI:10.1016/j.jas.2012.10.036 (accessed Jan 13, 2017).

Airstar. Airstar Aerospace Tethered Blimps Range. Airstar Aerospace, France, 2017. http://airstar.aero/en/tethered-balloons-range/. p 18 (accessed Dec 10, 2017).

Allsopp Helikites Ltd. GIS, Geomatics, Surveying and Inspection Helikites Balloons, 2017. http://www.allsopp.co.uk/index.php?mod=page&id_pag=63. pp 1–15 (accessed Nov 21, 2017).

Amoroso, L.; Arrowsmith, R. Balloon Photography of Brush Fire Scars East of Carefree, Arizona, 2012. http://activetectonics.su.edu/Fires_and_Floods/10_24_00_photod. pp 109–111 (accessed Sept 12, 2017).

Associated Press. Blimp, Infrared Camera Used to Fine-Tune Cotton Irrigation. Augusta Chronicle, 2004. http://chronicle.augusta.com/stories/2004/07/13/liv_422010.shtml. pp 1–5 (accessed Apr 29, 2017).

Bellis, M. History of the Parachute. Thought Co., 2017. https://www.thoughtco.com/history-of-the-parachute-1992334. pp 1–14 (accessed Nov 13, 2017).

Benton, C. C. Notes on Kite Aerial Photography. Berkeley, CA, 2016. http://kap.ced.berkeley.edu/kantoc.html. pp 1–4 (accessed Apr 28, 2017).

Cameron Balloons. Airships, 2017. http://www.cameronballoons.com/products/airships. pp 1–7 (accessed July 29, 2017).

CIMMYT. Obregon Blimp Airborne and Eyeing Plots. http://www.cimmyt.org/obregon-blimp-irborne-and-eyeing-plots/. pp 1–4 (accessed Apr 29, 2017).

Conrad, K. Kite Aerial Photography. Brooxes.com, 2017. http://www.brooxes.com/newsite/HOME.html. pp 1–4 (accessed Apr 28, 2017).

Digital Earth Watch. *Getting Images: Kite and Balloon Aerial Photography*. University of California, 2012. http://dew.globalsystemsscience.org/about/allaboutdew/03-kite-photography. pp 1–14 (accessed Aug 2, 2017).

Gardner, G. What Is the Etymology of the Word Blimp, 2015. https://www.quora.com/What-is-the-etymology-of- the-word-blimp. The Lawrence Hall of Science. p 1 (accessed Apr 29, 2017).

Gasser, W. R. Aerial Photography for Agriculture. CIA Historical Review Program, 1993. http://www.cia.gov/library/center-for-the-study (accessed Sept 12, 2017).

Gerard, B. It's a Bird, It's a Plane No, It's a Super Scientist. SAT Trends, ICRISAT Newsletter, 2016. http://www.icrisat.org/what-we-do/satrends/01dec/1.htm. p 14 (accessed Aug 3, 2017).

Grothaus, M. Hybrid Blimps Could Soon Take the Skies. https://www.fastcompany.com/3058428/hybrid-blimps-could-soon-take-to-the skies. pp 1–3 (accessed July 29, 2017).

Guglieri, G.; Ristorto, G. Safety Assessment for Light Remotely Piloted Aircraft Systems. International Conference on Air Transport- INAIR. pp 1–7.

Hailey, T. I. The Powered Parachute as an Archaeological Aerial Reconnaissance Vehicle. Archaeol. Prospec. 2005, 12, 69–78. DOI 10.1002/arp.247 (accessed Apr 24, 2017).

Hendrix, S. From Blimp Pilot to Successful Farmer: Montgomery County's New Crop Tillers. Washington Post, 2014. http://www.washingtonpost.com/local/. pp 1–3 (accessed Apr 25, 2017).

Hunt, E. R.; Horneck, D. A.; Gadler, D. J.; Bruce, A. F.; Turner, R. W.; Spinelli, C. B.; Brungardt, J. J. Detection of Nitrogen Deficiency in Potatoes Using Small Unmanned Aircraft Systems, 2015. https://www.ars.usda.gov/research/publications/publication/?seqNo115=301000. pp 1–4 (accessed Oct 27, 2017).

Jowit, J. Blimps Could Replace Aircraft in Freight Transport, Say Scientists. *Guardian*, 2010. http://www.thegurdian.com/environment/2010/jun/30/blimps-ircraft-freight (accessed July 29, 2017).

Kerman, B. A Brief History of the Parachute, 2012. http://www.popularmechanics.com/flight/g815/a-brief-history-of-the-parachute/. pp 1–17 (accessed Nov 13, 2017).

Khot, L. R.; Zhang, Q.; Karkee, M.; Sankaran, S.; Lewis, K. Unmanned Aerial Systems in Agriculture: Part 1 (systems). Washington State University Extension Service. Pulmann, Washington, DC, 2016, pp 1–12.

Kipemed-10. Kite Aerial Photography. International Conference on Kite Aerial Photography, 2015. https: www.wokipi.com/Kpined/rt-va.html. pp 1–3 (accessed Apr 26, 2107).

Koizumi, T. M.; Murai, M.; Manabe, H. An Automated System and Its Application for Aerial Photography Using Kite Balloon. *J. Japan Soc. Photogramm. Remote Sens.* **1986,** *25,* 12–23.

Lelong, C. C.; Burger, P.; Jubein, G.; Roux, B.; Labbe, S.; Baret, F. Assessment of Unmanned Aerial Vehicles Imagery for Quantitative Monitoring of Wheat Crop in Small Plots. *Sensors* 2016, 8, 3557–3585.

Meyer, J. An Introduction to Deployable Recovery Systems, 1985. http://www.parachutehistory.com/eng/drs.html. p 187 (accessed Nov 21, 2017).

Miyamoto, M.; Yoshino, K.; Nagano, T. *Wetlands*. Society of Wetland Scientists: Springer Verlag, 2004. http://doi.org/10.1672/0277-5212(2004)024[0701: UOBAPF]2.0.CO:2 (accessed Dec 27, 2017).

Mothership Aeronautics. 100 Applications of Solar-Powered Blimps, 2016. http://www.mothership.aero/100-applications. pp 1–5 (accessed Apr 4, 2017).

Mulders, M. A. Aerial Photography. *Develop. Soil Sci.* 1987, *15*, 155–180.
https://doi.org/10.1016/S0166-2481(08)70033-3 (accessed Aug 9, 2017).

Nagano, T. Studies on Mangrove in Thailand-an Aerial Photographic Survey Using Kite Balloons. *Japanese Soc. Environ. Cont. Biol.* **1990,** *28,* 119–124.

NASA. Kites: A Background. *Aeronautics Research Mission Directorate-NASA*. National Aeronautics and Space Agency: Washington DC, 2016. http://www.nasa.gov/sites/default/files/atoms/files/kites_t4.pdf. pp 1–65 (accessed Apr 28, 2017).

NASA STAFF. A Canopy of Confidence: Orion's Parachutes, 2017. http://www.nasa.gov/exploration/systems/mpcv/canopy_of_confidence.html. pp 1–8 (accessed Nov 20, 2017).

Pat Works. Early History of Parachutes. Skydiving Museum and Hall of Fame, 2014. http://works-words.com/NSM-WIKI/WP/wordpress/wiki/skydiving/early-history/history/early-history-of-parachutes/.

Prentice, B. E.; Beilock, R. P.; Philipps, A. J. Economics of Airships for Perishable Food Trade. 5th International Airship Convention and Exhibition, 2017. https://umanitoba.ca/faculties/management/ti/media/docs/AA04_airship_large1.pdf. pp 1–12 (accessed Nov 27, 2017).

Purcell, L. *Kites, Balloons Collect Aerial Data for Soybean Drought Tolerance Research. APS Crop Protection an Management Collection: Plant Management Network.* University of Arkansas: Fayetteville, Arkansas, 2014. http://www.plantmangementnetwork.org/pub/crop/news/2014/erialData/. pp 1–2 (accessed Apr 28, 2017).

Robinson, M. Meteorological Kites: Scientific Kites of the Industrial Revolution. http://kitehistory.com/Miscellneous/meteorological_kites.htm. pp. 1–5 (accessed Apr 29, 2017).

Sankaran, S.; Khot, L.; Espinoza, C. Z.; Jarolmasjed, S.; Sathuvalli, V. R.; Vandermerk, G. J.; Mikkls, P. N.; Carter, A. H.; Pumphrey, M. O.; Knowles, N. R.; Pavek, M. J. Low-Altitude,

High Resolution Aerial Imaging Systems for Row and Field Crop Phenotyping: Review. *EU. J. Agron.* **2015,** *70,* 112–123.

Shi, Y. Unmanned Aerial Vehicles for Height Throughput Phenotyping and Agronomic Research. *PLOSone* **2016,** *11.* DOI: 10.1371/journal.pone.0159781 (accessed Dec 27, 2017).

Sudhofff, S. Balloons, Drones and Satellite Images, 2012. Http://cartong.org/2012/08/01/ballons-drones-and-satellite-images/ (accessed Jan 5, 2018).

The McGraw-Hill Companies Inc. Unmanned Free Balloon. An Illustrated Dictionary of Aviation, 2005. http://encyclopedi2.the freedictonary.com/unmanned+free+ balloon. p 1 (accessed Nov 25, 2017).

Themistocleous, K. In *The Use of UAV Platforms for Remote Sensing Applications: Case Studies in Cyprus*. Proceedings of SPIE-The International Society for Optical Engineering DOI: 10.1117/12.2069514. pp 1–12 (accessed Nov 16, 2017).

Thamm, H. P. SUSI 62 A Robust and Safe Parachute UAV with Long Flight Time and Good Payload. *Int. Arch. Photogramm. Remote Sens. Spatial Inform. Sci.* **2011,** *38,* 19–24.

Thamm, H. P.; Judex, M. The Low-Cost Drone. An Interesting Tool for Process Monitoring in a High Spatial and Temporal Resolution. *Int. Arch. Photogramm. Remote Sens. Spatial Inform. Sci.* ISPs Commission 7th Mid-Term Symposium. Remote Sensing: From Pixels to Process. Enchede. The Netherlands 2006, 36, 140–144.

Thamm, H. P.; Menz, G.; Becker, M.; Kuria, D. N.; Misana, S.; Kohn, D. The Use of UAS for Assessing Agricultural Systems in AN Wetland in Tanzania, in the Wet-Season, for Sustainable Agriculture and Providing Ground Truth for Terra Sar X Data. ISPRS Int. Arch. Photogramm. *Remote Sens. Spatial Inform. Sci.* 2013, XL-1/w2, 401–406. DOI: 10.5194/isprsarchives-XL-1-W2-401-2013 (accessed Oct. 10, 2017).

Vericat, D.; Brasington, J.; Wheaton, J.; Cowic, M. Accuracy Assessment of Aerial Photographs Acquired Using Lighter Than Air Blimps: Low Cost Tools for Mapping River Corridors. *River Res. App.* 2008, *25*, 985–1000.

Wikipedi. Kite Types, 2017. https://en.wikipedia.org/wiki/Kite_types. pp 1–16 (accessed Apr 28, 2017).

Wundram, D.; Lofflier, J. Kite Aerial Photography in High Mountain Ecosystem Research. Institute of Geography, University of Bonn, Germany, 2014. https://www.researchgate.net/publication/255620360_Kite_Arial_Photography_in_Mountain_Ecosystem_Research. pp 1–18 (accessed Apr 28, 2017).

Zhou, J.; Reynolds, D.; Websdale, D.; Cornu, T. L.; Gonzalez-Nevarrro, O.; Lister, C.; Orford, S.; Laycock, S.; Finlayson, G.; Stitt, T.; Clark, M.; Bevan, M.; Griffiths, S. CropQuant: An Automated and Scalable Fie and Other Products to Different Locations. ls Phenotyping Platform for Crop Monitoring and Trait Measurements to Facilitate Breeding and Digital Agriculture, 2017. DOI: http://dx.doi.org/10.1101/161547. ppl–14 (accessed July 10, 2017).

CHAPTER 5

Sensors and Image Processing Computer Software Relevant to Unmanned Aerial Vehicle-Based Technology in Agriculture

ABSTRACT

Sensors that provide aerial imagery of agricultural terrain and crops form the centre piece of UAV technology as applied to precision farming. While UAVs offer the necessary vantage points, transit over land resources, vegetation and crops, it is the sensors that offer the crucial data to farmers. Sensors used in agriculture are mostly electro-optical, such as visual bandwidth (red, green and blue), near infra-red (NIR), infra-red (IR), red-edge, and LIDAR (light detection and ranging). A few UAVs also possess physico chemical electrodes aimed at analysing atmospheric gases and pollutants. There are indeed several companies that offer sensors (camera lenses) of wide range of capabilities. Farmers should select sensors that are apt for their purpose. In this chapter, a few examples, that are currently popular brands of sensors adopted on agricultural UAVs have been listed and discussed. Information about manufacturers, specifications, and uses in farming are provided, for each brand of sensors. A few examples are Canon sensors, Mapir sensors, Micasense Red-Edge sensor, Parrot Sequoia multi-spectral sensors, Sentera series, Tetracam RGB sensor, Rikola Hyperspectral sensor, Zenmuse-FLIR series of sensors, WIRIS Thermal cameras, and Leica LIDAR sensor. This list is only indicative not exhaustive. In practical agriculture, farmers really have very wide choice of UAVs, set of sensors and computer programs to process aerial imagery, acquire digital data and utilize them on variable rate vehicles.

It is rightly said that once we have opted for apt UAVS model and set of sensors, it is the computer programs that decide the flight path that influence the accuracy of data collection. Next, computer programs that help us in enhancing and sharpening the aerial imagery are important. They provide the farmers with accurate data. Salient features of flight path related computer programs such as "ATLAS flight plan," "Pixhawk," and "*E-Motion*" are listed. It is followed by features about image processing software such as PIX4D Mapper, Agisoft's Photoscan, and Trimble's Inphos.

5.1 INTRODUCTION

Sensors on unmanned aerial vehicles (UAVs) that scout the agricultural farms, surveillance farm machinery, and monitor agronomic procedures may, in fact, be the most important driving force. In future, sensors could be revolutionizing food generating systems all over the world. The UAV-aided farming, of course depends to a great extent, on the sensors that UAVs carry. Salami et al. (2014) state that miniaturization of sensors, electronic gadgets, and computer has allowed us to use UAV-aided technology more frequently and easily. In the near future, the images (visual range), spectral and physicochemical data gathered by sensors should be able to play a vital role. In the farm land, precision agricultural procedures could become easier to conduct using data collected by sensors placed on different kinds of UAVs.

It is rightly said that learning how to operate an UAV above crops and to collect data may be easier. We may use preprogrammed flight paths or manually control UAVs. Spectral data from a 50 ha plot could be obtained, in say, 30 min of an UAV flight. However, creating ortho-images of the crop field and assessing variations in normalized difference vegetation index (NDVI), nitrogen (chlorophyll), or water stress index (thermal imagery) needs good computer skills. The data capture, their storing, retrieval, and analysis using specific programs need skilled computer-savvy technicians. There are now several agricultural UAV-related companies that offer to manage spectral data derived from UAVs. They offer quick suggestions to farmers. Such companies thoroughly analyze spectral data, using different software (Miller et al., 2017). Right now, sensor technology engulfs about 40% of the total UAV market. The UAV market itself is forecasted to increase to US$ 3.69 billion by 2022. In an agricultural UAV, sensors, GPS, and inertial navigation system (INS) form the major component of its payload. Here, in this chapter, we are concerned more with the "sensors" and their utility in the UAV technology, as applied to agriculture. Sensors could be broadly classified into two main types. They are:

(1) Passive sensors: These passive sensors need an external source of energy, such as sunlight, to observe the crop fields. The sensors capture spectral and spatial data of an object on the ground (e.g., crops, weeds, soil type, etc.). These sensors are useful in the electromagnetic spectral range of 300–3000 nm. Most visual sensors operate

in the bandwidths: red (650 nm), green (510 nm), and blue (450 nm). These are the spectral ranges most frequently used to assess crop growth. Also, to assess influence of drought or diseases on the crop maturity stages. Near infrared (NIR) and infrared (IR) bands are used to assess thermal properties of crops. They are used to get values for water stress index and judge drought effects on crops.

(2) Active sensors: Active sensors depend on their own source of energy to locate the target and obtain images. Light detection and ranging (LiDAR) is an example of active sensor. Often, LiDAR is integrated with a light source, such as IR light. The reflected energy (light) is measured at the sensors. For example, LiDAR sensor is an active optical sensor that impinges light beam on the target, that is, terrain/crops (Khot, 2016; Howard, 2015).

SENSORS ON UAVS

There are indeed innumerable companies that deal with optical sensors (cameras). Each brand has certain highlights, advantages, and disadvantages to consider prior to their adoption on the UAVs. Here, we are considering UAVs utilized to photograph and assess crop growth. Such UAVs used above croplands often possess a set of sensors such as: visual (red, green, and blue bandwidth), NIR, and IR. We may note that, globally, certain brands and models of sensors (cameras) are used more frequently. A few could be custom made for a specific purpose. A few could be preferred because they are low cost, easily available, and replaceable if they get damaged during UAV's landing. Some sensor brands could be preferred because they are sturdy, last for a longer duration, and need no replacements. Sensors are usually protected inside a gimbal. A few sensors could be preferred because they are compatible with certain models of UAVs and not all (Cardinal, 2017). A recent report by Marinov (2017) aims at highlighting the relatively frequently adopted models of sensors. Considering a series of specifications and utility, three top multispectral sensors have been picked. They are: MicaSense RedEdge-M, Parrot Sequoia, and Sentera Double 4K.

As stated earlier, UAV technology revolves around the sensors that gather useful data about terrain, crops, and the surroundings in a farm. The choice of cameras and the wavelength band are crucial. In the general course, an agricultural UAV carries at least five different cameras (sensors) to gather data. They are visual, NIR, IR (thermal), red edge, and LiDAR. No doubt, the sensor technology we discuss here is not new. The cameras just need to be suitably adapted to perform while the UAV (aircraft or blimp or parachute) is on the move above crop fields. These sensors are based on electro-optical principles. There are also UAVs fitted with physicochemical sensors (electrodes). They collect data about atmospheric pollution, gaseous emissions, and contaminations, if any (Scentroid Inc., 2017).

The set of sensors used most commonly operate at visual range bandwidths. Sensors with multiple bandwidth are also used. They are known as multispectral cameras. The most frequently obtained data using such sensors on UAVs is NDVI. It needs a visual and NIR bandwidth imagery. It means generally two flights are required. However, a multispectral camera with four to five bandwidths can overcome repeated flights and aerial photography. The multispectral camera obtains data at different wavelengths in one go. A multispectral camera may have even more than four lenses. A recent report by a company, Light Inc. (Palo Alto, California, USA), suggests that there is a sensor model with up to 16 lenses. They operate simultaneously and collect useful data. Of course, the UAV operator has option to select any of the few apt lenses, to collect images of crops (Light Inc., 2017). In some cases, the required bandwidth of sensors is obtained by adding or removing optical filters to the commonly utilized R, G, and B cameras. A few examples of cameras are Sony DSC VI, Canon Powershot SX100, FinePix 10 *FD*, Canon PowerShot A 480, etc. (Salami et al., 2014). According to Corrigan (2017b), multispectral sensors on UAVs are already popular with agriculturalists. Farmers could reap benefits in terms of finding out about crop growth, soil surface conditions, fertilizer requirements, and water status.

The UAVs carry yet another type of cameras called "hyperspectral cameras." These are sensors that operate at wider wavelength and at a finer resolution. We can obtain useful data about soil, crop, crop's vigor, pest/disease pressure, and weed infestation. They really offer quite an amount of aerial data from which farmers have to pick. There are now small-sized, light-weight, low-cost hyperspectral sensors available for use on UAVs. They employ several (hundreds) of sequential bandwidths and offer a wide spectral range. There are also microhyperspectral cameras (e.g., Micro-Hyperspec VNIR). It operates at 260 bandwidths within spectral range of 400–1000 nm (Salami et al., 2014; Adao et al., 2017; Bareth et al., 2015; Burkart et al., 2014; Kalisperakis et al., 2015).

Thermal cameras operate at IR range bandwidth in the electromagnetic spectrum. They collect data about variations in heat (temperature) within the canopy and ambient air. They can also detect nighttime movement of insect pests and animals in the farm (Miller et al., 2017). Many brands of IR and NIR range bandwidth sensors are available. They are often used to assess temperature of the crop's canopy, soil, and ambient air. These parameters obtained by thermal sensors are used to calculate the crop's water status, influence of drought, and then to prescribe irrigation appropriately. They are used often to prepare water status maps of crops in fields (Hoffman et al., 2016; Gago et al., 2015). A good example is MicaSense IR and red edge sensors (Plate 5.3.1). A few other examples are Gobi 384 (Gago et al., 2016) and FLIR B20 (Erena et al., 2012; Plate 5.10). Thermal cameras are used in conjunction with visual and LiDAR to assess water status of large areas of crops or natural vegetation (DeBell et al., 2016; Dunn, 2016).

LiDAR is a laser-based sensor. LiDAR is an active sensor by classification. It emits light and captures its reflection. The time elapsed from the moment pulse left the system and to gather its reflection is used to calculate the distance between the crop and sensor. The distance is usually calculated accurately using GPS and INS. LiDAR can be used to prepare accurate maps of terrain and crops. It can even measure crop height and depict it. LiDAR sensors offer good resolution and elevation of terrain and objects on the farm. It is often used in conjunction with three-dimensional (3D) imagery of crop fields. LiDAR sensors have found application in measuring crop growth. Also, it is used to measure tree height and biomass of forest stands (Salami et al., 2014; Miller et al., 2017; Corrigan, 2017a).

GreenSeeker is a sensor carried most commonly by ground vehicles such as tractors, etc. A GreenSeeker emits light at NIR and red wavelength bands. It offers data about NDVI. A GreenSeeker could be adapted and fitted on UAVs for photography.

The set of sensors (R, G, B, NIR, IR, and LiDAR) discussed above are most commonly carried by agricultural UAVs. They are light based and operate at different wavelength bands of the electromagnetic spectrum. They offer detailed images of crops and wide range of spectral data pertaining to crop growth. We should be alert to the fact that recent trends in crop production involve high-input farming systems. Fertilizer-based nutrients, particularly, nitrogen is added in different forms at different intervals to the crop. Soil accumulates N in different forms. It also emits a sizeable fraction of N into atmosphere. Agricultural fields do emit greenhouse gases such as NH_3, NO, and NO_2. Repeated plowing and burning of crop residue results in loss of soil C as CO_2, CO, etc. A few crops emit CH_4 during the growing season (e.g., rice), in quantities, that can be construed as unwanted C-emissions. Also, there is a general need to assess atmospheric quality over crop production zones, particularly, if the fields are close to urban or industrial zones. We know that, at present, farming companies and farmers are always inquisitive about atmospheric conditions, emissions, and soil and air quality. Farmers depend on data about their crops, soil, and atmosphere. The agronomic procedures are mostly based on data and computer-aided decision support systems. Hence, UAVs with a range of sensors too may become common. Sensors (electrodes) that measure the physicochemical parameters such as atmospheric moisture (relative humidity), N-emissions (NH_3, NO, and NO_2), C-emissions (CO, CO_2, CH_4, and ethylene), and sulfur dioxide emissions could be carried on board by the UAVs. There are examples of UAV models that carry electrodes and conduct rapid electrochemical determination. They relay the data to the ground station computers. The ground control station (GCS) may map the fluctuations in emissions. The GCS records their concentrations and provides a map depicting it to farmers much rapidly. Balloons, blimps, and parachutes are often used by meteorological stations to conduct analysis of atmospheric gases and determine air quality. The UAVs could be rapid in data collection. They can pick samples and conduct electrochemical analysis of ambient air above the crop canopy.

Next class of sensors carried by UAVs on them are the microbiological sensors. They are not the typical sensors, like optical or physicochemical sensors discussed above. They are at best, containers or traps, to collect microbial propagules (spores, hyphae, microbial cells, dust particles with microbes on them, etc.). The UAVs carry petri plates or suitable traps (sterile felt pad) with nutrient medium to collect microbes. Farmers could be alerted about impending disease by noting the microbial diversity and density above the crop canopy. For example, "Scentroid UAV" attached with petri plates containing nutrient medium agar collects plant pathogens (Scentroid Inc., 2017). Microbiological sensors are a different kind of sensors on UAVs indeed (Aylor and Fernandino, 2008; Aylor et al., 2011; West et al., 2010). Payload for microbiological analysis of atmosphere above crops usually contains petri plates that can be opened and closed, like a small flat box, with a hinge or a shell. The petri plates are opened to trap microbial propagules. In the laboratory, microbiologists use several different common media to identify and enumerate the microbial diversity. Often, this technique is adopted to assess density of plant pathogen's propagules in the atmosphere. Such data provide an early warning about pathogen's buildup in the atmosphere (Salami et al., 2014; Brown and Hovmoller, 2002; West et al., 2010).

5.2 SENSORS ADOPTED ON AGRICULTURAL UAVS

As stated above, there are several companies that deal with sensors that agricultural UAVs often carry, to conduct spectral imagery. Most often, it is the optical sensor's brands that are of concern to UAV technologists. These optical sensors are located on the UAV in a gimbal. Gimbal helps to withstand vibrations, so that aerial images are not hazy. Gimbals also help in providing correct angles for the sensors. Khot (2016) states, that UAV technology could be adopted rampantly across agrarian regions in different continents. Since, sensors are the vital aspects of UAV technology, we need to pick and use most appropriate ones, at accurate bandwidths. Sensors may have to be improved and adopted to suit a crop species or location. Overall:

a) sensor should be small or even minute, to suit the payload size of the small UAV;
b) sensor should be light in weight;
c) sensor should provide a large view of the crop field and its details;
d) sensors should be integrated well with shutter triggering mechanisms;
e) sensors should allow sufficient overlap of images. This helps in image processing and stitching;
f) sensors should be easy to calibrate;
g) sensors should be integrated sufficiently to relay images/spectral data, to the ground station computers;
h) sensors should consume low amount of power to operate; and
i) sensors should be rugged enough to withstand the rough weather, landing and usage.

The following is a list of sensor brands, companies that produce them, and a few salient features:

5.2.1 VISUAL (RED, GREEN, AND BLUE) AND MULTISPECTRAL SENSORS

Name: Canon Sensors

Company and address: Several locations in USA and in other nations.

E-mail: sensor_info@cusa.canon.com; Phone: 1-800-652-2666; website: https://www.usa.canon.com/internet/portal/us/home/explore/product-showcases/sensors/

Canon Inc., USA is a 60 years old company specializing in production of cameras and sensors (Plate 5.1). Their products are used in wide range of applications, including in UAVs (SenseFly Inc., 2015a, 2015b). These digital sensors offer spectral data pertaining to agricultural terrain, crops, and their growth status. The thermal cameras provide images depicting water status of crops.

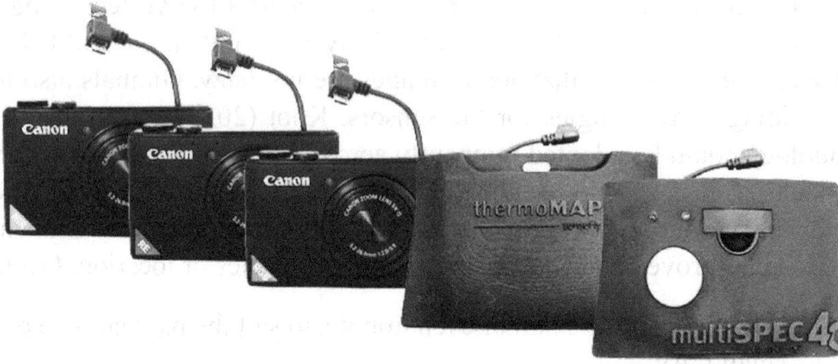

PLATE 5.1 A series of Canon sensors that could be fitted to UAV to obtain digitized maps. *Source*: Dr. Mathew Wade, SenseFly—A Parrot Company, Cheseaux-sur-Lausanne, Switzerland; SenseFly, 2015a, 2015b.
Note: From left to right, the cameras used are *SR110* for NIR; *S100 RE* for red edge bandwidth photography; *S110RGB* for photography at visible bandwidths red, green, and blue; *multiSPEC 4c* for multispectral imagery; and *thermoMAP* is used to record crop canopy temperature and moisture stress, if any.

References, Websites, and YouTube Addresses

SenseFly Inc. *Drones for Agriculture*; 2015a; pp 1–9. http://www.sensefly.com/applications/agriculture.html (accessed Aug 16, 2015).

SenseFly Inc. *eBee by SenseFly*; 2015b; pp 1–4. http://www.sensefly.com (*accessed* Sept 7, 2015).

https://www.sensefly.com/drones/accessories.html (*accessed* Aug 25, 2017).

https://www.sensefly.com/applications/agriculture.html (*accessed* Aug 25, 2017).

http://www.aerialmediapros.com/store/thermal-imaging (*accessed* Aug 25, 2017).

https://www.youtube.com/watch?v=ic3fDd2ykP4 (*accessed* Sept 3, 2017).

https://www.youtube.com/watch?v=_PIQtvP8uWY (*accessed* Sept 3, 2017).

https://www.youtube.com/watch?v=NQAHrlD4OGg (*accessed* Sept 3, 2017).

https://www.youtube.com/watch?v=iuQvZu7PGN4 (*accessed* Sept 3, 2017).

Name: MAPIR Sensors

Company and address: 7592 Metropolitan Drive, Suite 401, San Diego, CA, USA; Phone: (877) 949-1684; e-mail: info@peauproductions.com, cody@peauproductions.com, ethan@peauproductions.com; website: https://www.mapir.camera/pages/examples

Peau Productions Inc. has developed a 12-megapixel UAV camera. It captures images every 3 s. Images taken at 400 ft altitude produce an accuracy of 6.83 cm/pixel. The MAPIR camera is about the size and weight of a "GoPro Hero" camera. It weighs 65.3 g. Peau Productions has also developed a series of six cameras with different specifications. A different lens type distinguishes each camera. Camera one (Plate 5.2.1; top extreme left) has a standard lens that operates at visible RGB light.

Camera two works at blue and IR bandwidth light. Cameras three through six are IR, red, green, and blue light. We may note that multiple devices can be mounted on an UAV (Plate 5.2.2). The captured images can be used for crop scouting, assessing plant health (NDVI), identifying different areas of vegetation, surveying, detecting chlorophyll content, and creating photomosaic maps.

PLATE 5.2.1 A set of six MAPIR sensors.
Source: Dr. Nolan Ramseyer, CEO, Peau Productions, San Diego, USA.
Note: The above sensors operate at visual bandwidth range, blue, and NIR to obtain NDVI values, and at IR range to get thermal imagery. The sensors with individual bandwidths for red, green, and blue are also available (three sensors on bottom row).

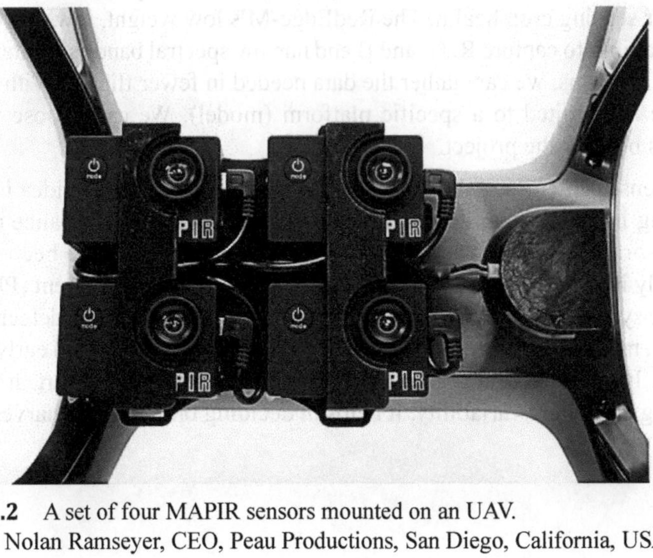

PLATE 5.2.2 A set of four MAPIR sensors mounted on an UAV.
Source: Dr. Nolan Ramseyer, CEO, Peau Productions, San Diego, California, USA.

References, Websites, and YouTube Address

http://www.mapir.camera/ *(accessed June 20, 2017).*
https://www.youtube.com/watch?v=PJVLXAK_sRQ *(accessed June 20, 2017).*
https://www.youtube.com/watch?v=s5J_CJkw2a8 *(accessed June 20, 2017).*
https://www.youtube.com/watch?v=vAjrDWKq4Lw *(accessed June 20, 2017).*
https://www.youtube.com/watch?v=Uny9nTr22go *(accessed June 20, 2017).*

Name: MicaSense RedEdge Camera

Company and address: MicaSense Inc., 1300 N Northlake Way, Suite 100, Seattle, Washington, WA 98103, USA; website: https://www.micasense.com/contact/

MicaSense is a Seattle-based start-up company specializing in sensors. It has developed the RedEdge multispectral camera for UAV operations. The Parrot Inc. (France) has backed MicaSense Inc. by investing in the company. This relationship enabled Parrot Inc. to use RedEdge technology to build Sequoia sensor. It is a more affordable and smaller camera. Both, the RedEdge and Sequoia cameras which are fitted to UAVs, now use "MicaSense's ATLAS" software to analyze and measure plant reflectance. RedEdge is priced at $5900 and has been shipping for about a year. Both Parrot Inc. and MicaSense Inc. intend to license these cameras to other UAV companies (MicaSense Inc., 2017).

RedEdge-M is a reliable, accurate, high-precision sensor. Usually, the MicaSense RedEdge-M sensor is action ready. It integrates with small UAVs. It is a compact camera. Since, MicaSense RedEdge is a light-weight sensor, it integrates even with small fixed-winged or copter UAVs. The sensor is sold with manual and kits to aid easy integration with wide range of UAV models (Plate 5.3.1; Fig. 5.1).

RedEdge-M has five bands—blue, green, red, red edge, and NIR. These are optimal for sensing crop health. The RedEdge-M's low weight, low power requirement, and ability to capture R, G, and B and narrow spectral bands simultaneously, is important. It means, we can gather the data needed in fewer flights. With RedEdge-M, we are not limited to a specific platform (model). We can choose whichever platform is best for the project.

MicaSense cameras are designed for precision agriculture. The idea is to collect overlapping images. Then, merge them together to create a reflectance map, from which all sorts of analysis can be done. MicaSense products have been used most successfully in the detection of spatial variation and farm management (Plate 5.3.2). MicaSense system can help with a wide variety of issues. It includes detection of pest infestation, monitoring crop health, monitoring invasive species, and early detection of disease. It provides data for fertilizer application decision support. It also helps in detecting crop yield variability. It helps in deciding optimal crop harvesting date.

SPECIFICATIONS OF MICASENSE REDEDGE CAMERA

Weight: 173 g;

Dimensions: 9.4 cm × 6.3 cm × 4.6 cm;

Power source: 4.2–15.6 V DC;

Spectral bands: blue, green, red, red edge, and NIR (global shutter, narrow band);

Ground sample distance: 8 cm per pixel at 120 m above the ground;

Capture rate: 1 image per s;

Interfaces: it interfaces with ethernet, removable Wi-Fi, Global Positioning System (GPS), and external trigger;

Field of vision: 47.2;

It has timer mode, overlap mode, serial ethernet options, and manual capture mode.

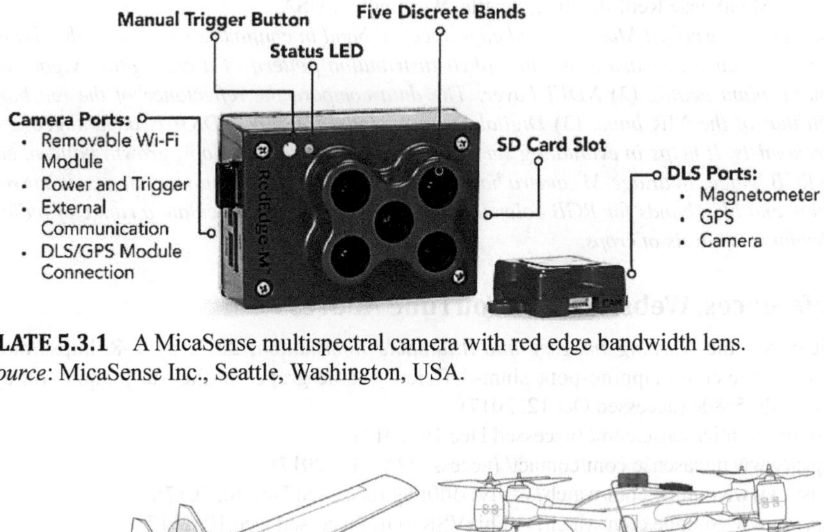

PLATE 5.3.1 A MicaSense multispectral camera with red edge bandwidth lens.
Source: MicaSense Inc., Seattle, Washington, USA.

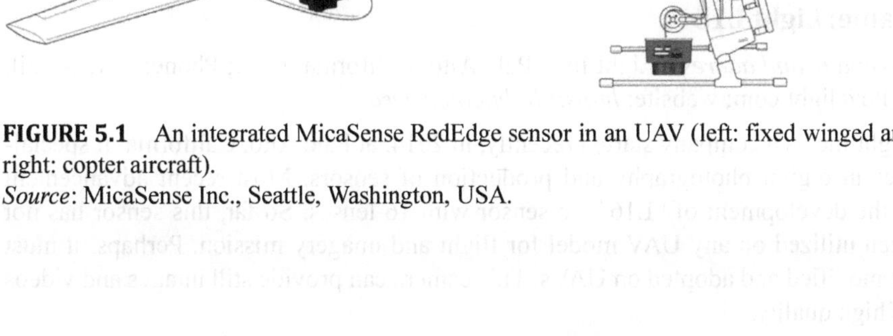

FIGURE 5.1 An integrated MicaSense RedEdge sensor in an UAV (left: fixed winged and right: copter aircraft).
Source: MicaSense Inc., Seattle, Washington, USA.

PLATE 5.3.2 A set of four different spectral maps offered by MicaSense multispectral sensors.

Source: MicaSense RedEdge Inc., Seattle, Washington, USA.

Note: (1) *Chlorophyll Map: The red edge spectral band in conjunction with the other bands provides accurate measure of chlorophyll distribution pattern of a crop, plant vigor, and general plant health.* (2) *NDVI Layer: The data compare the reflectance of the red band with that of the NIR band.* (3) *Digital Surface Model (DSM): A DSM is advantageous to agronomists. It helps in evaluating surface properties of crop's canopy, growth pattern, etc.* (4) *RGB Image: RedEdge-M camera has global shutters for distortion-free images. It has red, green, and blue bands for RGB color images. They allow us to calculate a range of indices relevant to analysis of crops.*

References, Websites, and YouTube Addresses

MicaSense Inc. *Turning Imagery into Attainable Information*; 2017; pp 1–8. https://blog.micasense.com/mapping-potassium-deficiency-table-grapes-in-the-san-joaquin-valley-434a4b85a80e (accessed Oct 12, 2017).

http://www.micasense.com/ (accessed Dec 10, 2017).

https://www.micasense.com/contact/ (accessed Dec 10, 2017).

https://www.youtube.com/watch?v=vIvOJtt0P9g (accessed Dec 10, 2017).

https://www.youtube.com/watch?v=FhixVSKfrDE (accessed Dec 10, 2017).

https://www.youtube.com/watch?v=yLdbMRD9wyc (accessed Dec 10, 2017).

Name: Light L16

Company and address: Light Inc., Palo Alto, California, USA; Phone: NA; e-mail: hello@light.com; website: *https://light.co/camera*

Light Inc. is a company started recently, in 2014, at Palo Alto, California. It specializes in digital photography and production of sensors. Most recent advancement is the development of "L16"—a sensor with 16 lenses. So far, this sensor has not been utilized on any UAV model for flight and imagery mission. Perhaps, it must be modified and adopted on UAVs. This camera can provide still images and videos of high quality.

The L16 is a point and shoot-sized camera. It houses 16 different lenses. It is not available in the market yet. The company expects shipments to begin again in 2017. "Light L16" is the world's first multiaperture computational camera. Firmware on the camera merges images from the lenses to form a 52-megapixel image. Depth of field, focus, and optical zoom (35–150 mm) can be adjusted, after the photo is taken. The L16 is not currently being promoted as a drone (UAV) camera. However, considering the image resolution and optical zoom capabilities, it is hard to believe that this device will not be used as an aerial mapping and surveying tool (Light Inc., 2017).

SPECIFICATIONS

Dimensions: H × W × D, 6.5 in. × 3.3 in. × 0.94 in. (or 165 mm × 84.5 mm × 24.05 mm); Weight: 15.3 oz (435 g); Chassis: Die-cast aluminum alloy; Grip: rubberized nonslip grip with lanyard connection; Screen: 5" FHD touchscreen

Tripod mount: Standard 1/4"–20 tpi; USB: USB 3.0 Super Speed, Type-C connector

Lenses: Sixteen individual modules of 5 mm × 28 mm *f*/2.0, 5 mm × 70 mm *f*/2.0, and 6 mm × 150 mm *f*/2.4 focal lengths, at full frame equivalent; Sensor: 16 individual 13 MP sensors; Zoom: 5 × Optical zoom

ISO sensitivity: 100–3200; Focus: minimum is 10 cm at 28 mm, 40 cm at 70 mm, 1 m at 150 mm, tap to focus, half-shutter press to focus; Shutter speed: 1/8000 to 15 s; Image formats captured: JPEG; Effective pixels (megapixels): up to 52 million+ (52+ MP); Video (coming soon): 4K video; 28, 70, or 150 mm focal lengths at full frame equivalent

Operating system: Android; Connectivity: GPS, Wi-Fi, and Bluetooth; Security: full disk encryption; Chipset: Qualcomm Snapdragon 820 + Light ASIC; Battery: 4120 mAh lithium-ion polymer (up to 8 h of life).

References, Websites, and YouTube Addresses

Light Inc. Light L16: A Multispectral Camera; Light Inc.: Palo Alto, CA, USA, 2017; pp 1–6. https://light.co/camera (accessed Dec 26, 2017).
https://light.co/ (accessed Dec 27, 2017).
https://light.co/technology (accessed Dec 27, 2017).
https://light.co/camera (accessed Dec 27, 2017).
https://www.youtube.com/watch?v=cW3rx7jiX8E (accessed Jan 12, 2018).
https://www.youtube.com/watch?v=F6VHol7aJxk (accessed Jan 12, 2018).
https://www.youtube.com/watch?v=bUXoj-0ba8k (accessed Jan 12, 2018).
https://www.youtube.com/watch?v=L9CIO-eOuck (accessed Jan 12, 2018).
https://www.youtube.com/watch?v=8n1wTJ9UjnQ (accessed Jan 12, 2018).
https://www.youtube.com/watch?v=RpXNyWPfs8E (accessed Jan 12, 2018).

Name: Parrot's Sequoia

Company and address: Parrot Inc., Paris, France; website: http://www.parrot.com/usa

The Parrot Inc. located in Paris, France was founded by Mr. Henri Seydoux, in the year 1994. It has worldwide collaborations and outlets. The company has made a few acquisitions to become a key player in the aspect of sensors. Sensor companies such as SenseFly, Airinov, MicaSense, and Pix4D are all owned by Parrot Inc.

The Parrot Sequoia multispectral sensor is a popular sensor among UAV technologists (Plate 5.4). The company is offering a comprehensive and adaptable solution. This sensor is compatible with all types of UAV models. With its two sensors, multispectral and sunshine, Sequoia analyzes plants' vigor by capturing the amount of light they absorb and reflect. The sensors offer NDVI data about crop, chlorophyll content, and water status. Such data means that farmers can do what is best for their crop fields. The Sequoia camera has two types of sensors. The multispectral sensor has visible bandwidth sensors that operate at green and blue range. Plus, it has two sensors that operate two at IR bands, red edge, and IR. Sequoia offers advanced aerial imagery facility to farm scientists and farmers.

SPECIFICATIONS OF PARROT'S SEQUOIA MULTISPECTRAL SENSOR

Four bands (green, red, red edge, and NIR);

RGB color outputs: Yes (rolling shutter);

RGB color outputs aligned across all bands: No;

Resolution GSD @ 100 m (400 ft): 13 cm per pixel;

Light sensor: Yes;

Calibration: Yes (for accurate and repeatable data);

Bit depth (discrete bands): 10 Bit (1024 values);

Weight: 135 g (with Sunshine Sensor and cable);

Power usage: 8 W nominal, 12 W peak;

IMU: Yes;

USB: No;

External trigger: No;

Ethernet: No;

Custom bands: No;

PPK–RTK capability: No;

Image/photos: 16 MPIX RGB camera;

Definition: 4608 × 3456 pixels;

Bandwidths: Green (550 BP 40); red (660 BP 40), red edge (735 BP 10), and NIR (790 BP 40)

PLATE 5.4 Parrot's Sequoia multispectral sensors.
Source: Parrot Inc., Paris, France.

References, Websites, and YouTube Addresses

https://www.riseabove.com.au/industrial-agriculture-drones/cameras-sensors-modules/ (accessed Dec 10, 2017).

https://www.riseabove.com.au/parrot-sequoia-multispectral-camera-for-ndvi-crop (*accessed Dec* 10, 2017).

https://www.youtube.com/watch?v=vY1eFp3Z5XU (accessed July 22, 2017).

https://www.youtube.com/watch?v=SaztuWuDEsg (accessed July 22, 2017).

https://www.youtube.com/watch?v=YJOl8VFC6so (accessed July 22, 2017).

https://www.youtube.com/watch?v=QnVywZirkLU (accessed July 22, 2017).

https://www.youtube.com/watch?v=_zuS7KrdZlI (accessed July 22, 2017).

https://www.youtube.com/watch?v=f9gC-TQjopg (accessed July 22, 2017).

https://www.youtube.com/watch?v=pBrhFsyxoO0 (accessed July 22, 2017).

https://www.youtube.com/watch?v=cIKJLxT9dvQ (accessed July 22, 2017).

https://www.youtube.com/watch?v=SaztuWuDEsg (accessed Dec 20, 2017).

https://www.youtube.com/watch?v=J-4yoGnNOoo (accessed Sept 10, 2017).

Name: Sentera Sensors

Company and address: Sentera LLC., 6636 Cedar Avenue S., Minneapolis, Minnesota, MN 55423, USA; Phone: 844 736 8372; e-mail: info@setara.com; website: https://sentara.com

Sentera LLC. offers wide range of products related to UAVS, sensors, and image processing software. They offer multispectral and IR sensors for use on agricultural UAVs. Sentera's sensors offer consistently good-quality images. These sensors produce high-quality, context-rich color and NIR images. They also provide NDVI data to farm companies and growers. Sentera's sensors are usually sold to customers along with "AgVault" software license. This helps in rapid, accurate storage and retrieval, plus processing of data as needed. They also provide data sheet and information about necessary calculations, if any.

Following are the sensors produced by Sentera LLC. that are relevant to adoption of UAV technology, in agrarian regions:

Name: Sentera's Double 4K dual sensor

The Sentera Double 4K is a small, fully customizable twin-imager sensor (Plate 5.5.1). It is universally compatible with any UAV. Fitting in the footprint of a "GoPro® HERO 4," the rugged, high-throughput Double 4K Sensor is designed for use in harsh environments. The configuration makes it ideal for use in agriculture and infrastructure inspection applications. This camera is capable of capturing high-megapixel color stills, NIR images, NDVI data, and 4K video.

The multispectral Double 4K sensor variant offers five spectral bands: blue, green, red, red edge, and NIR. Sentera's Multispectral Double 4K sensor helps agronomists, crop consultants, and growers, to identify issues with their crops earlier and with greater precision. The intelligence provided by this sensor makes it ideal for universities, researchers, large growers, and crop advisors.

PLATE 5.5.1 Sentera's Double 4K sensors.
Source: Ms. Sarah Ritzen, Sentera LLC., Minneapolis, Minnesota, USA.
Note: Sentera 4K is a twin imager.

Name: Sentera's Quad Sensor

The Sentera Quad Sensor is the latest among the cameras used on UAV models. Particularly, UAVs engaged in precision agriculture (Plate 5.5.2). With four fully customizable multispectral imagers, the innovative sensor offers standard NDVI, green NDVI, normalized difference red edge (NDRE), and high-resolution color capture—all in a single flight. The Quad sensor can measure key indicators of chlorophyll in crops. The sensor package includes Sentera Quad data sheet.

PLATE 5.5.2 The Sentera Quad sensor.
Source: Ms. Sarah Ritzen, Sentera LLC., Minneapolis, Minnesota, USA.

Name: Sentera's High Precision NDVI and NDRE (red-edge) Single Sensor

This is an elegantly engineered, high-precision single sensor (Plate 5.5.3). It is designed specifically to monitor crop health through NIR, NDVI, and NDRE (red edge) data collection. The High-Precision NDVI Single sensor and High-Precision NDRE Single sensor utilize advanced technology. They improve spectral band separation. They generate more accurate vegetation index measurements. The High-Precision Single sensor is available in two variants: NDVI and NDRE.

The "single sensor" is unmatched in its performance due to its global shutter, tiny size, and lightweight. It is a good tool for those who want to collect highest quality NDVI or NDRE data. The High-Precision Single sensors integrate onto any UAV model. It helps agronomists, consultants, and producers in identifying crop growth issues, earlier and with greater precision. The sensors are available preinstalled on DJI Phantom 4, Mavic, and Inspire 1 equipment.

PLATE 5.5.3 The Sentera High Precision NDVI and NDRE Single sensors.
Source: Ms. Sarah Ritzen, Sentera LLC., Minneapolis, Minnesota, USA.

Name: The Sentera Incident Light Sensor

For several years, Sentera's research customers have been using Sentera's Incident Light Sensor (ILS) (Plate 5.5.4). It is useful to measure the color spectrum of incident light from the sun. Now, these capabilities have been seamlessly integrated into Sentera's sensors, making it easy for agronomists, crop consultants, and growers, to accurately compare images of the same area, captured over time.

PLATE 5.5.4 The Sentera Incident Light Sensor.
Source: Ms. Sarah Ritzen, Sentera LLC., Minneapolis, Minnesota, USA.

Name: Sentera Q

The Sentera Q sensor provides highly accurate aerial imagery and data in a single pass. Therefore, it is ideal for surveying, mapping, surveillance, and reconnaissance operations. It is one of the most advanced cameras placed on UAVs (Plate

5.5.5). The Sentera Q captures highly accurate, 18 MP, and 10× optical zoom color imagery. The high-resolution images can be transformed into precise 3D models. It assists public safety officials with preplanning activities. Geospatial analysts use it for analyzing surface data.

PLATE 5.5.5 Sentera Q.
Source: Ms. Sarah Ritzen, Sentera LLC., Minneapolis, Minnesota, USA.

References, Websites, and YouTube Addresses

https://sentera.com/sensors/ (accessed Oct 14, 2017).
https://www.youtube.com/watch?v=4b4fsezjHnQ (accessed Jan 14, 2018).
https://www.youtube.com/watch?v=0wkCbz1zu3o (accessed Jan 14, 2018).
https://www.youtube.com/watch?v=UeZUBa1vaOY (accessed Jan 14, 2018).
https://www.youtube.com/watch?v=LjJdtIAXLIE (accessed Jan 14, 2018).

Name: Tetracam Multispectral Sensor

Company and address: Tetracam, Inc., Devonshire Street, Chatsworth, California, USA; Phone: 818-288-4489; Fax: 818-718-7103; e-mail: info@tetracm.com; website: www.tetracam.com

Tetracam Inc. is a sensor producing company headquartered in California, USA. Tetracam's units are also located in Hong Kong and Shenzhen, China. Tetracam Inc. has manufactured over 100 different models of sensors useful to agriculturists, foresters, and public. Tetracam offers multispectral sensors, NIR and IR (thermal) sensors (Plates 5.6.1–5.6.3). Tetracam Inc. has sold over million units of multispectral and special types of sensors. They are useful to farmers and farming companies adopting modern methods. Tetracam's cameras are among most frequently fitted (chosen) on UAVs, used in agriculture. Tetracam systems monitor subtle changes in the visible and NIR radiation that crops reflect. Growers use this data to spot plants under stress, monitor plant growth, or to perform dozens of other functions (Wehrhan et al., 2016; Plates 5.6.1 and 5.6.2). Tetracam sensors have been used to estimate chlorophyll content (i.e., crop's N status) (Caturegli et al., 2016), also potassium status of crops (Severston et al., 2016), and to study weed infestation in fields (Penia et al., 2015).

PLATE 5.6.1 Tetracam's multispectral sensor.
Source: Mr. John Edling, Tetracam Inc., Chatsworth, California, USA.

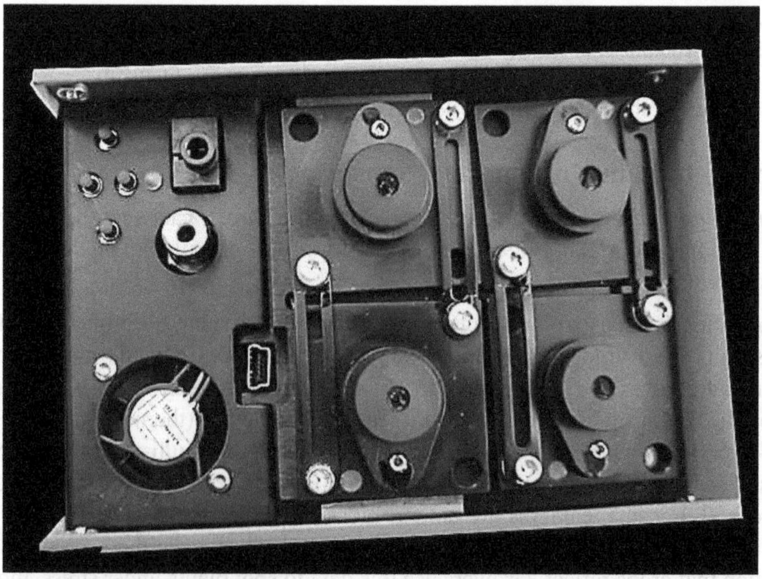

PLATE 5.6.2 Tetracam RGB +3 sensor.
Source: Mr. John Edling, Tetracam Inc., Chatsworth, California, USA.

PLATE 5.6.3 Tetracam's Thermal IR sensor.
Source: Tetracam Inc., Chatsworth, California, USA.

References, Websites, and YouTube Addresses

Caturegli, L.; Comeglia, M.; Gaetani, M.; Grossi, N.; Magni, M.; Miglizzi, M.; Angelini, L.; Mazzoncini, M.; Silvestri, N.; Fontanelli, M.; Raffaelli, M.; Peruzzi, A.; Volterrani, M. Unmanned Aerial Vehicle to Estimate Nitrogen Status of Turf Grass. *PLoS One* **2016,** 1–9. https://doi.org/10.1371/journal.pone0158268 (*accessed* July 27, 2017)

Penia, J. M.; Torres-Sanchez, J.; Serrano-Perez, A.; de Castro, A. I.; Lopez-Granados, F. Quantifying Efficacy and Limits of Unmanned (UAV) Technology for Weed Seedling Detection as Affected by Sensor Resolution. *Sensors* **2015,** *15*, 5609–5626. DOI: 10.3390/s150305609 (*accessed* Sept 23, 2017).

Severtson, D.; Callow, N.; Ken Flower, K.; Neuhaus, A.; Olejnik, M.; Nansen, C. Unmanned Aerial Vehicle Canopy Reflectance Data Detects Potassium Deficiency and Green Peach Aphid in Canola. *Precis. Agric.* **2016,** *17*, 659–677. http://link.springer.com/article/10.1007/s11119-016-9442-0 (*accessed* Aug 24, 2017).

Wehrhan, M.; Rauneker, P.; Sommer, M. UAV-based Estimation of Carbon Exports from Heterogenous Soil Landscapes: A Case Study from CarboZALF Experimental Area. *Sensors* **2016,** *16*, 255. DOI: 10.3390/s16020255 (*accessed* July 27, 2017).

http://www.tetracam.com/ImagingSystems.htm (*accessed* July 24, 2017).

http://www.tetracam.com/Services.htm (*accessed* July 24, 2017).

https://www.youtube.com/watch?v=fVt_BFSu2SM (*accessed Dec* 30, 2017).

https://www.youtube.com/watch?v=B62Xf0E-vtc (*accessed Dec* 30, 2017).

5.2.2 HYPERSPECTRAL SENSORS

Name: AISA Hyperspectral Sensors for Galileo Group

Company and address: Galileo Group, Research triangle park, North Carolina, USA; website: http://galileo-gp.com/

The Galileo Group Inc. is a company initiated some two decades ago. It specializes in remote sensing appliances and services. They offer highly specialized hyperspectral imaging services and processing. They are known to supply hyperspectral imagers to major users such as National Aeronautics and Space Agency of the United States, United States Air Force, water management agencies, natural resource monitoring agencies, etc. They offer large-scale imaging with options, for automatic processing of images derived. Additional expertise includes spectral and spatial analysis, and 3D modeling. They develop software for specific purposes in agriculture and related aspects. Galileo Group Inc. uses state-of-the-art AISA hyperspectral sensors. The sources offer high spectral and spatial resolution data in the very near infrared (VNIR) (400–1000 nm) and/or shortwave infrared (SWIR) (1000–2500 nm) spectral range. The AISA sensors are radiometrically calibrated to a National Institute for Standards and Technology (NIST) traceable radiance source. The hyperspectral imaging system also consists of a high-precision IMU/GPS unit. In addition, it has a ruggedized flight computer equipped with professional imaging software. The rapidly collected raw data is easily radiometrically and geometrically calibrated for further analysis. The result is a cost-effective, end-to-end, airborne data collection capability. The camera acquires research grade spectral imaging data for use in military and commercial remote sensing programs.

The AISA Eagle is a high-performance hyperspectral imaging system operated by Galileo Group Inc. (Plate 5.7). The system has been used in a number of operational scenarios. It has achieved success in detecting and discriminating low-observable targets. The sensor operates across the VIS/NIR portion of the spectrum (400–1000 nm). It resolves spectral differences as fine as 2–5 nm. With around 1000 pixels per scanning line, the system can image at submeter resolution. The spectral range is 1000–2500 nm.

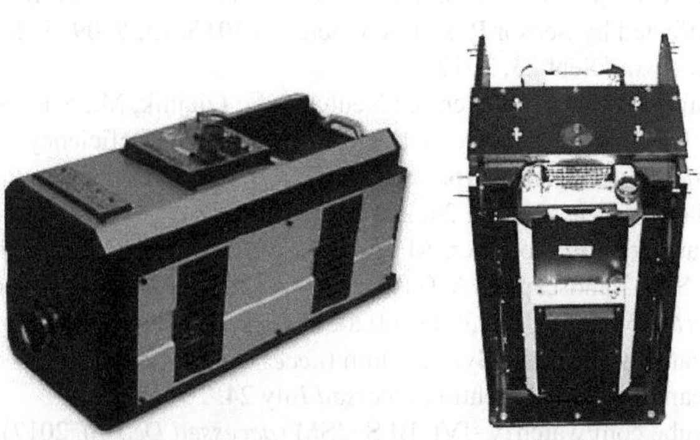

PLATE 5.7 Left: AISA Eagle hyperspectral sensor; right: AISA Hawk hyperspectral imaging sensor.

Source: Mr. Michael Barnes, Galileo Group, Research Triangle Park, North Carolina, USA.

References, Website, and YouTube Addresses

http://galileo-gp.com/applications/agriscience/ (accessed Apr 23, 2017).
http://galileo-gp.com/technologies/airborne-hyperspectral-imaging/ (accessed Apr 23, 2017).
http://galileo-gp.com/?gclid=EAIaIQobChMIvaS4qKOi2AIV0worCh0-tgSEEAMYAyAAEgJ_XPD_BwE (accessed Apr 23, 2017).
https://www.youtube.com/watch?v=xAlQDKj7MCU (accessed Dec 24, 2017).
https://www.youtube.com/watch?v=xwT1fPJJt9Q (accessed Dec 24, 2017).
https://www.youtube.com/watch?v=e6hbmSfPnMQ (accessed Dec 24, 2017).

Name: OCI UAV Airborne Hyperspectral Camera

Company and address: BaySpec Inc.; 1101, MacKay Drive, San Jose, California, CA 95131, USA. Phone +1(408)512-5928; Support@bayspec.com; sales@bayspec.com; website: http://www.bayspec.com/contact

BaySpec Inc. is a company that deals with spectral imagery, sensors, and spectrometers. It is located at San Jose in California, USA. They design and manufacture advanced spectral instruments. The list includes UV–VIS–NIR–SWIR spectrometers, benchtop and portable NIR/SWIR, Raman analyzers, Confocal Raman microscopes, hyperspectral imagers, mass spectrometers, and OEM spectral engines. In addition, they produce components for the R&D, biomedical, pharmaceuticals, chemical, food, semiconductor, health monitoring, human and animal medical devices, and the optical telecommunications industries.

Hyperspectral cameras are designed specifically for use on UAV/unmanned aerial systems (UAS), or remotely operated vehicles (ROV). They are used to obtain VIS-NIR hyperspectral data with continuous spectral and spatial coverage. Generally, hyperspectral cameras used on UAVs, such as OCI-UAV, are automatic. They require least human intervention. The OCI-UAV is small, compact, and weighs less than 300 g. The OCI-UAV is a straightforward system. It integrates with UAVs effortlessly, particularly, for applications such as in precision agriculture and remote sensing (Plate 5.8).

BaySpec's Compact OCI™-OEM cameras are vital optical components of the OCI-1000 and OCI-2000 hyperspectral imagers (Plate 5.8). These sensors acquire full, VIS-NIR hyperspectral/multispectral data with high spectral resolution and fast speed. Continuous hyperspectral data capturing can happen at video rates. The OCI-UAV hyperspectral cameras are designed, specifically, for use on UAV/UAS or ROV. The system is packed with a high-performance, miniature single-board computer. Sensors acquire VIS-NIR hyperspectral data with continuous spectral and spatial coverage. OCI-UAV is automatic and requires minimal human intervention. Conventional hyperspectral imagers rely on intensive software effort on orthorectification. However, the innovative design of the OCI-UAV-1000 features "true push-broom" imagers. They can move to scan at random speeds. OCI-UAV-2000 is a snapshot multispectral imager. It eliminates artifacts caused by vibrations in flight.

SPECIFICATIONS

Model: OCI-UAV-1000

Size: 8 cm × 3 cm × 3 cm (with head only) and 10 cm × 7.5 cm × 3 cm with computer chip;

Weight: 180 g (including standard lenses);

Operation mode: push broom;

Data interface: USB 3.0 (up to 120 bps)

Model: OCI-UAV-2000

Size: 8 cm × 3 cm × 3 cm (with head only) and 10 cm × 7.5 cm × 3 cm with computer chip;

Weight: 272 g (including standard lenses);

Operation mode: snap shot;

Data interface: USB 3.0 (up to 120 bps).

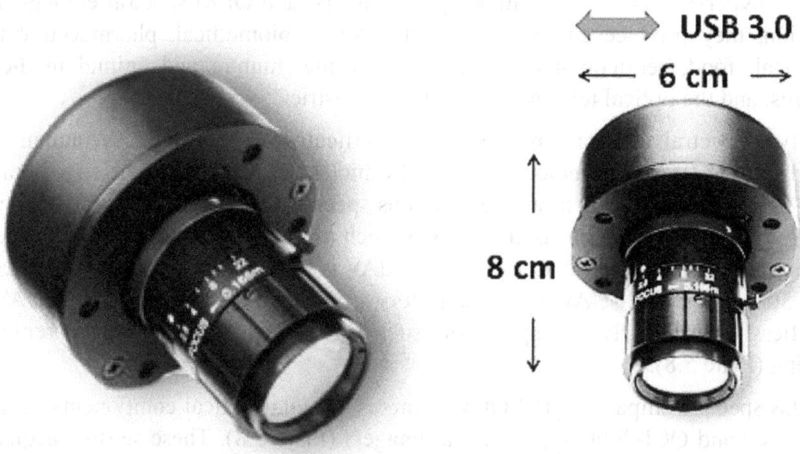

PLATE 5.8 OCI-UAV Ultra-Compact hyperspectral camera.
Source: Mr. Michael Zemlan, Sensors Division, BaySpec. Inc., San Jose, California, USA.

References, Websites, and YouTube Addresses

http://www.bayspec.com/spectroscopy/oci-uav-hyperspectral-camera/ (accessed Apr 20, 2017).
http://www.bayspec.com/spectroscopy/oci-uav-hyperspectral-camera/ (accessed Jan 14, 2018).
http://www.bayspec.com/spectroscopy/oci-uav-airborne-vis-nir-hyperspectral-imager/ (accessed Jan 14, 2018).

https://www.youtube.com/watch?v=OQF6GXT4p7k (accessed Jan 13, 2018).
https://www.youtube.com/user/BaySpecInc (accessed Jan 13, 2018).
https://www.youtube.com/watch?v=EmhgSt743N0 (accessed Jan 13, 2018).
https://www.youtube.com/watch?v=EmhgSt743N0 (accessed Jan 13, 2018).

Name: Rikola Hyperspectral Sensor

Company and address: Rikola Ltd., Kaitoväylä 1 F 2, FIN-90950 Oulu, Finland; Phone: +358 50 358 3516; e-mail: info@rikola.fi www.rikola.fi

The spectral resolution of sensor used on the UAV is important. The standard blue, green, red, and CIR bands cannot always provide the required details of the target. As a response to this shortcoming, the Finnish Technical Research Center and Rikola Ltd. have developed a camera. It collects hyperspectral narrow-band images with single exposure. The camera can record up to 50 bands in the range of 500–950 nm. This camera is available since 2013. It may create new business opportunities in fertilizer and pesticide optimization, forest inventory, and environmental monitoring. For example, MosaicMill Inc. is currently developing EnsoMOSAIC software to process these hyperspectral images into hyperspectral mosaics and hyperspectral 3D models.

Rikola Ltd. has developed the sensor, together with VTT, the world's smallest and most lightweight hyperspectral camera. It provides full two-dimensional (2D) images at every exposure, enabling hyperspectral stereophotogrammetry, using UAVs (Makelainen et al., 2013; Plate 5.9).

The Rikola hyperspectral camera has been tested in several locations. They have used it on various platforms such as: Microdrone, Mikrokopter, Infrotron, and C-Astral Bramor. The camera has proven its high performance with several benefits. The remote sensing applications include estimating chlorophyll content, leaf area index of forest stands, identifying species, or determining reflectance of forest canopies. In agriculture, Rikola sensor has been used for getting crop indices such as NDVI, GNDVI, CWSI, yield estimates, detection of crop diseases, and flux indices.

SPECIFICATIONS:

Weight: <700 g;

Max 24 bands by data cube in-flight or 400 bands in the office;

Spectral resolution: 10 nm, FWHM;

Image resolution: max. 1024 × 1024;

Spectral range: 500–900 nm (spectral step 1 nm);

Total time, for example, data cube of 16 bands—1.6 s;

Optical system: focal length 6 mm,

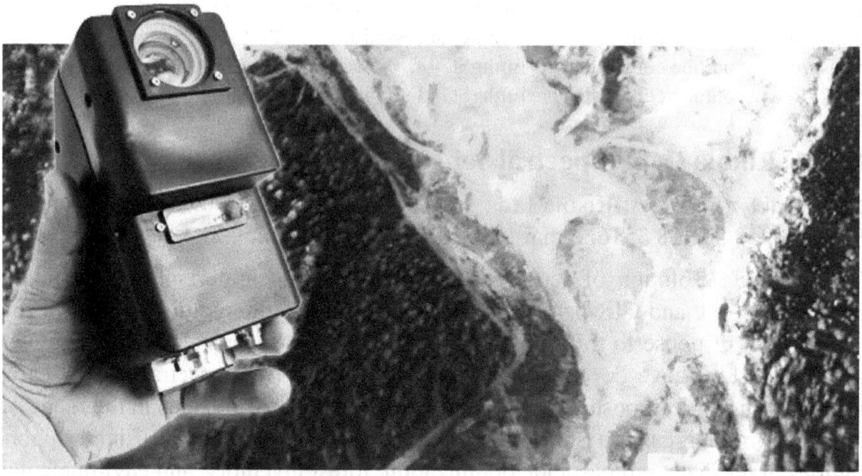

PLATE 5.9 Rikola hyperspectral camera.
Source: Rikola Ltd., Oulu, Finland.
Note: Rikola hyperspectral camera offers frame-based solution. It is light in weight. It offers accurate mosaics and output data cubes.

References, Websites, and YouTube Addresses

Makelainen, A.; Saari, H.; Hippi, I.; Sarkeala, J.; Soukkamäki, C. In *2D Hyperspectral Frame Imager Camera Data in Photogrammetric Mosaicking*, International Archives of the Photogrammetry, Remote Sensing and Spatial Information Sciences, 2014; Vol. 40 (1), 2013. UAV-g2013, Sept 4–6, 2013, Rostock, Germany; pp 263–268.

http://www.mosaicmill.com/uas/cropdrone.html (*accessed Oct* 2, 2017).

https://www.youtube.com/watch?v=npFjRT-vJsQ (*accessed Oct* 2, 2017).

https://www.youtube.com/watch?v=e6hbmSfPnMQ (*accessed Oct* 2, 2017).

https://www.youtube.com/watch?v=fAG1jSi7Fa4 (*accessed Oct* 2, 2017).

https://www.youtube.com/watch?v=icjvpOWijRo (*accessed Oct* 2, 2017).

https://www.youtube.com/watch?v=xsm3bogEhYY (*accessed Oct* 2, 2017).

5.2.3 THERMAL SENSORS (IR)

Name: FLIR and Zenmuse XT Thermal Camera

Company and address: FLIR Systems Inc. FLIR Corporate Office, 27700 SW Parkway Ave, Wilsonville, Oregon, OR 97070, USA

Phone: 1-503-498-3547; website: http://www.flir.com

DJI Inc. is a globally reputed and popular UAV company located in Shenzhen province of China. It offers UAVs to technologists, farmers, and public organizations. It offers wide range of UAV models, particularly, copter UAVs. They also offer several different types of cameras capable of visual, multispectral, hyperspectral, and thermal imagery of ground conditions, installations, crop fields, etc. FLIR is a large, established, public company that has been making thermal imaging cameras for almost 40 years. They have recently introduced three new lightweight cameras designed specifically for UAV operations.

It seems, DJI's gimbal technology and knowledge of image transmissions was combined with FLIR's thermal imaging expertise to build the "Zenmuse XT Thermal Camera" (Plate 5.10). The sensor provides aerial IR scanning at 640/30 fps or 336/60 fps. It is usually fitted to DJI's Inspire 1 or M100 UAVs. The cameras are available with four different lens options. The "DJI Go" app provides real-time camera controls for color palette selection isotherms, zoom levels, and the selection of either video recording or still-image capture. Applications include fire-fighting, tower and substation inspections, and precision agriculture (FLIR, 2017). FLIR camera has been used to detect water status of crops (Park et al., 2015).

FLIR offers two additional thermal devices. They are "FLIR Vue" and the "FLIR Vue Pro." Both versions have optional GoPro mounting holes, but, neither is hardware integrated like the Zenmuse XT. In other words, users will need to independently mount the FLIR Vue on their UAV. The FLIR Vue and the FLIR Vue Pro have thermal video recording capability.

PLATE 5.10 Infrared cameras for thermal imagery of crops. Top: Zenmuse gimbal and XT thermal IR camera; below left: FLIR Vue; and below right: FLIR Vue Pro
Source: Adam Najberg, DJI Inc., Shenzhen, China; (Images: http://www.flir.com/suas/content/?id=70733*).*

References, Websites, and YouTube Addresses

FLIR. *Zenmuse XT Powered by FLIR: Unlock the Possibilities of Sight*; 2017; pp 1–15. http://www.flir.com/suas/content/?id=73063 (*accessed Jan* 7, 2017).

Park, S.; Hernandez, E.; Chung, H.; O'Connell, M. O. *Estimations of Crop Water Stress in a Nectarine Orchard Using High Resolution Imagery from UAV*, 21st International Congress on Modelling and Simulation, Gold Coast, Australia, 2015; pp 1413–1419.

https://www.youtube.com/watch?v=IOgekgSaaG4 (*accessed Feb* 12, 2017).

http://www.flir.com/suas/content/?id=70733 (*accessed Feb* 12, 2017).

https://www.youtube.com/watch?v=zGNCi52UgEo (*accessed Feb* 12, 2017).

https://www.youtube.com/watch?v=QFt4hJak5SE (*accessed Feb* 12, 2017).

https://www.youtube.com/watch?v=B9gv6N1bnm4 (*accessed Feb* 12, 2017).

https://www.youtube.com/watch?v=tYUTBe5c_xM (*accessed Feb* 12, 2017).

https://www.youtube.com/watch?v=4wj2wj3XkoE (*accessed Feb* 12, 2017).

Name: WIRIS 2nd Gen Thermal Camera

Company and address: Workswell s.r.o.; Keller HCW Gmbh; Division MSR, TCS; Carl-KellerStrasse, D-49479 Ibbenburen-Lagenbeck, Germany, 2010; Czeck HQ address: Workswell CZ, Libocká 653/51b, 161 00 Praha 6, Czech Republic; Phone: +420725877063; e-mail: info@workswell.eu, support@workswell.cu; website: https://www.workswell-thermal-camera.com/contact/

Workswell s.r.o. is an European company with headquarters in the Czech Republic. They specialize in the advancement of machine vision and touch-free temperature measurement. They have offered innovative thermographic products. Their products include thermal WIC cameras, NDT thermal imaging system ThermoInspector, and the Workswell WIRIS thermal imaging system for UAVs (Plate 5.11). Usually, Workswell "CorePlayer software" is delivered together with the sensor system. Workswell "ThermoFormat" software is used for batch editing and to export pictures. For example, images are exported to GCS to create 3D models. WIRIS 2nd gen is a state-of-the-art thermal imaging system. It is designed for commercial UAVs. It is a compact system that combines a thermal camera and a digital camera. The processor unit has possibility to record radiometric data with a digital HDMI output, all in one case. WIRIS thermal sensors have been utilized to prepare thermal maps of variety of locations, such as buildings, industries, and agricultural crop fields (Hale, 2017).

WIRIS 2nd gen thermal imaging system is supplied with either 336 × 256 (WIRIS 336) or 640 × 512 (WIRIS 640) resolution. We may also choose from different objective lenses. The specifications for WIRIS thermal camera include:

Resolution: up to 640 × 512 pixels;

Temperature scale: −25°C to +150°C, −40°C to +550°C, +1°C to 500 °C on request;

Temperature sensitivity: up to 30 mK;

Accuracy: ±2% or ±2°C;

Weight: 400 g; External memory: USB flash disk; Internal memory: 32 GB;

Calibration: Yes Inc. certificate;

Sbus; CAN bus

PLATE 5.11 WIRIS 2nd Gen thermal camera for UAVs.
Source: Mr. Adam Svestka, Workswell s.r.o., Praha, Czeck Republic; and Mr. Jeroslav Bohm, Workswell Gmb, Ibbenbüren-Laggenbeck, Germany.

References, Websites, and YouTube Addresses

Hale, K. Thermal Drones/Infrared Imaging. *Dronefly*, 2017; pp 1–3. https://www.dronefly.com/pages/thermal-drones (*accessed Dec* 15, 2017).

https://www.workswell-thermal-camera.com/workswell-wiris-en/ (*accessed Jan* 3, 2017).

https://www.dropbox.com/sh/gx4mmara35nehnn/AACw0gjsedifGfVzcuIYnTeKa?dl=0 (*accessed Dec* 15, 2017).

https://www.youtube.com/watch?v=_gdbeH5fFFo (*accessed Dec* 15, 2017).

https://www.youtube.com/watch?v=ivI_6il2PF0 (*accessed Jan* 3, 2017).

https://www.youtube.com/watch?v=OZhEddn_jKU (*accessed Jan* 3, 2017).

https://www.youtube.com/watch?v=z5-YD7oBrZ8 (*accessed Jan* 3, 2017).

5.2.4 LIDAR SENSORS ON AGRICULTURAL UAVS

LiDAR is a precise remote sensing technology. It measures distance by illuminating a target with a laser and analyzing the reflected light. LiDAR is commonly used to measure buildings and land masses with precision, as well as, to develop an accurate 3D model of an area. LiDAR helps farmers to find areas where costly fertilizer is being overused. It helps to create elevation maps of farmland. Such data are converted to create slope and sunlight exposure area maps. LiDAR imaging can detect elevation changes and drainage issues.

Reports suggest that Airborne LiDAR is the fastest growing area within the geospatial technology market. The UAV-aided technique covers large areas of land, safely and accurately, in a matter of minutes. LiDAR sensors have been taking to the sky. An Airborne LiDAR System fixed to a helicopter or aircraft is capable of scanning land covering many miles.

Airborne LiDAR sensors are typically housed in a lightweight pod. The pod can accommodate additional equipment, such as an IMU. It helps in more accurate data capture. Airborne LiDAR has number of applications such as coastal monitoring, mining, and forestry where dense vegetation can easily be penetrated.

Name: Leica SPL100 (Single Photon LiDAR) Sensor

Company and address: *Leica Geosystems B.V.,* Turfschipper 39, Wateringen, 2292 JC, Netherlands; Telephone: +31 88 00 18 000; e-mail: customercare.nl@leica-geosystems.com; website: leicageosystems.nl

Leica Inc. is a company with production units and distribution centers spread in different nations. It specializes in several different sophisticated optical and electronic gadgets. Among them, sensors and cameras are prominent items.

The Leica SPL100 is the Single Photon LiDAR airborne sensor. It is meant for terrain mapping projects. It includes mapping land resources, water, and crops. The Leica SPL100 system is accompanied by an 80 MP RGBN camera. This camera captures 6 million measurements per second and delivers reliable results, during day or night, leaf-on or leaf-off conditions, and in dense vegetation (Leica Inc., 2017).

The SPL100 helps in creating the highest density point clouds by collecting 12–30 points per m^2 (depending on flying height). It penetrates semiporous obscurations, such as: vegetation, ground fog, and thin clouds. The information collected with the SPL100 Single Photon LiDAR has several applications. They are: country and state-wide mapping, disaster risk planning, emergency management, forest inventory, flood mapping, and control of soil erosion. Reports state that Leica SPL100 Single Photon LiDAR is part of *Leica RealTerrain,* the latest airborne reality capture solution developed by Leica Geosystems. It combines innovative sensor technology with the intuitive, high-performance, multisensor, processing workflow *HxMap* (Leica Inc., 2017).

The SPL100 sensor reaches the highest efficiency for large area mapping. It acquires data at the lowest cost per data point. Collecting 6 million points per second using 100 output beams makes the SPL100 sensor 10 times more efficient than many other conventional LiDAR sensors.

Image Processing is Fast

The SPL100 has high-performance multisensor LiDAR and imaging postprocessing workflow *HxMap*. This software (HxMap) offers the highest data throughput; eliminates the limitations of single workstation processing; accelerates data delivery; and reduces training costs.

"Leica RealTerrain" software offers the technical basis for the LiDAR mapping projects. We should note that, in crop fields, success of making informed decisions depends on readily available terrain and elevation data. Here, the "Leica SPL100" sensor seems useful. It captures and visualizes large terrain areas and ecosystems. It is a high-performance airborne sensor for farm, crop fields, and utility corridor surveys.

There are several other models of LiDAR sensors produced by Leica Geosystems Inc. For example, the Leica AL-S80 is a single family of Airborne LiDAR sensors. It offers speed and flexibility without compromising on quality and accuracy. Leica AL-S80 LiDAR system offers high-density point clouds for general purpose mapping, in a fraction of the time. It helps to perform high-AGL mapping with nearly 8 km swath. Leica Al-S80 is a widely used topographic LiDAR sensor. Leica AL-S80 is field proven in every corner of the world. Its major applications are in county and state-wide mapping, damage assessment, forestry, and terrain mapping. Leica AL-S80 offers high flexibility for every application. Leica AL-S80-CM is apt for city and corridor mapping applications from lower flying heights. Otherwise, Leica AL-S80-HP supports general purpose mapping at the most widely used flying heights. Leica AL-S80-UP is a high-altitude variant. It allows wide-area mapping on a state or national level. Usually, Leica AL-S80 systems are sold along with peripheral products and software. They provide a seamless workflow from mission planning through point cloud generation.

References, Websites, and YouTube Addresses

Leica Inc. *New Leica SPL100 Brings up to 10× more Efficiency to Airborne LiDAR Mapping*; 2017; pp 1–12. https://leica-geosystems.com/about-us/news-room/news-overview/2017/02/2017-02-13-new-leica-spl100-brings-up-to-10x-more-efficiency-to-airborne-lidar-mapping (*accessed Dec* 31, 2017).

https://leica-geosystems.com/products (*accessed Jan* 2, 2018).

https://leica-geosystems.com/en-in/products/airborne-systems/lidar-sensors/leica-als80-airborne-laser-scanner (*accessed Jan* 2, 2018).

https://leica-geosystems.com/en-in/products/airborne-systems/lidar-sensors (*accessed Jan* 2, 2018).

https://www.youtube.com/watch?v=s8Adhf1FxRc (*accessed Dec* 31, 2017).

https://www.youtube.com/watch?v=UB3vfZTnDYg (*accessed Dec* 31, 2017).

https://www.youtube.com/watch?v=HfV7jJgrw4Q (*accessed Dec* 31, 2017).

https://www.youtube.com/watch?v=QE9_enLWRqs (*accessed Dec* 31, 2017).

https://www.youtube.com/watch?v=U1ohTX2YPkU (*accessed Dec* 31, 2017).

https://www.youtube.com/watch?v=ssCRuTY4SPM (*accessed Dec* 31, 2017).

Name: Robin Airborne LiDAR System

Company and address: 3D Laser Mapping Ltd., Ranch House, Chapel Lane, Bingham, Nottingham, NG13 8GF, UK; Phone: +44 (0) 1949 838 004, e-mail: info@3dlasermapping.com; website: https://www.3dlasermapping.com/contact/

The company "3D Laser Mapping Ltd." was initiated in 1999 in United Kingdom. It has a multinational production unit in South Africa, Australia, and the United States. They specialize in laser scanners and 3D technology.

3D Laser Mapping Ltd. is a geospatial technology supplier and innovator. It serves customers in mining companies, governments, universities, blue-chip firms, operators of highways, power lines, and railways. 3D Laser Mapping Ltd. offers mapping and monitoring benefits. LiDAR is useful for a range of environmental applications, including agriculture and precision forestry. LiDAR sensor has vegetation penetrating capabilities. Therefore, it is useful when scanning through dense vegetation or crops. ROBIN +WINGS can assist organizations to scan forests and fields. It covers many acres with ease from the air (3D Laser Mapping Ltd., 2017).

The UAV LiDAR can be fast and effective. The UAVs can take to sky repeatedly with ROBIN +WINGS airborne laser scanner (Plate 5.12). Robin laser scanner integrates well with several different models of UAVs. The pod (gimbal) supports up to three industrial cameras to assist with point cloud visualization and ortho mapping.

Specifications of ROBIN +WINGS Airborne Systems as quoted by the company are:

a) Designed for ROBIN standard sensors (VUX scanner and IGI IMU);
b) Up to three Phase One camera support (one oblique, two downwards);
c) Helicopter pod suitable for: AFSP-1 and AF200 single pole mount for: Eurocopter AS-350, Eurocopter AS-355;
d) GV-V2 Nose Mount for: Bell 206A/B/B1, Bell 407;
e) Certification in compliance with the STC for the AFSP-1 pole mount and Bell 407;
f) FAR-27 Compliance Checklist;
g) Aircraft antenna;
h) Total weight: 15 kg (Pod only) + 7 kg (Scanner and IMU) = 22 kg

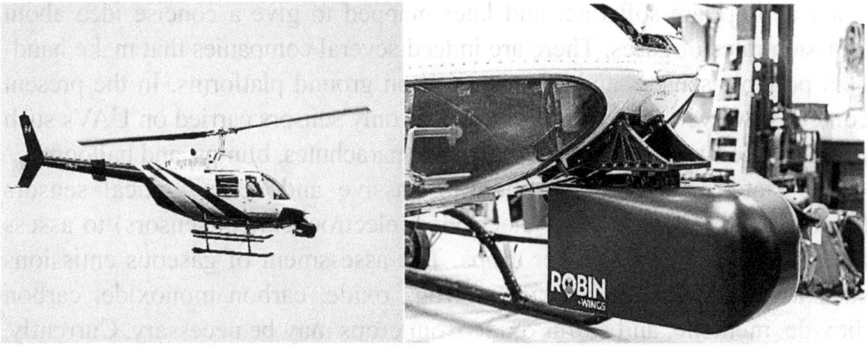

PLATE 5.12 Left: a manned helicopter (*not UAV*) attached with ROBIN +WINGS airborne laser scanner. Right: a close-up view of the ROBIN +WINGS laser sensor.
Source: Ms. Harriet Brewitt, Marketing Division, 3D Laser Mapping Ltd., Nottinghamshire, United Kingdom.

References, Websites, and YouTube Addresses

3D Laser Mapping Ltd. *UAV LiDAR*; 2017; pp 1–7. https://www.3dlasermapping.com/uav-lidar/ (*accessed Dec* 30, 2017).

https://www.3dlasermapping.com/ (*accessed Jan* 5, 2017).

https://www.3dlasermapping.com/uav-lidar/ (*accessed Jan* 5, 2017).

https://www.3dlasermapping.com/robin-mobile-mapping-system/robin-wings-airborne-laser-scanner/ (*accessed Jan* 5, 2017).

https://www.3dlasermapping.com/agriculture-precision-forestry/ (*accessed Jan* 5, 2017).

https://www.youtube.com/watch?v=HfV7jJgrw4Q (*accessed Jan* 5, 2017).

https://www.youtube.com/watch?v=eBUCGxZq_xg (*accessed Jan* 5, 2017).

https://www.youtube.com/watch?v=2-wNeIn0D0o&list=PLZAXXOpJVlw-q1bmaDm8kAhN1SUpCXSF8 (accessed Jan 5, 2017).

https://www.youtube.com/watch?v=BhHro_rcgHo (*accessed Jan* 5, 2017).

https://www.youtube.com/watch?v=0ZwuwGgfPX4&list=PLZAXXOpJVlw87POOWKFJQ1G0q_C3jsqwp (accessed Jan 5, 2017).

https://www.youtube.com/channel/UCpb9POt_xNsq5VQUYwy7jvA (accessed Jan 5, 2017).

https://www.youtube.com/watch?v=IbuJB5y5Ai0 (accessed Jan 5, 2017).

https://www.youtube.com/watch?v=e2OhaV5OveQ (accessed Jan 5, 2017).

https://www.youtube.com/watch?v=dkfolP-e6Uo (accessed Jan 5, 2017).

5.2.5 SENSORS TO ASSESS ATMOSPHERIC QUALITY AND GASEOUS EMISSION FROM AGRARIAN ZONE

There are hand-held sensors that can detect range of gaseous compounds in the atmosphere above crop fields, marshes, mines, etc. If hand-held gadgets are used, it is tedious to collect samples/readings at innumerable spots over a crop field of say, 10,000 ha. The data has to be logged, stored, and then processed using appropriate software, and later mapped to give a concise idea about emission rates of gases. There are indeed several companies that make hand-held portable sensors and those carried on ground platforms. In the present context, however, we are concerned with only sensors carried on UAVs such as small fixed-winged or copter aircrafts, parachutes, blimps, and balloons.

As stated above, in addition to passive and active optical sensors (cameras), UAVs may carry electrodes (electrochemical sensors) to assess the atmospheric quality over crops. The assessment of gaseous emissions such as ammonia, nitric oxide, nitrous oxide, carbon monoxide, carbon dioxide, methane, and sulfur oxide from crops may be necessary. Currently, there is a general trend to avoid excessive inputs of fertilizer-N and avoid burning crop residues. Agricultural agencies may often require data about the extent of greenhouse gas emitted from large farms. In this regard, air sampling using aircrafts could be costly. Then, the samples may have to be analyzed at a laboratory, using tedious analytical chemistry procedures. Preservation, shipment, and accurate analysis of gas samples is not easy. Balloons and blimps are of course useful in collecting air samples. In the present case, we are concerned more with rapid sampling and analysis of gaseous and particulate chemical components in the atmosphere, using UAVs. The UAVs meant for sampling and analysis of atmospheric gases usually carry a sampling device. For example, an UAV such as "Scentroid" carries a sampler and physicochemical sensors (electrodes). The electrodes may measure gaseous components such as NH_3, NO, NO_2, CO, CO_2, CH_4, H_2S, and SO_2. At a time, the UAV may carry up to two different electrodes. Then, relay data about the specific gases to the GCS.

The data derived using electrode sensors can help farm experts/technicians to thwart gaseous emissions by regulating fertilizer-N inputs. Undue accumulation of N which is vulnerable for emission could be avoided by noting the current levels of N-emissions. Similarly, C-emissions could be avoided by reducing plowing and disturbance to soil and also, by recycling organic matter without resorting to burning. Farms near industries may encounter excessive SO_2 in the atmosphere above crops. UAVs with sensors (electrodes) for SO_2 detection and estimation could be very useful (Scentroid Inc., 2017). UAVs with electrode for methane detection may easily assess its emissions from crop fields (e.g., rice), industries, mines, marshes, etc.

In fact, reports suggest that researchers at University of Bristol, United Kingdom have developed UAVs that carry methane sensors, exclusively. They detect methane above variety of terrain conditions, installations, and farm land (Phillips, 2017). They have used octocopters that sample atmospheric air from above crop canopy, till 9000 ft altitude above ground level. Then, they have detected methane emission levels in the samples. The sensors on the UAV are controlled by a GCS computer (laptop). The values for methane are marked for each location with GPS coordinates. Accompanying data about relative humidity and temperature are also given. Such UAVs may serve useful purposes. We may detect and map methane emissions by different rice varieties across vast stretches of low wetland paddy. Dairies that emit methane could also be monitored using UAVs carrying methane sensors (Miller, 2016).

Let us consider another example of an UAV system deployed exclusively to detect methane emissions from farm land, dairies, industries, and marshland. The UAV and sensor technology were developed and offered by VentusGeospatial Inc. of Canada and Boreal Laser Inc., Canada. Reports by Hugenholtz and Barchyn (2016) of University of Calgary suggest development of a small, battery-operated, fixed-winged UAV (3.8 kg). It has 90 min endurance in flight and carries a CH_4 sensor, that is, a methane gas detecting electrode. The UAV could take off repeatedly in 15–20 min. The CH_4 sensor is a miniaturized version (Boreal Laser Inc., Canada). It is a laser-based gas monitor and is integrated with UAV's electronic circuit. The changes in electronic signals are calibrated. They indicate the CH_4 gas concentration at each spot in the atmosphere. The location where CH_4 is measured is also provided with GPS tags plus temperature and relative humidity values.

In the following paragraph, a detailed discussion about a company that produces aerial electrochemical sensors is provided. These UAVs carry such sensors and detect levels of atmospheric gases. The UAV models Scentroid DR300 and Scentroid DR1000 plus the sensors they carry are discussed. Sensors that are pertinent to agricultural settings are only mentioned.

Name: Scentroid Sensors (deployed on Copter UAVs)

Company and address: Scentroid Inc., 70 Innovator Avenue, Unit 7, Whitchurch-Stouffville, Ontario, L4a Oy2, Canada. Phone: 416.479.0078; e-mail: support@idescanada.com, sales@idescanada.com; website: http://scentroid.com/scentroid-sampling-drone/

The company Scentroid Inc. produces a range of physicochemical sensors useful to industries and weather monitoring agencies. Their product range includes olfactometers that detect trace gases. They produce electrode-based chemical analyzers for rapid detection of contaminants. They also offer sensors that track pollutants, etc. In the present context, we may note that Scentroid Inc. offers DR1000 flying laboratory and DR300 flying laboratory. These sensors can assess gaseous components of atmosphere. They can assess a range of different chemicals using an UAV fitted with correct electrodes. Usually, about two to three electrodes could be fitted into UAV.

It helps to sample, assess, and relay the data to GCS computers (Scentroid Inc., 2017; Plate 5.13.1). The list of gases that Scentroid UAV plus sensors (electrode) combination can assess swiftly while flying at a definite path is still small. However, many of them are in the list of agriculturally more important emissions.

The sensors offered by Scentroid Inc., Canada are used on hand-held gadgets (Plate 5.13.2), on ground plant forms, and on UAVs. They are of utility in agriculture, food industry, wastewater treatment, refinery, mining industry, material testing, and environmental protection agency. Scentroid Inc. also provides training to technicians who wish to work with sensors (electrodes) on UAVs.

According to the company, "Scentroid DR300 Sampling UAV" is a versatile air sampler (5.13.3). The DR300 sampling UAV can be used to sample ambient air at heights of up to 125 m above ground level. In addition to flying above agricultural terrain, it can also be flown very close or above the hazardous industrial chimneys or plumes (5.13.3). With the DR300 sampling UAV, an operator can stay safely away from hazardous sources (pesticide sprays), while acquiring environmental samples, while flying above crop fields sprayed with hazardous chemicals or above industries. The UAV usually records relative humidity, temperature, and wind speed.

Continuous monitoring of chemicals is the most important aspect of Scentroid UAVs fitted with electrochemical sensors. For example, Scentroid DR300 air sampling UAV can monitor more than 50 chemicals in the atmosphere above crops/industries. Data from on-board sensors are transmitted to operator's tablet. Such data could be viewed live and logged.

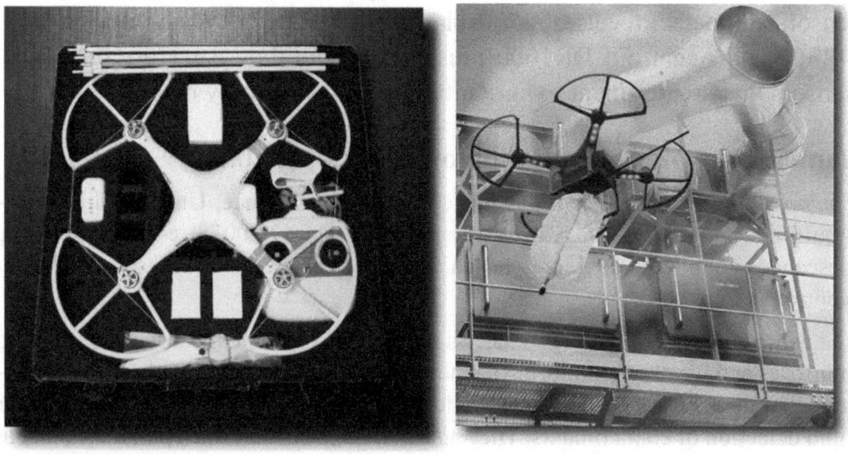

PLATE 5.13.1 Left: a Scentroid unmanned aerial vehicle for sampling; right: an UAV collecting samples from an industry, right at the plume of emission.
Source: Dr. Ardevan Bakhtari, President, Scentroid Inc., Stouffville, Ontario, Canada.

PLATE 5.13.2 A hand-held remote controller cum gaseous emission detector cum mapper.
Source: Dr. Ardevan Bakhtari, President, Scentroid Inc., Stouffville, Ontario, Canada.
Note: One operator can handle everything.

The Scentroid DR300 sampling UAV is a complete package. A single person can fly the UAV and monitor the exact location of the UAV including longitude, latitude, altitude, and height. Then, take samples without being exposed to any physical or chemical hazards. The dedicated remote and Drone App allow the operator to initiate pump, monitor H_2S and ammonia, to take temperature, altitude readings, and GPS position. Scentroid DR300 sampling UAV comes as a complete sampling solution. The kit includes the UAV, remote controller with the device holder (Plate 5.13.2), standing legs, sampling probe, three custom-made sampling bags, set of spare rotor blades, battery charger, spare battery pack, safety string rolls, etc. (Plate 5.13.3).

PLATE 5.13.3 Chemical monitoring in lakes, reservoirs, and water treatment plants.
Source: Dr. Ardevan Bakhtari, President, Scentroid Inc., Stouffville, Ontario, Canada.
Note: The small copter UAV that hovers above the water surface collects samples. The sampled locations are tagged with GPS coordinates, then, data is relayed. If one wants data from water bodies at different depths, then, perhaps regular amphibious UAVs should be developed! The UAV may have to dip into water or polluted water bodies, pick sample, and fly out. Perhaps, we have to mimic certain bird species that dip into water body, pick a fish, and fly away.

Scentroid DR300 air sampling UAV provides continuous monitoring of multiple chemicals. While in flight, the built-in sensors can provide remote monitoring of H_2S, ammonia, VOC, and more than 30 other chemicals. Data is transmitted to the operator's tablet for live monitoring and recording. Readings of chemical concentration along with GPS position and altitude can provide 3D mapping of ambient pollution and odor levels (Table 5.1). As stated earlier, Scentroid's DR300 UAV adopted to detect gaseous emissions and pollutants above crop land can be equipped with only two or three sensors, at a time. No doubt, it needs improvement to hold more sensors. It can then detect as many gaseous emissions and contaminants, above the crop fields. Perhaps, it is a matter of time before it could be achieved.

TABLE 5.1 Sensors Placed on UAVs that Are of Utility in Agrarian Regions.

Sensors on UAV	Gaseous emission detected	Range	Resolution
CD1 and CD2	Carbon dioxide	0–5000 ppm	15 ppm
CO1 and CO2	Carbon monoxide	500–10,000 ppm	20 ppm
MT1	Methane	0–100%	1%
AM1	Ammonia	100 ppm	1 ppm
ON1	Nitrogen dioxide	0–20 ppm	15 ppb

TABLE 5.1 *(Continued)*

Sensors on UAV	Gaseous emission detected	Range	Resolution
NO1	Nitrogen oxide	100 ppm	0.1 ppm
NC1	Nitric oxide	20 ppm	80 ppb
SD1 and SD2[1]	Sulfur dioxide	2000 ppm	2 ppm
PD2	Volatile chemicals (e.g., butylene)	50 ppm	1 ppb
PM1	Particulate dust	0–10,000 particles/s	NA
O1 and O2	Oxygen	0–100%	0.10–1%

Source: Mainly excerpted from Scentroid Inc., Canada; Environmental Protection Agency, Washington D.C., USA.; MacDonnel et al., 2013.

Note: The sensors listed above could be used on UAVs (e.g., Scentroid DR300; DR1000) that fly above crop fields. The sensors record emission levels (concentration) along with temperature and relative humidity values for the very spot. Each sample and value recorded is tagged with GPS coordinates.

References, Websites, and YouTube Addresses

MacDonell, M.; Raymond, M.; Wyker, D.; Finster, M.; Chang, Y.; Raymond, T.; Temples, D.; Scofield, M. *Mobile Sensors and Applications for Detection of Air Pollution*; Argonne National Laboratory and Environmental Protection Agency: USA, 2013; pp 278.

Scentroid Inc. *Scentroid: The Future of Sensory Technology*; 2017; pp 1–12. http://scentroid.com/scentroid-sampling-drone/ (*accessed Dec* 21, 2017).

https://cfpub.epa.gov/si/si_public_file_download.cfm?p_download_id=518291 (*accessed Dec* 20, 2017).

http://www.scentroid.com/scentroid-sampling-drone/ (*accessed Dec* 20, 2017).

http://www3.epa.gov/ttnamti1/files/2014conference/wedngambaxter.pdf (*accessed Dec* 20, 2017).

http://www.ncbi.nlm.nih.gov/pmc/articles/PMC4969839/ (*accessed Dec* 20, 2017).

http://www.mdpi.com/2073-4433/8/10/206/pdf (*accessed Dec* 20, 2017).

http://www.cen.acs.org/articles/94/i9/drones-help-us-study-climate.html (*accessed Dec* 20, 2017).

http://www.spectrum.ieee.org/energywise/energy/environment/drones-take-to-the-skies-to-screen-for-methane-emissions (*accessed Dec* 20, 2017).

http://www.adsabs.harvard.edu/abs/2017EGUGA..1911068C (*accessed Dec* 20, 2017).

http://www.alberta.ca/climate-methane-emissions.cfm (*accessed Dec* 20, 2017).

http://www.theengineer.co.uk/european-drone-project-toxic-gases (*accessed Dec* 20, 2017).

http:www.nist.gov/news-events/news/2017/06/nistcu-team-launches-comb-and-copter-system-map-atmospheric-gases (*accessed Dec* 20, 2017).

5.3 FLIGHT MISSION SOFTWARE FOR AGRICULTURAL UAVS

The flight plan of an agricultural UAV has direct impact on efficiency with which we gather aerial imagery. The accuracy, quality, resolution, and the ease with which images can be processed or analyzed depend to a certain extent, on flight plan. The overall flight mission and waypoints may decide the usefulness of the images that sensors gather. So, software used for flight planning has to be apt. Flight mission should be accurate to the purpose. Focusing entire crop field or only selected areas is important. Therefore, during aerial photography, flight plan decides the quality and usefulness of the spectral images gathered. There are several companies that deal with software for deciding flight plan for UAVs. A few of the flight plans can even be simulated on a computer screen, prior to actual use. For example, let us consider "*eMotion*" a flight plan software developed and offered by SenseFly Inc., Switzerland. "*eMotion 3*" enables us to plan, simulate, and manage the flights of every professional SenseFly UAV, including fixed-wing *eBee* UAV and copter UAVs (SenseFly Inc., 2017). "Pixhawk" is a flight plan software developed by PrecisionHawk Inc., Indiana, USA. ATLAS is another flight mission designing software. Robota's "Goose Autopilot" is another example of software that allows us to prepare flight plans and execute. It interfaces the UAV with GCS and helps in directing or changing flight plans mid-way, if needed. We may cover up to 40 acres in 50 min of UAV flight, using the flight plans (Robota, 2016).

Name: ATLAS Flight Plan

Company and address: MicaSense Inc., 1300 N Northlake Way, Suite 100, Seattle, WA 98103, USA

MicaSense Inc. is a sensor and computer software developing company. It specializes in UAV technology and related products. They offer products such as MicaSense and Parrot Sequoia cameras for UAVs. They also specialize in flight plan development software for UAVs and image processing software.

"Atlas Flight Plan" is a computer software that helps UAV technicians, to mark out a flight plan for the UAV, to fly while capturing useful images. It is a simple flight mission software. Several other softwares are available, but they need elaborate preparations and instructions. Atlas Flight Plan allows quick and easy "planned missions," for the DJI company's UAV models fitted with RedEdge-M, RedEdge, or Parrot Sequoia multispectral camera. With just a few taps, we can collect useful data. We should set desired altitude, speed, and overlap.

The basic features quoted for this flight plan developing software as follows: We must create a simple rectangular flight plan or log into atlas and choose a field.

The flight plan will be drawn over the field, automatically. Parameters needed for flight plan software are desired speed, altitude, and overlap. It configures camera's settings automatically, via Wi-Fi. It has facility for pause, stop, and resume flights. For example, we can change the batteries and resume flights.

Specifications include: it is compatible with iPad and iPhone devise only. It works with DJI company's UAV models only, for example, Matrice 100, Phantom 3, Phantom 4, and Inspire 1. We may note that, to device a flight plan, there is no need to fix a sensor.

References, Websites, and YouTube Addresses

http://www.micasense.com/ *(accessed Dec 16, 2017).*
https://www.micasense.com/contact/ *(accessed Dec 16, 2017).*
https://www.micasense.com/atlasflight/ *(accessed Dec 16, 2017).*
https://support.micasense.com/hc/en-us *(accessed Dec 16, 2017).*
https://www.youtube.com/watch?v=yNVRSh0GsCM *(accessed Dec 16, 2017).*
https://www.youtube.com/watch?v=ZogbFbLLWtA *(accessed Dec 16, 2017).*

Name: "*eMotion*" software for flight mission of UAVs

Company and address: SenseFly—A Parrot Company, 1033 Route Geneva, (Z.I. Châtelard Sud), Cheseaux-sur-Lausanne, Switzerland; Phone: +41 21 552 04 20; e-mail: info@sensefly.com; website: www.sensefly.com

eMotion is a flight mission software offered by SenseFly Inc., Switzerland. This software is utilized by ground station computers to plan, simulate, monitor, and control an UAV's flight. Some common steps adopted while deciding the flight mission, for an agricultural UAV are:

(a) Import the background map of choice and draw a rectangle (or create a more complex polygon) over the area, you want to map. Then, define your required ground resolution and image overlap. *eMotion* will automatically generate a full flight plan. It calculates the eBee's (fixed-winged UAV) required altitude and displays its projected trajectory.
(b) If the flight plan is on undulated terrain, then, we have to use *eMotion*'s 3D mission planning feature. It considers elevation data when setting the altitude and mission waypoints. Then, it offers the resulting flight lines. Using Simulator mode, we can check the performance of the flight mission even before actually flying and imaging the terrain/crop field. We can monitor the UAV at the GCS computer (SenseFly, 2017a, 2017b).

eMotion displays all its flight parameters, battery level, and image acquisition progress in real time.

The main "Map Area" displays the UAV's current position updated live as it executes its flight. A small arrow and info box show the wind speed and direction, as measured by the UAV. The "Status Panel" that follows the moving UAV icon displays its current waypoint status, battery charge, flight time, and altitude. In the event of a problem, such as extreme winds or low battery, a warning message instantly appears. With *eMotion,* we can reconfigure UAV's flight plan and landing point while in flight. *eMotion 3* minimizes the time we spend on planning flights and managing the data collected. *eMotion 3*'s full 3D environment adds a new dimension to UAV flight management. *eMotion 3* software can be used for both fixed-winged and multirotor UAVs. A built-in Flight Data Manager handles the georeferencing and preparation of images (SenseFly, 2017b).

Following are a few different types of fight missions commonly adopted by UAV experts. They are suitable during aerial imagery and processing. We have to decide the flight mission based on specific field locations and information needed about farm, terrain, and crops.

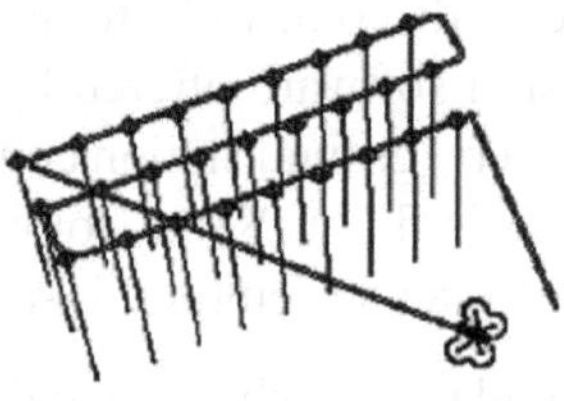

Horizontal Mapping

We may use this flight mission block to fly a "bird's eye" horizontal mapping mission. We have to set a few key mission parameters, such as preferred ground resolution. The *eMotion 3* software does the rest—creating flight lines and setting GPS waypoints automatically.

Corridor

This type of flight mission is used to map linear infrastructure and sites, such as roads and rail lines, pipelines, rivers, and coastlines.

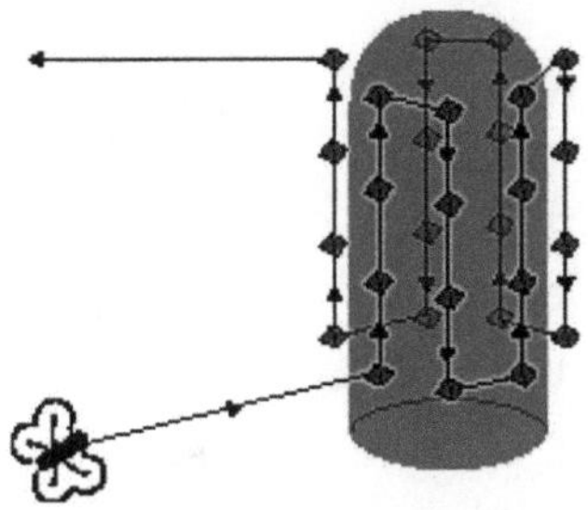

Cylinder

This type of flight plan is utilized to inspect digitally model structures such as wind turbines and towers using an UAV. The *eMotion 3* software generates the UAV's flight path, places waypoints, and sets the UAV's orientation/ head angle to capture the photos required.

Around Point of Interest

This flight helps in imaging vertical structures, towers, mines, etc. *eMotion 3* then automatically generates a circular flight path on a single horizontal plane and around the object of interest. It programs the UAVs' image capture points.

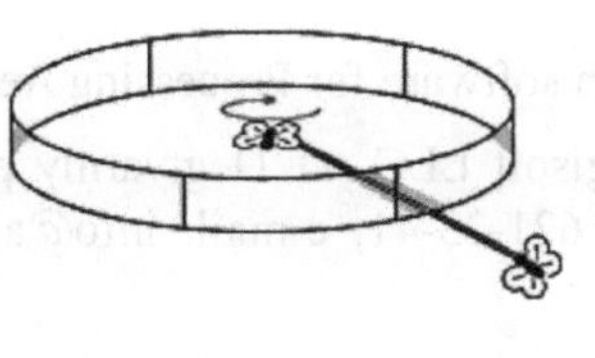

Panoramic

This type of flight mission provides flexibility to autonomous mapping missions. We may use this mission block to more accurately map and model concave environments, as well as to create new impressive marketing assets.

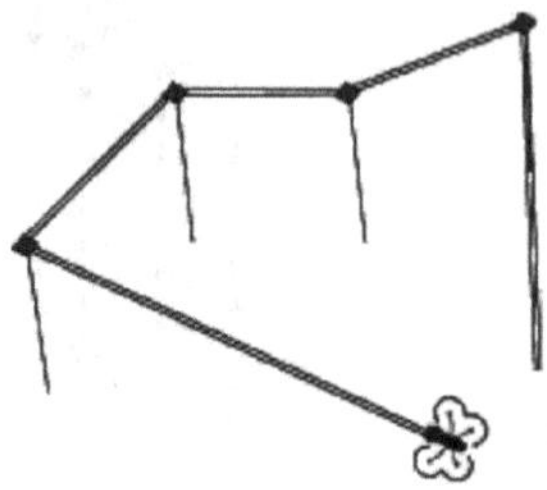

Custom Route

This flight mission is perfect for guiding UAVs through the most complex of environments.

eMotion is a highly efficient 2D flight planning approach. However, *eMotion 3* goes a step further by including a full 3D environment for flight planning, simulation, and management. The 3D approach is naturally intuitive. It enables us to plan and visualize the UAVs' flight path more accurately. It is especially useful in areas of varied elevation, such as mountainous regions.

References, Websites, and YouTube Addresses

Sense Fly. *Flight Plan and Control Software*; SenseFly—A Parrot Company: Cheseaux-Lausanne, Switzerland, 2017a. https://www.sensefly.com/software/emotion-2.html (*accessed Jan* 7, 2017).

SenseFly Inc. *eMotion 3*: *Drone Flight and Data Management*; SenseFly—A Parrot Company: Cheseaux-Lausanne, Switzerland; 2017b; pp 1–10. https://www.sensefly.com/software/emotion-3.html (*accessed Jan* 9, 2017).

https://support.micasense.com/hc/en-us/articles/115003789194 (*accessed Jan* 9, 2017).

https://support.micasense.com/hc/en-us/sections/205705068-Guides-and-Documentation (*accessed Jan* 9, 2017).

5.4 IMAGE PROCESSING SOFTWARE

Name: Agisoft's PhotoScan software for Processing Aerial Imagery

Company and address: Agisoft LLC, 11 Degtyarniy per., St. Petersburg, Russia, 191144; Phone: +7 (812) 621-33-41; e-mail: info@agisoft.com; website: http://www.agisoft.com/

Agisoft LLC is a software company initiated in 2006 at St. Petersgrad, Russia. This company focuses on computer vision technology. It also conducts intensive R&D work. Agisoft LLC has gained expertise in image processing algorithms, and with digital photogrammetry techniques. Agisoft is among the pioneers of digital photogrammetry solution developers. "Agisoft PhotoScan" is a software that performs photogrammetric processing of digital images obtained by agricultural UAVs. It even generates 3D spatial data. Agisoft PhotoScan is currently a highly popular and competitive photogrammetric software on the market. It helps in solving problems related to 3D reconstruction, visualization, surveying, and mapping tasks.

Agisoft's PhotoScan can be handled by a newcomer to image processing technology. It has simple steps to learn and conduct. It also caters to advanced image processing aspects for specialists. PhotoScan software produces accurate and good quality results. There are at least two editions of Agisoft's PhotoScan available for use. They are: Agisoft PhotoScan-Professional Edition and Agisoft PhotoScan-Standard Edition.

Important features touted by the company for this Agisoft's Image Processing software are as follows:

a) Photogrammetric triangulation
b) Dense point cloud: editing and classification
c) Digital elevation model: DSM/DTM export
d) Georeferenced orthomosaic export
e) Measurements: distances, areas, and volumes
f) Ground control points: high accuracy surveying
g) Python scripts: customize processing workflow
h) Multispectral imagery processing
i) 3D model: generation and texturing
j) 4D modeling for dynamic scenes
k) Panorama stitching

References, Websites, and YouTube Addresses

http://www.agisoft.com/ (*accessed Dec* 16, 2017).

https://www.questuav.com/services/questuav-external-services/image-processing-services/agisoft-photoscan/ (*accessed Dec* 16, 2017).

https://www.youtube.com/watch?v=doXm3pThz3s (*accessed Dec* 16, 2017).

https://www.youtube.com/watch?v=doXm3pThz3s&list=PLpMzaMqd-8eK4yLKWGWVW4oeHkV8USxIj (*accessed Dec* 16, 2017).

https://www.youtube.com/watch?v=Uny9nTr22go (*accessed Dec* 30, 2017).

https://www.youtube.com/watch?v=EELoTXJxoiA (*accessed Dec* 30, 2017).

https://www.youtube.com/watch?v=-LRbmaUAHR4 (*accessed Dec* 30, 2017).

https://www.youtube.com/watch?v=kEcUPc55Efc (*accessed Dec* 30, 2017).

Name: PIX4D MAPPER (Professional Photogrammetry software)

Company and address: PIX4D SA, EPFL Innovation Park, Building F, 1015 Lausanne, Switzerland; Phone: +41 21 552 0596; e-mail: info@pix4d.com, sales@pix4d.com, pr@pix4d.com; website: https://pix4d.com/industry/agriculture/#

"Pix4D Mapper" is a computer software that converts aerial images obtained by UAV's cameras into highly precise images. The images are processed, then, georeferenced maps and/or 3D models are offered. These maps are utilized in wide range of applications in farming situations. Such digital maps are used in planters, fertilizer applicators, and combine harvesters. The imagery is also used in yield forecasting.

Pix4DAG is a specialized software that converts aerial imagery into digital data and maps. It is useful during precision farming.

Some of the highlights of Pix4D Mapper software are:

a) It is amenable for both desktop and cloud processing. It allows farmers to adjust and chalk out clear workflow, in the field (Stein, 2017).

b) The multispectral data is thoroughly analyzed using Pix4DAG. It helps in generating NDVI, GNDVI, and CWSI maps for farmers. It also helps in classifying and offering prescriptions to famers, particularly, about planting schedules, fertilizer application, and harvesting.

c) Pix4DAG actually allows us to import or export processed maps and useful digital data to different locations (GCS), where it is used by technicians. At present, Pix4DAG maps can be accessed by several persons at a time. Also, useful prescriptions can be generated.

A few examples: Pix4DAG has been used in field management of an organic farm in Netherlands. The multispectral imagery obtained was processed using Pix4DAG Mapper. It helped in optimizing organic manure and irrigation inputs (Stein, 2017). Pix4DAG Mapper helped in preparing prescriptions for large areas of plantations suffering from rampant soil erosion. UAV-aided mapping and processing using Pix4D Mapper, it seems, gave dramatic results in reducing loss of soil and soil fertility. In California, it helped in assessing and mapping sequoia forests. It has helped in mapping soil conditions and emission rates of greenhouse gases above tree canopy (Stein, 2017).

References, Websites, and YouTube Addresses

Stein, N. *Agriculture: Aerial Crop Analysis*; Pix4D: Switzerland, 2017; pp 1–8. https://pix4d.com/industry/agriculture/# (accessed Dec 30, 2017).

https://pix4d.com/product/pix4dmapper-photogrammetry-software/ (accessed Dec 30, 2017).

https://pix4d.com/industry/agriculture/ (accessed Dec 30, 2017).

https://www.youtube.com/watch?v=XoPV-eT-1Ds (accessed Dec 30, 2017).

https://www.youtube.com/watch?v=p30Rt_aO0os (accessed Dec 30, 2017).

https://www.youtube.com/watch?v=p30Rt_aO0os (accessed Dec 16, 2017).

https://www.youtube.com/watch?v=bCWvR2KCOr8 (accessed Dec 16, 2017).

https://www.youtube.com/watch?v=Pt_miByA05U (accessed Dec 30, 2017).
https://www.youtube.com/watch?v=_7vMkbzrMYY (accessed Dec 16, 2017).
https://www.youtube.com/watch?v=RJW1UEwu2-8 (accessed Dec 16, 2017).

Name: Trimble's "Inpho" Image Processing Software

Company and address: Trimble Inc., Agricultural Division, Sunnyvale, CA 94085; San Jose, California, USA. Phone: +1 (408) 481-8000; e-mail: pr@trimble.com; website: http://www.trimble.com/Corporate/About_at_Glance.aspx

Trimble Inc. is a company that specializes in very wide range of agricultural products and services. It is located at San Jose in California, USA and elsewhere in other locations. They assess crop production trends in different regions and individual farms, analyze the aerial data, and offer prescriptions. They produce UAVs of great utility to agricultural activity. They also produce software that helps us in processing the images derived using UAVs. In the present context, we are interested in an image processing software called *Inpho*.

The Inpho® software is designed to precisely transform aerial images into consistent and accurate point clouds and surface models, orthophoto mosaics, and digitized 3D features. The modules can be used as a complete system or as individual components. They integrate easily into any photogrammetric production workflow using third party products. The portfolio also includes point cloud and terrain modeling for massive data. It also caters to Airborne LiDAR scanning and processing. Inpho software update service keeps the process up-to-date. Software maintenance agreements are renewed on an annual basis. Maintenance includes the update service, as well as access to Trimble's technical support. Inpho modules are offered in educational or research packages. Inpho also offers regular consultancy for farmers and farm agencies.

Inpho is compatible with most photogrammetric camera systems from Trimble, Leica, Vexcel, IGI, Phase One, or other companies. Inpho includes Geo-referencing, Geo-Imaging, Geo-Capturing, Geo-Modeling, Support, and Training.

A few other softwares relevant to aerial imagery and produced by Trimble Inc. are "OrthoMaster," "OrthoVista," and MATCH-T DSM.

OrthoMaster: Professional software for high-quality orthorectification of digital aerial frames, push broom, or satellite imagery generating true orthophotos, for both single images and complete image blocks. Trimble's OrthoMaster adopts classic rectification methods to generate classic orthos or true orthos.

OrthoVista: OrthoVista is a well-known software in the world of photogrammetry. It is used for creating orthomosaics with perfect homogeneity. It could be adopted even for multichannel imagery. It helps to adjust and combine thousands of orthophotos from any source. It converts them into perfect, seamless, color-balanced, and geometrically correct orthomosaics without any necessary subdivision.

MATCH-T DSM: It automatically creates digital terrain models, surface models, and dense point clouds from aerial image blocks or satellite image blocks.

DTMaster: It is a point cloud and LiDAR data editing, refining, and basic CAD/GIS mapping module. It is suited for large volumes of airborne data.

References, Websites, and YouTube Addresses

http://www.terraspatium.gr/inpho--trimble.html (accessed Dec 30, 2017).

https://community.trimble.com/docs/DOC-10806-get-to-know-inpho-trimble-aerial-photogrammetry-software (accessed Dec 10, 2017).

http://geospatial-solutions.com/new-trimble-products-focus-on-geospatial-imagery/ (accessed Dec 30, 2017).

https://www.youtube.com/watch?v=WEEg5c43hrw (accessed Dec 30, 2017).

https://www.youtube.com/watch?v=_LcO_bdcZjg (accessed Dec 10, 2017).

https://www.youtube.com/watch?v=_LcO_bdcZjghttps://www.youtube.com/watch?v=_LcO_bdcZjg (accessed Dec 30, 2017).

https://www.youtube.com/watch?v=Udeyc4mT-U0 (accessed Dec 30, 2017).

https://www.youtube.com/watch?v=xDEavl5Avzg (accessed Dec 30, 2017).

http://geo-matching.com/products/id1819- (accessed Dec 10, 2017).

inpho.html?gclid=EAIaIQobChMI0PX6gbex2AIVAiQrCh3BqwNcEAAYASAAEgKeBPD_BwE (accessed Dec 30, 2017).

KEYWORDS

- **sensors**
- **multi-spectral sensors**
- **hyperspectral sensors**
- **infra-red sensors**
- **LIDAR sensors**
- **ortho-images**
- **image processing software**
- **flight mission software**
- **Pix4D Mapper**
- **Agisoft's Photoscan**
- **physico-chemical sensors**
- **gaseous emissions**

GENERAL REFERENCES

Adao, T.; Hruska, J.; Padua, L.; Peres, E.; Morais, R.; Sousa, J. J. Hyperspectral Imaging: A review on UAV-based Sensors, Data Processing and Applications for Agriculture and forestry. *Remote Sens*. **2017,** *9,* 1110. DOI: 10.3390/rs9111110 (*accessed Dec* 10, 2017).

Aylor, D. E.; Fernandino, F. J. In Prospects for Precision Agriculture to Manage Aerially Dispersed Pathogens in Patchy Landscapes, Proceedings of 9th International Congress of Plant Pathology, Healthy and Safe Food for Everybody, Torino, Italy. *Plant Pathol*. **2008,** *90,* 59.

Aylor, D. E.; Schmale, D. G.; Shields, E. J.; Newcombe, M.; Nappo, C. J. Tracking the Potato Late Blight Pathogen in the Atmosphere Using Unmanned Aerial Vehicles and Lagrangian Modelling. *Agric. For. Meteorol.* **2011,** *151*, 251–260.

Bareth, G.; Helge, A.; Bendig, J.; Gnyp, K. M.; Bolten, D. A.; Jung, K. A.; Michels, L. R.; Soukkamaki, J. *Low-weight and UAV-based Hyperspectral Full-frame Cameras for Monitoring Crops: Spectral Comparison with Portable Spectroradiometer Measurements*; Schweizerbart'sche Verlagsbuchhandlung: Stuttgart, Germany, 2015. www.schweizerbart.de. DOI: 10.1127/pfg/2015/0256 1432-8364/15/0256 (*accessed Jan* 8, 2018)

Brown, J. K. M.; Hovmoller, M. S. Aerial Dispersal of Plant Pathogens on the Global and Continental Scales and Its Impact on Plant Disease. *Science* **2002,** *297*, 537–541

Burkart, A.; Cogliati, S.; Schickling, A.; Rascher, U. A Novel UAV-based Ultra-light Weight Spectrometer for Field Spectroscopy. *IEEE Sens. J.* 2014, *14*, 1462–1468. DOI: 10.1109/JSEN.2013.2279720 (March 10, 2017)

Cardinal, D. DxOMark Benchmarks for Popular Drone Camera Sensors. *Camera Sensor Review*, 2017; pp 1–11. http://www.dxomark.com/dxomarks-for-popular-drone-camra-sensors/ (*accessed Dec* 15, 2017).

Corrigan, F. Introduction to UAV Photogrammetry and LiDAR Mapping Basics. *DroneZon*, 2017a; pp 1–13. http://www.dronezon.com/learn-about-drones-quadcopters/introduction-to-uav-photogrammmetry-and-lidar-mapping-basics (*accessed Dec* 16, 2017).

Corrigan, F. Multispectral Imaging Camera Drones in Farming Yield Big. *DroneZon*, 2017b; pp 1–10. https://www.dronzon.com/learn-about-drones-quadcopters/multispectral-sensor-drone (*accessed Dec* 12, 2017).

DeBell, L.; Anderson, K.; Brazier, R. E.; King, N.; Jones, L. Water Resource Management at Catchment Scales Using Light Weight UAVs: Current Capabilities and Future Perspectives. *J. Unmanned Vehicle Syst.* **2016,** *4* (1), 7–30. https://doi.org/10.1139/juvs-2015-0026 (*accessed Jan* 8, 2018).

Dunn, M. *Applications of UAVs in Agriculture*; 2016; pp 1–15. https://www.linkedin.com/pulse/applications-uavs-agriculture-michael-dunn-cca/ (*accessed Jan* 7, 2017).

Gago, J.; Douthe, C.; Coopman, R. F.; Gallego, P. P.; Ribas-Carbo, M.; Flexas, J.; Esclona, J.; Medrano, H. UAVs Challenge to Assess Water Stress for Sustainable Agriculture. *Agric. Water Manag.* 2015, *153*, 1–14. DOI: 10.1016/j.agwat.2015.01.020 (March 27, 2017).

Gago, J.; Martorell, S.; Douthe, C.; Fuentes, S.; Tomais, M.; Hernandez, E.; Mir, L.; Carriqui, M. Escolana, J. M.; Medrano, H. *Up-scaling Levels for Drought Assessment in Agriculture: From Leaf to the Whole-vineyard*; I Jornadas del Grupo de Viticultura y Enologia de la SECH No. 7, 2016; pp 114–119.

Erena, M.; Lopez-Francos, A.; Montesinos, S.; Berthoumieu, J. P. *The Use of Remote Sensing and Geographic Information Systems for Irrigation Management in Southwest Europe*; CIHEAM Intereg IVB Etudes de Researches No. 67, 2012; pp 175–184.

Hoffmann, H.; Jensen, R.; Thomsen, A.; Nieto, H.; Rasmussen, J.; Friborg, T. Crop Water Stress Maps for Entire Growing Seasons from Visible and Thermal UAV Imagery. *J. Biogeosci.* 2016. DOI: 10.5194/bg-2016-316-2016 (*accessed Jan* 8, 2018).

Howard, B. LiDAR and Its Use in Agriculture. *Agric. Innov.* **2015,** *27*, 13. www.gesmapping.com.au (*accessed Jan* 7, 2018).

Hugenholtz, C.; Barchyn, S. C. *A Drone in Search of Methane*; 2016; pp 1–6. http://www.ventusgeo.com/wp-content/uploads/2016/04/Project_outline_no_watermark-1.pdf (*accessed Dec* 20, 2017).

Kalisperakis, I.; Stentoumis, C.; Grammatikopoulis, L.; Karantzalos, K. *Leaf are Index Estimation in Vineyards from UAV Hyperspectral Data, 2D Image Mosaics and 3D Canopy*

Surface Models. In The International Archives of the Photogrammetry, Remote Sensing and Spatial Information Sciences; 2015; Vol. 40 (1), pp 299–303.

Khot, L. R. *Unmanned Aerial Systems in Agriculture: Part 2 (Sensors)*; Washington State University, Extension Service: Pullman, WA, USA, 2016; pp 1–7. extension.wsu.edu (*accessed Dec* 10, 2017).

Light Inc. *Light L16: A Multispectral Camera*; Light Inc.: Palo Alto, CA, USA, 2017; pp 1–6. https://light.co/camera *(accessed Dec 26, 2017)*.

Marinov, M. The Top 3 High-end Near Infra-red Drone Sensors. *AgroHelper*, 2017; pp 1–6. https://medium.com/agrohelper/the-top-3-highend-nnear-infrared-drone-sensors fd95/ (*accessed Dec* 15, 2017).

Miller D. J. Ammonia and Methane Dairy Emission Plumes in the San Joaquin Valley of California from Individual Feedlot to Regional Scales. *J. Geophys. Res. Atmos*. **2016,** *120*, 9718–9738.

Miller, J. O.; Adkins, J.; Tully, K. *Providing Aerial Images Thorough UAVs*; University of Maryland Extension Service, 2017; pp 1–4. www.extension.umd.edu (*accessed Dec* 20, 2017).

Phillips, A. *University of Bristol Uses UAV Technology for Atmospheric Research*; 2017; pp 1–4. http://www.bristol.ac.uk (*accessed Dec* 20, 2017).

Robota Inc. The Complete Mapping Solution; 2016; pp 1–4. http://www.robota.us (*accessed Oct* 30, 2016).

Salami, E.; Barrodo, C.; Pastor, E. UAV Flight Experiments Applied to the Remote Sensing of Vegetated Areas. *Remote Sens*. **2014,** *6*, 11051–11081. DOI: 10.3390/rs61111051 (March 27, 2017).

Scentroid Inc. *Scentroid: The Future of Sensory Technology*; 2017; pp 1–12. http://scentroid.com/scentroid-sampling-drone/ (*accessed Dec* 21, 2017).

West, J. S.; Bravo, C.; Moshou, D.; Ramon, H.; Alastair McCartney, H. Detection of Fungal Diseases Optically and Pathogen Inoculum by Air Sampling. In *Precision Crop Production: The Challenges and Use of Heterogeneity*; 2010; pp 135–149. http://lnk.springer.com/chapter/10.1007/978-90-481-9277-9_9?no-a (*accessed Oct* 16, 2015).

CHAPTER 6

Applications of Unmanned Aerial Vehicle Systems in Agriculture: A Few Examples

ABSTRACT

This chapter is about unmanned aerial vehicles systems (UAVS) that are gaining acceptance in farming. The UAVs (aerial robots) are being examined for application during crop production. Firstly, UAVS are efficient in providing farmers with excellent aerial imagery of land resources, soils, water, and crops. A fixed-winged UAV covers about 250–300 acres of crop land in a single flight of 40 min. A quadcopter, which is slightly slower covers 100–150 acres per flight of 30 min. A tethered aerostat or helikite with sensors can relay ortho-images and digital data continuously. The aerial imagery is shrewdly utilized while preparing the "management blocks" and managing large farms. In fact, aerial monitoring of fields helps farmers to make decision regarding agronomic procedures and their timing. The spectral data derived using sensors in payload area of UAVs are of great utility to farmers. The data helps in calculating a very wide range of crop-related indices. They are indictive of crops' status. Such data forms the basis for farmers decisions regarding several farm procedures starting from land preparation till harvest and transport of grains/forage. A few examples of indices offered are Normalized Difference Vegetative Index (NDVI), Soil Adjusted Vegetative Index (SAVI), Normalized Green Vegetative Index (NGDRI), Normalized green red vegetative index (NGDRI), Green Leaf Index (GRI), Excess Red Vegetation (ExR), Excess green-red Index (Ex G-R), etc. During past 5 years, UAVS are being utilized to develop crop surface models (CSMs). The CSMs are utilized by farmers to decide about agronomic procedures. The other utility of significant value to crop researchers, plant breeders, and farmers in general is the ability of UAVs with sensors, to obtain large posse of phenomics data of crops. UAVs are known to provide detailed data from seed germination, its emergence, growth rate at seedling stage, plant height, plant biomass accumulation trends, variations in grain formation, etc. Data derived using near infra-red (NIR) sensors help farmers in knowing the variation in crops' water status. Farmers practicing precision farming methods will be at advantage since UAVs provide detailed digital data about growth, nutrients (N), and water stress index of the crop. Such data could be directly utilized in the GPS-guided fertilizer applicators and irrigation equipment. Currently, UAVs are adopted to assess and manage crops such as maize, wheat, rice, sorghum, potato, apple, grapes forest plantations, etc.

UAVs have also been applied by researchers to assess damage caused by weeds, pests, and disease. At present, there is a strong need to collect spectral signatures (data) of healthy and affected crops/fields. Data banks with spectral signatures of healthy or diseased crops would be essential, in future. There are several models of helicopter and multi-copter sprayers available. Sprayer drones have gained popularity in China, Japan, and other nations in Far East (e.g., Yamaha's RMX, DJI's MJ-Agras, HSE's Hercules, ZHNY, ProDrone 6B, and AGStar X8). UAVs spray pesticides at a rapid rate and low cost. They keep skilled technicians away from contact with harmful pesticides.

UAVS provide digital data that depicts variations in soil fertility, crop growth, disease/pest attack, and weed intensity. Such digital data is highly relevant to farmers adopting precision farming. UAVs have been utilized to surveillance agricultural experimental stations. They are also effective in collecting data about large number of crop genotypes that get evaluated in experimental farms.

Over all, we have to evaluate UAVs in many regions. Their economic efficiency needs to be examined. They say, as number of models of UAVs sensors and computer programs manufactured increases, their cost may reduce. Farmers may find them remunerative and highly efficient in managing crops with least drudgery. It is perhaps a matter of time before UAVs get examined and adopted by farmers, in different agrarian regions. They would affect the "agricultural skyline." They have the greatest potential to reduce farm related drudgery and improve efficiency of agronomic procedures.

6.1 INTRODUCTION

Unmanned aerial vehicles (UAVs), more frequently termed as "drones" are gaining acceptance in agrarian regions. They are finding useful applications in variety of ways. UAVs utilized above agricultural fields are usually small fixed-winged, or single rotor helicopters, or multirotor UAVs. Their application has been well standardized and is now a routine procedure in some areas.

Broadly, UAVs have found applications in many aspects, such as in aerial surveillance of land and natural resources, vegetation, infrastructure, and industrial installations; policing city events and traffic; sampling atmosphere, lakes, rivers, and ponds; and monitoring polar caps, rivers, sanctuaries, and wild life. There are still many aspects where UAVs could be applied. But, a degree of standardization, repeated evaluation, and streamlining of procedures are needed. In fact, there are several recent publications that list a series of applications for UAVs (Krishna, 2016; 2018; Gogarty and Robinson, 2012; Frey, 2015; Johnson et al., 2017). Here, in this chapter, we are concerned only with applications of UAVs relevant to monitoring and surveillance of farms, farm structures, and most importantly crop fields. We are also concerned with various agronomic procedures, wherein UAVs can be effectively utilized. The ability of UAVs to fly over crop fields rapidly or hover at vantage locations and offer multispectral images of crops is important. Most of the applications depend on the aerial images and spectral data that UAVs offer. UAVs, particularly, helicopters and multirotor types fitted with tanks (containers) and spray nozzles are being effectively adopted, to spray the crops with plant protection chemicals. Aerial spraying is indeed another important application of UAVs in crop production. It has gained in popularity among several farmers.

The theme of this chapter is to explore and list various UAV-based methods that reduce human drudgery, in open crop fields. UAVs add to accuracy of data collection, its storage, and its processing. UAVs help us in conducting agronomic procedures rapidly and "in time," to match the crop's needs. They, finally, improve crop productivity (Krishna, 2018). Several types of UAVs and models have been described. Their specifications have been listed in Chapters 2–4. Most of them could be applied very effectively in conducting certain farm activities. Agricultural UAVs have found their way into almost all continents and their agrarian regions. UAVs have been adopted during production of several different crop species. They are useful in accomplishing wide range of tasks, relevant to the conduct of various soil and crop management procedures (Krishna, 2016; 2018).

Deering (2014) states that, UAVs are quickly moving from battle fields to agricultural fields and pastures. They have now found utility in farm world. There are indeed several reports that have compiled the military history of UAVs and their range of applications. A few of them have clearly traced the transition of application of UAVs, from predominantly military to agricultural crop fields (Krishna, 2018). It has taken about a decade for UAVs, to gain acceptability in farms. No doubt, UAVs could be utilized in several different ways within agricultural landscapes. Yet, according to McKinnon and Hoff (2017), an UAV is a new gadget in the hands of farmers, technicians, and crop scientists. They need to get acquainted with its handling (remote piloting), fixing the pathways and waypoints, matching apt platforms (models), sensors, computer programs, and data preservation methods. Computer analysis of data/maps and arriving at accurate prescription need to be mastered by farmers and farm companies.

UAVs are supposedly most efficient during landscape survey, capture of digital data, preparation of 2D and 3D maps of terrain, in marking plots and the "management blocks." UAVs are said to offer most accurate maps and details of fields with crops and those in fallow. A close-up can provide a good deal of details about soil color, texture, and surface features. Usually, crop production in any location begins with land survey. UAVs are cost effective compared to skilled technicians, while conducting survey and preparing the maps. So, application of UAVs during farming begins with land survey. Timing of land preparation events such as deep ploughs, clod crushing, and ridging could be managed effectively using aerial images derived from UAVs. Ramaswamy (2016) has reviewed the status of UAVs, particularly, their usage in agricultural belts. She states that, if used with sufficient shrewdness and prudence, UAVs could become a very useful gadget in the hands of rural community and farmers. UAVs could positively improve crop productivity, even in locations with meager resources. We should note that, agricultural UAVs are rapid in flight over crops and in collecting useful data. Torres-Rua (2017) states that, a quadcopter could cover 100–150 acres of cropped land per flight of 30 min. A fixed-winged UAV may cover 300–4000 acres of cropped land, in about 30–90 min flight. Obviously, this is a quantum jump in the efficiency of surveillance, detection of crop's nutrient, water status, and maladies, if any. UAVs are supposed to give agriculture a boost in terms of technology, accuracy, and productivity. Mazur (2016) states that, there are at least six ways, that is, different applications in farming sector, through which UAVs can revolutionize crop production trends. The list includes: (1) aerial survey of soil and field, (2) seeding, (3) spraying crops with pesticides/fungicides, (4) crop monitoring and collecting data on vigor, biomass accumulation, and nutrient status (crops N status), and (5) collecting data and deciphering variations in soil/plant water status and crop health assessment, that is, detecting pest and disease attacks. UAVs can help us in forecasting crop yield. A few other ways by which UAVs can improve farming is by the use of "swarms of UAVs." Crop monitoring and supply of inputs are done rapidly and in time, if swarms are used. UAV swarms can cover a large farm in a matter of minutes.

A few other reports emanating from agricultural UAV companies suggest that, UAVs are versatile. UAVs observe fields/soils and crop plants from above, that is, locations at 300–1000 m above ground level. They are also capable of observing crop plants and the happenings in the field, from a close range of few inches, above crop canopy. Hence, they offer a great advantage to farmers. They offer digital data and maps that could be used, in at least seven different ways, during crop production (Drone Deploy, 2017). They are: (1) crop scouting for growth and detecting pest/disease; (2) observing seed germination and establishment of plant stand, from close range; (3) analyzing stand establishment, plant population, and identifying location with gap; (4) obtaining data on leaf chlorophyll, that is, crop's N status, and crop water stress index (CWSI), that is, crop's irrigation needs and offering most appropriate decisions; (5) UAVs aid in developing variable-rate prescription; (6) UAVs can be utilized in developing "crop surface models (CSM)," forecasting yields, and detecting yield loss; (7) UAVs can quickly identify locations affected by floods, drought, soil erosion, weeds, etc.

Agricultural experts believe that, in future, UAVs may have several important applications during crop production. Many more applications may be discovered and standardized. At present, the need for digital data showing variations in soil/crop productivity obtained using UAVs are important. Economic consequences will also decide the acceptance of these flying gadgets and related computer software (Snow, 2016; Krishna 2016; 2018). Right now, there are only a very few attempts, to assess economic feasibility and profitability of UAVs, in different agrarian regions. The common expectation is that, large-scale production of UAVs (i.e., platforms, sensors, computer software, etc.) and liberalization of rules governing use of UAVs in agriculture are needed. Farmers may then adopt UAVs regularly. Most common applications listed by several different experts are as follows: (1) simple crop scouting; (2) pivot irrigation nozzle inspection; (3) precision spraying of liquid fertilizer or pesticide/fungicide; (4) development of maps of individual fields or large cropping zones; (5) collecting data showing variations in soil fertility, crop productivity, disease/pest-attacked zones, variability in soil moisture, and CWSI; and (6) in providing insurance claims reports (Snow, 2016; Krishna 2018).

6.2 UNMANNED AERIAL VEHICLES AND AGRICULTURAL WEATHER PATTERNS

Experts dealing with UAVs state that UAVs may be small gadgets. Also, they could be infrequently flying over the crop fields. However, their utility in assessing various parameters of weather over farms and agrarian regions seems very important. It may have a significant impact on farmer's decision-making (Wharton, 2013; Richardson, 2014). UAVs, no doubt, have a role in land and natural resource management. They are utilized to obtain aerial images depicting the terrain, topography, land forms, soil types, vegetation, water resources, and general atmospheric conditions. Johnson et al. (2017) have reviewed the present status of usage of UAVs in different aspects of environment. Some of the applications are in (1) monitoring ambient atmosphere, using aerial samples and sensor data, (2) judging environmental conditions in natural vegetation reserves, (3) in developing terrain and weather models, (4) monitoring water resources, and (5) vegetation assessments, including crops in relation to local weather patterns. They have clearly suggested that, there is a need for greater details, and accurate data about utility of UAVs in monitoring environment over crops. Air sampling by UAVs can even allow us to estimate CO_2, CO, semivolatile organic compounds, volatile organic compounds, pesticides, particulate contaminants, etc. We require standardization and repeated trials using UAVs. UAVs could also be used to monitor several parameters related to agricultural weather. UAVs with appropriate sensors provide the data about surface soil temperature, surface soil moisture, crop canopy temperature, ambient temperature, and atmospheric composition (CO_2, N_2O, NO_2, moisture, etc.). Incidentally, National Aeronautics and Space Agency, USA has found application of UAVs, to collect data about ozone layer above agrarian belts. They are studying effect of ozone on weather parameters (Oskin, 2013).

UAVs could be effectively used to assess uncongenial weather patterns and their effects on crop growth and productivity. Effect of flood, drought, or soil erosion could be assessed, by analyzing aerial images picked, by sensors on UAVs. Shi et al. (2016) have reported that, aerial imagery collected using a fixed-winged UAV (e.g., Precision Hawk's Lancaster II) could clearly show the effects of flooding on maize/wheat growth, in a field. Further, phenomics data such as NDVI, LAI, plant height, and biomass were collected periodically. This allowed experts to take note of depression of growth, specifically, in patches affected by flooding.

UAVs have a role in the collection of weather data and in forecasting impending thunder, storms, typhoons, dusts storms, etc. Reports by FAO of the United Nations suggest that, UAVs have an important application in the reduction of loss due to agriculture-related disasters. They are usually part of "early warning" and "first reaction" systems. Agricultural UAVs have been tested thoroughly for their utility, in obtaining data about impending storms; uncongenial weather, particularly wind patterns; droughts; and dust storms.

In Philippines and Bangladesh, fluctuations of water flow into rivers, lakes, tanks, and floods have been detected relatively quickly. So, it allows us more time for initial reaction by farmers (Krishna 2018; FAO, 2016; Galimberti 2014 a, b; NASA, 2013). UAVs could be useful in detecting and mapping areas affected by excessive precipitation. Aerial images of soil erosion, loss of embankments, and loss of seedlings in field are useful. UAVs are used to map locations affected by soil erosion, in cropfields (Terraluma, 2014). Postflood damages could be quickly mapped. Based on which remedial measures could be adopted (Towler et al. 2012). In a few agrarian regions, manual methods for detection of flood-prone regions and mapping could be obsolete. They may need updating. Methods adopting satellite imagery too may need improvement (Shan, et al. 2009). Therefore, UAVs that offer high-resolution images are being suggested.

In addition to their role in collecting useful data about the weather above the farms, UAVs could be utilized in cloud seeding programs (Galimberti, 2014b). Cloud seeding is done using solid CO_2 or silver iodide. For example, weather department in China, it seems, regularly uses cloud seeding. This way, UAVs could have a role in avoiding impending scarcity of soil moisture and droughts. UAVs could be used in micromanagement of cloud seeding.

6.3 SOIL FERTILITY AND CROP MANAGEMENT USING AERIAL IMAGERY VIA UAVS

We can adopt different types of UAVs and their improvised models, to obtain accurate images of a crop. For example, in case of rice, low-altitude remote sensing using small UAV aircraft has shown certain advantages. We can obtain accurate images at visual and NIR bandwidth. We can compute a series of vegetative indices. We can evaluate the UAV-derived data for greater accuracy and correlation with ground data from laboratory analysis. There are indeed innumerable indices that could be adopted to evaluate crops. Following are a few examples:

NDVI (normalized difference vegetation index) = (NIR − r)/(NIR + r);

GNDVI (green normalized vegetation index) = (NIR − g)/(NIR + g);

NGRDI (normalized green red vegetative index) = (g − r)/(g + r);

GRVI (green–red vegetation index) = (G − R)/(G + R);

RGRI (red–green ratio index) = (R/G);

I_{pca} (principal component analysis index) = 0.994/(R − B) + 0.96 (G − B) + 0.914 (G − R);

GLI (green leaf index) = (2G − R − B)/(2G + R + B);

ExR (excess red vegetation index) = (1.4r − g);
ExB (excess blue vegetation index) = (1.4b − g);
ExG (excess green vegetation index) = (2g − r − g);
ExGR (excess green − excess red) = (ExG − Exr)
(Excerpted Saberioon and Gholizadeh, 2016).

At this juncture, it is useful to note that, there are innumerable indices developed. Their mathematical formulae have been listed in literature. For example, Wojtowicz et al. (2016) have compiled a list of over 22 different types of indices that relate to crop growth, its nutrient status, water status and irrigation needs, pest attack, disease, destruction due to soil erosion, flood damage, drought, etc. We should note that, these indices and formulae are relevant for measurements done, using ground vehicles, airplanes, or satellite imagery. The indices may have to be modified to suit the imagery derived from UAVs. Also, such indices derived via UAVs may have to be authenticated thoroughly, before, using them in agricultural farms.

The most commonly utilized crop growth index is NDVI. It has been in vogue, to study general vegetation and crop growth. A set of visual, NIR, and IR cameras are used in most cases. Beyond NDVI, there are other derivatives of vegetation index. For example, SAVI is soil-adjusted vegetative index. It removes the interference caused by soil while obtaining reflectance data (Huete, 1988). There are several other variations to NDVI that are in vogue. Most of them are derived using different formulae. To quote, recently, the trend has been to develop a single or couple of vegetative indices that could be calculated, easily and rapidly. Also, indices offer farmers with reliable information on variations in crop growth and nutritional status. For example, visible atmospheric resistance index (VARI) = $(R_{green} - R_{red})/(R_{green} + R_{red} - R_{blue})$ and triangulation greenness index (TGI) = $(R_{green} - 0.39 \times R_{red} - 0.61 \times R_{blue})$ are the indices being tested by some companies such as Agribotix (McKinnon and Hoff, 2017; Saberioon and Gholizadeh, 2016; Argyrotic, 2017). As a caution, any new method based just on R, G, and B visual sensors needs to be thoroughly compared with the established methods. The basic thrust is to develop vegetation maps or values that offer farmers with data, which is good enough, to judge crop health and vigor. Farmers often need data almost instantaneously. In fact, they like to get data and digital maps immediately after a UAV flight, mainly because, NDVI and other indices could help in deciding remedial measures. Most farm companies in the United State, Europe, and Far East, currently, offer crop analysis data. It includes NDVI, canopy chlorophyll, crops N status, vigor, and disease/pest infection, if any. A few companies also offer CWSI and recommendations for quantity and timing of irrigation.

Rogers (2013), points out that, UAV usage to study crop growth (NDVI) and assess variations in nutrient content has become feasible. It has been attributed to the fact that UAV technology is becoming easier and cheaper. UAVs are useful in assessing the crop's response to fertilizer application. In most cases, the data accrued about NDVI and leaf chlorophyll are accurate, within a margin of 5%. Primicerio et al. (2012) have utilized UAV (VIPtero), to assess crop vigor and NDVI periodically, in grape vine. The idea was to standardize procedures for getting maps showing variations in vigor, NDVI, and biomass accumulation. Based on it, growers could judge the need for nutrient application. UAVs seem to have a wide range of application, during management of grapevines, particularly in managing soil fertility and irrigation. A few other plantation crops such as citrus, apples, bananas, etc., have also been monitored for growth, NDVI, nutrient, and water requirements (Table 6.1).

6.3.1 CROP PHENOTYPING USING AGRICULTURAL UAVS

Phenotypic expression of a particular cultivar may vary in response to environmental factors. Factors such as soil type, its fertility status, fertilizer inputs, irrigation, abiotic stresses, and agronomic procedures adopted may all affect phenomics (Araus and Cairns, 2014; Burkart, 2015; Shi et al., 2016; Zaman-Allah et al., 2015). Studying morphogenetics and phenotyping is currently a very important aspect of crop improvement effort, in agricultural experimental stations. UAVs with their ability for rapid collection of data about crop's traits, right there in field, are gaining in acceptance. Plant breeders, agronomists, and soil fertility specialists seem to prefer UAVs. Primarily, UAVs are nondestructive while collecting phenomics data. Traditional methods of collecting phenomics data such as plant height, leaf area index, canopy size, leaf chlorophyll content, biomass, and yield are time consuming. Usually, farm scouts (skilled technicians) take days, if not weeks, to collect, collate, and analyze phenomics data. UAVs are rapid, with possibility for frequent turnaround and relay of raw or processed data, to agricultural experts; hence, Yang et al. (2017) state, sooner or later, UAVs are going to be very powerful farm gadgets. The performance of breeding programs may hinge on phenotyping by UAVs. Nowadays phenotyping is essential during selection of crop genotypes.

Agricultural researchers have mostly adopted a series of phenotypic traits of a crop species to assess the growth pattern; also, to quantify extent of crop's response to factors such as fertilizers, pesticide sprays, irrigation, etc. Until recently, manual methods that are tedious and costly were adopted. It continues to be so in many experimental stations. Sankaran et al. (2015) have compiled a series of phenotypic traits and standard methods used prior to advent of UAVs. According to them, most commonly used phenotypes are:

a) Plant height that was measured previously using a scale could now be assessed, using LiDAR system. The elevation of terrain and crop, and ability of the UAV, to direct its sensors in oblique fashion is important.
b) Plant biomass is estimated using destructive sampling and weighing, either in fresh or dried condition. Visual rating by experts is also common. UAVs with sensors that operate at visible to NIR are utilized, to get NDVI. We can then compute biomass.
c) Seed germination, emergence, and establishment are measured using visual scores, by trained technicians and actual plant counting. UAVs can offer images of canopy in a few minutes. Using hyper-spectral cameras even each individual seedling could be marked (using GPS co-ordinates). Then, plant number could be counted at ground station.
d) Flowering stage: Flowering is usually visually scored for 50% flowering. UAVs with visual range sensors (R, G, and B) could be adopted, to assess flowering. UAVs usually cover larger areas and rapidly, compared to manual methods.
e) Estimation of plant water status: Visual rating, soil moisture determination, and porometers are used, if done manually. UAVs fitted with visual, NIR, and IR (thermal sensors) offer data about crop's canopy temperature and ambient air temperature. We can use them to calculate CWSI quickly.
f) Plant nutrition and fertilizer requirement is usually estimated using destructive sampling of leaf tissue or entire plants. Samples from several locations within a field are used. Hand-held leaf-N meters are common. UAVs with NIR sensors allow us to estimate leaf chlorophyll and canopy N status. So, fertilizer-N requirement could be calculated rapidly, using appropriate computer programs. We can obtain crop canopy digital maps showing variations in crop-N distribution. This data is utilized in variable-rate applicators.

g) Leaf area is measured using leaf area meter, destructive sampling and plant canopy analyzer, if manual methods are adopted. If UAVs are employed, we derive images using visual and NIR cameras. It helps to estimate the leaf area and biomass.
h) Crops disease susceptibility and incidence scores are usually done manually, on a relative scale. UAVs with sensors can judge the disease incidence better. They adopt spectral signatures of healthy and disease-afflicted fields.
i) Crop's senescence is assessed using visual score, if, manual methods are adopted. Whereas, UAVs with visual and NIR sensors offer plant greenness images, rapidly.

As a bottom line, we should realize that manual methods are tedious and time consuming. Also, trained technicians or experts are needed. Technicians must roam around the entire field to chart out or map the areas. Human fatigue-related errors creep in during data collection; whereas, UAVs offer data in a matter of minutes with better accuracy. Using appropriate computer programs, the digital data could be analyzed rapidly or even instantaneously. Based on it, prescription could be developed (Plate 6.1).

6.3.2 CROP SURFACE MODELS

CSM is a novel approach that utilizes aerial images and digital data derived, using low flying UAVs. The ability of UAVs to offer high-resolution close-up images and accurate digital data about LAI, leaf chlorophyll, canopy structure, and NDVI is utilized, in developing CSMs. In fact, UAVs are being utilized frequently, to obtain plant height data, using oblique shots. This helps in forecasting biomass accumulation trends of crops (Anthony et al., 2014).

Anderson et al. (2014) believe that in the era of "smart farming" that is at the door step, it is most important, to acquire spectral imagery (digital data) of crop surface, at a rapid pace. We may have to cover large farms at short notice, rapidly, repeatedly and then analyze data. We may be asked to offer prescriptions about nutrient and irrigation needs, at rather short notice. So, we need UAVs and CSMs, to accomplish such tasks. We can then forecast final yield accurately. We can then offer most tangible prescriptions, to farmers. Experts at IFPRI state that, using UAVs, we can develop CSM and forecast biomass/grain yield from small plots (micro), farms, large acreages of farm companies, or even a large patch of agrarian region (macro) (IFPRI, 2016). They say, CSMs allow us to forecast yield and in detecting yield gaps accurately.

Forecasting crop growth, biomass accumulation patterns, and grain yield levels attainable is an essential aspect of crop husbandry. Most, if not all, of the agronomic procedures adopted, timing of inputs and quantity depend on yield forecasts. Farmers do revise their yield forecasts and inputs. So, developing CSM quickly and accurately, using UAV-derived imagery is essential. In future, some of the agronomic procedures may entirely depend on CSMs developed, using sensor data from UAVs. Computer software that helps in developing CSM and one that compares them to arrive at prescriptions is very important. For example, Agisoft PhotoScan 0.90 and ArcGIS help in comparing quantitative data.

Maize is an important grain and fodder crop in European plains. Farmers apply fertilizers, water, and plant protection chemicals, after periodic scouting. The data collected is analyzed and recommendations are derived, using various computer programs. Recently, Geipel et al. (2014) have assessed the utility of high-resolution images of maize canopy and vegetative indices, in developing CSM. They state that along with plant height, the spectral data could be useful, in developing CSMs. Farmers may compare data acquired by UAVs, periodically, with known CSMs. Then, rapidly arrive at a prescription. Bendig et al. (2013a, b) have reported that, forecasts done about foliage and grain productivity have often correlated highly (R^2 = 0.74). Further, Bendig et al. (2015) have assessed barley genotypes grown an experimental station. They have used hyperspectral data and CSM. A combination of CSM, plant height, and vegetative indices were used, to assess barley performance. Actual effects of fertilizer inputs were assessed. Predictions about biomass accumulation were accurate, if done using CSM, particularly, at early stages of crop. A few UAV companies believe that use of UAVs to develop CSM may enthuse farmers in German plains to use smart farming. Farmers could rapidly acquire data, analyze it, and arrive at prescriptions, using CSM (Ascending Technologies Gmbh, 2016).

The concept of CSM has also been evaluated for plantation crops. Kalisperakis et al. (2015) have developed CSM for vineyard grapes. They used hyperspectral data derived through a multicopter UAV (model: ONYX-STAR BAT-F8). They acquired hyperspectral data, 2D RGB mosaic imagery, and 3D CSM, then, compared them for accuracy of biomass and fruit yield forecasts. They say spectral data tended to over-estimate LAI. Best correlations were obtained for hyperspectral data, particularly, canopy greenness index with 3D CSM (R^2 = 90%).

PLATE 6.1 A fixed-winged UAV being ready for launch above a soybean cropfield.
Note: Soybean is grown in rotation with wheat. Sensors attached to the UAV help in evaluating crop's performance, in response to application of nutrients and water. CSM could be adopted to evaluate crop's response to fertilizer and water inputs. Disease/pest incidence too could be demarcated.
Source: Clyde Beaver, Creative Services Manager, International Centre for Maize and Wheat, Obregon Experimental Station, Mexico, Photo credit: CIMMYT Archives, Mexico.

6.3.3 LEAF CHLOROPHYLL AND CROP NITROGEN STATUS ESTIMATED, USING AERIAL IMAGERY BY UAVS

Agricultural UAVs are used to detect variations in the leaf chlorophyll status. The UAVs fitted with multispectral sensors are used, to conduct spectral analysis of crop canopy. Survey of literature indicates that such studies on leaf chlorophyll encompass a range of crop species (Table 6.1). The basic principle is that leaf chlorophyll content varies depending on soil fertility and crop-related factors. Leaf chlorophyll content is directly proportional to leaf N status. Therefore, it is indicative of crop's N status and requirements. Recommendations about fertilizer-N supply often depend on chlorophyll estimations, done using chlorophyll meters (Francis and Piekielek, 2012; Rambo et al., 2010). Leaf-chlorophyll estimation is done at different stages of the crop, mainly to assess fertilizer-N requirements. There are innumerable examples that prove the fact that spectral measurement of crop canopy done, using sensors could be utilized, to assess crop's N status (Table 6.1). The basic concept is that canopy chlorophyll index is directly related to leaf greenness multiplied by biomass or ground cover (Muharam et al., 2017). Leaf chlorophyll estimated using UAVs are often correlated with those measured via leaf chlorophyll meter and/or laboratory chemical extraction and estimation (Krishna 2016, 2018; Kokila et al., 2017). Further, canopy/leaf chlorophyll is directly related to leaf/canopy N content. The multiplication factors, correlation coefficients, and equations need to be standardized for a specific crop.

Nitrogen is a key input to paddy fields. Knowledge about crop's N status and fertilizer-N needs are necessary, almost on a weekly basis or at even shorter intervals. Crop's N status is determined periodically, to decide the in-season, split-N dosages. Fertilizer-N supplies are often dependent on yield goals and anticipated losses to soil and environment. Rapid detection of N deficiency in the rice crop is difficult if tedious hand-held leaf color meter methods are adopted. The recent trend is to adopt UAVs with a series of sensors that operate at visual and NIR bandwidth. We measure the leaf chlorophyll content and compute the crop's N status. In case of rice crop, Saberioon and Gholizadeh (2016) have examined the possibility of using low altitude remote sensing aided by a small fixed-winged UAV such as "Swinglet Cam" (Chapter 2). They have indeed evaluated a series of vegetative indices derived using sensors.

Hunt et al. (2015) conducted a field trial with potatoes exposed to different levels of fertilizer-N. They found that aerial images, plus NDVI and GNDVI data could easily detect the effects on low-N. The canopy reflectance could indicate N deficiency/sufficiency. UAV-aided spectral images clearly depicted the growth effects of different levels of fertilizer-N. Lowest levels of fertilizer (control plots) developed N-deficiency rather rapidly. The canopy reflectance pattern clearly depicted N deficiency. A Tetracam Hawkeye UAV system fitted with visual and NIR sensors were used, to assess NDVI and GNDVI.

Perry et al. (2016) aptly state that, UAVs and sensor technology could be utilized effectively, to assess nitrogen status of crop canopy. We can then arrive at suitable fertilizer recommendations. Such techniques are said to be accurate and cost effective. So, they provide definite savings on fertilizer-N supply. They have proved that estimation of canopy chlorophyll index on apples and pears using UAVs, helps in assessing crop's N status.

We should note that, on turf grasses and pastures, the spectral reflectance data is a useful parameter. It helps us in assessing plant nutritional status and needs. Let us consider an example. Caturegli et al. (2016) grew three different turf grass species, namely, Cynodon dactylon, Zoysia matrella, and Paspalum vaginatum. These species were grown at different levels of fertilizer-N supply (input). They determined NDVI using leaf meter, green seeker hand-held leaf colorimeter, and UAV-derived spectral values. The N% in grass clippings could be assessed using leaf color meter, while UAV-derived data could be used to get NDVI values. It is said that for a relatively small area, hand-held leaf color meter may suffice. However, for large areas, UAVs with multispectral sensors should be used, to obtain NDVI and canopy/leaf chlorophyll data (Caturegli et al., 2016). Generally, UAVs could be used to assess sod farms, golf courses, turf stretches, and pastures for growth, vigor, and nitrogen status. In fact, in North America, agricultural consultancy firms are now offering to monitor pastures and turf grass zones, using small UAVs or even long-endurance blimps.

TABLE 6.1 Agricultural UAVs Are Utilized to Measure NDVI, GNDVI, Leaf Chlorophyll, Leaf N, and Crop N Status: A Few Examples.

Crop species	Parameter estimated using UAVs	References
Maiz	Leaf chlorophyll (R/R + G + B)	Kokila et al. 2017
	NDVI, leaf chlorophyll, LAI	Shi et al. 2016; Bendig et al. 2013a, 2013b
Wheat	NDVI, Leaf chlorophyll, crop N status	Lelong et al. 2016
	LAI, plant height, leaf chlorophyll	Shi et al. 2016; Hunt et al. 2015; Croft, 2013; Song and Wang, 2016
Rice	NDVI, canopy vigor, and leaf chlorophyll	Swain et al. 2010
	Several Vis-NIR band derived indices and leaf chlorophyll index	Saberioon and Gholizdeh, 2016
Sorghum	Plant height, NDVI, LAI, CWSI	Shi et al. 2016; Van der Staay, 2017
Potato	NDVI and GNDVI were estimated to assess variation	Stevenson, 2015; Oregon State
	in leaf area and ground cover, also leaf chlorophyll	University, 2014; Hunt et al. 2015;
Apple	Canopy chlorophyll concentration index (CCCI)	Perry et al. 2016
Grapes	Leaf chlorophyll and carotenoid pigments	Zarco-Tejada et al. 2013;

TABLE 6.1 *(Continued)*

Crop species	Parameter estimated using UAVs	References
	Vigor, NDVI maps, LAI, Yellowing maps, N and K+	Primicerio et al. 2012; MicaSense, 2017
Citrus	Plant vigor, NDVI, leaf chlorophyll, leaf-N, tree water status	Bouffard, 2015; Colaco and Moulin, 2014; Colaco et al. 2014; Macarther et al. 2005
Banana	Spatial variations in NDVI, leaf chlorophyll and CWSI	Machovina et al. 2016
Pears	Canopy chlorophyll concentration index (CCCI)	Perry et al. 2016
Turf grass	NDVI	Caturegli et al. 2016
Pastures/ forages		Von Bueren et al. 2015

6.3.4 APPLICATION OF FOLIAR FERTILIZER-N, USING UAVS

Crop productivity depends immensely on timely supply of nutrients to the rooting zone. The nutrients are supplied usually via chemical fertilizers or organic manures. The form of fertilizer (granules or fluid), timing, and concentration of nutrients released to roots, in available form, are crucial. Fertilizers are also supplied as "foliar spray," so that, crops can absorb nutrients via leaf surface (hydathodes). They are supplied as solution or emulsion. Most commonly, nitrogen fertilizers such as urea are supplied as foliar sprays. Foliar sprays of liquid fertilizer-N are done as 0.2–0.3% urea solution. That means fertilizer needs decrease enormously, if, it is supplied as foliar sprays when compared to soil application. Also, soil processes that affect the "fertilizer utilization efficiency" such as chemical fixation, loss via percolation and erosion are avoided, if farmers adopt foliar sprays. Farmers often resort to foliar sprays to supply the in-season split application of fertilizer-N. Foliar sprays could be swift. UAVs fitted with liquid tanks and nozzles/atomizers are used. During precision farming, UAVs are supplied with digital information depicting variation in soil-N fertility and crop's need for nutrients. Such data is utilized by CPU of the UAV, to control the variable-rate nozzles and release foliar fertilizer-N, accurately, at each spot. More than 30 different models of UAVs, mostly copters with tanks and aerial spray bars/nozzles have been listed, in Chapter 3. Such UAVs often carry 5–10 L of nutrient solution. The UAVs spray it uniformly when the recommendation is for blanket distribution. UAVs could be fitted with variable-rate nozzle system, if precision farming is prescribed. UAV models such as RMAX (RMAX, 2015) and DJI MG-1 AGRAS are apt (DJI, 2017; Real Agriculture, 2017; XM2Aerial, 2017). There are several popular sprayer UAVs. A few of them with slightly larger tanks of 15–20 L are sought (Tiltuli, 2016; Media Al, 2016). For example, Unzerbrechlich X820, Trump Sprayer UAV (Model S4-1400), SZM-X4-1200 Agricultural Sprayer UAV, and Pulverizador Sprayer Drone (15 kg) are the copter UAVs with 10–15 L tanks. These agricultural UAVs could be best suited to distribute foliar fertilizer-N, that is, in-season split dosages. Of course, they could also be utilized to distribute granular fertilizers, uniformly on soil surface.

6.3.5 ESTIMATION OF POTASSIUM STATUS OF CROPS USING AERIAL IMAGERY BY UAVS

Since several decades, crop's K status has been measured by obtaining leaf/plant samples methodically, from entire fields, then, resorting to chemical extraction and estimation. These procedures require skilled technicians proficient in chemical estimations and cartography. They cannot be repeated often. However, a farmer may like to assess his fields for soil/plant K status and trace the variations in K, frequently. Further, accurate digital data depicting soil K fertility variations are required, to guide variable-rate K applicators, during precision farming. It is believed that if we successfully standardize UAV-aided sensor techniques, then, it could be rapid and accurate. It can offer us data on soil/plant K variation in one go (Severston et al., 2016).

Recently, UAVs have been adopted to assess crop's potassium status. The canopy reflectance data obtained by sensors on UAVs are useful. They offer adequately accurate assessment of crop's K status. This concept has been tested on crops like canola (Brassica sp.) (Severston et al., 2016). Higher resolution images of the canola crop could be obtained, by flying the UAV at low height above the crop canopy. Interestingly, such UAV images revealed greater details about the effect of K deficiency. They could clearly detect several symptoms caused by K deficiency such as reduced growth, accumulation of N in leaves, greater density of infection by aphids, etc. Many of the images confirmed the variations in K in the field soils, particularly, spots with deficiency of K.

Recently, MMC Hydrogen-powered UAV which is an octocopter has been used, to assess plant stress of different intensities (Fulwood, 2016). Maps generated by UAVs showed an excellent correlation spatially regarding the K^+ stressed patches of canola. The hydrogen-powered UAVs were used for 4 months, that is, during the entire crop season. This was done to verify the accuracy of aerial imagery, particularly, in identifying the K^+ deficiency in soils. This information could be used in digital form in the variable-rate fertilizer applicators. Incidentally, crop growth (NDVI) was relatively low in fields with K deficiency. Further, crop became susceptible to pests such as aphids (Fulwood, 2016). We may have to study and accumulate data about spectral signatures of K-deficient crops, as observed by UAVs. We should also pick data at different stages of crops. Such data banks are essential for decision-support computers to refer. In a different study, grapevines in San Joaquin area of California have been assessed for K^+ deficiency. They have used MicaSense Red-edge cameras. Researchers dealing with UAVs have obtained K^+ atlas and digital data that could be utilized, during fertilizer supply (MicaSense, 2017).

6.4 UNMANNED AERIAL VEHICLES IN STUDYING WATER RESOURCES, THEIR DYNAMICS, AND MANAGEMENT IN CROPPING ZONES

UAVs have been evaluated repeatedly, for their utility in conducting various tasks, such as collection of data relevant to water resource management. They are used in judging crop's water needs at various stages of crop growth. UAVs also help us in allocating and monitoring irrigation systems, in a farm. Recent literature suggests that, UAVs help in studying aspects of water shed hydrology. Sensors offer data about dynamics of water, at fine scale, in the terrestrial and aquatic ecosystems (Anderson et al., 2012; Vivoni, 2012). Further, Vivoni et al. (2014) point out that, in an agroecosystem, UAVs are utilized to study interactions of soil and water. UAVs are versatile and offer greater accuracy. Hence, they are preferred. However, tethered balloons and blimps too could be adopted, to study water resources (McGrey and Saripalli, 2013; Johnson et al., 2014). UAVs have been effectively utilized to study the dynamics of water, during and after a monsoon. UAVs have been effectively used to surveillance dams, lakes, ponds, and irrigation channels, periodically (Jimmenez-Bello et al., 2013; Achtilik, 2015, McCabe, 2014, Detweiler and Elbum, 2013). UAVs with appropriate sensors have offered data about spatial variations of water, in water shed. UAVs have been utilized to estimate evapotranspiration trends, stream and lake water temperature regimes, and groundwater discharge (Templeton et al., 2014). Recently, Krishna (2018) has reviewed and identified several important roles for UAVs in assessment and allocation of water resources, and in irrigation of crops. UAVs are involved in keeping vigil of dams, reservoirs, and irrigation channels, round the clock. They are utilized to obtain data about CWSI. Such data helps to decide the quantum of irrigation. UAVs are utilized effectively to monitor the irrigation equipment and their functioning (Grassi, 2014). Centre-pivot sprinklers and their accurate movement in a field and unhindered working can be monitored effectively. Clogs in sprinkler jets are easily detected (Heck, 2016). UAVs, of course, are among the best bets to assess drought or flood damage, to an agroecosystem or farm or a crop field (Lewis, 2014). UAVs provide data on extent of soil and water loss, due to erosion. We can derive the data regarding flooding/drought almost instantaneously. The digital maps that UAVs offer can be utilized directly in the field, to correct the water status or control erosion. Manual assessment of drought/flood or mapping irrigation requirements of a crop field or farm is tedious. Manual methods are time consuming and several types of inaccuracies creep into the data sets. DeBell et al. (2016) state that, selecting appropriate UAV platform (model) and sensors are crucial; particularly, if we intend to obtain best data sets, then, utilize them during assessment of water resources and irrigation of crops. Lightweight platform in ready-to-use condition and fitted with visual, NIR, and IR cameras are most preferred for such tasks.

Global agriculture depends immensely on water resources. We produce a large share of total food grains, using irrigation. Further, irrigation efficiency is indeed an important aspect. At present, there is a great interest in adopting UAV-aided techniques that provide data about crop phenology, biophysical parameters, and plant's physiological disposition to water (Jones and Vaughan, 2010; Gago et al., 2015a, b; Krishna, 2018). In a recent review about wide-ranging applications of agricultural UAVs, Gago et al. (2015a and b) point out that, UAVs offer exciting opportunities regarding studying water requirements of crops, particularly, in monitoring changes in CWSI. Also, in deciding quantum of irrigation, that crops may require, at different growth stages. It involves measurement of ambient temperature, canopy/leaf temperature reflectance, estimation of stomatal conductance (Gs), and leaf water potential (Ψ). UAVs are rapidly and repeatedly deployable. They are also quick to offer useful data about crop's water status. The computer programs and decision-support systems are already in place for some crops. They are used to decide quantum and frequency of irrigation. Thermal imagery of crop fields is the centerpiece of the methods. Thermal sensors help in obtaining data about crop's canopy temperature, ambient temperature, and soil temperature. These parameters help us to calculate CWSI. They further state that leaf chlorophyll content, that is, GNDVI could be mapped for the same crop or spot, in a cropfield. They could be overlayered while analyzing the data. So, it adds to the accuracy while judging crop vigor, CWSI, and irrigation needs.

Basically, the canopy temperatures recorded by the IR cameras located on UAVs are related to air and soil water content, actual transpiration, and crop water stress (Jackson et al., 1981; Idso et al, 1986). If the crop experiences dearth for water, then stomatal regulation limits the loss of water and conserves it, in leaf tissue. So, water deficit leads to a situation, wherein, stomata close and conserve energy and water. It means such canopies will show up higher temperature than those supplied with ample water. Often in practical field studies, we find difficulties in recording canopy temperature, alone. Land surface temperatures may confound the data. Hoffman et al. (2016) point out that, due to such interference, researchers have tried to select crops with large canopies such as grapes (Baluja et al., 2012; Gago et al., 2013) or olive trees (Berni et al., 2009). No doubt, we should develop methods to correct interference by soil surface temperature reflectance.

According to DeBell et al. (2016), knowledge about water resources in the vicinity of farms and its shrewd management, during crop production is important. Water resource management is adopted at different scales. It ranges from a small plot, to a field of a few acres, to a big farm of several 1000 ha, or even a small agrobelt. The data accrual about water resources in the catchment must be timely and apt. They say, UAVs (or drone) are among the most recent techniques available to farmers. UAVs can offer data at appropriate scale and at short intervals. DeBell et al. (2016) further suggest that, UAVs are becoming important components of water resource management, during crop production. UAVs are offering the fine scale data about crop's water status. In an agrarian region, UAVs with visual, NIR, and IR (thermal) cameras can provide data about the water resources and their dynamics, in the catchment area. They can help us to allocate water accurately within a farm. Such fine scale accurate data about canopy water status cannot be obtained, using satellite or aircraft mounted sensors. However, such fine scale assessment of CWSI is required due to heterogeneous distribution of soil moisture, in crop fields. Thermal imagery by UAVs helps in marking the spatiotemporal variations in CWSI. They point out that regular use of UAVs with multispectral and IR sensors is still feeble. There is a strong need to integrate UAV technology, during water resource management, in a cropfield.

Barley is an important crop in the temperate regions of Europe. It includes Scandinavian countries such as Denmark, Sweden, and Finland. Barley does encounter uncongenial weather conditions. Its need for water in growing season differs, based on climate. Investigation using UAVs has shown that CWSI for barley can be estimated with optimum accuracy. Usually, the thermal imagery provides maps showing canopy temperature, ambient air temperature, and land surface temperature. However, often, canopy temperature may be a composite of land canopy/temperature. Soil reflectance confounds data, particularly, during early seedling stages. However, as the crop canopy gets established, in the mid and late season, we can derive accurate maps of barley canopy temperate and air temperature. This allows us to calculate CWSI. In Germany, Bendig et al. (2015) have utilized CSM to compare the digital data obtained, using UAVs, with previous models. They have prescribed irrigation water based on CSM.

Labbe et al. (2012) state that UAVs are useful in estimation of water stress experienced by crops. Field experimentation with soybean at Montpellier (France) suggests that, canopy temperature derived from UAVs and ambient air temperature known from the nearest weather station could be utilized, to calculate CWSI. The difference between surface air temperature and canopy temperature is a good indicator of crop's water status. During periods of stress, if crop limits its transpiration via stomatal regulation, then, canopy temperature increases.

Torres-Rua (2017) states that sensors attached to UAVs are crucial components, in terms of datasets that farmers can get, per flight. Usually, R, G, B, NIR, and IR sensors are loaded on to UAVs. These sensors could be effectively utilized to monitor soil moisture, crop's evapotranspiration trends and crop's water stress index (thermal sensors), etc. Using such data, we can calculate irrigation needs of the crops. Ultimately, we may gain in accuracy regarding timing, frequency, and quantity of irrigation required by crops. Aerial imagery also collects data about water movement in canals (Torres-Rua, 2017). In fact, in Utah, USA, agricultural researchers have recommended the use of UAVs to monitor county-wise irrigation canal systems.

No doubt, agricultural UAVs that are attached with thermal sensors are useful in getting thermal images. Such images could help farmers in ascertaining water stress that tree crops suffer. Gomez-Candon et al. (2017) state that datasets could be obtained using computation of spectral indices and image classification. It helps to assess drought stress experienced by individual apple trees and entire orchard. Entire apple orchard was subjected to different levels of drought stress, by withholding irrigation. Subsequently, datasets were obtained at noon during the summer season. The thermal sensors offered data that depicted spatiotemporal variations of water stress, both at inter- and intra-canopy level. Generally, water stressed trees showed higher canopy temperature compared to well-irrigated normal tree canopy. The variations in temperature within canopy could be observed periodically and mapped. The datasets from UAVs clearly showed that, tolerance to drought stress differed between genotypes of apple.

6.5 UNMANNED AERIAL VEHICLES HAVE A ROLE TO PLAY IN DETECTION AND CONTROL OF INSECT PESTS IN CROP FIELDS

We are still learning the rudiments of UAV technology as applied to detection and control of insect pests. Our knowledge about spectral signatures of pest-attacked and healthy genotype is feeble. Spectral signature banks need to be developed. They should at least cover most, if not all, of field crops and plantation species. Sprayer UAVs are being manufactured and used in good number, particularly, single-rotor helicopters and multicopters. Yet, their economic efficiency needs to be evaluated in farmers' fields. Right now, there is a great interest evinced on UAV technology as applied to plant protection. Let us consider a few examples in the following paragraphs.

A project involving UAVs aims at monitoring crops such as wheat grown worldwide. UAVs are utilized to surveillance wheat crop for disease, pests, and any other maladies. The idea is to rapidly scout the crop in any location. Later, inform the farmers about the size of the pest or fungal damage and remedies required. This is a global wheat-related project run currently by Kansas State University and Plant Biosecurity Projects of Australia and New Zealand. This project aims to standardize UAV-aided surveillance of wheat disease/pest. The aim is to disseminate the knowledge rapidly. McComack (2015) states that, early detection of pest/fungal infestation is essential. Small UAVs are efficient in detecting and offering aerial images of infestation. The suggestions about appropriate UAV model, sprayer accessories, and sprays could also be disseminated to researchers/farmers in different parts of the world.

Soybean production in USA involves intensive production practices. This vast stretch of 75 million acres is experiencing a definite change in the way pests, particularly aphid infestation is detected. Farmers are rapidly changing over to UAV-aided aerial survey and spectral analysis of crops to assess aphid damage (Kansas State University, 2015). UAVs with ability to spray the fields rapidly are a boon, to soybean cultivators. UAV models efficient in aerial images are available. We may simulate, compare, and select the most appropriate UAV model, for the said purpose. UAV characters such as wing type, its size, number of rotors, cameras, size of sprayer bar, number of nozzles, and batteries needed may all have to be weighed out. Spectral data in banks that help in identification of soybean patches with pest/disease attack and healthy ones is needed. Appropriate spectral data must be accrued rapidly, for computers to refer and arrive at accurate decisions.

UAVs seem to have definite application in tracing the infestation by Spodoptera frugiperda (Army worm) and related species. Army worms affect several crop species. Usually, they would have caused several large patches of devastation, by the time satellite imagery can help in tracing the insect attack. A significant amount of destruction of crops would have occurred, even before it is identified, using satellite imagery. However, with the advent of UAVs, we can repeatedly fly and trace out the Spodoptera buildup, at early stages. Early detection and remedy helps in avoiding crop's destruction (Zhang et al., 2015). In Northeastern USA, UAVs have been examined for their utility in the control of insect pest attack and build up on soybean stretches (Vogel, 2014). In Minnesota, they are trying to detect Spodoptera attack, by studying the spectral signatures of healthy and insect-attacked soybeans. They are imaging crops with distorted leaf chlorophyll formation and leaf fall, using UAVs (Tigue, 2014, United Soybean Board, 2013).

Severston et al. (2016) have shown that, low-flying UAVs with ability (sensors) for high-resolution images are useful. They can clearly prove the occurrence of "green peach aphid" on canola. The density of aphid population can be assessed, using aerial images. Usually, images are drawn from a height of 15 m above crop's canopy. It provided a resolution of 8.1 mm and that was enough to judge the aphid attack, accurately.

Canola grown in some parts of Australia is prone to damage, due to slug infestation. General prophylactic methods include burning stubbles, culling, and spraying after tracing the slug-caused damage. The time between first sighting and build of pest and damage is, it seems, small. Therefore, it needs constant monitoring, quick detection, and application of pesticides. Researchers at GRDC, Victoria, have adopted UAVs to detect slug attack. They are using spectral signatures of healthy and slug-attacked patches. They report that, time required to fly the UAVs (UX5 model, Trimble Inc. CA, USA; (Trimble, 2015)) over the canola fields, retrieve the digital data into GCS computers and analyze the images is optimum. We may still need a couple of days before deciding on a spray strategy and quantum. This time lapse may eventually get smaller. Therefore, pesticide treatment could be hastened (Beveridge and Russel, 2015).

Let's consider a pest that attacks plantations. Cao et al. (2012) have shown that spectral signatures of peach tree canopy can be monitored and studied, using small UAVs. They further state that UAVs could be used to detect, spatially mark crop zones/trees and spray pesticides, only at spots that require attention. Again, it has been suggested that spectral data banks and computer software, to assess disease/pest damage on peaches are required. The development of efficient UAV models is also essential.

Phenotyping crops exposed to insect damage and estimating yield loss due to insect invasion is important. Clearly, in future, agricultural UAVs may play a vital role in detecting insect pests and their damage, at an early stage. UAVs offer greater accuracy. They could be flown over crops, in-season, to detect buildup of population and destructive behavior of pests on crops. Pests-devouring leaves, attacking panicles, stem, fruits, etc., may be detected, using UAVs. Spectral signatures of healthy and pest-attacked crop species/genotypes are utilized to detect and quantify the pest damage. This obviously helps agricultural agencies to arrive at the most appropriate pesticide (chemical brand) and quantity to be applied on to crops. Agricultural UAVs could be most useful in phenotyping crop genotypes and their reactions to insect attack. They say, resistance to pests or tolerance depends on the mechanisms by which a host plant suppresses insect and its destructive activity. So, a crop tolerant to pest grows proportionately better than the susceptible cultivar. Phenotyping susceptible and resistant genotypes will be helpful (Goggin et al., 2015). In fact, regular phenotyping of crops is essential during crop breeding. In future, agricultural experimental stations evaluating crop genotypes for pest resistance may use UAVs, regularly, to collect data on phenotypic expression of crops.

6.5.1 UAVS TO SPRAY PESTICIDES

Effective pest control involves both rapid detection and application of pesticides, to a crop field. As stated earlier, UAVs would have a major role in detecting the pest attack, using spectral signatures of healthy and pest-inflicted crop. There are indeed several examples wherein rapid scouting, by using UAVs, has helped farmers to control insect attack. Such examples pertain to rice, wheat, and soybean (Krishna, 2018). In Japan, UAVs have been consistently utilized, to spray the crop with pesticides. For example, RMAX, a helicopter UAV by Yamaha Motor Company is a preferred helicopter model. It is used to spray the crops. Similarly, MJ-AGRAS1 by DJI Company, China is a preferred multirotor sprayer UAV. Several different models of helicopters and multirotor (VTOL) sprayer UAVs have been designed, tested, and used (Chapter 3; Krishna 2018). They all possess ability to follow instructions from either a remote controller or those prescribed by decision-support computers based on spectral maps of insect spread. The UAVs used for application (spraying) of pesticides (or even fertilizers) usually have tanks, to hold the formulation. The tanks could hold 5–20 L of pesticide formulation. Several different types of nozzles and atomizers have been evaluated, for dispensing liquid/granular pesticide formulation (Huang et al., 2009; Ru et al., 2011; Aerdron, 2017). Pesticides could be dispensed as uniform blanket spray, all cross the large fields. Precision application is possible, if there are computer decision-support and variable-rate nozzles fitted to UAVs. Normally, metallic spray tracers are used to quantify the spray deposition. Spray rates are also affected by the swatch width, number of passes, and flight path strategy. For example, Giles and Billing (2015) report that, work rate ranged from 2–7 ha h^{-1} depending on swatch width and number of passes. They say, on an average 2–2.5 ha h^{-1} of grapevines could be dusted or sprayed with pesticides, using RMAX helicopters.

In the Far East China and Japan, farmers are now in the early stages of adopting multicopter known as MJ AGRAS-1. They are utilizing it to spray pesticides as blanket spray all over the fields. The UAV keeps the farmers safe without getting exposed to pesticide. It completes spraying at a much rapid pace than human skilled sprayers. In Southern Japan, researchers at Saga University have tested a quadcopter with multispectral cameras and sprayer bar, plus variable-rate nozzles. It first identifies spots attacked by pests, using their spectral signatures and then lowers its altitude to just a couple of meters above rice crop's canopy. It sprays pesticide at only spots affected. This way, it reduces on pesticide usage enormously. This UAV model has also been tested on crops such as sweet potato and soybean, again, to detect and control (spray) pesticides (Atherton, 2017).

Yue et al. (2012) have opined that efficient pest control involves rapid detection and forthright immediate adoption of spray schedules. Time lapses between onset of pest attack and detection must be small. Remote-sensing satellites do allow us to analyze and detect pests, but only when it occurs in large patches. By then, considerable crop loss would have occurred. Further, turnaround time of satellites, haziness, and low resolution are the limitations. Hence, UAVs with ability for high-resolution images and repeated flight capability are preferable. Immediate analysis of aerial images seems the best option, right now. They say, in China, adoption of UAVs in large scale seems a good idea, particularly, to detect and control pests. Already, farmers in China do use helicopter sprayers. UAVs spray pesticides to rice, wheat, and soybean crops (Krishna, 2018).

Let us consider another example. Giles and Billing (2015) have utilized a helicopter UAV (Yamaha's RMX) to spray grapevines with pesticides. They have tested the spray efficacy on "Cabernet Sauvignon" wine grapes located in University of California, Davis, Experimental Station. Using UAV, they could spray 2.0–4.5 ha h^{-1}. The UAV dispensed about 14–39 L of pesticide formulation per hour. The spray volume could be controlled by manipulating the nozzles using GCS computers. UAV-aided spray offers great advantages. Particularly, in terms of the total quantity of pesticides needed. The need for pesticide decreases enormously, if precision spraying is adopted. UAVs dispense pesticide rapidly in a matter minutes. So, it is highly efficient in terms of cost and need for human labor. UAVs keep farmers away from the harmful chemicals.

In another study, researchers at University of California, Davis have attempted to utilize low-flying quadcopters. The idea is to detect and control "spider mite" attack on strawberries. They are using spectral data from UAVs, to first identify the spots affected by spider mite infestation. Then, they use a different UAV machine carrying "predator mite," to control the pest. It is a biological control method (Kruetz, 2017).

Bolton (2016) says that UAVs have utility in patrolling crop fields. Also, in providing aerial imagery, nutrient distribution maps, and those depicting pest/disease-attacked spots. In addition, it is interesting to note that UAVs can deliver "glow light." It is an electrical payload. It points to the location where insects are congregating. It is comparable to light traps with black or white light.

6.6 UNMANNED AERIAL VEHICLES IN THE RAPID DETECTION AND CONTROL OF CROP DISEASE

The UAV-aided surveillance of crops is getting more frequent in large farms. Next, detection and mapping of disease incidence could become a routine procedure. Most farms may adopt spectral analysis as a tool, to decipher the type of disease and its spread across different locations, in a crop field. The use of satellite imagery to detect large patches of disease-affected crops is not new. However, satellite imagery does not offer high-resolution images of healthy and disease-attacked crops. The images could be hazy and turnaround time to reassess crop loss is uncongenial, to most farmers. Hence, in future, UAVs will be sought more often to keep a regular vigil and judge the crop's health.

The collection of spectral signatures of healthy and disease-afflicted crops is the central piece of this technique. In a crop season, periodic flights above the crop's canopy may also provide data about temporal changes in disease occurrence, its spread rate, and intensity. Upon obtaining high-resolution images and maps, the digital data could be channeled to decision-support computers. Such computers send commands into variable-rate applicators fitted on sprayer UAV. So, UAVs keep farm labor away from contact with harmful chemicals.

The UAVs could be applied at two stages. Firstly, to make aerial survey of cropped fields, to detect disease-affected areas and map them accurately. Secondly, multirotor UAVs with an ability to spray plant protection chemicals could be utilized. We may note that, UAV-aided crop disease control is getting more commercial and streamlined. There are now many private UAV companies that offer famers with crop's spectral data and disease situation, for a fee. This is done periodically and at short intervals, so that, disease buildup is avoided. The private agency usually offers accurate suggestions about the disease, its spread, and potential damage, it may cause. The plant protection chemical (brand and active ingredient) to be sprayed using UAVs are also prescribed. So, UAVs could reduce farm drudgery required while scouting the crop for disease attack, also, while spraying the crop with appropriate chemicals (HSE, 2015; Trimble, 2015; 2017; Lacewell and Harrington, 2015). Krishna (2018) has reviewed usage of UAVs to detect and control various diseases affecting field and plantation crops. Let us now consider a few examples of crop diseases that could be controlled, using UAVs.

Several UAV models/sensors have been screened above wheat fields. The aim is to select one that offers accurate imagery of wheat crop affected by yellow rust (Puccini recondita). UAVs offer images showing patches of crops affected by rust (Huang et al., 2008b). The spectral reflectance of rust-affected zone is usually higher. The crop is usually imaged at 560–570 nm band.

Maize is an important crop in tropical East Africa. Here, it is severely exposed to attack by rust fungus, Puccinia sorghi (Agape Palilo, 2014). The disease reduces photosynthetic efficiency of crops. The rust pustules are thick and red or black. Spectral data obtained clearly depicts rust-affected and healthy zones. Reports suggest that small UAVs could be used routinely, to monitor rust occurrence on wheat. Early detection is highly beneficial, in reducing crop loss. UAVs fitted with sensors (R, G, B, NIR, and IR) are utilized, to get spectral signatures of rust-affected and healthy crop. Regions with rust disease are then sprayed with fungicides.

Potatoes grown in North American plains are affected by fungal pathogens, pests, and environmental vagaries. In each case, the crop suffers retardation in growth, loss of foliage, and reduced tuber yield. There have been attempts to identify the cause of retarded growth, using spectral data from UAVs and comparing it with ground truth data. A correlation up to R^2 =

0.91 have been obtained. It shows the accuracy of judgment of crop's health, using UAV imagery (Khot et al., 2014). In Louisiana, USA, UAVs have been used to detect various diseases of sugarcane (Schultz, 2013).

UAVs could be adopted to detect disease-causing organism in environment and on crops, at an early stage. Then, prophylactic measures could be considered, accordingly. UAVs have also been flown above crops periodically. This is to collect spores or any other propagules of disease-causing fungi or other detrimental microbes in the air. Therefore, we can judge the buildup of inoculum in the air above canopy. It helps to thwart the disease, before it manifests in a big way (Aylor and Fernandino, 2008; West et al., 2003; 2009; 2010). For example, Aylor et al. (2011) have shown that, sporangia and other fungal propagules of Phytophthora sp. could be detected in the atmosphere, using UAVs fitted with traps. This application, that is, aerobiological sampling of atmosphere above crops is possible, only when spores of disease-causing organism is airborne or is disseminated via storms or sand particles and atmospheric dusts. In addition, this procedure needs microscopic examination of samples or media to grow microbes held in air samples. Appropriate selective medium that captures microbes in petri plates is required.

UAVs have been utilized to detect spectral differences between healthy and disease-affected field crops, for example, wheat, maize, rice, soybean, etc. (Krishna et al., 2018). UAVs are being evaluated for use in cropfields affected by leaf diseases. The foliage showing discolorations, pustules, mosaics, or curls have been picked more accurately, by the sensors. Loss of chlorophyll pigment is among the most easily detected leaf trait by the sensors. However, we may note that UAVs are also useful, to detect diseases like seed rot, loss of germination, seedling wilt, malformations, stunting, and lack of productivity. High-resolution close up shots are possible using visible and IR sensors. So, they can reveal the details of disease spread. Loss of seedlings and gaps can be easily detected, by visual bandwidth images. Panicle disease such as head blight, neck blast, smuts, and ergot can also be imaged using UAVs.

6.6.1 PLANTATION DISEASES AND UAVS

Let us consider a few reports about experimental evaluation of UAVs in plantations. Banana plantations are a common feature of Philippines. They are grown intensely. Its productivity is relatively high. Like any other plantation crop, bananas are afflicted by several different fungal, bacterial, and viral diseases. "Black sigatoka" caused by *Mycosphaerella fijiensis* var. difformis and Panama disease caused by *Fusarium oxysporum vas* cubense are endemic. They reduce foliage and fruit productivity significantly. Recently, a Dutch farm research group has introduced UAVs into banana-cultivating zones, to detect diseases. They adopt periodic aerial scouting by UAVs. Disease control measure, that is, spraying fungicide has also been accomplished, using UAVs (Triple 20, 2015).

Reports suggest that UAVs could effectively trace and map a disease like "flavescence doree" on grape vines. The UAVs imagery correlated with ground truth data, to an extent of 97%. Such an UAV-aided detection of grapevine diseases at an early stage is said to thwart the damage. Otherwise, the disease potentially reduces grapevine yield, by 35% annually. Simultaneously, farmers could reduce on water usage and irrigation requirements, if UAV-aided surveillance was adopted (Cornell, 2015).

Apple orchards are susceptible to a wide range of diseases. Such diseases reduce tree growth, leaf formation, and fruit production. Sometimes, diseases reduce the economic value of the product. Apple scab is a disease caused by a fungus known as Venturia inaequalis. The scabs and lesions reduce the value of apple fruits. A few apple orchard owners have adopted UAVs, to detect the occurrence and intensity of scabs on fruits. They could adopt suitable sprays, in time (Kara, 2013; Modern Farmer, 2013).

Avocados grown in southeastern USA are being attacked periodically, by "Laurel wilt." Laurel wilt is caused by a fungal complex. The fungal complex is disseminated by Ambrosia beetles (vector). Rapid detection of beetles and fungal infection on stems is crucial. Since, the disease spreads rapidly, timely spraying and culling is essential. Disease symptoms should be identified at the earliest. In a large farm, human scouts may find it difficult, to trace the Laurel wilt symptoms quickly. Usually, damage might have already taken place, by the time skilled labors detect and map the disease spread, accurately. So, UAVs seem to be the best bet. UAVs can be flown periodically over the trees. It could be done at short intervals to bring home highly accurate pictures of trees/canopy. Specialized thermal cameras are said to be useful, in detecting the Laurel wilt, early. In future, we may find more farmers in Florida and Georgia, using UAVs with IR sensors. This is to detect Laurel wilt on avocados (Associated Press 2015; Buck, 2015).

Olive tree plantations in Cordoba, Spain are affected by Verticillium wilt. The causal agent is a fungus—Verticillium dahliae. The disease results in water stress symptoms and low fruit bearing. Calderon et al. (2013; 2015) have shown that UAVs with visual, NIR, and IR sensors could be used, to scout the olive trees, from a very close height. UAVs offer hyperspectral images with greater details. High-resolution images allow farmers, to detect and identify wilt and water stress symptoms, quickly. Of course, UAVs could also be used to spray fungicides. Fungicide sprays are based on digital data obtained via UAVs.

During past decade, Huanglongbing (HLB) disease also called "citrus greening" has affected Florida's citrus belt. It is spread by vector, a psyllid insect named Diaphorina citri. Florida state in USA supports a large citrus belt of 220 thousand ha. It is said that HLB has reached devastating proportions reducing fruit yield (Garcia-Ruiz et al. 2013). Citrus experts at CREC, Lake Alfred in Florida have experimented with UAV-aided surveillance and data procurement, about HLB disease. This is to adopt remedial sprays. The high-resolution cameras on UAVs allow researchers to examine each tree, rather closely, from vantage points. They can mark the affected trees accurately, using GPS co-ordinates. Results indicate that UAVs offer digital maps of HLB that coincide with ground reality data, to an extent of 85% (Ehsani and Sankaran, 2010; Garcia-Ruiz et al., 2013; Sankaran and Ehsani, 2012). UAVs were rapid. They were also low cost and accurate when compared to skilled scouts judging the citrus orchards. It may be a matter of time before a range of different data such as growth, foliage, canopy, water and nutrient status, disease/pest attacks, and fruit-bearing pattern of citrus orchards are scouted, using UAVs. No doubt, digital data could be retrieved rapidly and analyzed, using appropriate computer programs.

6.7 UNMANNED AERIAL VEHICLES IN WEED CONTROL

Weed control is a major preoccupation during crop production, especially, when fields are large. Weeds of diverse species could be traced in most of the cropfields. Weeds are almost endemic to certain agrarian patches/regions. Weeds, no doubt, are detrimental to crop growth. They compete for soil moisture, nutrients, and photosynthetic space. If left unattended, weeds may often outgrow the canopy of the major crop and suppress photosynthetic activity, growth, and yield formation by crop species. Weeds also act as collateral host for many disease-causing agents and pests. During the past two decades, farmers have opted for no-till systems rather too frequently. No-till systems often induce large and intense buildup of weed patches. Weed control during early period, that is, immediately after seeding and at early seedling stage is almost mandatory. In high input farms, farmers use pre-emergent application of herbicides to soil. It could be followed by postemergent sprays, at seedling stage. High-input farms have accepted the use of "herbicide-tolerant" cultivars of crop species. This induces them to adopt high-intensity herbicide spray to control weeds. However, herbicides could be harmful, if it accumulates in soil or persists in crop tissue or contaminates atmosphere. Therefore, at present, most procedures and prescriptions about weed control try to reduce the application of herbicides. Most recent method that is being standardized, tested, and repeatedly evaluated in experimental farms is the use of UAV-aided techniques. The UAV-aided technique is supposed to adopt precision agricultural principles and reduce the use of herbicides, to a very great extent (Krishna, 2018; Glen, 2014; Shi et al., 2016; Samseemoung et al., 2012). UAVs cover larger areas in short flight and bring home data about weed infestation, if any, in the interrow and even intra-row spaces. The sensors on UAVs provide the data based on spectral signatures of crops and weeds that are different. Of course, if we use UAVs, there is need for data banks possessing spectral signatures. This allows GCS computers to consult the data banks, then, identify and verify different weed species and crops. Digital maps depicting weed infestation and intensity could then be obtained. Lottes et al. (2017) point out that, accurate identification, classification, and location is essential, when we use ground robots or UAVs or spray herbicides. Most accurate detection, identification, and classification of weeds in cropfields (e.g., in sugar beet) are possible, using UAVs. UAV-aided aerial images and maps could be 99% accurate when compared with the ground data (Kooistra, 2014). Weed mapping is to be done very early in the growing season, using spectral signatures. Pena et al. (2013; 2015) have described a method that involves object-based image analysis (OBIA) along with spectral signatures, to distinguish weed species and crops. It supposedly leads us to maps and digital data with greater accuracy. The OBIA procedure involves classification of crop rows and weeds, based on dynamic and autoadaptive classification approach. It discriminates weeds and crops based on their relative positions, in a crop field. Then, it generates weed infestation map depicting species diversity and intensity on a grid map. Such weed maps are then used during precision sprays of herbicides, on crop fields. They say, OBIA procedure computes data from multiple sources to classify weed species. It permits calculation of herbicide requirement, accurately (Tropnevad, 2013). Hence, it could reduce on herbicide usage.

Let us consider an example from southern plains of USA, where maize, sorghum, and wheat are prominent cereals. Here, weed management in cereal fields is a constant challenge, to farmers. They need rapid and accurate information about infestation, dominant weed, size of weed species, weed species diversity, and growth stage. Weedy spots along with their size and location should be identified. Routine scouting is almost essential, if site-specific methods are to be adopted. Incidentally, here, Shi et al. (2016) have conducted experiments that examined the utility of UAVs and UAV-derived images, in controlling weeds. Major weed species recorded were Palmer amaranth (Amaranthus palmen), Barnyard grass (Echinochloa crus-galli), Texas Panicum (Panicum texanum), and Morning glory (Ipomea sp.). Spectral data was used to identity and classify the weed-affected patches. Classification produced maps showing soil and general vegetation. It compared aerial data about weeds with ground truth data. Spectral data was also used to study the effect of different treatments on weeds in plots. Shi et al. (2016) have reported a high degree of correlation ($R^2 = 0.89$) between UAV-aided detection of weed species and ground truth data. However, they point out that UAV-derived images did not depict 3D or volume parameters of weeds, in fields. Weed maps were all 2D depiction. Also, rapid transit of UAVs above cereal fields may result in blurs and reduced resolution.

It is interesting to note that, we have now achieved a certain degree of standardization, about use of UAVs, to spray herbicides. There are several models of UAVs that could be branded as apt for conducting aerial spray of herbicides. UAVs, mostly single-rotor copters (i.e., RMAX, Autocopter, Yintong's UX5, etc.) or multirotor types (e.g., MJ-AGRAS-1; Hercules; Chapter 3) are suited. These UAVs could be adopted for both, blanket sprays of herbicides or to supply exact quantities only, on spots infested with weeds. The precision methods involve variable-rate nozzles, digital data, and computer guidance.

According to Rice (2015), the data collected by UAVs is processed, by adopting the following steps:

(1) we must identify the plant characters using spectral data, then match the spectral signatures and classify using algorithms; (2) We have to cluster the like-pixels to estimate the extent of weed spread, and (3) create a product that shows land cover, including crop and weed distribution. This digital output showing weed-infested zones is then used, in variable-rate herbicide applicators. Weed detection, its mapping, and herbicide application procedures are currently conducted, by several private UAV companies. There are now, agricultural UAV companies that specialize in periodic surveillance of cropfields. They detect weeds species, map their intensities, and offer timely prescriptions, to farmers, for a fee. It is very rapid, since UAVs cover large fields of 50 ha in 30 min. The data is accurate. The quantity of herbicide prescribed is very small compared to blanket sprays, if precision techniques

are adopted. Let us consider an example. A farming related company (GAIA Data Collection, 2017) operates UAVs, to collect data on weed infestation in wheat fields. They firstly monitor cropfields using UAVs, then, identify weed sprouting and growth in fields, on a regular basis (weekly). Firstly, they monitor fields to decide postemergent spraying based on weed appearance and weed growth stage. They offer geospatial weed maps to farmers. Farmers may apply herbicides, either manually or using ground robots or UAVs. Maps are often colored showing weedy spots brightly and conspicuously. The farm companies ensure that weed control measures are effective, by periodic flights of UAVs over cropfields. They deliver weed maps and information about field conditions in 24 h, after a UAV's flight over the farmer's fields. Usually, they offer contracts with farmers that cover an entire crop season, for 32–35 weeks. Farm companies have reported that quantity of herbicide required reduces enormously, if UAV-derived maps are adopted. The spraying cost also reduces. Most importantly, weed control is timely, so, it thwarts weed development right at early stages. There are now innumerable agricultural private companies in North America and Europe that offer timely maps of weed infestation and prescriptions for herbicide sprays. They together reduce cost on skilled farm scouts. To quote an example, reports by the UAV company suggested that, 23% of cropped area was weed-free and 47% area had feeble weed infestation, that is, not needing immediate attention (spraying). Hence, if precision technique based on UAV imagery is adopted, herbicide requirement reduces appreciably. Blanket spray would have meant a large dosage of herbicide applied into entire cropped zone. Blanket sprays do not consider weed-free area or low-intensity weed zone (Tropnevad, 2013).

The spectral characteristics of crops noted using sensors on UAVs should lead us to the accurate identification, classification, and clustering of weed species. In most cases, we may identify crop species accurately. For example, Zhenkun et al. (2017) state that, UAV-aided reflectance measurements could identify winter wheat crop grown around Beijing with great accuracy of 96.1%. For maize, it was 90% accurate. Spectral data also helps in accurate discrimination of intercrops. It can also separate weed patches. We may note that discrimination of weeds and crop seedlings is not always easy, using specific spectral signatures. The stage of the crop species, weed species encountered, and their stages also affect accurate identification and mapping. So, agronomists at Oregon State University first wanted to use UAVs, to study reflectance pattern of Brassica crop and weeds that infest the field. Then, they opted to spray the field with weedicides, using sprayer

UAVs (Plaven, 2016). Accuracy of spectral data seems essential. Herbicide spraying using UAVs is gaining greater acceptance, particularly, in large farms. UAVs could be efficient and accurate. They replace a sizeable requirement of human skilled technicians (Krishna, 2018; Cornett, 2013; CREC, 2015; University of California, Davis, 2014). For example, autonomous helicopters such as Yamaha's RMAX cover 1.0 ha cropped land in 8 min of flight. There is clear possibility for improvement of efficiency of UAV-aided herbicide spray systems. The swath width, nozzles, and digital data may all affect rapid release and accurate distribution of herbicides (Zhu et al., 2014). Krishna (2018) lists at least six different most popular copter models utilized during herbicide spray, they are, Yamaha's RMAX, Yintong's YT P5, Crop duster AG-RHCD, Rotomotion's SR 200, AG-V8A Octocopter, and DJI's MJ-AGRAS.

6.8 UNMANNED AERIAL VEHICLES AND PRECISION AGRICULTURE

"Precision farming" is now a decade-old farming method. It is adopted more frequently in developed nations, wherein, farmers or private farming companies own large farms of a few thousand ha. Precision farming involves formation of management blocks of smaller size. Management blocks are based on soil traits such as topography, soil type, fertility/productivity levels, cropping systems adopted, soil moisture levels, input levels, etc. (Krishna, 2013; Zhang and Kovacs, 2012, Zhang et al., 2014; Sharma, 2017; Raymond Hunt et al., 2014; Primicerio et al., 2012; Hafsal, 2016; Abdullahi et al., 2015; Al-Arab et al., 2013; Mulla, 2013). Precision farming is essentially a geospatial technique. It involves use of GPS connected farm vehicles such as planters, variable-rate fertilizer applicators, variable-rate pesticide/fungicide applicators, combine harvesters with a facility for yield maps depicting variable grain yield, etc. Blanket applications could lead us to apply larger inputs that do not match the crop's requirement. This aspect is entirely avoided if precision technique is adopted. Hence, it may lead us to greater efficiency and reduced cost on inputs (Krishna, 2013). Precision farming needs accurate maps showing variable soil productivity, insect/pest attack, or soil moisture status. Agricultural UAVs that fly over the cropfields can swiftly obtain digital data about crop's vigor, NDVI, biomass, leaf chlorophyll (i.e., leaf-N), and water stress index (i.e., water status and requirement). They are currently deemed, as best bets. UAVs are extremely swift, accurate, and low-cost methods compared to human skilled technicians. UAVs avoid human fatigue related errors. The digital data is stored and retrieved in a matter seconds and analyzed, to prescribe the inputs. Inputs can be channeled at variable rates, if digital data is channeled to decision-support computers. Hence, UAVs are among most sought after gadgets in farms adopting precision farming.

One of the advantages related to the use of UAVs in crop fields is the early detection of crop diseases, pests, weeds, or any other malady. The UAVs could be flown above crop fields, say daily, or on alternate days or even once a week. It then allows farmers to get the much-needed detailed surveillance images of crops/soil. It tells about the condition of crops. Just, early detection of pest or disease makes it easy for farmers, to take control measures, in time. This aspect thwarts disease/pest quickly. It also avoids undue damage to crops. Farmers may have to apply much less quantity of pesticides/fungicides. In fact, early detection of nutritional deficiencies, using high-resolution aerial images and NDVI mapping helps farmers, to correct the soil fertility, using much less quantity of fertilizers. Fertilizers are applied only at spots that need nutrients and in apt quantities. It is said precision agriculture measures, if adopted with full-scale support of UAV imagery, reduces cost. The return on investment on UAVs and inputs amounted to 12 USD acre-1 in case of corn, 2 USD with soybean, and 3 USD with wheat (Hampton Technical Associates, 2016). Selection of UAV, its model, and various salient features, all influence economic advantages.

Forecasts by UAV companies suggest that the multibillion agricultural enterprises may soon depend immensely on precision farming methods. Aerial imagery depicting variations in soil fertility, moisture, and crop growth pattern is an essential aspect of precision farming. This will mean that UAVs will be adopted in large number, particularly, to obtain digital data useful for variable-rate applicators. Agricultural UAVs may after all be necessary for regular analysis of crops, in future, if precision techniques are adopted. The inter- and intrafield variations could be detected, then, data relayed more efficiently (Precisionhawk, 2014, 2017; Trimble, 2015; 2017; Aeroscraft, 2017). In addition to small UAV aircrafts, aerostats are also amenable during precision farming of large acreages of cropland. Aerostats could be hovering over fields, to collect accurate data about crop's progress. Aerostats provide us with detailed digital data about soil fertility, moisture, weeds, pests/disease, and growth indices (NDVI, GNDVI, GVI, etc.) (Aeroscraft, 2017).

UAVs adopted for use during precision agriculture could be low-cost models, such as Sky hunter, eBee, Precision Hawk Lancaster, or Trimble's UX5. For example, Skyhunter, which is a small UAV is available at 2000

Euros. It rapidly takes to sky and offers digital maps of over 150 ha in 30 min (Danovich, 2014). A few hours later, the images could be processed at GCS computer. Decision-support computers could then offer accurate suggestions. In all, it is a matter of less than an hour in flight for collecting data. Later, it takes about 24–48 h to arrive at prescriptions. On the go prescriptions are also possible, if we opt for sophisticated UAV with facility for variable-rate applicators (e.g., multirotor copter). A few multicopters are capable of on the go data analysis and prescriptions, at a very rapid pace.

6.9 UNMANNED AERIAL VEHICLES IN AGRICULTURAL EXPERIMENTATION

The UAVs are forecasted to revolutionize data collection, storage, retrieval, and analysis method during experimental evaluation of crops. The maintenance of experimental blocks and entire station may get highly streamlined. Plant breeding and agronomic field trials may become much easier to conduct. The cost of maintenance of experimental farms and conducting a field trial may reduce, perceptibly, if UAVs are adopted.

Forecasts by agricultural scientists at University of Georgia, USA, suggests that, it is a matter of time, before ground robots and UAVs could become as common as tractors or combine harvesters in farms. Further, UAVs may gain prominence in agricultural experimental farms. They could be utilized regularly to collect data about crop growth, phenomics, and tolerance to disease/pest and drought. UAVs are relatively rapid in collecting large amount of data. They do so in a matter of minutes. They also store the data for retrieval and careful analysis, using different statistical programs. The process of collection of data about genetic nature of genotypes of crop species is, right now, tedious. It is believed that, advent of UAVs could hasten crop breeding procedures (Wooten, 2017).

Agricultural researchers in Wisconsin, USA, have adopted UAVs to monitor and assess the performance of several hundreds of wheat genotypes grown in experimental stations. They have collected the data about phenomics, periodically, to evaluate the wheat genotypes. Research staff could offer details of any genotype about which farmers evinced interested, right through the crop season. This was done using internet and mobile networks (DMZ Aerial Inc., 2013).

Researchers at Texas A&M University have recently evaluated genotypes of wheat and maize, for their performance in experimental fields. They have also collected useful agronomic data of crops (wheat, maize, and sorghum) exposed to different soil fertility levels. Shi et al. (2016) opine that, recent trends in crop breeding involve elaborate collection of field data (at short intervals) and analysis. In fact, crop breeding programs spend enormous amount of time and funds. They hire skilled technicians to manually collect routine phenotypic data of segregating populations. Usually traits such as plant height, LAI, plant population, canopy growth rate, biomass accumulation pattern, boot leaf initiation, flowering time, yield attributes such as panicle initiation, size, etc., are measured with great care. Yet, human error may affect data collection. Clearly, manual methods are tedious and costly. Hence, automated data collection and analysis systems are preferred. Ground-based robotic methods are available, but they too could be tedious. They could be slow and cumbersome. At present, UAVs hold maximum promise to offer excellent and accurate phenomics data, to crop experts. UAVs are fast paced and low cost (Shi et al., 2016). Further, if precision farming is practiced it also requires periodic data about crop's performance. We require digital data showing spatial and temporal variations of crop's traits. This is to optimize application of fertilizers, water, and pesticides. No doubt, UAV-aided aerial imagery can offer such accurate data. Also, if experimental fields are large, high throughput data collection at rapid pace is essential. Agricultural UAVs with GCS computers capable of rapid collection and analysis of data become highly relevant. Reports by Shi et al. (2016) indicate that, UAV-collected data on phenomics are excellently correlated with ground data accrued using skilled technicians. So, UAVs may ultimately find their way into experimental stations and large farms.

At Kearney Agricultural Experimental Station (KARE), of the University of California, researchers specializing on sorghum have evaluated and adopted UAVs, to collect crop data rapidly and at low cost. The Idea is to automate data collection in experimental stations. Van der Staay (2017) has collected a series of phenomics data of sorghum plots at the KARE station. She reports that sorghum drought nurseries with over 100 genotypes could be evaluated, using UAVs. UAVs with appropriate sensors were also effective, in obtaining heat stress and soil moisture data within sorghum fields. Researchers aimed specifically at scoring sorghum genotypes for rapid recovery trait, once heat/soil moisture stress was alleviated using irrigation. The recovery symptoms on sorghum crop canopy could be assessed very accurately, using sensors on UAVs. We may note that most of the procedures involved in scoring crop genotypes, for drought tolerance and recovery, are similar to many crop species. In future, it could be possible to evaluate drought nursery of several crop species grown in drought-prone regions.

Sankaran et al. (2015) have reviewed the status about low-altitude, high-resolution aerial imaging systems, for phenotyping of row crops. They believe that UAVs may find more common use during evaluation of field crops, at experimental stations. It is attributable to their ability for regular and rapid monitoring of crops/genotypes. UAVs offer flexibility during data collection. The data collection, its storage, retrieval, and analysis are much easier, if UAVs are utilized. The UAV-based methods cost less compared to manual data collection and analysis. UAVs offer accuracy up to subcentimeter level, during aerial imagery. They can be used to asses a range of physiological parameters of each crop genotype sown in the experimental field (Chapman, 2015; Zarco-Tejada et al., 2013). However, most of the digital data and imagery obtained using UAVs, pertain to canopy level traits (Sankaran et al., 2015). UAVs have been utilized to collect data for traits such as plant height, canopy size, leaf chlorophyll, biomass, NDVI, disease and pest tolerance (i.e., incidence scores and spread), CWSI, yield attributes such as panicle initiation, panicle number, length of ripening, yield forecast, etc. UAVs have also been adopted, to collect data relevant to specific field or plantation species. Table 6.2 depicts a few examples of crop species evaluated in experimental farms, using UAVs.

UAVs offer definite advantages while assessing crops/genotypes at canopy level. They are currently highly useful in obtaining data and digital imagery, to develop CSM. These CSMs are utilized while comparing the performance of different genotypes. The influence of crop husbandry procedures and inputs could be assessed, using CSMs. The CSMs are highly useful to farmers while assessing agronomic progress of their crops, particularly, in large fields. They can adopt CSMs and ascertain crop's progress almost daily. However, ability of UAVs, to assess individual plants in greater detail is limited or even nil. There may be situations when plant breeders need data from each plant in a row. However, a multicopter with ability to hover over a single plant in the row could be tried, to assess a single plant's growth progress. Parachutes and blimps are now being evaluated in experimental farms. They could be used to assess crops/genotypes, in greater detail. We may be able to focus and assess a single plant, if blimps can be made to stay at a particular altitude and a point above the crop field. Blimps with long endurance can be a good suggestion, if experts intend to monitor genotypes continuously and throughout the season.

At Toulouse, France, Lelong et al. (2016) have evaluated the performance of 17 genotypes of durum wheat and 2 of bread wheat. The genotypes were sown at different planting densities. These wheat genotypes were also

TABLE 6.2 Utilization of UAVs in Agricultural Experimental Stations to Evaluate Performance of Crop Species and Their Genotypes.

Crop species/location	Remarks	References
Wheat		
CIMMYT, El Baton, Mexico	Evaluate performance of wheat genotypes in different seasons using blimps. Collect data about CWSI, leaf chlorophyll (leaf-N status, disease/pest attack on each genotype of wheat, using sensors on blimps.	Tattaris and Reynolds, 2015; CIMMYT 2012; Lumpkin, 2012
Oklahoma State University Stillwater, OK, USA	Identification of new germplasm lines with tolerance to low soil fertility and disease.	Reynolds 2012; Babar, 2006; Prasad, 2007; Pinto, 2010
DMZ Aerial Inc. Whitewater, Wisconsin, USA	Evaluation of wheat genotypes using phenomics data collected periodically via UAVs.	Feine, 2017
Washington State University Pullman, WA, USA	Evaluation of wheat seed germination in response to soil moisture and crusting.	Khot et al. 2014 Zhang et al. 2014
	Influence of organic manure on wheat crop productivity.	
Oklahoma State University Stillwater, OK, USA	Estimation of genetic variation in wheat genotypes for in-season biomass, leaf chlorophyll, and canopy temperature.	Babar, 2006
Oklahoma State University Stillwater, OK, USA	Genetic evaluation, analysis, and indirect selection of winter wheat grain yield using spectral reflectance indices.	Prasad, 2007
Texas A and M University Ag Experimental Station, Brazos County, Texas, USA	Evaluation of growth and yield performance of wheat genotypes, using NDVI, LAI, and corp surface models.	Shi et al. 2016
Rothamsted Experimental Station Harpenden, England	Screen large number genotypes of wheat, barley and lentil in experimental blocks. They collect data pertaining to all the genotypes in a matter of minutes using visible, NIR, and IR (thermal) sensors. The data are collated, analysed, and released for farmers to assess crop material.	Farmingonline, 2014; Case, 2013
INRA, Auzeville, Toulouse, France	Evaluation of wheat crop's response to fertilizer-N inputs at four different levels, using UAVs fitted with multi-spectral high-resolution sensors.	Lelong et al. 2016

TABLE 6.2 (Continued)

Crop species/location	Remarks	References
Institute of Soil Science, Pulway, Poland	Experimental station's installations and crop fields were monitored, using Parachute (UAV) "Paraglider-Moto' fitted with visible, IR, and NIR sensors. Experimental fields with wheat and maize were assesses for eye spot disease and couch grass weed infestation.	Pudelko et al. 2012
CIMMYT, Harare, Zimbabwe	Evaluation wheat/soybean rotation using NDVI, leaf chlorophyll, and thermal data (IR sensors).	CIMMYT, 2012; Mortimer 2013
CSIRO, Brisbane, Australia	Monitor agronomic fields and evaluate genotypes using NDVI values.	Duan et al. 2017
Maize		
Agriculture-Canada	Field evaluation of maize crop's response to fertilizer-N application. Estimations of biomass (NDVI) and leaf chlorophyll (canopy-N)	Trembley et al. 2014
Experimental station,		
Montague, Quebec		
CIMMYT, Zimbabwe.	Evaluation of maize genotypes using phenomics and CSMs.	CIMMYT, 2012
Rice		
Locations in China	Experimental evaluation of fertilizer-N effects on rice.	Zhu et al. 2014; RMAX, 2015; Yamaha, 2014
Barley		
Get address	Study the response of cereal crop to fertilizer-N supply using R, G, B, and IR sensors. Arriving at appropriate fertilizer-N dosages using aerial imagery and crop canopy-N status.	Inen et al. 2013
University of Cologne, Cologne, Germany,	UAV-derived imagery has been used develop 'Crop Surface Models (CSM)' and prescribe fertilizers, water, and pesticides based on such CSM.	Bendig et al. 2015
Klein-Altendorf, Koln, Germany	Hyperspectral data (ground truth) and UAV-derived 'Digiital Surface Models' were compared for their efficacy, in judging the influence of different agronomic procedures and fertilizer-N inputs, on productivity of barley	Aasen, 2016

TABLE 6.2 (Continued)

Crop species/location	Remarks	References
Sorghum		
Kearney Agricultural Research and Extension Centre University of California, USA	Aerial survey of sorghum drought nurseries; collection of data about soil moisture and heat stress tolerance and scoring drought tolerant genotypes; scoring for drought traits of sorghum; and collection of routine phenomics data of hundreds of sorghum genotypes.	Van der Staay, 2017
Rye		
USDA ARS, Beltsville Agricultural Experimental Station Maryland, USA	Fields with Secale cereale were assessed for NDVI, Greenness index, and leaf chlorophyll content using UAVs. The UAV derived data and those from ground (manual) correlated ($r^2 = 0.73$)	Raymond Hunt et al. 2014
Sunflower		
Dept. of Rural Engineering, Escuela Superior de Ingeniería, University of Almería, Almeria, Spain	Estimating soil-N fertility using leaf chlorophyll from plots exposed to different treatments.	Aguera et al. 2011
Potato		
Centro de La Papa, Lima, Peru	UAVs have been used to obtain spectral data of various genotypes of potato. Healthy and pest-attacked potato crop has been scored in experimental plots, using UAV data.	Allen, 2016
Hermiston Agricultural Experimental Station, Columbia River Basin, Oregon, USA	To detect nitrogen deficiency and response to fertilizer-N by Potato genotypes.	Hunt et al. 2015
Canola		
GRDC, South Australia, Australia	To detect spectral difference of slug attacked and health patches	Beveridge and Russell, 2015

TABLE 6.2 (Continued)

Crop species/location	Remarks	References
Cotton		
Agricultural Experimental Station Knoxville, Tennessee	Evaluation of water status and drought tolerance of cotton using thermal camera (IR sensors) at Tennessee Valley Agricultural Experimental Station.	Sullivan et al. 2007
Sugarcane		
CIRAD, Re Union Islands, France	Evaluation of sugarcane germplasm using aerial imagery by UAVs Sugarcane response to fertilizer-N application.	Lebourgeois et al. 2012
Grapes		
University of California, Agricultural Experimental Station, Davis, CA, USA	Spraying pesticides to grapevine (Cabernet Sauvignon) at Oakville, California, using different swatch width (spray bars length) and number of passes.	Giles and Billing, 2015
Vineyard Experimental Station Central Italy,	To assess grapevine growth using vigour maps, NDVI, and to compare it with ground truth data.	Primicerio et al. 2012
Agricultural Farms, The University of Melbourne, Australia	Periodic flights to assess experimental grape vineyards for health, pest infestation, and disease.	Crys, 2017
Citrus		
Lake Alfred, Florida, USA	To assess citrus tree growth and orchard soil moisture variation, detect Huanglongbing (greening) disease	Sankaran and Ehsani, 2012; Sankaran et al. 2010

Note: The above list is only indicative of number of agricultural experimental stations that have explored the use of Unmanned Aerial Vehicles.

supplied with four different levels of fertilizer-N. A free-flying parachute (aerial robot) guided using a remote controller or via pre-programmed flight path was used, to collect data. Sony DSC F828 and Canon EOS 350 sensors were used to collect data. The data pertained to canopy growth, NDVI, leaf chlorophyll, canopy temperature, and disease incidence, if any. The data collected by UAV was crosschecked, using ground reality data. The rapid collection of data and ability for repeated flights are major advantages, if UAVs are used in experimental stations. The cost of collecting accurate data at short intervals reduces enormously, if experimental stations adopt parachute or blimp UAVs.

Duan et al. (2017) point out that, NDVI is an excellent indicator of crop vigor, biomass production trend, and leaf chlorophyll. The NDVI data could be efficiently and accurately derived, using UAVs. So, suitable modifications and computer-based programs should be prepared, to utilize NDVI data, particularly, during experimental evaluation of crops. Agronomic performance of crop genotypes could be judged better, using NDVI data. Further, we should be able to use NDVI data effectively during selection of parent lines and elite genotypes. The UAVs offer NDVI data of each genotype rather rapidly. UAVs can cover a large area of wheat breeding material, say in a few acres, in just 30–45 min and offer stored digital data. Usually, noise from soil reflectance could be measured using high-resolution sensor. The NDVI values could be corrected appropriately. Duan et al. have reported that NDVI measured using UAV and those derived by ground vehicles were highly correlated ($R^2 = 0.85$). Strong correlation between UAV and ground-vehicle-derived NDVI occurred during flowering stage ($R^2 = 0.87$).

Parachutes with their ability for low altitude slow movement over crop fields have also been explored for use, during agricultural experimentation. Crop fields supporting large number of genotypes of a crop species could be monitored regularly with greater accuracy, by a low-flying parachute. Parachutes have longer endurance. They can obtain high-resolution images of large number of crop genotypes and experimental plots, using visual, NIR, and IR cameras. Let us consider an example. Pudelko et al. (2012) have adopted parachute (UAV) fitted with "Sony DSC F828" 8-megapixel photo camera, to obtain data such as NDVI, leaf chlorophyll content, and CWSI maps of wheat and maize genotypes. Further, aerial images were used to detect and mark patches with "eye spot disease" and "couch grass (weed)" infestation. Such paraglider UAVs can stay afloat for long duration, monitor experimental fields incessantly, and collect useful data. They are invariably much cheaper to adopt compared to human skilled scouts utilized to collect data.

6.10 UNMANNED AERIAL VEHICLES ARE UTILIZED TO DEVELOP AND MAINTAIN PASTURES

Pastures and turfgrass stretches could be effectively monitored, using low-flying UAVs (Caturegli et al., 2016; Watts et al., 2012). The UAVs fitted with multispectral sensors offer digital data about NDVI, leaf chlorophyll, and nitrogen status. Aerial maps of pastures depicting variations in growth and canopy N status are also obtained, using UAVs. Such data helps in judging response of pasture grass/legume species, to fertilizer. If an IR sensor is also used, then water stress index of the pasture too could be assessed.

Agricultural UAVs seem most apt to surveillance and manage large pastures and forage fields. They could be flown repeatedly, at short intervals to collect data about the farm animals. UAVs could be utilized to decipher grazing trends. Most importantly, UAVs aid in judging the growth and health of pasture grass/legume mixtures. In fact, precision techniques involving application of fertilizers and plant protection chemicals could be adopted. The digital maps of pastures have to be supplied. Now, let us consider an example from pasture producing regions of New Zealand. Von Bueren et al. (2015) have standardized procedures for detecting NDVI, greenness index, and water stress index of pastures using UAVs. They have used SC-tech's "Falcon-8" as UAV platform with cameras, such as MCA6 (Multiple Camera Array)—a six-band multispectral sensor, to collect the data about pastures. They report that data from UAVs were highly correlated with ground reality data. The accuracy of the data depended on strict adherence to protocols related to flight path, way points, and computer programs adopted to analyze the data. So, it may not be very long before a good share of pasture surveillance is accomplished, using aerial imagery drawn by UAVs.

Reports emanating from Southern Plains of USA suggest that, UAVs are gaining in popularity with ranchers, also forage and pasture owners. UAVs, both fixed-winged and multicopter models are being utilized, to manage land and soil resources. UAVs are being adopted mainly in geospatial survey of forage and pasture land. UAVs offer digital maps that help in marking management blocks. Such maps also depict nutrient deficiencies, diseases, and erosion, if any (Santa Ana, 2016).

6.11 UNMANNED AERIAL VEHICLE SYSTEMS IN PLANTATIONS AND FORESTS

Small UAVs with ability for aerial imagery, foliar fertilizer application, and spraying plant protection chemicals are expected to throng the fruit plantations. Agricultural UAVs have now gained acceptance in many plantations. They have been adopted to conduct an aerial survey of plantation. The aims are to assess canopy growth, leaf chlorophyll content, pests, diseases, and weeds. Also, to assess water stress index and decide about timing and quantum of irrigation. More specifically, UAVs have now invaded grape vines in many regions of the world. They have been adopted in the grape orchards of North America (Sky Squirrel, 2017). France, Italy, and Germany, in the European grapevine region have examined UAVs, to obtain aerial imagery and detect grapevine vigor and disease (Krishna, 2018). Australian farmers too have tested UAVs to monitor and assess pest damage. UAVs are also utilized to assess need for inputs of grape vines. Deployment of UAVs offers certain advantages to grape growers. They are as follows: (1) UAVs offer high precision well-calibrated data about NDVI and leaf chlorophyll content (crop N status); (2) they offer GPS-tagged maps of grape orchard; (3) we can use optical filters to enhance imagery; and (4) UAVs can also collect thermal data, using FLIR, that is, infrared cameras. Maps obtained using thermal sensors help in calculating crop's water stress index (Krishna, 2018).

In Spain, vineyards are currently experiencing introduction of UAV-aided assessment of crops' water status. The UAVs are flown periodically above vineyards to collect data about NDVI, leaf chlorophyll, canopy temperature, ambient temperature, soil temperature, and CWSI. The maps depicting variations in CWSI are utilized, to calculate requirements of water and to apply it, using variable-rate methods (i.e., precision farming) (Gago et al., 2015a, b). Further, Gago et al. (2015b) state that, UAVs could play an important role in assessing grapevine water status. The data helps in prescribing frequency and quantum of irrigation. Obviously, UAVs with IR (thermal) sensors can provide the data pertaining to grape canopy temperature and ambient temperature. With suitable correction for soil-related reflectance (heat), we can assess crop's water stress index and prepare a map (Gago et al., 2015b). They have reported that, data from an experimental station in Spanish grape-producing tract is encouraging. The procedure needs to be scaled up for large tracts of grape vine, particularly, in the areas prone to drought. In a drought-prone zone, rapid detection of water stress is essential. UAVs are best suited here because they could be flown rapidly and repeatedly. Also, data collected could be confirmed, using ground proof, if needed.

Agricultural UAVs help in overall management of canopy growth and its size. An UAV offers data that is required in planning cultivation procedures. Aerial imagery by UAVs is destined to help grape growers in monitoring growth and nutrient (plant-N) status. The UAV technology offers to optimize fertilizer supply to grape vines with greater accuracy and economic efficiency. Agricultural UAVs offer disease/pest infestation maps that are of great value to farmers who adopt autonomous sprayers. Such robotic sprayers could be either on ground vehicles or UAVs. The data procured by UAVs is helpful to develop forecasts about fruit yield. Again, UAV models are now available in large number in the market. UAVs apt for each agronomic procedure could be searched, selected, and used. For example, small, flat-winged UAVs (e.g., eBee or Precision Hawk's Lancaster) are sought, when aerial imagery is the only requirement. An UAV with four or six rotors, a payload that allows pesticide tank and has spray bar and nozzles suits best, if the need is to spray the grape crop. Selection and use of the most appropriate UAV model, for each or a group of agronomic procedures, is the need of the hour. Computer simulations, ready reckoners and software that help farmers, to arrive at few best decisions are needed.

In Canada, private UAV-based technology companies such as VineView—an image-processing company and SkySquirrel Technology Inc. (UAV producer) are collaborating. Their aim is to apply UAVs, to assess the crops and prescribe. They are picking images using visual, IR (thermal), and NIR sensors. So far, they have found that UAV-aided techniques are useful to horticulturists with small- and medium-sized vineyards, say <100 acres (Quackenbush, 2017). UAVs obtain images of vineyard at 1 acre min^{-1}. So, they cover 20–25 acres per flight. They fly at low altitudes of 120 ft. above the grape canopy. UAVs offer high-resolution images of canopy, soil conditions, also disease/pest attack, if any.

Reports suggest that we can fly a UAV fitted with MicaSense RedEdge cameras, to obtain NDVI values. NDVI maps are used by farmers, to judge crop health and decide on fertilizer and water supply. In California's San Joaquin Valley, grapes are being regularly surveyed, using Sequoia or MicaSense RedEdge cameras (MicaSense, 2017). UAVs provide digital data and maps of various vegetation indices. We can detect stresses in nutrients and water, using such maps of vegetation indices. Sometimes, grape farmers conduct tedious leaf and twig sampling, then, get them chemically analyzed in laboratories, for nutrient deficiencies, if any. This is done at seedling stage, so that, corrections to soil fertility could be made early in the season. We should note that a swift flight of UAV with MicaSense cameras can reveal equally good details, in one flight of a few minutes. Usually vegetation and leaf chlorophyll maps are utilized. Leaf yellowing shows deficiency of N, K, and other elements. Low K^+ levels are also indicated by retarded bunch growth, and small-sized berries. Grape fruits photographed, using close up shots by UAV, indicate loss of color in leaves and fruits. Fruits become brittle and break. They are low in quality, if K^+ deficiency occurs. UAVs can aid in detection of K^+ deficiency and map it in detail, in a single flight (MicaSense, 2017).

UAV-aided citrus plantation management is gaining acceptance in Florida, USA. UAVs help planters to monitor growth, disease/pest incidence, and water status. Planters are using spectral signatures collected via low-flying UAVs. Several UAV models have been adopted to study citrus groves. Some of them are efficient and easy to operate. Farmers may reap economic advantages. A few other UAV models may not be apt for certain reasons, yet, farmers may be using them. In such cases, a computer with ability to reach data banks and select most suitable UAV models is a necessity. In fact, there are possibilities to develop software that compares performance of various UAV models, in terms of agronomic and economic efficiency.

Now, let us consider a different plantation species. Prunus (Nectarine) are cultivated in Australia. Park et al. (2015) have examined the utility of UAVs during Prunus production in farms, in the state of Victoria, Australia. They have conducted a field trial using a multirotor (DJI's 900) fitted with thermal infrared camera (TIR) (A65, FLIR Systems Inc.). The data obtained related to assessing CWSI. They generated maps showing variation in CWSI, for Prunus trees planted on fine sandy loam. They also collected ground proof data such as stem water potential and g_c exchange (stomatal activity). They have reported that estimation of CWSI, using UAV technology is useful. It correlates with ground data. Therefore, nectarine (Prunus) orchards could be exposed to UAV-based techniques, more often, to accomplish several tasks. A few of the tasks are estimation of NDVI, monitoring growth, and detecting disease/pest attack, if any. UAVs could also be used to spray the orchards with fertilizers (liquid formulations), pesticides, and fungicides.

The UAVs have also been adopted in the tropical plantations, such as banana, sugarcane, or spices. There are only very few examples. Agricultural UAVs are being evaluated to detect spatial variations in growth (NDVI), leaf chlorophyll (plant-N), water stress index (thermal imagery), and disease/pest status, in Costa Rica (Machovina et al., 2016). UAVs have been flown over banana plantations to assess spatial variation in photosynthetic activity (NDVI) and fruit production. They have also collected ground data about soil characteristics such as texture, soil moisture, fertility, etc. They reported a good correlation between UAV-derived data about photosynthetic activity and fruit production efficiency, also fruit quality. Soil physical conditions did not correlate well with fruit quality or the production efficiency. They have concluded that UAVs could be regularly utilized to assess NDVI, greenness index, water stress index, fruit quality, and production efficiency, in a banana plantation.

Agricultural UAVs could be applied to conduct several different agronomic procedures relevant to forestry. UAVs could be drafted for use, right from landscape imaging, seeding, nursery development, transplantation, monitoring growth, and biomass accumulation rates of trees, till harvest and logging out. She et al. (2017) believe that, UAVs could offer solutions to several small- and medium-scale forest planters. Their use begins with counting containers in nursery via low-altitude flight over nursery regions. Tree counting and aerial imagery of seedling size and greenness (NDVI) is important to farmers. Using such data, we can forecast density and biomass of forest tree stands. UAVs may be used to supply fertilizers and irrigation. Using computer software such as "Microsoft ICE," the aerial images could be processed. It provides an idea of tree count, canopy size, and biomass accumulation trends. Satellites could be utilized, but the resolution, turn-around time, and cost may not be congenial, to foresters. Hence, at present, they prefer to utilize copters flying at low altitude, to conduct forest nursery analysis.

6.12 FUTURE OF UNMANNED AERIAL VEHICLES IN FARM WORLD

UAV-aided agriculture is said to gain in popularity and cover the most agrarian regions, if not entirely, all over in the high-intensity farming zones. They seem to be equally well suited and currently sought, by medium- and low-input farms. Subsistence farms too may derive benefit from UAV technology, if they operate in collective cooperative. Forecasts based on current trends plus the interest evinced by farmers worldwide suggests that, sooner or later UAVs could become essential aspect of global crop-production zones. We may be depending on them too frequently, to accomplish every other agronomic procedure. Now, if UAVs could become important, then, "sensors" placed on it are definitely the centre piece of UAV technology itself. So, sensors would play an important role in the global food production tactics. The spectral data that UAV's sensors collect and offer at any time, as required by farm companies and farmer will be crucial. Data from sensors will help to decide many of the crop husbandry procedures, also their timing and intensity. Most importantly, procedures such as fertilizer-N supply, irrigation, and water stress management may entirely be decided, by the aerial imagery that UAV's sensors offer to farm experts. Computers, computer software, computing skills will have to match the sensor-based UAV technology that farmers may prefer to adopt, in future. Offering training to farmers and other personnel about UAV technology may be important.

There are many improvements required before UAV technology as applied to farm world becomes routine. We may encounter difficulties in its usage from time to time. UAVs must suit the farmers' requirements. Specific crops may need certain modifications to UAV machine, sensors, software, and their integration. Clearly, there is a lot for aviation engineers, computer software developers and most importantly the farm scientists/companies to toil. They could get the UAV technology streamlined and make it easy to operate. Economic feasibility may dictate terms during the selection of UAV models. The returns from its application have to be consistent to the farmers. Otherwise, it may fade away just like several contraptions that were previously introduced into agrarian regions. Ultimately, they were withdrawn when found not so efficient or remunerative. Although it has not been pointed out clearly the UAV-based techniques remove farm drudgery. This is a reason enough for farmers and farming agencies to popularize UAVs. The other overriding advantage is in the ease and safe application of pesticides, fungicides, and herbicides to large areas of farmland. Human skilled workers may lose out to UAVs (e.g., RMAX, Hercules) miserably as time lapses. The cost incurred on farm scouting and spraying plant protection chemicals is expected decrease perceptibly.

In due course, we may find agricultural service agencies employing UAVs, computers, and technical experts to monitor and regulate crop production trends, in many parts of the world. Over all, investment on UAV technology must be commensurate with the returns. Removal of farm drudgery by UAVs needs its due consideration while assessing UAVs and their advantages.

KEYWORDS

- **aerial imagery**
- **crop surface models**
- **foliar fertilizer**
- **leaf nitrogen**
- **NDVI**
- **pest control**
- **plantation**
- **precision agriculture**
- **unmanned aerial vehicles**

REFERENCES

Aasen, H. The Acquisition of Hyper-spectral Digital Surface Models of Crops from UAV Snapshot Cameras. University of Koln, Germany, Doctoral Dissertation, 2016, p 156.

Abdullahi, H.; Mahiedddine, F.; Sheriff, R. Technology Impact on Agricultural Productivity: A Review of Precision Agriculture Using Unmanned Aerial Vehicles Wireless and Satellite Systems. *Wirel. Satell. Syst.* 2015. DOI: 10.1007/978-3-319-25479-1_29

Achtilik, M. UAV Inspection and Survey of Germany's Highest Dam. Ascending Technologies, 2015, pp 1–3 http://www.asctec.de/en/uav-inspection-survey-of-germanys-highest-dam. (accessed June 26, 2016).

Aerdron. *Pest Control and Crop Protection, Pesticide Spraying*, 2017, pp 1–7. http://aerdron.com/agriculture/ (accessed Nov 9, 2017).

Aeroscraft. *Aeroscraft Corporation in Brief: Precision Agriculture*, 2017, pp 1–6. http://aeroscraft.com/precision-agriculture/4584020239 (accessed Oct 30, 2017).

Agape Palilo. *Monitoring and Management of Maize Rust (Puccinia sorghi) by a Drone Prototype in Southern Highlands, Tanzania. Sokoine University of Agriculture, Morogoro, Tanzania,* 2014, pp 1–15. Http://Www.Academia.Edu/8063999/Monitoring_and_Management of maize-rust disease. (accessed May 30, 2015).

Aguera, F.; Carvajal, F.; Perez, M. Measuring Sunflower Nitrogen Status from an Unmanned Aerial Vehicle-based System and an on the Ground Device. *Int. Arch. Photogram. Remote Sens. Sp. Inform. Sci.* **2011,** *38,* 1–5.

Al-Arab, M.; Torres-Rua, Ticlavilca, A.; Jensen, A.; McKee, M. Uses of High Resolution Multi-spectral Imagery from an Unmanned Aerial Vehicle in Precision Agriculture. Paper Presented at the Geo Science and Remote Sensing Program IGARSS)—2013 pp 1–8 Conference: IGARSS 2013–2013 IEEE International Geoscience and Remote Sensing Symposium. 10.1109/IGARSS.2013.6723419 (October 30th, 2017)

Allen, W. Drones Detect Crop Stresses More Effectively. ICT Update 2016, 82: 10–11 http://ictupdate.cta.Int (accessed Apr 2, 2017).

Anderson, C. A.; Vivoni, E. R.; Pierini, N.; Robles-Morua, A.; Rango, A.; Laliberte, A.; Saripalli, A. 2012, Characterization of Shrubland-Atmosphere Interactions Through Use of the Eddy Covariance Method, Distributed Footprint Sampling and Imagery from Unmanned Aerial Vehicles, Poster presentation, American Geophysical Union Fall Meeting, San Francisco, California, USA. 3-7 December 2012.

Anderson, W.; You, L.; Anisimova, E. Mapping Crops to Improve Food Security. International Food Policy Research Institute, Washington, DC. 2014, pp 1–5. http://www.ifpri.org/blog/mapping-crops-improve-food-security?print. (accessed May 10, 2016).

Anthony, D.; Elbaum, S.; Lorenz, A.; Detweiler, C. On Crop Height Estimation with UAVs, 2014, pp 1–8. *cse.unl.edu/~carrick/papers/AnthonyELD2014* (accessed Sept 18, 2016).

Araus, J. L.; Cairns, J. E. Field-high Through-put Phenotyping: The New Crop Breeding Frontier. *Trends Plant Sci.* **2014,** *19,* 52–61.

Argyrotic. Comparing RGB-Based Vegetation Indices with NDVI for Agricultural Drone Imagery, 2017, pp 1–7. http//Agribotix.com/blog/2017/04/30/compring-rgb-based-vegetation-indices-with-NDVI-for-agricultural-drone-imgery. (accessed Aug 1, 2017).

Ascending Technologies GMBH. *UAVs- Drone-based Precision Agriculture and Smart farming*, 2016, pp 1–4. http://www.asctec.de/en/uav-drone-based-precision-agriculture--smart-farming.htm (accessed Feb 5, 2016).

Associated Press. Drones, Dogs Deployed to Save Avocados from Deadly Fungus in Florida. The Associated Press, 2015, pp 1–6. http://www.nydailynews.com/life-style/eats/drones-dogs-deployed-save-avocados-deadly-fungus-in-floirda (accessed Oct 18, 2015).

Atherton, K. The Drone Sprays Pesticides Around Crops, 2017, pp 1–3. https://www.popsci.com/agri-drone-is-precision-pesticide-machine (Nov 9, 2017).

Aylor, D. E.; Fernandino, F. J. In *Prospects for Precision Agriculture to Manage Aerially Dispersed Pathogens in Patchy Landscapes*. Proceedings of 9th International Congress of Plant pathology, Healthy and Safe Food for Everybody. Torino, Italy. Plant Pathology, 2008, 90, p 59.

Aylor, D. E.; Schmale, D. G.; Shields, E. J.; Newcombe, M.; Nappo, C. J. Tracking the Potato Late Blight Pathogen in the Atmosphere, Using Unmanned Aerial Vehicles and Lagrangian modelling. *Agric. Forest Meteorol.* **2011,** *151*, 251–260

Babar, S. A. Spectral Reflectance to Estimate Genetic Variation for In-Season Biomass, Leaf Chlorophyll, and Canopy Temperature in Wheat. *Crop Sci.* **2006,** *46* (3), 1046–1049.

Baluja, J.; Diago, M. P.; Balda, P.; Zorer, R.; Meggio, F.; Morales, F.; Tardaguilla, J. Assessment of Vineyard Water Status Variability by Thermal and Multispectral Imagery Using Unmanned Aerial Imagery. *Irrigat. Sci.* **2012,** *30*, 511–522. DOI: 10.1007/s00271-012-0382-9

Bendig, J.; Bolten, A.; Bareth, G. UAV-based Imaging for Multi-temporal, Very High-resolution Crop Surface Models to Monitor Crop Growth Variability. *Photogramm, Fernerkund Geo-Inform.* **2013a,** *13*, 551–562.

Bendig, J.; Willkomm, M.; Tilly, N.; Gnyp, M. L.; Bennertz, S. Qiang, C.; Miao, Y.; Lenz-Wiedmann, V. L. S.; Bareth, G. Very High-resolution Crop Surface Models (CSMs) from UAB-based Stereo Images of Rice Growth Monitoring, in Northeast China. *Int. Arch. Photogramm Remote Sens., Spatial Inform. Sci.* **2013b,** *40*, 45–50.

Bendig, J.; Yu, K.; Assen, H.; Bolten, A. Bennertz, S.; Broscheit, J.; Gnyp, M.; Bareth, G. Combining UAV Based Plant Height from Crop Surface Models, Visible and Infrared Vegetation Indices for Biomass Monitoring in Barley. *Int. J. Appl. Earth Observ. Geo-Inform.* **2015,** *39*, 79–87.

Berni, J. A. J.; Zarco-Tejada, P. J.; Sepulcre-Cnto, G.; Ferrees, E.; Villalobis, F. Mapping Canopy Conductance and CWSI in Olive Orchards, Using High Resolution Thermal Remote Sensing Imagery. *Remote Sens. Environ. Sci.* **2009,** *113*, 2380–2388.

Beveridge, M.; Russell, A. Using Drones for the Detection of Crop Pest Damage in Canola. Grain Research and Development Centre Southern Farming Systems: A Report, 2015, pp 1–8 http://www.farmtrials.com.au/trial/18829 (accessed July 27, 2017).

Bolton, A. *Japan Develops Drone to Patrol Farmland and Destroy Insect Pests*. CBS Interactive Inc., 2016, pp 1–3. https://www.cnet.com/news/japan-develops-a-drone-to-patrol-farmland-and-destry-insect-pests. (accessed Sep 23, 2017).

Bouffard, K. Drones Find Uses on Farms, 2015, pp 1–7. http://www.heraldtribune.com/article/20150326/ARTICLE/303269992. htm (accessed June 20, 2015).

Buck, B. Low Altitude Aerial Images Allow Early Detection of Devastating Avocado Diseases. Growing Florida, 2015, pp 1–3. http://growingfl.com/news/2015/05/low-altitude-aerial-images-allow-early-detection-devastating-avaacado-disease.htm (accessed June 20, 2015).

Burkart, A. Multi-temporal Assessment of Crop Parameters Using Multisensorial Flying Platforms. Frederich-Williams Universitat Bonn, Bonn, Germany Dissertation, 2015, pp 133.

Calderon, R.; Naves, H.; J. A.; Lucena, C.; Zarco-Tajeda, P. J. High-resolution Airborne Hyper-spectral and Thermal Imagery for Early Detection of Verticillium Wilt of Olive Using Fluorescence, Temperature and Narrow-band Spectral Indices. *Remote Sens. Environ.* **2013,** *139*, 231–245.

Calderon, R., Naves, J. A.; Zarco-Tejada, P. J. Early Detection and Quantification of Verticillium Wilt in Olive Using Hyperspectral and Thermal Imagery Over Large Areas. *Remote Sens.* **2015,** *7*, 5584–5610.

Cao, H.; Yang, Y.; Pei, Z.; Zhang, W.; Ge, D.; Sha, Y.; Zhang, W.; Fu, K.; Liu, Y.; Chen, Y.; Dai, H.; Zhang, H. Intellectualized Identifying and Precision Control System for Horticultural Crop Disease Based on Small Unmanned Aerial Vehicle. Jiangsu Province Scientific Technology Support Program, 2012, pp 1–14. http:// ifip.org/db/conf/ifip12/cta2012-2/CaoYPZGSFLCDZ12.pdf (accessed Mar 28, 2017).

Case, P. Rothamsted Unveils Octocopter Crop-monitoring Drone, 2013, pp 1–2. http://www.fwi.co.uk.arable/rothamsted-unveils-octocopter-crop-monitoring-drone.htm (accessed June 25, 2013).

Caturegli, L.; Comeglia, M.; Gaetani, M.; Grossi, N.; Magni, M.; Miglizzi, M.; Angelini, L.; Mazzoncini, M.; Silvestri, N.; Fontanelli, M.; Raffaelli, Peruzzi, Volterrani, M. Unmanned Aerial Vehicle to Estimate Nitrogen Status of Turf Grass. *PLOS One* 2016, 1–9. https://doi.org/10.1371/journal.pone0158268 (accessed July 27, 2017).

Chapman, A. Types of Drones: Multi-rotor vs Fixed Winged vs Single Rotor vs Fixed Winged VTOL Drones. Australian UAV, 2015, pp 1–9. http://www.auav.com.au/articles/drone-types (accessed Aug 12, 2016).

CIMMYT. *Obregon Blimp Airborne and Eyeing Plots*. International Maize and Wheat Centre, Mexico, 2012, pp 1–4 (accessed Apr 4, 2017).

Cornell, C. *Farmers Use Drones and Data to Boost Crop Yields*. The Globe and Mail, 2015. http://www.theglobend mail.com/report-on-business/sb-growth/farm/ (accessed Aug 26, 2017).

Cornett, R. Drones and Pesticide Spraying: A Promising Partnership. Western Plant Health Association. Western Farm Press, 2013, pp 1–3. http://westernfarmpress.com//grapes/drones-and-pesticides-spraying-partnership (accessed Aug 30, 2014).

Colaco, A. F.; Molin, J. P. In *A Five-year Study of Variable Rate Fertilization in Citrus*. Proceedings of 12th International Conference on Precision Agriculture, Sacramento, California, USA, 2014, p 127.

Colaco, A. F.; Ruiz, M. A.; Yida, D. Y.; Molin, J. P. In *Management Zones Delineation in Brazilian Citrus Orchards*. Proceedings of 12th International Conference on Precision Agriculture, Sacramento, California, USA, 2014, p 126.

CREC. UAV Application in Agriculture. Citrus Research and Education Centre, Lake Alfred, Florida, 2015, pp 1–3. http://www.crec.ufl.ifas.edu/publications/news/PDF/UAVwsflyer3-pdf.pdf (accessed Sept 30, 2015).

Croft, J. Sowing the Seeds for Agricultural Drones. Aviation Week and Space Technology, 2013, pp 1–3. http://www. Aviationweek.com/awin/sowing-seeds-agriculture-uavs (accessed June 21, 2015).

Crys J. Digital Vineyards. The University of Melbourne, Melbourne, Australia, 2017. https://pursuit.unimelb.edu.au/features/digital-vineyards pp 1–8 (Nov 7, 2017).

Danovich, A. Actual and Potential Use of Drones in Precision Agriculture. Hot Wires, 2014, pp 1–3. http://circuitsassembly.com (accessed Mar 25, 2017).

DeBell, L.; Anderson, K.; Brazier, R. E.; King, N.; Jones, L. 2016 Water Resource Management at Catchment Scales Using Light Weight UAVs: Current Capabilities and Future perspectives. *J. Unmanned Veh. Syst.* 2016, *4* (1): 7–30. https://doi.org/10.1139/juvs-2015-0026

Deering, C. *Growing Use of Drones Poised to Transform Agriculture*. USA Today, 2014, pp 1–4. http://www.usatoday.com/story/money/business/2014/03/23/drones-agriculture-growth/6665561/ (accessed Jan 9, 2017).

Detweiler, C.; Elbaum, S. UNL Developing Water-collecting Drones for Tests, Remote Locales-UNL Crop Watch Sept 4, 2013- Archives. University of Nebraska-Lincoln Crop Watch, 2013, pp 1–4. http://cropwatch.unl.edu/archive/-asset_publisher/VHeSptv0Agju/content/unl-developmeng...html (accessed Oct 9, 2015).

DJI. *DJ1 MG-1S Agricultural Wonder Drone*, 2017. https://www.youtube.com/watch?v=P2YPG8PO9JU (accessed Aug 28, 2017).

DMZ Aerial Inc. Unmanned Aerial Vehicles and Scouting, 2013, pp 10. http://www.dmzaerial.com/uavscouting.html (accessed Aug 15, 2014).

Drone Deploy. *Seven Ways to Use Drone Mapping on Farm this Season*, 2017, pp 1–5. https://blog.dronedeploy.com/seven-ways-to-use-drone-mapping-on-the-farm-this-season-5d62e66e1c78 (accessed Oct 30, 2017).

Duan, T.; Chapman, S. C.; Guo, Y.; Zheg, B.; Zheng, B. Dynamic Monitoring of NDVI in Wheat Agronomy and Plant Breeding Trials Using an Unmanned Aerial Vehicles. The Commonwealth Scientific and Industrial Research Organization, Brisbane, Australia, 2017, pp 1–12. DOI: 10.1016/j.fcr.2017.05.025 (accessed Sept 20, 2017).

Ehsani, R.; Sankaran, S. Sensors and Sensing Technologies for Disease Detection. Citrus Industry, 2010, pp 1–5. http://www.crec.ifas.ufl.edu/.../2010junesensoringtechnology.pdf (accessed Oct 14, 2015).

FAO. *FAO Philippines Newsletter*, 2016, pp 1–3. http://www.fao.org/resilience/resources/resources-detail/en/c/409967/ (accessed Jan 15, 2017).

Farming Online Using Drones to Monitor Crops. Rothamsted Experimental Agricultural Station, United Kingdom, 2014, pp 1–2. http://farming.co.news/article/9243 (accessed May 22, 2015).

Feine, M. Inspire Ag. DMZ Aerial Inc. Whitewater, Wisconsin, USA, 2017, pp 1–4 http://www.dmzaerial.com/?page_id=1467 (accessed Sept 29, 2017).

Francis, D. D.; Piekielek, W. P. Assessing Crop Nitrogen Needs with Chlorophyll Meters. Site-Specific Management Guidelines SSMG-12, 2012, pp 1–9. www.ppi-far.org/ssmg (accessed Aug 29, 2017).

Frey, T. *Future Uses for Flying Drones*, 2015, pp 1–37. http://www.futuristsspeaker.com?2014/09/192-future-uses-for-flying-drones/ (accessed May 5, 2015).

Fulwood, J. *Drones Diagnose Plant Health and Decrease Insecticide Use*, 2016, pp 1–3. http://phys.org/news/2016-06-drones-health-decrease-insecticide.html (accessed Mar 28, 2017).

Gago, J.; Douthe, C.; Coopman, R. F; Gallego, P. P; Ribas-Carbo, M.; Flexas, J.; Esclona, J.; Medrano, H. UAVs challenge to assess water stress for sustainable agriculture. *Agric. Water Manag.* **2015a,** *153*, 1–14. DOI: 10.1016/j.agwat.2015.01.020 (accessed Mar 27, 2017).

Gago, J.; Martorell, S.; Douthe, C.; Fuentes, S.; Toms, M.; Hernandez, E.; Mir, Li.; Carriqui, M. Escalona, J. M.; Gallego, P. P.; Medrano, H. Upscaling Levels for Drought Assessment in Agriculture: from Leaf to the Whole-vineyard. 1 Journadas del Grupo de Viticultura y Enologia de la SECH- Reos Actuales de I +D en Viticultura. Agricultural Water Management, 2015b, Vol. 153, C, pp 9–19.

Gago, J.; Martorell, S.; Tomas, M.; M.; Pou, A.; Millan, B.; Ramon, J.; Ruiz, M.; Sanchez, R.; Galmes, J.; Conesa, M.; High Resolution Aerial Thermal Imagery for Plant Water Status Assessment in Vineyards, Using Multi-copter RPAS 2013, pp 1–7. http://sechaging-madrid2013.org/geystiona/adjs/communications/272/C00790001.pdf (accessed Sept 6, 2017).

GAIA Data Collection. *Weed Identification*, 2017, pp 1–5. http://www.gaiadatacollection.com/new-page/ (accessed Mar 28, 2017).

Galimberti, K. *Can Drones Offer New Ways to Predict Storms, Save Lives*? 2014a, pp 1–4, AccuWeather.com (accessed Sept 23, 2015).

Galimberti, K. *Drones Offer New Horizon, Solutions for Weather Modification*, 2014b, pp 1–3. AccuWeathr.com (accessed Oct 20, 2015).

Garcia-Ruiz, F.; S. Sankaran, J. M. Maja, W. S. Lee, W. S.; Rasmussen, J.; Ehsani, R. Comparison of Two Aerial Imaging Platforms for Identification of Huanglongbing Infected Citrus Trees. Comput. Electron. Agric. 2013, *91*, 106–115. DOI: http://dx.doi.org/10.1016/j.compag.2012.12.002 (accessed Sept 7, 2016).

Geipel, J.; Link, J.; Claupein, W. Combined Spectral and Spatial Modeling of Corn Yield Based on Aerial images and Crop Surface Models acquired with an Unmanned Aircraft System. *Remote Sens*. **2014,** *11*, 10335–10355. DOI:10.3390/rs61110335 (accessed June 23, 2015).

Giles, D. K.; Billing, R. C. Deployment and Performance of a UAV for Crop Spraying Chemical Engineering. *Transactions* **2015,** *44*, 307–312.

Glen, B. *Drones Put to Work Hunting Weeds.* The Western Producer, 2014, pp 1–3. http://www.producer.com/2014/07/drones-put-to-work-hunting-weeds/ (accessed Mar 28, 2017).

Gogarty, B.; Robinson, I. Unmanned Vehicles: A (Rebooted) History, Background and Current State of the Art. *J. Law, Informat. Sci.* **2012,** *21*, 1–18.

Goggin, F. Lorence, A.; Topp, N. C. Applying High-throughput Phenotyping to Plant-insect Interactions: Picturing More Resistant Cops. *Curr. Opin. Insect Sci.* **2015,** *9*: 69–76.

Gomez-Candon, J.; D.; Torres-Sanchez, J.; Labbe, S.; Jolivot, S.; Martinez, S.; Renard, S. L. Water Stress Assessment at Tree Scale: High Resolution Thermal UAV Imagery Acquisition and Processing. *Acta Hortic. (ISHS)* **2017,** *1150*, 159–166. http://www.actahort.org/members/showpdf?booknramr=1150_23 (accessed Sept 13, 2017).

Grassi, M. *5 Actual Uses for Drones in Precision Agriculture Today. Drone Life*, 2014, pp 1–4. http://dronelife.com/2014/12/30/5-actual-uses-drones-precision-agriculture-today/ (accessed May 13, 2015).

Hafsel, L. P. Precision Agriculture with Unmanned Aerial Vehicles for SMC Estimations-Towards a More Sustainable Agriculture. Department of Applied Ecology and Agriculture, Hedmark University of Applied Sciences. Masters Dissertation, 2016, p 32.

Hampton Technical Associates Inc. *UAS/Drone-Agriculture*, 2016, pp 1–2. http://www.hampton-technical.com/uas/droneuasagriculture/ (accessed Mar 28, 2017).

Heck, K. *Is that a Hummingbird Outside my Window. Heck Land Company*, 2016, pp 1–2. http://hecklandco.com/agricultural-commodity/helicopter-drone/ (accessed Apr 28, 2016).

Hoffmann, H.; Jensen, R.; Thomsen, A.; Nieto, H.; Rasmussen, J.; Friborg, T. Crop Water Stress Maps for Entire Growing Seasons from Visible and Thermal UAV Imagery. Biogeosciences Discussions, 2016. DOI: 10.5194/bg-2016-316

HSE, *Intelligent Imaging*, 2015, pp 1–3. http://www.uavcropdustersprayers.com/agriculture_delta_fw_70_fixed_wing_uav.htm (accessed May 19, 2015).

Huang, Y.; Hoffman, W. C.; Lan, Y.; Wu, W.; Fritz, B. K. Development of a Spray System for an Unmanned Aerial Vehicle Platform. *Appl. Eng. Agric*. **2009,** *25*, 803–809.

Huang, W.; Lin, L.; Haung, M.; Wang, J.; Wan, H. Monitoring Wheat Yellow Rust with Dynamic Hyperspectral Data, 2008b, pp 1–12. http://www.researchgate.net/c/nveacx/javascript/lib/pdfjs/web/ viewer.html?file=http (accessed Sept 29, 2015).

Huete, A. R. A. A Soil Adjusted Vegetation Index (SAVI). Remote Sensing Environment, New York, USA 1988; pp 205–309ie.

Hunt, E. R.; Horneck, D. A.; Gadler, D. J.; Bruce, A. F.; Turner, R. W.; Spinelli, C. B.; Brungardt, J. J. Detection of Nitrogen Deficiency in Potatoes Using Small Unmanned Aircraft Systems, 2015, pp 1–4. https://www.ars.usda.gov/research/publications/publication/?seqNo115=301000 (accessed Oct 27, 2017).

Idso, S. B.; Clawson, K. L.; Anderson, M. G. Foliage Temperature: Effects of Environmental Factors with Implications for Plant Water Stress Assessment and the CO_2/Climate Connections. *Water Resour. Res*. 1986, *22*, 1702–1716.

IFPRI. *Yield analysis. Harvest Choice*, 2016, pp 1–2. http://harvestchoice.org/topics/yield-analysis. (accessed May 3, 2016).

I'nen, I. P.; Saan, H.; Kaivosoja, J.; Honkavara, E.; Pesnonen, L. Hyperspectral Imaging Based Biomass and Nitrogen Content Estimations from Light-weight UAV. *Remote Sens. Agric. Ecosyst. Hydrol*. **2013,** *15*, 87–88 http://dx.doi.org/10.1117/12.2028624 (accessed May 23, 2015).

Jackson, R. D.; Idso, S. B.; Reginato, R. J.; Pinter, P. J. Canopy Temperature as a Water Stress Indictor. Water Resour. Res. **1981,** *17*, 1133–1138 DOI: 10.1029/WR017i004p01133

Jiménez-Bello, M. A.; Royuela, A.; Manzano, J.; Zarco-Tejada, P. J.; Intrigliolo, D.. Assessment of Drip Irrigation Sub-units Using Airborne Thermal Imagery Acquired with an Unmanned Aerial Vehicle (UAV). In *Precision Agriculture '13*; Wageningen Academic Publishers: The Netherlands, 2013; pp 705–711.

Johnson, K.; Nissen, E.; Saripalli, S.; Arrowsmith, J. R.; McGarey, P.; Scharer, K.; Williams, P.; Blisniuk, K... Rapid Mapping of Ultrafine Fault Zone Topography with Structure from Motion. *Geosphere* **2014,** *10*, 969–986.

Johnson, R.; Smith, K.; Wescott, K. Unmanned Aircraft System (UAS) Application to Land and Natural Resource Management, **2017,** pp 170–177. https://www.cambridge.org/core DOI: 10.1017/S1466046615000216

Jones, H. G.; Vaughan, R. A. *Remote Sensing of Vegetation, Principles, Techniques, and Applications;* Oxford Press University: Oxford, UK, 2010; p 384.

Kalisperakis, I.; Stentoumis, C.; Grammatikopoulis, L.; Karantzalos, K. *Leaf Are Index Estimation in Vineyards from UAV Hyperspectral Data, 2D Image Mosaics and 3D Canopy Surface Models. The International Archives of the Photogrammetry, Remote Sensing and Spatial Information Sciences,* **2015;** Vol. riXLI/W4, pp 299–303.

Kansas State University. Project Using Drones to Detect Emerging Pest Insects, Disease in Crops. Department of Entomology, Kansas State University at Salina and Manhattan: USA, 2015; p 104. http://www.agprofessional.com/news/project-using-drones-detect-emerging-pest-insects-diseases-in-crops/ (accessed Apr 5, 2017).

Kara, A. Low Cost UAV Disease Devastating Apple Crops. Aris Plex, 2013, pp 1–3.

Khot, L.; Sankaran, S.; Cummings, T.; Johnson, D.; Carter, A.; Serra, S. and Musacchi, S. *Unmanned Aerial Systems Applications in Washington State Agriculture*. Proceedings of 12th International Conference on Precision Agriculture, Sacramento, California, USA, 2014, pp 129.

Kooistra, L. Wageningen University and Research, Netherlands, 2014, pp 1–3. http://www.wur.nl/en/activity/Weed_detection_with_Unmanned_Aeril_Vehicles-inagricultural_fields/ (accessed Mar 28, 2017).

Kokila, M.; Karti, J.; Madhuvsaki, E.; Sathya, S.; Vignesh, B. In *Estimation of Chlorophyll Content in Maize Leaf: A Review*. International Conference on Emerging Trends in Engineering, Science and Sustainable Technology, 2017, pp 73–82.

Krishna, K. R. Precision Farming: Soil Fertility and Productivity Aspects. Apple Academic Press Inc.: Waretown, New Jersey, USA, 2013; pp 185.

Krishna, K. R. Push Button Agriculture. Apple Academic Press Inc.: Waretown, New Jersey, USA, 2016; pp 499.

Krishna, K. R. Agricultural Drones: A Peaceful Pursuit. Apple Academic Press Inc.: Waretown, New Jersey, USA, 2018; p 425.

Kruetz, L. Drones Helping to Fight Mites in Strawberry Fields and Almonds, 2017, pp 1–5. http://www.abc10.com/news/local/davis/drones-helping-to-fight-mites-in-strawberry-almond-fields/446347500 (accessed Nov 9, 2017).

Labbe, S.; Lebourgeois, Jolivot, A.; Marti, R. Thermal Infra-red Remote Sensing for Water Stress Estimation in Agriculture. Options Mediterraneennes: The Use of Remote Sensing and Geographic Information Systems for Irrigation in Southwest Europe. Options Mediterraneenes: Serie B. *Etudes et Researches*. 2012, *67*, 175–184

Lacewell, R. D.; Harrington, P. *Potential Cropping Benefits of Unmanned Aerial Vehicles (UAVs) Applications*. Texas A and M Agrilife, College Station, Texas, Texas Water Resource Institute Technical Report, 2015, 477, pp 1–7.

Lebourgeois, V.; Begue, A.; Labbe, S.; Houles, M.; Martine, J. F. A Light Weight Multispectral Aerial Imaging System for Nitrogen Crop Monitoring. *Precis. Agric.* 2012, *13*, 525–541.

Lelong, C. C.; Burger, P.; Jubein, G.; Roux, B.; Labbe, S.; Baret, F. Assessment of Unmanned Aerial Vehicles Imagery for Quantitative Monitoring of Wheat Crop in Small Plots. *Sensors* 2016, *8*, 3557–3585.

Lewis, K. *Remote Aerial Surveys Given the Go-ahead.* The Marketing Eye, Press Release, 2014. http://www.themarketingeye.com/client_news/remote_aerial_survey_drones.html (accessed Aug 20, 2014).

Lottes, P.; Khanna, R.; Pfeiffer.; Siegwart, R.; Stachniss, C. UAV-based Crop and Weed Classification for Smart Farming, 2017, pp 1–23. http://www.ipb.uni-bonn.de/wp-content/papercite-data/pdf/lottes17icra.pdf/

Lumpkin, T. CGIAR Research Programs on Wheat and Maize: Addressing Global Hunger. International Centre for Maize and Wheat (CIMMYT), Mexico. DG's Report, 2012, pp 1–8.

Macarther, D.; Scheller, J. K.; Crane, C. D. Remotely Piloted Mini-helicopter Imaging of Citrus. American Society of Agricultural Engineering Paper No. 051055 ASAE St. Joseph. Michigan, USA, 2005, pp 1–7.

Machovina, B. L.; Feeley, K. J.; Machovina, B. J. UAV Remote Sensing of Spatial Variation in Banana Production. *Crop Pasture Sci.* **2016,** *67*, 1281–1287 https//doi.org/10.1071/CP16135 (accessed Aug 3, 2017).

Mazur, M. Six Ways Drones Are Revolutionizing Agriculture. MIT Technical Review, 2016, pp 1–4. https://www.technologyreview.com/s/601935/six-ways-drones-are-revolutionizing-agriculture/ (accessed Oct 3, 2017).

McComack, S. Multimillion-dollar Project Using Unmanned Aerial Systems to Detect Emerging Pests, Diseases in Food Crops, 2015, pp 1–14. http://www.k-state.edu/media/newsreleases/mar15/uasinsect31815.html (accessed Oct 14, 2017).

McCabe, D. Sky is the Limit for UAVs in Agriculture. *Nebraska Farmer* 2014, *75,* 34.

McGrey, P.; Saripalli, S. *Autokite: Experimental Use of a Low Cost Autonomous Kite Plane for Aerial Photography and Reconnaissance.* International Conference on Unmanned Aircraft Systems, 2013, 208–213.

McKinnon, T.; Hoff, L Agricultural Drones: What Farmers Need to Know. Agribotix LLC, 2017, pp 1–12. https://agribotix.com/whitepapers/farmers-need-know-agricultural-drones/ (accessed Sept 27, 2017).

Media Al. *The 20 lts drone*, 2017. https://www.youtube.com/watch?v=IuDj6om2eJc (accessed Aug 28, 2017).

MicaSense. *Turning Imagery into Attainable Information*, 2017, pp 1–8. https://blog.micasense.com/mapping-potassium-deficiency-table-grapes-in-the-san-joaquin-valley-434a4b85a80e (accessed Oct 12, 2017).

Modern Farmer. Using Drones in the Fight Against Apple Scab, 2013, p 1. http://huffingtonpost.com/modern-farmer/using-drones-in-the-fight_b_4171110. (accessed Oct 16, 2015).

Mortimer, G. Skywalker: Aeronautical Technology to Improve Maize Yields in Zimbabwe. International Maize and Wheat Centre, Mexico, DIY Drones, 2013 pp 1–6. http;//www.ubedu/web/ub/en/menu_eines/notices/2013/04/006.html (accessed Feb, 10, 2016).

Muharam, F. M.; Ms, S. J.; Bronson, K. F.; Delhunty, T. Estimating Cotton Nitrogen Status Using Leaf Greenness and Ground Cover Information. *Remote Sens.* 2017, *7*, 7007–7028 http://dx.doi.org/10.3390/rs70607007 (accessed Aug 24, 2017).

Mulla, D. J. Twenty-five Years of Remote Sensing in Precision Agriculture: Key Advances and Remaining Knowledge Gaps. *Biosyst. Eng.* 2013, *114*, 358–371.

NASA. NASA Weather Drones Used to Determine How Tropical Storms Strengthen, 2013, pp 1–3. http://www.theepochtimes.com/n3/author/associated-press/ (accessed Sept 23, 2015).

Oregon State University. *Drones to Check Out Acres of Potato*, 2014, pp 1–2. http://www.cropand soil.oregonstate.edu/context/drones-check-out-acres-potato. (accessed July 22, 2014).

Oskin, B. NASA Drone to probe ozone layer loss, 2013, pp 1–5. http://www.livescience.com/26161-nasa-drones-, ozone-study.html (accessed Nov 25, 2015).

Park, S.; Nolan, A.; Ryu, D.; Fuentes, S.; Hernandez, E.; Chung, H. N. O'Connell, M. O. *Estimation of Crop Water Stress in a Nectarine Orchard Using High Resolution Imagery from Unmanned Aerial Vehicle (UAV).* In Proceedings of the 21st International Congress on Modelling and Simulation, Gold Coast, Australia, 29 November–4 December 2015, 2015, pp 1–12 (Oct 30, 2017)

Pena, J. M.; Torres-Sanchez, J.; deCastro, A.; Kelly, M.; Granados, F. Weed Mapping in Early-season Maize Fields Using Object Based Image Analysis of Unmanned Aerial Vehicle Images. *PLoS One* 2013. DOI: 10.1371/journal.pone.oo77151

Pena, J. M., Torres-Sanchez, J., Serrano-Perez, A., deCastro, A.; Lopez-Granados, F. Quantifying Efficacy and Limits of Unmanned Aerial Vehicle (UAV) Technology for Weed Seedling Detection as Affected by Sensor Resolution. *Sensors* **2015,** *15*, 5609–5626.

Perry, E. M., Bluml, M., Goodwin, D.; Swarts, N. D. Remote Sensing of N Deficiencies in Apple and Pear Orchards. International Society of Horticultural Science. *Acta Hortic.* **2016,** *1130*, 575–580 http://www.ishs.org/ishs-article/1130_86 (accessed Aug 24, 2017).

Pinto R. F. Heat and Drought Adaptive QTL in a Wheat Population Designed to Minimize Confounding Agronomic Effects. *Theor. Appl. Genet.* **2010,** *121* (6), 1001–1021.

Plaven, G. Drones to Quinoa, Field Day Showcase Research Center, 2016, pp 1–3. http://www.opb.org/news/articles/dsrones-to-quinoa-field-day-show-cases-research-center. (accessed Apr 12, 2016).

Prasad M. M. Genetic Analysis of Indirect Selection for Winter Wheat Grain Yield Using Spectral Reflectance Indices. *Crop Sci.* **2007,** *47* (4), 1416.

Precisionhawk, The Fundamentals of UAVs. Precision Hawk Media, 2014, pp 1–8. http://www.precision hawk.com/media/topic/the-fundamentals-of-uavs/ (accessed Mar 22, 2017).

Precisionhawk. An Enterprise Drone Platform for Better Business Intelligence. Precision Hawk Inc.: Indiana, USA, 2017, pp 1–12. http://www.precisionhawk.com/ (accessed May 5, 2017).

Primicerio, J.; Fillipo Di Gannaro, S.; Fiolrillo, E.; Lorenzo, G.; Lugata, E.; Matese, A.; Vaccari, F. P. A Flexible Unmanned Aerial Vehicle for Precision Agriculture, 2012. DOI: 10.1007/s1119-012-9257-6 (accessed Feb 20, 2015).

Pudelko, R.; Stuczynski, T.; Borzecka-Walker. The Suitability of an Unmanned Arial Vehicle (UAV) for the Evaluation of Experimental Fields and Crops. *Zemdirbyste-Agric.* **2012,** *99*, 431–436 UDK 631.5.001.4:629.734

Quackenbush, J. North Coast Vineyards See More Drone Use as Agriculture Market Soars. *North Bay Bus. J.* **2017,** 104. https://www.vineview.com/single-post/2017/01/17/VineView-Featured-in-North-Bay-Business-Journal-Flight-of-the-Vineyard-Skybots pp 1-7 (September 25th, 2017)

Ramasamy, S. 2016 Why drones are the latest buzz in Agriculture. The WIRE. http://thewire.in/70610/drones-agriculture-slantrange-ndvi/ (accessed Jan 9, 2017).

Rambo, L.; Ma, L.; Xionh, Y.; Silvia, P. R. F. Leaf and Canopy Characteristics as Crop-N Status Indicators for Field Nitrogen Management in Corn. *J. Plant Nutr. Soil Sci. Temuco* **2010,** *173*, 434–443.

Raymond Hunt, E.; Daughtry, C. S. T.; Mirsky, S. B.; Hively, W. D. Remote Sensing with Simulated Unmanned Aircraft Imagery for Precision Agriculture Applications. IEEE J. Select. Topics Appl. Earth Observ. *Remote Sens.* **2014,** *7*, 1–12. DOI: 10.1109/JSTRS.2014.2317876 (accessed Sept 20, 2017).

Real Agriculture. DJI Spraying Drone MG-1 Agras, 2017. https://www.youtube.com/watch?v=K3zGVi08wAw (accessed Aug 28, 2017).

Reynolds, M. Phenotyping Approached for Physiological Breeding and Gene Discovery in Wheat. *Ann. Appl. Biol. Precision Agriculture* **2012,** *155* (3), 309–320.

Rice, C. Weed Detection from Drones… Finally Here. *Spectrabotics* 2015, 1–4. http://www.spectrabotics.com/blog/221-weed-detection-from-drones-finally-here/ (accessed Mar 28, 2017).

Richardson, B. *Drones Could Revolutionize Weather Forecasts, but Must Overcome Safety Concerns*. Washington Post, 2014, pp 1–2, https://www.washingtonpost.com/news/capital-weather-gang/wp/2014/04/25/drones-could-revolutionize-weather-forecasts-but-must-overcome-safety-concerns/ (accessed June 26, 2016).

RMAX. *RMAX Specifications*. Yamaha Motor Company, Japan, 2015, pp 1–4. http://www.max.yamaha-motor.Drone.au/specifications. (accessed Sept 8, 2015).

Rogers, D. R. *Unmanned Aerial System for Monitoring Crop Status*. Virginia Polytechnic Institute and State University, Blacksburg, VA, USA, MSc Thesis, 2013, pp 1–12.

Ru, Y.; Zhou, H.; Fan, Q.; Wu, X. In *Design and Investigation of Ultra-low Volume Centrifugal Spraying System on Aerial Plant Protection.* Proceedings of American Society of Agricultural and Biological Engineering. Paper No. 11-10663, 2011, pp 231–236.

Saberioon, M. M.; Gholizadeh, A. *Novel Approach for Estimating Nitrogen Content in Paddy Fields Using Low Altitude Remote Sensing System.* The International Archives of the Photogrammetry, Remote sensing and Spatial Information Sciences XLI-B1, 2016, pp 12–19. DOI:10.5194/isprsarchives-XLI-B1-1011-2016 (accessed Aug 23, 2017).

Samseemoung, G.; Soni, P.; Jayasuriya, H. P. W.; Salokhe, V. M. Application of Low Altitude Remote Sensing (LARS) Platform for Monitoring Crop Growth and Weed Infestation in a Soybean Plantation. Precision Agric **2012,** *13,* 611–627.

Sankaran, S.; Khot, L. R.; Espinoza, C. Z.; Jarolmasjed, S.; Pavek, M. J. Low altitude, high resolution aerial imagery system for row and field crop phenotyping. *Eur. J. Agron.* **2015,** *70,* 112–123.

Sankaran, S.; Ehsani, R. A Detection of Huanglongbing Disease in Citrus Using Fluorescence Spectroscopy. Trans. ASABE 2012, *55,* 313–320.

Sankaran, S, Mishra, A.; Ehsani, R.; Davis, C. A Review of Advance Techniques for Detecting Plant Disease. *Comput. Electron. Agric.* **2010,** 1–13. http://www.science direct.com/science/article/pii/S0168169910000438 (accessed July 7, 2015).

Santa Ana. Drone-based Land Management Meeting in Corpus Chrisiti. AgriLife Today, 2016, pp 1–4. https://today.agrilife.org/2016/01/12/drone-land-management-meet/ (accessed Apr 22, 2017).

Schultz, B. Louisiana: Researchers Study Use of Drones in Crop Monitoring, 2013, pp 1–4. Agfax.com http://www.lsuagcenter.com/news_archive/2013/deceber/headline_news/AgCenter-researchers-study-use-of-drones-in-crop-monitoring (accessed Jan 1, 2016).

Severtson, D.; Callow, N.; Ken Flower, K.; Neuhaus, A.; Olejnik, M.; Nansen, C.; M. Unmanned Aerial Vehicle Canopy Reflectance Data Detects Potassium Deficiency and Green Peach Aphid in Canola. *Precision Agric.* **2016,** *17,* 659–677. http://link.springer.com/article/10.1007/s11119-016-9442-0 (accessed Aug 24, 2017).

Shan, J.; Hussain, E.; Kim, K.; Biehl, Flood Mapping and Damage Assessment: A Case Study in the State of Indian. *Geospat. Technol. Earth Observ.* **2009,** 473–495.

Sharma, P. Precision Farming. Gene-Tech Books.: New Delhi, India, 2017; p 256.

She, Y.; Ehsani, R.; Robbins, J.; Leiva, J. N.; Owen J. *Application of Small UAV Systems for Tree and Nursery Inventory Management.* Proceedings of 14th, International Conference on Precision Agriculture Montreal, Quebec, Canada, 2017, Pp 1–7 (accessed June 30, 2017).

Shi, Y.; Thomasson.; Murray, S. C.; Puch, N. A.; Rooney, W. L.; Shafian, S.; Rajan, N.; Rouze, G.; Morgan, C. L. S.; Neely, H. L.; Rana, A.; Bagvthiannan, M. V.; Herrickson, J.; Bowden, E.; Vilsack, J.; Olsenholler, J.; Bishop, M. P.; Sheridan, R.; Putman, E. B.; Popescu, S.; Burks, T.; Cope, D.; Ibrahim, A.; McCutchen, B. F.; Baltensperger, D. D.; Vent, R.; Vidrine, M.; Yang, C. Unmanned Aerial Vehicles for High Throughput Phenotyping and Agronomic Research. *PLOS One* **2016,** 1–15. http://dx.doi.org/10.1371/journal.pone.0159781 (accessed Nov 9, 2017).

Snow, C. The Truth About Drones in Precision Agriculture, They Are Great Scouting Tools, but Can They Unseat Incumbents, 2016, pp 1–12. https://www.angeleyesuav.com/crop-content/uplands/2016/08/the truthAboutDrones_g.pdf (accessed Sept 20, 2017).

Song, Y. ;Wang, J. Evaluation of the UAV-based Multispectral Imagery and Its Application for Crop Intra-field Nitrogen Monitoring and Yield Prediction. The University of Western

Ontario Electronic Thesis and Dissertation Repository No. 4085, 2016, pp 112 (accessed Aug 24, 2017).

Sullivan, D. G., Fulton, J. P. Shaw, J. and Bland, G. 2007 Evaluating the sensitivity of an unmanned thermal infrared aerial system to detect water stress in a cotton canopy. Transactions of ASABE 50: 1955-1962

Swain, K. C., Thomson, S. J., Jayasuriya, H. P. Adoption of Unmanned Helicopter for Low Altitude Remote Sensing to Estimate Yield and Total Biomass of a Rice Crop. *Trans. Am. Soc. Agric. Eng.* **2010,** *53*, 21–27.

Sky Squirrel. Vineyard Drone Research: Applications, 2017, pp 1–2. https://www.skysquirrel.ca/applictions.html (accessed Mar 28, 2017).

Stevenson, A. Drones and the Potential for Precision Agriculture. Altech Inc., 2015, pp 1–2. http://www.altech.com/blogposts/drones-and-potential-precision-agriculture.htm (accessed June 26, 2015).

Tattaris, M.; Reynolds, M. *Applications of an Aerial Remote Sensing Platform*. Proceedings of the International TRIGO (Wheat) Yield Workshop. Reynolds, M., Mollero, G., Mollins, J., Braun, H., Eds.; International Maize and Wheat Centre (CIMMYT): Mexico, 2015; pp 1–5.

Templeton, R. C.; Vivoni, E. R., Méndez-Barroso, L. A.; Pierini, N. A.; Anderson, C. A.; Rango, A.; Laliberte, A.; Scott, R. L. High-resolution Characterization of a Semiarid Watershed: Implications on Evapotranspiration Estimates. *J. Hydrol.* **2014,** *509*, 306–319. DOI: 10.1016/j.jhydrol.2013.11.047.

Terraluma. Applications Selected Case Studies, 2014, pp 1–8. http://www.terraluma.net/showcases.html (accessed Aug 28, 2016).

Tigue, K. University of Minnesota Research Group Pushes for Ag. Minnesota Daily. Precision Farming Dealer, 2014, pp 1–3. http://www.mndaily.com/news/campus/2014/04/29/u-research-pushes-agriculture-drones.

Tiltuli. *New UAV Fertilizer and Pesticide Spraying and Small Fire Extinguishing*, 2016. https://www.youtube.com/watch?v=IuDj6om2eJc (accessed Aug 28, 2017).

Torres-Rua, A. *Drones in Agriculture: An Overview of Current Capabilities and Future Directions*, 2017, pp 1–5. *https://conference.usu.edu/uwuw/includes/AlfonsoTorres_DronesinAgriculture.pdf?* (accessed Oct 30, 2017).

Towler, J.; Krawiec, B.; Kochersberger, K. Terrain and Radiation Mapping in Post-disaster Environments Using an Autonomous Helicopter. *Remote Sens*. 2012, *4*, 1995–2010.

Tremblay, N.; Vigneault, P.; Belec, C.; Fallon, E.; Bouroubi, M. Y. *A Comparison of Performance Between UAV and Satellite Imagery for N Status Assessments in Corn.* Proceedings of 12th International Conference on Precision Agriculture, Sacramento, California, USA, 2014, pp 19.

Trimble. *Trimble UX5 Aerial Imaging Solution for Agriculture*, 2015, pp 1–3. http://www.trimble.com/Agriculture/UX5.aspx (accessed May 20, 2015).

Trimble Inc. The Complete Unmanned System. Trimble Inc. San Jose, California, USA, 2017, pp 1–10. http://www.us.trimble/ux5 (accessed May 5, 2017).

Triple20. *Using Drones to Detect Crop Diseases*, 2015, pp 1–2. http://www.foodvalleyupdate.com/news/using-drones-to-detect-crop-diseases/ (accessed Oct 18, 2015).

Tropnevad. *Object-based Weed Analysis of Unmanned Aerial Vehicle (UAV) Images*, 2013, pp 1–4. http://www.dronetrest.com/t/object-based-weed-analysis-of-unmanned-aerial-vehicle-images/ (accessed Mar 28, 2013).

United Soybean Board. *Farming's Newest Precision Agriculture Tool Takes Data to on New Horizon*, 2013, pp 1–5. http://www.unitedsoybean.org/article/new-precision-agriculture-could-revoluitonize-farming. (Oct 26, 2015).

University of California, Davis, UC Davis Investigates Using Helicopter Drones for Crop Dusting. Agriculture-UAV Drones, 2014, pp 1–7. http://www.agricultureuavs.com/uc_davis_uav_crop_dusting.htm (accessed June 21, 2015).

Van der Staay, L. J. Sorghum research at the UC Agriculture and Natural Resources Research & Extension Centers Available to the Public. Kearney News Updates, 2017, pp 1–12. http://ucanr.edu/blogs/kearney/index.cfm?tagname=sorghum. (accessed Nov 7, 2017).

Vivoni, E. R. Spatial Patterns, Processes and Predictions in Ecohydrology: Integrating Technologies to Meet the Challenge. *Ecohydrology* **2012,** *5* (3), 235–241. DOI:10.1002/eco.1248.

Vivoni, E. R.; Rango, A.; Anderson, C. A.; Pierini, N. A.; Schreiner-McGraw, A. P.; Sripalli, S.; Lalberte, A. S. Ecohydrology with Unmanned Aerial Vehicles. *Ecosphere* 2014, *5*, 1–14. DOI: 10.1890/ES14-00217.1 (accessed Sept 15, 2017).

Vogel, J. 2014 Ready to Fly a UAV Drone Over Your Fields. Prairie Farmer Magazine http://farmprogress.com/story-ready-fly-uav-drone-fields-9-120813 pp 1-3 (accessed Oct 9, 2015).

Von Bueren, S. K. Burkart, A.; Hueni, A.; Rascher, U.; Touhy, M. P.; Yule I. J. Deploying Four Optical UAV-based Sensors Over Grassland: Challenges and Limitations. *Bio-Geosciences* 2015, *12*, 163–175 DOI: 10.5194/bg-12-163-2015 (Sept 26, 2017).

Watts, A. C.; Ambrosia, V. G.; Hinkeley Unmanned Aircraft Systems in Remote Sensing and Scientific Research: Classification and Considerations. *Remote Sens*. 2012, *4*; 1671–1692.

West, J. S.; Atkins, S. D.; Fitt, B. D. L. Detection of Airborne Plant Pathogens: Halting Epidemics Before They Start. *Outlooks Pest Manag*. **2009,** *20*, 111–113.

West, J. S.; Bravo, C. and Oberti, R. The Potential of Optical Canopy Measurement for Targeted Control of Crop Diseases. *Ann. Rev. Phytopathol*. 2003, *41*, 593–614.

West, J. S.; Bravo, C.; Moshou, D.; Ramon, H.; Alastair McCartney, H. Detection of Fungal Diseases Optically and Pathogen Inoculum by Air Sampling. In *Precision Crop Production: The Challenges and Use of Heterogeneity,* 2010, pp 135–149. http://lnk.springer.com/chapter/10.1007/978-90-481-9277-9_9?no-a (accessed Oct 16, 2015).

Wharton, C. Nevada Looks at Dronesfor Economic Development and Natural Resources, 2013, pp 1–4. http://www.unce.unr.edu/news/article.asp?ID=1871 (Sept 9, 2014).

Wooten, M. UGA Scientists Use Robots and Drones to Accelerate Plant Genetic Research, Improve Crop Yield. UGA Today, 2017, pp 1–3. http://news.uga.edu/releases/articles/robots-and -drones-improve-crop-yield/ (accessed Aug 26, 2017).

Wojtowicz, M.; Wojtowicz, Piekarczyk, J. Application of Remote sensing methods in Agriculture. International Journal of the Faculty of Agriculture and Biology, Warsaw University of Life Sciences-SGGW, Poland. *Commun. Biomet.Crop Sci*. **2016,** *11*, 31–50.

XM2 Aerial. *DJI AGRAS Review-Spraying Drone*, 2017. https://www.youtube.com/watch?v=dCHvICOJ7mY (accessed Aug 28, 2017).

Yang, G.; Liu, J.; Zhao, C.; Li, Z.; Huang, Y.; Yu, H.; Xu, B.; Yang, X.; Zhu, D.; Zhang, X.; Feng, H.; Zhao, X.; Li, Z.; Li, H.; Yang, H. Unmanned Aerial Vehicle Remote Sensing for Field-based Crop Phenotyping: Current Status and Perspectives. *Front. Plant Sci*. **2017,** *8*, 1111–1140.

Yamaha, *RMAX-History*, 2014, pp 1-4. http://www.rmax.yamaha-motor.com.all/history (accessed Sept 20, 2015).

Yue, J.; Lei, T.; Li, C.; Zhu, J. The Application of Unmanned Aerial Vehicle Remote Sensing in Quickly Monitoring Crop Pests. *Intell. Automat. Soft Comput.* **2012,** *18*, 1043–1052.

Zaman-Allah, M.; Vergara, O.; Araus, J. L. Tarekegne, A.; Magarokosho, C.; Zarco-Tejada, P. J.; Hornero, A.; Alba, A. H.; Das, B.; Craufurd, P.; Olsen, M, Prasanna, B. M.; Cairns, J. Unmanned Aerial Platforms-based Multi-spectral Imaging for Field Phenotyping of Maize. *Plant Methods* **2015,** 123. https//doi.org/10.1186/s13007-015-0078-2 (accessed Aug 24, 2017).

Zarco-Tejada, P. J.; Guillen-Climent, M. L.; Hernandez-Clement, R.; Catalina, A.; Gonzalez, M. R.; Martin, P. Estimating Leaf Carotenoid Content in Vineyards Using High Resolution Imagery Acquired from an Unmanned Aerial Vehicle (UAV). *Agric. Forest Meteorol.* **2013,** *171*, 281–294.

Zhang, W.; Kovacs, J. M. The Application of Small Unmanned Aerial Systems for Precision Agriculture: A Review. *Precis. Agric.* **2012,** *13*, 693–712. DOI: 10.1007/s11119-012-9274-5.

Zhang, J.; Huang, Y.; Yuan, L.; Yang, G.; Chen, I.; Zhao, C. Using Satellite Multispectral Imagery for Damage Mapping of Armyworm (*Spodoptera frugidera*) in Maize at a Regional Scale. *Pest Manag. Sci.* **2015,** 1–3. DOI: 10.1002/ps.4003 (accessed Oct 18, 2015).

Zhang, C.; Walters, D.; Kovacs, J. M. Application of Low Altitude Remote Sensing in Agriculture Upon Farmer's Request: A Case Study in North-eastern Ontario, Canada. *PLOS One* 2014, 1–9. http://journals.plos.org/plosone/article?id=10.137/jurnal.pone.0112894 (accessed June 25, 2015).

Zhenkun, T.; Yingying, F.; Suhong, L.; Liufeng, N. Rapid Crops Classification Based on UAV Low-altitude Remote Sensing. *Trans. Chinese Soc. Agric. Eng.* **2017,** 1–14. DOI: 10.3969/j.issn.1002-6819.2013.7.014 (accessed June 30, 2017).

Zhu, H.; Lan, Y. Wu, W.; Hoffman, C.; Huang, Y.; Xue, X.; Liang, J.; Fritz, B. Development of a Precision Spraying Controller for Unmanned Aerial Vehicles, 2014, pp 1–8. DOI:10.1016/S1672-66529(10)60251-X (accessed May 25, 2015).

Index

G

H

I

U